# Lecture Notes in Civil Engineering 435

**Lecture Notes in Civil Engineering** (LNCE) publishes the latest developments in Civil Engineering—quickly, informally and in top quality. Though original research reported in proceedings and post-proceedings represents the core of LNCE, edited volumes of exceptionally high quality and interest may also be considered for publication. Volumes published in LNCE embrace all aspects and subfields of, as well as new challenges in, Civil Engineering. Topics in the series include:

- Construction and Structural Mechanics
- Building Materials
- Concrete, Steel and Timber Structures
- Geotechnical Engineering
- Earthquake Engineering
- Coastal Engineering
- Ocean and Offshore Engineering; Ships and Floating Structures
- Hydraulics, Hydrology and Water Resources Engineering
- Environmental Engineering and Sustainability
- Structural Health and Monitoring
- Surveying and Geographical Information Systems
- Indoor Environments
- Transportation and Traffic
- Risk Analysis
- Safety and Security

To submit a proposal or request further information, please contact the appropriate Springer Editor:

- Pierpaolo Riva at pierpaolo.riva@springer.com (Europe and Americas);
- Swati Meherishi at swati.meherishi@springer.com (Asia—except China, Australia, and New Zealand);
- Wayne Hu at wayne.hu@springer.com (China).

**All books in the series now indexed by Scopus and EI Compendex database!**

Maria Antonietta Aiello · Antonio Bilotta
Editors

# Proceedings of Italian Concrete Conference 2022

Springer

*Editors*
Maria Antonietta Aiello
Department of Engineering for Innovation
University of Salento
Lecce, Italy

Antonio Bilotta
Department of Structures for Engineering and Architecture
University of Naples Federico II
Naples, Napoli, Italy

ISSN 2366-2557 ISSN 2366-2565 (electronic)
Lecture Notes in Civil Engineering
ISBN 978-3-031-43101-2 ISBN 978-3-031-43102-9 (eBook)
https://doi.org/10.1007/978-3-031-43102-9



This Springer imprint is published by the registered company Springer Nature Switzerland AG
The registered company address is: Gewerbestrasse 11, 6330 Cham, Switzerland

Paper in this product is recyclable.

# Topics and Participants

Requirements for structural concrete cover sustainability, durability, and adaptability to reuse (of materials, elements, or entire structure). On the one hand, these aspects are treated during construction processes with new ideas from conceptual design and technologies or analytical and experimental tools, as well as innovative assessment criteria and controls. On the other hand, the material, too, undergoes remarkable development, differentiating itself for specific applications, up to the multiple today's concretes, binders, additives, and reinforcements (metallic or not).

The Italian Concrete Conference ICC2022 held in Naples during the autumn of the 2022 offered to all players—such as industrialists of construction, producers of components, technology and materials, professionals of architecture and civil engineering, field technicians, researchers, and academics—an opportunity for updating their cultural sight, establishing contacts, sharing information and opinions, discussing applications, proposing solutions, and consolidating their own role awareness. A round table at the beginning of the conference on the topic of the existing bridges pointed out one of the leitmotifs of the conference.

## Presentation by Invited Speakers

The conference featured three leading figures in the world of reinforced concrete, Karen Scrivener, for material development, Andrew W. Taylor, for a view on regulatory aspects, and Hugo Corres Peiretti, for an overview of realizations.

Karen Scrivener

Andrew W. Taylor

Hugo Corres Peiretti

Karen Scrivener has been Professor and Director of the Laboratory of Construction Materials in the Department of Materials of Ecole Polytechnique Federale de Lausanne (EPFL) for the last 20 years. She is Fellow of the UK Royal Academy of Engineering and Author of over 200 journal papers. Her research focuses on understanding the chemistry and microstructure of cement-based materials and improving their sustainability. In her presentation titled: "*Realistic options to reduce CO2 emissions from Cement and*

*Concrete*", Karen provided a review on the realistic options to reduce CO2 emissions associated with cement and concrete, to underline that there are no magic solutions, despite the frequent hypes appearing in the media. During her speech, Karen emphasized that substantial reductions are possible if all parts of the construction chain are considered: cement level, concrete level, and structure level. Karen also highlighted some particularly promising solutions at all these levels, notably LC3 (limestone calcined clay cement) which can reduce the CO2 emissions of cement by up to 40%.

Andrew W. Taylor is Technical Director with KPFF Consulting Engineers in Seattle, Washington, and leads for design of special seismic systems such as seismic isolation and seismic damping. Dr. Taylor serves as Affiliate Professor in the Department of Civil and Environmental Engineering at the University of Washington, where he has enjoyed teaching courses in structural analysis and reinforced concrete design. His professional experience spans 30 years of structural engineering research and practice, including seven years with the Building and Fire Research Laboratory at the National Institute of Standards and Technology. He is currently Chair of the American Concrete Institute 318 Building Code Committee, which will complete the next edition of the U.S. national concrete building code in 2025. His presentation titled "*Sustainable Structural Design and the American Concrete Institute Building Code Requirements for Structural Concrete, ACI 318–25*" explored areas of the ACI 318 Building Code that could be examined for opportunities to directly or indirectly reduce greenhouse gas emissions. Andrew illustrated the new Appendix providing guidelines for improving the sustainable design of reinforced concrete structures, written by ACI 318 Subcommittee N, which includes structural engineers, concrete producers, and experts on the environmental impacts of the life cycle of structural concrete. His contribution also highlighted that producers continue to search for ways to reduce greenhouse gas emissions through developing engineered structural materials without adversely affecting their engineering design properties, quality, and economy.

Hugo Corres Peiretti is Professor of structural concrete and conceptual design of structures at the School of Civil Engineering at the Technical University of Madrid, Honorary President of fib, and Member of the Project Team that prepared Eurocode 2 (Structural Concrete), authored several papers and books, and founded the FHECOR Consulting Engineers. His keynote titled "*Interventions in existing structures. A sustainable option*" showed different interventions on existing structures: extension of a bridge built by successive cantilevers, from 12 to 24 m, extension of a mixed cable-stayed bridge, from 21 to 41 m, replacement of cables for a cable-stayed bridge with a prefabricated concrete deck, and expansion of a tall building in Madrid. All these examples show different aspects of the state of the construction before the intervention, the assessment process, the conception, and the construction of the interventions. During the speech, Hugo highlighted that intervention in existing structures, to extend the useful life, is a sustainable option, and these projects are of great responsibility, involving a reusable assessment and a complex design.

## Conference's Overview and Proceedings

The event has been organized in the beautiful setting of museum site of "Donnaregina" in a city, Naples, that is undergoing a cultural rebirth. The conference was among the first in its field to be completely in the presence after a long period plagued by the problems of the pandemic and it re-energized all participants. More than one hundred technical and scientific papers on theory and modelling, applications and realizations, materials and investigations, and technology and construction techniques were presented during the conference. Applications concern not only new constructions, but more and more analyses, rehabilitation, conservation, strengthening, and upgrading, of the existing structures, particularly degraded bridges, and buildings prone to seismic action. This volume represents a synthesis of the proceedings, aimed at an international promotion of the event, collecting a set of contributions, presented therein in English, which may deserve a wider international acquaintance.

We would like to take this opportunity to thank all the colleagues of the scientific committee, particularly the international experts, for reviewing and finalizing the papers in a tight time. Our sincere appreciation is also extended to the speaker that provided a summary of all the papers presented during the conference, divided in four main topics: *Sustainability* shown by Prof. Enzo Martinelli; *Infrastructures development and upgrading* shown by Prof. Nino Recupero; *Durability* shown by Prof. Luigi Coppola; and *Road and industrial pavements* shown by Prof. Giovanni Plizzari.

A special acknowledgement is addressed to the International Association of Structural Concrete fib, for having supported this event in the perspective of an International Federation aimed at concrete structures' improvement worldwide.

We hope that this volume can contribute to favouring the dialogue among the stakeholders acting in the construction field, in the new era of globalization and digitalization, and to further enhance in it the international cooperation.

Maria Antonietta Aiello

Antonio Bilotta

# Contents

## Durability

# Sustainability

# Structural Performances of Recycled Aggregates Concrete Foundation Plinths

Luisa Pani, Lorena Francesconi, Monica Valdes, and Flavio Stochino(✉)

Department of Civil Environmental Engineering and Architecture, University of Cagliari, Cagliari, Italy

{lpani,m.valdes,fstochino}@unica.it, lorenafrancesconi@libero.it

**Abstract.** The recycling of concrete construction and demolition wastes to obtain coarse recycled aggregates for structural concrete production represents an interesting strategy to develop circular economy in the construction sector. It reduces landfill waste and raw material exploitation. In this applied research, the recycled aggregate, obtained from construction and demolition waste, was used in partial replacement of coarse natural aggregate for casting a set of real scale foundation plinths. These structures have been tested under lateral forces applied on the top of precast reinforced concrete columns connected to the plinths. The mechanism and force characterizing the collapse have been determined showing how concrete structures with recycled aggregates have similar performances to the ordinary ones.

**Keywords:** sustainability · recycled concrete · plinths

## 1 Introduction

Current concrete technology contributes to the exploitation of non-renewable natural resources. For this reason, recycling concrete demolition waste promotes a circular economy strategy and reduces the environmental impact. The use of recycled aggregates represents a valid alternative to natural ones for concrete production helping natural resources preservation and reducing landfill disposal (Kovler and Roussel, 2011; Pepe et al. 2014).

In Italy DM 11/10/17 defines, for all public contracts, the Minimum Environmental Criteria (CAM). They are the environmental requirements aimed at identifying the best design solutions, products, and services from an environmental point of view throughout the entire life cycle.

Currently, applied researches show that there is no technical or scientific limit to the use of recycled concrete aggregate for structural and non-structural concretes (González-Fonteboa & Martínez-Abella 2008; Pani et al. 2020; Pepe et al. 2014).

Experimental data concerning concrete made with Recycled Concrete Aggregate (RCA) show that recycled concrete having medium compressive strength can be produced regardless of the parent concrete quality (Pacheco et al. 2019; Francesconi et al. 2016; Stochino et al. 2017; Tabsh and Abdelfatah 2009, Pani et al. 2020).

L'autore di riferimento per questo paper è: Flavio Stochino


M. A. Aiello and A. Bilotta (Eds.): ICC 2022, LNCE 435, pp. 3–13, 2024.
https://doi.org/10.1007/978-3-031-43102-9_1

Recently, RCA have been used for full scale reinforced concrete elements (see Stochino et al. 2017; Xu et al. 2017; Xu et al. 2018; Xu et al. 2019).

In addition to the clear environmental benefits, the use of RCA can bring economic advantages also to the precast concrete industry where scraps of processing waste can be transformed in RCA, reducing land-filling costs and raw materials.

In this work the performance of precast recycled concrete casted with coarse RCA has been analyzed.

An experimental campaign was carried out at the University of Cagliari in collaboration with a company that produces and sells precast reinforced concrete elements (beams, plinths, and hollow core slabs). With more accuracy we focused our attention on prefabricated plinths, made with recycled concrete, in which the coarse aggregates, were obtained from the crushing of processing waste. The plinths have been tested under service loads condition.

## 2 Experimental Campaign

Two recycled concrete mixes having respectively 30% and 50% substitution percentage, by weight, of coarse recycled aggregates were used. Precast reinforced concrete plinths made with these mixes were produced and tested.

In order to compare the structural performance, a reinforced ordinary concrete plinth, made with natural aggregates, was also casted and tested.

The test aim is to evaluate the capacity and the collapse mechanism of the plinths. The load on the structure is applied on the top section of a precast reinforced concrete column connected to the plinth (Fig. 1).

**Fig. 1.** Plinth column element.

The precast elements were manufactured by "Vibrocemento Srl", a precast concrete company, the geometrical details of the specimens and the reinforcements arrangements are presented in Fig. 2.

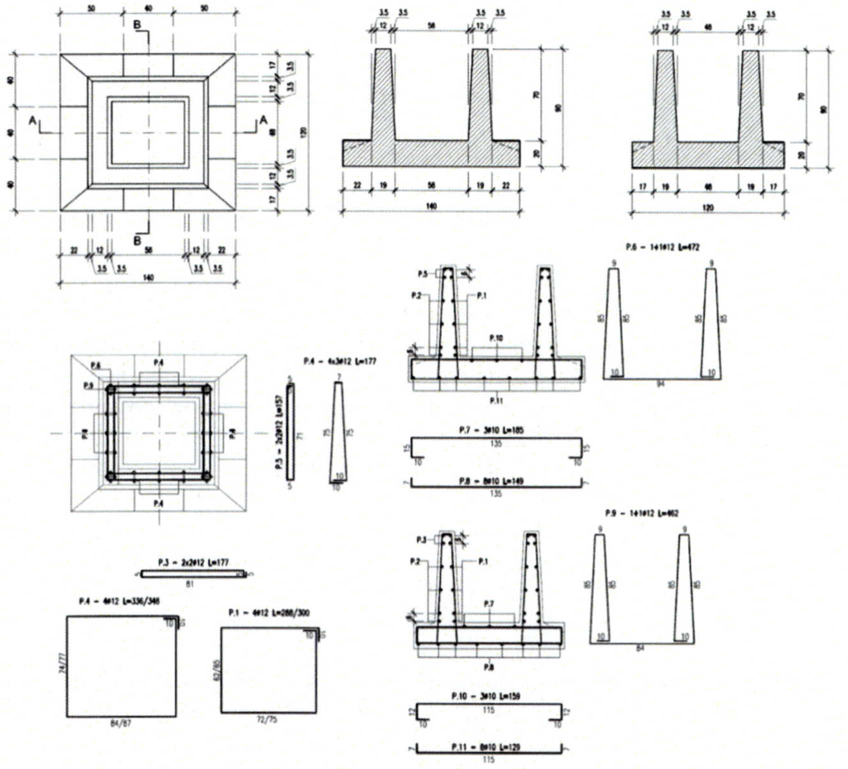

**Fig. 2.** Plinth details and reinforcements distribution. Measures in cm. Materials

Three different concrete mixes were casted varying the recycled aggregate replacement percentage: RC0% with fine and coarse natural aggregates, RC30% with fine natural aggregates and natural and recycled coarse aggregates (30% by weight of recycled aggregates and 70% by weight of natural aggregates), RC50% with fine natural aggregates and natural and recycled coarse aggregates (50% by weight of recycled aggregates and 50% by weight of natural aggregates). Precast columns were made with a mix labelled RC0%. Details of the concrete mixes are shown in Table 1.

In the present investigations the recycled aggregates come from the concrete structures of the old Cagliari football stadium (built between 1965 and 1970). Recycled Concrete Aggregates (RCA) are obtained from the foundation blocks demolition waste by dry crushing and consequent sieving in order to obtain the required particle size. The recycled aggregates were subjected to all the tests complying with UNI EN 12620: 2008 and UNI 8520-1: 2015. The obtained results are summarized in Table 2, while Table 3 presents water absorption, density and fines content of filler aggregate (see UNI EN

**Table 1.** Mix design of concrete (1 $m^3$)

| Concrete | | RC0% | RC30% | RC50% |
|---|---|---|---|---|
| Cement CEM 52,5 RI (kg) | | 350 | 356 | 352 |
| Water (kg) | | 160 | 186 | 215 |
| Fine natural aggregates (kg) | 0–2 mm | 120 | 130 | 120 |
| | 0–6 mm | 600 | 595 | 600 |
| Coarse natural aggregates (kg) | 4–10 mm | 520 | 365 | 255 |
| | 8–16 mm | 600 | 415 | 300 |
| Coarse recycled aggregates (kg) | 4–16 mm | 0 | 340 | 560 |

12620: 2008). With more details looking at column 4 of Table 3: $f_x$ means that in the given particle size range the percentage passing by mass of fine content is less than x%, so for example f1.5 denotes fine content below 1.5%.

Six cubic specimens (150 × 150 × 150 mm) and 1 prismatic specimen (150 × 150 × 300 mm) were made for each concrete mix. Table 4 presents the results of the experimental compressive strength ($R_{c,28d}$), tensile strength ($f_{ct,}$), and elastic modulus $E_c$ tests. Reference standards are UNI EN 12390-3: 2019, UNI EN 12390-6: 2010, and UNI EN 12390-13: 2013. B450C Steel reinforcing bars were used, their characteristics yield strength is 450 MPa, while their characteristics tensile strength is 540 MPa.

**Table 2.** Recycled aggregates characteristics

| Identification | Recycled Aggregates |
|---|---|
| Size designation | 4/16 |
| Cat. Grading | GC 90/15, GT 17.5 |
| Flakiness Index | 4 |
| Shape Index | 59 |
| Saturated surface-dried particle density | 2.39 Mg/$m^3$ |
| Loose bulk density and voids | $\rho_b = 1.23$ Mg/$m^3$<br>v% = 45 |
| Percentage of fines | 0.15% |
| Percentage of shells | absent |
| Resistance to fragmentation | 39 |

*(continued)*

**Table 2.** (*continued*)

| Identification | Recycled Aggregates |
|---|---|
| Constituents of Coarse RCA | X = 0;<br>Rc = 74%;<br>Ru = 27%;<br>Rb = 0;<br>Ra = 0;<br>Rg = 0 |
| Content of water-soluble chloride salts | 0.005% |
| Content of acid-soluble chloride salts | 0.325% |
| Content of acid-soluble sulphate | 0.43% |
| Content of total sulfur | S < 0.1% |
| Content of water-soluble | SS = 0.148% |
| Water absorption | WA24 = 7.0 |

**Table 3.** Parameters of recycled aggregates

| Aggregate particle size (mm) | water absorption (%) | density (kg/m$^3$) | fine content |
|---|---|---|---|
| 0–2 (natural aggregate) | 0.80 | 2562 | f3 |
| 0–6 (natural aggregate) | 1.31 | 2600 | f3 |
| 4–10 (natural aggregate) | 1.54 | 2673 | f1.5 |
| 8–16 (natural aggregate) | 1.06 | 2673 | f1.5 |
| 4–16 (recycled aggregate) | 4.48 | 2498 | f4 |

**Table 4.** Mechanical properties of concrete

<table>
<tr><th>Concrete mix design</th><th>$R_{c,28d}$ (MPa)</th><th>$R_{c,28d\ average}$ (MPa)</th><th>fct, (MPa)</th><th>$f_{ct,average}$ (MPa)</th><th>$E_c$ (MPa)</th></tr>
<tr><td rowspan="3">RC0%</td><td>37.5</td><td rowspan="3">37.87</td><td>3.55</td><td rowspan="3">3.44</td><td rowspan="3">28640</td></tr>
<tr><td>38.2</td><td>3.41</td></tr>
<tr><td>37.9</td><td>3.35</td></tr>
<tr><td>RC30%</td><td>37.1</td><td>38.10</td><td>3.53</td><td>3.48</td><td>28814</td></tr>
</table>

(*continued*)

**Table 4.** (*continued*)

| Concrete mix design | $R_{c,28d}$ (MPa) | $R_{c,28d\ average}$ (MPa) | fct, (MPa) | $f_{ct,average}$ (MPa) | $E_c$ (MPa) |
|---|---|---|---|---|---|
| | 39.1 | | 3.49 | | |
| | 38.1 | | 3.43 | | |
| RC50% | 37.8 | 34.97 | 3.30 | 3.13 | 24678 |
| | 33.7 | | 2.96 | | |
| | 33.4 | | 3.13 | | |

## 2.1 Testing Methods

A horizontal load was applied at the top of the reinforced concrete column connected to the plinth (Fig. 2). The horizontal load is increased with a constant displacement rate of 0.10 mm/s until the precast plinth-column system reaches the collapse mechanism.

This loading system can apply a horizontal force up to 500 kN, with a maximum displacement at the top of the column equal to 200 mm.

The plinths constraints system is developed by two steel beams denoted by the red squares in Fig. 3. These two beams prevent the plinth from rigid motion due to the load system.

A set of displacement transducers have been applied to the column and to the plinths and are pointed by the blue arrows in Fig. 3.

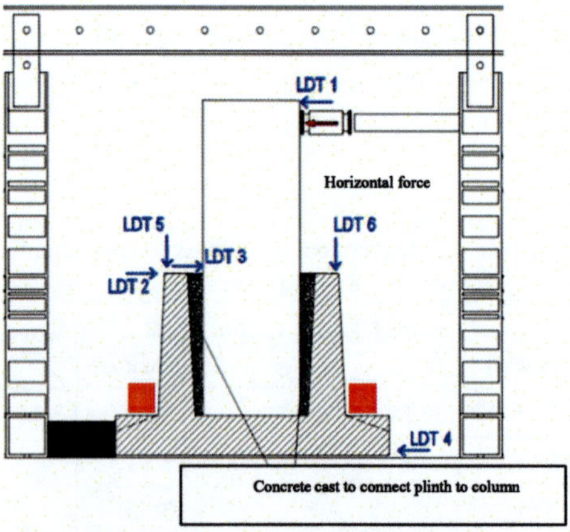


**Fig. 3.** Test set-up/Schema del test.

## 3 Results

With reference to LDT1 located on the top section of the column, see Fig. 3, it is possible to represent the load – displacement diagram for each specimen, see Fig. 4:

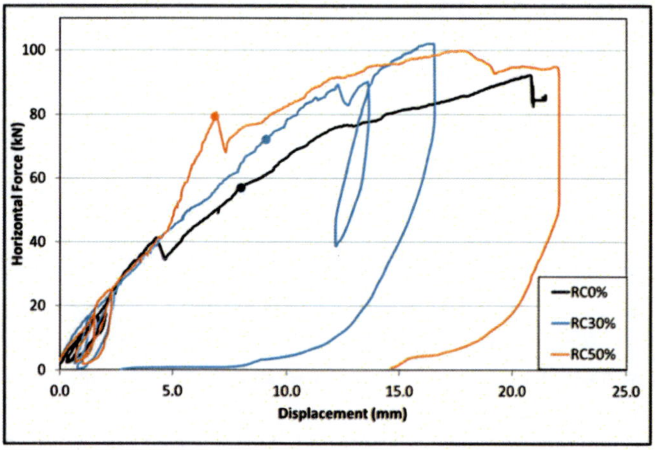


**Fig. 4.** Load – Displacement curves.

In addition, looking at the results of LDT6 and LDT5 located on the top of the plinths it is possible to calculate the plinth rotation, as shown in Fig. 5. The load-rotation curves are presented in Fig. 6.

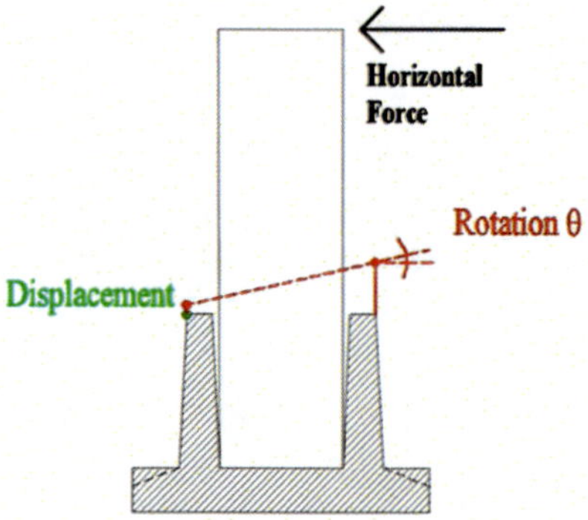


**Fig. 5.** Rotation calculation

The rounded spots present in Figs. 4 and 6 denote the first cracking for each specimen.

The collapse mechanism is similar for each specimen: a single crack at the base of the plinth was formed, highlighting that this is the weak point of the structures. A fragile shear collapse is obtained, as shown by Fig. 7. Table 5 presents the collapse forces of

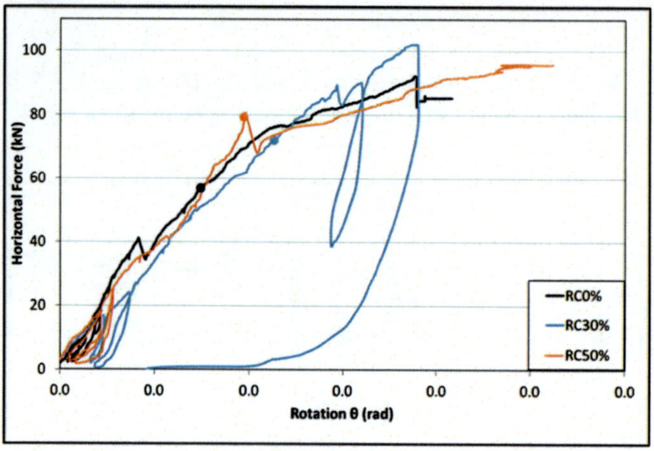


**Fig. 6.** Load Rotation curves.

each specimen. The inclination of the crack and the lack of shear reinforcements confirm the identification of a shear failure.

**Table 5.** Capacity of each specimen

| Specimen | Collapse Force (kN) |
|---|---|
| Plinth RC0%-Column RC0% | 98 |
| Plinth RC30%-Column RC0% | 102 |
| Plinth RC50%-Column RC0% | 100 |

RC 0%

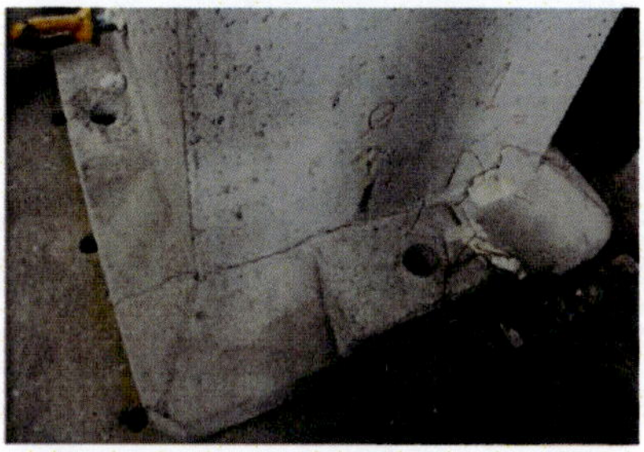

RC 30%

RC 50%

**Fig. 7.** Collapse Mechanism for different mixes

## 4 Discussion and Conclusions

In this paper the structural performance of precasted plinths have been tested. Three specimens obtained from 3 different concrete mixes having different RCA replacement percentage have been compared under the load test presented in Sect. 2.

Looking at the results shown in Sect. 3 it is clear that the small differences in the concrete mixes mechanical properties (see Table 3) correspond to similar structural performances of each specimen (see Fig. 4, Fig. 6 and Table 4). Indeed, the capacity of each plinth is almost the same with a very similar collapse load. In addition, the collapse mechanism and the first cracking load are very similar for the three specimens (see Fig. 7) too. Thus, tests prove that it is possible to obtain a structural concrete plinth using RCA and how its structural behaviour is almost similar to the one obtained by a plinth casted with natural aggregates. This can be explained taking into account the structural role of the reinforcement bars and of the plinths geometry.

The potential applications of including RCA in the production process of precast reinforced concrete elements can be seen as a good example of circular economy. There are various environmental and economic positive aspects coming from this action.

Indeed, the use of RCA limits the exploitation of non-renewable natural resources, decreases the concrete debris disposed in landfills or the concrete production scraps. It also reduces production and landfilling costs, due to the lower price of recycled aggregates if compared to natural ones.

Further developments are expected considering other structural elements or different replacement percentage.

**Acknowledgements.** Authors would like to acknowledge Sardegna Ricerche for the financial support of project: Materials for Sustainable Building and Infrastructure - Recycled Aggregates (MEISAR). POR FESR 2014/2020 - ASSE PRIORITARIO I "RICERCA SCIENTIFICA, SVILUPPO TECNOLOGICO E INNOVAZIONE.

## References

DM 11/10/17. Ministero dell'ambiente e della tutela del territorio e del mare 2017, Minimum environmental criteria for the award of design services and works for the new construction, renovation and maintenance of public buildings. In Italian

Francesconi, L., Pani, L., Stochino, F.: Punching shear strength of reinforced re-cycled concrete slabs. Constr. Build. Mater. **127**, 248–263 (2016)

González-Fonteboa, B., Martínez-Abella, F.: Concretes with aggregates from demolition waste and silica fume. Materials and mechanical properties. Build. Environ. **43**(4), 429–437 (2008)

Kovler, K., Roussel, N.: Properties of fresh and hardened concrete. Cem. Concr. Res. **41**, 775–792 (2011)

Pepe, M., Toledo Filho, R.D., Koenders, E.A., Martinelli, E.: Alternative processing procedures for recycled aggregates in structural concrete. Constr. Build. Mater. **69**, 124–132 (2014)

Pani, L., Francesconi, L., Rombi, J., Mistretta, F., Sassu, M., Stochino, F.: Effect of parent concrete on the performance of recycled aggregate concrete. Sustainability **12**(22), art. Num 9399, 1–17 (2020)

Pacheco, J., de Brito, J., Chastre, C., Evangelista, L.: Experimental investigation on the variability of the main mechanical properties of concrete produced with coarse recycled concrete aggregates. Constr. Build. Mater. **201**, 110–120 (2019)

Stochino, F., Pani, L., Francesconi, L., Mistretta, F.: Cracking of reinforced recycled concrete slabs. InT. J. Struct. Glass Adv. Mater. Res. **1**(1), 3–9 (2017)

Tabsh, S.W., Abdelfatah, A.S.: Influence of recycled concrete aggregates on strength properties of concrete. Constr. Build. Mater. **23**, 1163–1167 (2009)

UNI EN 12620: 2008 Aggregates for concrete

UNI 8520-1: 2015 Aggregates for concrete. Additional provisions for the application of EN 12620 Part 1: Designation and conformity criteria

UNI EN 12390-3: 2019, Testing hardened concrete. Part 3: Compressive strength of test specimens

UNI EN 12390-6: 2010 Tests on hardened concrete - Part 6: Splitting tensile strength of the specimens

UNI EN 12390-13: 2013, Test on hardened concrete - Part 13: Determination of the secant modulus of elasticity in com-pression

Xu, J.J., Chen, Z.P., Zhao, X.Y., Demartino, C., Ozbakkaloglu, T., Xue, J.Y.: SEIS-MIC performance of circular recycled aggregate concrete-filled steel tubular columns: FEM modelling and sensitivity analysis. Thin Walled Struct. **141**, 509–525 (2019)

Xu, J.J., Chen, Z.P., Ozbakkaloglu, T., Zhao, X.-Y., Demartino, C.: A critical assessment of the compressive behavior of reinforced recycled aggregate concrete columns. Eng. Struct. **161**, 161–175 (2018)

Xu, J.J., Chen, Z.P., Xiao, Y., Demartino, C., Wang, J.H.: Recycled aggregate concrete in FRP-confined columns: a review of experimental results. Compos. Struct. **174**, 277–291 (2017)

# Experimental and Analytical Study of a Welded Steel Wire Mesh for Anti-Falling Debris Protection System

Beatrice Belletti[1](✉), Simone Ravasini[1], Alice Sirico[1], Lorenzo Franceschini[1], Andrea Vitali[2], and Paolo De Berardinis[2]

[1] Department of Engineering and Architecture, University of Parma, Parma, Italy
beatrice.belletti@unipr.it
[2] Metallurgica Abruzzese Spa (Cavatorta Group), Teramo, Italy

**Abstract.** Prevention and safety of protection systems against debris falling from ceilings are gathering increasing interest in the engineering community. The design of an adequate protection system can prevent severe damage and human losses. In this paper, the use of welded steel wire meshes as system against debris falling from ceiling, in retrofitting and repairing interventions, is investigated. The behaviour of galvanized steel wire meshes as system against debris falling is experimentally and analytically analysed in the case of diaphragms realised with concrete joists and interposed bricks, which represent a typical solution for existing reinforced concrete or masonry buildings, for both public and residential use. Firstly, the main outcomes of an experimental test carried out at the University of Parma are described. Secondly, a simplified analytical method is presented to evaluate the resistance of such a ceiling protection system subjected to uniformly distributed load. Finally, the suitability of the analytical method for design and verification purposes is demonstrated.

**Keywords:** prevention and safety · welded steel wire mesh · anti-falling debris protection system · experimental campaign · simplified analytical method

## 1 Introduction

In Italy, floor diaphragms of existing reinforced concrete (RC) or masonry structures are often realised by adopting RC concrete joists with interposed hollow clay block elements.

The issue of debris falling from ceilings in the case of reinforced concrete-brick floors has been identified as one of the main risks for damage to contents and human safety in civil buildings, both in private and public ones, such as schools and offices (Belletti et al. 2018). The debris falling is associated with the detachment of portions of bricks or plaster in existing ceilings. This phenomenon is related to inadequate detailing, poor workmanship, excessive applied loads, poor quality of materials, and incorrect maintenance. In addition, the degradation processes, related to long-term effects of concrete and corrosion of steel rebars, can lead to structural damage and debris falling with


M. A. Aiello and A. Bilotta (Eds.): ICC 2022, LNCE 435, pp. 14–28, 2024.
https://doi.org/10.1007/978-3-031-43102-9_2

higher probability of injuries and human losses. In the recent two decades, increasing interest in the scientific community has arisen in the use of galvanized welded steel wire mesh as protection systems.

The studies available in the literature aim to the evaluation of the response of welded wire meshes for surface control in underground coal mines (Dolinar 1999; Kang et al. 2022; Morton et al. 2007; Player et al. 2007; Sengani, 2021; Shan et al., 2014). From the experimental tests, it was found that the most relevant parameters governing the wire mesh resistance are: (i) the dimension of wire diameters, (ii) the shape and position of the loading plate, (ii) the spacing of fixing elements and (iv) the orientation and the shape of the welded mesh. Several numerical techniques have been also used to estimate the response of the welded wire mesh (Baek et al. 2020; Gadde et al. 2006; Karampinos et al. 2018; Wang et al. 2021; Xu et al. 2019), which can be a powerful tool for the evaluation of test results as well as in establishing useful design parameters (Dolinar 2009). However, no studies are available in the scientific literature on the assessment of the performance of welded wire meshes aimed to anti-falling debris protection of ceilings.

This paper represents a further development of a previous work (Franceschini et al. 2021) and aims to assess the performance of a galvanized welded steel wire mesh under a uniformly distributed load caused by debris falling of brick elements. For this purpose, an ad-hoc experimental test was carried out at the University of Parma, Italy. In addition, a simplified analytical method is proposed and described to estimate the load-deflection response of the system and compared against experimental findings.

## 2 Experimental Test Setup

### 2.1 Geometry of the Tested System

The system is composed by a steel wire mesh fixed to the intrados of joists by means of steel washers and dowels, as shown in Fig. 1. Moreover, circular washers made of SBR (Styrene Butadiene Rubber) are introduced to prevent premature failures caused by the rupture of welds in the region of dowels.

The steel wire mesh is composed of welded longitudinal and transversal wires with different mechanical properties, as reported in the following paragraph. Both transversal and longitudinal spans (labelled as 1 and 2, respectively) are related to the spacing of dowels fixing the wire mesh. In this case, the span lengths in direction 1 and 2 are equal to 450 and 500 mm, respectively. Due to steel washer's diameter and wire's spacing the number of wires in transversal and longitudinal directions (labelled as "n") placed under the circular washers made of SBR results equal to 3 in this case. These wires in the following are indicated as "fixed wires".

### 2.2 Mechanical Properties and Loading Procedure

The mechanical properties of the steel wires along both transversal (1) and longitudinal (2) directions were derived from tensile tests and are reported in Table 1: the diameter, the yield and maximum strengths as well as the elastic modulus and ultimate strain are

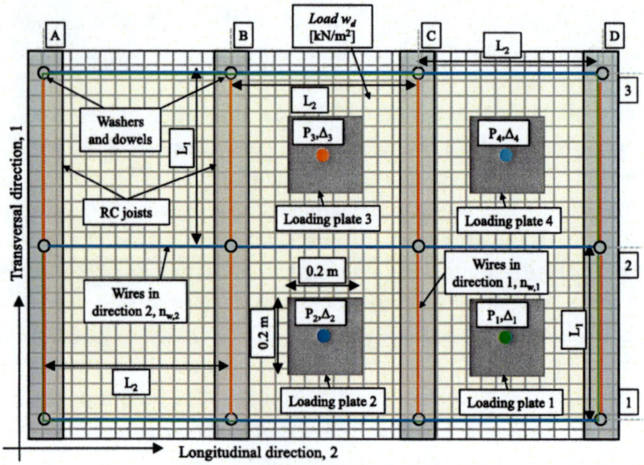


**Fig. 1.** Experimental test setup.

**Table 1.** Mechanical properties of steel wires.

| Wires' direction | $f_y$ (MPa) | $f_u$ (MPa) | $E_s$ (GPa) | $\varepsilon_u$ (%) | $\phi$ (mm) |
|---|---|---|---|---|---|
| Transversal - 1 | 763.6 | 791.5 | 203 | 1.01 | 1.75 |
| Longitudinal - 2 | 489.8 | 543.9 | 196 | 7.38 | 1.70 |

shown. Such parameters and the mentioned geometrical features are fundamental to characterize the system's capacity.

The response of the system in terms of load P vs displacement at central span, $\Delta$, for the different loading points (LP) labelled from 1 to 4, as shown in Fig. 1, were recorded. A quasi-static test procedure was adopted by applying a displacement-controlled action through hydraulic actuators placed at the four loading points shown in Fig. 2.

# 3 Simplified Analytical Method

## 3.1 Main Hypotheses

The schematic representation of the geometry of the system is reported in Fig. 3 together with the assumed tributary areas subjected to a distributed load corresponding to the design load combination (demand) $w_d$ [kN/m$^2$]. It is worth mentioning that the resulting triangular-shaped distributed load applied to the wire is equivalent to two concentrated loads, named $P_{d,i}/2$, where $P_{d,i}$ is calculated according to Eq. (1). In this case, for wires in directions 1 and 2:

$$P_{d,1} = P_{d,2} = 2\left(w_d \frac{L_1 L_2}{2}\frac{1}{2}\right) = w_d \frac{L_1 L_2}{2} \tag{1}$$

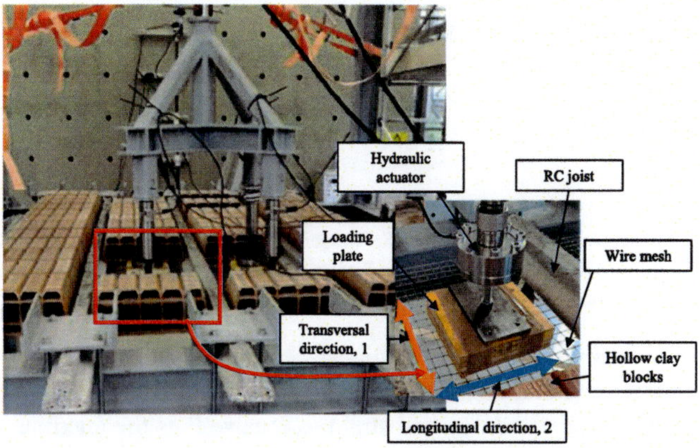


**Fig. 2.** Loading plates, hydraulic actuators and wire mesh.

where two times of the tributary area acting on the central wires is considered (refer to wires BC-2, CD-2 and C-12, C-23 in Fig. 3).

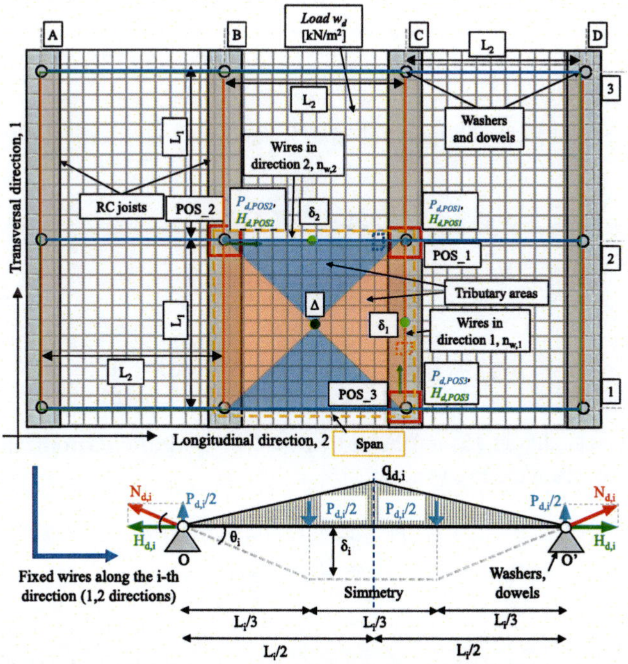


**Fig. 3.** Geometry of the system and n-fixed wires scheme along the i-th direction.

In this work, the load components of the n-fixed wires along the i-th direction ($N_i$, $H_i$, $P_i$) associated to demand are reported with the subscript "d", which is omitted when related to the capacity of the wires or the entire system.

### 3.2 Capacity of Wires

Concerning the capacity of n-fixed wires along the i-th direction, the equilibrium at the deformed state as well as the curves related to the load components from the equilibrium condition are reported in Fig. 4. The following assumptions are adopted:

1. Both mechanical and geometrical nonlinearities are considered.
2. A bilinear stress-strain relationship is assumed for n-fixed wires along 1 and 2 directions. The steel yield stress, $f_{y,i}$, and strain, $\varepsilon_{y,i}$, as well as the maximum stress, $f_{u,i}$, and the ultimate strain, $\varepsilon_{u,i}$, are considered. The elastic and hardening moduli are indicated as $E_i$ and $E_{h,i}$, respectively.
3. The wire's deformation is localized at the inclined part ($L_i/3$) and the elongation of the straight part ($L_i/6$) is neglected. The vertical displacement between two adjacent dowels of n-fixed wires, $\delta_i$, is function of the strain achieved in the wires, $\varepsilon_i$.
4. The load P – displacement δ response of the n-fixed wires is represented by a curve composed by two branches: (i) from O to A corresponding to the elastic response till yielding, (ii) from A to B corresponding to the wires' plastic behaviour till fracture. Note that the shape of the branches is obtained by discretizing the following equations into several points.

From the equilibrium configuration at deformed state, the vertical displacement is given by Eq. (2):

$$\delta_i = \frac{L_i}{3}\sqrt{\varepsilon_i^2 + 2\varepsilon_i} \tag{2}$$

The rotation of the wires, $\theta_i$, is given by Eq. (3):

$$\theta_i = arctg\left(\frac{\delta_i}{L_i/3}\right) = arctg\left(\sqrt{\varepsilon_i^2 + 2\varepsilon_i}\right) \tag{3}$$

The horizonal reaction, $H_i$, and the vertical load, $P_i$, can be found from the equilibrium condition by adopting the Eq. (4):

$$\begin{cases} H_i = N_i cos(\theta_i) \\ P_i = 2N_i sen(\theta_i) \end{cases} \tag{4}$$

where $N_i$ is the n-fixed wires' axial load, expressed as the product between the axial stress, $\sigma_i$, and the "n" wires' cross-sectional area, $A_i$, along the i-th direction.

The expressions of the axial load $N_i$ at elastic and plastic stages are reported in the Eq. (5):

$$\begin{cases} N_i = E_i A_i \varepsilon_i; & 0 \le \varepsilon_i \le \varepsilon_{y,i} \\ N_i = A_i\left[f_{y,i} + E_{h,i}\left(\varepsilon_i - \varepsilon_{y,i}\right)\right]; & \varepsilon_{y,i} < \varepsilon_i \le \varepsilon_{u,i} \end{cases} \tag{5}$$

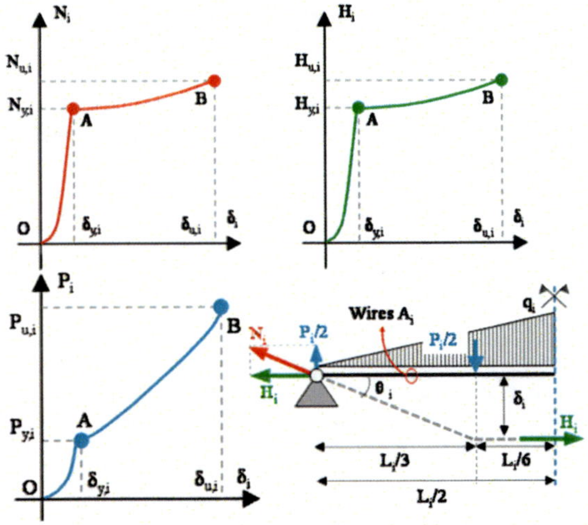


Fig. 4. Equilibrium condition at deformed state.

The previous expressions can be further simplified by highlighting the dependency from the wires' axial strain, $\varepsilon_i$, through the trigonometric expressions reported in the Eq. (6):

$$\begin{cases} cos\left[arctg\left(\sqrt{\varepsilon_i^2 + 2\varepsilon_i}\right)\right] = \frac{1}{1+\varepsilon_i} \\ sen\left[arctg\left(\sqrt{\varepsilon_i^2 + 2\varepsilon_i}\right)\right] = \frac{\sqrt{\varepsilon_i^2+2\varepsilon_i}}{1+\varepsilon_i} \end{cases} \tag{6}$$

Such simplified expressions are reported in the following paragraphs for both elastic and plastic stages.

### 3.2.1 Wires' Capacity at the Elastic Stage

At small deformation stage, the response of the system is governed by the elastic axial stiffness of the wires. The previous equations may be used to express the horizontal reaction and vertical load in correspondence of the achievement of yield strain (point A in Fig. 4) through the Eq. (7):

$$\begin{cases} H_{y,i} = \left(E_i A_i \varepsilon_{y,i}\right)\frac{1}{1+\varepsilon_{y,i}} \\ P_{y,i} = 2\left(E_i A_i \varepsilon_{y,i}\right)\frac{\sqrt{\varepsilon_{y,i}^2+2\varepsilon_{y,i}}}{1+\varepsilon_{y,i}} \end{cases} \tag{7}$$

The corresponding vertical displacement, $\delta_{y,i}$, can be obtained by substituting the yield strain, $\varepsilon_{y,i}$, in the Eq. (2). The expressions for $P_i$ in the Eq. (4) can be used to discretize the concavity of the curve from points O to A.

### 3.2.2 Wires' Capacity at the Inelastic Stage

At large deformation stage, the response of the system is governed by the hardening axial stiffness of the wires. The horizontal reaction and vertical load in correspondence of the achievement of ultimate strain (point B in Fig. 4) can be obtained by using the Eq. (8):

$$\begin{cases} H_{u,i} = A_i\left[f_{y,i} + E_{h,i}\left(\varepsilon_{u,i} - \varepsilon_{y,i}\right)\right]\frac{1}{1+\varepsilon_{u,i}} \\ P_{u,i} = 2A_i\left[f_{y,i} + E_{h,i}\left(\varepsilon_{u,i} - \varepsilon_{y,i}\right)\right]\frac{\sqrt{\varepsilon_{u,i}^2+2\varepsilon_{u,i}}}{1+\varepsilon_{u,i}} \end{cases} \quad (8)$$

The corresponding vertical displacement, $\delta_{u,i}$, can be obtained by substituting the yield strain, $\varepsilon_{u,i}$, in the Eq. (2). Finally, the expressions for $H_i$ and $P_i$ in the Eq. (4) can be used to discretize the concavity of the curve from A to B. The displacements at point A and B are calculated by using the Eq. (9):

$$\begin{cases} \delta_{A,i} = \delta_{y,i} = \frac{L_i}{3}\sqrt{\varepsilon_{y,i}^2 + 2\varepsilon_{y,i}} \\ \delta_{B,i} = \delta_{u,i} = \frac{L_i}{3}\sqrt{\varepsilon_{u,i}^2 + 2\varepsilon_{u,i}} \end{cases} \quad (9)$$

## 3.3 Exact Solution of the Wires' Behaviour Subjected to the Demand $w_d$

In this paragraph, the procedure to obtain the vertical displacement of n-fixed wires along the i-th direction ($\delta_{d,i}$) subjected to the load combination $w_d$ is described. Both elastic and inelastic stages may be encountered depending on the system demand, as shown in Fig. 5. Note that the corresponding load $P_{d,i}$ derived from the demand $w_d$ is calculated by adopting the Eq. (1).

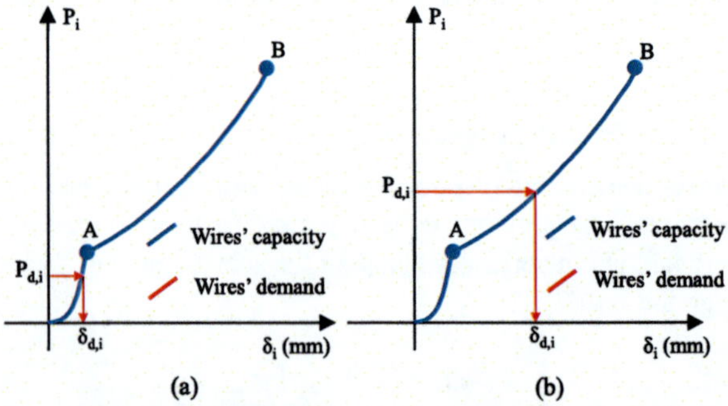


**Fig. 5.** Displacement demand at (a) elastic and (b) inelastic stages.

### 3.3.1 Demand at the Elastic Stage

The calculation of the displacement demand of wires at the elastic stage is reported herein. The solving expressions are reported in the Eq. (10) based on equilibrium and stress-strain relations:

$$\begin{cases} N_{d,i}sen(\theta_{d,i}) = \frac{P_{d,i}}{2} \text{ (equilibrium)} \\ H_{d,i}tg(\theta_{d,i}) = \frac{P_{d,i}}{2} \text{ (equilibrium)} \\ N_{d,i}cos(\theta_{d,i}) = H_{d,i} \text{ (equilibrium)} \\ N_{d,i} = \varepsilon_{d,i}\frac{N_{y,i}}{\varepsilon_{y,i}} \text{ (elastic } \sigma - \varepsilon) \\ H_{d,i} = \varepsilon_{d,i}\frac{H_{y,i}}{\varepsilon_{y,i}} \text{(elastic } \sigma - \varepsilon) \end{cases} \tag{10}$$

By substituting the Eq. (6) in the second expression of the Eq. (10), the Eq. (11) can be derived:

$$H_{d,i}tg(\theta_{d,i}) = \varepsilon_{d,i}\frac{H_{y,i}}{\varepsilon_{y,i}}\sqrt{\varepsilon_{d,i}^2 + 2\varepsilon_{d,i}} = \frac{P_{d,i}}{2} \tag{11}$$

Rearranging the terms, the solving expression is reported in the Eq. (12):

$$\varepsilon_{d,i}^4 + 2\varepsilon_{d,i}^3 + k_{d,i} = 0 \tag{12}$$

Such relationship is a fourth-order equation $ax^4 + bx^3 + cx^2 + dx + e = 0$. Thus, given the unknown variable equal to the wire strain, $\varepsilon_{d,i}$, the coefficients to solve the Eq. (12) are reported as follows:

$$\begin{cases} a = 1 \\ b = 2 \\ c = d = 0 \\ e = k_{d,i} = -\frac{P_{d,i}^2\varepsilon_{y,i}^2}{4H_{y,i}^2} \end{cases} \tag{13}$$

By solving the previous expression, the wire strain, $\varepsilon_{d,i}$, can be obtained and the corresponding vertical displacement of fixed wires, $\delta_{d,i}$, can be derived from the Eq. (2).

### 3.3.2 Demand at the Inelastic Stage

The calculation of the displacement demand of wires at the inelastic stage is reported herein. The solving expressions are reported in the Eq. (14) based on equilibrium and stress-strain relations:

$$\begin{cases} N_{d,i}sen(\theta_{d,i}) = \frac{P_{d,i}}{2} \text{ (equilibrium)} \\ H_{d,i}tg(\theta_{d,i}) = \frac{P_{d,i}}{2} \text{ (equilibrium)} \\ N_{d,i}cos(\theta_{d,i}) = H_{d,i} \text{ (equilibrium)} \\ N_{d,i} = N_{y,i} + A_iE_{h,i}(\varepsilon_{d,i} - \varepsilon_{y,i}) \text{ (plastic } \sigma - \varepsilon) \\ H_{d,i} = H_{y,i} + k_{H,i}(\varepsilon_{d,i} - \varepsilon_{y,i}) \text{ (plastic } \sigma - \varepsilon) \end{cases} \tag{14}$$

Similarly to the previous case, the solving expression is reported in the Eq. (15):

$$k_{H,i}^2\varepsilon_{d,i}^4 + 2\left(k_{d,i,2} + k_{H,i}^2\right)\varepsilon_{d,i}^3 + \left(k_{d,i,1} + 4k_{d,i,2}\right)\varepsilon_{d,i}^2 + 2k_{d,i,1}\varepsilon_{d,i} + k_{d,i,3} = 0 \tag{15}$$

where the following constants are used:

$$\begin{cases} k_{d,i,1} = H_{y,i}^2 + k_{H,i}^2 \varepsilon_{y,i}^2 - 2H_{y,i} k_{H,i} \varepsilon_{y,i} \\ k_{d,i,2} = H_{y,i} k_{H,i} - k_{H,i}^2 \varepsilon_{y,i} \\ k_{d,i,3} = -\frac{P_{d,i}^2}{4} \end{cases} \quad (16)$$

Therefore, the coefficients to solve the fourth-order equation are reported in the Eq. (17):

$$\begin{cases} a = k_{H,i}^2 \\ b = 2\left(k_{d,i,2} + k_{H,i}^2\right) \\ c = \left(k_{d,i,1} + 4k_{d,i,2}\right) \\ d = 2k_{d,i,1} \\ e = k_{d,i,3} = -\frac{P_{d,i}^2}{4} \end{cases} \quad (17)$$

### 3.4 Resistance of the System

#### 3.4.1 Extension of the Method to Wire Mesh

The expressions reported in the previous paragraphs may be used for the analysis of an entire wire mesh system by adopting the following assumptions:

1. The fixed wires with lower ductility govern the resistance of the system, which is attributed to the achievement of the ultimate strain corresponding to the first wires' fracture.
2. The central span deflection of the system, Δ, is calculated from the displacements in correspondence of the fixed wires in the two orthogonal directions, labelled as $\delta_1$ and $\delta_2$, respectively, as shown in the Fig. 3.
3. In the case of geometry spans investigated in this work, the ratio between the displacement Δ and $\delta_1$ (and $\delta_2$) was experimentally found equal to 3. Different values may be obtained for different configurations from experimental and numerical findings.
4. Only the n-fixed wires restrained along the transversal and longitudinal directions contribute to the resistance of the system.
5. All the spans of the system were assumed to be loaded with the distributed design load combination, $w_d$, (refer to Fig. 3) to be compared with the system capacity.

The total resistance of the system can be expressed as the sum of the contributions by considering the n-fixed orthogonal wires (in this case, n = 3) in Eq. (18) and shown in Fig. 6 based on the equilibrium conditions:

$$\begin{cases} P_\alpha = 2\frac{P_{\alpha,1}}{2} + 2\frac{P_{\alpha,2}}{2} = P_{\alpha,1} + P_{\alpha,2} \\ P_\beta = 2\frac{P_{\beta,1}}{2} + 2\frac{P_{\beta,2}}{2} = P_{\beta,1} + P_{\beta,2} \\ P_\gamma = 2\frac{P_{\gamma,1}}{2} + 2\frac{P_{\gamma,2}}{2} = P_{\gamma,1} + P_{\gamma,2} \end{cases} \quad (18)$$

The points α, β and γ are referred to the main events occurring in the system: (i) the first yielding of the wires along the direction 1 or 2, (ii) the second yielding of the wires along the direction 1 or 2 and (iii) the first fracture of the wires along the direction 1 or 2. Additional wires' failure sequences are conservatively omitted in this analytical method. As shown in Fig. 6, the total resistance of the system can be expressed in terms of surface distributed load, w, which is calculated by considering the tributary area with the following Eq. (19):

$$\begin{cases} w_\alpha = \frac{P_\alpha}{L_1 L_2} \\ w_\beta = \frac{P_\beta}{L_1 L_2} \\ w_\gamma = \frac{P_\gamma}{L_1 L_2} \end{cases} \tag{19}$$

It should be noted that the previous expressions are equivalent to the vertical reaction exhibited by the central dowel in the position-1 reported in Fig. 3. It can be also demonstrated that such resistance is representative of the capacity of a single span $L_1$ x $L_2$ and can be compared to the experimental test results at each loading point described in the paragraph 2. In addition, the demand $w_d$, can be intersected with the capacity curve, as shown in Fig. 6, leading to three possible scenarios:

1. If the demand $w_d$ is lower than $w_\alpha$, the system response is at the elastic state.
2. If the demand $w_d$ is between $w_\alpha$ and $w_\gamma$, the system enters the inelastic state.
3. If the demand $w_d$ is greater than $w_\gamma$, the failure of wires fixed along longitudinal or transversal directions is achieved, corresponding to the failure of the system as assumed by the model, although the experimental test results demonstrated that the steel wire mesh system can sustain higher load.

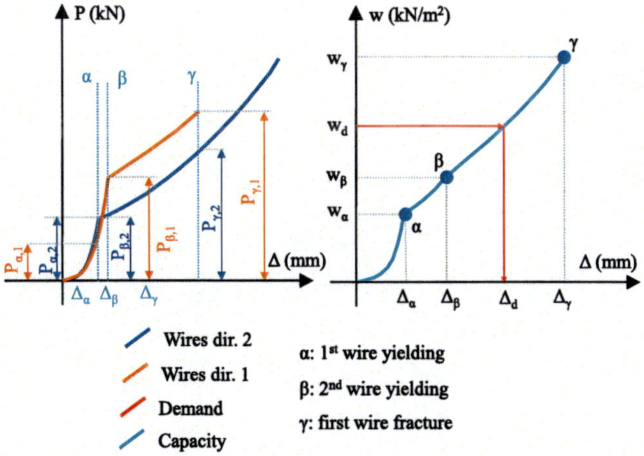


**Fig. 6.** Superimposition of effects from the two orthogonal directions.

It is worth mentioning that the exact value of the vertical displacement of the n-fixed wires attributed to the demand, $\delta_d$, could be obtained by inverting the described formulations previously described in the paragraph 3.3 both for elastic and inelastic stages. Alternatively, a linear interpolation may be adopted.

### 3.5 Predicted Actions on Dowels

In the context of the design and/or assessment of steel wire mesh, the role of washers and dowels is fundamental to ensure the load redistribution capacity of the system. For this purpose, the scheme reported in Fig. 7 illustrates the axial and shear actions on dowels (labelled as P and H, respectively) at different positions. From the equilibrium conditions, the axial loads on dowels in positions 1, 2 and 3 can be calculated by using the Eq. (20):

$$\begin{cases} P_{d,POS\_1} = P_{d,1} + P_{d,2} \\ P_{d,POS\_2} = P_{d,2} \\ P_{d,POS\_3} = P_{d,1} \end{cases} \tag{20}$$

In a similar way, the shear loads on dowels in positions 1, 2 and 3 are reported in the Eq. (21):

$$\begin{cases} H_{d,POS\_1} = 0 \\ H_{d,POS\_2} = H_{d,2} \\ H_{d,POS\_3} = H_{d,1} \end{cases} \tag{21}$$

Due to the loading symmetry of the system, the shear load on the dowel in position 1 is null, conversely to the dowels in position 2 and 3. Note that the same relations are valid to estimate the axial and shear loads on dowels for the main events related to the system capacity or to calculate the actions derived from the demand, $w_d$.

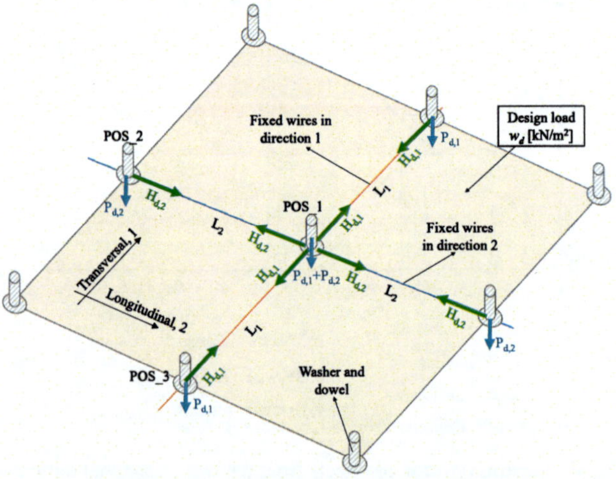


**Fig. 7.** Position and actions on washers and dowels.

# 4 Results and Discussion

## 4.1 Comparisons with Experimental Tests

In this paragraph are reported the comparisons between the analytical provisions and the experimental results in terms of load vs displacement curves from the different loading points from 1 to 4. The comparisons are expressed in terms of load P vs displacement Δ curves and are shown in Fig. 8. It can be observed that the experimental response is characterized by an initial increasing stiffness at the elastic stage followed by the yield and sequential fractures of the fixed wires in longitudinal or transversal directions. The experimental data recorded from the different loading points were similar. On average, the maximum load achieved in the experimental tests was between 5.5 and 7.2 kN in correspondence of a central span displacement between 130 and 150 mm. The values of load P (and the corresponding distributed load w) and displacements Δ in correspondence of the main events occurring in the system are reported in Table 2 both from the analytical provisions and the experimental results. By comparing the experimental load-displacement curves, the lower load and the corresponding displacement at the achievement of the wires' yielding and first fracture were used.

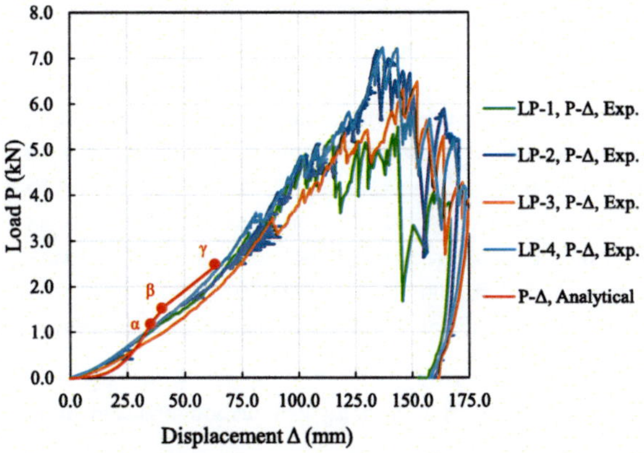


**Fig. 8.** Comparison between analytical and experimental curves.

It can be observed that the analytical model well predicts the experimental loads and the occurrence of the main events. At both yield and fracture stages, the analytical method tends to underestimate both the vertical displacement and load, which is attributed to the assumption of neglecting the elongation of the wires' straight portion mentioned in paragraph 3.1 and shown in Fig. 4.

In particular, the yield of longitudinal wires firstly occurred immediately followed by the yield of transversal wires. Finally, the fracture is associated to the achievement of the ultimate strain in the transversal wires, which show lower ductility. It is worth to mention that the analytical approach provides a conservative estimation of the system capacity, stopping at the first wires' fracture. Indeed, the actual maximum load achieved in the experimental test is associated to sequential wire fractures in both directions with

**Table 2.** Results from the analytical method and experimental test.

| Event | Analytical | | | Experimental | |
|---|---|---|---|---|---|
| | P (kN) | w (kN/m$^2$) | Δ (mm) | P (kN) | Δ (mm) |
| Yielding, α | 1.18 | 5.24 | 35.37 | 1.67 | 50.64 |
| Yielding, β | 1.47 | 6.53 | 39.07 | | |
| Fracture, γ | 2.47 | 10.98 | 64.15 | 3.17 | 77.80 |
| Failure mode | Transversal wires fracture | | | Transversal wires fracture | |

a final drop corresponding to the exhausted system resistance. Such sequential events are very difficult to be considered in the analytical model and hence, further events may be detected by performing ad-hoc nonlinear finite element analyses.

### 4.2 Check of Loads Acting on Dowels

Although the horizontal actions on dowels were not recorded during the experimental tests, the results derived from the analytical method are reported by following the expressions in paragraph 3.5 for the main events of the system's capacity curve. The axial and shear loads (labelled as P and H, respectively) on steel dowels are reported in Table 3. It can be observed that the shear loads are much greater than axial loads, which is attributed to the equilibrium conditions of the fixed wires. Indeed, the achievement of a considerable strain in wires is related to a high axial load, N, whose horizontal component is the shear load acting on dowels, H, as shown in Fig. 4. Therefore, careful design and/or verification of steel dowels must be performed to avoid the premature failure of the system due to interaction between axial and shear forces. The axial and shear load on dowels in the different positions may be calculated from the loading demand by adopting the expressions in paragraph 3.3 or, alternatively, a linear interpolation may be used.

**Table 3.** Results from the analytical method - loads on dowels.

| Dowel position | $P_\alpha$ (kN) | $H_\alpha$ (kN) | $P_\beta$ (kN) | $H_\beta$ (kN) | $P_\gamma$ (kN) | $H_\gamma$ (kN) |
|---|---|---|---|---|---|---|
| 1 | 1.18 | 0.00 | 1.47 | 0.00 | 2.47 | 0.00 |
| 2 | 0.47 | 3.33 | 0.52 | 3.33 | 0.86 | 3.34 |
| 3 | 0.71 | 4.50 | 0.95 | 5.49 | 1.61 | 5.65 |

## 5 Conclusions

In the present work, the capacity of a galvanized welded steel wire mesh in terms of P-Δ response has been analytically investigated against experimental findings. In the first part, a brief description of the experimental setup is provided. In the second section, a simplified analytical method is proposed to describe the capacity of the tested system. Finally, the comparison between experimental results and analytical prediction is pointed out.

The following conclusions can be drawn:

- The investigated system is considered suitable for in-situ application as anti-falling debris protection system.
- Good agreement between the experimental and analytical results is observed. The main events of the system are well predicted by the analytical approach till the first wires' fracture. Therefore, the proposed analytical approach tends to provide a conservative system's resistance.
- The analytical approach could be used for both design and assessment purposes. The verification of the system may be efficiently carried out by constructing the system capacity curve to be compared with the demand.
- Special attention must be paid to the design of the washers and dowels, which are demonstrated to be subjected to considerable shear actions. If not adequately designed, the rupture of the dowels may lead to the premature failure of the system.

## References

Baek, B., Karampinos, E., Hadjigeorgiou, J.: Understanding the impact of test confguration on welded-wire mesh labor test results. Rock Mech. Rock Eng. **53**(11), 4873–4892 (2020)

Belletti, B., Berardengo, M., Collini, L., Foresti, R., Garziera, R.: Design of an instrumentation for the automated damage detection in ceilings. NDT E Int. **94**(October 2017), 31–37 (2018)

CEN-EC0. EN 1990 Eurocode 0: basis of structural design. European Committee for Standardization (2002)

Dolinar, D.R.: Performance characteristics for welded wire screen used for surface control in underground coal mines. In: SME Annual Meeting and Exhibit and CMA's 111th National Western Mining Conference, vol. 2, pp. 659–666 (2009)

Dolinar, D.R.: Load capacity and stiffness characteristics of screen materials used for surface control in underground coal mines. In: 25th International Conference on Ground Control in Mining, pp. 152–158 (1999)

Franceschini, L., Sirico, A., Ravasini, S., Belletti, B., Vitali, A., De Berardinis, P.: Numerical and analytical assessment of a galvanized welded steel wire mesh : an anti-falling debris protection system for ceiling. In: Proceedings of the 2nd fib Symposium on Concrete and Concrete Structures, November 22, 2021, Sapienza University, Rome, Italy, pp. 1–8 (2021)

Gadde, M.M., Rusnak, J.A., Honse, J.W.: Behavior of welded wire mesh used for skin control in underground coal mines. In: 25th International Conference on Ground Control in Mining, pp. 142–151 (2006)

Kang, H., Yuan, G., Gao, F., Lou, J.: experimental study on the performance of different meshes under quasi-static loading. In: Rock Mechanics and Rock Engineering, vol. 55, issue 1, pp. 249–258 (2022)

Karampinos, E., Baek, S., Hadjigeorgiou, J.: Discrete element modelling of a laboratory static test on welded wire mesh. In: Caving 2018 - Proceedings of the Fourth International Symposium on Block and Sublevel Caving, vol. 1981, pp. 735–746 (2018)

Morton, E.C., Thompson, A.G., Villaescusa, E., Roth, A.: Testing and analysis of steel wire mesh for mining applications of rock surface support. ISRM Congress, Lisbon (2007)

NTC-2018. Norme Tecniche per le Costruzioni. Gazzetta Ufficiale Della Repubblica Italiana (2018)

Player, J.R., Morton, E.C., Thompson, A.G., Villaescusa, E.: Static and dynamic testing of steel wire mesh for mining applications of rock surface support. In: 6th International Symposium on Ground Support in Mining and Civil Engineering Construction, pp. 693–706 (2007)

Sengani, F.: A short note on the performance of steel-welded mesh under static loading. In: Innovative Infrastructure Solutions, Vol. 6, Issue 2 (2021)

Shan, Z., Porter, I., Nemcik, J.: Performance of full scale welded steel mesh for surface control in underground coal mines. In: 31st Annual International Pittsburgh Coal Conference: Coal - Energy, Environment and Sustainable Development, PCC 2014, pp. 1–10 (2014)

Wang, C., Wang, H., Shankar, K., Morozov, E.V., Hazell, P.J.: On the mechanical behaviour of steel wire mesh subjected to low-velocity impact. Thin-Walled Struct. **159**(November 2019), 107281 (2021)

Xu, C., Tannant, D.D., Zheng, W.: Discrete element analysis of the influence of bolt pattern and spacing on the force-displacement response of bolted steel mesh. Int. J. Numer. Anal. Meth. Geomech. **43**(12), 2106–2125 (2019)

# Carbon-Doped Eco-Earth Concretes for Sustainable Monitoring of Structures

Antonella D'Alessandro(✉), Andrea Meoni, and Filippo Ubertini

Department of Civil and Environmental Engineering, University of Perugia, Perugia, Italy
antonella.dalessandro@unipg.it

**Abstract.** The construction sector has a great impact on the environment. The growing attention to more sustainable alternatives to traditional construction materials is stimulating the development of novel eco-friendly composites. This work presents the first investigations on novel concretes realized with earth and carbon fillers. In particular, electrical and mechanical tests have been carried out in order to evaluate the conductive and smart sensing properties of the novel composites. Cube samples doped with different amounts of carbon microfibers have been tested under compression, with the final aim of investigating the characteristics of the materials and identifying the most performant mixes. The tests demonstrated the good electrical and electromechanical properties of these new eco-earth concretes which appear promising to fabricate embedded sensors or structural smart elements in sustainable structures.

**Keywords:** Carbon-based fillers · Cement-based composites · Carbon microfibers · Sustainable concrete · Structural health monitoring · Earth materials

## 1 Introduction

The field of constructions demonstrated a great progress with the development of concrete technology (Metha & Monteiro 2014). Such a composite material appears very versatile, both in terms of tailoring of composition and in the manufacturing of complex structures not achievable with other construction techniques.

Many buildings and infrastructures could be designed and formed by exploiting the physical and mechanical capabilities of such a construction material (Collepardi 2010). However, the environmental impact of the production, the use, the maintenance, and the disposal of concrete results critical (Meyer 2009; D'Alessandro et al. 2017). As a matter of fact, the concrete technologies use raw materials and a great amount of energy, producing resource waste, $CO_2$, and other greenhouse gas emissions, with a great environmental footprint (Flower & Sanjayan 2007; Purnel 2012).

In this work, for the enhancement of concrete sustainability, a novel cementitious construction material, made of the addition of clay and carbon fillers, is proposed, and investigated. This eco-friendlier composite can decrease the environmental impact of constructions and provide multifunctional properties of self-sensing.


M. A. Aiello and A. Bilotta (Eds.): ICC 2022, LNCE 435, pp. 29–39, 2024.
https://doi.org/10.1007/978-3-031-43102-9_3

Once properly characterized, also in terms of mechanical properties, the smart-earth can be adopted to construct structures and structural elements capable of monitoring their state of deformation and performance, thus aiding the identification of possible critical states and losses of structural integrity.

The optimized maintenance during the service life could enhance the durability of the elements, thus further increasing their sustainability. In this sense, further investigations at laboratory and real scale are needed. The presented work is part of a research project on the development and characterization of an eco-friendly earth-based material (Curto et al. 2020; D'Alessandro et al. 2020a).

## 2 State-of-Art

Nowadays the sensitivity to environmental issues is increasing in several fields of science and technology. The sustainability represents a good choice for limiting the use of raw resources, waste, and air pollution, soil, and water (Pacheco-Torgal & Jalali 2012). In engineering, many researches are developing for producing eco-friendly materials and devices, with less energy consumption and promoting the use of recycled materials (Tam et al. 2018; Zamora-Castro et al. 2021). In particular, the cement industry in the recent years has started to experiment new strategies for reducing its environmental footprint (D'Alessandro et al. 2020b). Possible solutions are in the development of less energy-consuming production processes, the use of recycled aggregates, the addition of additives that enhance the performance, and the adoption of natural binders. Among possible natural materials, earth is an interesting alternative. It is one of the first construction materials, utilized both in dried and burned forms. The use of clay and bricks is now increasing in civil engineering due to their sustainable nature. If used in dried state, clay could be taken from recycled earth from excavation during construction works (Guettala et al. 2006; Xiao et al.2005). The presence of earth modifies the resistance properties of the composite, so that fiber additions or tailored production systems are needed for guaranteeing a suitable performance (Birgin et al. 2021; Van Damme & Houben 2017). In particular, the use of carbon fillers could confer multifunctional capabilities to the construction material, as demonstrated in various cement-matrix composites (Azhari & Banthia 2012; Meoni et al. 2018; Metaxa et al. 2009). Among possible enhanced properties, the self-sensing ability could be provided to structural elements, thus originating structures able to monitor their state of strain and performance in their service life (Han & Ou 2015; Meoni et al. 2021).

## 3 Aim of the Research

In the light of developing a sustainable cementitious construction material, this research is aimed at developing and characterizing an earth-cementitious composite with carbon microfibers. For this purpose, various types of composites, with different levels of carbon inclusions have been produced, for evaluating their physical and sensing properties. Promising mixes were then in depth investigated. Cubic samples with copper electrical setups were tested. This research represents the first attempt to develop such novel fiber-reinforced composite for evaluating its feasibility and applicability in civil engineering.

Future steps of the research will concern the production and performance analysis of higher-scale elements.

# 4 Materials and Devices

## 4.1 Matrix Material and Preparation Procedure

The material investigated in this paper was formed by cement, dry earth, carbon microfibers, and water. The cement was Portland type 42.5R, fibers were SIGRAFIL®, provided by SGL Carbon, with a diameter of 5 μm and length of 6 mm. The earth is pure clay taken directly from a brick factory placed in Bevagna (Italy). Table 1 reports the mix designs for a cubic meter of normal composite and smart-earth, with various additions of carbon microfibers. The table also shows the water/binder ratio of the mixes, where the binder represents the sum of cement and earth. The clay blocks were dried and mechanically pulverized up to a maximum dimension of 2 mm. Smart-earth is a cementitious construction material in which more than half of the weight of the binder (cement + earth) consists of earth. Consequently, it can be produced by using a high rate of excavated soil in field, thus increasing the sustainability and circularity of the construction market, also by reducing the logistics and the supply of cement. Possessing sensing capabilities, this new smart material also eliminates the need to install external sensors on structures and structural elements, with obvious benefits in terms of sustainability and safety. All the components were manually mixed until a homogeneous dough was obtained (Fig. 1a). Then, the composite was poured into oiled steel molds, cubic samples were formed, and two copper wires of 0.8 mm of diameter were embedded linearly in the central part, as electrodes (Fig. 1b).

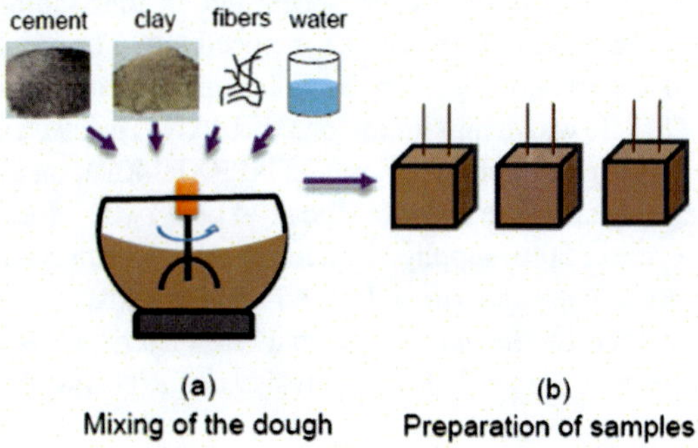


**Fig. 1.** Preparation procedure of smart-earth samples: (a) mechanical mixing of the components, (b) formation of the samples.

**Table 1.** Mix designs (kg/$m^3$) of novel composites produced with matrices made of earth and cement, and smart-earth composites with addition of increasing amounts of carbon fillers.

| Components | Normal | Carbon-added |
|---|---|---|
| | kg | kg |
| Earth | 560 | 560 |
| Cement | 240 | 240 |
| Sand | 643 | 642 |
| Water | 400 | 400 |
| CMF | - | 0.08-8 |
| Water/Binder | 0.5 | 0.5 |

### 4.2 Samples and Procedures

The samples were kept in a controlled environment for some days, hence unmolded and cured at laboratory conditions. The carbon microfibers were added at increasing amounts, with respect to the weight of the binder (cement + earth): 0.01, 0.025, 0.05, 0.1, and 0.25. The samples analyzed in this research were cubes with sides of 5 cm. Figure 2 shows an example of a series of samples investigated in this work, where are also visible the aligned copper electrodes. The cubic samples were first analyzed without applying loads, for evaluating the conductive properties, and then under cyclical compressive loads in the direction of the electrodes, to investigate the sensitivity of the different mixes and identify the most promising ones. Electrical tests were performed on six samples produced with a filler content of 0.00%, 0.01, 0.025, 0.05, 0.1, and 0.25, respectively. Compression tests were performed on four samples produced with a filler content of 0.00%, 0.025, 0.05, and 0.1, respectively. The mechanical tests were performed by use of a servo-controlled pneumatic universal dynamic testing machine type IPC Global UTM14P, with a maximum load of 14 kN. The electrical measurements were carried out with a digital multimeter, model NI PXIe-4071, on a chassis, model NI PXIe-1073. A voltage square wave of 3V (6V peak-to peak) with a frequency of 1 Hz was applied with a duty cycle of 50%, supplied by a function generator type RIGOL DG1022. The current was recorded at approximately 80% of the positive wave, as described in previous works developed by the authors on concrete materials (Downey et al. 2017). From the acquired current, the total electrical resistance R(t) was obtained through the Ohm's law:

$$R(t) = V/(I(t)) \tag{1}$$

where V is the applied constant voltage in the positive part of the square wave input signal, and I is the current intensity value sampled at the time instant t. The electrical properties of the composites strongly depend on the homogeneity of the conductive carbon fillers. For investigating the quality of the fibers' dispersions, micrographs by use of an optical microscope (Fig. 3a) and a Scanning Electron Microscope (Fig. 3b) were carried out. Figure 3 reports the inspections carried out on a fragment of smart-earth with 0.05% of carbon microfibers. The picture demonstrates that the filler is clearly visible, not damaged, and quite dispersed into the matrix.

**Fig. 2.** Cubic samples of smart-earth produced with addition of carbon microfibers with embedded copper wires.

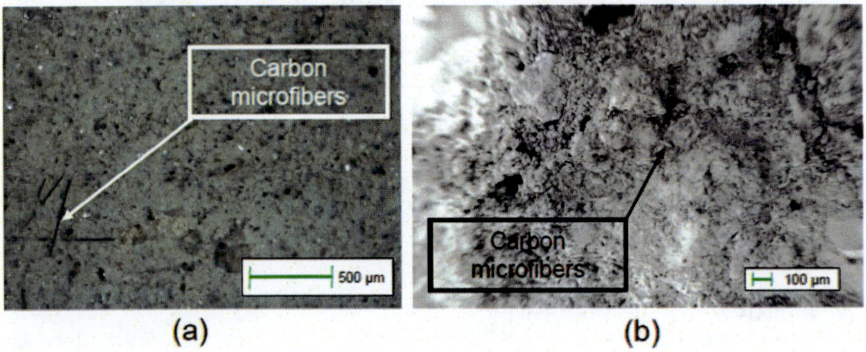


**Fig. 3.** Micrographs of concrete with carbon microfibers and graphite using (a) an optical microscope; (b) SEM.

## 5 Experimental Methodology

The experimental tests are aiming at investigating the electrical and sensitive properties of the novel material. For this purpose, electrical tests without applied load are first conducted. Then, compression tests were carried out to evaluate the strain monitoring capabilities of the various carbon-modified composites.

### 5.1 Percolation

All the carbon-doped smart-earth cubes were subjected to a maximum voltage of 3V on the two copper electrodes embedded in the center of the sample (Fig. 4). The acquired current was converted into electrical resistance to identify the jump in conductivity, which represents the percolation threshold of the filler content. This value represents the optimal content in filler for enhancing the electrical properties of the conductive composite, due to the synergic effects of the fibers, the matrix, and of the contact surface.

**Fig. 4.** Setup of electrical tests on cubes in smart-earth with increasing inclusions of carbon fillers.

## 5.2 Compressive Tests on Cubes

Compressive tests were conducted for evaluating the sensitivity of the different types of carbon composites to the axial strain (Fig. 5). The figure represents the sketch of the data acquisition systems and the detailed view of a sample subjected to electromechanical tests. The investigated percentages were whose ones resulted more promising and performant from the electrical tests. The function generator applied a square wave input on the internal copper electrodes, while the loads and the electrical current were acquired from specific systems. The applied loads were triangular and cyclical loads, from 0.5 to 4 kN and a speed of 2kN/s. Figure 6 shows the time history of the applied loads for compressive tests.

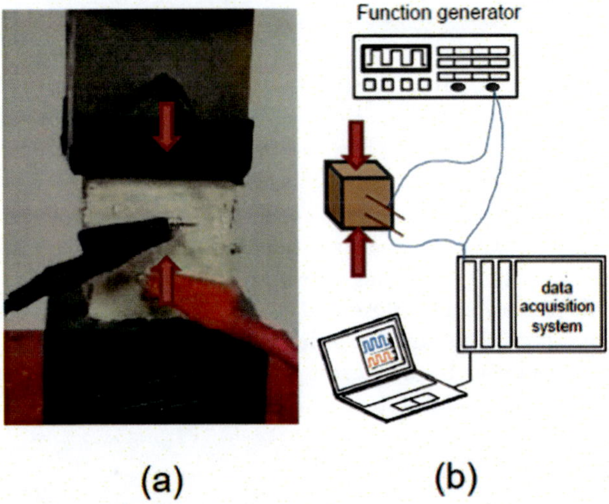


**Fig. 5.** Setup for the electromechanical compressive test on cubic samples: (a) picture of the tests, (b) sketch of the electrical setup.

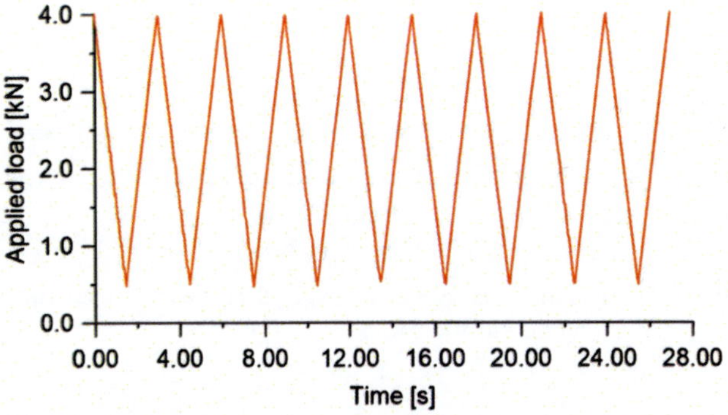


**Fig. 6.** Time history of the cyclical applied loads during the compressive tests on cubic samples.

### 5.3 Evaluation of Sensitivity

From the sensitivity tests, Gauge factors GF were evaluated by use of the following equation:

$$GF = (\Delta R/R_0)/\Delta\varepsilon \tag{2}$$

where R0 and ΔR are the initial and changing electrical resistance, respectively, and Δε is the variation of maximum strain during the tests. The sensitivity properties of different samples and test repetitions were then compared, also considering the feasibility of the results.

# 6 Results

The outputs were recorded, post-processed, and organized for the different types of tests.

## 6.1 Electrical Tests

Figure 7 shows the value of the electrical conductivity obtained for each cube sample subjected to electrical tests. From the graph it is clear that over 0.1% of fillers the resistance decreases highly, thus identifying the percolation threshold. The percentages from 0.025 to 0.1%, staying beyond but close to such a threshold represent good candidates for monitoring applications. This range of inclusions could be considered as a percolation zone for effective materials. For this reason, such types of carbon-modified composites have been selected for the evaluation of the sensitivity tests through compressive cyclical load application.

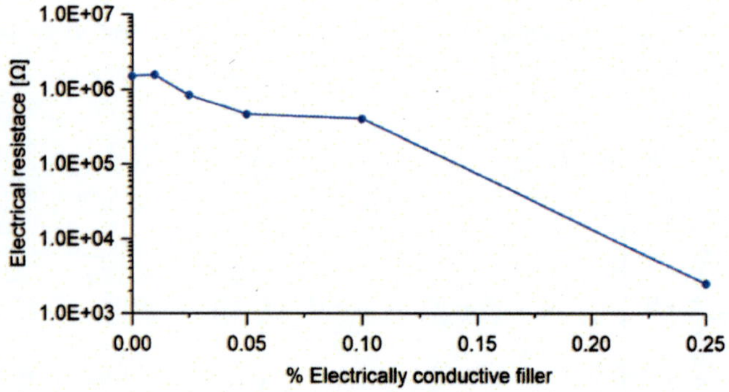


**Fig. 7.** Variation in the electrical resistance measured on the cubes with increasing fillers' content up to 0.25% in weight, during the electrical tests.

## 6.2 Compression Tests on Cubes

Compression tests were carried out on three types of smart cubes with 0.025, 0.05, and 0.1 % of carbon fillers. Eight load test repetitions were performed in total for each type of sample. The results were compared with those obtained by testing a plain sample. An example of outputs, concerning the sample with 0.05% of fillers, is shown in Fig. 8. The graphs in the figure represent the measured variation in strain, collected by strain gauges applied on the lateral sides of the sample in the direction of the load, and the normalized variation in electrical resistance attained from the electrical measurements on the embedded copper electrodes. The negative sign of the strains represents the compression state. The nine load cycles are clearly visible in the outputs of both strain gauges and smart-earth sensor.

The electrical signals, which were acquired at 1 Hz, appear with a good quality. The electrical resistance decreases with the increase of the applied compression strain, due to the increase of the contact points among conductive carbon fibers, and the decrease of the distance among the various internal conductive paths. Similar results were found when testing the other samples involved in this experimental campaign.

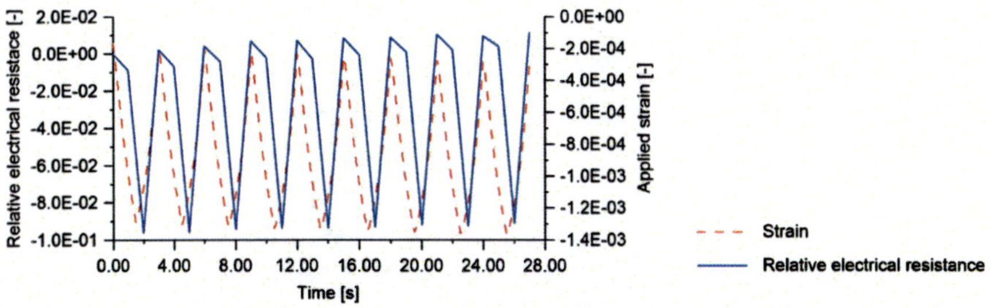


**Fig. 8.** An example of time histories of applied strain and relative electrical resistance obtained during the sensitivity compressive tests on smart-earth cubes with 0.05% of carbon microfibers.

### 6.3 Sensitivity Evaluation

The electromechanical tests allowed the evaluation of the sensitivity properties of the composites doped with carbon microfibers. In particular, from the cyclical compression loads, the change in electrical resistivity was measured and compared with the variation in strain, in order to determine the GF, as in Eq. 2. Table 2 reports the average values of the normalized variation in electrical resistance, strain variation, and GF obtained for each sample typology. The coefficient of variation (CoV) computed for each GF was also specified. Considering the obtained values of the GF, it is evident that the best performance was obtained from the carbon-doped samples. Accordingly, the addition of the filler improved the strain-sensing capability of the plain material. In particular, the sample with 0.025% in weight of CMF shows the highest GF, while the sample with 0.05% of fillers exhibits the best CoV, and a GF higher than that obtained for the plain sample. The sample with 0.05% of CMF therefore represents a good compromise between improved strain-sensitivity and repeatability of measurement under compressive loads. The electromechanical tests demonstrated that the proposed material could be effectively utilized as a multifunctional material for constructions.

## 7 Discussion

The physical, electrical, and electromechanical tests carried out on this new type of smart material, which contains both earth and cement as binder, were carried out to investigate the feasibility of the use of this composite as a sensing construction material. The workability, the electrical, and sensing properties of the analyzed and optimized composites appear promising for the development of those materials in the field of civil engineering. Further investigations, to allow the adoption of such materials in real structures and infrastructures will concern mechanical and durability analyses.

**Table 2.** Results from electrical tests on cubes with 0.000, 0.025, 0.050 and 0.100% of carbon fillers with respect to the weight of the binder. $R_0$ is the initial resistance, $\Delta R$ is the change in electrical resistance, GF the Gauge Factor, CoV the coefficient of variation.

| CMF %w | $\Delta R/R_0$ [-] | $\Delta\varepsilon \cdot 10^{-4}$ | GF | CoV |
|---|---|---|---|---|
| 0,000 | -0,0366 | 4,8 | 77,78 | 0,46 |
| 0,025 | -0,0870 | 5,2 | 172,73 | 0,62 |
| 0,050 | -0,0833 | 9,0 | 94,88 | 0,31 |
| 0,100 | -0,0480 | 5,0 | 94,52 | 0,72 |

## 8 Conclusion

The paper presents the first results of an experimental investigation on a novel multifunctional and self-sensing composite with a matrix made of cement and earth, and with addition of carbon microfibers as fillers. The composite was tailored for obtaining a workability suitable for construction materials. Various amounts of fillers were investigated for identifying the most performing mix. The electrical and sensing tests demonstrated that the proposed composite is promising for application on structures. The novel construction material could represent a more sustainable solution for reducing the environmental impact of cementitious building composites. Further investigations will concern samples of higher dimensions or specific applications.

**Acknowledgements.** This work was financially supported by the Italian Ministry of University and Research (MUR) through the Project FISR 2019: "Eco Earth" (code 00245). The authors would like to acknowledge funding by the European Union – Next Generation EU under the Italian Ministry of University and Research (MUR) National Innovation Ecosystem Grant ECS 00000041 – Vitality.

## References

Azhari, F., Banthia, N.: Cement-based sensors with carbon fibers and carbon nanotubes for piezoresistive sensing. Cem. Concr. Comp. **34**, 866–873 (2012)

Birgin, H.B., D'Alessandro, A., Laflamme, S., Ubertini, F.: Hybrid carbon microfibers-graphite fillers for piezoresistive cementitious composites. Sensors **21**(2), 518 (2021). https://doi.org/10.3390/s21020518

Collepardi, M.: The new Concrete, 436 pp. Editor: Tintoretto ISBN: 8890377720 (2010)

Curto, A., Lanzoni, L., Tarantino, A.M., Viviani, M.: Shot-earth for sustainable constructions. Construct. Build. Mater. **239** (2020)

D'Alessandro, A., Materazzi, A.L., Ubertini, F.: Nanotechnology in Cement-Based Construction, 424 pp., Jenny Stanford Publishing (2020a)

D'Alessandro, A., Coffetti, D., Crotti, E., Coppola, L., Meoni, A., Ubertini, F.: Self-sensing properties of green Alkali-activated binders with carbon-based nanoinclusions. Sustainability **12**(23), 1–13 (2020)

D'Alessandro, A., Fabiani, C., Pisello, A.L., Ubertini, F., Materazzi, A.L., Cotana, F.: Innovative concretes for low-carbon constructions: a review. Int. J. Low-Carbon Technol. **12**(3), 289–309 (2017)

Downey, A., D'Alessandro, A., Ubertini, F., Laflamme, S., Geiger, R.: Biphasic DC measurement approach for enhanced measurement stability and multi-channel sampling of self-sensing multi-functional structural materials doped with carbon-based additives. Smart Mater. Struct. **26**(6), 065008 (2017)

Flower, D., Sanjayan, J.: Greenhouse gas emissions due to concrete manufacture. Int. J. Life Cycle Assess. **2007**(12), 282–288 (2007)

Guettala, A., Abibsi, A., Houari, H.: Durability study of stabilized earth concrete under both laboratory and climatic conditions exposure. Constr. Build. Mater. **20**(3), 119–127 (2006)

Han, B., Yu, X., Ou, J.: Self-Sensing Concrete in Smart Structures. Butterworth-Heinemann 2015 ISBN 978-0-12-800517-0, Elsevier Inc. (2015)

Meoni, A., et al.: An experimental study on static and dynamic strain sensitivity of embeddable smart concrete sensors doped with carbon nanotubes for SHM of large structures. Sensors **18**(3), 831 (2018). https://doi.org/10.3390/s18030831

Meoni, A., D'Alessandro, A., Mancinelli, M., Ubertini, F.: A multichannel strain measurement technique for nanomodified smart cement-based sensors in reinforced concrete structures. Sensors **21**(16), 5633 (2021). https://doi.org/10.3390/s21165633

Metaxa, Z.S., Konsta-Gdoutos, M.S., Shah S.P.: Carbon nanotubes reinforced concrete. ACI Symposium Publication, vol. 267, pp. 11–20 (2009)

Meyer, C.: The greening of the concrete industry. Cement Concr. Compos. **31**, 601–605 (2009)

Purnell, P.: Material nature versus structural nurture: the embodied carbon of fundamental structural elements. Environ. Sci. Technol. **46**(1), 454–461 (2012)

Mehta, P.K., Monteiro, P.J.M.: Concrete: Microstructure, Properties, and Materials, McGraw Hill, 704 pp. (2014). ISBN-10: 0071797874

Pacheco-Torgal, F., Jalali, S.: Earth construction: Lessons from the past for future eco-efficient construction. Construct. Build. Mater. **29**, 512–519 (2012)

Tam, V.W.Y., Soomro, M., Evangelista, A.C.J.: A review of recycled aggregate in concrete applications (2000–2017). Constr. Build. Mater. **172**, 272–292 (2018)

Van Damme, H., Houben, H.: Earth concrete. Stabilization revisited. Cem. Concr. Res. **114**, 90–102 (2017)

Xiao, J., Lia, J., Zhang, C.: Mechanical properties of recycled aggregate concrete under uniaxial loading. Cem. Concr. Res. **35**, 1187–1194 (2005)

Zamora-Castro, S.A., et al.: Sustainable development of concrete through aggregates and innovative materials: a review. Appl. Sci. **11**, 629 (2021)

# A Preliminary Study on the Use of Recycled Asphalt Pavement (RAP) in Mortars

Salma Jaawani, Annalisa Franco, Giuseppina De Luca(✉), Orsola Coppola, and Antonio Bonati

Construction Technologies Institute of the Italian National Research Council, ITC-CNR, Via Lombardia 49, 20098 San Giuliano Milanese, Milan, Italy
{jaawani,frranco,deluca,coppola,bonati}@itc.cnr.it

**Abstract.** The use of different types of recycled materials in the building sector became of central importance in recent years as one of the main goals worldwide is to reduce the depletion of natural resources and favour the exploitation of already existing materials in a more sustainable manner. The aim of this paper is to present a recycled material deriving from the milling of asphalt pavement (Recycled Asphalt Pavement, RAP) and its possible uses as aggregate in mortars. More specifically, a preliminary review on the influence of RAP on the mechanical characteristics of mortars, such as grading, compressive strength, flexural strength, splitting tensile strength and durability, is presented in relation to the various percentages of substitution of natural aggregates with this recycled material. Lastly, an overview on the codes and standards regulating the use of recycled materials in mortars is given, together with the respective imposed restrictions.

**Keywords:** Sustainability · Recycled Asphalt Pavement (RAP) · cement mortar · mechanical properties · standards

## 1 Introduction

The construction sector is facing big issues related to the depletion of natural resources, the need of raw materials, and the increasing amount of waste to be disposed. In this sense, great attention has been given also by the European Union, which aims to move towards a more sustainable construction sector (COM/2018/656 final 2018). As a result, in the last few years, "circular economy" has become the target of the whole scientific world as a strategy to reduce natural resource consumption and waste production. Concrete and mortar are two of the most used construction materials, and many attempts have been made to substitute totally or partially natural aggregates in concrete or mortar by aggregates deriving from construction and demolition waste (C&D) (Vitale et al. 2021). As a matter of fact, nowadays the most accumulated waste material on earth is from C&D and it represents 25–30% of the total amount of produced waste (Pavlů 2018). In Europe, the amount of C&D waste is about 45% of the total European waste (Galvez-Martos et al. 2018, Reis et al. 2021) and its possible reuse will constitute an alternative solution to their disposal on landfills, also because of the lack of available spaces (Vitale et al. 2021).


M. A. Aiello and A. Bilotta (Eds.): ICC 2022, LNCE 435, pp. 40–51, 2024.
https://doi.org/10.1007/978-3-031-43102-9_4

Research on the use of recycled aggregates (RA) in cementitious materials is still ongoing, and the aim of this work is to investigate the possible substitution of fine natural aggregates with fine RA in mortars. A first overview on the different RA available on landfills and their suitability to be used in mortars, which include their effect on the mix design, on the mechanical characteristics of the recycled mortars, and the durability aspects, will be given. For the reasons mentioned above, the percentage of replacement plays a major role. The focus will be then moved to the description of a specific type of RA coming from the milling of road surfaces, referred to as Recycled (or reclaimed) Asphalt Pavement (RAP). In fact, the studies on the use of RAP in cementitious products, such as concrete and mortars, are still in progress. The main characteristics of RAP will be presented, and the effects on its use in mortars are discussed by reporting the experimental results found in literature and discussion on the limitations that needs to be faced in its application.

## 2 Use of Recycled Aggregates in Mortars

In the production of coarse RA, the fine fraction is involuntarily produced and it represents a large amount of the weight of the crushed C&D waste (Restuccia et al. 2016). According to Angulo et al. (2009), the produced fine fraction is about 50%, therefore, there is a strong need to find a suitable application for it (Restuccia et al. 2016). For this reason, the incorporation of fine RA as a replacement to natural aggregates in mortars has been investigated. A classification of mortars is given by the harmonized European Standards EN 998-1 (EN 998-1:2016) and EN 998-2 (EN 998-2:2016), which distinguish between rendering/plastering and masonry mortars, respectively. While rendering mortars are coating mortars, which are used to cover external and internal walls and ceilings, masonry mortars are used to lay and bind ceramic bricks or concrete blocks. The characteristics of the mortars depend therefore on their intended use. In the following paragraphs, the different types of RA, which are used in mortars, will be presented.

### 2.1 Fine Recycled Concrete Aggregates (FRCA)

The analysis of literature evidenced that the most studied RA used in mortars, as a replacement to natural aggregates, are those coming from concrete. Several authors have investigated the properties of mortars containing fine recycled concrete aggregate (FRCA), evidencing a reduction of properties as the replacement percentage increases (Restuccia et al. 2016). In Roque et al. (2020), rendering mortars including two types of fine RA coming from C&D waste are studied: Recycled Concrete Aggregate (RCA) and Mixed Recycled Aggregate (MRA, i.e. with various constituents such as concrete, mortar, ceramic and other contaminants). The percentages of substitution used were 0%, 20%, 50% and 100% (by volume). The results were generally positive, with a strong influence of the percentages of substitution and the type of RA used, as it can be seen in Fig. 1, where the values of both compressive (Fig. 1a) and flexural strength (Fig. 1b) at 28 days of curing have been reported for each type of mortar with RA mentioned in this paper as a function of the percentages of substitution. Values are expressed in terms of ratio with the reference mortar.

It can be seen that by increasing the incorporation of RA, the mechanical strength of the mortars after 28 days of curing increased (see Fig. 1). But in a longer time span (90 days), both RCA and MRA mortars showed a decrease in mechanical strength, even though RCA showed a lower decrease than MRA mortars (Roque et al. 2020). Indeed, the mixes containing 100% MRA showed the greatest decrease in terms of flexural and compressive strength, respectively of 44% and 49% at 90 days.

Among the other properties, it is noteworthy mentioning the positive effect of the inclusion of FRCA on the dynamic modulus of elasticity. Indeed, a reduction in this property (about 62% for MRA and 42% for RCA at 100% replacement) may be favourable to rendering mortars, thanks to the lower cracking susceptibility and improved deformation capability.

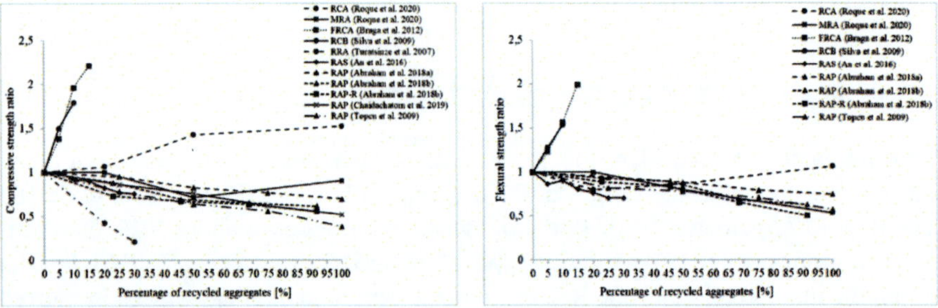


(a) (b)

**Fig. 1.** Mechanical properties of mortars with RA at 28 days: (a) Compressive strength; (b) Flexural strength.

In a more specific research on the incorporation of FRCA, Braga et al. (2012) also evidenced positive results (see Fig. 1). In this study, the size of the RCA did not exceed 0.150 mm and the percentages of substitution went from 0% to 15%. As it can be seen from Fig. 1a, the compressive strength of the mixes containing 5%, 10%, and 15% increased by 38%, 96% and 121%, respectively (Braga et al. 2012). In a similar way, flexural strength also showed an increment of about 23%, 56% and 99%, respectively (Fig. 1b). The mix containing 15% was also studied in terms of drying shrinkage. The results showed an increase in shrinkage of about 44%, compared to the reference mortar. This could be due, according to another author (Hansen 1986), to the high content of hardened mortar bonded to the recycled aggregates surface.

In Zhao et al. (2015), the saturation state of FRCA (dried or saturated) played a major role on the properties of the mortars. In fact, it was reported that the compressive strength of mortars made with dried FRCA was better than that of mortars with saturated FRCA. However, similarly to the other studies, the compressive strength of recycled mortars decreases as the percentage of recycled fine aggregates increases. In fact, for a w/c equal to 0.5 and for 100% FRCA, mortar with dried FRCA showed a decrease of the compressive strength of about 30.8% compared to the conventional mortar, while for the same w/c mortar with saturated FRCA showed a decrease of about 32.3%.

Some studies related to the durability can also be found in literature. The resistance of mortars containing FRCA to the magnesium sulfate attack was investigated in Lee

(2009). The results of visual appearance, the compressive strength loss and expansion tests indicated that replacement levels by recycled fines beyond 50% exhibited a better resistance to magnesium sulfate attack compared to the reference mortar, especially at the later exposure duration. However, this improvement in resistance was not proved (Lee 2009).

### 2.2 Fine Recycled Red Brick (RCB)

Some research has been performed to include aggregates coming from C&D wastes that contain fine crushed red clay brick (RCB) with the aim to improve the performance of cementitious mortars (Martínez et al. 2013; Reis et al. 2021; Silva et al. 2009). In Silva et al. (2009), incorporation of brick powder in the different mortars was performed in various percentages: 0%, 5% and 10%. In terms of compressive strength, mortars containing 10% of RAP showed an increase of 79% when compared with the conventional mortar, as it can be seen in Fig. 1. Meanwhile, the shrinkage and water vapour permeability did not significantly improve (Silva et al. 2009).

In a similar study, masonry mortars have been investigated with a percentage of substitution of 100%, and various types of filler (Martínez et al. 2013). After 28 days' period of curing in a humidity room, an increase was observed in terms of compressive strength and flexural strength for all the mixes, between 10% and 30% and between 10% and 35%, respectively.

### 2.3 Fine Recycled Rubber (RRA)

RA coming from shredded non-reusable tyres were also used to study the behaviour of recycled mortars (Turatsinze et al. 2007). Two recycled mixes, containing 20% and 30% of rubber aggregates (RRA), were compared with a reference mortar. The results showed that rubberized mortars exhibit a large decrease in tensile and compressive strengths: 67% and 79% respectively for the maximum percentage of substitution (see Fig. 1a). Furthermore, the presence of less stiff aggregates reduces the internal restraint, increasing therefore the length change resulting from free shrinkage. Despite these drawbacks, restrained shrinkage cracking tests showed instead an enhanced strain capacity, evidencing therefore that the balance between enhanced strain capacity and increased free shrinkage results in a clear benefit from rubber aggregate incorporation in mortars (Turatsinze et al. 2007).

### 2.4 Fine Recycled Shingle (RSA)

In the US, tear-off roofing shingle, referred to as Reclaimed Asphalt Shingle (RAS) are considered a product of the C&D waste. An et al. (2016) investigated the use of RAS in the production of recycled mortars. The percentages of inclusion used are up to 30%. The results showed that the inclusion of RAS in mortars causes a reduction in compressive and flexural strengths in the order of 30% and 25% respectively at 100% of replacement (see Fig. 1). Under flexural loading, RAS mortars showed a better ductility thanks to the presence of the asphalt film layer that impedes the crack propagation. However, this improvement was considered slight because the increase of toughness between the reference mortar and the mortar with the highest toughness value, which was the one containing 10% of RAS, was of about 4.7% (An et al. 2016).

### 2.5 Discussion

From the results found in literature, it is evident that the behaviour of mortars containing different types of waste strongly depends on various factors, such us the type of recycled material used, the gradation of the aggregate, the percentage of replacement, etc. Fig. 1 sums up the strength behaviour of the mortars analysed: all the recycled materials causes a decrease in compressive and flexural strength as the content increases, except for the aggregates coming from RCA and RCB. This trend is certainly due to the higher compatibility of such waste materials in cementitious mixes compared to other types of RA, which involve instead the presence of aggregates with lower stiffness (rubber aggregates) or an asphalt film layer (in RSA but also in RAP, see Sect. 3), which, in both cases, cause a "bond defect" with the cementitious matrix (Turatsinze et al. 2007).

However, there is an issue concerning the quality of RA especially coming from C&D waste for sand production that limits their uses and it relates to the presence of the porous and low strength phases (Reis et al. 2021). This produces sand with low quality which affects workability, mechanical and durability performance at hardened state (Evangelista et al. 2010). For these reasons, in most countries, the use of recycled sands is not allowed in structural cementitious material production (concrete or mortar). A possible solution could be the removal of adhered mortar, which will allow the improvement of the aggregate performances. However, this is not a simple task due to the high costs of the necessary treatments (Reis et al. 2021).

## 3 Mortars Containg Recycled Asphalt Pavement (RAP)

### 3.1 Description, Origin and Main Characteristics of RAP

Another C&D waste containing asphalt but with a different origin than RAS is Recycled Asphalt Pavement (RAP), which is a product of the milling of road/motorway surface. Large amounts of recycled asphalt are still deposited on landfills and mostly remain unused (Pavlů 2018). The majority of RAP is re-used in asphalt in a sort of "closed cycle", while the studies on the use of RAP in concrete is still ongoing and only few results are available (Jaawani et al. 2021a, b). Recycled (or reclaimed) asphalt pavement is the name given to crushed, milled, pulverized, processed or/and unprocessed pavement materials containing asphalt and aggregates. These materials are generated when asphalt pavements are removed for reconstruction by milling from road/motorway surface. RAP consists of two components:

1. RAP aggregate, which is the aggregate part of the reclaimed asphalt;
2. RAP binder that consists of the asphalt cement of the reclaimed asphalt.

In Italy, RAP is still not considered as a valuable resource: according to SITEB, in Italy, only 25% of the available milled asphalt is destined to be reused (SITEB 2020).

The mechanical properties of RAP also depend on the original asphalt pavement type. Several studies affirm that RAP has its own grading curve, which is mainly due to the different origin of each RAP (Jaawani et al. 2021a, b). Similarly to the other types of aggregates coming from C&D waste, the percentage of substitution is another aspect that highly influences the mechanical characteristics. The fine fraction was also studied in a possible inclusion in mortars.

### 3.2 The Use of RAP in Mortars

Most of the studies found in literature are focused on the effect of coarse fraction of RAP aggregates (aggregate size >4.75 mm) rather than the fine fraction. Only a few studies (Abraham and Ransinchung 2018a, b; Chaidachatorn et al. 2019; Jaawani et al. 2021a, b; Qiang et al. 2011; Topçu and Isikdag 2009) have been conducted on cement mortar to study the effect of fine fraction of RAP as a replacement to the natural aggregates.

### 3.3 Mechanical Properties

In a study of Abraham and Ransinchung (2018a), natural aggregates have been replaced by the fine fraction of RAP aggregates, in order to obtain different mixes with different percentages of substitution: 25%, 50%, 75% and 100% by volume. The cement-to-sand ratio and w/c ratio were maintained constant at 1:3 and 0.44 respectively for all mixes (Abraham and Ransinchung 2018a). Some of the main mechanical characteristics of cement mortar such as compressive strength, flexural strength and splitting tensile strength, were investigated, for different curing times: 7, 28 and 90 days (see Fig. 1a and b for compressive strength and flexural strength at 28 days, respectively). With the increase in RAP aggregate content, compressive strength of cement mortar decreases. The authors found that this reduction is maintained for 7, 28 and 90-day specimens. The rate of gain of strength for the mixes was more in early ages until 28 days in comparison to later ages. When compared to the reference mix, a reduction of 4.85%, 17.48%, 23.30% and 30.10% was observed for the mixes containing 25%, 50%, 75%, and 100% of RAP, respectively (see Fig. 1). Similarly to compressive strength, the flexural strength decreases at all ages of curing (7, 28 and 90 days) with the increase in RAP aggregate content in cement mortar and the maximum reduction of strength (about 25%) was registered at 100% of replacement. The same trend was observed for the splitting tensile strength, where the increase of RAP content leads to a decrease in splitting tensile strength reaching about 45% for 100% replacement by RAP (Abraham and Ransinchung 2018a).

From the studies found in literature, gradation is one of RAP characteristics that strongly influences the mechanical behaviour of mortars. In (Abraham and Ransinchung 2018b), different mixes of cement mortar were prepared with various percentages of replacement of natural aggregates by RAP (25%, 50%, 75% and 100%) by weight. Mortar mixes were prepared with fine aggregates with their specific gradation (RAP mix) as well as with gradation falling in the Average Gradation Line (AGL), which is a line drawn through the average values of upper and lower limits of passing percentage of each designated Indian Standard (IS 383 2016) (R mix). The results showed that hardened mortar properties of RAP and R mixes, like density, compressive strength, flexural strength and splitting tensile strength, were lower compared to reference mix with natural fine aggregates and decreased with increase in fine RAP aggregate content (Abraham and Ransinchung 2018b). A maximum decrease of 41.7%, 39.3% and 51.5% in compressive strength, flexural strength and splitting tensile strength, respectively, was observed for R mixes with 100% RAP content as compared to the reference mix (Abraham and Ransinchung 2018b). According to the authors, the decrease in strength of cement mortar may be due to the increase in the porosity of the Interfacial Transition Zone (ITZ). When focusing on the effect of gradation, significant difference was observed between the strength parameters of RAP and R mixes with 25%, 50% and 75%, except for 100% mixes. RAP mixes were observed to have higher values of strength than R mixes.

Another mix design factor that influences the mechanical characteristics is w/c. In Chaidachatorn et al. (2019) this aspect has been investigated on different mixes of RAP mortars where the percentage of substitution were of 0%, 25%, 50%, 75% and 100% (see Fig. 1). The variation of w/c chosen for this study was between 0.40 and 0.70. From the characterization of the aggregates, it was observed that RAP had higher water absorption at saturated surface dry (SSD) state and slower rate of absorption than natural sand. This causes the cumulus of additional water to be compensated for the SSD state as a free water in the mix after hardening. At low w/c ratio (w/c $< 0.5$), which is insufficient in terms of hydration of mortar, the RAP replacement increased the compressive strength of RAP mortars. For high w/c ratio mixes, where the water is sufficient for complete cement hydration, the after-hardening unabsorbed water increased the amount of water to be reacted with cement. This high amount of water causes higher porosity and lower compressive strength: the lowest value of compressive strength was found at w/c $=$ 0.75 and for 100% of RAP and was about 8 MPa, while the reference mix at the same w/c ratio was about 19 MPa. Thanks to the experimental analyses (Chaidachatorn et al. 2019), it was possible to define an optimal percentage of RAP of 25%, where at w/c $=$ 0.45 the compressive strength reached 28 MPa, which was higher than the value of the control mix compressive strength (about 26 MPa). The role of RAP replacement on the reduction in the degree of hydration was considered for the development of extended water to cement ratio law. In Chaidachatorn et al. (2019) strength development over time has been represented by a logarithmic function and it was observed that the normalized compressive strength development by 28-day strength was practically unique. Based on this law and the equations developed, it was possible to predict the values of compressive strength in relation to various w/c ratios, RAP replacement ratios and curing times. Furthermore, this formulation of the strength prediction equations can be applied to various RAP containing different types of bitumen and aggregate.

### 3.4 Porosity and Sorptivity Parameters

The percentage of replacement plays a major role also when porosity of mortars is investigated: in Abraham and Ransinchung (2018a), porosity has been studied with the Mercury Intrusion Porosimetry (MIP). From the results of porosity, it was observed that an increase in RAP aggregate content, causes an increase in porosity: from the values of porosity, it was noticed that a similar result to that of the mix with 0% of RAP is registered for the RAP mix which contained 25% of RAP aggregates. While for higher concentration of RAP, such as 50%, 75%, and 100%, the porosity values increased by

71.5%, 73.8% and 89.8%, respectively. The increase in porosity may be due to a weak ITZ in RAP mortar (Brand and Roesler 2017). The plausible reason for a porous ITZ may include the hydrophobic nature of asphalt, repelling the water gradient formed around the aggregate in fresh cementitious mix, preventing an efficient hydration of the mix (Brand and Roesler 2017). In fact, in another study (Qiang et al. 2011), which focused on the mechanisms that regulate the strength in cement-asphalt mortar, the results showed that the asphalt emulsion delayed the early hydration of the cement and the asphalt membrane had a negative effect on the further hydration of the cement, to some extent (Qiang et al. 2011). For this reason, the production of a porous ITZ might be due to the poisoning effect on cement hydration by asphalt (Brand and Roesler 2017).

Sorptivity parameters like initial rate of absorption and capillary absorption coefficient were decreasing with increase in RAP aggregate content (Abraham and Ransinchung 2018b). This behaviour was a contradiction due to the higher porosity of RAP and R mortar, and lower density of fine RAP aggregates. Decrease in sorptivity parameters was due to the clogging of pores with molten asphalt during oven drying before the absorption process, thus hindering and blinding the actual capillary action. The trend of sorptivity of RAP cement mortar were in contradiction to the porosity nature of the same, which suggest the unsuitability of the conventional method used (Abraham and Ransinchung 2018b). For this reason, a modified method should be developed to study mortar and concrete mixes with RAP aggregates is necessary (Abraham and Ransinchung 2018b)). In general, other studies evidenced the need to study RAP and its effect by modifying the now existing testing methods (Jaawani et al. 2021a, b). Indeed, in previous studies (Berry et al. 2010; Brand et al. 2012; Okafor 2010; Spreadbury et al. 2021), it was found that the leaching test, the resistance to crushing and the alkali-silica activity are among the characteristics that need to be investigated with a proper method, which is suitable for RAP.

### 3.5 Durability

In terms of durability aspects, it is important to observe the behaviour of RAP mortar under sulfate attack. From the experimental data of Abraham and Ransinchung (2018a), an increase in mass reduction was observed with an increase in RAP content. Interesting was the correlation between the effect of sulfate environment on the compressive strength of cement mortar mixes (Abraham and Ransinchung 2018a). When comparing mixes under moist curing, reduction in strength was observed for 0%, 75% and 100% RAP mixes in the order of 21.6%, 9%, and 40% respectively, whereas for 25% and 50% RAP mixes, an increase in strength was observed after 90-day period, in the order of 13.21% and 1.04%, respectively. This may be due to the difference in rate of formation and filling of pores by expansive products (gypsum and ettringite) between the mixes (Abraham and Ransinchung 2018a). In order to maintain a safe margin to strength parameters and resistance to sulfate attack, according to Abraham and Ransinchung (2018b), replacement of natural fine aggregates by RAP aggregates at 25% by volume shall be suggested in cementitious mixtures as the optimal percentage of substitution.

Another durability characteristic that needs to be investigated, because of its strong correlation with the main mechanical characteristics, is drying shrinkage. Drying shrinkage has been investigated as percent shrinkage and shrinkage strain of RAP inclusive cement mortar specimens at various inspection periods (Abraham and Ransinchung 2018a). According to Abraham and Ransinchung (2018a), RAP mortar mixes undergone more shrinkage in comparison to mixes with natural aggregates with rate of shrinkage to be higher at the initial days of air drying which causes an increase in elongation. Furthermore, drying shrinkage behaviour was generally uniform with time as the content of RAP increases (Abraham and Ransinchung 2018a). Further studies need to be taken up in order to see whether this behaviour is being maintained in longer periods of drying. In Topcu et al. (2009), a more focused research on free and restrained shrinkage is given to predict fractural behaviour of mortars, with different cement and replacement ratios. Satisfactory test results of the restrained shrinkage were obtained at higher RAP content despite the lower strengths. The shrinkage in mortars influenced also the other mechanical properties. From the results of shrinkage, it was showed that the optimum RAP content, in order to limit the influence on mechanical properties, was obtained for percentages of replacement between 25% and 50% (Topçu and Isikdag 2009). Furthermore, destructive and non-destructive tests were conducted to delay crack formation and improve strain capacity of mortars before cracking (Topçu and Isikdag 2009). A measurement of the crack widths was also registered and, from the results, they could be considered in an acceptable range for all the mixes. Furthermore, crack widths were reduced with the increase in RAP content: at 100% RAP replacement the maximum crack opening was about 0.17 mm, while the reference mix showed a maximum crack opening of 0.70 mm. In general, the results showed that incorporation of RAP in mortars is adequate for delaying crack formation and reducing crack opening. The RAP enhances the strain capacity of mortars despite the decrease in mechanical properties (Topçu and Isikdag 2009).

### 3.6 Standards

The inclusion of RA such as RAP needs to be investigated also in terms of applicability related to the requirements of current standards. In Jaawani et al. (2021a, b), a research has been done to highlight the limitations that could be faced by RAP in its incorporation in concrete, with a slight look also at the use in mortars according to current harmonized standards. In fact, the harmonized standard EN 13139 (2004), that defines the properties and the characteristics of aggregates and fillers obtained from natural, artificial or recycled materials to be used for mortars, has been analysed and some limitations have been found. According to Jaawani et al. (2021a, b), the heterogeneity of RAP could constitute a possible issue in its use in mortars, also in regards of its own grading curve. Furthermore, its constantly different chemical composition could be a limit, as the content of organic material in the aggregates shall not influence the setting speed of the mortar by extending its time by more than 120 min (EN 13139:2004). In relation to the content of organic material in the aggregates, EN 13139 requires also that the compressive strength of the specimens do not decrease by more than 20% after 28 days (EN 13139:2004).

## 4 Discussion and Conclusions

In this paper, an overview on the use of different RA, especially coming from C&D waste in both rendering and masonry mortars, is given. From the results found in literature and analysed in this paper, it was observed that:

- Among the different RA, FRCA and RCB positively influence the mechanical characteristics of the mortars, for their higher compatibility with the cementitious paste, while RRA and RSA negatively affect the strength parameters of mortars due to the poor rubber (or asphalt) –cementitious matrix bond. Mortars with 50% of FRCA exhibited a better resistance to magnesium sulfate attack.
- The low quality of RA from C&D waste for sand production limits their uses. A possible solution could be the removal of adhered mortar on RA, which is economically not convenient.
- Few studies on RAP incorporation in structural cementitious products are present. A deeper investigation is needed.
- The strength characteristics decreases as the content of RAP in mortar increases. Grading of RAP, w/c ratio, and the percentages of substitution strongly influence the strength behaviour of mortars.
- A high content of RAP produces high values of porosity, in contradiction with a decrease of sorptivity: the conventional method used for sorptivity and other properties might be unsuitable when RAP is included.
- The optimum percentage of RAP in cementitious mixtures should be around 25% by volume, to maintain a safe margin on strength parameters and resistance to sulfate attack.
- Mortars containing RAP showed more shrinkage but cracking phenomena was acceptable, as crack width reduces with the increase in RAP content.
- A deeper study of the current harmonised standards on mortars and RA should be performed in order to understand the possible limitations that could be faced by RAP in its incorporation in mortars.

**Acknowledgements.** Authors wish to acknowledge the Fondazione Cariplo for financial support under the research project RAPCON ("Sustainable concrete made with recycled asphalt pavement") in the framework of the national Call for the Scientific Research Area "Circular Economy: Research for a Sustainable Future".

## References

Abraham, S.M., Ransinchung, G.D.R.N.: Influence of RAP aggregates on strength, durability and porosity of cement mortar. Constr. Build. Mater. **189**, 1105–1112 (2018)

Abraham, S.M., Ransinchung, G.D.R.N.: Strength and permeation characteristics of cement mortar with Reclaimed Asphalt Pavement Aggregates. Constr. Build. Mater. **167**, 700–706 (2018)

An, J., Nam, B.H., Youn, H.: Investigation on the Effect of Recycled Asphalt Shingle (RAS) in Portland Cement Mortar. Sustainability **8** (2016)

Berry, M., Stephens, J., Cross, D.: Feasibility of Reclaimed Asphalt Pavement as Aggregate in Portland Cement Concrete Pavements (2010)
Braga, M., de Brito, J., Veiga, R.: Incorporation of fine concrete aggregates in mortars. Constr. Build. Mater. **36**, 960–968 (2012)
Brand, A.S., Roesler, J.R.: Bonding in cementitious materials with asphalt-coated particles: part I - The interfacial transition zone. Constr. Build. Mater. **130**, 171–181 (2017)
Brand, A.S., Roesler, J.R., Al-Qadi, I.L., Shangguan, P.: Fractionated Reclaimed Asphalt Pavement (FRAP) as a Coarse Aggregate Replacement in a Ternary Blended Concrete Pavement. Illinois Center for Transportation, Urbana, IL: 1–129 (2012)
Chaidachatorn, K., Suebsuk, J., Horpibulsuk, S., Arulrajah, A.: Extended water/cement ratio law for cement mortar containing recycled asphalt pavement. Constr. Build. Mater. **196**, 457–467 (2019)
COM/2018/656 final. Report from the commission to the European Parliament, the council, the European economic and social committee and the committee of the regions on the implementation of EU waste legislation, including the early warning report for member states at risk of missing the 2020 preparation for re-use/recycling target on municipal waste
EN 998-1:2016. Specification for mortar for masonry - Part 1: Rendering and plastering mortar. European Committee for Standardization (CEN), Brussels, Belgium
EN 998-2:2016. Specification for mortar for masonry - Part 2: Masonry mortar. European Committee for Standardization (CEN), Brussels, Belgium
EN 13139:2004. Aggregates for mortar. European Committee for Standardization (CEN), Brussels, Belgium
Evangelista, L., de Brito, J.: Durability performance of concrete made with fine recycled concrete aggregates. Cement Concr. Compos. **32**, 9–14 (2010)
Galvez-Martos, J.-L., Styles, D., Schoenberger, H., Zeschmar-Lahl, B.: Construction and Demolition Waste Best Management Practice in Europe. Resourc. Conserv. Recycl. **136** (2018)
Hansen, T.C.: Recycled aggregates and recycled aggregate concrete second state-of-the-art report developments 1945–1985. Mater. Struct. **19**, 201–246 (1986)
IS 383. Coarse and Fine Aggregate for Concrete - Specification. Bureau of Indian Standards (BIS)
Jaawani, S., Franco, A., De Luca, G., Coppola, O., Bonati, A.: Durability of concrete containing Recycled Asphalt Pavement (RAP). In: Proceedings of the 2nd fib Symposium on Concrete and Concrete Structures "The future of concrete structures: towards resilience and technological innovation", Roma, 18–19 November 2021
Jaawani, S., Franco, A., De Luca, G., Coppola, O., Bonati, A.: Limitations on the Use of Recycled Asphalt Pavement in Structural Concrete. Appl. Sci. **11** (2021b)
Lee, S.-T.: Influence of recycled fine aggregates on the resistance of mortars to magnesium sulfate attack. Waste Manage. **29**, 2385–2391 (2009)
Martínez, I., Etxeberria, M., Pavón, E., Díaz, N.: A comparative analysis of the properties of recycled and natural aggregate in masonry mortars. Constr. Build. Mater. **49**, 384–392 (2013)
Okafor, F.O.: Performance of recycled asphalt pavement as coarse aggregate in concrete. Leonardo Electron. J. Pract. Technol. **9**, 47–58 (2010)
Pavlů, T.: The utilization of recycled materials for concrete and cement production- a review. In: Proceedings of the IOP Conference Series: Materials Science and Engineering, Malta, April 10 (2018)
Qiang, W., Peiyu, Y., Ruhan, A., Jinbo, Y., Xiangming, K.: Strength mechanism of cement-asphalt mortar. J. Mater. Civ. Eng. **23**, 1353–1359 (2011)
Reis, G.S., Quattrone, M., Ambrós, W.M., Grigore Cazacliu, B., Hoffmann Sampaio, C.: Current applications of recycled aggregates from construction and demolition: a review. Materials **14** (2021)
Restuccia, L., Spoto, C., Ferro, G.A., Tulliani, J.-M.: Recycled mortars with C&D waste. Procedia Struct. Integrity **2**, 2896–2904 (2016)

Roque, S., Maia Pederneiras, C., Brazão Farinha, C., de Brito, J., Veiga, R.: Concrete-based and mixed waste aggregates in rendering mortars. Materials **13** (2020)

Silva, J., de Brito, J., Veiga, R.: Incorporation of fine ceramics in mortars. Construct. Build. Mater. **23**(1), 556–564 (2009). https://doi.org/10.1016/j.conbuildmat.2007.10.014

SITEB. Strade: in Italia, solo il 25% del fresato d'asfalto viene avviato al recupero. https://www.ingenio-web.it/26047-strade-in-italia-solo-il-25-del-fresato-dasfalto-viene-avviato-al-recupero

Spreadbury, C.J., Clavier, K.A., Lin, A.M., Townsend, T.G.: A critical analysis of leaching and environmental risk assessment for reclaimed asphalt pavement management. Sci. Total Environ. **775**, 1–16 (2021)

Topçu, İ, Isikdag, B.: Effects of crushed RAP on free and restrained shrinkage of mortars. Int. J. Concrete Struct. Mater. **3**, 91–95 (2009)

Turatsinze, A., Bonnet, S., Granju, J.L.: Potential of rubber aggregates to modify properties of cement based-mortars: improvement in cracking shrinkage resistance. Constr. Build. Mater. **21**, 176–181 (2007)

Vitale, F., Nicolella, M.: Mortars with recycled aggregates from building-related processes: a 'four-step' methodological proposal for a review. Sustainability 13 (2021)

# Seismic Fragility Curves: A Comparison Among Nonlinear Static and Dynamic Analysis Procedures

Carlotta Pia Contiguglia[1(✉)], Angelo Pelle[1], Davide Lavorato[1], Bruno Briseghella[2], and Camillo Nuti[1]

[1] Department of Architecture, Roma Tre University, 00153 Rome, Italy
carlottapia.contiguglia@uniroma3.it

[2] College of Civil Engineering, Fuzhou University, Fuzhou 350108, China

**Abstract.** Nonlinear response time-history analyses are used in well-known seismic performance assessment methods such as Incremental Dynamic Analysis, Multi-Stripes Analysis and Cloud Method to describe the relationship between chosen Damage Measure versus Intensity Measure. Numerous researches have presented simplified procedures or nonlinear static procedures to develop fragility over the last two decades. In NSPs, pushover analysis is used to assess the system capacity, and response spectra quantify the seismic demand. In addition to the known ones, Incremental Modal Pushover Analysis (IMPA) is a novel nonlinear static procedure that has been proposed in recent years and is used in this study to advance an IM-based fragility estimation. The accuracy and effectiveness of various vulnerability assessment methods are investigated by comparing fragility curves obtained from MPA-based Cloud Analysis, IMPA, Cloud Analysis to IDA fragility curves. The results from two relatively small bins of record motions differing by ranges of Joyner-Boore Distance and are scattered across a range of Magnitude are presented.

**Keywords:** IDA · IMPA · modal pushover · nonlinear analysis · fragility curves · seismic vulnerability

## 1 Introduction

### 1.1 Main Info

The first generation of PBEE assessment and design methodologies for buildings (Krawinkler 1999; ATC-20 1989; FEMA-273 1997; SEOAC 2000) made great progress toward performance-based earthquake engineering. Ever since, the Pacific Earthquake Engineering Research Center (PEER) has been developing a more robust methodology that consists of four stages: hazard analysis, structural analysis, damage analysis, and loss analysis (Porter 2003). In the damage analysis, the fragility function describes the conditional probability for a given intensity level of a system, element or component to be damaged. The Seismic Design Decision Analysis (SDDA) approach, proposed in the


M. A. Aiello and A. Bilotta (Eds.): ICC 2022, LNCE 435, pp. 52–65, 2024.
https://doi.org/10.1007/978-3-031-43102-9_5

United States in 1975, was the first attempt to determine fragility curves (Whitman et al. 1975). Further advances (Kennedy et al. 1980, 1984) were first used in the nuclear sector to construct a probabilistic relationship between the seismic input and the seismic failure of a nuclear power plant component. Ever since, researchers have developed a variety of ways to evaluate fragility, but due to various shortcomings, including the highly subjective nature of expert-based, empirical and experimental methods, the common practice has garnered attention on analytical and hybrid methods over the last two decades.

Incremental Dynamic Analysis (IDA) is one among several analytical approaches to derive fragility curves and is a parametric analysis method developed in 1998 (Luco & Cornell 1998), extensively discussed in 2002 (Vamvatsikos and Cornell 2002) and widely used still today. Numerous researchers have discussed that one of the IDA's fundamental drawbacks, along with its high computing demand, is the introduction of uncertainties as a result of an overly coarse description of seismic input (Han & Chopra 2006; Mackie & Stojadinovi 2005; Kiani & Khanmohammadi 2015, Jalayer et al. 2017). In IDA in fact, a given number of inputs are amplitude scaled to define IM = im, NL-THA is executed and the distribution $DCR_{LS}|IM = im$ is defined using the results. This procedure is repeated by changing the scale factor to describe the seismic response over a wide range of earthquake intensities. Because a single IM is a slightly simpler characterization of ground-motion severity, the value of $DCR_{LS}$ from different ground-motion with IM = im may vary, making the relationship $DCR_{LS} = f(IM)$ probabilistic. The lack of strong ground motion records across high-intensity intervals at specified periods of a structure is a common limitation in current databases (Vargas-Alzate & Hurtado 2021; Bradley 2013). As a result, excessive scaling to fit high-intensity intervals could bias the structural response (Luco & Bazzurro 2007), resulting in an erroneous correlation between IMs and EDPs and enhancing structural response uncertainties.

Unlike IDA and MSA (Multiple-Stripe Analysis), to cover a wide range of seismic intensities, the Cloud Method (Shome 1999; Bazzurro et al. 1998; Jalayer 2003) require nonlinear analysis of a structure given a sample of different distance/intensity pairs of unscaled as-recorded ground motion and is based on a regression in the logarithmic space of structural response vs seismic intensity. Accordingly to Miano et al. (2019), Cloud Method can minimize the number of analyses, biases in seismic input with intensity and computation cost in developing seismic fragility curves.

Many studies on the use of pushover analysis methodologies to assess seismic vulnerability, especially on bridges, have been published since the late 1990s in response to the need for simplified, faster, and approximate methods. In these researches, Nonlinear Static Procedures (NSPs) (i.e. the Capacity Spectrum Method (Mander 1999; Shinozuka et al. 2000; Banerjee & Shinozuka 2007; Rossetto et al. 36), the N2 method (Dolšek & Fajfar 2004; Faella et al. 2008; Jalayer et al. 2019) and Modal Pushover Analysis (Han & Chopra 2006)) are used to assess the system's capacity, while response spectra are used to estimate demand. Some authors have compared developed fragility curves to those obtained by nonlinear time history analysis to assess the effectiveness of various analytical approaches (Shinozuka et al 2000; Rossetta et al. 2016; Faella et al. 2008).

In recent years Bergami et al. have suggested a new nonlinear static procedure called Incremental Modal Pushover Analysis (IMPA) (Bergami et al. 2015) to produce a multimodal capacity curve in terms of base shear vs top displacement. IMPA requires the execution of Modal Pushover Analysis (MPA) using scaled response spectra to assess the structural performance across a variety of seismic intensity levels. Although this method is suitable for performing a displacement-based design procedure and structural analysis of existing structures, the authors did not suggest an analytical estimation of fragility yet.

The authors claim that scaling response spectra over a wider range of intensities, may introduce fewer uncertainties than simple amplitude scaling of ground motions. Further steps of this research will investigate how seismic input uncertainties affect the accuracy of IMPA vs IDA seismic fragility for different ground motion bins, in particular, pulse-like records. This work focuses on determining the reliability of structural fragility derived using aforementioned methods by comparing the curves obtained for an actual RC frame by Cloud Method, the accuracy of pushover-based seismic vulnerability estimation is validated.

## 2 Methodology

### 2.1 Record Selection

A core database of 210 ground motions was extracted from the PEER Next Generation Attenuation (NGA) - West 2 Database (Ancheta et al. 2014). In addition, two ground motion records, with epicentral distances of 4.6 km and 26.9 km respectively, representing the 2016 Norcia earthquake, were collected from the Italian Accelerometric Archive (Luzi et al. 2008) and included in the previous database.

The following Table 1 gives some details on the two sub-sets of records considered divided into two groups of 16 accelerograms by epicentral distance, from 0 to 10 km and 10 to 50 km respectively. A wide range of IM and dispersed $DCR_{LS}$ values are included in record selection, with at least one-third of the $DCR_{LS}$ values being greater than 1. (Jalayer et al. 2017). The selection of near-fault records excludes pulse-like ground motions which are well known to differ from common ground motions (Somerville 2002). The Shahi and Baker procedure was used to identify impulsive signals (Shahi & Baker 2014).

### 2.2 Chosen EDP and IM

Various Engineering Demand Parameters (EDPs) have been presented in the literature (Shome et al. 1998) and each of them is meant to describe the structure's local or global damaged condition. This study assumes the EDP to be the critical demand to capacity ratio for the intended Limit State (LS), abbreviated as $DCR_{LS}$ (Jalayer et al. 2015, 2017). It denotes the demand-to-capacity ratio that brings the system closer to the limit state (herein, the life-safety limit state). The weakest-link formulation is used to assess the $DCR_{LS}$ (Eq. 1). The structure achieves the desired limit state for the $l^{th}$ mechanism if the

**Table 1.** Details of the two subsets of ground motion data deepened for the study based on the NGA—West 2 database.

| RSN | Earthquake Name | Year | Mw | $R_{jb}$ (km) |
|---|---|---|---|---|
| 1 | "Helena_ Montana-01" | 1935 | 6 | 2.07 |
| 112 | "Oroville-03" | 1975 | 4.7 | 5.95 |
| 134 | "Izmir_ Turkey" | 1977 | 5.3 | 0.74 |
| 139 | "Tabas_ Iran" | 1978 | 7.35 | 0 |
| 144 | "Dursunbey_ Turkey" | 1979 | 5.34 | 5.57 |
| 145 | "Coyote Lake" | 1979 | 5.74 | 5.3 |
| 150 | "Coyote Lake" | 1979 | 5.74 | 0.42 |
| 226 | "Anza (Horse Canyon)-01" | 1980 | 5.19 | 5.85 |
| 240 | "Mammoth Lakes-04" | 1980 | 5.7 | 1.37 |
| 251 | "Mammoth Lakes-07" | 1980 | 4.73 | 3.33 |
| 6 | Imperial Valley-02 | 1940 | 6.95 | 6.09 |
| 563 | Chalfant Valley-04 | 1986 | 5.44 | 8.88 |
| 564 | Kalamata, Greece-01 | 1986 | 6.2 | 6.45 |
| 565 | Kalamata, Greece-02 | 1986 | 5.4 | 4 |
| 752 | Loma Prieta | 1989 | 6.93 | 8.65 |
| 1 | "Helena_ Montana-01" | 1935 | 6 | 2.07 |
| 112 | "Oroville-03" | 1975 | 4.7 | 5.95 |
| 134 | "Izmir_ Turkey" | 1977 | 5.3 | 0.74 |
| n.a. | Central Italy | 2016 | 6.5 | 4.6 |
| 15 | "Kern County" | 1952 | 7.36 | 38.42 |
| 33 | "Parkfield" | 1966 | 6.19 | 15.96 |
| 70 | "San Fernando" | 1971 | 6.61 | 22.23 |
| 72 | "San Fernando" | 1971 | 6.61 | 19.45 |
| 101 | "Northern Calif-07" | 1975 | 5.2 | 28.73 |
| 107 | "Oroville-02" | 1975 | 4.79 | 13.55 |
| 125 | "Friuli_ Italy-01" | 1976 | 6.5 | 14.97 |
| 152 | "Coyote Lake" | 1979 | 5.74 | 20.44 |
| 155 | "Norcia_ Italy" | 1979 | 5.9 | 31.43 |
| 225 | "Anza (Horse Canyon)-01" | 1980 | 5.19 | 12.24 |
| 231 | "Mammoth Lakes-01" | 1980 | 6.06 | 12.56 |
| 239 | "Mammoth Lakes-03" | 1980 | 5.91 | 10.31 |
| 265 | "Victoria_ Mexico" | 1980 | 6.33 | 13.8 |
| 274 | "Mammoth Lakes-09" | 1980 | 4.85 | 10.96 |
| 442 | "Borah Peak_ ID-02" | 1983 | 5.1 | 16.31 |
| n.a. | Central Italy | 2016 | 6.5 | 26.9 |

demand-to-capacity ratio $D_{jl}/C_{jl}$ is equal to or higher than unity in just one component.

$$DCR_{LS} = \max_{l}^{N_{mech}} \max_{j}^{N_e}\left(\frac{D_{jl}}{C_{jl}(LS)}\right) \tag{1}$$

where $N_{mech}$ = the number of the considered potential mechanism of failure and N = the number of the elements taking part in the lth mechanism, respectively. $D_{jl}$ and $C_{jl}(LS)$ are the demand and limit state capacity for the $j^{th}$ element of the $l^{th}$ mechanism, respectively.

In this study, only a ductile failure mechanism in columns and beams is considered a potential failure mechanism ($N_{mech} = 1$). As a result, the demand D is defined in terms of maximum chord rotation in the $j^{th}$ component, while the chord rotation capacity C is assessed according to Equation C8.7.2.5 given in the NTC 2018 Commentary (Circ. 2009). For the life-safety limit state, the chord rotation is defined as 3/4 of that corresponding to the near-collapse limit state $\theta_u$ (Eq. 2).

$$\theta_u = \left(\theta_y + (\phi_u - \phi_y)L_{pl}\left(1 - \frac{0,5L_{pl}}{L_v}\right)\right) \tag{2}$$

where the $\theta_y$ = yielding chord rotation, $\phi_u$ = ultimate curvature, $\phi_y$ = yielding curvature, respectively, $L_{pl}$ = plastic hinge length, $L_v$ = shear length.

Regarding chosen IM, the first-mode spectral acceleration $S_a(T_1, \xi = 5\%)$ is commonly used as an Intensity Measure (IM) parameter (Han & Chopra 2006; Mackie & Stojadinović 2005). As the studied frame is dominated by the first mode of vibration (structure's first-mode period of vibration $T_1 = 0.62$ s and the modal mass participation at first mode is 82%), the 5% damped spectral acceleration at the structure's fundamental period $Sa(T_1, \xi = 5\%)$ or simply Sa is adopted as the IM in this work. This option is consistent with what Shome et al. (1998) previously stated about the nonlinear response of an MDOF structure where the fundamental period is dominant.

### 2.3 Cloud-Based Analysis

Because of the simplicity of its formulation and the low computational effort required, the Cloud-based Analysis (CA) is especially well suited to assess structural fragility. Conversely, it is highly sensitive to record selections and is based on a few simplifying assumptions, such as the regression's fixed standard error (Jalayer et al. 2015, 2017, 2019; Jalayer 2003; Miano et al. 2018, 2019). To fit the pairs of demand to capacity ratio ($DCR_{LS}$) and IM, CA uses a linear regression model in the logarithm space, with $DCR_{LS}$ derived using nonlinear dynamic analysis. The $DCR_{LS}$ for a given IM level is described by a regression-based probability model, which may be assessed using the following equations:

$$E[\ln DCR_{LS}|IM] = \ln \eta_{DCR_{LS}|IM} = \ln a + b \ln IM \tag{3}$$

$$\sigma_{\ln \beta_{DCR_{LS}|IM}} \cong \beta_{DCR_{LS}|IM} = \sqrt{\sum_{i=1}^{N} \left(\ln, DCR_{LS,i} - \ln \eta_{DCR_{LS}|IM_i}\right)^2/(N-2)} \tag{4}$$

where $E[\ln DCR_{LS}|IM]$ = the expected value for the natural logarithm of $DCR_{LS}$ given IM, $\eta DCR_{LS}|IM$ = median and $\sigma \ln DCR_{LS}|IM$ = logarithmic standard deviation for $DCR_{LS}$ given IM. The constants ln a and b are the linear least square regression coefficients. Finally, the structural fragility obtained based on the CA is:

$$P(DCR_{LS} > 1|IM) = P(\ln DCR_{LS} > 0|IM) = \Phi\left(\frac{\ln \eta_{DCR_{LS}|IM}}{\beta_{DCR_{LS}|IM}}\right) \quad (5)$$

where $\Phi$ = standard Gaussian cumulative distribution function.

To determine the relationships of IM versus $DCR_{LS}$ for the structure under examination, two alternative approaches are used in this study. Time-history analyses are used in one scenario to evaluate the demand $D_{jl}$ (demand of the $j^{th}$ element of the $l^{th}$ mechanism) at each time step. This method is referred to as Dynamic Cloud Analysis (D-CA) in this research. In the second scenario, the demand $D_{jl}$ is calculated using the modal pushover analysis (MPA), so MPA-CA is the name given to this method. The modal pushover analysis (Chopra & Goel 1999, 2002) is a nonlinear static approach based on a static analysis of a structure subjected to lateral forces distributed over the building height following the nth modal shape. The Capacity Spectrum Method (CSM) is implemented in the MPA procedure: a response spectrum (representing structure demand) and a pushover curve (representing building capacity) are comparing by converting them in the ADRS format to find a correlation between earthquake ground motions and building performance (ATC-40 1982). This seeks to define the Performance Point (PP), which represents the demand of a building in terms of maximum inelastic displacement for a given seismic event.

## 2.4 Incremental Dynamic Analysis (IDA)

In IDA, a nonlinear model is subjected to a set of accelerogram $a_\lambda$ that are each scaled to multiple levels of a monotonic scalable intensity measure like $S_a$, PGV, or PGA (herein $IM = S_a(T_1, \xi = 5\%)$). Unscaled "as-recorded" time histories are scaled with a non-negative scale factor ($\lambda$) to produce a scaled accelerogram $a_\lambda$ in which amplitudes are scaled without affecting signal frequency content. The analysis' output is represented by a set of IDA curves, which are plots of the recorded DM ($DCR_{LS}$) against IM ($S_a(T_1, \xi = 5\%)$), all parameterized with the identical IMs and DMs (Vamvatsikos and Cornell 2002). The following is one of the simplest suggested analytical methods for developing fragility based on IDA (Porter et al. 2007):

$$P(LS|IM = x) = P(DCR_{LS} \geq 1|IM = x) = P\left(IM^{DCR=1} \leq x\right) \quad (6)$$

In an EDP-based interpretation of fragility (Eq. 6), the conditional probability of exceeding a limit state given an IM is equal to the probability of the $DCR_{LS}$ reaching 1 for a given $S_a$. Yet, the fragility can also be expressed as a complementary cumulative distribution function or as "IM-based fragility" (Eq. 7). The IM-based derivation of fragility is well suited to incremental dynamic analysis. This interpretation describes seismic fragility as the probability of spectral acceleration values being smaller than a particular value, indicated as $Sa_{DCR} = 1$ and defined by intersecting IDA curves with the $DCR_{LS} = 1$.

$$P\left(IM^{DCR=1} \leq x\right) = \phi\left(\frac{\ln x - \ln \eta_{S_a|DCR=1}}{\beta_{S_a|DCR=1}}\right) \tag{7}$$

In Eq. 7, the standard normal (Gaussian) cumulative distribution (CDF) of two parameters (median or log of mean and standard deviation) is estimated by the method or "METHOD A", suggested by Porter (Eqs. 8 and 9) (Porter et al. 2007).

$$\ln \eta_{S_a|DRC=1} \cong \frac{\sum_{i=1}^{n} \ln S_a^{DCR=1}}{n} \tag{8}$$

$$\beta_{S_a|DRC=1} \cong \sqrt{\frac{\sum_{i=1}^{n}\left(\ln S_{a,1}^{DCR=1} - \ln \eta_{S_a|DRC=1}\right)^2}{n-1}} \tag{9}$$

### 2.5 Incremental Modal Pushover Analysis (IMPA)

Bergami and co-authors presented IMPA, a novel nonlinear static technique, initially for buildings (Bergami et al. 2015a, b, 2017) and later for bridges (Bergami et al. 2020a, b, 2021). By scaling down response spectra, this method takes advantage of the simplicity of nonlinear static analysis while allowing for the definition of seismic demand for a certain range of intensity levels. Although the procedure for determining the maximum expected demand for the $j^{th}$ element of the $l^{th}$ mechanism $D_{jl}$ is conceptually similar to that described for MPA-CA, response spectra in the IMPA procedure are scaled to multiple levels of a chosen monotonic scalable intensity measure such as $S_a$, PGA, or PGV (herein $IM = S_a(T_1, \xi = 5\%)$).

If two or more modes of vibration are considered, the performance points (P.P.) can be obtained for each intensity level, and the demand measure $D_{jl}$ can be combined to define a multimodal $DCR_{LS}$. The analysis' outcome is expressed as a set of "IMPA curves", which are plots of $DCR_{LS}$ against IM that are very similar to IDA curves. The empirical distribution of the random variable (IM) for the probabilistic model of fragility is given by the intersection of IMPA curves and the chosen limit state threshold ($DCR_{LS} = 1$) (Eqs. 6 and 7).

## 3 Numerical Analysis

### 3.1 Frame and FE Model Description

The 2D frame used as a case study in this research was taken from a school facility built in 1962 in Norcia (Italy, 42.7941° N, 13.0963° E). It is considered to be fixed at the base (Fig. 1a) and it is a two-bay (5.65 m and 5.56 m span) regular cross frame with an inter-story height of 3.50 m for the ground floor, 3.30 m for the other stories and 2.5 m for the attic (Fig. 1). The fundamental and second periods of the frame are equal to 0.62 s and 0.21 s, respectively. The geometry of the frame, beams and columns geometry and reinforcement details are shown in Figs. 1b.

The nonlinear FE model of the investigated frame was developed in the OpeenSEES platform (McKenna et al. 2000). The "Beam With Hinges Element" from the OpenSEES library is used to model beams and columns to account for the nonlinearity. The length of the plastic hinge $L_p$ is assumed to be equal to the cross-section height, which is in good agreement with the $L_p$ assessed by the equation proposed by Priestley and Park (1987). A four-point moment-curvature law accounting for crack, yield, ultimate failure and collapse state (80% of ultimate failure) has been evaluated through the software Response 2000 and assigned to the plastic hinges. Shear failure is not considered in the model.

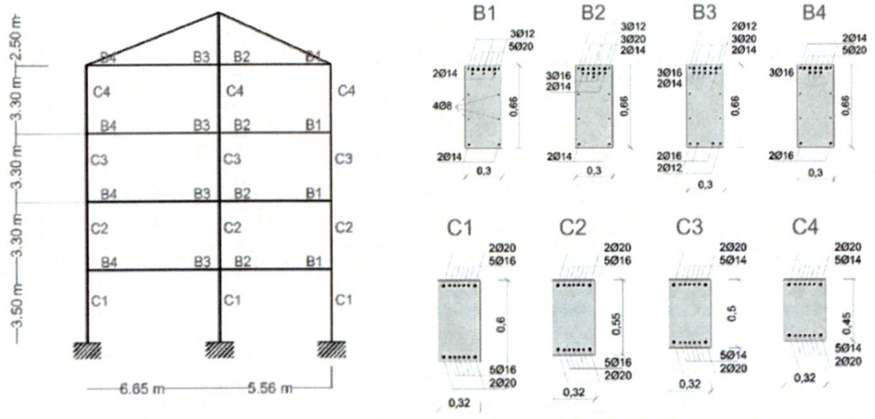


(a) (b)

**Fig. 1.** (a) Geometry of the frame, (b) Columns "C" and beams "B" cross-sections and longitudinal reinforcement details Captions should be written both in English and in Italian.

Two-point Gauss integration is used in elements' interior, while two-point Gauss-Radau integration is used over lengths of 4Lp at the element ends. Six integration points are used in total (Scott & Fenves 2006, Scott & Ryan). The Newton with Line Search method is the solution algorithm used for time-history analyses, Newmark has been used as the integrator and the convergence test is the Normal Displacement Increment. OpenSees' default tolerance (10–7) and a maximum number of iterations (50) are used. To account for energy dissipation, Rayleigh damping is adopted.

Accelerograms and response spectra, in IDA and IMPA analysis respectively, are scaled by a scale factor $\lambda$ chosen to scale the spectral acceleration at the fundamental period $S_a(T_1,\xi = 5\%)$ to $IM = a_\lambda \in [0.1\text{ g}, 0.8\text{ g}]$ with $\Delta a_\lambda = 0.1$ g. The mass damping coefficient and the stiffness damping coefficient of the Rayleigh damping are evaluated by considering the first and the second natural frequency of the case study. The percentage of critical damping is equal to 5%.

# 4 Results

## 4.1 Nonlinear Static Analysis

The pushover analysis has been performed in displacement control to reach a target displacement of 350 mm, with each computation step incrementing by 1 mm. The capacity curves, derived using two load distributions proportional to the first and second modal forms, respectively, are shown in Fig. 2a. The first load distribution's capacity curve reaches a maximum base shear of 492 kN, which corresponds to a top displacement of around 220 mm. Figure 2b depicts the activation sequence of the plastic hinges at different steps of the analysis. Plastic hinges were activated first in the columns rather than the beams, especially in the upper stories which, according to Eurocode 8, have weak column-strong beams.

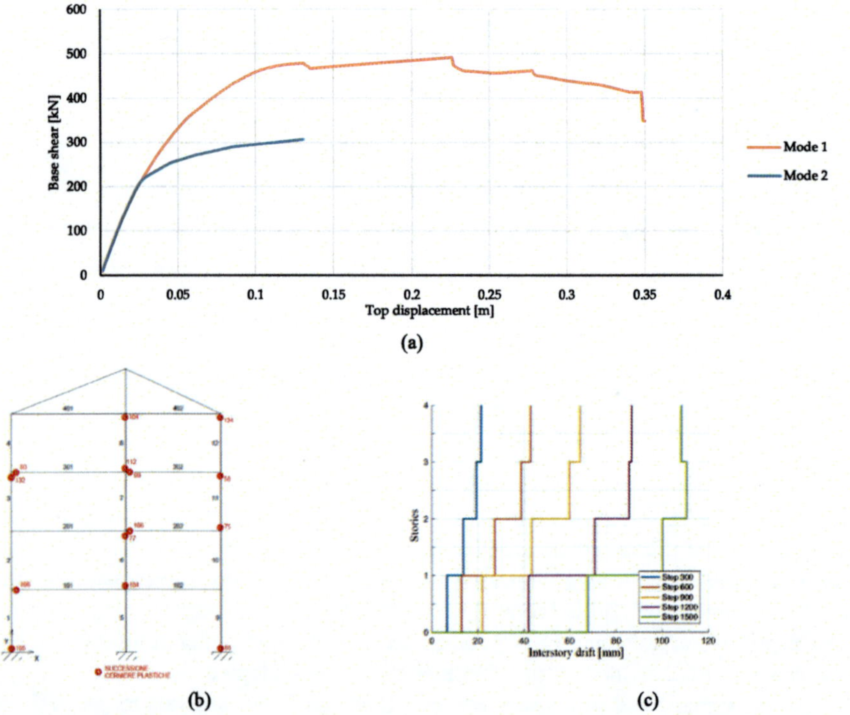


**Fig. 2.** (a) Capacity curve of the frame, (b) plastic hinges sequence, (c) interstory drift at different steps of the pushover analysis.

## 4.2 Nonlinear Dynamic Analysis

This paper analysed in total 12 subsets of records generated arbitrarily from the main set of 210 records in order to follow the general record selection guidelines outlined in

Jalayer et al. 2017. As expected, IDA and IMPA are less sensitive on record selection, with mean fragility curve values ranging from 0.463 g to 0.525 g and 0.387 to 0.432 g, respectively, and small values of standard deviation equal to 0.02 and 0.01, respectively. D-CA and MPA-CA, on the contrary, are more dependent on record selection, with higher values of standard deviation for the $16^{th}$, $50^{th}$ and $84^{th}$ fractiles of the 12 fragility curves (Table 3). To verify the solidity of each method, the derived fragility curves are compared to those obtained using D-CA. The accuracy of various fragility models' with respect to Cloud Analysis predictions is measured using normalised root-mean-square deviation (RMSD) (Eq. 10), assuming the values obtained by Eq. 5 are the reference values:

$$RMSD(\%) = \sum_{i=1}^{n} \sqrt{\frac{(\hat{y}_i - y_i)^2}{\hat{y}_i^{\,2}}} \tag{10}$$

where n = number of points, $y_i$ = predicted probability of exceeding the considered limit state and $\hat{y}_i$ reference probability of exceeding the considered limit state.

The following Table 2 compares results arised from two record selection for each described method to Cloud Analysis in terms of percentage variation of median values, $16^{th}$ and $84^{th}$ fractiles of fragility curves, and absolute values of β. When compared to D-CA, which IDA appears to be more accurate in calculating the $50^{th}$, $16^{th}$ and $84^{th}$ fractiles and has the smallest normalised root-mean-square deviation, equal to 3%,. However, when it comes to evaluating fragility in Far Field recordings, IMPA and IDA leads to comparable results. The inclusion of two or more vibration modes in determining the maximum multimodal DCRLS ($MPA_2$-CA and $IMPA_2$) seemed to be unnecessary, as it yielded high conservative results. Table 1. Details of the two subsets of ground motion data deepened for the study based on the NGA—West 2 database.

**Table 2.** $16^{th}$, $50^{th}$ $84^{th}$ fractiles and β parameter of fragility curves derived via different methods for two selection of records. The table shows also the normalised root-mean-square deviation with respect to D-CA.

| Methodology | Near-Fault Record | | | | | Far-Field Record | | | | |
|---|---|---|---|---|---|---|---|---|---|---|
| | $\eta_{16\%}$ [g] | $\eta_{50\%}$ [g] | $\eta_{84\%}$ [g] | β | RMSD | $\eta_{16\%}$ [g] | $\eta_{50\%}$ [g] | $\eta_{84\%}$ [g] | β | RMSD |
| D-CA | 0.38 | 0.50 | 0.66 | 0.26 | - | 0.53 | 0.66 | 0.81 | 0.19 | - |
| IDA | 0.42 | 0.51 | 0.63 | 0.2 | 3% | 0.38 | 0.48 | 0.60 | 0.23 | 14% |
| $IMPA_1$ | 0.31 | 0.41 | 0.55 | 0.29 | 10% | 0.27 | 0.38 | 0.52 | 0.32 | 17% |
| $IMPA_2$ | 0.23 | 0.31 | 0.42 | 0.28 | 16% | 0.21 | 0.28 | 0.38 | 0.3 | 19% |
| $MPA_1$-CA | 0.26 | 0.35 | 0.47 | 0.32 | 14% | 0.37 | 0.54 | 0.77 | 0.29 | 8% |
| $MPA_2$-CA | 0.21 | 0.27 | 0.36 | 0.32 | 17% | 0.26 | 0.37 | 0.54 | 0.29 | 17% |

**Table 3.** Percentage variation of the 16th percentile, mean and 84th percentile of fragility curves for different methods with respect to D-CA for other record selections

| | D-CA | IDA | $IMPA_1$ | $MPA_1$-CA | D-CA | IDA | $IMPA_1$ | $MPA_1$-CA | D-CA | IDA | $IMPA_1$ | $MPA_1$-CA |
|---|---|---|---|---|---|---|---|---|---|---|---|---|
| Fractile | 0.16 | 0.16 | 0.16 | 0.16 | 0.5 | 0.5 | 0.5 | 0.5 | 0.84 | 0.84 | 0.84 | 0.84 |
| | [g] | % | % | % | [g] | % | % | % | [g] | % | % | % |
| Sel. 1 $_{NF}$ | 0.393 | −1% | −21% | −35% | 0.525 | −7% | −21% | −35% | 0.701 | −13% | −20% | −34% |
| Sel. 2 $_{NF}$ | 0.361 | −2% | −15% | −34% | 0.521 | −11% | −17% | −34% | 0.751 | −19% | −19% | −33% |
| Sel. 3 $_{NF}$ | 0.422 | −6% | −26% | −41% | 0.539 | −7% | −22% | −34% | 0.688 | −8% | −18% | −26% |
| Sel. 4 $_{NF}$ | 0.372 | −5% | −19% | −49% | 0.502 | −7% | −18% | −36% | 0.677 | −9% | −18% | −21% |
| Sel. 5 $_{NF}$ | 0.355 | 13% | −21% | −34% | 0.480 | 5% | −16% | −27% | 0.649 | −1% | −11% | −20% |
| Sel. 6 $_{NF}$ | 0.367 | 4% | −15% | −38% | 0.488 | −1% | −15% | −33% | 0.648 | −6% | −15% | −28% |
| Sel. 7 $_{FF}$ | 0.574 | −30% | −49% | −24% | 0.724 | −34% | −45% | −19% | 0.914 | −37% | −41% | −14% |
| Sel. 8 $_{FF}$ | 0.496 | −24% | −44% | −45% | 0.658 | −26% | −40% | −32% | 0.871 | −27% | −36% | −16% |
| Sel. 9 $_{FF}$ | 0.573 | −31% | −47% | −29% | 0.687 | −27% | −39% | −17% | 0.822 | −23% | −31% | −4% |
| Sel. 10 $_{FF}$ | 0.589 | −32% | −52% | −33% | 0.708 | −27% | −42% | −12% | 0.849 | −22% | −30% | 16% |
| Sel. 11 $_{FF}$ | 0.574 | −25% | −45% | −27% | 0.687 | −24% | −38% | −14% | 0.822 | −22% | −31% | 1% |
| Sel. 12 $_{FF}$ | 0.570 | −32% | −47% | −33% | 0.668 | −28% | −40% | −22% | 0.784 | −23% | −32% | −8% |
| μ [g] | 0.471 | 0.389 | 0.300 | 0.308 | 0.599 | 0.492 | 0.411 | 0.448 | 0.765 | 0.622 | 0.565 | 0.653 |
| σ | 0.10 | 0.02 | 0.01 | 0.09 | 0.10 | 0.02 | 0.01 | 0.12 | 0.09 | 0.02 | 0.02 | 0.17 |

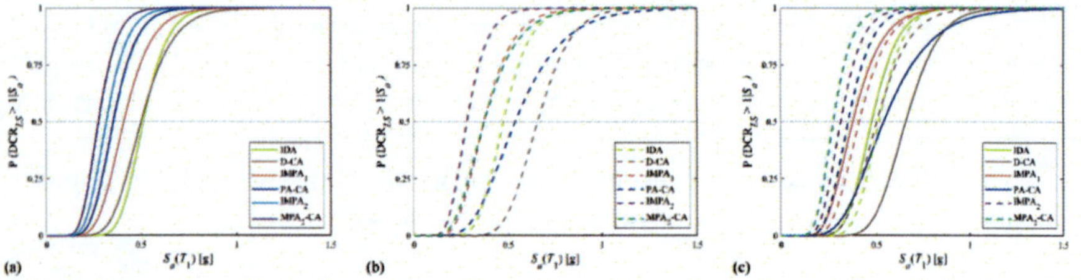

**Fig. 3.** Comparison among fragility curves for all aforementioned methods: (a) comparison among fragility curves for near-fault (NF) selection (b) far-field (FF) selection, and (c) total.

# 5 Conclusions

Fragility curves generated by different known static and dynamic nonlinear techniques are compared in this work. To determine IMPA curves and hence seismic fragility, Incremental Modal Pushover Analysis (IMPA) is offered as an alternative to IDA, which is currently regarded as the most reliable method. In the well-known Cloud Method, both MPA and NL-THA are employed to determine capacity. From this investigation, which is limited to a simplified 2D frame model and a small set of data, the following findings can be drawn:

in the case of methodologies distinguished by scaling, IDA and IMPA, a comparison of fragility curves reveals that near-fault non-pulse like and far-field record selections produced essentially identical results. In contrast, when using records without scaling, such as in D-CA and MPA-CA, the results in terms of fragility reveal clear variations over the whole range of intensities. The addition of higher mode contributions does not appear to be necessary for low to medium buildings (up to 9 stories (Han & Chopra 2006)); nonetheless, fragility curves that just evaluate the contribution of the first mode in calculating $DCR_{LS}$ have led to more accurate results in respect to IDA. In all subsets and over the whole range of intensities, D-CA leads to a greater vulnerability than MPA-CA and IMPA. Pushover analysis-based methodologies, on the other hand, have produced more conservative results, particularly for the 16 and 50 fractiles IDA appears less sensitive to record-to-record variability but IMPA, however, has the advantage of a considerable reduction in the computational time required to complete the structural analysis, despite its slightly higher sensitivity. The total time required in IMPA is primarily related to the post-processing phase, which is the same for small 2D frames as it is for more complex 3D buildings. To confirm the observed data and derive more general conclusions, a more thorough validation is required (Fig. 3).

## References


Ancheta, T.D., et al.: NGA-West2 database. Earthq. Spectra **30**(3), 989–1005 (2014)

Applied Technology Council. (ATC-20) Procedures for Post-earthquake Safety Evaluation of Buildings. Redwood City, Calif.: Applied Technology Council (1989)

Applied Technology Council. Seismic evaluation and retrofit of concrete buildings. Report ATC 40. November 1996

Banerjee, S., Shinozuka, M.: Nonlinear static procedure for seismic vulnerability assessment of bridges. Comput. Aided Civil Infrastruct. Eng. **22**(4), 293–305 (2007)

Bazzurro, P., Cornell, C.A., Shome, N., Carballo, J.E.: Three proposals for characterizing MDOF nonlinear seismic response. J. Struct. Eng. **124**(11), 1281–1289 (1998)

Bergami, A.V., Fiorentino, G., Lavorato, D., Briseghella, B., Nuti, C.: Application of the incremental modal pushover analysis to bridges subjected to near-fault ground motions. Appl. Sci. **10**(19), 6738 (2020)

Bergami, A.V., Forte, A., Lavorato, D., Nuti, C.: The incremental modal pushover analysis (IMPA): proposal and application. 16WCEE, Santiago del Chile, Chile (2017)

Bergami, A.V., Liu, X., Nuti, C.: Evaluation of a modal pushover based incremental analysis. In: Proceedings of ACE, Vietri sul mare, Italy, pp. 12–13 (2015)

Bergami, A.V., Nuti, C., Liu, X.: Proposal and application of the incremental modal pushover analysis (IMPA). In: Proceedings of the IABSE Conference, Geneva (2015)

Bergami, A.V., Nuti, C., Lavorato, D., Fiorentino, G., Briseghella, B.: IMPAβ: incremental modal pushover analysis for bridges. Appl. Sci. **10**(12), 4287 (2020)

Bergami, A.V., et al.: Seismic assessment of corroded concrete bridges using incremental modal pushover analysis. In: Proceedings of the Institution of Civil Engineers-Bridge Engineering, pp. 1–29. Thomas Telford Ltd (2021)

Bradley, B.A.: A critical examination of seismic response uncertainty analysis in earthquake engineering. Earthquake Eng. Struct. Dynam. **42**(11), 1717–1729 (2013)

Chopra, A.K., Goel, R.K.: Capacity-demand-diagram methods for estimating seismic deformation of inelastic structures: SDF systems. Report No. PEER1999/02 (1999)

Chopra, A.K., Goel, R.K.: A modal pushover analysis procedure for estimating seismic demands for buildings. Earthquake Eng. Struct. Dynam. **31**(3), 561–582 (2002)

Code, P.: Eurocode 8: Design of structures for earthquake resistance-part 1: general rules, seismic actions and rules for buildings. European Committee for Standardization, Brussels (2005)

Contiguglia, C.P., Pelle, A., Briseghella, B., Nuti, C.: IMPA versus cloud analysis and IDA: different methods to evaluate structural seismic fragility. Appl. Sci. **12**(7), 3687 (2022)

Dolšek, M., Fajfar, P.: IN2-A simple alternative for IDA. In: 13th World Conference on Earthquake Engineering, pp. 1–6 (2004)
Faella, C., Lima, C., Martinelli, E.: Non-linear static methods for seismic fragility analysis and reliability evaluation of existing structures. In: Proceedings of the 14th World Conference on Earthquake Engineering, pp. 12–17. Beijing, China (2008)
Freeman, S.A.: The capacity spectrum method as a tool for seismic design. In Proceedings of the 11th European Conference on Earthquake Engineering, pp. 6–11, Paris, France (1998)
Han, S.W., Chopra, A.K.: Approximate incremental dynamic analysis using the modal pushover analysis procedure. Earthquake Eng. Struct. Dynam. **35**(15), 1853–1873 (2006)
Iervolino, I., Cornell, C.A.: Record selection for nonlinear seismic analysis of structures. Earthq. Spectra **21**(3), 685–713 (2005)
Jalayer, F.: Direct Probabilistic Seismic Analysis: Implementing Non-Linear Dynamic Assessments. Stanford University (2003)
Jalayer, F., De Risi, R., Manfredi, G.: Bayesian cloud analysis: efficient structural fragility assessment using linear regression. Bull. Earthq. Eng. **13**(4), 1183–1203 (2015)
Jalayer, F., Ebrahimian, H., Miano, A.: N2 with cloud: a non-linear dynamic analysis procedure for the equivalent SDOF system. N2 with Cloud: A Non-Linear Dynamic Analysis Procedure for the Equivalent SDOF System, pp. 215–225 (2019)
Jalayer, F., Ebrahimian, H., Miano, A., Manfredi, G., Sezen, H.: Analytical fragility assessment using unscaled ground motion records. Earthquake Eng. Struct. Dynam. **46**(15), 2639–2663 (2017)
Kennedy, R.P., Ravindra, M.K.: Seismic fragilities for nuclear power plant risk studies. Nucl. Eng. Des. **79**(1), 47–68 (1984)
Kennedy, R.P., Cornell, C.A., Campbell, R.D., Kaplan, S., Perla, H.F.: Probabilistic seismic safety study of an existing nuclear power plant. Nucl. Eng. Des. **59**(2), 315–338 (1980)
Kiani, J., Khanmohammadi, M.: New approach for selection of real input ground motion records for incremental dynamic analysis (IDA). J. Earthquake Eng. **19**(4), 592–623 (2015)
Krawinkler, H.: Challenges and progress in performance-based earthquake engineering. In: International Seminar on Seismic Engineering for Tomorrow–In Honor of Professor Hiroshi Akiyama, vol. 26 (1999)
Luco, N., Bazzurro, P.: Does amplitude scaling of ground motion records result in biased nonlinear structural drift responses? Earthquake Eng. Struct. Dynam. **36**(13), 1813–1835 (2007)
Luco, N., Cornell, C.A.: Effects of random connection fractures on the demands and reliability for a 3-story pre-Northridge SMRF structure. In: Proceedings of the 6th US National Conference on Earthquake Engineering, vol. 244, pp. 1–12. El Cerrito, CA, USA: EERI (1998)
Luzi, L., Hailemikael, S., Bindi, D., Pacor, F., Mele, F., Sabetta, F.: ITACA (ITalian ACcelerometric Archive): a web portal for the dissemination of the Italian strong motion data. Seismol. Res. Let. (2008)
Mackie, K.R., Stojadinović, B.: Comparison of incremental dynamic, cloud, and stripe methods for computing probabilistic seismic demand models. In: Structures Congress 2005: Metropolis and Beyond, pp. 1–11 (2005)
Mander, J.B.: Fragility curve development for assessing the seismic vulnerability of highway bridges. Res. Progress **89** (1999)
McKenna, F., Fenves, G.L., Scott, M.H., Jeremir, B.: Open System for Earthquake Engineering Simulation. University of Berkeley, OpenSEES (2000)
Miano, A., Jalayer, F., Ebrahimian, H., Prota, A.: Cloud to IDA: efficient fragility assessment with limited scaling. Earthquake Eng. Struct. Dynam. **47**(5), 1124–1147 (2018)
Miano, A., Jalayer, F., Ebrahimian, H., Prota, A.: Nonlinear dynamic analysis procedure with limited number of analyses and scaling. In: Proceedings of 7th ECCOMAS Thematic Conference on Computational Methods in Structural Dynamics and Earthquake Engineering (COMPDYN), Crete, Greece (2019)

MIT Ministry of Infrastructures and Transportation. Circ. C.S.Ll.Pp. No. 617 of 2/2/2009 "Istruzioni per l'applicazione delle nuove norme tecniche per le costruzioni di cui al Decreto Ministeriale 14 Gennaio 2008," Consiglio superiore dei lavori pubblici. S.O. n.27 alla G.U. del 26.02.2009, No. 47 (2009)

Muntasir, A.H.M., Shahria, M.: Seismic fragility assessment of highway bridges: a state-of-the-art review. Struct. Infrastruct. Eng. **11**(6), 804–832 (2015)

Porter, K.A.: An overview of PEER's performance-based earthquake engineering methodology. In: Proceedings of Ninth International Conference on Applications of Statistics and Probability in Civil Engineering, pp. 1–8 (2003)

Porter, K., Kennedy, R., Bachman, R.: Creating fragility functions for performance-based earthquake engineering. Earthq. Spectra **23**(2), 471–489 (2007)

Priestley, M.J.N., Park, R.: Strength and ductility of concrete bridge columns under seismic loading. Struct. J. **84**(1), 61–76 (1987)

Rossetto, T., et al.: FRACAS: a capacity spectrum approach for seismic fragility assessment including record-to-record variability. Eng. Struct. **125**, 337–348 (2016)

Scott, M.H., Fenves, G.L.: Plastic hinge integration methods for force-based beam–column elements. J. Struct. Eng. **132**(2), 244–252 (2006)

Scott, M.H., Ryan, K.L.: Moment-rotation behavior of force-based plastic hinge elements. Earthq. Spectra **29**(2), 597–607 (2013)

Shahi, S.K., Baker, J.W.: An efficient algorithm to identify strong-velocity pulses in multicomponent ground motions. Bull. Seismol. Soc. Am. **104**(5), 2456–2466 (2014)

Shinozuka, M., Feng, M.Q., Kim, H.K., Kim, S.H.: Nonlinear static procedure for fragility curve development. J. Eng. Mech. **126**(12), 1287–1295 (2000)

Shome, N.: Probabilistic Seismic Demand Analysis of Nonlinear Structures. Stanford University (1999)

Shome, N., Cornell, C.A., Bazzurro, P., Carballo, J.E.: Earthquakes, records, and nonlinear responses. Earthq. Spectra **14**(3), 469–500 (1998)

Somerville, P.G.: Characterizing near fault ground motion for the design and evaluation of bridges. In: Proceedings of the 3rd National Seismic Conference and Workshop on Bridges and Highways, vol. 28, pp. 137–148 (2002)

Vamvatsikos, D., Cornell, C.A.: Incremental dynamic analysis. Earthquake Eng. Struct. Dynam. **31**(3), 491–514 (2002)

Vargas-Alzate, Y.F., Hurtado, J.E.: Efficiency of intensity measures considering near-and far-fault ground motion records. Geosciences **11**(6), 234 (2021)

Whitman, R.V., Biggs, J.M., Brennan, J.E., III., Cornell, C.A., de Neufville, R.L., Vanmarcke, E.H.: Seismic design decision analysis. J. Struct. Div. **101**(5), 1067–1084 (1975)

# Recycled Aggregates and Circular Economy: The Case of Centrifuged Reinforced Concrete Poles for Electric Power Lines

Marco Pepe[1], Bianca Maria Mennini[2], Silvio Di Cesare[3], Jean Pierre Goossens Alayon[3], Enrico Valigi[3], Fabrizio Gasbarri[3], Carmine Lima[1], and Enzo Martinelli[4(✉)]

[1] TESIS srl, Fisciano, Italy
{m.pepe,c.lima}@tesis-srl.eu
[2] SIPA spa, Benevento, Italy
b.mennini@sipapali.it
[3] Enel Global Infrastructure and Networks Srl, Roma, Italy
{silvio.dicesare,jeanpierre.goossensalayon,enrico.valigi, fabrizio.gasbarri}@enel.com
[4] Department of Civil Engineering, University of Salerno, Fisciano (SA), Italy
e.martinelli@unisa.it

**Abstract.** Several solutions are currently under investigation with the aim to reduce environmental impact of concrete production processes. These solutions often consist of partially/totally replacing ordinary "natural" constituents with recycled ones, in view of the twofold objective of both reducing the raw materials demand and the amount of waste to be disposed in landfills. This paper summarizes the results of an industrial R&D project aimed at evaluating the feasibility of recycling dismissed concrete poles for the production of new precast centrifuged reinforced concrete elements made of Recycled Concrete Aggregates. The results show that it is possible to obtain recycled concrete poles with mechanical performances compatible to the ordinary ones, provided that an adequate processing procedure are adopted for transforming concrete debris in recycled aggregates and a specific design method is required for the structural concrete mixture composition. Finally, a specific Life-Cycle Analysis with the aim to quantify the environmental impacts of the proposed recycling procedure.

**Keywords:** Recycled Aggregates · Structural concrete · Circular Economy

## 1 Introduction

The construction sector is one of the most energy-intensive and raw-material demanding human activities (Lima et al. 2021) and, hence, it contributes a significant share of greenhouse gas emissions (Wieser et al. 2021). Therefore, making the construction sector "greener" is one of the main challenges for policy makers, private companies and the scientific community (Udomsap & Hallinger 2020). To this aim, one of the most


M. A. Aiello and A. Bilotta (Eds.): ICC 2022, LNCE 435, pp. 66–82, 2024.
https://doi.org/10.1007/978-3-031-43102-9_6

promising actions is based on recycling Construction and Demolition Waste (CDW) and converting them into secondary raw materials for the construction sector itself (Wu et al. 2019). On the other hand, since concrete is one of the most widely used construction materials, enhancing sustainability in its production processes would contribute to reducing the aforementioned environmental impacts (Menegaki & Damigos 2018). To do so, the use of recycled aggregates in new concrete mixtures represents a technically feasible and environmentally effective solution. The latter is often referred to as Recycled Aggregate Concrete (RAC) and the most up-to-date codes and guidelines al-ready provide indications about its production and application (NTC 2018, *EN 206*). One of the main issues related to the diffusion of recycling processes for CDW is the lack of confidence of consumers and builders regarding the quality and heterogeneity of the generated aggregates. CDW can be classified in relation to their original source: construction, demolition, renovation of buildings and natural disasters. In general, CDW are a mixed residue, but despite the wide heterogeneity of their composition, they have a high potential for recycling, since around 40% to 85% of them are composed of concrete, mortar, rocks and ceramic materials (Chen et al. 2018). In the last decades, many studies (Amario et al. 2017) demonstrated the potential of using fine and coarse Recycled Concrete Aggregates (RCAs, presenting at least 90% by mass of Portland cement-based fragments and rocks) also for high-strength structural RAC, even though most of these studies have been performed at a laboratory scale without demonstrating the real possibility of upscaling the RCAs production at an industrial level.

In this context, this paper summarizes the results of an industrial R&D project aimed at evaluating the feasibility of recycling dismissed concrete poles for the production of new precast centrifuged reinforced concrete elements made of RCAs. The study is organized as follow: (i) *Processing procedure* for RCAs from dismissed concrete poles; (ii) *Properties* of RCAs; (iii) *Mixture proportioning* of RACs; (iv) *Prototyping* and *Life Cycle Analysis.*

## 2 Recycled Aggregates: Processing Procedures and Properties

### 2.1 Production of Recycled Concrete Aggregates

The recycled aggregates used in the present study were obtained from dismissed "old" centrifuged reinforced concrete poles. As a matter of principle, since the "original" waste concrete source (i.e., dismissed concrete poles in Fig. 1) was already homogenized and separated from possible impurities, during the RCAs processing procedure, the following phases were executed:

- *demolition and preliminary size reduction*: this allows to generate concrete debris from dismissed concrete poles;
- *size reduction, sieving and cleaning*: these were executed in one step since the machine available in the powerplant allow to fill the crushing mill with concrete debris, which are later cleaned (by washing) and sieved; during this phase, the steel remaining are also removed as well as the fine fraction (i.e., nominal diameter lower than 4 mm) is discharged.

The RCAs were produced by processing two batches of dismissed concrete poles and, moreover, the process was set to obtain two different class sizes: 4 mm to 8 mm and 8 mm to 16 mm.

**Fig. 1.** Production of Recycled Aggregates

Dismissed poles for an overall amount of around 23 tons of concrete were processed and, from them, around 8.2 tons of coarse RCAs were obtained. It is worth highlighting that, although a significant amount (around 60%) of processed debris was lost during processing, the removal of the fine particles certainly contributes to improve the quality of the produced coarse recycled aggregates.

### 2.2 Properties of Recycled Concrete Aggregates

This section compares the results obtained on the physical and geometric properties from tests performed on fine and coarse aggregates that have been employed for producing centrifuged concrete poles prototypes. Specifically, the following fractions are considered:

- Natural sand: fine aggregate, obtained from natural source and characterized by a maximum nominal diameter equal to 4 mm;
- Natural 4–8 mm: coarse aggregate, obtained from natural sources and characterized by a nominal diameter ranging between 4 mm and 8 mm;
- Natural 8–16 mm: coarse aggregate, obtained from natural sources and characterized by a nominal diameter ranging between 8 mm and 16 mm;
- Recycled 4–8 mm: coarse aggregate, obtained from old centrifuged concrete poles and characterized by a nominal diameter ranging between 4 mm and 8 mm;
- Recycled 8–16 mm: coarse aggregate, obtained from old centrifuged concrete poles and characterized by a nominal diameter ranging between 8 mm and 16 mm;

Figure 2 summarizes the results of the grain-size distribution tests performed on both natural and recycled aggregates in accordance with the *EN 933-1*.

The curves plotted in Fig. 2 show the fine nature of the employed sand fraction and highlight the low scatter registered on the grain size distribution for coarse aggregates of the same class size (4–8 mm and 8–16 mm) derived from either natural or recycled original sources. This can be attributed to the fact that both natural and recycled aggregates were produced by the same plant: in fact, the granulometric distribution of aggregates for concrete is strongly influenced by the processing procedure (sequence and adopted machineries) adopted for transforming crushed rocks and/or CDW in the form of aggregates. In addition, the similar grain size distribution of natural and recycled aggregates also characterized by a neglectable content of fines represents a first indicator of quality for produced RCAs (Rangel et al. 2019).

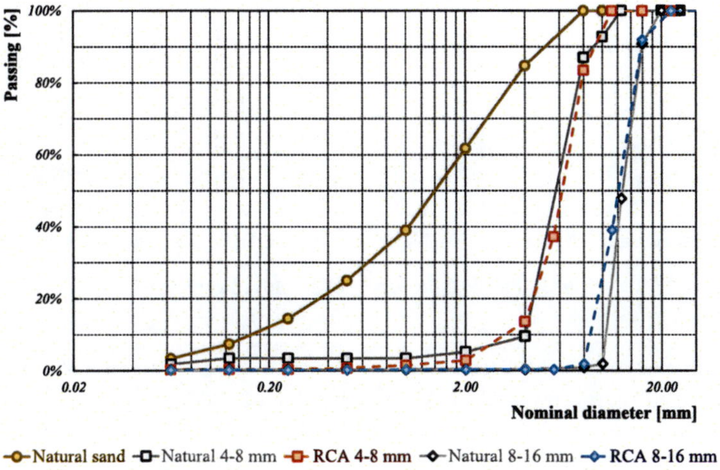


**Fig. 2.** Grain-size distribution for employed aggregates

Another fundamental aspect to be considered in the production of aggregates for structural concrete is related to their shape and geometry. In accordance with the *EN 933-3* and *EN 933-4*, the geometric characteristics of the coarse particles can be estimated through the definition of the Flakiness (FI) and Shape (SI) Indices (*EN 12620*), respectively: the produced coarse recycled aggregates pertain to the $FI_{15}$ and $SI_{15}$ category (*EN 12620*) and it is worth to mention that both fractions stay within the limits proposed by the *EN 206* for the use of recycled aggregates in concrete *(i.e.,* $FI < FI_{50}$ and $SI < SI_{15}$*).*

The physical properties of the particles employed herein were evaluated in accordance with the *EN 1097-6*: water absorption capacity at 24h (A) and particle density at both saturated with dry surface ($g_{SSD}$) and oven dried ($g_{DRY}$) conditions (Fig. 3).

As expected, the natural sand is characterized by a higher water absorption capacity than the natural coarse fraction. On the other hand, the recycled aggregates present a significantly higher water absorption capacity at 24h moving from 1.31% to 4.14% in the case of 4–8 mm fraction and from 1.03% to 4.33% for the bigger (8–16 mm) fraction. Conversely, an examination of the RCA particle density reveals a slight difference in comparison with the corresponding natural aggregate. As will be further discussed in the coming section, the higher water absorption capacity and slightly lower particle density is due to the composite nature of the RCA which contain natural rocks and "old" mortar derived from the previous concrete (the latter can be defined as Attached Mortar). Also

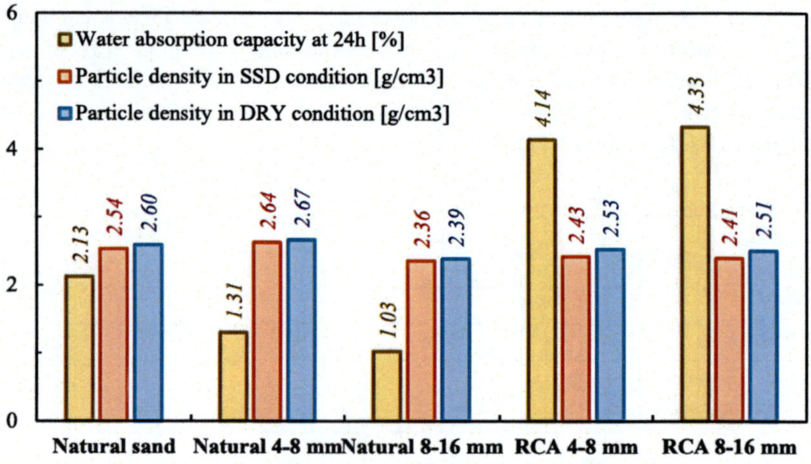


**Fig. 3.** Physical properties of employed aggregates

in this case, the produced RCAs stay within the limits proposed by the *EN 206* for the use of recycled aggregates (i.e., particle density above 2100 kg/m$^3$).

Nevertheless, even though the value of water absorption after 24h of soaking is generally assumed as a reference, the time evolution of the absorbed water can be a further relevant aspect for qualifying the key physical properties of the aggregates and, particularly, of RCAs. For this reason, the absorption rate was measured for both classes (i.e., 4–8 mm and 8–16 mm) of recycled aggregates. The obtained results show that the RCAs particles are not able to absorb the whole amount of water during the first minutes: this aspect will be fundamental for the adequate mixing of the resulting Recycled Aggregate Concrete produced with them since it could affect the effective water-to-cement ratio being detrimental for the resulting mixtures (Pepe 2015).

## 2.3 Quality Control Parameter and Proposed Classification for RCAs

Based on the results obtained herein, this subsection presents a comparison of the experimental data of this project with data available in the scientific literature, also in the view of a possible performance-based classification for Recycled Concrete Aggregates. Certainly, the first quality indicator for coarse aggregates is represented by their porosity (Rangel et al. 2019). On the other hand, the particle density (*EN 1097-6*) represents a fundamental parameter to be determined for the mix design of RAC mixtures. The results reported in the Sect. 2.2 highlight the particle density and show that, as expected, this property tends to decrease in aggregates with higher open porosity. As also reported in the literature (Pepe et al. 2016) for a fundamental understanding of RCAs properties, their characterization should be complemented with the evaluation of the attached mortar content. Specifically, a clear correlation can be detected between the AM content and both the water absorption capacity and the oven dried particle density (Pepe et al. 2016):

$$A = A_{NAT} \cdot (1 - AM) + A_{AM} \cdot AM \quad (1)$$

where $A_{NAT}$ (equal to 1%) and $A_{AM}$ (equal to 15%) represent the porosity of the two phases present in the RCAs: natural aggregates and attached mortar, respectively. It is worth to highlight that the above correlations demonstrate that the evaluation of the AM content allows a fundamental interpretation of key aggregate properties independently of the RCAs source and their granulometric class. This approach unveils that the Recycled Aggregates produced herein are characterized by an Attached Mortar (AM) contend below 25%.

Moreover, since the AM content defines the values of both the particle density and the water absorption capacity of coarse RCAs, a clear existing relationship can be identified between these two parameters as a well as a generalized coarse RCAs classification can be proposed (Rangel et al. 2019) (Fig. 4).

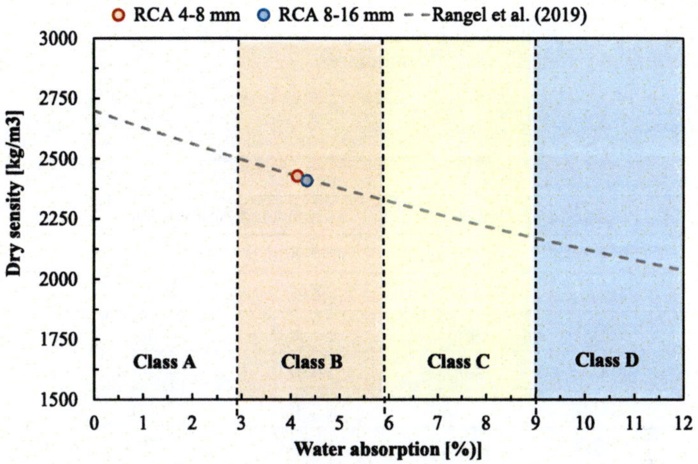


**Fig. 4.** Classification of RCAs

In accordance with this classification, the RCA produced in the present study can be classified on Class B (RCAs coarse particles characterized by a A values ranging between 3% and 6% and an AM content between 10% and 35%) clearly indicating a good quality of the produced aggregates.

## 3 Mixture Composition and Mechanical Performance

### 3.1 Selection of Trial Batches and Mix-Design

The selection of the trial batches aims at analysing the influence of the two different fractions of the RCAs, produced during the R&D activities, on the resulting Recycled Aggregate Concrete performances at both fresh and hardened state. In this aim, the following seven mixtures were designed and produced (see also Table 1):

- *NAT*, the reference mixture made with all natural components: it represents the typical mixture realized by S.I.P.A. spa for producing centrifuged reinforced concrete poles;

- *RAC-P1*, where a direct volume replacement of the 4–8 mm fraction (*Pietrisco 1*- P1) was adopted;
- *RAC-P2*, where a direct volume replacement of the 8–16 mm fraction (*Pietrisco 2*-P2) was adopted;
- *RAC20*, in which the coarse aggregates were composed by 80% of natural aggregates and by 20% of recycled ones: this mixture complies the limits proposed by the IT regulation (NTC 2018);
- *RAC30*, in which the coarse aggregates were composed by 70% of natural aggregates and by 30% of recycled ones: this mixture complies the limits proposed by the EU standards (*EN 206*);
- *RAC50*, in which the coarse aggregates were composed by 50% of natural aggregates and by 50% of recycled ones: this mixture exceeds the limits proposed by the IT and EU standards;
- *RAC100*, made with all coarse recycled aggregates: also this mixture exceeds the limits proposed by the IT and EU standards.

**Table 1.** Mixtures composition..

| Mix | Natural aggregates | | Recycled Aggregates | | $w_{abs}$ |
|---|---|---|---|---|---|
| | 4–8 mm | 8–16 mm | 4–8 mm | 8–16 mm | |
| | kg | kg | Kg | kg | kg |
| *NAT* | 700.0 | 400.0 | 0.0 | 0.0 | 30.3 |
| *RAC-P1* | 0.0 | 400.0 | 630.0 | 0.0 | 34.2 |
| *RAC-P2* | 700.0 | 0.0 | 0.0 | 360.0 | 34.0 |
| *RAC20* | 560.0 | 320.0 | 126.0 | 72.0 | 31.8 |
| *RAC30* | 490.0 | 280.0 | 189.0 | 108.0 | 32.6 |
| *RAC50* | 350.0 | 200.0 | 315.0 | 180.0 | 34.1 |
| *RAC100* | 0.0 | 0.0 | 630.0 | 360.0 | 37.9 |

All batches were produced with 800 kg/m$^3$ of natural sand, 360 kg/m$^3$ cement, a water-to-cement ratio (free water equal to 120 kg/m$^3$) of 0.33 and 3 kg/m$^3$ of super-plasticizer. To compensate for the water absorption of both dried recycled and natural aggregates, additional water was poured during mixing ($w_{abs}$ in Table 1). In accordance with the mix-design method proposed by Pepe (2015) the evaluation of $w_{abs}$ for RCAs was based on the measurement of the time evolution of the water absorption for coarse RCAs. Consequently, in the mixture proportioning of RACs proposed herein, the RCAs are employed in fully dried condition and a 50% of their water absorption capacity at 24 h is considered for saturation during mixing, since this amount of additional water do not affect the free water available in the mixture (Pepe et al. 2016).

The different mixtures were produced according to the following procedure. First, sand, fine and coarse aggregates were added to the mixer and mixed for 90 s; then, cement was added and mixed for another 90 s. In the following step, 70% of the total water was added to the aggregate blend before adding it to the mixture, then, this blend was mixed for one minute, and then added to the mixer. Finally, the superplasticizer and the rest of water was added to the mixer, and mixing proceeded for another 8 min.

### 3.2 Experimental Methods

For each mixture, slump tests were performed (*EN 12350–2, 2019*) and six cubic specimens (150 mm x 150 mm x 150 mm) were prepared for the mechanical characterization at 28 days of the compressive strength performed in accordance with the European standards (*EN 12350–3/4, 2019*) as well as the hardened concrete density (*EN 12350–7, 2019*). The samples were demoulded after 24 h and the cure was carried out in a water bath and temperature of 20 °C up to 28-days.

### 3.3 Results and Analysis

The results of workability performed through slump tests for the concrete mixtures under investigation are presented in Fig. 5.

The RAC workability, as in the case of ordinary concrete, is mainly governed by the aggregate shape, grain size distribution and the amount of free water available in the mixture. In the present study the amount of free water has been kept constant (equal to 120 l/m$^3$) and the grain size distribution is similar for natural and recycled aggregates as shown in Fig. 2 (only in the case of 4–8 fraction a lower amount of fine particles is observed for RCA in comparison with the companion natural aggregate). For these reasons, the higher slump values observed as the RCA content increases can be mainly attributed to the more rounded shape observed for both RCA fractions considered herein.

The results of physical properties (Fig. 6) show that the presence of RCAs is not affecting the RAC density at 28-days in comparison with natural reference mixtures (i.e., NAT). This result is certainly attributed to the specific mixture proportion adopted

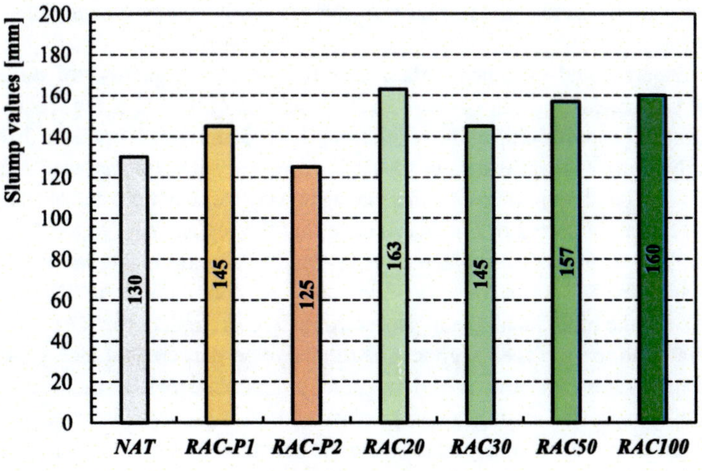


**Fig. 5.** Slump test results

herein (i.e., based on the approach proposed by *Pepe et al.* in *2016*) as well as to the "high quality" of the employed RCAs. The latter confirms the conclusion above remarked, in which a performance-based classification for recycled aggregates was presented. As a matter the principle, generally, the presence of recycled aggregates tends to decrease the concrete density at the hardened state: this is due to the higher porosity characterizing the coarse particles and, this phenomenon is more pronounced when mixed recycled aggregates (i.e., not derived only from concrete waste) are used.

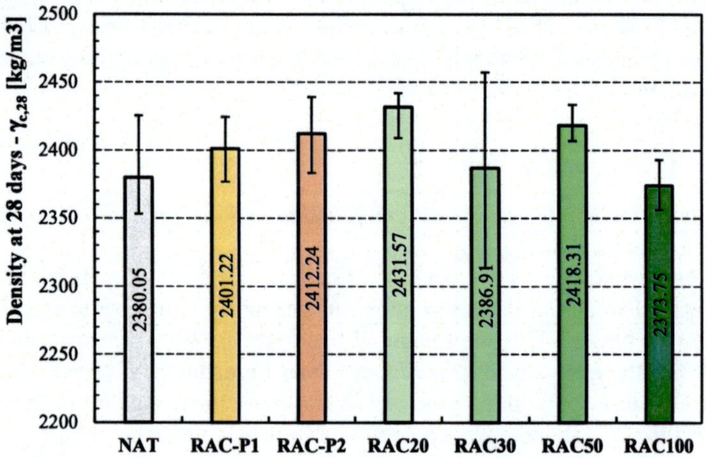


**Fig. 6.** Hardened concrete density at 28 days

The results obtained from cubic compressive strength at 28 days ($R_{c,28}$), summarized in Fig. 7 in terms of average values, show that all the RACs obtained slightly higher values in comparison with the reference natural mixture (i.e., NAT). Moreover, it is worth to mention that also the registered scatter is similar for both natural and recycled concrete.

This experimental evidence further confirms the "high-quality" of the produced Recycled Concrete Aggregates as well as the efficiency of the use of the specific method for the mixture proportioning of the high-strength RAC mixtures (Pepe et al. 2016). More specifically it proves that the data of the RCAs used are correctly analyzed and the consideration of their absorption capacities during mixing (50% of the total 24 h absorption) is adequate. Data reported in Fig. 7 evidence that the replacement of the finer fraction (*Pietrisco 1*, 4–8 mm) of the coarse particles lead to slightly higher values than in the case in which the 8–16 mm fraction is replaced (i.e., comparing RAC-P1 and RAC-P2 mixtures). On the other hand, the higher compressive strength of the RAC mixtures can be associated to the surface roughness of the recycled particles. In fact, generally, the recycled aggregates are characterized by a more pronounced surface roughness and when "high-quality" recycled aggregates are used, it leads to improve the Interfacial Transition Zone (ITZ) linking the coarse particles and the surrounding cementitious matrix.

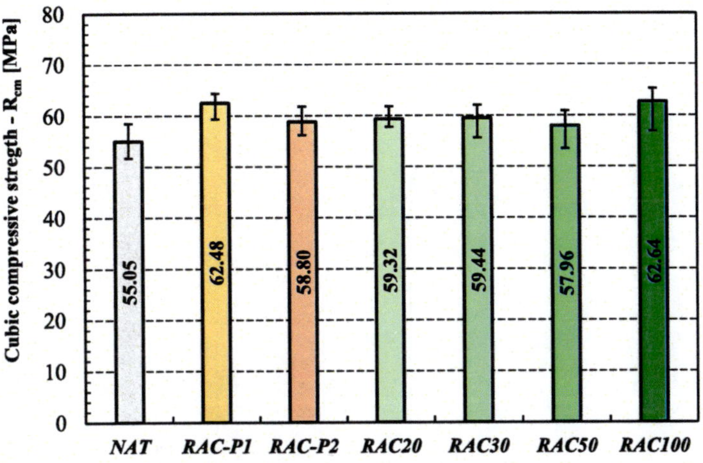


**Fig. 7.** Compressive strength results at 28 days

# 4 Production of Prototypes

## 4.1 Selection of Optimized Concrete Mixture Produced and Representative Prototypes

Based on the results obtained in the previous phases of the present R&D project, several RAC mixtures were selected for the production of the poles' prototypes. More specifically, the following three optimized mixtures were selected (Sect. 3.1): *RAC20, RAC50* and *RAC100.* Once defined the composition of the RAC mixtures, some representative concrete poles (defined by *GSS002:*Fig. 8) were selected for the production of the prototypes: *Type D* (length 10 m, d 20 cm, D 35 cm label D/10/20) and *Type E* (length 10 m, d 24 cm, D 35 cm label E/10/24).

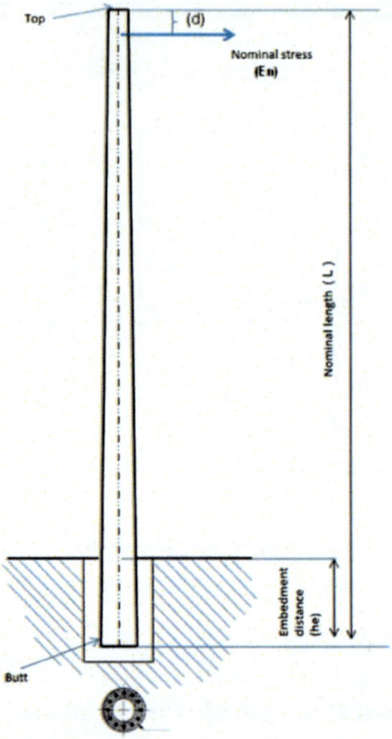


**Fig. 8.** Schematic of concrete pole in accordance with *GSS002*

## 4.2 Production Process of Selected Prototypes

The following Fig. 9 summarizes the several steps for the production of the centrifuged reinforced concrete poles adopted by *S.I.P.A. s.p.a.*: (a) preparation of steel reinforcement, (b) concrete mixing and casting; (c) centrifugation; (d) curing and demolding and (e) storage.

A total of 6 prototypes were produced, considering the variation of the pole typology (i.e., *D/10/20* and *E/10/24*) and the concrete mixture (i.e., *RAC 20*, *RAC 50* and *RAC 100*). During the production of the centrifuged reinforced concrete poles prototypes it was verified that the produced RAC mixtures did not affect the centrifugation process, and, in addition, curing time was comparable with ordinary (produced with 100% of natural aggregates) concrete poles. In addition, at the end of the curing process, it was verified that the presence of the recycled compound did not influence the visual inspection test by "checking the presence of eventual erroneous characteristics on pole" (Fig. 10).

(a)
(b)
(c)
(d)
(e)

**Fig. 9.** Production process for the poles 'prototype

**Fig. 10.** Representative prototype produced with RAC100

## 4.3 Tensile Tests Performed on RAC Prototypes

The produced RCA poles were then submitted to the tensile test defined by the *GSS002:* Breaking Strength Test as schematized in Fig. 11 (T1 and T2 represent tensile forces applied on two specific points of the pole: 0.1 and $h_2$ from the top of the pole). During the test execution it was ensured that the interlocking length was continuous and enough to ensure that every stress shows no buckling.

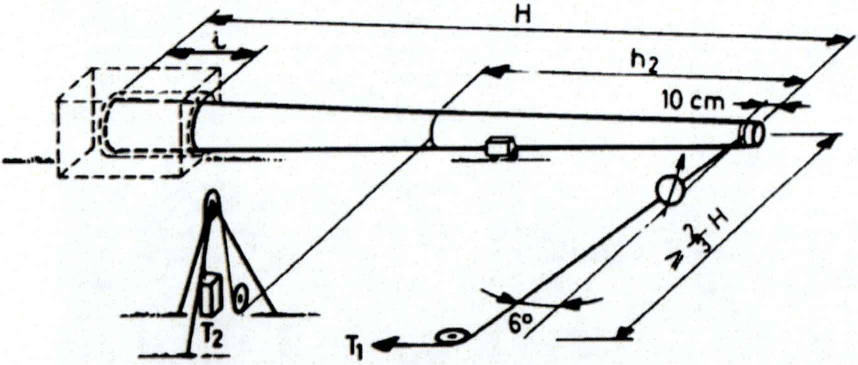


**Fig. 11.** Representative scheme of the Breaking Strength Test in accordance (*GSS002*)

## 4.4 Properties of Produced Prototypes

During the execution of the tests, for each produced prototype, it was measured the pole deflection at each meter length at both the nominal stress force (i.e., 40% of T1) and the tensile force (i.e., T1). The results of the measurement of the pole deflection are reported in Fig. 12.

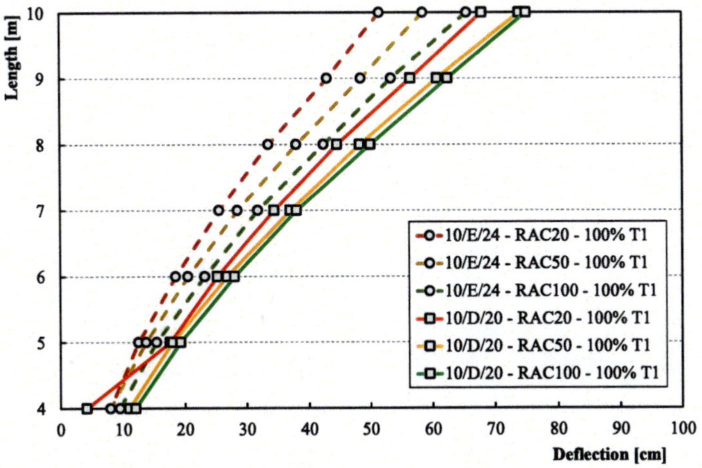


**Fig. 12.** Poles' deflection at the tensile force (100% T1)

It is worth mentioning that, no cracks were detected during the test, for all the tested prototypes, at the load level equal to the nominal stress. In addition, the poles did not register any residual deformation. The graphs show that the presence of recycled aggregates in the produced mixtures lead to a more "flexible" behavior of the resulting pole. As a matter of fact, when the overall amount of recycled particles increases (i.e., moving from RAC 20 to RAC 100 mixture) a higher deflection is registered by keeping constant the load level. This trend is confirmed at both the nominal stress and tensile stress levels. Moreover, as expected, the type E pole present a more "rigid" behavior in comparison with the companion D/10/20 prototypes. Furthermore, all the produced prototypes showed a regular deflection curve for the whole length up to the failure of the tested poles. The tests were executed up to the failure of the HC poles (Fig. 13).

Only at this stage some cracks were observed. The registered breaking force was always higher than the tensile force (T1).

**Fig. 13.** Failure of the poles during the tensile test

## 5 Life Cycle Analysis

The environmental impact resulting from the production process under consideration was evaluated by considering a representative 10/D/20 pole under the two following alternative scenarios:

- the first scenario, referred to as "NAT", considers natural coarse aggregates as raw constituents for the concrete production;
- the second scenario, referred to as "RAC100", assumes the use of recycled aggregates (as a replacement of 100% of coarse natural aggregates) derived from the demolition and recovery processes of dismissed concrete poles.

The study examines the entire supply chain at different stages of the life cycle using a cradle-to-cradle approach. Table 2 shows the overall environmental impacts (for the defined functional unit equal to 1 cubic meter of concrete pole) relating to the impact categories from the "General Program instructions" of the EPD® system and those indicated in the "Construction products and Construction services" PCRs: *Global Warming Potential* (GWP), *Eutrophication Potential* (EP) and *Acidification Potential* (AP).

**Table 2.** Results of the LCA

| Scenario | GWP | EP | AP |
|---|---|---|---|
| | kg $CO_{2\text{-eq}}$ | kg $PO4^{3}{}_{\text{-eq}}$ | kg $SO_{2\text{-eq}}$ |
| *NAT* | 419.025 | 0.427 | 1.839 |
| *RAC100* | 393.341 | 0.404 | 1.651 |

It can be recognized that, in various respects, RAC100 is less impactful than NAT. Specifically, a reduction of 6.1% in GWP category, 5.4% in the EP category and 10.2% is the AP category has been quantified via LCA. As it is apparent, the decrease in impacts is mainly due to the reuse of aggregates deriving from demolition processes of dismissed poles.

## 6 Conclusions

This paper presented the results of an industrial R&D project aimed at evaluating the feasibility of recycling dismissed concrete poles to produce new precast centrifuged reinforced concrete elements made of RCAs. The results shown that by adopting an adequate processing procedure for transforming the concrete debris in secondary raw materials, as well as by using a specific mixture proportion method for the structural concrete production, it is possible to obtain RAC poles without affecting the resulting mechanical performances. In addition, the Life-Cycle Analysis quantified the environmental impacts of the proposed recycling procedure. It can be concluded that for the next step of the present R&D project the prototypes can be certainly designed by complying the IT and/or EU regulations but it would be possible to produce representative elements using 50% or 100% of coarse recycled aggregates.

**Acknowledgements.** Il Gruppo Enel si è impegnato a sviluppare un modello di business in linea con gli obbiettivi della COP21 di Parigi per contenere l'aumento della temperatura globale al di sotto di 2 °C rispetto ai livelli preindustriali e per continuare a limitare tale aumento a 1,5 °C al 2050. Net Zero Ambition è l'impegno della divisione infrastrutture e reti di Enel per raggiungere tale obiettivo e l'iniziativa di innovazione tecnologica sui materiali da costruzione rientra in questo contesto: in particolare Enel con questo progetto mira a ridurre le proprie emissioni indirette di gas serra, Scope 3, per le quali ha fissato l'obiettivo di riduzione dell'55% entro il 2030 rispetto ai livelli del 2017.

## References

Amario, M., Rangel, C.S., Pepe, M., Toledo Filho, R.D.: Optimization of normal and high strength recycled aggregate concrete mixtures by using packing model. Cem. Concr. Compos. **84**, 83–92 (2017)

Chen, J., Su, Y., Si, H., Chen, J.: Managerial areas of construction and demolition waste: A scientometric review. Int. J. Environ. Res. Public Health **15**(11), 2350 (2018)

EN 1097-6:2013. Tests for mechanical and physical properties of aggregates. Determination of particle density and water absorption

EN 12350-2:2019. Testing fresh concrete - Part 2: Slump test

EN 12390-3:2019. Testing hardened concrete - Part 3: Compressive strength of test specimens

EN 12390-4:2019. Testing hardened concrete - Part 4: Compressive strength - Specification for testing machines

EN 12390-7:2019. Testing hardened concrete - Part 7: Density of hardened concrete

EN 12620:2013. Aggregates for concrete

EN 206:2013+A1:2016. Concrete. Specification, performance, production and conformity

EN 933-1:2012. Tests for geometrical properties of aggregates. Determination of particle size distribution. Sieving method

EN 933-3:2012. Tests for geometrical properties of aggregates. Determination of particle shape. Flakiness index

EN 933-4:2008. Tests for geometrical properties of aggregates. Determination of particle shape. Shape index

Global Standard GSS002. Concrete poles for distribution networks. Rev. 04 of 21/09/2018

Lima, L., Trindade, E., Alencar, L., Alencar, M., Silva, L.: Sustainability in the construction industry: a systematic review of the literature. J. Clean. Prod. **289**, 125730 (2021)

Menegaki, M., Damigos, D.: A review on current situation and challenges of construction and demolition waste management. Current Opin. Green Sustain. Chem. **13**, 8–15 (2018)

NTC (2018) DM 17.01. 2018: Nuove Norme Tecniche per le Costruzioni. Italian Ministry of Infrastructures and Transportation (in Italian)

Pepe, M.: A Conceptual Model for Designing Recycled Aggregate Concrete for Structural Applications. Springer, Springer Theses (2015)

Pepe, M., Toledo Filho, R.D., Koenders, E.A., Martinelli, E.: A novel mix design methodology for Recycled Aggregate Concrete. Constr. Build. Mater. **122**, 362–372 (2016)

Rangel, C.S., Toledo Filho, R.D., Amario, M., Pepe, M., de Castro Polisseni, G., de Andrade, G.P.: Generalized quality control parameter for heterogenous recycled concrete aggregates: a pilot scale case study. J. Clean. Prod. **208**, 589–601 (2019)

Udomsap, A.D., Hallinger, P.: A bibliometric review of research on sustainable construction, 1994–2018. J. Clean. Prod. **254**, 120073 (2020)

Wieser, A.A., Scherz, M., Passer, A., Kreiner, H.: Challenges of a healthy built environment: air pollution in construction industry. Sustainability **13**(18), 10469 (2021)

Wu, H., Zuo, J., Zillante, G., Wang, J., Yuan, H.: Status quo and future directions of construction and demolition waste research: a critical review. J. Clean. Prod. **240**, 118163 (2019)

# Advances on the Use of Geopolymer Recycled Aggregate Concrete in Construction

Muhammad Ahmed, Piero Colajanni, and Salvatore Pagnotta(✉)

Department of Engineering, University of Palermo, Palermo, Italy
salvatore.pagnotta@unipa.it

**Abstract.** Geopolymer cement is made up of waste materials, an efficient alternative to ordinary cement in order to reduce the emission of carbon dioxide. The use of geopolymer recycled aggregate concrete (GRAC) in structural elements has been an object of recent several experimental research in which the behaviour of either the material or structural element was investigated. A review paper will be done critically to compile the present research on the use of GRAC in construction. It will cover the effect of different materials composition in GRAC and the factors affecting its strength focusing on recycled aggregate's replacement ratio. The problems related to the strength of GRAC structural members, their tensile and flexural behaviour will also be addressed. It will also help in finding the future prospects of the use of GRAC in construction.

**Keywords:** geopolymer recycled aggregate concrete · recycled aggregate · structural members

## 1 Introduction

Ordinary Portland Cement (OPC) concrete is the world's most common and widely used binding constructional material. It has a number of advantages as it is easily available, has low cost and is durable. But this cement is criticized due to the emission of carbon dioxide during its manufacturing process. The production of OPC involves the world's 5% $CO_2$ emission, and one tonne of ordinary Portland cement releases approximately 0.9 tons of $CO_2$ (Malhotra & Mehta 2002 and Imbabi et al. 2012).

In the normal OPC, there are almost 75–80% fine and coarse aggregates which are obtained through mining. This mining is causing a negative effect on the environment, and it is also causing a lot of waste of energy. It is in the utmost favour of the environment to develop a way which ultimately benefits both environment and the construction sector (Lamond & Pielert 2006).

At the same time, a massive amount of construction waste is produced every year due to the demolition of buildings and other concrete structures (Akhtar & Sarmah 2018). The depletion of natural resources to produce OPC urges a need to develop some alternative

The original version of the chapter has been revised. A correction to this chapter can be found at https://doi.org/10.1007/978-3-031-43102-9_55


M. A. Aiello and A. Bilotta (Eds.): ICC 2022, LNCE 435, pp. 83–93, 2024.
https://doi.org/10.1007/978-3-031-43102-9_7

to ordinary Portland cement so that it can be helpful in protecting the environment as well as it will provide an efficient alternative in terms of cost and characteristics Verian et al. (2018).

Geo Polymer Cement (GPC) is an alternative to OPC which can be widely used in construction. Geopolymers are inorganic materials which have a polymeric structure of molecules. It is called a "Geopolymer" because the material used for its manufacturing is mainly of geological origin. Geopolymer cement is produced when waste material such as fly ash, Ground Granulated Blast Furnace Slag (GGBS), and clay containing aluminosilicate minerals are treated with an alkali solution such as sodium hydroxide. This cement with aggregates produces geopolymer concrete which can be used for various constructional applications.

Almutairi et al. (2021) did comprehensive research and found that the use of geopolymer cement in the construction industry will reduce 80% of carbon dioxide emissions associated with the production of concrete. It will also be helpful in reducing the cost of raw materials.

Le et al. (2021) evaluated that, the old concrete obtained from demolished construction waste can be mechanically (crushed, sieved and cleaned) and sometimes also chemically treated in order to obtain Recycled Aggregates (RA) for structural concrete. These recycled aggregates can be used in the concrete as a partial substitute or full substitute depending upon the requirement.

The use of recycled aggregates for the concrete structural elements is hindered by the attached mortar since it has some negative effects on the strength of the mixture. It increases both the porosity of the recycled aggregate and develops two different Interfacial Transition Zones (ITZ)s between the recycled aggregate and new mortar, and between the new mortar and attached mortar. This ITZ between new and old mortar is a weak zone that causes the reduction of strength.

To take benefit both from the GPC and recycled aggregate, it's a practical solution to use both together, i.e., use Geopolymer Concrete with Recycled Aggregate Concrete (GRAC). There are already several kinds of research going on for the use of GPC with Natural Aggregate (NA) and OPC with RA, but there are very few research on the use of GPC with RA. Besides that, there is less availability of research on the structural assessment of the use of GRAC. The available data also provide contradicting results and assessments of the efficiency of GRAC.

This review paper briefly describes the past and current research development on the use of GRAC, especially from the structural point of view.

## 2 Research Development on the Mechanical Properties of Grac

Uddin Ahmed Shaikh (2016) did an experimental study to find out the durability and mechanical properties of GRAC. The RA was used as a partial replacement (15%, 30% and 50%) of NA. The GPC with 100% NA was used as a reference for comparison. The test results showed that with the increase of RA content, the compressive strength, indirect tensile strength and elastic modulus of geopolymer concrete decreased whether the test was done after 7 or 28 days (Fig. 1).

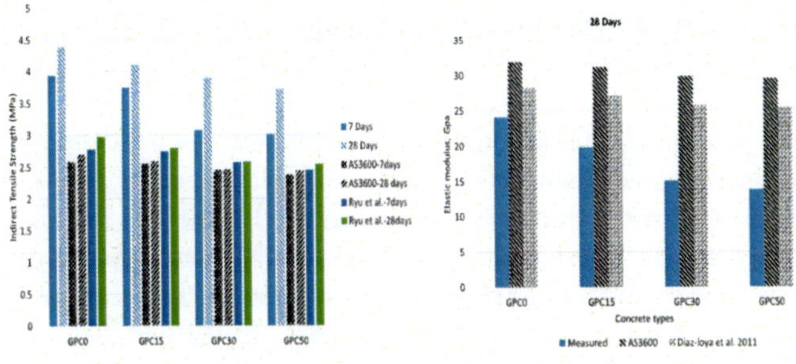


(a) (b)

**Fig. 1.** (a): Indirect tensile strength of GPC containing different percentage of NA and RA; and (b) Elastic modulus of GPC containing different percentage of NA and RA (U.A.Shaikh 2016)

Moreover, it was found that the existing empirical models for OPC (AS3600 (2009)) and GPC (Ryu et al. 2013 & Diaz-Loya et al. 2011) containing natural aggregate underestimate the indirect tensile strength and overestimate the elastic modulus of GRAC (Fig. 1b). Currently, no empirical model is available for reliable prediction of the compressive and tensile strength of GRAC, since they depend on several factors such as binder types, aggregate types, the molarity of alkaline solutions, mixing procedure, casting temperature and environmental conditions.

Nuaklong et al. (2016) studied the effect of recycled aggregate on the strength and durability of flyash-based geopolymer concrete having sodium silicate solution and sodium hydroxide solution. Two types of aggregates were used; recycled concrete aggregate and crushed limestone coarse aggregate. The results revealed that recycled concrete aggregate can be used as a coarse aggregate as its compressive strength is just slightly lower than the geopolymer concrete prepared with NA.

More recently, Nuaklong et al. (2018) did an experimental study using Metakaolin (MK) with fly ash-based geopolymer concrete and did a comparative assessment between the use of NA and 100% RA concrete. It was found that the partial replacement of High Calcium Fly Ash (HCFA) by metakaolin in geopolymer binders increases the strength of GRAC. Two different schemes were adopted using: 1. Limestone as natural aggregate in geopolymer concrete; 2. 100% recycled aggregate in geopolymer concrete. It was found that when the metakaolin amount was increased the compressive strengths of GRAC with metakaolin (0, 10, 20, and 30%) were 32.9, 40.4, 45.0, and 47.2 MPa, respectively. GRAC mixtures with metakaolin achieved approximately 15–34% higher compressive strength than the concrete without metakaolin. This is because of enhanced compressive strength due to the increased geo-polymerization and denseness of microstructure. Also increasing metakaolin from 10% to 30% led to an increase in the splitting tensile strength from 2.9 to 5.4 MPa for GPC with NA and from 2.7 to 3.5 MPa for GRAC. The strength of geopolymer concrete also increased in both types (1 and 2) since the compressive strength of (1) was almost 7%–19% more than (2).

Moreover, researchers stressed that usually, the formation of geopolymer is done by casting geopolymer slurry in the mould with a significant amount of alkali solution, increasing the chances of high porosity. These pores can act as a point of stress concentration and mechanical failure. In this context, the application of pressure reduces porosity.

Thus, Wongsa et al. (2020) did comparative research to find out the physical properties of Pressed Geopolymer Concrete (PGC) with recycled aggregate, Recycled Concrete Block Aggregate (RB), and limestone dust. The RA and RB meshed into fine aggregates having 4.75 mm diameter. The results showed that: a) the pressed geopolymer concrete made up of limestone dust exhibited more compressive strength, less porosity and water absorption than the concrete made up of RB and RA; b) the compressive strength of pressed GPC made up of RA and RB was nearly equal to the strength of moderate strength lightweight concrete prepared according to the ACI 213; c) it was also recommended that the pressed GPC with RA can be used, besides the structural application, to make hollow geopolymer-based concrete blocks with better thermal insulation than cement-based concrete blocks.

Le et al. (2021) studied the use of GRAC. The alkali-activated binder (geopolymer binder) was made using low calcium Fly Ash (FA), sodium silicate solution, sodium hydroxide solution, and lignosulfonate superplasticizer. Specimen were prepared to cure both at 60 °C and at ambient temperature.

The results showed an encouraging behaviour of the GRAC specimen as a very less decrease in strength was observed if the use of low calcium fly ash was joined with curing at 60 °C, even if a 100% replacement NA with RA was done.

Berhanul et al. (2021) find out the effect of metakaolin as a cement replacement on the properties of fresh and hardened recycled aggregate concrete and natural aggregate concrete. The recycled aggregates were obtained from first-hand cast laboratory cubes whose compressive strength was already known. Concrete mixtures were prepared and tested with different percentages of recycled aggregates. The results showed that the addition/use of metakaolin as a cement replacement improved the strength of concrete made up of recycled aggregates.

Xu et al. (2021) reviewed the current research on the mechanical properties of GRAC reconfirming that geopolymer in GRAC is an ideal substitute for cement and similarly RA is also a substitute of NA, because of their environment-friendly effects; The strength of GRAC depends on many factors such as the type of geopolymer, the casting temperature, type of aggregates etc. But these strength influencing factors are somehow the same as GPC; there is a lack of research on the use of GRAC so the focus must be done on the practical use of GRAC.

Srinivas and Abhignya (2020), from the literature observed that by using fly ash and GGBS as a replacement for cement and RA as a replacement for NA, geopolymer concrete beams and columns with compressive and flexural strengths perform much better than conventional reinforced concrete beams and columns. The optimum replacement percentage of RA was found to be 30% because at this replacement ductility nature of both Geopolymer and conventional concrete beams were almost the same.

To mitigate the vulnerability of failure due to the ITZ of recycled aggregates, in the past Liang et al. (2013) proposed two different mixing procedures, named as Mortar Mixing Approach (MMA), and the Sand Enveloped Mixing Approach (SEMA). The schematic diagrams of both methods are given in (Fig. 2). The results showed that using

MMA improvement in compressive strength on concrete made with 100% coarse RA was obtained. In the SEMA method, RA underwent pre-surface treatment 7 days before the mixing, obtaining higher 28-day compressive strength, as compared to that of recycled aggregate concrete made with the MMA method. Due to these mixing processes an additional layer of cement formed on the aggregates surface which decreased the porosity as well as reduced high-water absorption which ultimately resulted in the improvement of strength. However, these methods produce an increase in cost and casting time.

Recently, Alqarni et al. (2021) proposed a new two-stage mixing approach with silica fume and cement. In this process, a cement-silica fume slurry solution was prepared by mixing the cement and silica fume in different percentages with water by weight. RA dried in an oven for a day and then cooled. After that, RA was mixed with cement-silica fume slurry solution for about 30 min. It was found that it increased the compressive strength of concrete.

The bond strength between the concrete and steel reinforcement is essential for the ultimate strength of structural members. Romanazzi et al. (2021) did an experimental investigation to find out the bond behaviour of GPC with steel bars and sand coated (Glass Fiber Reinforced Polymer) GFRP bars using the pull-out test.

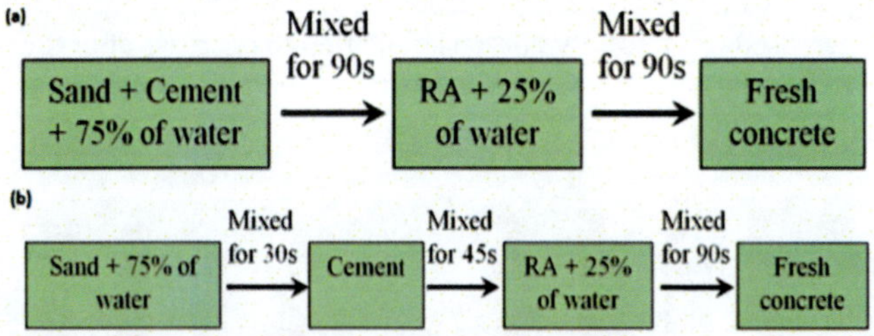


**Fig. 2.** Schematic diagram of (a): Mortar Mixing Approach MMA method; and (b) Sand Enveloped Mixing Approach SEMA method (Liang et al. 2013)

It was evaluated from the experimental investigation that the ultimate bond strength results of GPC with steel bars were 2 to 3 times higher than sand coated GFRP bars. It proved that there is adequate mechanical interlocking and good bond strength between the GPC and steel bars.

## 3 Research Development on the Structural Elements Made up of Grac

Thangamanibindhu & Murthy (2015) carried research on the behaviour of ambient cured GRAC beams. GPC was prepared using GGBS, fly ash, and sodium silicate solution. Sodium hydroxide was used as an alkali activator, and a superplasticizer was used to improve the characteristics. Total 9 beams were tested in flexure. Three beams were prepared with conventional concrete mixes while 6 other beams had varying proportions of fly ash, GGBS and recycled coarse aggregates.

Two-point loading was applied to the beams. It was found that the cracking load of GRAC beams on average 30% more than that of conventional reinforced concrete beams. The load-carrying capacity of all the beams decreased when replaced with more quantity of recycled aggregates. The same load-deflection characteristics were obtained for the ordinary reinforced cement concrete beams and geopolymer concrete for 0% (without replacement) and 10% replacement of RA. The failure behaviour of geopolymer concrete beams was found to be similar to the cement concrete beams, as both types of beams failed initially due to the yielding of the tensile steel; then the crushing of concrete occurred.

Kathirvel & Kaliyaperumal (2016) did an experimental study to investigate the influence of recycled aggregates obtained from demolished construction waste on the mechanical properties of GRAC. The casting of GRAC was done at room temperature. GGBS, alkaline solution and superplasticizers were used to achieve high strength. Beams with dimensions of 1.5 m × 0.1 m × 0.15 m were tested under a four-point flexure load scheme. RA was pre-wetted to mitigate the consequence of the rapid reduction of concrete workability. The results revealed that with the increase of RA content, there was a slight decrease in the initial stiffness. Due to pre-wetting and inclusion of plasticizers up to 50% RA replacement, the compressive strength and water absorption characteristics improved (Fig. 3a), while up to 75% replacement of NA with RA, the load-bearing capacity of beams increased (Fig. 3b). By contrast, it started to decrease after the replacement exceeds 75%.

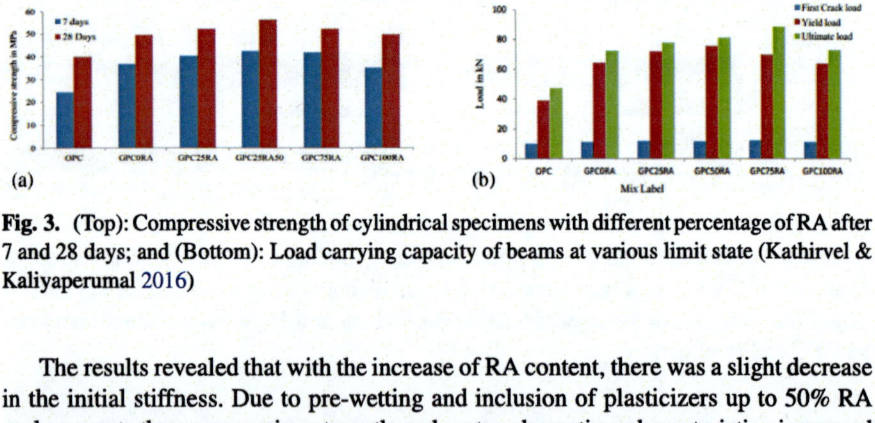


**Fig. 3.** (Top): Compressive strength of cylindrical specimens with different percentage of RA after 7 and 28 days; and (Bottom): Load carrying capacity of beams at various limit state (Kathirvel & Kaliyaperumal 2016)

The results revealed that with the increase of RA content, there was a slight decrease in the initial stiffness. Due to pre-wetting and inclusion of plasticizers up to 50% RA replacement, the compressive strength and water absorption characteristics improved (Fig. 3a), while up to 75% replacement of NA with RA, the load-bearing capacity of beams increased (Fig. 3b). By contrast, it started to decrease after the replacement exceeds 75%.

Raj S & Bhoopesh (2017) did an experimental study to find the strength and behaviour of GRAC beams. Recycled aggregates taken from demolition waste were used as coarse aggregate in geopolymer concrete. The ingredients of GPC were low calcium fly ash (Class F), sodium silicate alkaline solutions, Sodium Hydroxide solution. Coarse aggregate, fine aggregate, and superplasticizer were used together with GPC to prepare GRAC. NA was replaced with RA with the replacement percentage as 20%, 30%, 40%, 50%

and 60%. The optimum ratio was found to be 40%. Beams were casted of dimensions 175 mm × 150 mm × 1200 mm and subjected to bending tests. From the experimental study, it was concluded that with the addition of RA there was a slight reduction in its strength and durability (Fig. 4b). It was also observed that the GRAC beams showed a larger size and number of cracks as compared to normal geopolymer concrete beams. This is due to the porous structure of recycled aggregates, which produces a reduction in the tensile strength of GRAC.

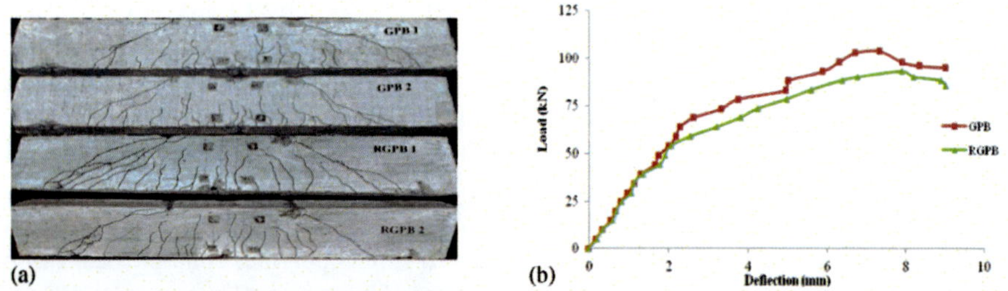


**Fig. 4.** (Top): Crack pattern of geopolymer concrete beams and GRAC beams; and (bottom): Load deflection curves of geopolymer concrete beams and GRAC beams (Raj S & Bhoopesh 2017)

The main aim of the experimental study performed by Srinivas & Abhignya (2021) was to determine the optimum percentage of RA for GRAC and to find out the behaviour of structural members like beams and columns when subjected to axial compression or bending. Recycled aggregates were obtained from demolition waste and mixed with geopolymer cement to produce GRAC. The geopolymer cement was prepared using fly ash, ground granulated furnace slag, and alkaline solution (sodium silicate and sodium hydroxide). Naphthalene sulphonate formaldehyde and superplasticizers were used for better strength of GPC. In this experimental study, 20%, 30%, 40%, 50%, and 60% replacement of recycled aggregates was done. Beams and columns of dimensions (150 mm × 150 mm × 1200 mm) were prepared. The beams were tested under a two-point loading test and columns were tested under axial loading. The results revealed that based on mechanical and workability, the optimum replacement percentage of RA was 40%. Comparison is done between GRAC and GPC with NA beams. The crack pattern and failure mode of GRAC beams and GNAC beams were the same. GRAC column with 40% of RA behaved in axial compression as same as of GPC with NA column. Due to the inclusion of naphthalene sulphonate formaldehyde, it is also found that almost the same ultimate load strength is obtained for both GRAC and GPC with NA, so it is suggested that GRAC is a practical and environment-friendly solution.

Zhang et al. (2021) did a comparative study on GRAC and ordinary recycled aggregate concrete beams. Static loading tests were conducted on three ordinary recycled aggregate concrete beams and seven GRAC beams. Metakaolin-based fly ash geopolymer and alkaline solution were used in the preparation of GPC. The test variables included the RA replacement ratio, the replacement pattern, and the reinforcement ratio.

Three replacement ratios (30%, 70%, and 100%) of RA were taken. The conventional aggregate replacement pattern was to replace the same percentage of all particle sizes.

But in a new replacement pattern, only larger (up to 19mm) NA particles were replaced with RA; a 70% replacement ratio was taken in both replacement patterns. Ten reinforced concrete beams with dimensions (1800 mm (L) × 100 mm (W) × 250 mm (H)) having the same geometry, but different concrete types and replacement ratios were made.

From the tests, it was revealed that the geopolymer concrete has the same compressive strength as ordinary concrete but with a smaller elastic modulus (e.g., 28.9 GPa for ordinary and 10.2 GPa for geopolymer) because the Young's modulus of geopolymer concrete is affected by microstructure based on speciation of the alkali silicate activating solutions as well as properties of the aggregates. Due to this, GRAC beams have less height of neutral axis and more deflection than ordinary recycled aggregate concrete beams at the same loading (Fig. 5a), depending on replacement patterns. The ultimate deflection was found to be 17.9 mm for CC14-100 (Ordinary concrete beam with longitudinal bars of 14mm ø with 100% RA), 19 mm for GC14-100 (Geopolymer concrete beam with longitudinal bars of 14mm ø with 100% RA), 12.7 mm for CC14-70-L (Ordinary concrete beam with longitudinal bars of 14 mm ø with 70% replacement of large natural aggregates), 21.9 mm for GC14-70-L (Geopolymer concrete beam with longitudinal bars of 14 mm ø with 70% replacement of large natural aggregates). GRAC beams also have slightly less cracking load, ductility and bending capacity. It was also found that the cracking load and cracking moment of the GRAC with 100% RA were found to be approximately 23% less as compared to the ordinary concrete beam with 100% NA. In this study, high alkali solution was used which reduced the elastic modulus and when the concentration was reduced the geopolymer concrete showed better results.

Aldemir et al. (2022) did research to find out the shear behaviour and structural performance of geopolymer concrete beams in detail. A new type of geopolymer concrete was prepared from demolition wastes. Roof tiles, red clay and hollow bricks, concrete rubble along with slag, fly ash, sodium hydroxide and sodium silicate were used to prepare GPC. In that study, GRAC, geopolymer natural aggregate concrete, ordinary concrete with RA, and ordinary concrete with NA were used. In previous studies the authors found that with the addition of RA both in ordinary concrete and GPC, the workability and compressive strength decrease but, in this study, it was assessed that, when the same w/c ratio is used, in GRAC the porosity of the concrete decrease because part of the water is absorbed by the RA, thus increasing the strength of the concrete.

Three shear span-to-effective depth ratios, namely (a/d = 0.5,1,1.65) were used to examine the different failure modes. Bending tests were performed to determine the shear behaviour of beams. Parameters including load-deflection curves, moment curves and crack propagation were used to assess the mechanical performance of the beams. It was also found that the compressive strength of members made of this GRAC was 3% more than conventional concrete members. From the results, it was assessed that the beams made up of geopolymer concrete exhibited similar performance as normal concrete beams of the same grade. But when recycled aggregates were used, then the failure mechanism shifted from flexure dominated to shear dominated. This shifting was more common in the beams with a larger span to effective depth ratio.

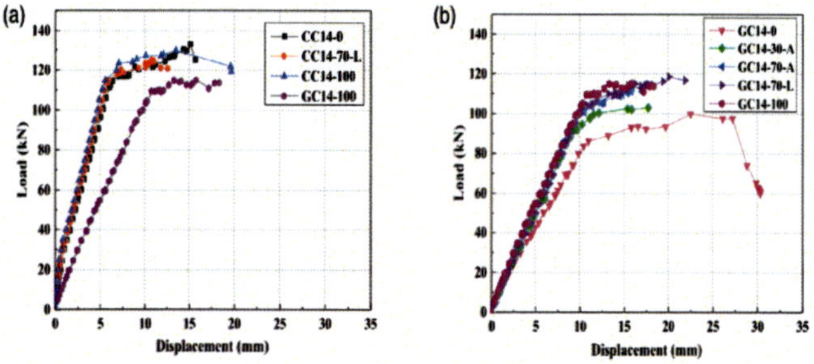


**Fig. 5.** (a): Load-displacement curves of geopolymer and ordinary Portland cement beams; and (b): Load- displacement curves of GRAC with different ratio and replacement patterns (Zhang et al. 2021)

## 4 Conclusions

The use of geopolymer concrete with recycled aggregate is a complex topic. When recycled aggregates are used with (Geo Polymer Cement) GPC or with ordinary cement, interfacial transition zone develops between the attached mortar and (Recycled Aggregate) RA. This is one of the weakest zones, and proper analysis is necessary before using the RA. In the latest research it is also analyzed that addition of fillers and fly ash is helpful to fill the pores of RA and thus it somehow reduces the vulnerability of failure along (Interfacial Transition Zones) ITZ; to this aim Two Stage Mixing Technique (TSMA) can be also adopted, in which first the recycled aggregates are mixed with cement paste which forms a coating on the surface of the aggregate and thus it fills up the cracks before actual mixing of concrete.

The type of geopolymer cement is one of the most important aspects, as different amount and type of chemicals used in the production of GPC. To choose proper type and quantity of materials for GPC production is a bit tricky task. In the older studies, the GPC is prepared without using metakaolin and the results were not so much encouraging, limiting the recycled aggregate replacement ratio up to 30–40%. In recent research, it is found that when metakaolin based GPC were used, larger values of replacement ratio, sometimes up to a full replacement, can be used without significant reduction of strength of the element. When more amount of RA is used, it causes workability issues, but from the latest research, it is also analyzed that the incorporation of superplasticizers up to 2% in volume is very useful.

It was also found that excess use of alkali activators causes the reduction in elastic modulus of concrete. So, it is very important to pay attention to the accurate quantity of alkali activators with reference to the other materials used in the production of GPC. There is not unanimous consensus yet about the exact quantity and type of the material to be used. Depending on the characteristics of available materials, the quantity and type of GPC should be chosen.

From all the literature review it is also analyzed that there are no general limits on the use of coarse RA in a concrete mixture. Some of the former researchers recommend 30% replacement of NA (Natural Aggregate) with RA as maximum, while recently researchers analyzed that RA replacement can go up to 50% or 100% if the mix design, batching methodology, and the moisture condition of the RA are properly handled.

## GLOSSARY

| | |
|---|---|
| GRAC | Geopolymer Recycled Aggregate Concrete |
| OPC | Ordinary Portland Cement |
| GPC | Geo Polymer Cement |
| GGBS | Ground Granulated Blast Furnace Slag |
| RA | Recycled Aggregates |
| ITZs | Interfacial Transition Zones |
| NA | Natural Aggregate |
| MK | Metakaolin |
| HCFA | High Calcium Fly Ash |
| PGC | Pressed Geopolymer Concrete |
| RB | Recycled Concrete Block Aggregate |
| FA | Fly ash |
| MMA | Mortar Mixing Approach |
| SEMA | Sand Enveloped Mixing Approach |
| GFRP | Glass Fiber Reinforced Polymer |
| TSMA | Two Stage Mixing Technique |

## References

Akhtar, A., Sarmah, A.K.: Construction and demolition waste generation and properties of recycled aggregate concrete: a global perspective. J. Clean. Prod. **186**, 262–281 (2018)

Aldemir, A., et al.: Shear behaviour of reinforced construction and demolition waste-based geopolymer concrete beams. J. Build. Eng. **47**, 103861 (2022)

Almutairi, A.L., Tayeh, B.A., Adesina, A., Haytham, F.I., Abdullah, M.Z.: Potential applications of geopolymer concrete in construction: a review. Construct. Mater. **15**, e00733 (2021)

Alqarni, A.S., Abbas, H., Al-Shwikh, K.M., Al-Salloum, Yousef, A.: Treatment of recycled concrete aggregate to enhance concrete performance. Construct. Build. Mater. **307**, 124960 (2021)

Berhanu1, A., Gebreyouhannes, E., Zerayohannes, G., Zeleke, E.: The effect of metakaolin on the properties of recycled aggregate concrete, concrete structures: new trends for eco-efficiency and performance. In: Proceedings of the fib Symposium 2021, Lisbon, Portugal (14–16 June 2021) (2021)

Diaz-loya, E.I., Allouche, E.N., Vaidya, S.: Mechanical properties of fly ash based geopolymer concrete. ACI Mater. J. **108**(3), 300–306 (2011)

Imbabi, M.S., Carrigan, C., McKenna, S.: Trends and developments in green cement and concrete technology. Int. J. Sustain. Built Environ. **1**(2), 194–216 (2012)

Kathirvel, P., Kaliyaperumal, S.M.: Influence of recycled concrete aggregates on the flexural properties of reinforced alkali activated slag concrete. Constr. Build. Mater. **102**(1), 51–58 (2016)

Lamond, J.F., Pielert, J.H.: Significance of Tests and Properties of Concrete and Concrete-Making Materials. ASTM, Philadelphia, PA (2006)

Le, H.-B., Bui, Q.-B., Tang, L.: Geopolymer recycled aggregate concrete: from experiments to empirical models. Materials **14**(5), 1180 (2021)

Liang, Y.-C., Ye, Z.-M., Vernerey, F., Xi, Y.: Development of processing methods to improve strength of concrete with 100% recycled coarse aggregate. J. Mater. Civ. Eng. **27**(5), 1–9 (2013)

Malhotra, V.M., Mehta, P.K.: High-performance, high-volume fly ash concrete: materials, mixture proportioning, properties, construction practice, and case histories, Book, Canada (2002)

Nuaklong and Sata, and V., Chindaprasirt, P. , 2016.Nuaklong, P., Sata, V., Chindaprasirt, P.: Influence of recycled aggregate on fly ash geopolymer concrete properties. J. Clean. Product. **112**(4), 2300–2307 (2016)

Nuaklong, P., Sata, V., Chindaprasirt, P.: Properties of metakaolin-high calcium fly ash geopolymer concrete containing recycled aggregate from crushed concrete specimens. Constr. Build. Mater. **161**, 365–373 (2018)

Raj, S.D., Bhoopesh, J.: Strength and behaviour of recycled aggregate geopolymer concrete beams. Adv. Concrete Construct. **5**(2), 145–154 (2017). https://doi.org/10.12989/acc.2017.5.2.145

Romanazzi, V., Leone, M., Aiello, M., Maddaloni, G., Pecce, M.: Bond strength of geopolymer concrete with steel and GFRP bars. Italian Concrete Days 2020, Italy (14–17 April 2021) (2021)

Ryu, G.S., Lee, Y.K., Koh, K.T., Chung, Y.S.: The mechanical properties of fly ash-based geopolymer concrete with alkaline activators. Constr. Build. Mater. **47**, 409–418 (2013)

Srinivas, T., Abhignya, G.: A review on geopolymer RCC beams made with recycled coarse aggregate. In: 2nd International Conference on Design and Manufacturing Aspects for Sustainable Energy, E3S Web of Conferences, vol. 184, p. 01095 (19 Aug 2020) (2020)

Srinivas, T., Abhignya, G.: Behaviour of structural elements made of geopolymer concrete with recycled aggregates. In: IOP Conference Series: Materials Science and Engineering, vol. 1091, Coimbatore, India (22–23 Jan 2021) (2021)

Thangamanibindhu, M.K., Murthy, D.R.: Flexural behavior of reinforced geopolymer concrete beams partially replaced with recycled coarse aggregates. Int. J. Civil Eng. Technol. **6**(7), 13–23 (2015)

Uddin Ahmed Shaikh, F.: Mechanical and durability properties of fly ash geopolymer concrete containing recycled coarse aggregates. Int. J. Sustain. Built Environ. **5**(2), 277–287 (2016)

Verian, K.P., Ashraf, W., Cao, Y.: Properties of recycled concrete aggregate and their influence in new concrete production. Resources Conserv. Recycl. **133**, 30–49 (2018). https://doi.org/10.1016/j.resconrec.2018.02.005

Wongsa, A., Siriwattanakarn, A., Nuaklong, P., Sata, V., Sukontasukkul, P., Chindaprasirt, P.: Use of recycled aggregates in pressed fly ash geopolymer concrete. Environ. Progress Sustain. Energy **39**(2), e13327 (2020)

Xu, Z., Huang, Z., Liu, C., Deng, X., Hui, D., Deng, S.: Research progress on mechanical properties of geopolymer recycled aggregate concrete. Rev. Adv. Mater. Sci. **60**(1), 158–172 (2021)

Zhang, H., Wan, K., Wu, B., Hu, Z.: Flexural behavior of reinforced geopolymer concrete beams with recycled coarse aggregates. Adv. Struct. Eng. **24**(14), 3281–3298 (2021)

# Bond Behavior of Geopolymer Concrete with CFRP and GFRP Bars

Maria Antonietta Aiello[1], Riccardo Angiuli[2], Ilaria Ingrosso[2], Marianovella Leone[1], Vincenzo Romanazzi[1(✉)], and Vito Tarantino[2]

[1] Department of Innovation Engineering, University of Salento, Lecce, Italy
vincenzo.romanazzi@unisalento.it
[2] CETMA, Brindisi, Italy

**Abstract.** The high mechanical performances showed by geopolymer concrete led several researchers to investigate about possibilities of using this material in reinforced structural elements. Since geopolymer binder has a different microstructure from ordinary Portland Cement (OPC) it is necessary to investigate on its bonding behavior with steel bar that as well-known influences the service and ultimate conditions. For this reason, in the last decades both direct pull-out and beam-end tests were carried out with this material. Generally, it has been observed that geopolymer concrete (GPC) has higher bond strength than OPC due to the higher compression strength and the dense and compact microstructure of GPC. This means that the existing design equation for bond strength prediction of ordinary concrete can be conservatively used also for GPC. In this paper the mechanical properties of GPC will be analyzed and the bond-slip behavior of GPC and both CFRP and GFRP bars has been studied. The experimental data are analyzed and discussed in terms of bond stress transfer mechanisms and mode of failure.

**Keywords:** Geopolymer concrete · Bond behavior · CFRP bar · GFRP bar

## 1 Introduction

The recent develops in construction material technology have showed the great potential of Geopolymer Concrete (GPC). The necessity of this new concrete material arose from the environmental issues related to the Portland Cement production. It is estimated that more than 4 billion tonnes of cement are produced each year, leading to the production of 8% of the global carbon dioxide ($CO_2$) emissions (Lehne & Preston 2018). At this purpose, using geopolymer binder instead of OPC allow to reduce the $CO_2$ emissions up to the 80% since the activation process of geopolymer cement requires no heat. Moreover, GPC have been confirmed to have high mechanical properties so that several authors consider the use of this material also in structural elements in place of OPC concrete (Singh et al. 2015, Mo et al. 2016, Ding et al. 2016, Ma et al. 2018).

The bond behaviour between GPC and deformed steel bar have been investigated by means of both direct pull-out and beam tests. The results obtained from these tests


M. A. Aiello and A. Bilotta (Eds.): ICC 2022, LNCE 435, pp. 94–103, 2024.
https://doi.org/10.1007/978-3-031-43102-9_8

showed that GPC has higher bond strength with deformed steel bar than OPC concrete (Sarker 2010).

In addition to steel bar, some researchers focused on bond-slip behaviour between GPC and glass-fiber reinforced polymer (GFRP) bar. In particular, they observed that the bond strength of GFRP reinforced GPC was similar to that of OPC concrete (Tekle et al. 2015).

In this paper the bond slip of GPC with sand-coated CFRP reinforcing bar is analysed by performing direct pull-out test and compared the results obtained with that of GFRP bar reported in (Romanazzi et al. 2021). The direct pull-out test results on GPC with GFRP and CFRP underlined that the bond strength values obtained with the two different bar material are comparable and that in both cases the maximum load increases increasing the bond length from 2.5ϕ to 5ϕ.

## 2 Experimnetal Investigation

### 2.1 Experimental Program

The experimental test discussed in the following refers to material characterization of GPC and bond tests between GPC and FRP bars. As better specify in the following sections the obtained results are compared with those already discussed in (Romanazzi et al. 2021, Attanasio et al. 2021).

The experimental program is resumed in Table 1. In particular, it includes a total of 19 pull-out tests with both GFRP and CFRP bars and two different bond lengths of which 5 (marked with an asterisk) belong to experimental tests reported in (Romanazzi et al. 2021). The chosen bond length was 2.5 and 5 times the bar diameter (Table 2).

**Table 1.** Pull-out samples casted and tested

| Bar material | Bar diameter | Bond length | Number of specimens |
|---|---|---|---|
| GFRP | ϕ12 mm | 2.5ϕ | 5 |
| | | 5ϕ | 5* |
| CFRP | ϕ8 mm | 5ϕ | 5 |
| | | 2.5ϕ | 4 |

* (Romanazzi et al. 2021)

### 2.2 Materials

The binder of the cement-free concrete mix investigated in this work were obtained by mixing: Ground Granulated Blast Furnace Slag (GGBFS); expanded glass and water-glass. The geopolymer mixtures GPC_1, defined in (Romanazzi et al. 2021, Attanasio et al. 2021), and GPC_2 are reported in Table 2.

**Table 2.** Geopolymer concrete mixes design

| GPC_1** | |
|---|---|
| Components | Quantity |
| GGBFS | 224 kg/m$^3$ |
| Limestone and gypsum | 128 kg/m$^3$ |
| Silica Fume | 48 kg/m$^3$ |
| Ingessil (Activator solution) | 170 kg/m$^3$ |
| Water | 140 kg/m$^3$ |
| Additive (Plasticizer) | 8 kg/m$^3$ |
| Sand | 1092 kg/m$^3$ |
| Gravel | 471 kg/m$^3$ |
| GPC_2 | |
| Components | Quantity |
| GGBFS | 222.2 kg/m$^3$ |
| Expanded Glass | 55.6 kg/m$^3$ |
| Waterglass | 77.8 kg/m$^3$ |
| Activator solution (NaOH) | 25.9 kg/m$^3$ |
| Water | 122.2 kg/m$^3$ |
| Natural Sand (0-4 mm) | 707.8 kg/m$^3$ |
| Gravel (4-8 mm) | 354.0 kg/m$^3$ |
| Magnetite (0-2 mm) | 689.0 kg/m$^3$ |

**(Romanazzi et al. 2021, Attanasio et al. 2021)

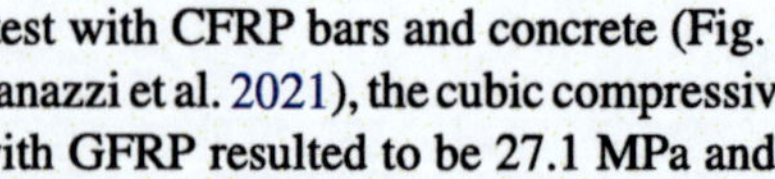

**Fig. 1.** GFRP (up) and CFRP (down) bars detail

GPC_1 mix design was used to realize the bond test between GFRP bars and concrete while GPC_2 for bond test with CFRP bars and concrete (Fig. 1).

As reported in (Romanazzi et al. 2021), the cubic compressive strength of GPC_1 cast for pull-out specimen with GFRP resulted to be 27.1 MPa and 19.5 MPa for specimen with bond length equal to 5$\phi$ and 2.5$\phi$, respectively. The cubic compressive strength of the GPC_2 mix determined according to UNI EN 12390–3 (2001) and at 28 days of curing resulted to be equal to 42.3 MPa.

Two different types of reinforcement bar have been used in the present work: sand coated glass fiber-reinforced polymer (GFRP) and sand coated carbon fiber reinforced polymer (CFRP) bar. The GFRP bars, with 12 mm diameter, were made of chemically resistant glass fiber and polyester thermosetting resin with a nominal tensile strength equal to 800 MPa and modulus of elasticity of 3 5GPa. The CFRP bars with dimeter of 8 mm have a nominal tensile strength equal to 1700 MPa and modulus of elasticity of 130 GPa. For both bars the mechanical properties refers to the data sheet of the manufacture.

### 2.3 Test Setup

The direct pull-out tests setup and specimens were designed by following the RILEM Recommendations (RILEM TC 1983) as showed in Fig. 2.

Each GPC specimen was cubic with side equal to 10ϕ with a centred embedded bar in order to have a concrete cover equal to 4.5ϕ. The bonded zone, $L_b$, was 5ϕ and 2.5ϕ and it was realized by applying a plastic sheet on the embedded length of the bar according to the Fig. 2.

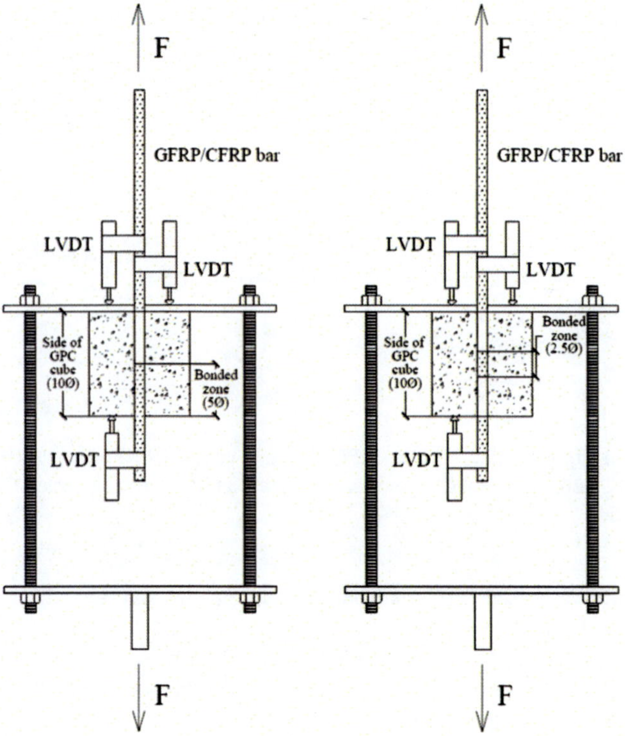


**Fig. 2.** Pull-out tests setup

As showed in Fig. 2, three Linear Variable Displacement Transducer (LVDT) have been used to measure the slip between bars and GPC: one anchored on the unloaded side of the bar and the others positioned on the loaded side. It has been assumed that the bond stress has a constant trend within the bond length $L_b$, so the bond stress could be calculated using Eq. 1:

$$\tau_b = \frac{F}{\pi \cdot \phi \cdot L_b} \tag{1}$$

where F is the applied tensile load.

# 3 Experimental Results

## 3.1 Pull-out Test with GFRP Bars

In Fig. 3 the bond stress versus slip curves obtained by means pull-out tests between GPC and GFRP bar with bond length equal to 5ϕ (Romanazzi et al. 2021) and those with bond length of 2.5ϕ are reported. The shape of the curves is almost similar varying the bond length: since the GFRP bars used had no transversal ribs, the resistance to slip was only due to chemical adhesion between GPC matrix and sand-coated bar as confirmed by the first branch of the curve that results characterized by very small slip values.

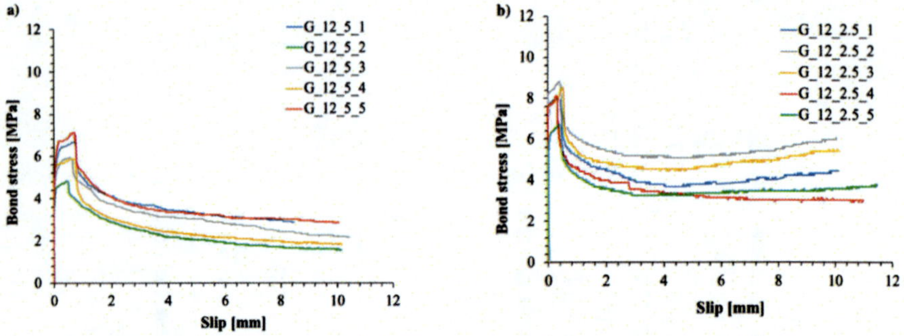


**Fig. 3.** a) Bond slip behaviour between GPC and sand-coated GFRP bar with bond length 5ϕ (Romanazzi et al. 2021); b) Bond slip behaviour between GPC and sand-coated GFRP bar with bond length 2.5ϕ

All the specimen exhibited a pull-out failure meaning that the reinforcing bar was pulled out from the concrete cube.

The experimental bond strength values obtained from pull-out tests between GPC and GFRP bar are reported in Table 3. As showed in (Romanazzi et al. 2021), in case of bond length equal to 5ϕ, the average value of GPC-GFRP bar bond strength resulted to be 6,1 MPa (CoV = 14%). On the other hand, when the bond length was set to 2.5ϕ the average bond strength resulted to be 8.1 MPa.

**Table 3.** Pull-out tests between GPC and GFRP results

| Bond Length | Sample | $\tau_{b,max}$ [MPa] | $\tau_{b,max,av}$ [MPa] |
|---|---|---|---|
| 5ϕ | G_12_5_1 | 6.7 | 6.1 (CoV = 14%) |
| | G_12_5_2 | 4.8 | |
| | G_12_5_3 | 6.0 | |
| | G_12_5_4 | 5.9 | |
| | G_12_5_5 | 7.2 | |
| 2.5ϕ | G_12_2.5_1 | 8.0 | 8.1 (CoV = 10%) |
| | G_12_2.5_2 | 8.8 | |
| | G_12_2.5_3 | 8.6 | |
| | G_12_2.5_4 | 8.1 | |
| | G_12_2.5_5 | 6.7 | |

### 3.2 CFRP Bars

The results of bond behaviour of GPC with CFRP bars is reported in Fig. 4.

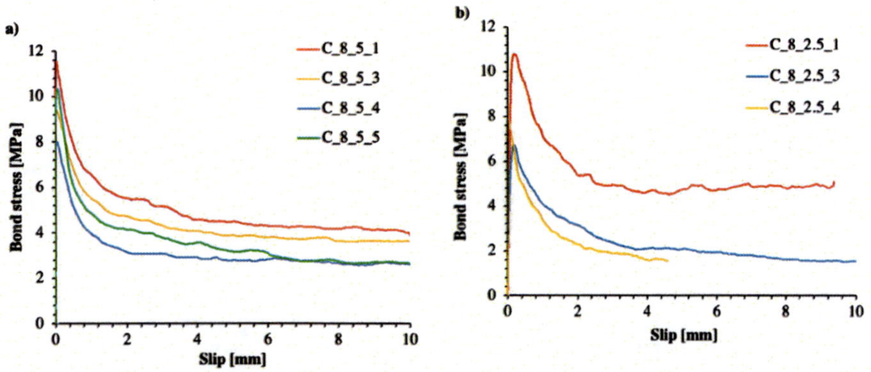


**Fig. 4.** a) Bond slip behaviour between GPC and sand-coated CFRP bar with bond length 5ϕ; b) Bond slip behaviour between GPC and sand-coated CFRP bar with bond length 2.5ϕ

Again, the finishing of the CFRP bars and the small value of the slip in the ascending branch of the curve suggest that the stress transfer behaviour between CFRP bars and GPC is manly governed by the chemical friction. In fact, from Fig. 3 and Fig. 4 it can be noted that a sharp bond stress decay occur once the peak has been reached.

As in case of GFRP bar, all the specimen's failures were pull-out type. The results obtained on specimen C_8_5_2 and C_8_2.5_2 have been excluded from the respective averages values since they were largely out from the statistical ranges. However, the results are still under analysis to better understand the cause of these scattered data.

The bond strength results are resumed in Table 4. In particular, the average bond strength of GPC with CFRP bars resulted to be 9.8 MPa and 8.5 MPa for bond length equal to 5ϕ and 2.5ϕ, respectively.

**Table 4.** Pull-out tests between GPC and CFRP results

| Bond Length | Sample | $\tau_{b,max}$ [MPa] | $\tau_{b,max,av}$ [MPa] |
|---|---|---|---|
| 5ϕ | C_8_5_1 | 11.6 | 9.8 (CoV = 15%) |
| | C_8_5_3 | 9.4 | |
| | C_8_5_4 | 8.0 | |
| | C_8_5_5 | 10.3 | |
| 2.5ϕ | C_8_2.5_1 | 10.8 | 8.5 (CoV = 25%) |
| | C_8_2.5_3 | 6.7 | |
| | C_8_2.5_4 | 7.8 | |

Finally, in order to compare the maximum bond stress reached by specimen with GFRP and CFRP bar, the results of specimens with GFRP bars have been corrected ($\tau_{b,max,cor}$) by using the coefficient

$$\left(\frac{f_c}{f_{cREF}}\right)^{1/2}$$

in which $f_{cREF}$ has been assumed equal to the compressive strength of the second GPC mix (used to realize the pull-out specimen with CFRP bar). This procedure, applied also by other previous research works (Aiello, Leone, and Pecce 2007; CEB-FIP 2010), has become necessary since the two mixes has a different compressive strength. The adjusted results are showed in Table 5.

From a direct comparison between the results reported in Table 4 and those of Table 5 it is not evident a clear effect of stiffness of the bars on bond performance: taking into account the data variability the recorded bond strength are almost similar. Probably the stress transfer mechanism activated for this type of bar (i.e. chemical adhesion and friction) are not linked to the type of FRP rebars. However, the data are under in-depth analysis to better understand the bond mechanisms taking into account also the different diameter of the bars.

**Table 5.** Bond strength corrected results between GPC and GFRP results

| Bond Length | Sample | $\tau_{b,max,adj}$ [MPa] | $\tau_{b,max,adj,av}$ [MPa] |
|---|---|---|---|
| 5ϕ | G_12_5_1 | 8.3 | 7.6 (CoV = 14%) |
| | G_12_5_2 | 6.0 | |
| | G_12_5_3 | 7.5 | |
| | G_12_5_4 | 7.4 | |
| | G_12_5_5 | 8.9 | |
| 2.5ϕ | G_12_2.5_1 | 11.8 | 11.9 (CoV = 10%) |
| | G_12_2.5_2 | 13.1 | |
| | G_12_2.5_3 | 12.6 | |
| | G_12_2.5_4 | 12.0 | |
| | G_12_2.5_5 | 9.9 | |

## 3.3 Effect of Bond Length

The graph of Fig. 5 puts in evidence the effect of the bond length on maximum load for both GFRP and CFRP bar.

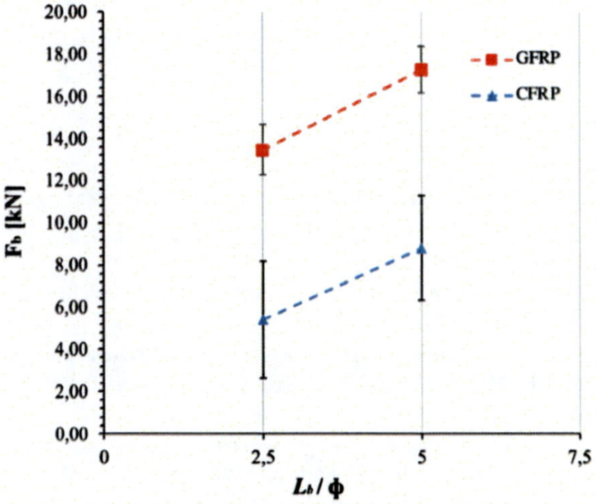


**Fig. 5.** Effect on bond length on the maximum load

Analyzing the Fig. 5 it possible to observed that the maximum load sharply increases for both bars materials when the bond length passes from 2.5ϕ to 5ϕ. This observation underlines that, in both cases, the transfer length (i.e. the length above which maximum load doesn't increase anymore) is equal or higher than 5ϕ.

## 4 Discussion and Conclusions

In the present work the bond strength of GPC with both sand-coated GFRP and CFRP bar has been investigated. A total of 14 pull-out specimens have been casted and tested: five were realized with ϕ12 mm GFRP bar (with bond length of 2.5ϕ) and nine were realized with ϕ8 mm CFRP bar (with bond length of 5.0ϕ and 2.5ϕ). According to the results illustrated in the previous section, the following consideration can be taken:

- In case of both sanded coated GFRP and CFRP bars the bond mechanism was mainly due to chemical adhesion and friction with GPC matrix.
- The ultimate bond strength obtained seems to be independent of the stiffness of the FRP bars: results of GPC with sand-coated GFRP bar resulted to be comparable with those obtained with sand-coated CFRP bar.
- The maximum load sharply increases when the bond length passes from 2.5ϕ to 5ϕ, meaning that the transfer length is equal or higher than 5ϕ.

The experimental campaign introduced in this work will be extended with future research activities in order to validate the results obtained and develop useful analytical relationship useful from design point of view.

**Acknowledgements.** This research was carried out in the framework of the research project MAREWIND "MAterials solutions for cost Reduction and Extended service life on WIND off-shore facilities" funded by the European Union's Horizon 2020 research and innovative programme.

## References

Aiello, M.A., Leone, M., Pecce, M.: Bond performances of FRP rebars-reinforced concrete. J. Mater. Civ. Eng. **19**, 205–213 (2007). https://doi.org/10.1061/(ASCE)0899-1561(2007)19

Attanasio, A., et al.: Secondary Raw Materials for a Circular Economy in Concretes. Italian Concrete Days – Napoli (2021)

CEB-FIP. 2010. Fib Model Code for Concrete Structures. https://doi.org/10.1007/s13398-014-0173-7.2

Ding, Y., Dai, J.G., Shi, C.J.: Mechanical properties of alkali-activated concrete: a state-of-the-art review. Constr. Build. Mater. **127**, 68–79 (2016). https://doi.org/10.1016/j.conbuildmat.2016.09.121

Lehne, J., Preston, F.: Making concrete change innovation in low-carbon cement and concrete Chatham House Report 1–122 https://doi.org/10.1088/1742-6596/1015/3/032163

Ma, C.K., Awang, A.Z., Omar, W.: Structural and material performance of geopolymer concrete: a review. Constr. Build. Mater. **186**, 90–102 (2018). https://doi.org/10.1016/j.conbuildmat.2018.07.111

Mo, K.H., Johnson Alengaram, U., Jumaat, M.Z.: Structural performance of reinforced geopolymer concrete members: a review. Constr. Build. Mater. **120**, 251–264 (2016). https://doi.org/10.1016/j.conbuildmat.2016.05.088

Rilem, T.C.: RC 6 Bond Test for Reinforcement Steel. 2. Pull-out Test. Edited by RILEM (1983)

Romanazzi, V., Leone, M., Aiello, M.A., Maddaloni, G., Pecce, M.: Bond Strength of Geopolymer Concrete with Steel and GFRP Bars. Italian Concrete Days - Napoli, no. April (2021)

Sarker, P.: Bond strengths of geopolymer and cement concretes. Adv. Sci. Technol. **69**, 143–151 (2010). https://doi.org/10.4028/www.scientific.net/ast.69.143

Singh, B., Ishwarya, G., Gupta, M., Bhattacharyya, S.K.: Geopolymer concrete: a review of some recent developments. Constr. Build. Mater. **85**, 78–90 (2015). https://doi.org/10.1016/j.conbuildmat.2015.03.036

Tekle, B.H., Amar, K., Obada, K.: Bond properties of glass fibre reinforced polymer bars with fly-ash based geopolymer concrete. In: 10th International Conference on Composite Science and Technology, no. October: 1–8 (2015)

UNI EN 12390–3.: Testing Hardened Concrete - Part 3: Compressive Strength of Test Specimens (2001)

# Ultra High-Performance Fibre-Reinforced Cementitious Composites as the Link Between Structural Durability and Sustainability: The Experience of the H2020 Project ReSHEALience

Francesco Lo Monte(✉), Salam Al-Obaidi, and Liberato Ferrara

Department of Civil and Environmental Engineering, Politecnico di Milano, Milan, Italy
{francesco.lo,salammaytham.alobaidi,liberato.ferrara}@polimi.it

**Abstract.** The need for more environmentally and economically sustainable structures pushes towards ahead the formulation of new advanced cementitious materials which, on the one hand, can reduce the carbon footprint in the production phase, and, on the other hand, can significantly improve the structural durability, thus resulting into a longer service life. In this context, the ReSHEALience project has been launched in 2018 within the Programme Horizon 2020. The project aims at developing a new approach for the design of structures exposed to extremely aggressive environments, based on durability and life cycle analysis. The starting point is represented by a novel concept of Ultra-High Performance Fibre-Reinforced Cementitious Composites (UHPFRCCs), which has been called Ultra-High Durability Concretes (UHDC), characterized by multiple cracking behavior and enhanced self-healing. In a design perspective, this makes it necessary to develop an effective approach for identifying the main parameters describing the overall behavior in tension. In the paper an overall description of the concepts and of the structure behind ReSHEALience project is provided. The core is represented by Sustainability of structures and infrastructures, intended as optimization of materials performance and extension of the service life.

**Keywords:** ReSHEALience · Ultra-High Durability Concrete – UHDC · tensile constitutive behaviour · structural validation · self-healing · Durability Assessment-based Design (DAD) · Life Cycle Analysis (LCA)

## 1 Introduction

### 1.1 Durability and Sustainability

In the last two decades, the sustainability issue has received more and more attention, thus becoming a primary goal in many fields of human activities. To this end, in particular in the last decade, several efforts have been made in order to decrease the overall environmental impact of the construction sector, often acting in the production processes of the materials with the aim of reducing the consumption of raw materials and/or the overall carbon footprint. Alternatively, sustainability in construction industry can be pursued by extending the service life of structures and infrastructures, consequently decreasing maintenance and restoration interventions.


M. A. Aiello and A. Bilotta (Eds.): ICC 2022, LNCE 435, pp. 104–115, 2024.
https://doi.org/10.1007/978-3-031-43102-9_9

This second approach, however, needs higher material and structural durability, hence, typically, a higher initial cost of construction. On the opposite it allows to reduce the final economical and environmental impact if the whole structural life cycle is considered (Bribián et al., 2011; Banthia 2008; Mackechnie and Alexander, 2009).

The clear definition of such approach represents the scope of the Research Project ReSHEALience (GA 760824), launched in January, 2018, within the framework of the Research and Innovation Programme Horizon 2020 and concluded in March, 2022. Such framework entails the development of a durability-oriented structural approach for both ordinary and extremely aggressive environments, based on the concepts of Durability Assessment-based Design (DAD) and Life Cycle Analysis (LCA).

The explicit evaluation of key durability parameters is then required, in the classic performance-based approach, this allowing to effectively take advantage of the great benefits (in terms of both mechanical performance and improved durability) linked to the use of advanced cementitious materials (Al-Obaidi et al., 2020; Serna et al., 2019).

In this context, Ultra-High Performance Fibre-Reinforced Cementitious Composites (UHPFRCC) rises on the stage, thanks to the powerful interaction of reduced crack-opening and engineered self-healing which, if properly calibrated, can significantly increase the long term performance in the cracked state, even in particular aggressive environments.

Extremely Aggressive Environments (EAE), in fact, represent a severe and challenging condition for Reinforced Concrete structures which experience durability time-dependent problems (including, for example, ageing and corrosion of reinforcement) even resulting into early and/or continuous repairs. Matthews (1997), for example, reports as almost half of repaired concrete structures fails again within the following 25 years. Despite the need for increasing durability, current solutions are often not taking into account new cement-based construction materials, such as UHPFRCC, neither new constituents specifically conceived to improve the concrete durability, because of the lack of standards and technical awareness by most designers and contractors.

It is worth remarking, in fact, that UHPFRCC represents the most significant innovation in concrete technology of the last thirty years, based on the balance between crack-tip toughness and fibre pull-out work in the matrix-fibre interaction (Li and Wu, 1992; Li et al., 1993, Li, 2003). If the mix design is properly designed so to have a hardening response in tension (or bending), once the first crack is formed under tension (or bending), the crack bridging action of the fibres is activated. The possibility of further increase the applied load, thanks to the hardening response, allows for the formation of new cracks, this resulting into the stable multiple cracking process up to the onset of the unstable localization of one single crack (Naaman and Reinhardt, 2006). Furthermore, hardening response in tension/bending requires mix compositions characterized by a high binder content and a low water/binder ratio (this being compensated by a high dosage of superplasticizer), this guaranteeing a far higher durability in the un-cracked state with respect to ordinary concretes, thanks to the high compactness of the matrix.

On the other hand, a superior durability is provided in the cracked state also, thanks to the above-mentioned multiple cracking response translating into an inherently effective crack-width control. Moreover, the synergy between crack tightness and material composition (characterized by a large quantity of un-hydrated particles) also results into a high propensity to autogenous self-healing, with the following enhancement of the material and structural durability (Li and Li, 2011; Özbay et al., 2013; Zhang et al., 2014). As it is well-known, healing products allow the partial or total closure of cracks and can also improve the fibre-matrix bond. This latter aspect makes it possible that, when reloading a healed specimen, new cracks are formed instead of re-opening the healed ones (Ferrara et al., 2015).

In this framework, the H2020 project ReSHEALience proposes to upgrade the concept of UHPFRC through the incorporation of tailored functionalities to a "metamaterial" concept, named Ultra High Durability Concretes (UHDCs), which will share the signature tensile "strain hardening" behaviour of UHPFRC. The combined effects of inherent higher durability at the un-cracked state, multiple cracking under tension/bending and autogenous self-healing represent the key point for improving the durability and extending the service life of structures exposed to ordinary conditions or EAEs, subsequently reducing maintenance needs and the overall economic impact.

## 1.2 ReSHEALience Project

ReSHEALience Projects involves 13 partners all around Europe, consisting of 5 universities (Politecnico di Milano – Project Coordinator, Universitat Politècnica de València, Technische Universität Dresden, University of Malta, Ben-Gurion University of the Negev), 2 research centers (CSIC and STRESS Scarl) and 6 industrial partners (API Europe, Banagher Precast Concrete, Enel Green Power, Penetron Italia, RDC, Rover Maritime,) as illustrated in Fig. 1.

As previously described, the final goal of extending the service life of structures is based on the overall improvement of durability performance. As shown in Fig. 2, this approach starts from the definition of UHPFRCCs with improved durability, thus called UHDC. It is worth noting as this was not intended as the development of patented mix designs, but as the definition of guidelines for the design of cementitious mixes fulfilling the requirements of UHDC mixes. This is instrumental for the implementation of more sustainable materials which must make use of local available constituents.

These mixes have been firstly characterized at the material level from both the mechanical and durability point of view according to a "multi test/multi parameter approach" (Lo Monte and Ferrara, 2020: Lo Monte and Ferrara, 2021, Lo Monte et al., 2022; Davolio et al., 2023). In parallel, a Durability Assessment-based Design (DAD) Approach is defined, so to implement a Performance-Based approach able to make "fully exploitable" the great benefits linked to UHDC in terms of durability performance. This entails the definition of durability indexes linked to key performance points.

Secondly, full-scale demonstrators (also called pilots, see Fig. 3) have been constructed, in order to assess the overall performance at the structural scale, this being important for checking the whole procedure: production process and technologies, simplified or advanced design approaches and durability performance monitoring with time.

The 6 demonstrators consist of: a precast break water element made of UHDC-based textile, an aquaculture mussel raft consisting of pre-stressed UHDC slender beams, an offshore wind floater made of hollow core UHDC members, the restoration of a R/C historical water tower, and, finally, a mud collection basin and a geothermal power plant tower basin made of UHDC cast-in-situ or prefabricated walls/panels.

The former 4 pilots are exposed to XS (chlorides) aggressive environments, while the latter two to XA (sulphate) exposure. The design and the following monitoring of the full-scale demonstrators have been based on the implementation of both advanced (such as finite element modelling) and simplified (namely, analytical practitioner-oriented) calculations (Al-Obaidi et al., 2020; Al-Obaidi et al., 2021; Al-Obaidi et al., 2022). In parallel, advanced numerical simulations have been formulated taking into account the coupled chemo-physical-mechanical problem, able to simulate mechanical cracking and damage, self-healing kinetics and material properties recovery (Cibelli et al., 2019). Such advanced numerical simulations aim at understanding the main mechanisms behind self-healing and material properties recovery, thus generalizing the experimental results.

This paved the way to the last step of the project, namely the formulation of the Durability Assessment-based Design (DAD) via practitioner- and standards-oriented approaches. In turn, DAD is developed together with a Life-Cycle Assessment with a cradle-to-cradle approach, in order to estimate the overall economic and environmental impact, on the basis of a set of key performance durability parameters.

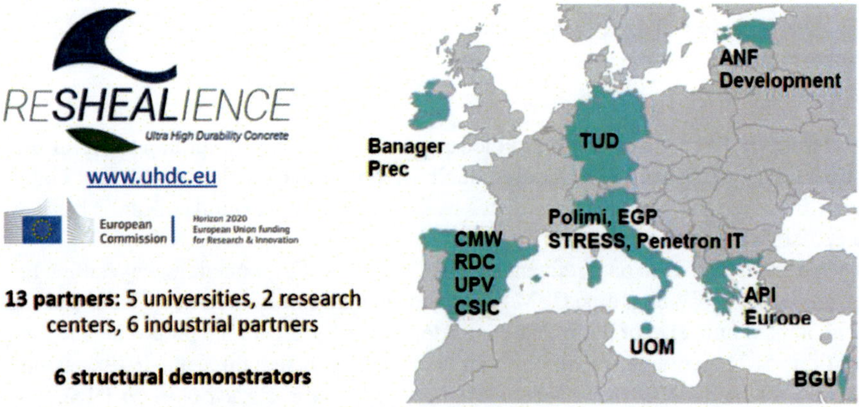


**Fig. 1.** ReSHEALience Consortium and geographical collocation of the partners.

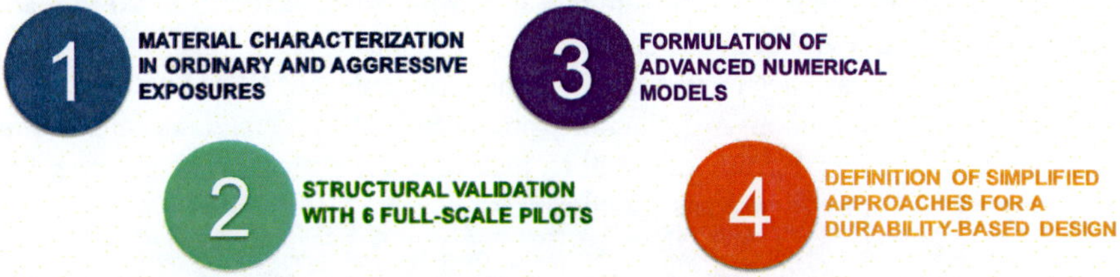


**Fig. 2.** Conceptual scheme of the ReSHEALience Project development.

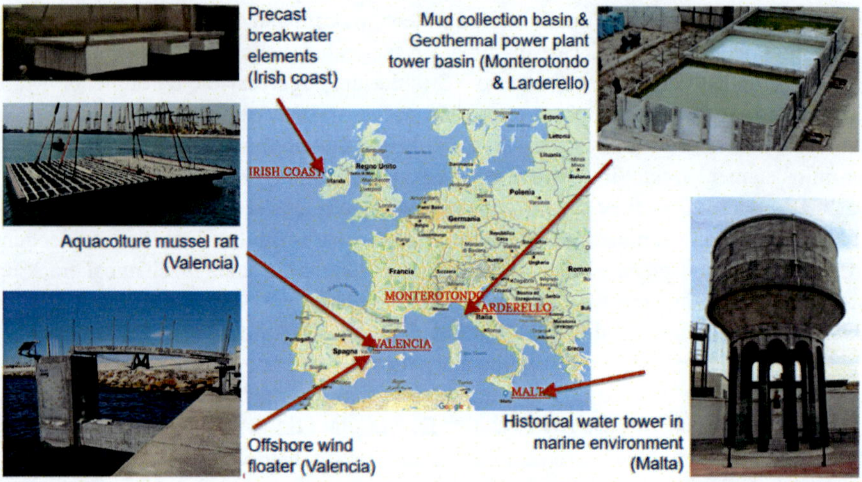


**Fig. 3.** Full-scale demonstrators of ReSHEALience Project for the validation at the structural level.

## 2 Materials Characterization

### 2.1 Mix Design

Different UHPFRCCs have been studied as part of the Project, according to the different structural applications and exposures. Table 1 reports the typical mix studied at Politecnico di Milano for the pilots constructed by EGP in Tuscany (a mud collection basin and a geothermal power plant tower basin).

The study is based on a reference UHDC mix (XA-CA) containing crystalline Penetron Admix®. Cement type CEM I 52.5 and slag have been used as a binder and sand with a maximum size of 2 mm has been adopted, according to the proportions reported in Table 1. The volume fraction of both cement (c) and slag (s) is about 19%, while it is 38% for sand (a), 20% for water (w) (thus, with a ratio c:s:a:w close to 1:1:2:1) and 1.5% for steel fibre (almost minimum amount of fibre for attaining a strain-hardening behaviour). An important role is obviously played by superplasticizer, added to get the correct rheology during casting, studied for fostering fibre alignment with pouring flow. Straight brass-plated fibres (tensile strength $f_t \geq 2400$ MPa, length $l_f = 20$ mm and diameter $d_f = 0.22$ mm) have been used.

Further aspects have been investigated, as the replacement of CEM I with CEM III, or the adoption of amorphous metallic fibre (produced with a special procedure leading to an amorphous microstructure of the fibre, instead of the typical crystalline one, this providing far higher resistance against corrosion).

### 2.2 Mechanical Characterization

Mechanical characterization has been firstly performed via different indirect tension tests, namely 4-Point Bending Test (4PBT) on 100x100x500 $mm^3$ and 25 x 100 x

**Table 1.** UHDCs' mix compositions.

| Constituents [kg/m$^3$] | XA-CA | XA-CA-Amf | XA-CA-CEMIII |
|---|---|---|---|
| CEM I 52.5 | 600 | 600 | - |
| CEM III 52.5 | - | - | 600 |
| Slag | 500 | 500 | 500 |
| Water | 200 | 200 | 200 |
| Steel fibres | 120 | - | 120 |
| Amorph. Fibres | - | 111 | - |
| Sand (0–2 mm) | 982 | 982 | 982 |
| Superplasticizer | 33 | 33 | 33 |
| Cryst. Adm | 4.8 | 4.8 | 4.8 |

500 mm$^3$ specimens and Double-Edge Wedge Splitting (DEWS) on 100x100x25 mm$^3$ specimens (see Fig. 4).

The following calibration of the material constitutive law in tension require the implementation of identification procedures (Figs. 5 and 6, Lo Monte et al., 2020) or even inverse analysis, (Lo Monte et al., 2022; Lo Monte et al., 2023a) due to the inherent redundancy of indirect tests.

Material behaviour has been assessed in ordinary conditions and after 6 months' exposure to aggressive environment, and the calibrated constitutive laws have been then adopted for the design of the related full-scale demonstrator.

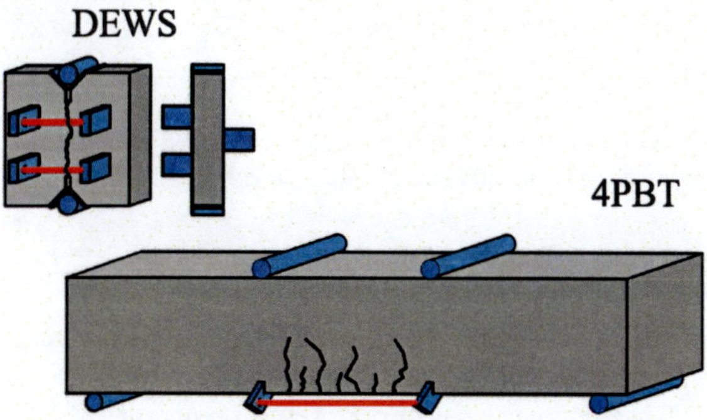


**Fig. 4.** Conceptual scheme of Double-Edge Wedge Splitting (DEWS) and 4-Point Bending Test (4PBT).

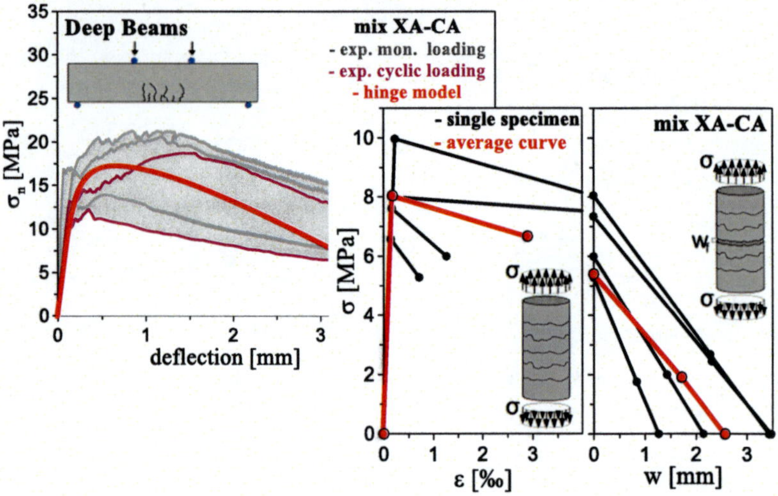


**Fig. 5.** Calibrated constitutive laws in tension for the reference mix (adapted from Lo Monte et al., 2022).

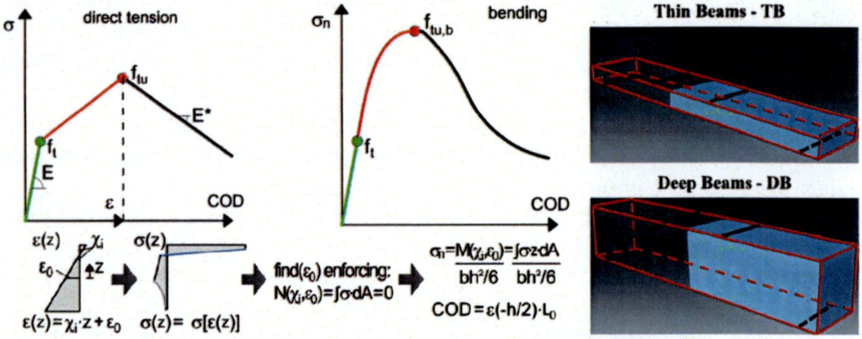


**Fig. 6.** Calibration of constitutive law in tension based on DEWS and back simulation of bending test via sectional integration (adapted from Lo Monte and Ferrara, 2020).

## 2.3 Self-healing Assessment

Durability performance at material level has been also investigating implementing a multi-test/multi-parameter approach (Lo Monte and Ferrara, 2021), in order to assess the benefits of self-healing in the recovery of several properties of UHDCs.

Three different test setups have been adopted, addressing the recovery brought in cracked samples by self-healing in terms of water permeability, flexural stiffness and crack-sealing:

- water permeability test on disks (diameter of 100 mm and thickness of 60 mm);
- 4-point-bending test on Thin Beams – TB (25 x 100 x 500 $mm^3$);
- 4-point-bending test on Deep Beams - DB (100 x 100 x 500 $mm^3$).

In all the three tests, crack sealing triggered by self-healing phenomenon has been assessed (Fig. 7a) as a function of the healing time. Then, in the first test, water permeability is assessed, while in the other two setups (based of flexural tests), stiffness (Fig. 7b) and strength recovery are investigated.

It is worth noting as the experimental assessment of self-healing still not see any established standard test. In this regards the authors have been involved in the Cost Action - European Cooperation in Science and Technology, CA15202 - Self-healing As preventive Repair of COncrete Structures (SARCOS). Under the umbrella of SARCOS Cost Action, in fact, several Round Robin Tests have been organized to study the effectiveness and sensitivity of reference tests for the assessment of self-healing regarding permeability, chlorides diffusivity, stiffness and strength recovery (Cuenca et al., 2021; Lo Monte et al., 2023b).

Focusing on the UHDCs studied within the ReSHEALience Project, the inherent crack-width control allowed by the multi-cracking response, allowed for keeping limited crack opening even for high average tensile strain. For strain at the tension side (under bending) of around 2‰ (this being the yielding strain of common steel rebar!) cracks with opening lower than 100 μm has been observed. (Typically, crack width was in the range 0–50 μm). Such limited crack opening fostered a significant recovery of material

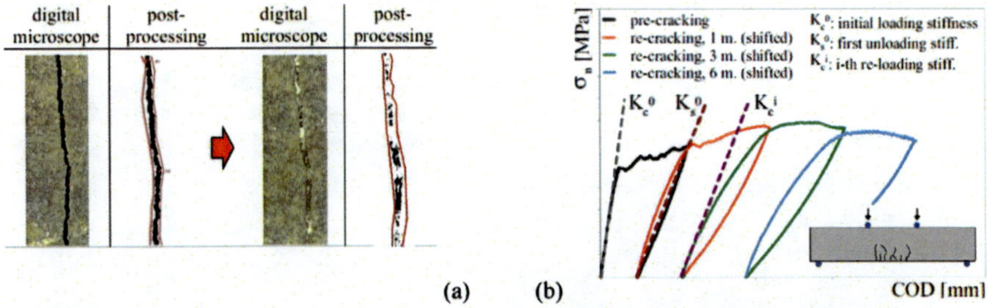


**Fig. 7.** Evaluation of crack-sealing recovery via digital image analysis (a) and evaluation of stiffness recovery via multiple cracking at different healing time durations (b) (adapted from Lo Monte et al., 2021).

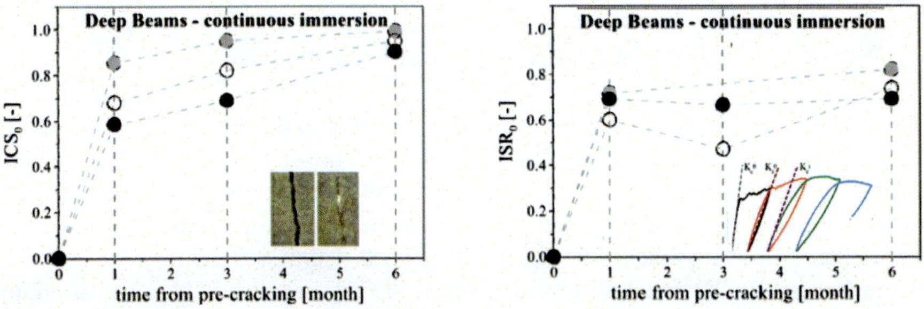


**Fig. 8.** Evolution of Index of Crack-Sealing (ICS) and Index of Stiffness Recovery (ISR) with time for the reference mixes under continuous immersion in geothermal water (adapted from Lo Monte et al., 2021).

parameters already after 1 month of healing, as shown in Fig. 8 for crack-sealing and stiffness.

## 3 Structural Validation

On the basis of the calibrated constitutive laws and of the assessment of self-healing capability, full-scale demonstrators have been designed. In Figs. 9 and 10, the geothermal power plant tower basin (*Politecnico di Milano-Enel Green Power* pilot) is shown together with simplified design methods.

Both more advanced methods (based on linear/non-linear finite element modelling) and simple analytical approaches (effective cantilever and yielding line approaches, as sketched in Fig. 9) have been implemented in a designer- and standards-oriented framework.

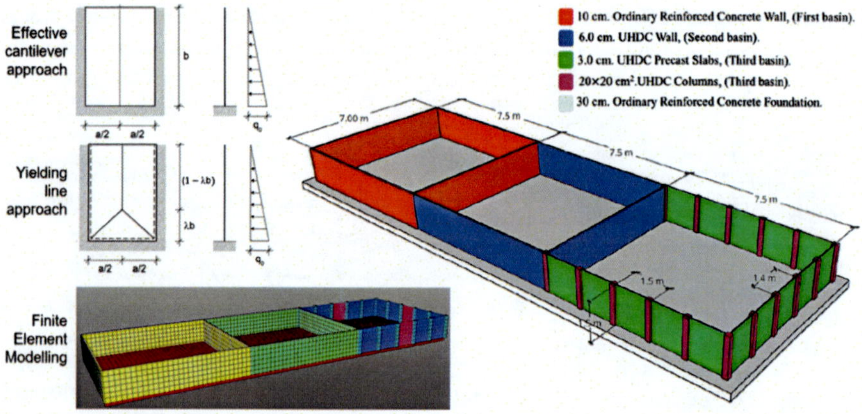


**Fig. 9.** *Politecnico di Milano-Enel Green Power* pilot: geothermal power plant tower basin (adapted from Al Obaidi et al., 2021).

## 4 Numerical Modelling and Simplified Approaches

The simplified approaches have been also implemented in the monitoring phase of the durability and mechanical response of the pilot in the following two years of survey. During this period, different water filling-emptying tests have been performed measuring the displacements (Fig. 10) and comparing them with the results of the numerical/analytical simulations. Such approaches have been developed within a Durability Assessment-based Design (DAD) and Life Cycle Analysis (LCA) framework. It is worth noting that experimental characterization at the material level, regarding both mechanical and durability performance, has been accompanied by advanced multi-physics numerical simulation (see Cibelli et al., 2019) with the aim of highlighting the main mechanism behind self-healing and recovery of material properties.

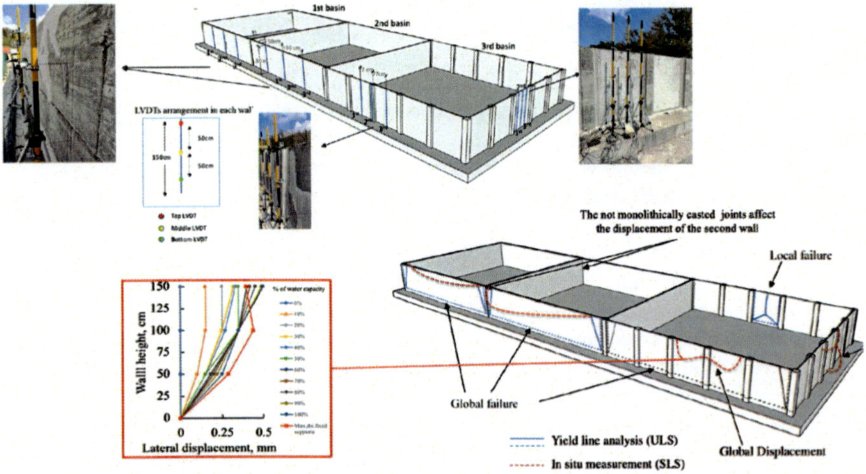


**Fig. 10.** Structural monitoring of the full-scale demonstrator (adapted from Al Obaidi et al., 2022).

## 5 Conclusions

In the paper an overall description of the concepts and of the structure behind ReSHEALience project has been provided. The core is represented by Sustainability of structures and infrastructures, intended as optimization of materials performance and extension of the service life. In this regard, UHPFRCCs with improved durability, thus called Ultra-High Durability Concretes (UHDCs), represented the key of success, thanks to the effective crack-width control and engineered self-healing. After a comprehensive material characterization of both mechanical and durability performance, structural validation has been carried out (and it is still ongoing) on full-scale demonstrators realized within the project and studied according to a Durability Assessment-based Design (DAD) and Life Cycle Analysis (LCA) framework.

**Acknowledgements.** The research activity reported in this paper has been performed in the framework of the ReSHEALience project (Rethinking coastal defence and Green-energy Service infrastructures through enHancEd-durAbiLity high-performance cement-based materials) which has received funding from the European Union's Horizon 2020 research and innovation program under grant agreement No 760824. The information and views set out in this publication do not necessarily reflect the official opinion of the European Commission.

The authors acknowledge the cooperation of MEng. Lorenzo Papa, Stefano Passoni, Angelo Alferi, Nicola Borgioni, Andrea Cervini and Luca Famiani in performing experimental tests, in partial fulfilment of the requirements for the MEng in Civil Engineering and Building Engineering, respectively.

The kind collaboration of ReSHEALience partner Penetron Italia (MArch. EnricoMaria Gastaldo Brac) in supplying the crystalline self-healing promoter is also acknowledged.

The authors also thank Mr. Marco Francini (BuzziUnicem) for supplying of cement, Mr. Michele Gadioli and Roberto Rosignoli (Azichem ltd) for supplying of steel fibres, dr Florian Bernard (Saint-Gobain Seva) for supplying of amorphous fibres and Mr. Sandro Moro (BASF Italia) for supplying the superplasticizer employed for casting the investigated UHDC mix.

## References

Al-Obaidi S., et al.: Innovative design concept of cooling water tanks/basins in geothermal power plants using ultra-high-performance fiber-reinforced concrete with enhanced durability, Sustainability (2021)

Al-Obaidi S., et al.: Structural Validation of Geothermal Water Basins constructed with Durability Enhanced Ultra High-Performance Fiber Reinforced Concrete, Case Studies in Construction Materials (2022)

Al-Obaidi, S., Bamonte, P., Luchini, M., Mazzantini, I., Ferrara, L.: Durability-based design of structures made with UHP/UHDC in extremely aggressive scenarios: application to a geothermal water basin case study. MDPI Infrastructures **5**(11), 1–44 (2020)

Banthia, N.: Fiber reinforced concrete for sustainable and intelligent infrastructure. In: Proceedings of First International Conference on Microstructure Related Durability of Cementitious Composites. Rilem, Nanjing (2008)

Bribián, I.Z., Capilla, A.V., Usón, A.A.: Life cycle assessment of building materials: comparative analysis of energy and environmental impacts and evaluation of the eco-efficiency improvement potential. Build. Environ. **46**(5), 1133–1140 (2011)

Cibelli, A., Di Luzio, G., Ferrara, L.: "Towards the numerical modelling of autogenous healing for ultra high-performance fibre-reinforced cementitious composites". In: Proceedings of the 1st fib Italy YMG Symposium on Concrete and Concrete Structures, FIBPRO 2019, pp. 1–8 (2019)

Cuenca, E., Lo Monte, F., Moro, M., Schiona, A., Ferrara, L.: Effects of autogenous and stimulated self-healing on durability and mechanical performance of uhpfrc: validation of tailored test method through multi-performance healing-induced recovery indices. Sustainability **13**(20), 11386 (2021)

Davolio, M., Al-Obaidi, S., Altomare, M.Y., Lo Monte, F., Ferrara, L.: A methodology to assess the evolution of mechanical performance of UHPC as affected by autogenous healing under sustained loadings and aggressive exposure conditions. Cement Concr. Compos. **139**, art. n. 105058 (2023)

di Prisco, M., Ferrara, L., Lamperti, M.G.L.: 'Double Edge Wedge Splitting (DEWS): an indirect tension test to identify post-cracking behaviour of fibre reinforced cementitious composites. Mats. Structs. **46**(11), 1893–1918 (2013)

Ferrara, L., Faifer, M., Muhaxheri, M., Toscani, S.: A magnetic method for non-destructive monitoring of fiber dispersion and orientation in steel fiber reinforced cementitious composites part 2: correlation to tensile fracture toughness. Mats. Structs. **45**(4), 591–598 (2012)

Ferrara, L., Ferreira, S.R., Krelani, V., Della Torre, M., Silva, F., Toledo, R.: Natural fibers as promoters of autogenous healing in HPFRCCs: Results from on-going Brazil-Italy cooperation. In: M.A. Chiorino et al., eds., ACI SP-305. Proc. DSCS, Bologna, Italy, Oct. 1–3 (2015)

Ferrara, L., Ozyurt, N., di Prisco, M.: High mechanical performance of fiber reinforced cementitious composites: the role of "casting-flow" induced fiber orientation. Mats. Structs. **44**(1), 109–128 (2011)

Lepech, M.D., Li, V.: Water permeability of engineered cementitious composites. Cem. Concr. Comp. **31**, 744–753 (2009)

Li, M., Li, V.: Cracking and healing of engineered cementitious composites under chloride environment. ACI Mats. J. **108**(3), 333–340 (2011)

Li, V.C., Wu, H.C.: Conditions for pseudo strain-hardening in fiber reinforced brittle matrix composites. Appl. Mechs. Rev. **45**(8), 390–398 (1992)

Li, V.C.: On engineered cementitious composites. A review of the material and its applications. J. Adv. Concr. Tech. **1**(3), 215–230 (2003)

Li, V.C., Stang, H., Krenchel, H.: Micromechanics of crack bridging in fibre-reinforced concrete. Mats. Structs. **26**(8), 486–494 (1993)

Lo Monte, F., Ferrara, L.: Self-healing characterization of UHPFRCC with crystalline admixture: Experimental assessment via multi-test/multi-parameter approach. Construct. Build. Mater. **283**, 122579 (2021)

Monte, F.L., Ferrara, L.: Tensile behaviour identification in Ultra-High Performance Fibre Reinforced Cementitious Composites: indirect tension tests and back analysis of flexural test results. Mater. Struct. **53**, 145 (2020). https://doi.org/10.1617/s11527-020-01576-8

Lo Monte F., Mezquida-Alcaraz E. J., Navarro-Gregori J., Serna P. and Ferrara L.: "Experimental Characterization of the Tensile Constitutive Behaviour of Ultra-High Performance Concretes: Effect of Cement and Fibre Type". In: Proceedings of RILEM-fib X International Symposium on Fibre Reinforced Concrete BEFIB2020, 20–22 September 2021, Valencia (Spain). In RILEM BOOKSERIES - ISSN:2211–0844 vol. 36 (2022)

Lo Monte F., Mezquida-Alcaraz, E. J., Navarro-Gregori, J., Serna, P., Ferrara, L.: Mechanical Characterization in Tension of UHPFRCs with Steel or Amorphous Metallic Fibres via a Combined Experimental/Numerical Identification Procedure. submitted to Structural Concrete (2023a)

Lo Monte F.: Multi-Performance Experimental Assessment of Autogenous and Crystalline Admixture-Stimulated Self-Healing in UHPFRCCs: Validation and Reliability Analysis through an Inter-Laboratory Study. submitted to Cement and Concrete Composites (2023b)

Mackechnie, J.R., Alexander, M.G.: Using durability to enhance concrete sustainability. J. Green Build. **4**(3), 52–60 (2009). https://doi.org/10.3992/jgb.4.3.52

Matthews, S.: Conrepnet: performance-based approach to the remediation of reinforced concrete structures: achieving durable repaired concrete structures. J. Bldng. Appr. **3**(1), 6–20 (1997)

Naaman, A.E., Reinhardt, H.W.: 'Proposed classification of HPFRC composites based on their tensile response. Mats. Structs. **39**(5), 547–555 (2006)

Özbay, E., Şahmaran, M., Lachemi, M., Yücel, H.: Self-healing of microcracks in high volume fly ash incorporated engineered cementitious composites. ACI Mat. J. **110**, 33–44 (2013)

Plagué, T., Desmettre, C., Charron, J.P.: Influence of fiber type and fiber orientation on cracking and permeability of reinforced concrete under tensile loading. Cem. Concr. Res. **94**, 59–70 (2017)

Serna, P.; et al.: Upgrading the Concept of UHPFRC for High Durability in the Cracked State: the Concept of Ultra High Durability Concrete (UHDC) in the Approach of the H2020 Project ReSHEALience. In: Proceedings of the International Conference on Sustainable Materials Systems and Structures SMSS 2019, Rovinj (Croatia), March, 20–22, 2019, pp. 764–771 (2019)

Zhang, Z., Qian, S., Ma, S.: Investigating mechanical properties and self-healing behavior of micro-cracked ECC with different volume of fly ash. Constr. Build. Mats **52**, 17–23 (2014)

# A New Concept of Additive Manufacturing for the Regeneration of Existing Tunnels

Stefano Guanziroli[1], Andrea Marcucci[2](✉), Alberto Negrini[1], Liberato Ferrara[2], and Bernardino Chiaia[3]

[1] Hinfra, Casale Monferrato, Italy
[2] Department of Civil and Environmental Engineering, Politecnico di Milano, Milan, Italy
andrea.marcucci@polimi.it
[3] Department of Structural Geotechnical and Building Engineering, Politecnico di Torino, Turin, Italy

**Abstract.** Among the new technologies driving the fourth industrial revolution in the construction industry, 3D Concrete Printing (3DCP) is playing a key role. The typical process is made through robotic arms or gantries equipped with nozzles, similarly to contour crafting in other industries. Despite 3DCP is appealing when applied to complex architectural shapes, the structural behaviour and geometrical size are limitations difficult to overcome. Upscaling the extrusion process to full sections, introducing a new concept of ultrafast adaptable slip-forming, is the access key to different sectors of the construction industry, as infrastructures. This paper will present the Extruded Tunnel Lining Regeneration (ETLR) technology developed by HINFRA with the scope to automatically regenerate the lining of existing damaged tunnels. "Tailored" features and issues of the aforesaid technology will be discussed in this paper, together with a design validation related to a Fibre Reinforced Concrete (FRC) tunnel lining.

**Keywords:** Fibre-reinforced concrete · Structural robustness · Material heterogeneity

## 1 Introduction

### 1.1 The Italian Problem of Existing Tunnels

Italy, because of its geomorphological features, is one of the countries with the largest highway and roadway network worldwide. In detail, Italy, with 1500 km of tunnels, has the first largest tunnel network in Europe and the second in the world. This provides lustre to the country, but requires nowadays a dedicated maintenance plan, where most of these infrastructures, which have been built between the late 50s and 60s, are in deep need of retrofitting and refurbishment. As a matter of fact, 1000 km of existing tunnels were built more than 20 years ago and must be subjected to heavy refurbishment in the next years; among them 500 km are currently under strict observation due to risk of collapse.


M. A. Aiello and A. Bilotta (Eds.): ICC 2022, LNCE 435, pp. 116–124, 2024.
https://doi.org/10.1007/978-3-031-43102-9_10

It is estimated that the impact of a eventual lack of critical infrastructures in heavy populated and industrialized areas in Italy generates a loss in the GDP of the same area of 36,000 Eur/km/day.

The degradation of the material over time, together with the accidental presence of defects, can drastically reduce the load carrying capacity of the tunnel linings, leading to a possible collapse of the structure (Zude et al. 2019). As a matter of fact, defects originating from an imperfect casting can promote the formation of cracks that can speed up degradation phenomena thanks to an easier ingress of aggressive substances, thus shortening the service life, as anticipated in the design stage.

In view of this, given the importance of the safety of a tunnel in the framework of transportation infrastructure networks, and the impact related to its collapse or out-servicing, it is necessary to reduce the probability of failure, intended as the attainment of a limit state for which the tunnel cannot be any longer used. The most recent decisions undertaken in this respect moved along this path, through the introduction of national standards that aim to regulate the construction of new tunnels, describing the minimum requirements that must be satisfied.

### 1.2 Advanced Technology for the Refurbishment

In these last years, the concrete construction industry has been walking along a steep innovation uptake curve also through the development, validation and implementation of new materials and processes, mutually enabling each other. Digital Fabrication with Concrete (DFC) is becoming appealing as it allows to combine the digital design to physical products, thanks to automated processes, as Additive Manufacturing (AM). DFC can provide multiple advantages for the construction industry, in different ways. As a matter of fact, thanks to the digitalization it is possible to design more complex structures that are topologically optimized (Menna et al. 2020), thus adopting lower amount of material, resulting in both costs and emissions savings. Another advantage is the implementation of automated processes that on the one hand increase the safety thanks to the reduction of the number of workers present during the construction process, and, on the other, promote an increment of the productivity tied to a faster construction, also characterized by a higher quality of the final product (García de Soto et al. 2018; De Schutter et al. 2018), including a better mechanical and, likely, durability performance.

3D Concrete Printing (3DCP) technology is emerging in this context as it has been employed in several start-up companies and academic research centres. It is often claimed to have great potential as almost everything can be built using it (Lloret-Fritschi et al. 2020). As a matter of fact, the material requirements become more demanding, starting from the rheological point of view (Buchli et al. 2018), as concrete must be pumped through a pipe, then extruded by a nozzle and finally it has also to harden until it reaches the target mechanical properties to maintain at least itself standing, while progressively being able to withstand the stresses arising in the structure from the ongoing construction. This makes 3DCP technologies a cutting-edge field of development both in the robotics and materials/structural engineering fields.

Tunnel engineering, and the management of existing tunnels in transportation infrastructures can represent a challenging field-test for digital fabrication with concrete. As remarked above, the problem of obsolete tunnels represents a huge societal and technical-economic burden, which requires large scale maintenance and regeneration programmes to be implemented nationwide in several countries. The maintenance techniques that are available nowadays have been shown able to extend the service life of the tunnels merely of a few years, providing localized and temporarily limited solutions. In this framework the slipforming 3DCP method, that consists of dynamic casting of concrete based on the continuously slipping formworks and the "feeding" of a concrete with tailored fast-setting and hardening properties, can represent an interesting solution for an overall regeneration of existing tunnel linings.

HINFRA has developed a signature holistic approach to the problem, encompassing the material concept development together with the design of the machinery and the construction process, to deliver this kind of technology for the refurbishment of existing tunnels. The final aim is to provide a solution for maintenance interventions able to deliver a final product (the refurbished tunnel lining) that has a service life comparable (if not higher) than the one of a newly built structure, designed according to the most recent standards.

This paper will summarize the main details of this new technologies together with an example of application to a real-case design. This is a key aspect for the validation and acceptance by the overall construction industry (Menna et al. 2020; Bos et al. 2022) in order to promote product/process standardization and codified design approaches also able to quantify the benefits digitally fabricated/3D printed concrete structures in terms of structural, durability, economic and environmental performance.

## 2 The "HINFRA" System

HINFRA core content consists in the development of a technological platform to handle concrete as a dynamic and constantly evolving medium in fast extrusion processes, that allow to boost the productivity in the construction of critical infrastructures as tunnels, bridges, dams, offshore wind power structures. Further to R&D and industrialization of advanced concretes, fit for the intended technology, the HINFRA business model is based on the design and building of the machines, the automation, and the application process, implementing the "Holistic approach to INFRAstructures Additive manufacturing".

The HINFRA team is currently working on its first development project, named ETLR (Extruded Tunnel Lining Regeneration), with the final scope to automatically regenerate the lining of existing damaged tunnels. The project is based on the reconstruction of the tunnel linings directly at site, through an ultrafast horizontal slip forming. The ETRL processing train is a machinery consisting of several modular units, each addressing a specific issue and performing the related function. The incremental industrialization of the intended set of operations, typically the demolition of the damaged tunnel lining, the subgrade surface preparation, and the new lining phases, combined with the performances of the special concrete, allow to target productivity rates far from the traditional methods in use in the tunnel refurbishment sector.

A solid step into the technology implementation has consisted in the construction of the real scale extrusion facility, where the main components of the ETLR train have been designed, built/assembled, and tested on a typical section of single lane roadway tunnel.

The main module tested is the tunnel lining extruder, a special horizontal formwork, designed for fast slip forming process. In this framework, the behaviour of a fibre reinforced extrudable concrete has also been evaluated at industrial scale through production cycles that will be then implemented in the real case scenario of the first tunnel pilot project.

The permanent extrusion facility is resting over an outdoor concrete platform and consists of three main items:

1. the fixed external formwork, directly connected to the mat-foundation, made by bolted steel panel that can be easily removed and relocated for longitudinal advancement. This macro element has the scope to simulate the confinement of an "equivalent" existing tunnel, also providing a constraint to stabilize the pressure during concrete injection;
2. the contrast and advancement system, made of two custom designed rails, provided with web holes, allowing to couple with two contrast structures connected with the extruder through hydraulic cylinders;
3. the tunnel lining extruder prototype, consisting of two main items: the "universal" frame, supported by N°4 steel wheels, and the adaptable formwork, transferring the actions arising from the different phases of the concrete slip forming process (injection, setting, advancement).

The same steel frame has been also used for demolition tests, by replacing the formwork panels with the mechanical asportation module. Thanks to the above-described facility and equipment, it is possible to produce equivalent tunnel sections were produced with the following geometrical characteristics: internal net width of 4,50 m; internal net height of 3,82 m, internal radius of 2,25 m; vertical abutments 1,57 m high; equivalent lining thickness of 0,40 m. A frontal view of this system is provided in Fig. 1a.

## 3 The Extrudable FRC: Performance Based Concept and Design Parameters

Lining segments are made with extrudable fiber reinforced concrete without traditional steel reinforcements. This is a key aspect of the technology since the absence of steel reinforcement bars has two major impacts: primarily environmental, since the weight saving of more than 50% due to the use of fibers as compared to steel bars lead to significant reduction in $CO_2$ emission, both because the production method of the small diameters fibers from primary coils is widely less consuming than one of bars, and due to fewer trucks on the road, also thanks to the optimization of the transportable quantities for each truck. Secondly, on productivity, since the application of traditional reinforcements requires a set of ancillary operations that strongly limit the potential advancement rate theoretically allowed by the fiber reinforced extrudable concrete.

The engineered fiber-reinforced concrete at issue is characterized by high early-age strength parameters, that allow a rapid slipform of the system at a certain predetermined time. To this purpose it is worth study the evolution of the fresh state properties of concrete to predict its open time with the aim of starting at the right time the slipping process. A methodology to assess the fresh state of cementitious mortars was introduced by Marcucci et al. (2022). Its properties, in particular initial setting and hardening are well managed through the correct mix design and can be customized over time according to the environmental conditions of the specific construction site. At the time of slipforming, which can take place from the first moments after casting, the material therefore has not only the ability to self-sustain but also the ability to withstand external loads, ability that increases speedly in the first hours of curing.

The production process of the lining segments is semi-continuous, in the sense that defined volume batches of concrete are produced in a truck-mixer and injected in the formwork through pumping operations. The advancement of the system is then generated by finite steps, coordinated with the setting and workability windows of the injected batch. The system can also be pushed to a continuous advancement, but it will be part of a further implementation stage.

The typical length of the lining segment is 1m along longitudinal direction. These dimensions can be adjusted, as parameters of the extrusion process, within the range 0,50 m – 1,50 m. The semi-continuous production advancement results in joints between subsequent segments; however, the behavior of the material at this interface should not be treated as “cold-joint”, because the timeframe between two injections is very limited. As a matter of fact, in the initial stages just after the workability window, the hydration process of the binder is very fast, especially in the first hours from the beginning of the setting phase, therefore this allows two successive castings to create a monolithic “hot-joint”. Moreover, during the slipforming phase, the adhesion that is created between concrete and formwork causes an important stress on the joint between the two successive castings, however, tests carried out on small-scale prototypes have already shown that the hybrid joint is able to produce a bonding resistance compatible with the strength of the base concrete material. A representation of an intersection between two segments is given in Fig. 1b. This effect is of primary importance, since could result in substantial technological breakdown in the management of joints, considering also that waterproofing is a crucial performance requirement.

## 4 Example of Application

As a case study, an extruded tunnel lining at the HINFRA test field is considered, the lining segment being 1 m long with the internal radius of 2,25 m and having a thickness of 40 cm (Fig. 1c).

Once the material properties and the structure geometry are known, it is possible to build the bending moment vs axial force (M-N) interaction diagram with the aim of defining the maximum bearing capacity of the sections and consequently the ultimate failure domain of the lining. Its computation starts form the plane cross section assumption and the knowledge of the strength parameters of the material, including the compressive strength and the flexural toughness parameters $f_{R1}$ and $f_{R3}$, as per Model Code 2010 prescription. Strain and stress cross sectional distribution for specific point of the M-N diagram are reported in Fig. 2.

These have been respectively obtained by testing cube specimens, for the compressive strength, and standard EN 14651 beam specimens in three-point bending, for the residual flexural strength parameters.

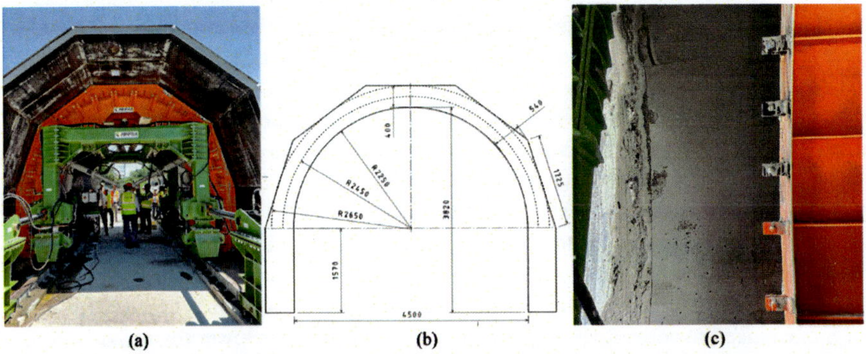

**Fig. 1.** (a): Front view of a real scale prototype extrusion tunnel lining through horizontal slipforming; (b) Intersection between two different tunnel segments extruded consecutively; (c) scheme of the cross-section of a real scale tunnel segment extruded by HINFRA.

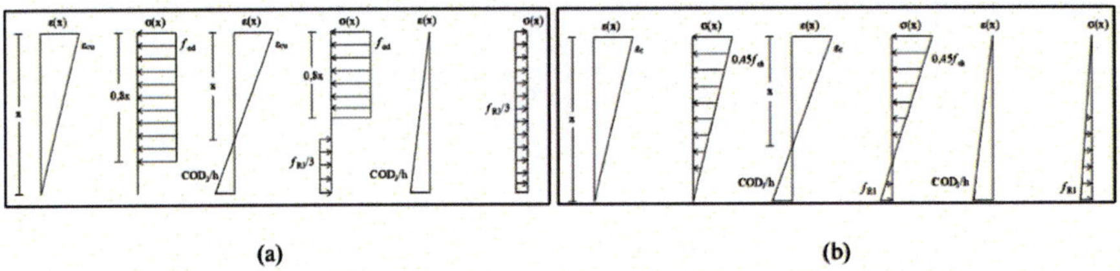

**Fig. 2.** Strain and stress cross sectional distribution at the ULS (a) and at the SLS (b) for the calculation of the M-N domain.

From the latter, a class "5b" for the Fibre Reinforced Concrete (FRC) has been achieved (Di Prisco et al. 2009), while a characteristic value of the cubic compressive strength of 60 MPa has been obtained for the former.

For what concerns the external forces, different loading and constraint conditions have been assumed, in order to obtain by means of a linear elastic finite element analysis the stresses acting in a tunnel segment 1m long. In particular, the following boundary conditions were considered: rigid constraints to maximize the rigidity of the structural system, and as load conditions: a symmetrical load in the centre of the shell (Fig. 3a) and a non-symmetrical lateral load (Figs. 3b, c).

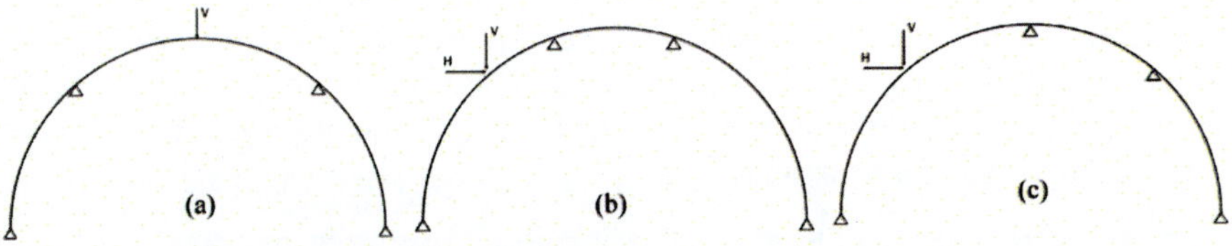

**Fig. 3.** Different loading and restraint conditions schemes applied to the tunnel lining.

Under the aforesaid assumptions, the M-N diagram has been computed for the Ultimate Limit State (ULS) and the Serviceability Limit State (SLS), in both cases adopting the design values of the strengths, following an approach similar to the one by Chiaia et al. (2007). The diagrams are reported in Fig. 4. It can be observed that the performance demand is contained into the domain for each of the considered lining segment points, thus providing positive check for the coupled bending moment - axial force values resulting from an external concentrated load of 1000 kN.

It is worth also noting that the SLS domain intersects the ULS one when approaching the tensile region of the axial force (that has a negative value for the reference adopted). As a matter of fact, the residual tensile strength when in the SLS is higher than to the ULS one since the concrete belongs to class "5b", thus $f_{R3}$ is lower than $f_{R1}$.

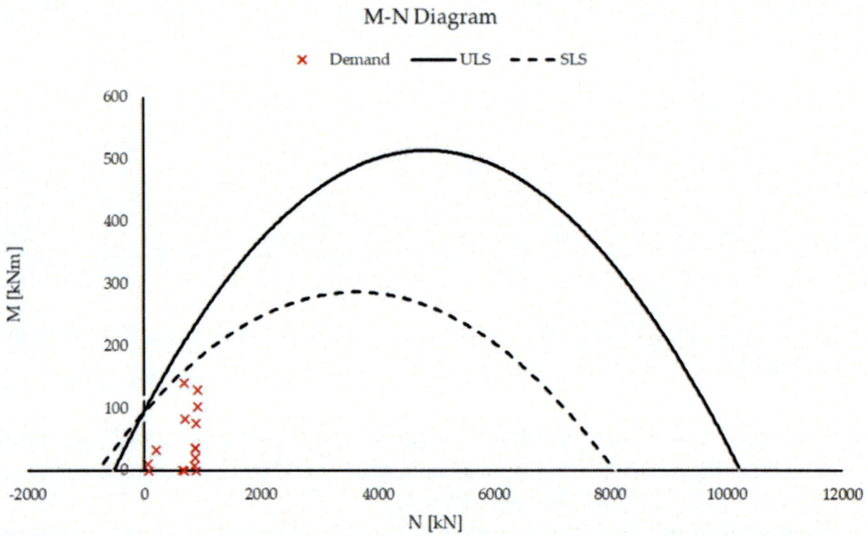


**Fig. 4.** M-N domain of the lining both at the ULS and SLS. The demand is always lower than the resistance.

## 5 Conclusions

In this paper an innovative horizontal slip-forming system for the retrofitting of existing tunnels, based on the use of extrudable fibre reinforced concrete and developed by the start-up company HINFRA has been presented. Along with this project, there are two main area of interest for its development:

1. on the material side, since the technology imposes stringent requirements for the fresh state and very early age properties of concrete. As a matter of fact, the material is first pumped through a hose and then is casted into the space between the existing subdgade tunnel and the slip-forming moulds. This means that it is necessary to start

with a self-compacting concrete consistency for the pumping and rapidly evolving towards a fast setting to allow a rapid advance of the slip-forming mould, as typical of the additive manufacturing/3dprinting concrete technology;
2. on the machinery side, because a customized machine is needed for the realization of the slipforming process.

Considering the dramatic condition of the Italian tunnel roadway and highway network, the proposed technology has already been proved successful on a medium/large scale, highlighting its great potential for the refurbishment capacity of existing damaged tunnels, disrupting the current productivity rates of the industry and guaranteeing a nominal life of the product longer than 35 years.

A preliminary assessment of the structural load-bearing capacity of the as built tunnel lining, as per the M-N resistance domain, has demonstrated that a 400 mm section made of the extrudable fibre-reinforced concrete at issue, without traditional steel reinforcement, fulfil the performance to which a tunnel lining is commonly subjected. The absence of traditional integrated reinforcements has in this context a beneficial effect, leading to a further increase in productivity, since all manual installation operations are avoided.

Finally, given the compressive and tensile properties of the intended C60 class 5b fibre reinforced concrete, which is integral part of the proposed technology, and the resulting M-N domain, it is clear that an as thin as 400 mm tunnel lining section is able to guarantee performances that go well beyond those of the tunnel linings existing nowadays on the Italian roadway network, which are in strong need of refurbishment.

**Acknowledgements.** The second author acknowledges the support of the Italian National Programme PON Ricerca e Innovazione in funding his PhD programme.

## References

Bos, F.: The realities of additively manufactured concrete structures in practice. Cement and Concrete Research, **156**, Article 106746 (2022)

Buchli, J., et al.: Digital in situ fabrication – challenges and opportunities for robotic in situ fabrication in architecture, construction, and beyond. Cem. Concr. Res. **112**, 66–75 (2018)

Chiaia, B., Fantilli, A.P., Vallini, P.: Evaluation of minimum reinforcement ratio in FRC members and application to tunnel linings. Mater. Struct. **50**, 593–604 (2007)

De Schutter, G., Lesage, K., Mechtcherine, V., Nerella, V.N., Habert, G., Augusti-Juan, I.: Vision of 3D printing with concrete – technical, economic and environmental potentials. Cem. Concr. Res. **112**, 25–36 (2018)

De Soto, B.G., et al.: Productivity of digital fabrication in construction: cost and time analysis of a robotically built wall. Autom. Constr. **92**, 297–311 (2018)

Di Prisco, M., Plizzari, G., Vandewalle, L.: Fibre reinforced concrete: new design perspectives. Mater. Struct. **42**, 1261–1281 (2009)

Lloret-Fritschi, E., et al.: From smart dynamic casting to a growing family of digital casting systems. Cem. Concr. Res. **134**, 106071 (2020)

Marcucci, A., Kompella, K.S., Lo Monte, F., Levi, M., Ferrara, L.: Early age shear and tensile fracture properties of 3d printable cementitious mortar to assess printability window. In: Third International RILEM Conference on Concrete and Digital Fabrication (DC2022), vol. 37, pp 337–342 (2022)

Menna, C., et al.: Opportunities and challenges for structural engineering of digitally fabricated concrete. Cement Concrete Res. **133**, Article 106079 (2020)

Zude, D., Xiafei, J., Xiaoqin, L., Zhihua, R., Sen, Z.: Influence of symmetric and asymmetric voids on mechanical behaviours of tunnel linings: model tests. Symmetry (Basel) **11**, 802 (2019)

# Influence of Infill-to-Frame Connection on the Seismic Response of RC Frames

Ciro Del Vecchio[1](✉), Marco Di Ludovico[2], Gerardo Mario Verderame[2], and Andrea Prota[2]

[1] Department of Engineering, University of Sannio, Benevento, Italy
cdelvecchio@unisannio.it

[2] Department of Structures for Engineering and Architecture, University of Napoli Federico II, Napoli, Italy
{diludovi,verderam,aprota}@unina.it

**Abstract.** Existing reinforced concrete buildings subjected to medium intensity earthquake loads often exhibit significant damage to structural and non-structural components that may compromise the structural safety and the building repairability. This damage is commonly concentrated on infills and partitions leading to relevant economic losses. This makes the research on the lateral response of infills and on the role of infill-to-frame connection of paramount importance to reliably assess the seismic performance of existing buildings. This paper discusses the results of an experimental program on full-scale multi-storey infilled RC frames tested under pseudo-dynamic loads. Emphasis is given to the variability of the results with the different infill-to-frame connection realized with classic mortar applied on four sides or with a gap below the beam. The lateral response and the observed damage at increasing earthquake intensities are discussed and compared.

**Keywords:** reinforced concrete · infill · experimental · pseudodynamic · full-scale/calcestruzzo armato · tamponature · prove sperimentali · pseudodinamiche

## 1 Introduction

Recent devastating earthquakes point out the high vulnerability of existing reinforced concrete (RC) buildings and the influence of infills on the global damage and related economic losses (Ricci et al. 2011; Cardone and Perrone 2017; Del Vecchio et al. 2020; Lan et al. 2020). Many literature studies focused on the seismic response of infilled RC frames remarking their influence on the global building response (Ricci et al. 2011), the interaction with the surrounding frame (Verderame et al. 2019), the influence of infill distribution and openings (Negro and Verzeletti 1996; Morandi et al. 2018), the damage quantification at increasing drift demand (Colangelo 2005; De Risi et al. 2018). Many other studies proposed macro-model to include the infill contribution in the non-linear analysis of RC buildings (Panagiotakos and Fardis 1996; Crisafulli 1997; Chrysostomou et al. 2002; Cavaleri and Di Trapani 2014; De Risi et al. 2018). However few studies are


M. A. Aiello and A. Bilotta (Eds.): ICC 2022, LNCE 435, pp. 125–133, 2024.
https://doi.org/10.1007/978-3-031-43102-9_11

available in literature on the influence of the degree of connection between the infills and the surrounding frame. This is an important aspect that may significantly affect the response of the entire building. Indeed, the distribution of shear forces transmitted to the surrounding RC members, significantly depends on the infill-to-frame connection (Bolis et al. 2020). Field reconnaissance in the aftermath of recent earthquakes showed that this interaction can be triggering for the shear failure at the top of the column with variable crack inclination and different infill-to-frame connection (see Fig. 1).

**Fig. 1.** Observed damage on the case study building with emphasis on infill-to-frame connection

In order to investigate the role of infill-to-frame connection in existing RC frame an experimental program of full-scale specimen was conducted at the DiST-CeSMA laboratory under pseudo-dynamic (PSD) loads. Different concrete strength and infill-to-frame connections were investigated.

This paper discusses the results of two specimens having the same concrete and a different connection between the top of the infills and the bottom of the beam. A non-linear FEM model is developed to capture the response of the two frames and the results are compared with experimental outcomes in terms of global and local (infill) response.

## 2 Frame Prototype

The prototype frames tested in this work were realized to be faithfully compliant with the perimetral frame of an existing building damaged by the 2009 L'Aquila earthquake. It is a two storeys RC frame 6.3 m height and 4.1 m long with reinforcement details typical existing RC buildings in the Mediterranean area (i.e. lack of stirrups in the joint panel). It consists of square columns 0,4 m side and rectangular beams 0,4 m width and 0,55 m height. The column is reinforced with 8$\phi$ 16 mm deformed bars uniformly distributed within the cross-section. The longitudinal reinforcement of the portion ends of the beams consists of 6$\phi$ 16 mm at the top and 4$\phi$ 16 mm at the bottom. The transverse reinforcement of beam and columns were $\phi$8 bars 250 mm spaced. The mean cylindrical concrete compressive strength is about 19.0 MPa in agreement with that found on the case study buildings by using in-situ destructive tests.

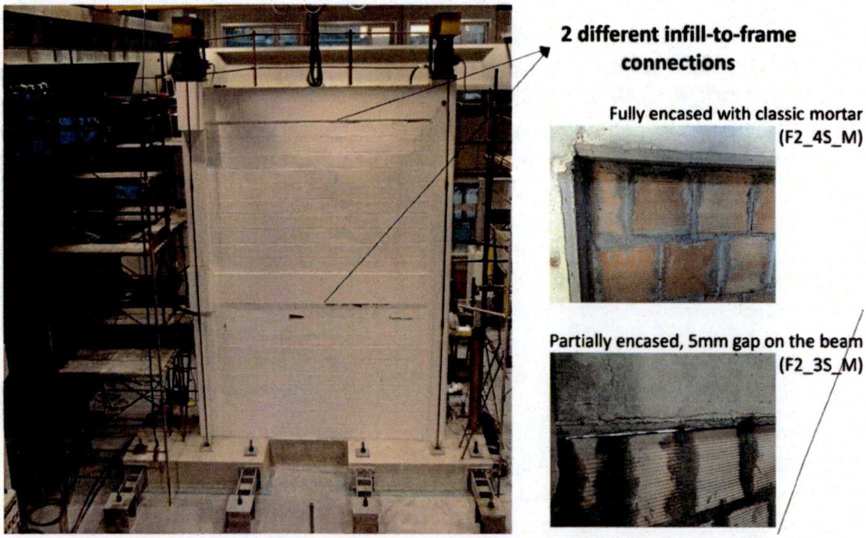


**Fig. 2.** Two-storey frame prototype with different infill-to-frame connection

The infills were fabricated using a M10 class mortar and hollow clay bricks with thickness of 200 mm. The mean shear strength determined by tests on masonry wallet is about 0.35 MPa.

More details on the reinforcement details and material properties can be found in Del Vecchio et al. (2022).

An overview of the tested frames is reported in Fig. 2. In this paper the results of two configurations are analysed. The specimen F2_4S_M was fabricated with the infills fully encased withing the RC frame by using a layer (1 cm thick) of classic mortar. The specimen F2_3S_M had the infill partially encased with 5 mm gap between the top of the infill and the bottom of the beam. The latter configuration can be representative of an infill with a degraded or weak top mortar layer.

These tests aim at quantifying the contribution of the infill transferred to the beam and the influence on the lateral response of the frame.

It is worth mentioning that in the real practice number of different infill to frame connections can be found and the results of this work are not intent to cover all the possible combination of materials and fabrication processes.

### 2.1 Test Setup and Loading Protocol

The specimen was fixed to the strong floor by using prestressed steel bars passing through the foundation. Additional connection systems made of steel profiles casted into the foundation block and connected to the strong floor by means of prestressed steel bars were used. The frame was loaded by means of two actuators connected at the beam level and contrasting on the reaction wall. Furthermore, a constant axial load (N) of 300 kN was applied on each column by means of two hydraulic jacks (see Fig. 2). The frames were subjected to PSD tests at increasing intensity. The AQG_NE record of the 2009 L'Aquila earthquake was selected and scaled from 10% to 150%. The adopted PSD testing framework uses the alpha-OS integration algorithm (Combescure and Pegon 1997) to solve the equation of motion. Mass and damping matrix of the selected frame are defined according to substructuring considerations. More details on the implementation of the PSD framework and the substructuring approach can be found in (Molitierno et al. 2021). The displacement control was set at the midspan of the beam and it was monitored by using high precision transducers. LVDTs, potentiometers and strain gauges were installed to monitor global and local deformations or strain on internal reinforcements.

## 3 Experimental Results

The experimental results of all the PSD tests at increasing intensity in terms of top displacement ($d_{top}$) demand vs base shear ($V_{base}$) are reported in Fig. 3a, while the envelope of the cyclic response is reported in Fig. 3b. The comparison outlines the influence of the infill-to-frame connection on the lateral response of infilled RC frames. The initial stiffness is very similar for both the specimens and no marked differences were observed until the infill-to-frame separation at about 2 mm of imposed top displacement. After this stage, during the run to 50% of the earthquake intensity, a significant stiffness degradation was observed in both the specimens. However, the response of the partially encased prototype (F2_3S_M) resulted less stiff than the one with infills fully encased (F2_4S_M). This led to a displacement demand in the F2_3S_M double than F2_4S_M during the runs to 75%, 100% and 125% of the earthquake intensity. This is related also to the significant differences in terms of strength (i.e. base shear) monitored during the tests. Indeed, following to the infill-to-frame separation the strength of the F2_4S_M specimen is significantly larger than F2_3S_M specimen.

This difference increases by increasing the displacement demand until the achievement of the peak strength (409 kN and 290 kN) around 18 mm and 20 mm in the positive load direction for the specimens F2_4S_M and F2_3S_M, respectively. At this stage a significant cracking of the infills at both the floors was found. In particular, significant

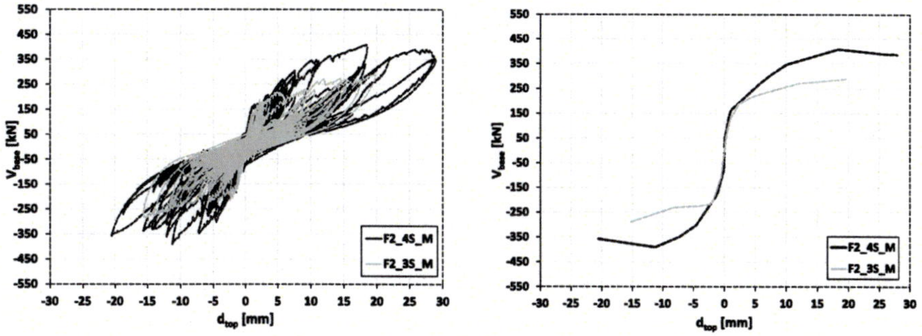


(a) (b)

**Fig. 3.** Comparison of the experimental results in terms base shear vs top displacement: hysteretic response (a); backbone curves (b)

diagonal cracking with multiple crack (i.e. development of multiple struts) were found on the specimen F2_4S_M along with a shear cracking at top of the ground floor column. The latter is because of the infill-to-structure interaction and the high shear force transmitted by the infill struct acting on the column. The test ended at 150% of the earthquake intensity when a significant degradation of the infills with crushing of some bricks was observed. At this stage the displacement demand at the top floor was about 29 mm and a base shear of about 389 kN was recorded. On the other hand the F2_3S_M specimen exhibited a different response. Because of the gap at the bottom of the beam, during the run to 100% and 125% of reference earthquake a slippage of the bricks along the column surface was observed. This is because of the gap at the top of the infill that did not retrain the vertical movement. This resulted in a significant reduction of the infill strut contribution and, in turn, of the action transmitted to the top of the column. Indeed, in this case the shear cracking of the column was not observed. The test ended at about 20 mm of displacement demand at the top floor where the crushing of corner bricks was observed. At this stage, the maximum infill contribution to the lateral strength was achieved corresponding to a base shear of about 290 kN.

## 4 Numerical Modelling

The available test results are of significant importance to calibrate refined numerical model capable of reproducing the seismic response of infilled RC frames. Indeed, the clear differences of the response of the two frames varying the degree of infill-to-structure connection allows to quantify the sensitivity of the model to this variable.

In order to reproduce the response of the tested frames a non-linear finite element model (FEM) was developed in the SAP 2000 (C.S.I. Computers and Structures Inc. 2004) environment. An overview of the proposed model is reported in Fig. 4.

The nonlinearities of beam and columns were considered lumping plastic hinges at the member's ends. Their mechanical behaviour was characterized according to the suggestion of the Eurocode 8 (CEN 2005) considering cracking, yielding and a perfectly plastic response up to the ultimate rotation. The response of beam-column joints was

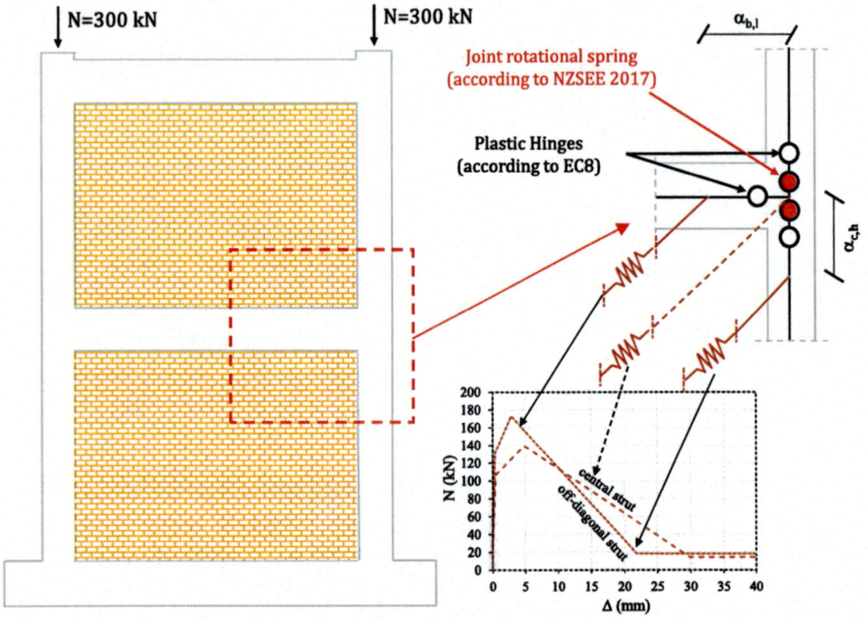


**Fig. 4.** Adopted numerical modelling

characterized according to Priestley et al. (1997) considering the cracking at a principal tensile stress of $0.3\sqrt{f_c}$ and peak strength at $0.42\sqrt{f_c}$, where $f_c$ is the mean cylindrical concrete compressive strength. The corresponding joint rotation were assumed as 0.0005 rad and 0.005 rad, respectively according to NZSEE (MBIE 2017). The joint ultimate rotation was set at 0.01 rad. The joint nonlinearties were included in two rotational springs located immediately above and belove the centre of the joint panel. The nonlinear response of the infills was characterized according to the model of Panagiotakos and Fardis (1996). The three-strut modelling approach suggested by Chrysostomou et al. (2002) was used to distribute the lateral capacity of the infills. This model may provide accurate estimations of the actual shear forces transmitted at the top of the column as discussed in Verderame et al. (2019).

The displacement histories recorded at the two floors during the experimental test were applied on the numerical model along with a constant axial load of 300 kN at the top of each column. The results and the comparison with the outcomes of the experimental tests are reported in Fig. 5 in terms of global (pushover) and local (infill) response.

The comparison shows the reliability of proposed model in reproducing the lateral response of infilled RC frames. In particular, the comparison in terms of pushover curves reported in Fig. 5a shows that the proposed model well approximates the experimental response both in terms of strength and stiffness for both the tested specimens with a full (F2_4S_M) or partial (F2_3S_M) connection. It is worth mentioning that the contribution of the top off-diagonal strut was omitted in the numerical model of the F2_3S_M frame to account for the 5 mm gap between the beam and the top of the infills.

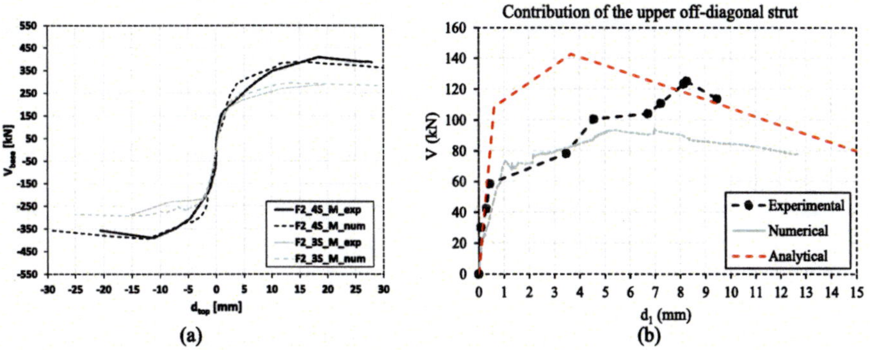


**Fig. 5.** Comparison of experimental and numerical results in terms of: pushover curves (a); contribution of the upper diagonal strut (b)

To further investigate the contribution of the infill-to-beam connection and how this phenomenon can be reproduced by an off-diagonal struct insisting on the beam, the analytical, numerical, and experimental results in terms of the contribution of the off-diagonal strut are compared in Fig. 5b in terms of shear (V) vs. floor displacement ($d_1$). The latter is the horizontal component of the curves reported in Fig. 4. In particular, the analytical contribution represents the portion of the total infill lateral response carried by the top diagonal strut. It is implemented in the numerical model. The experimental and numerical results are obtained by subtracting the shear strength of the specimen F2_3S_M to that of the specimen F2_4S_M at fixed displacement demands. The comparison shows a good agreement between the experimental and numerical results up to a displacement demand of about 7 mm. Beyond this point the experimental strength is significantly higher than the one predicted by the numerical model. This could be related to differences in the distribution of internal forces within the three struts that depends by number of variables that cannot be reproduced by a numerical model where the infill contribution is represented by microelements (i.e. the struts). It is worth noting that the maximum experimental strength achieved by the top off-diagonal strut well approximates (underestimation of about the 13%) the maximum strength of the analytical model. However, the maximum experimental strength is achieved at higher displacement demand and this can be related to the different distribution of internal actions during the test.

## 5 Conclusions

This paper reports the preliminary results of an experimental program involving multi-storey full-scale infilled RC frames tested with different infill-to-structure connection. The experimental tests were conducted by using a pseudodynamic testing framework under increasing earthquake intensity. The experimental results are then compared with those obtained by using a nonlinear numerical model. The main results can be summarized as follows:

- The experimental tests showed that the infill-to-frame connection has a significant influence on the lateral response of infilled RC frames;
- A full connection with classic mortar of all the four sides of the infill significantly increase the shear forces carried by the infill leading to a significant reduction of the lateral deformability respect to the specimen with a gap between the beam and the top of the infill;
- However, the increasing shear forces carried by the infill are transferred to the RC frame leading to the shear cracking at the top of the column;
- The proposed numerical model consisting of a three-strut model to reproduce the infill contribution well captures the lateral response of the tested frames, including the contribution of the off-diagonal infill strut insisting on the beam.

Further studies are needed to better quantify the influence of the infill-to-structure connection considering other type of materials and fabrication techniques.

**Acknowledgements.** This study was performed within the framework of the PE2022–2023 joint program DPC-ReLUIS, WP5: "Fast and Integrated Retrofit Interventions".

## References

Bolis, V., Paderno, A., Preti, M.: Experimental assessment of an innovative isolation technique for the seismic downgrade of existing masonry infills. In: Kubica, J., Kwiecień, A., Bednarz, Ł. (eds.) Brick and Block Masonry - From Historical to Sustainable Masonry, pp. 935–942. CRC Press (2020). https://doi.org/10.1201/9781003098508-133

C.S.I. Computers and Structures Inc.: SAP 2000, Static and Dynamic Finite Element Analysis of Structures (2004)

Cardone, D., Perrone, G.: Damage and loss assessment of Pre-70 RC frame buildings with FEMA P-58. J. Earthq. Eng. **21**, 23–61 (2017). https://doi.org/10.1080/13632469.2016.1149893

Cavaleri, L., Di Trapani, F.: Cyclic response of masonry infilled RC frames: experimental results and simplified modeling. Soil Dyn Earthq Eng **65**, 224–242 (2014). https://doi.org/10.1016/j.soildyn.2014.06.016

CEN: Design of structures for earthquake resistance - Part 3: Assessment and reofitting of buildings. EN-1998–3, Eurocode 8. European Committee for Standardization, Brussell (2005)

Chrysostomou, C.Z., Gergely, P., Abel, J.F.: A six-strut model for nonlinear dynamic analysis of steel infilled frames. Int. J. Struct. Stab. Dyn. **2**, 335–353 (2002). https://doi.org/10.1142/S0219455402000567

Colangelo, F.: Pseudo-dynamic seismic response of reinforced concrete frames infilled with non-structural brick masonry. Earthq. Eng. Struct. Dyn. **34**, 1219–1241 (2005)

Combescure, D., Pegon, P.: A-operator splitting time integration technique for pseudodynamic testing error propagation analysis. Soil Dyn. Earthq. Eng. **16**, 427–443 (1997). https://doi.org/10.1016/S0267-7261(97)00017-1

Crisafulli, F.J.: Seismic behaviour of reinforced concrete structures with Masonry Infills. University of Canterbury, New Zealand (1997)

De Risi, M.T., Del Gaudio, C., Ricci, P., Verderame, G.M.: In-plane behaviour and damage assessment of masonry infills with hollow clay bricks in RC frames. Eng. Struct. **168**, 257–275 (2018). https://doi.org/10.1016/j.engstruct.2018.04.065

Del Vecchio, C., Di Ludovico, M., Prota, A.: Repair costs of reinforced concrete building components: from actual data analysis to calibrated consequence functions. Earthq. Spectra **36**, 1–25 (2020). https://doi.org/10.1177/8755293019878194

Del Vecchio, C., Di Ludovico, M., Verderame, G.M., Prota, A.: Pseudo-dynamic tests on full-scale two storeys RC frames with different infill-to-structure connections. Eng. Struct. **266**, 114608 (2022). https://doi.org/10.1016/j.engstruct.2022.114608

Lan, Y.J., Stavridis, A., Kim, I., et al.: ATC Mw7.1 Puebla-Morelos earthquake reconnaissance observations: structural observations and instrumentation. Earthq. Spectra **36**, 31–48 (2020). https://doi.org/10.1177/8755293020977520

Molitierno, C,, Del Vecchio, C., Di Ludovico, M., Prota, A.: Validation of pseudo-dynamic testing protocol through full-scale tests on two-storey infilled reinforced concrete frame. In: 2nd FIB Italy YMG Symposium on Concrete and Concrete Structures. Rome, pp. 109–116 (2021)

Morandi, P., Hak, S., Magenes, G.: Performance-based interpretation of in-plane cyclic tests on RC frames with strong masonry infills. Eng. Struct. **156**, 503–521 (2018). https://doi.org/10.1016/j.engstruct.2017.11.058

Negro, P., Verzeletti, G.: Effect on infills on the global behaviour of R/C frames: energy considerations from pseudodynamic tests. Earthq. Eng. Struct. Dyn. **25**, 753–773 (1996). https://doi.org/10.1002/(sici)1096-9845(199608)

NZSEE/MBIE (2017) The Seismic Assessment of Existing Buildings: Technical Guidelines for Engineering Assessments. Part C–Detailed Seismic Assessment

Panagiotakos TB, Fardis MN (1996) Seismic response of infilled RC frame structures. In: Proceedings of the 11th World Conference on Earthquake Engineering. pp 1–8

Priestley, M.J.N.: Displacement-based seismic assessment of reinforced concrete buildings. J. Earthq. Eng. **1**(1), 157–192 (1997). https://doi.org/10.1080/13632469708962365

Ricci, P., Verderame, G.M., Manfredi, G.: Analytical investigation of elastic period of infilled RC MRF buildings. Eng. Struct. **33**, 308–319 (2011). https://doi.org/10.1016/j.engstruct.2010.10.009

Verderame, G.M., Ricci, P., De Risi, M.T., Del Gaudio, C.: Experimental assessment and numerical modelling of conforming and non-conforming RC frames with and without Infills. J. Earthq. Eng. **00**, 1–42 (2019). https://doi.org/10.1080/13632469.2019.1692098

# Green Geopolymer Mortars for Masonry Buildings: Effect of Additives on Their Workability and Mechanical Properties

Laura Bergamonti[1], Elena Michelini[2(✉)], Claudia Graiff[1], Daniele Ferretti[2], Marianna Potenza[1], Federico Pagliari[2], and Francesco Talento[3]

[1] Department of Chemistry, Life Sciences and Environmental Sustainability, University of Parma, Parma, Italy
[2] Department of Engineering and Architecture, University of Parma, Parma, Italy
elena.michelini@unipr.it
[3] Bacchi S.p.A., Boretto, RE, Italy

**Abstract.** The fight against climate changes has encouraged the development of most research on new binders, with the aim of progressively reducing the use of Ordinary Portland Cement, with its huge carbon footprint. Within this context, geopolymer-like binders can represent an eco-friendly alternative, also for the production of mortars for masonry buildings. In this work, a "one-pot" all in powder formulation was developed to the purpose, and special attention was given to explore the influence of different additives on mortar workability and mechanical properties. Different types of adhesives (like rice starch or maize starch, and alginate) were considered, which were separately applied at first, and finally mixed together with water retention additives and super-plasticizers. The obtained results highlighted the achievement of an adequate workability and adhesion to the support, at the cost of reduced mechanical performances.

**Keywords:** "one part" geopolymers · thin-layer mortar for masonry · sustainability · rice starch · maize starch

## 1 Introduction

In recent years, a new sensibility for environmental issues has promoted the research and development of sustainable building materials, with reduced carbon footprint. Alkali-activated binders represent a potential alternative to cement, whose production has a huge, no longer sustainable environmental impact. Manufacturing of Ordinary Portland Cement (OPC) produces indeed a large amount of $CO_2$ emissions and consumes huge quantities of unrenewable raw materials and energy (Ali et al. 2011; Miller et al. 2016). As reported in the literature (van Deventer et al. 2010; Nematollahi et al. 2015), the use of alkali-activated binders, often known as geopolymers, reduces $CO_2$ emission of about 80% with respect to OPC, with significant energy savings (up to 40% less). Alkali-activated binders can be synthetized starting from a multitude of reactive powders derived from natural sources (like for example aluminosilicate powders) or


M. A. Aiello and A. Bilotta (Eds.): ICC 2022, LNCE 435, pp. 134–149, 2024.
https://doi.org/10.1007/978-3-031-43102-9_12

industrial by-products, which are mixed with one or more alkaline activators. Usually, geopolymers are produced starting from a two-part mix, with the use of alkaline solutions (i.e. Bergamonti et al. 2018; Nawaz et al. 2020; Shehata et al. 2021; Michelini et al. 2022). More recently, the attention has shifted to one-part systems, which are more similar to premixed bagged materials for the construction industry, being formed by dry powders simply combined with water, which is added on site (Provis 2018). One-part solutions are very interesting in view of factory production and distribution, as well as of a feasible usage at the construction site, where highly corrosive alkaline solutions would be difficult and dangerous to use (Nematollahi et al. 2015; Provis 2018). The production of cementitious products satisfying engineering requirements is still under investigation, since one-part alkali-activated binders usually develop lower mechanical properties with respect to two-parts geopolymers. In one-part admixtures, the dissolution of the alkali activator only starts after water addition, and therefore a full reaction with solid aluminosilicate source could be hardly achieved in short time.

Being a recent technology, few studies have investigated the applicability of one-part alkali activated binders for the production of pre-mixed mortars up to date (e.g., Dong et al. 2020; Ouyang et al. 2020; Zhang et al. 2021). In this study, a one-part alkali-activated binder, previously developed by the Authors (Bergamonti et al. 2022), is adopted for obtaining a thin-layer mortar to be used with autoclaved aerated concrete blocks. Aim of the work is to investigate the effect of different additives on fresh and hardened properties of the mortar, with particular attention to workability, adhesion to the substrate and mechanical strengths. The main challenge of the study – also from a sustainability perspective - is to investigate the feasibility of using mainly natural additives (especially as adhesives and rheological modifiers), while keeping acceptable performances.

## 2 Experimental Program

### 2.1 Materials and Sample Preparation

The alkali-activated mortars analysed in this study were prepared starting from a geopolymeric binder previously developed by the Authors (Bergamonti et al. 2022). The binder, indicated in the following as $MK_{K1}$, was obtained using metakaolin ($Al_2Si_2O_5$) as aluminosilicate precursor, and adding calcium hydroxide ($Ca(OH)_2$) and potassium trisilicate ($K_2SiO_3$) in powder form as alkali activator. Mortar specimens were prepared according to EN 998–2 Standard, by adopting a binder to sand ratio equal to 1/3. The binder $MK_{K1}$ was first premixed with sand, and they were then homogenized into a Hobarth 5-Quart mixer at 139 rpm for 5 min with different water/binder ratios, ranging from 0.91 to 1.1, to obtain the reference mortar AAM (without any additive). To improve some key properties of the mortar, like workability, correction time, and adhesion, different natural and artificial polysaccharides additives were added to the admixture. First of all, the influence of adhesives was investigated, by preparing three different admixtures respectively containing alginate, rice starch and maize starch in powder form. The content of adhesives (in percentage with respect to the total mass of the mortar) were varied opportunely, so to obtain an acceptable workability. The final dosages added to the admixtures were consequently different for the three additives, as reported in Table 1.

The second step was the addition of water retention additives. Starting from the blend with maize starch, which seemed the most promising solution, different types of cellulose pulp were added to the admixture (medium viscosity cellulose, microfibrillated cellulose MF, nanocrystalline cellulose NC). Together with cellulose, an anionic surfactant was also added as superplasticizer. The analysed formulations are summarized in Table 1 for reading convenience.

**Table 1.** Main components of the developed mortar admixtures

| Admixture | w/b ratio | Additives(*) |
|---|---|---|
| AAM | 0.91, 0.93, 1.1 | - |
| AAM-ALG | 1 | alginate (0.25%) |
| AAM-AR | 0.97 | rice starch (3%) |
| AAM-M | 1.1 | maize starch (0.5%) |
| AAM-MC | 1.1 | maize starch (0.5%), medium viscosity cellulose (0.3%) |
| AAM-MCD | 1.1 | maize starch (0.5%), medium viscosity cellulose (0.3%), superplasticizer (0.05%) |
| AAM-$MC_{MF}D$ | 1.1 | maize starch (0.5%), microfibrillated cellulose MF (0.3%), superplasticizer (0.05%) |
| AAM-$MC_{NC}D$ | 1.1 | maize starch (0.5%), nanocrystalline NC cellulose (0.3%), superplasticizer (0.05%) |

(*) dosages referred to the total mass of the mortar

After the preparation of the different admixtures, part of the slurry was immediately used for the execution of tests on fresh mortar. The remaining part was instead destined to the casting of the specimens required for the determination of the properties at the hardened state. For the mechanical characterization of mortar under flexure and compression, the slurry was poured into 40 × 40 × 160 $mm^3$ steel moulds and vibrated for 2 min so to remove the entrapped air, according to EN 1015-11. The moulds were sealed by polyethylene plastic film and stored at room conditions (20 ± 2 °C, 65 ± 5% RH) for two days; subsequently, the specimens were demoulded and cured in laboratory, within closed plastic bags. The specimens were taken out of plastic bags two days before the execution of the tests. For the most promising admixtures, correction time of fresh mortar and adhesive strength of hardened mortar on substrate were also determined, according to relevant standards (EN 1015-9, EN 1015-12). In the performed experimental campaign, Autoclaved Aerated Concrete (AAC) blocks with density equal to 480 ± 50 $kg/m^3$ were chosen as substrate; for this reason, the mortar is intended to be laid in thin layers.

## 2.2 Methods

### 2.2.1 Chemical and Microstructural Characterisation

FTIR spectroscopy was used to characterize the formation of the geopolymer 3D network. The IR spectra were acquired by a Thermo-Nicolet Nexus equipped with a Thermo Smart Orbit ATR diamond accessory. The measurements were performed in the range 4000–400 $cm^{-1}$. The XRPD diffrattograms were collected on a Thermo ARL X'TRA X-ray diffractometer with Si-Li detector, using Cu-K$\alpha$ radiation (40 kV and 40 mA) at 0.2°/sec scan rate (in 2$\theta$) in the range 0–80° to get microstructural information on the alkali-activated binder $MK_{K1}$ and on the reference mortar AAM. For 2$\theta$ calibration, the powdered silicon reflections were used. The identification of crystalline phases was done for comparison with JCPDS (Joint Committee on Powder Diffraction Standard) cards. A Scanning Electron Microscope (SEM) JEOL 6400 equipped with Energy Dispersive X-Ray Microanalysis System (EDS) Oxford-INCA was used to analyse the surface morphology and the elemental distribution of the samples (binder and reference mortar).

### 2.2.2 Properties of Fresh Mortar

One of the most important requirements for a mortar is its workability, which is related to the effort required to manipulate the fresh mortar with minimum loss of homogeneity. Workability is in turn related to different mortar properties, like consistency, plasticity and cohesion (Panarese et al. 1991; Haach et al. 2011). Nevertheless, in current practice, consistency is often used as a measure of workability, due to the difficulties related to the determination of the other related properties.

In this work, consistency of fresh mortar was evaluated through flow tests, according to EN 1015-3. To this end, a mould was centrally placed on the flow table disc, and the mortar was introduced within the mould itself in two layers. Each layer was compacted to ensure uniform filling of the mould itself. After skimming off the excess of mortar, the mould was raised vertically. The flow table was jolted 15 times at a constant frequency, so to spread out the mortar on the disc. Consistency was finally determined as the average of diameters of the spread mortar in the disc measured in two perpendicular directions.

**Fig. 1.** Test for the determination of correction time.

Correction time was determined following EN 1015-9 recommendations. According to this Standard, correction time is defined as the time at which 50% of the contact surface of a cube, placed on a layer of mortar applied on a masonry substrate and then removed, is coved by adhering mortar. The cubes adopted in this work, with dimensions 50x50x50 $mm^3$, were cut from AAC masonry units, which also acted as substrate (Fig. 1). Both the cubes and the masonry substrate were preliminary dried in an oven at a temperature of 105 ± 5 °C, up to the reaching of constant mass. Each cube was kept on the mortar layer for 30 s, by applying to it a load equal to 0.5 kg, and then removed.

### 2.2.3 Properties of Hardened Mortar

At the hardened state, flexural and compressive tests were performed after 28 days from casting on each admixture, according to EN 1015-11. Flexural tests were carried out on prismatic samples under a three-point bending scheme, by using an MTS 2/M Universal testing machine (Fig. 2a). After failure, the two remaining halves were tested in compression, by using an INSTRON 5882 Universal machine working under loading control. Steel platens, with dimension 40 mm × 40 mm, were interposed between sample surfaces and the testing machine (Fig. 2b).

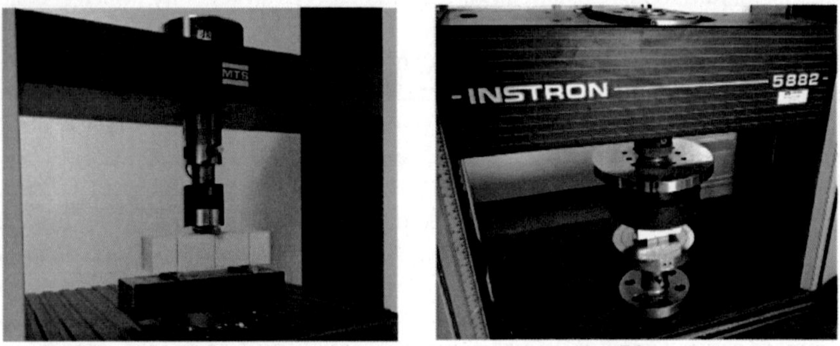


(a) (b)

**Fig. 2.** Test setup for the determination of (a) flexural and (b) compressive strength.

Adhesion to the support was determined for the most promising admixtures, according to EN 1015-12, with a MATEST pull-off tester (Fig. 3). A 5-mm thick mortar layer was first applied to the AAC block acting as substrate. Before test execution, specimens were conditioned for 28 days at laboratory conditions (20 ± 2 °C, 65 ± 5% RH). Adhesive strength was determined by using a detaching plate glued to the surface of the tested mortar specimen by means of epoxy resin. Circular plates with 50 mm diameter and 25 mm thickness were used in this study. Through the pull-off tester, a tensile load was applied to the plates at a constant speed, so that failure took place between 20 and 60 s. Adhesive strength was calculated as the maximum tensile strength caused by the detaching load perpendicular to mortar surface.

**Fig. 3.** Setup for the determination of adhesive strength.

# 3 Results and Discussion

## 3.1 Characterization of Alkali-Activated Binder and Reference Mortar

X-Ray diffrattograms of the alkaly activated binder $MK_{K1}$ and of the reference mortar AAM are reported in Fig. 4. Both the XRD patterns highlighted an amorphous halo at 2θ between 20° and 35°, which can be associated with the disordered structure of geopolymers (Provis et al. 2005). Several sharp peaks identified the different crystalline phases: quartz, calcium monocarboaluminate and calcium carbonate. Other peaks were attributed to K, Ca silicate and aluminum silicate, which are due to the reorganization of Al-O and Si-O network from metakaolinite after the reaction with the alkaline activator solution. Peaks attributable to the K-zeolite-like and piroxenes phases were also identified in the diffractogram of the mortar.

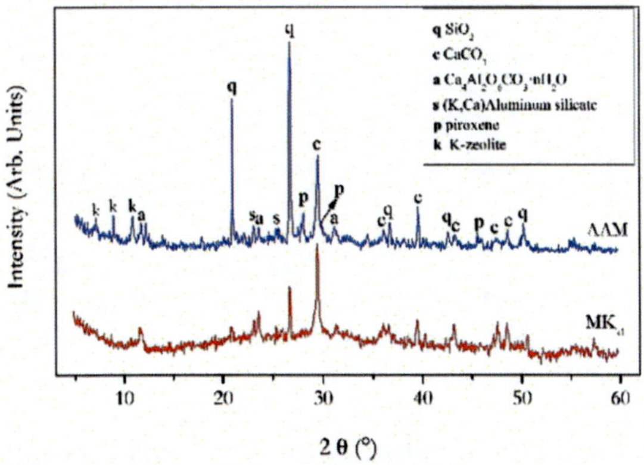


**Fig. 4.** XRD pattern of alkali activated binder $MK_{K1}$ and reference mortar AAM.

The FTIR spectra of the alkali activated binder ($MK_{K1}$) and reference mortar (AAM), shown in Fig. 5, were dominated by broad bands due to the stretching (at about 3300 $cm^{-1}$) and bending (at 1650 $cm^{-1}$) vibration of the OH groups (adsorbed water and/or silanol), and by the strong/broad band at about 1050 $cm^{-1}$ due to the stretching vibration of Si-O-T (T = Si or Al) groups. This band was shifted at lower wavenumber with respect raw metakaolin, from 1107 $cm^{-1}$ to 1074 $cm^{-1}$. This shift could be associated to the decrease of the Si-O-Si units following the silicates dissolution, and to the increase in Si-O-Al units during the formation of geopolymer lattice. Bands due to the vibrational modes (stretching in the 1500–1300 $cm^{-1}$ range, bending at 874 $cm^{-1}$, and rocking at 712 $cm^{-1}$) of the carbonate ion formed by the reaction between the metal hydroxide of the alkaline solution and atmospheric $CO_2$ were clearly visible in both spectra. Carbonate formation is indeed a frequent phenomenon (Yunsheng et al. 2010).

The morphological characterization and microanalysis of the binder $MK_{K1}$ and of the reference mortar AAM were performed by SEM-EDS measurements (Fig. 6 a,d).

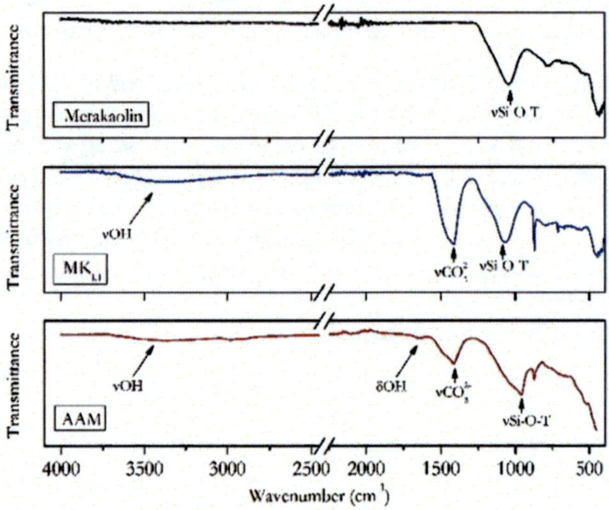


**Fig. 5.** FTIR spectra of metakaolin, geopolymers-like binder ($MK_{K1}$), and mortar (AAM).

The geopolymeric binder (Fig. 6a) had a homogeneous microstructure characterized by a widespread microporosity, with pores of about 20–30 μm in diameter. The main elements revealed by the EDS microanalysis (Fig. 6b) were Ca, Al and Si in a weight ratio of about 40:40:60, with a lower content of cation K. Due to sand addition, irregular submillimetric clasts, about 500 μm in size, and rounded micrometric grains, ≈ 50 μm in size, embedded in a very fine matrix were clearly visible in the reference mortar (Fig. 6c). The high Si content (Fig. 6d) was due to the quartz sand added to the mixture.

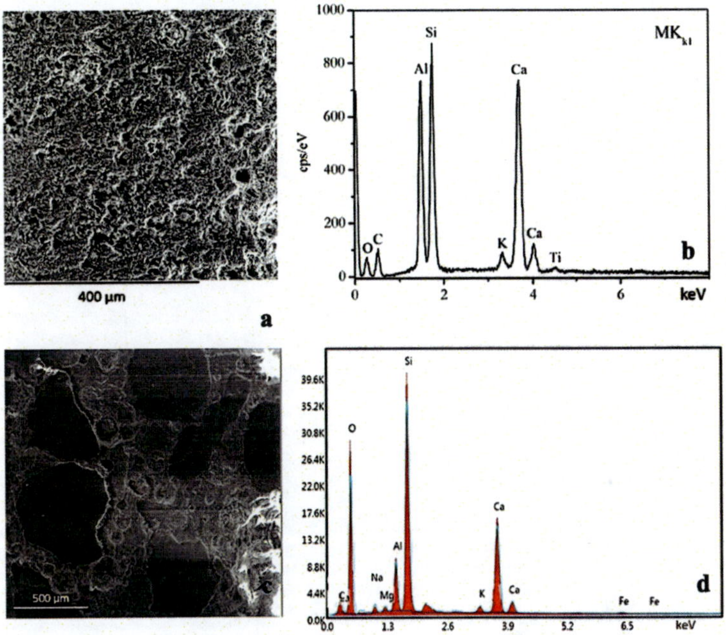


**Fig. 6.** Secondary electron SEM images and microanalysis EDS of (a,b) binder $MK_{K1}$, and (c,d) reference mortar AAM.

## 3.2 Influence of Additives on Consistency and Mechanical Strengths

### 3.2.1 Reference Mortar

The effect of water/binder ratio on mortar workability was preliminarily investigated for the reference admixture AAM, without any additive (Fig. 7). Acceptable values of consistency for the usage with AAC blocks were achieved for w/b ratios ranging between 0.97 and 1.06. However, since the addition of adhesives was expected to reduce the workability, it was decided to adopt the higher values of w/b ratios (ranging between 1 and 1.1) in the preparation of the other admixtures recalled in Table 1. This last choice seemed reasonable also with respect to mechanical performances. The corresponding variation in mechanical strengths for three different w/b ratios is depicted in Fig. 8.

It can be seen that the achieved compressive strength was acceptable also for w/b = 1.1, being higher than 10 MPa. Mortars belonging to strength class M5 and M10 (with $f_m$ respectively equal to 5 and 10 MPa) are indeed commonly used for AAC masonry assemblages.

### 3.2.2 Influence of Adhesive Type

The effect of polysaccharides additives to improve the adhesion between the mortar and the blocks is reported in Table 2 and Fig. 9, with reference to consistency and mechanical strengths respectively.

The selected dosages of the three adhesives provided similar consistency values for the mortar admixtures, while in terms of mechanical performances, the solution with the

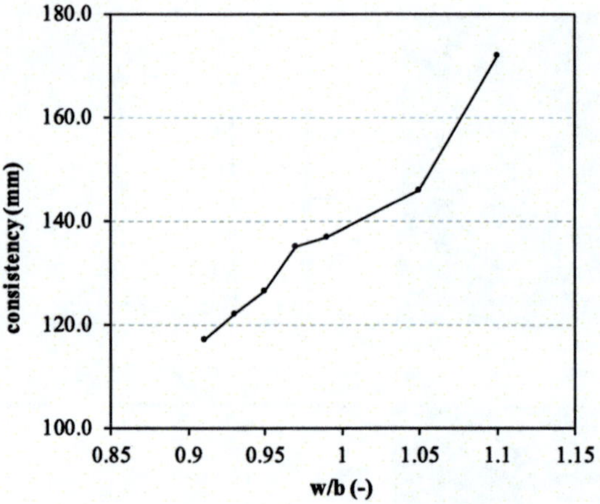


**Fig. 7.** Variation of consistency as a function of water/binder w/b ratio for mortar AAM without any additive.

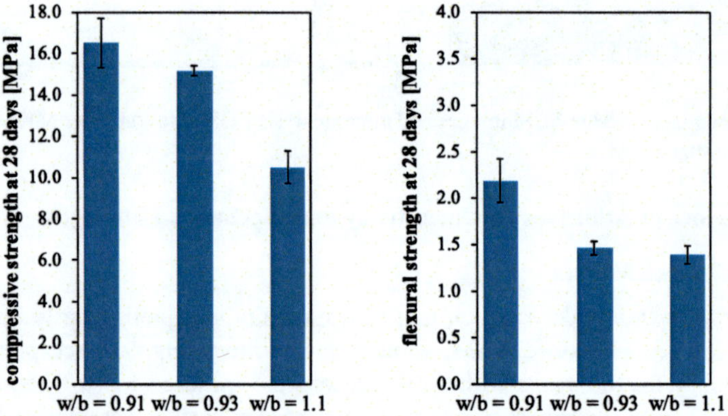


**Fig. 8.** Variation of mechanical strengths as a function of w/b ratio for mortar AAM without any additive.

**Table 2.** Influence of adhesive additives on mortar consistency.

| Admixture | w/b ratio | Consistency (mm) |
|---|---|---|
| AAM-ALG | 1 | 146 |
| AAM-AR | 0.97 | 153 |
| AAM-M | 1.1 | 148 |

addition of alginate was not satisfactory, since the achieved compressive strength was lower than 5 MPa. On the contrary, the two solutions with rice starch and maize starch provided very similar compressive and flexural tensile strengths, with values around 8 MPa and 1.3 MPa, respectively. The admixture with the addition of maize starch had

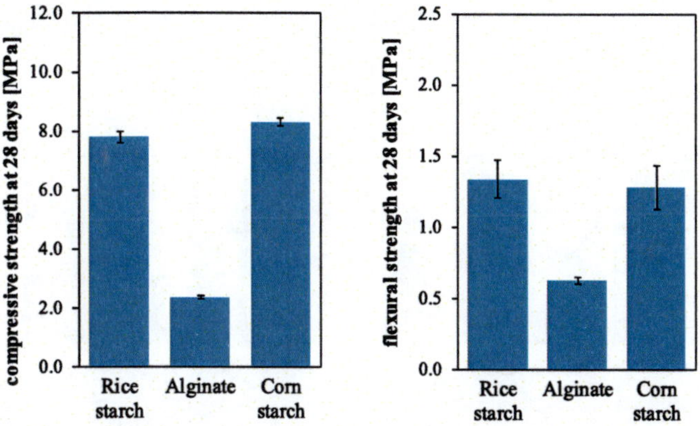


**Fig. 9.** Influence of adhesive additives on mortar mechanical strengths.

two main advantages: a lower required adhesive dosage, and the possibility of adding it to the admixture at ambient temperature. Instead, rice starch should be preliminary dissolved in warm water, at a temperature over 70 °C, which seemed not feasible for possible on-site applications.

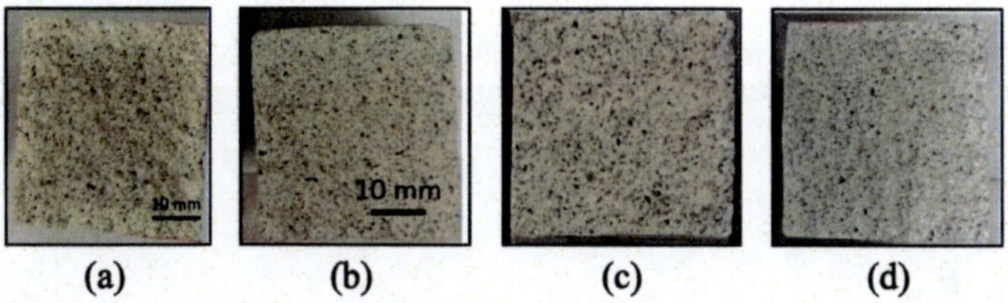


**Fig. 10.** Morphology of tested specimens: (a) reference mortar, and mortar with the addition of (b) rice starch, (c) alginate, (d) maize starch.

A general view of the specimens after the execution of three-point bending tests is reported in Fig. 10. It can be seen that in all cases the structure of the specimens was microporous, with evenly distributed submillimetre pores, and without the formation of evident air macro-bubbles.

### 3.2.3 Influence of Cellulose and Superplasticizer Addition

Cellulose is usually added in premixed mortars for its water retention properties and thickening effect, which improve the construction performance and workability of the final product. In this study, the addition of medium viscosity cellulose only in the admixture (MC blend, Fig. 11 and Tab. 3) determined however a significant reduction of mortar mechanical properties, as can be inferred from Fig. 11. On the contrary, the admixture containing both cellulose and superplasticizer (MCD) achieved an acceptable compressive strength, almost equal to 5 MPa.

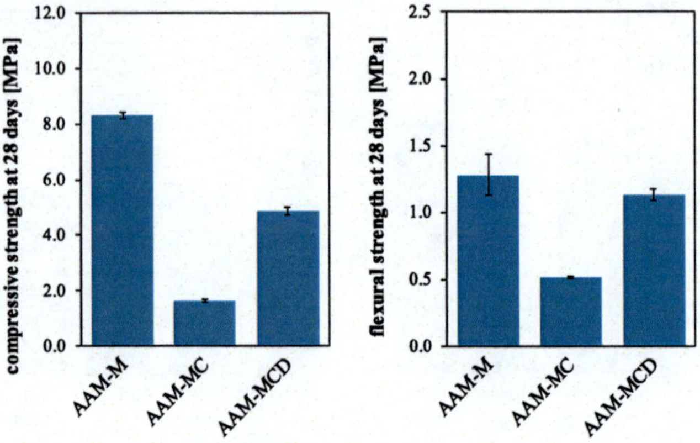


**Fig. 11.** Influence of cellulose and superplasticizer addition on mortar mechanical strengths.

The influence of cellulose type on mortar properties was also studied, by substituting the medium viscosity cellulose used in the first tests with microfibrillated (MF) or nanocrystalline (NC) cellulose. The type and amount of superplasticizer were kept the same for all the admixtures. The corresponding results are shown in Fig. 12 and Table 3. It can be seen that the influence of microfibrillated and nanocrystalline cellulose was remarkable even at low dosages, due to their high surface area. This effect was positive with regard to consistency measurements, but led to an excessive reduction of mechanical strengths, not compatible for the usage in bearing masonry structures.

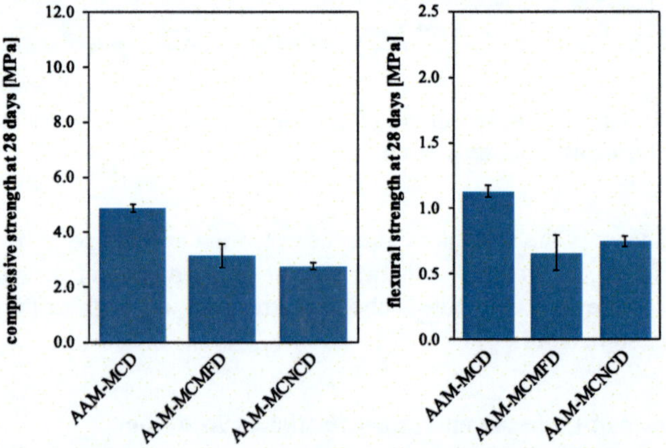


**Fig. 12.** Influence of cellulose type on mortar mechanical strengths.

### 3.3 Correction Time and Adhesion to the Support for Some Selected Admixtures

For three of the tested admixtures (that is the reference mortar AAM, and the two mortars AAM-MC and AAM-MCD, with cellulose only, or with cellulose and superplasticizer),

**Table 3.** Influence of different types of cellulose on mortar consistency.

| Admixture | w/b ratio | Consistency (mm) |
|---|---|---|
| AAM-MC | 1.1 | 146 |
| AAM-MCD | 1.1 | 153 |
| AAM-$MC_{MF}$D | 1.1 | 173 |
| AAM-$MC_{NC}$D | 1.1 | 167 |

correction time and adhesion to support were also determined. The measured correction time is reported in Table 4. As can be seen from Fig. 13a, the reference mortar and that with the addition of medium viscosity cellulose alone did not adhere to the support. With the addition of the superplasticizer, a correction time almost equal to 4 min was achieved (Fig. 13b). Commercial mortars used for the realization of AAC masonry assemblages often reach slightly higher correction times, which are however generally lower than 8 min, according to technical sheets from the producers.

**Table 4.** Correction time for three selected admixtures.

| Admixture | Correction time (min) |
|---|---|
| AAM | it did not adhere |
| AAM-MC | it did not adhere |
| AAM-MCD | ≈ 4 |

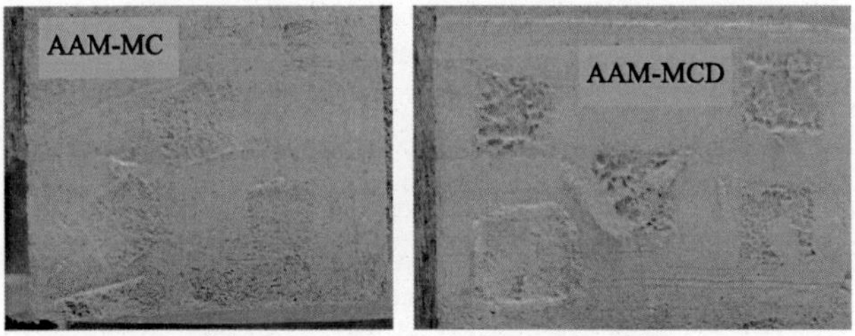


**Fig. 13.** Tests for the determination of correction time on mortar: (a) AAM-MC and (b) AAM-MCD.

In order to increase correction time, the possible addition of other commercial additives commonly used in mortar admixtures was also explored, although with not satisfactorily results.

Starting from AAM-MCD blend, tartaric acid was added as retarder, while polymeric gel was used to enhance the resistance to moisture and water seepage. In these attempts, high viscosity cellulose was mixed together with a small percentage (0.03%) of a different type of cellulose fiber, to improve water retention. The influence exerted by these further additives is shown in Table 5, in terms of consistency, correction time and mechanical strengths at 28 days.

**Table 5.** Admixtures with other commercial additives to improve correction time.

| Admixture | w/b ratio | Additives(*) | Consistency (mm) | Mean compressive strength (MPa) | Mean flexural strength (MPa) | Correction time (min) |
|---|---|---|---|---|---|---|
| AAM-MCTD | 1.3 | maize starch (0.5%), cellulose (0.3% + 0.03%), tartaric acid (0.025%), superplasticizer (0.03%) | 143 | 3.8 | 0.82 | ≈4 |
| AAM-MCTDp | 1.3 | maize starch (0.5%), cellulose (0.3% + 0.03%), tartaric acid (0.025%), superplasticizer (0.03%), polymer gel (0.05%) | 149 | 3.4 | 0.78 | it does not adhere |
| AAM-MCTDp-1 | 1.7 | maize starch (0.5%), cellulose (0.6% + 0.06%), tartaric acid (0.013%), superplasticizer (0.025%), polymer gel (0.05%) | 139 | 0.5 | 0.17 | ≥10 |

(*) dosages referred to the total mass of the mortar

As can be observed, correction time was not improved, but at the same time mechanical strengths were reduced. A correction time greater than 10 min was only obtained when doubling the cellulose dosage in the admixture. However, the corresponding decrease in compressive strength was unacceptable.

The results of adhesion tests are summarized in Table 6. The adhesive stress values obtained for mortar MCD are in line with those expected for commercial mortars for

**Table 6.** Adhesive strength for the three selected mortars.

| Admixture | Mean adhesive strength (MPa) | Failure mode |
|---|---|---|
| AAM | 0.03 | A |
| AAM-MC | 0.23 | B |
| AAM-MCD | 0.29 | B |

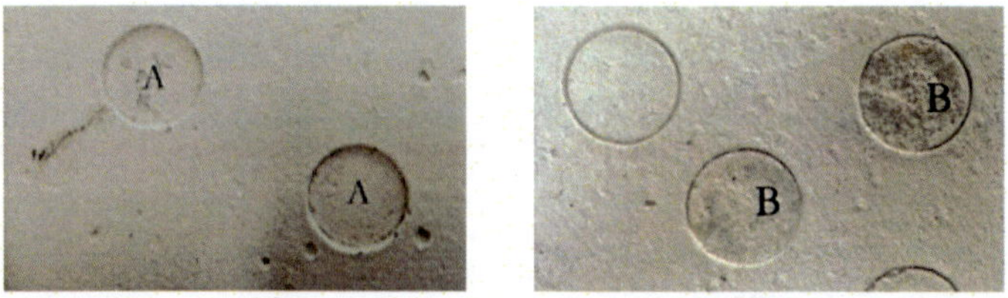


(a) (b)

**Fig. 14.** Failure modes at the end of adhesion tests: (a) mode A; (b) mode B.

AAC masonry. For the reference mortar, an adhesive fracture was observed, with failure at the interface between the mortar and the substrate (Fig. 14a). For MC and MCD mortars, a cohesive fracture took place, with failure within the mortar layer (Fig. 14b).

## 4 Conclusions

In this work, a new alkali-activated thin-layer mortar was developed for AAC masonry assemblages. While most of the published work in the literature are focused on mortars developed from two-parts geopolymeric binders, in this work a one-part "just-add-water" alkali activated binder was used. The effect of different additives on fresh and hardened state properties of the mortar was deepened, with special emphasis on the possible use of natural products, so to improve the environmental sustainability of the final products. Based on the experimental results, the following conclusions can be drawn:

- maize starch has good potential to be used as adhesive additive in alkali-activated mortars, since its addition did not produce an excessive decrease of mortar mechanical performances (compressive strength was reduced of about 20% with respect to the reference mortar, but its final value was suitable for bearing masonry). On the contrary, the use of alginate or rice starch was less feasible.
- Cellulose type had a limited effect on mortar workability, while it strongly influenced the mechanical performances. The most satisfactorily results were achieved by using medium viscosity cellulose. Mortar compressive strength was inversely proportional to cellulose content, while correction time increased with cellulose dosage.
- Superplasticizer addition improved both the correction time and the adhesion of the mortar to the substrate. The addition of other commercial additives, like tartaric acid and polymer gel, had instead a negligible influence on those properties.

The obtained results are promising, but further studies are required to develop a higher performing product to be used in the building sector. Possible applications of the final product would not only be limited to the case of thin-layer mortars, but also to the development of plasters or overlay coatings for new and existing masonry structures.

**Acknowledgements.** This work is part of a research founded by Ekoru s.r.l., within the PON national project n. F/090017/00/X36, CUP B18I17000450008.

## References

Ali, M.B., Saidur, R., Hossain, M.S.: A review on emission analysis in cement industries. Renew. Sustain. Energy Rev. **15**(5), 2252–2261 (2011)

Bergamonti, L., Taurino, R., Cattani, L., Ferretti, D., Bondioli, F.: Lightweight hybrid organic-inorganic geopolymers obtained using polyurethane waste. Constr. Build. Mater. **185**, 285–292 (2018)

Bergamonti, L., et al.: Influence of alkali cations on the mechanical properties of geopolymer-like binders based on solid alkali activator. In: Proceedings of fib International Congress 2022 on Concrete Innovation for Sustainability (2022)

Dong, M., Elchalakani, M., Karrech, A.: Development of high strength one-part geopolymer mortar using sodium metasilicate. Constr. Build Mater. **236**, 117611 (2020)

EN 998-2: Specification for mortar for masonry - Part 2: Masonry mortar (2016)

EN 1015-11: Methods of Test for Mortar for Masonry - Part 11: Determination of Flexural and Compressive Strength of Hardened Mortar (2019)

EN 1015-12: Methods of test for mortar for masonry - Part 12: Determination of adhesive strength of hardened rendering and plastering mortars on substrates (2016)

EN 1015-3, 1999/A2:2006: Methods of test for mortar for masonry – Part 3: Determination of consistence of fresh mortar (2006)

EN 1015-9, 1999/A1:2006. Methods of test for mortar for masonry - Part 9: Determination of workable life and correction time of fresh mortar (2006)

Haach, V.G., Vasconcelos, G., Lourenço, P.B.: Influence of aggregates grading and water/cement ratio in workability and hardened properties of mortars. Constr. Build. Mater. **25**(6), 2980–2987 (2011)

Michelini, E., et al.: Fracture energy of sustainable geopolymer composites with and without the addition of slaughterhouse by-products as fibre-reinforcement: an experimental investigation. Proc. Struct. Integr. **39**, 71–80 (2022)

Miller, S.A., Horvath, A., Monteiro, P.J.: Readily implementable techniques can cut annual $CO_2$ emissions from the production of concrete by over 20%. Environ. Res. Lett. **11**(7), 74029 (2016)

Nawaz, M., Heitor, A., Sivakumar, M.: Geopolymers in construction-recent developments. Constr. Build. Mater. **260**, 120472 (2020)

Nematollahi, B., Sanjayan, J., Shaikh, F.U.A.: Synthesis of heat and ambient cured one-part geopolymer mixes with different grades of sodium silicate. Ceram. Int. **41**(4), 5696–5704 (2015)

Ouyang, S., Chen, W., Zhang, Z., Li, X., Zhu, W.: Experimental study of one-part geopolymer using different alkali sources. J. Phys. Conf. Ser. **1605**(1), 012155 (2020)

Panarese W.C., Kosmatka S.H., Randall, F.A.: Concrete Masonry Handbook for Architects, Engineers, Builders, 5th edn. Portland Cement Association, USA (1991)

Provis, J.L., Lukey, G.C., van Deventer, J.S.: Do geopolymers actually contain nanocrystalline zeolites? A reexamination of existing results. Chem. Mater. **17**(12), 3075–3085 (2005)

Provis, J.L.: Alkali-activated materials. Cem. Concr. Res. **114**, 40–48 (2018)

Shehata, N., Sayed, E.T., Abdelkareem, M.A.: Recent progress in environmentally friendly geopolymers: a review. Sci. Total Environ. **762**, 143166 (2021)

van Deventer, J.S., Provis, J.L., Duxson, P., Brice, D.G.: Chemical research and climate change as drivers in the commercial adoption of alkali activated materials. Waste Biomass. Valoriz. **1**(1), 145–155 (2010)

Yunsheng, Z., Wei, S., Zongjin, L.: Composition design and microstructural characterization of calcined kaolin-based geopolymer cement. Appl. Clay Sci. **47**, 271–275 (2010)

Zhang, H.Y., Liu, J.C., Wu, B.: Mechanical properties and reaction mechanism of one-part geopolymer mortars. Constr. Build. Mater. **273**, 121973 (2021)

# FRCM Composites for the Structural Upgrading of Reinforced Concrete Shallow Beams

Marco C. Rampini(✉), Giulio Zani, Matteo Colombo, and Marco di Prisco

Department of Civil and Environmental Engineering, Politecnico Di Milano, Milan, Italy
{marcocarlo.rampini,giulio.zani,matteo.colombo,
marco.diprisco}@polimi.it

**Abstract.** In the last decades, there has been an increment in the use of Fabric-Reinforced Cementitious Matric (FRCM) composites, in both the restoring and the upgrading of the load carrying capacity of existing structural elements. At the beginning, FRCMs were mainly applied on masonry buildings, while the recent literature documents growing interventions on Reinforced Concrete members. In the present work, the results of a series of bending tests on RC shallow beams, alternatively strengthened or retrofitted with the application of FRCM composites at the intrados, are reported. The effect of predamage conditions is also discussed. At the end, the estimation of the mechanical response by means of a simplified plane-sectional approach and a proposal for the evaluation of the crack spacing are reported in case of the FRCM-strengthened RC beam.

**Keywords:** FRCM · retrofitting · strengthening · RC · shallow beams · crack spacing

## 1 Introduction

In Italy, there was a significant increase of the use of Fabric-Reinforced Cementitious Matrix (FRCM) composites for the structural reinforcement and rehabilitation, also due to the publication of both the Guidelines for the material qualification and control (Consiglio Superiore dei LL. PP. 2018) and of the one for the Design of FRCM interventions (CNR 2018). While they were initially used more for the reinforcement on masonry structures, today these composites are also being used for applications on reinforced concrete (RC) structures, thanks to the compatibility with irregular surfaces, the easier applicability, and the lower cost with respect to the alternative FRP-type intervention.

The goal of this paper was the experimental assessment of the effect of FRCM strengthening and retrofitting on the load-bearing capacity of reinforced concrete shallow beams and, in the view of understand the possible improving in durability, on the crack pattern evolution, studied both in terms of distance and opening.

At the end, the estimation of the mechanical response of the FRCM strengthened beam was performed with the aim to provide a simplified tool for designer to be used in parallel with the formulas proposed by the CNR Guidelines (CNR 2018). In addition, a simplified formulation for the evaluation of crack distance is proposed and critically assessed.


M. A. Aiello and A. Bilotta (Eds.): ICC 2022, LNCE 435, pp. 150–159, 2024.
https://doi.org/10.1007/978-3-031-43102-9_13

## 2 Materials Characterization

The shallow beams were realized with an ordinary concrete (average compressive cubic strength of 39.9 MPa and average tensile strength of 3.1 MPa), and they were reinforced at the intrados with three nominally identical $\phi$12 mm B450C bars (average yielding and ultimate strength of 530.2 MPa and 628.2 MPa), with a clear cover of 20 mm (see Fig. 2).

The FRCM system used were made by a commercial shrinkage-compensated cementitious thixotropic mortar (mean cubic compressive and tensile strengths of 58.9 MPa and 3.1 MPa) and a symmetric AR-glass fabric characterized by an equivalent thickness of 0.093 mm/m, a grid spacing of around 38 mm and an ultimate capacity of around 162.5 kN/m. The results of tensile tests on FRCM coupons and on fabric samples, considering only the warp direction, are reported in Fig. 1a. The complete description of specimen sizes and testing procedure are reported in (Rampini et al. 2019). For the selected FRCM composite an efficiency factor of 0.73 was obtained dividing the average ultimate strength of the composite by the one of the textiles, which are respectively 1268 MPa and 1744 MPa.

In addition, two shear single lap test were performed to assess the potential of the composite when applied on the substrate. The surface of the concrete blocks, made by the same concrete of the shallow beams, were prepared by means of a hydro-scarification with a water jet in the range of 1000–1400 atm prior the application of the FRCM strip on a 70 × 150 mm area. The water pressure was set in according to the experimental evidence reported in (Rampini et al. 2020) and was also used for the preparation of the shallow beam intrados. The shear tests were performed at a constant displacement rate of 0.01 mm/s and the same test setup of (Rampini et al. 2020) was used. The resulting curves, Fig. 1b, shown a good repeatability of the response and both the samples failed due to the rupture of the fabric not embedded in the mortar. No slip between mortar and substrate was observed. The average ultimate stress on the composite was equal to 1466 MPa, resulting higher than the one of the FRCM composite.

Despite the number of specimens, their sizes, and the test procedures were not equal to the ones requested for the FRCM qualification in according with the Italian Guidelines (Consiglio Superiore dei LL. PP. 2018), the results of the characterization tests were used to define the design law of the FRCM system, following the CNR instructions (CNR 2018). The minimum of the characteristic ultimate strengths obtained in the tests (composite, plain fabric, and shear) was compute and then used to define the ultimate design strength as:

$$\sigma_d = \eta \cdot \sigma_k / \gamma_M \tag{1}$$

in which $\eta$ is 0.9 in case of internal application, and $\gamma_{\mathrm{M}}$ is equal to 1.5. Up to the ultimate design strength, $\sigma_{\mathrm{d}}$, of 719.0 MPa, the design law coincided with the tensile curve of the plain fabric (a slope equal to $E_f$ was set).

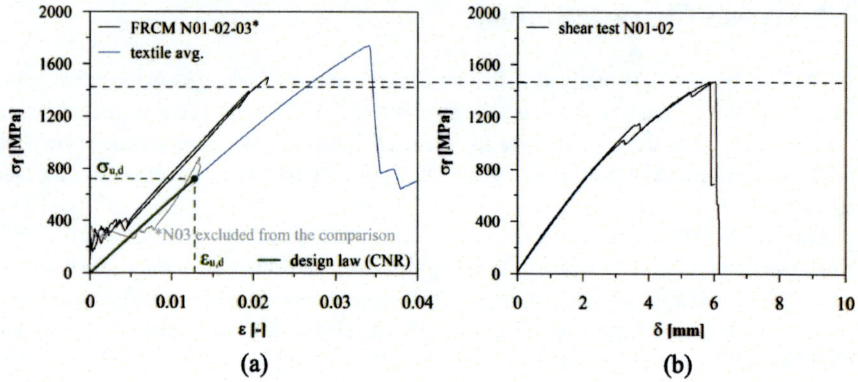


**Fig. 1.** Results of the FRCM composite experimental characterization (tensile tests on fabric (a) and composite coupon, and single lap shear test (b)). Definition of the composite design law.

## 3 Experimental Campaign

In this section the description and the results of the experimental tests are presented and discussed. The first beam was tested up to failure to define the reference response (REF), the second one was strengthened with the application at the intrados of the FRCM system previously presented (STR), and the last two beams were predamaged (PRE) at serviceability condition (SLS) and then retrofitted (RTF).

### 3.1 Setup Scheme and Test Procedure

The beams were tested in a symmetric four-point bending scheme, with a shear length of 900 mm, Fig. 2. All the tests were performed under displacement control; the rate of increasing of the actuator head displacement (stroke) was set to 20 μm/s.

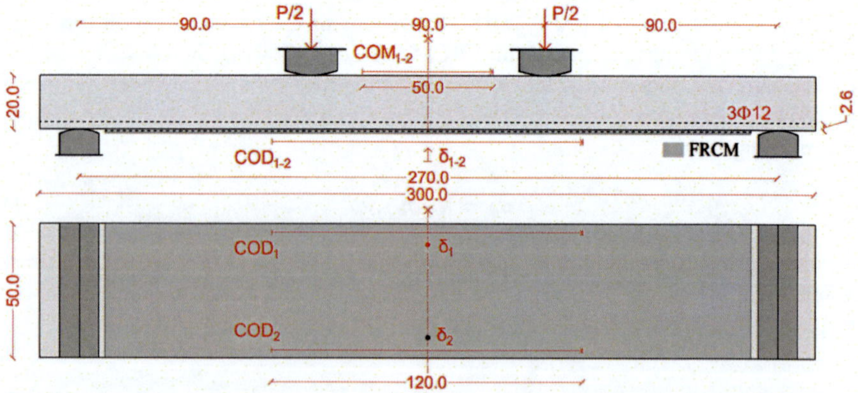


**Fig. 2.** Four-point bending test scheme and position of the measuring instruments (lateral and bottom view).

The samples were instrumented with two Potential Displacement Transducers (PDT) at the mid-span to measure the vertical deflection, $\delta_{1-2}$, and with other four PDTs applied on top and bottom surfaces to measure the integral crack opening displacement, $COD_{1-2}$, and the axial shortening, $COM_{1-2}$. A maximum measuring range of 20 mm and 10 mm were available respectively for COD and COM. Please note that, due to the moderate curvature of the two supports, the effective shear span length was equal to around 850 mm.

## 3.2 Experimental Results: RC Shallow Beams

The response of the REF and PRE beams are reported in Fig. 3, both in terms of vertical load vs. mid-span displacement and of moment vs. COD. The maximum load reached in the REF test was equal to 85.3 kN, corresponding to a bending moment of 36.3 kNm, and the failure was due to the rupture of one of the longitudinal bars. At a mid-span displacement of around 65 mm, to preserve the instruments from damage the mid-span PDTs were removed. It is possible to appreciate the ductility of the response observing the load vs. stroke curve; then, dividing the ultimate displacement with respect to the yielding one a ductility ratio, $\mu = \delta_u / \delta_y$, of around 7.5 was obtained.

The predamaged tests (PRE) were arrested at the reaching of a total load of 42 kN, corresponding to the bending moment at which both the top compressive and the steel tensile stresses remained lower than serviceability limits ($0.45f_{ck}$ and $0.8f_{yk}$).

An example of the crack patterns obtained in PRE tests is reported in Fig. 5. The mean distance, measured on both front and back sides on both the PRE beams, was equal to 145.5 mm and the average COD measurement read at the beginning and at the end of the unloading phases were respectively equal to 2.1 mm and 1.0 mm. Please note that for the evaluation of the crack distance, $s_{rm}$, only cracks formed between the two loading knives were considered and only a single main crack was counted in case of a local splitting in two parts at the bottom of the element.

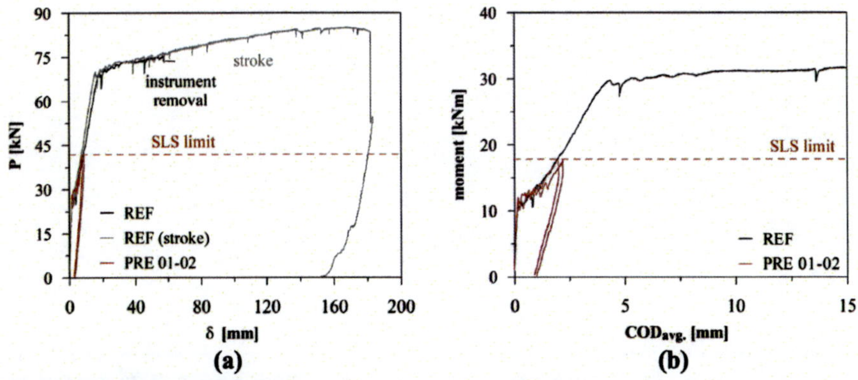


**Fig. 3.** Load vs. displacement (a) and moment vs. average COD in the center zone of the beam curves of the REF and PRE beams.

### 3.3 Experimental Results: FRCM Retrofitted/Strengthened Beams

The 15 mm thick FRCM layers were made by the typical hand lay-up technique, placing the fabric with the warp oriented in the longitudinal direction. To ensure the highest concrete to mortar compatibility, the intrados substrate was preliminary hydroscarificated, cleaned and saturated prior the cast of the FRCM system, and to simplify the application procedure the beams were tilted with the intrados facing upward.

The results of the experimental tests on the FRCM-reinforced beams are reported in Fig. 4. The maximum loads reached were equal to 99.2 kN and 100.5 kN (average between the two), respectively for the STR and the RTF beams, corresponding to an increment of around 16% with respect to the REF one. No relevant variation was observed between the maximum load reached in the STR and RTF tests. Please note that, if the comparison is done at the same mid-span displacement of the STR peak (around 40 mm), the increment reached a value around 30%.

After the reaching of the peak, at which the fabric failed, there was a sudden reduction in load up to the value corresponding to the nearly horizontal branch of the REF curve. Then, the failure was related to the rupture of the steel bars.

Considering the mid-span deflection at the curve peak, it is worth to notice a reduction of ductility of the response ($\mu$ equal to around 2.5). This reduction resulted lower if the longitudinal bar failure is considered as the ultimate displacement, $\delta_u$, leading to a $\mu$ in the 4.5 to 6.5 range.

Observing the moment vs. COD curves in Fig. 4b, it is interesting to notice the effect of the FRCM in the control of crack opening. In the STR beam, comparing the COD values recorded at the same applied bending moment, the FRCM has the effect to reduce the crack opening with respect to the REF case of more than 20%. This effect became negligible as soon as the value of 20 kNm is exceeded. Conversely, in case of RTF members, the controlling of the crack opening is visible only after around 15 kNm. The presence of pre-cracks caused a weakness of some sections of the beam in the initial phase of the response. Due to the consequent strain localization, the typical hardening response of the FRCM was activated in advance with respect to what observed in the

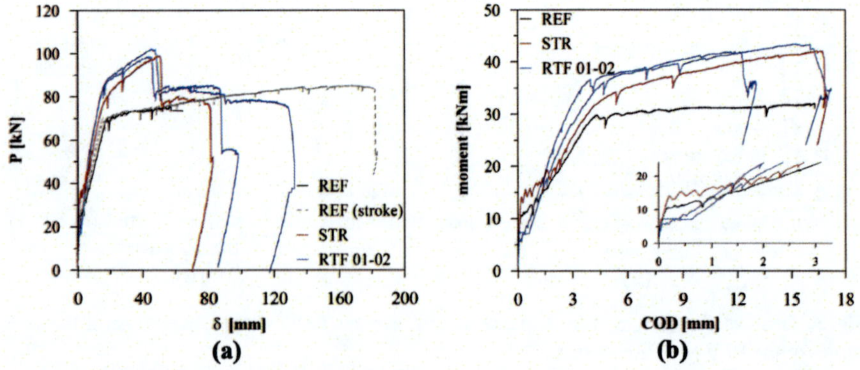


**Fig. 4.** Comparison of the load vs. displacement and of the moment vs. COD curves of the REF, STR and RTF beams.

STR case, inducing a consistent crack reduction (around 30–40% with respect to REF) up to the yielding of the longitudinal bars.

At the end of the tests, the distance between the cracks on the FRCM bottom face was in the range of 50–60 mm. This can be considered as the typical characteristic length of the composite.

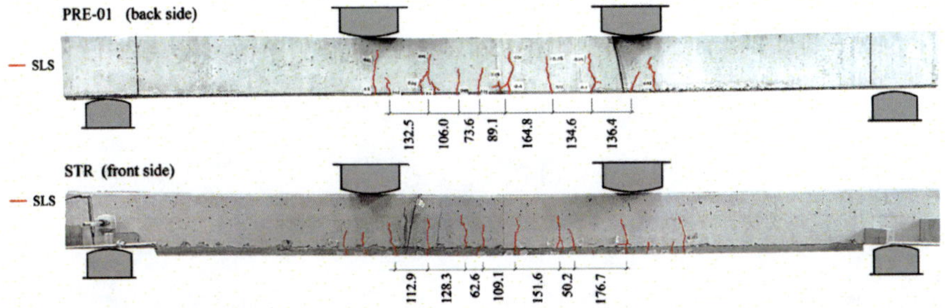


**Fig. 5.** Identification of the crack pattern at the chosen SLS limit for the PRE-01 and the STR beams. Measures in [mm]

## 4 Estimation of the Shallow Beams Responses

In this section, for both the REF and the STR beam, the analytical simulation of the moment vs. curvature curves and the evaluation of the design capacity are reported. In addition, a simplified estimation of the crack distances is proposed and performed.

### 4.1 Analytical Prediction of the Experimental Responses

Starting from the material mechanical properties previously reported, the compression and tension constitutive laws for both the materials (concrete and steel) were obtained in according with MC2010 (CEB-FIP 2013). Then, it was possible to estimate the moment vs. curvature response of the REF element performing a simplified analysis based on the plane-section approach and using the nominal sizes reported in Fig. 2. As visible in Fig. 6, the analytical curve adequately fit the experimental ones, especially regarding the estimation of the branch after the bar yielding.

In case of the STR beam, the same plane-section approach was used. In addition to the constitutive models of concrete and steel bars, a simplified stress-strain law of the FRCM was used. The latter was defined by means of the revised version of the stochastic cracking model (Rampini et al. 2019), originally presented in (Cuypers and Wastiels 2006), which has been proved to adequately replicate the FRCM behaviour in tension.

The most important input parameter of this model is the characteristic length, $l_{cs,FRCM}$, which represent the crack distance developed in tension in the FRCM layer. In order to validate the simulation performed with the aid of this model, three different lengths were taken into account: a) $l_{cs,min} = 38$ mm, corresponding to the fabric grid

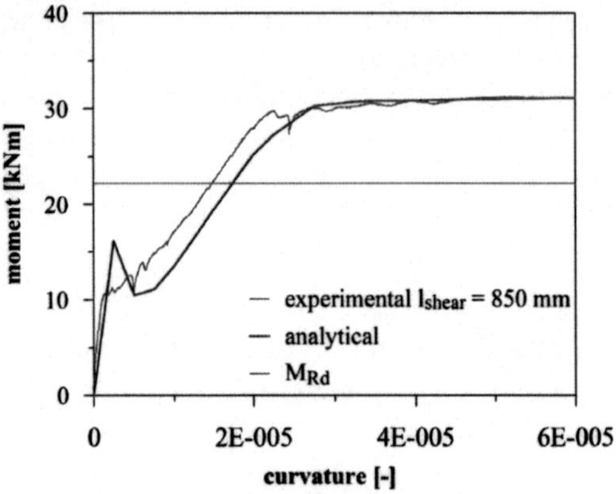


**Fig. 6.** Experimental and the analytical moment vs. curvature of the REF beam and value of the design bending resistance.

spacing; b) $l_{cs,max} = s_{rm,PRE}$, because the crack distance cannot be higher than the one obtained without the FRCM; and c) $l_{cs,exp} = 50$ mm, equal to the one experimentally observed on the intrados surfaces at the end of the tests on beams. In Fig. 7a the three different constitutive laws for the FRCM system are depicted in terms of stress on the composite vs. strain.

The experimental responses of the shear single lap tests have demonstrated the perfect adhesion of the composite to the concrete substrate, allowing to use the entire stress vs. strain law of the FRCM in the analytical estimation. The obtained moment vs. curvature behaviour is reported in Fig. 7b. It is worth to notice how the maximum and the minimum values of the $l_{cs,FRCM}$ led to upper and lower bound in the sectional response (respectively less and more deformable with respect to the real curve), while the use of the experimental characteristic length resulted in a proper approximation of the experimental behaviour.

## 4.2 Evaluation of the Design Bending Capacity

In addition to the simulation of the moment vs. curvature response, it is possible to assess the design formulation proposed in (CNR 2018) for FRCM-strengthened beams at the ultimate limit state (ULS), computing the safety coefficient corresponding to the ratio between the experimental and design capacity ($M_{exp}/M_{Rd}$).

In case of the REF beam, assuming the balanced failure of concrete and steel bars ($f_{cd} = 20.0$ MPa and $f_{yd} = 391.3$ MPa were used), the design resistant moment, $M_{Rd}$, resulted equal to 22.2 kNm (Fig. 6).

For the STR beam, the assumption of FRCM failure at already-yielded stress was initially verified (the design law of the composite defined in according with the CNR guidelines was used, Fig. 1a); then the bending resistance was computed and resulted equal to 28.7 kNm (Fig. 7b).

The ratio $M_{exp}/M_{Rd}$ passed from around 1.6, for the REF case, to around 1.5, in case of the STR beam. This limited variation did not pose any problem in terms of structural safety.

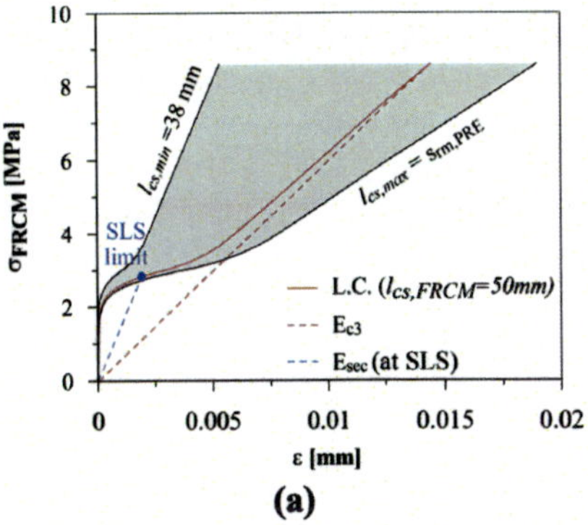


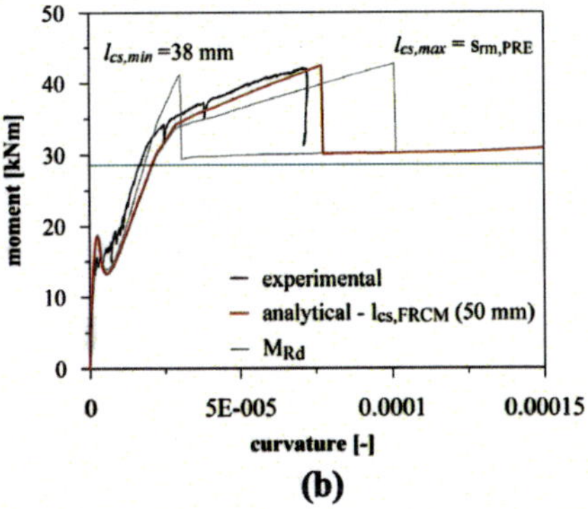


**Fig. 7.** Definition of the FRCM constitutive law at various $l_{cs,FRCM}$ values, and representation of the $E_{c3}$ and $E_{sec}$ stiffness(a), and experimental and the analytical moment vs. curvature of the STR beam and value of the design bending resistance (b).

## 4.3 Prediction of the Crack Distance

From the tests, it was observed that, in parallel to the increase of load-bearing capacity, another important effect of the FRCM was the help in crack control. In this section, a proposal for the estimation of crack spacing in FRCM strengthened beams is presented.

Initially, the well-known formulation of the EC2 (Standard 2004) was used to determine the crack spacing observed in the PRE beams at serviceability conditions (see the predamage phase in Fig. 3). The maximum crack distance, $s_{rm,max}$, was computed as:

$$s_{rm,max} = k_3 c + k_1 k_2 k_4 \frac{\phi}{\rho_{p,eff}} \quad (2)$$

where $\rho$ was the ratio between the total steel bars area and the effective concrete area in tension ($A_s/A_{c,eff}$), computed at the chosen SLS limit; $k_1, k_2, k_3, k_4$ equal to 0.8, 0.5, 3.4 and 0.425, as suggested from the code; $c$ was equal to the clear cover (20 mm); and $\phi$ was the bar diameter (12 mm). The average crack spacing, $s_{rm,avg}$, determined dividing the maximum one by 1.7, resulted equal to 135.2 mm, leading to an under-estimation of around 7% of the experimental value (see the crack pattern example Fig. 5).

For the evaluation of the crack distance in case of FRCM-strengthened beams it was decided to consider the composite as a reinforcement in parallel with the longitudinal rebar. In order to do this, the composite area was homogenized to the steel one, similarly to what already proposed by (Si Larbi et al. 2012; Zomorodian et al. 2016), computing the total reinforcing area as:

$$A_{tot} = A_s + A_{FRCM} \frac{E_{FRCM}}{E_s} \quad (3)$$

where the composite stiffness, $E_{FRCM}$, can be alternatively assumed equal to the one of the third branch of the multilinear response, $E_{c3}$, (close to the plain fabric one, $E_f$), or equal to the secant slope, $E_{sec}$, obtained connecting the stress point at SLS with the zero (see Fig. 7a). The SLS stresses on the FRCM were identified by means of the analytical simulation previously presented. As visible from Fig. 7b, the constitutive law which conduced to a better representation of the global response of the STR section behaviour in bending was the one obtained by using an $l_{cs,FRCM} = 50$ mm.

In presence of different reinforcements, FRCM and steel bar, it was necessary to compute an equivalent diameter, $\phi_{eq}$, assuming a fictitious equivalent circular section for the composite. This $\phi_{eq}$ can be computed as:

$$\phi_{eq} = \frac{n_s\phi_s^2 + n_f\phi_f^2}{n_s\phi_s + n_f\phi_f} \tag{4}$$

in which diameter and number of reinforcing elements are used ($n_s$ and $\phi_s$ for steel bar, $n_f$ and $\phi_f$ for the FRCM fabric).

Another important value to be set in the Eq. (2) is the clear concrete cover of the reinforcement, *c*. The first option was to consider the value of 35 mm, which corresponded to the sum of the composite layer thickness, 15 mm, and the clear cover already presence in the REF elements, 20 mm. In alternative, it was decided to set the clear cover referred to the position of the equivalent bar previously defined (depending on the different assumptions *c* was close to around 30 mm).

Varying these two parameters (stiffness and clear cover value), the results shown in Fig. 8 were obtained. The better agreement with the experimental crack spacing was obtained taking the secant stiffness of the composite at the set SLS limit, $E_{sec}$, and the clear cover related to the position of the equivalent reinforcement. As visible, the chosen stiffness has the main role on the estimation of the crack distance.

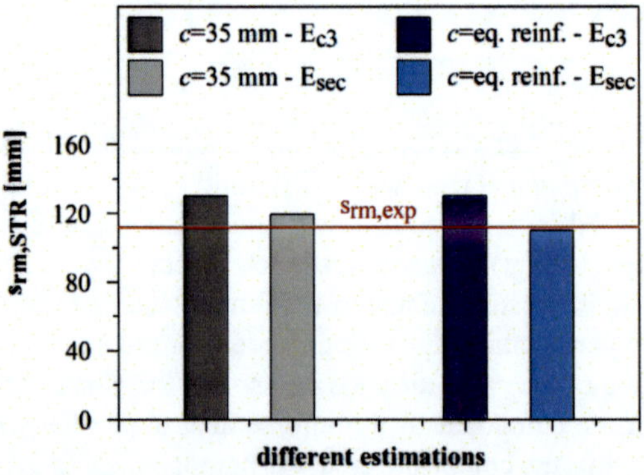


**Fig. 8.** Comparison between the estimated and the experimental $s_{rm}$ in case of the STR beam.

## 5 Conclusions

Based on the experimental results reported in this article, it is possible to draw some conclusions regarding the effect of applying FRCM composites for the strengthening and retrofitting of reinforced concrete slab beams subjected to bending actions. In particular, the following were observed: i) a modest increase in load-bearing capacity, independently to the presence of an initial pre-cracked state; ii) a reduction of both crack opening and distance, leading to a probable increase of durability; and iii) a decrease of the overall ductility. The latter aspect must be considered since, depending on the requirements to be guaranteed for the structural element, it could be a limitation of this type of reinforcement intervention.

The choice of a simplified trilinear constitutive law to be used in the classical plane-section approach allowed to obtain a good approximation of the experimental response, provided that the set characteristic length was carefully chosen. In addition, the use of this constitutive law, in which it was possible to determinate a secant stiffness at the serviceability limit state, was found to be critical in estimating the crack distance in the FRCM-reinforced beam. In fact, with the introduction of the equivalent area of the reinforcement (steel and FRCM composite) in the formulation for the calculation of the crack distance in EC2, typical for reinforced concrete structures, an adequate evaluation of the experimental results was found.

Finally, the estimation of the load-bearing capacity according to the CNR Guidelines lead to an on safe value with respect to the experimental behaviour.

**Acknowledgements.** The authors would like to acknowledge Gavazzi Tessuti Tecnici Spa and MBS Italia Spa. The research was financially supported by ReLUIS WP14 – 2019/2021.

## References

CEB-FIP: fib Model Code for Concrete Structures 2010 (2013). https://doi.org/10.1002/9783433604090

CNR: CNR-DT 215/2018 Istruzioni per La Progettazione, l'Esecuzione Ed Il Controllo Di Interventi Di Consolidamento Statico Mediante l'utilizzo Di Compositi Fibrorinforzati a Matrice Inorganica. CNR: Consiglio Nazionale delle Ricerche (2018)

Consiglio Superiore dei LL: Linea Guida per La Identificazione, La Qualificazione Ed Il Controllo Di Accettazione Di Compositi Fibrorinforzati a Matrice Inorganica (FRCM) Da Utilizzarsi per Il Consolidamento Strutturale Di Costruzioni Esistenti (2018)

Cuypers, H., Wastiels, J.: Stochastic matrix-cracking model for textile reinforced cementitious composites under tensile loading. Mater. Struct. **39**(8), 777–786 (2006). https://doi.org/10.1617/s11527-005-9053-0

Rampini, M.C., Zani, G., Colombo, M., di Prisco, M.: Mechanical behaviour of TRC composites: experimental and analytical approaches. Appl. Sci. **9**(7), 1492 (2019). https://doi.org/10.3390/app9071492

Rampini, M.C., Zani, G., Colombo, M., di Prisco, M.: The role of concrete substrate roughness on externally bonded fabric-reinforced cementitious matrix (FRCM) layers. In: First Fib Italy YMG Symposium on Concrete and Concrete Structures, pp. 56–63 (2020)

Si Larbi, A., Contamine, R., Hamelin, P.: TRC and hybrid solutions for repairing and/or strengthening reinforced concrete beams. Eng. Struct. **45**, 12–20 (2012)

Standard, British: Eurocode 2: Design of Concrete Structures. Part 1–1: General rules and rules for buildings (2004)

Zomorodian, M., Yang, G., Belarbi, A., Ayoub, A.: Cracking behavior and crack width predictions of FRP strengthened RC members under tension. Eng. Struct. **125**, 313–324 (2016)

# Indirect Identification of the Bond-Slip Model at SRP-Concrete Interface

Francesco Ascione[1], Marco Lamberti[2], Annalisa Napoli[1](✉), and Roberto Realfonzo[1]

[1] Department of Civil Engineering, University of Salerno, Fisciano, SA, Italy
annapoli@unisa.it

[2] ENEA, Italian National Agency for New Technologies, Energy and Sustainable Economic Development, Brasimone Research Centre, Camugnano, BO, Italy

**Abstract.** Steel Reinforced Polymer (SRP) systems represent an appealing solution for the external strengthening and repairing of existing reinforced concrete structures. However, similarly to what known for composite materials employing other fibers, such as carbon or glass ones, to assure an adequate bond between the reinforcement and the concrete substrate is fundamental for the success of many strengthening applications. In this paper, an analytical procedure recently published by the authors is applied for an indirect identification of the concrete-SRP interface law by varying some parameters, such as the compressive strength and the surface finish of the concrete. To this purpose, results from a wide experimental campaign performed by the authors have been considered. The bond-slip laws found case by case, implemented in the analytical procedure, allow for accurately fitting the experimental load-slip curves obtained from the performed direct shear tests.

**Keywords:** steel reinforced polymer (SRP) · bond-slip model · analytical procedure

## 1 Introduction and Background

In the civil engineering field, the growing interest in the development of cost-effective solutions for the external strengthening and repairing of reinforced concrete (RC) and masonry structures has recently led to the introduction of a very promising class of composites termed Steel Reinforced Systems (SRPs). In comparison to the more traditional carbon and glass Fiber Reinforced Polymers (FRPs), these systems make use of High Tensile Strength Steel (HTSS) micro wires twisted into small diameter cords or strands that are uniaxially embedded into an epoxy matrix.

By focusing on the SRP systems applied to concrete substrates, subject of this paper, it is highlighted that, even though there has been an even more increasing number of experimental investigations since 2004 (Wobbe et al. 2004, Prota et al. 2004), research is still limited if compared to that available for the more traditional FRP composites.

With the aim to acquire comprehensive and reliable data to use for the development of specific design guidelines, De Santis et al. (2016) published a state-of-the-art about


M. A. Aiello and A. Bilotta (Eds.): ICC 2022, LNCE 435, pp. 160–173, 2024.
https://doi.org/10.1007/978-3-031-43102-9_14

the experimental research performed on SRPs, in which they highlighted the need for a better understanding of the SRP-concrete interfacial behaviour. Indeed, it is known that controlling the parameters that govern such behavior is the key to the prevention of unexpected separation of the external reinforcement from the substrate, which leads to a premature failure of the strengthening system.

With the aim to fill some of the foregoing knowledge gaps, a wide experimental investigation was completed at the University of Salerno (Italy), consisting of 130 direct single-lap shear (DSLS) tests performed on SRP-concrete joints with different concrete surface finishes, i.e.: sandblasted (SB), bush hammered (BH) and grinded (GR). The results of these tests, widely documented in (Ascione et al. 2019) and briefly summarized herein, were used to develop a simplified analytical procedure simulating the DSLS tests in which closed-form solutions are derived to predict the entire debonding propagation process. This procedure, widely described in (Ascione et al. 2021) and summarized herein, requires the implementation of a proper local shear stress-slip ($\tau$-s) law from which specific expressions for the interfacial shear stress distribution, the axial stress profile and the concrete-SRP relative displacement (slip) function can be derived; analytical estimates of the SRP effective bonded length can be obtained as well as relationships for calculating the maximum axial stress (or peak force) at SRP laminate debonding.

In order to apply and validate the proposed analytical procedure, proper local bond-slip laws were directly calibrated in (Ascione et al. 2019, Ascione et al. 2020) by using the experimental data of DSLS tests with reference to the three concrete surface finishes (GR, BH and SB). Specifically, the sets of parameters defining such laws were found by applying an error minimization technique between the theoretical and the experimental shear stress ($\tau$) at given slip (s) values. The couples of experimental ($\tau$, s) values were evaluated by using the strain readings provided by the gauges bonded to the SRP strip and by applying the procedure described in (Ascione et al. 2019).

However, another interesting application of the proposed analytical modelling consists of indirectly identifying the relevant parameters of the local bond-slip interface laws through the so-called "back-analysis", i.e., by finding the interface law that yields the numerical load-slip global response fitting the corresponding experimental curve best.

The procedure for the indirect identification of the bond-slip model at SRP-concrete interface, similar to that used by Pecce et al. (2001) for the case of the bond between FRP reinforcing rebars and concrete, is deeply described herein, and the obtained results are compared with those obtained by applying the analytical procedure per "direct way", i.e., by employing the ($\tau$-s) law proposed in Ascione et al. (2020). In particular, the results are here discussed in terms of both ($\tau$-s) laws and load-slip (or axial stress-slip) curves with reference to DSLS tests performed on SRP systems bonded to concrete joints with SB finish.

## 2 Experimental Study: Outline

Figure 1 shows a schematic of the test specimen geometry.

The concrete blocks had 200 × 150 $mm^2$ cross-section and 400 mm length; they were manufactured from different concrete batches in order to have average values of the cylindrical compressive strength, $f_{cm}$, at 28 days spanning between about 13 and

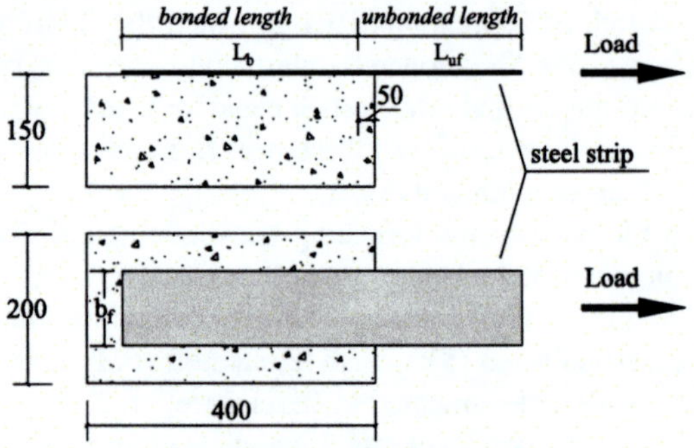


**Fig. 1.** Specimen geometry.

**Table 1.** Geometry and main mechanical properties of dry steel fabric (Kerakoll 2020).

| Parameter | Tape density | | | |
|---|---|---|---|---|
| | Low (L) | Low-Medium (LM) | Medium (M) | High (H) |
| $\rho$ [cords/mm] | 0.157 | 0.314 | 0.472 | 0.709 |
| $c_s$ [mm] | 6.35 | 3.18 | 2.12 | 1.41 |
| $\gamma$ [g/m$^2$] | 670 | 1200 | 2000 | 3300 |
| $t_f$ [mm] | 0.084 | 0.164 | 0.254 | 0.381 |
| $f_{f,u}$ [MPa] | > 3000* | | | |
| $E_f$ [GPa] | > 190** | | | |
| $\varepsilon_{f,u}$ [%] | > 2.00* | | | |

*Characteristic value ** Mean value

45 MPa. Among the 130 tests, 29, 48 and 53 were carried out on concrete blocks with sandblasted, grinded and bush hammered surface finish, respectively.

Within each typology of concrete surface finish, test specimens were divided in two main sub-groups termed "normal strength concrete (NSC)" and "high strength concrete (HSC)". In particular, the NSC group included concrete specimens with $f_{cm}$ values ranging between 13 and 26 MPa, whereas the HSC one collected specimens with $f_{cm}$ values ranging between 40 and 45 MPa. Only the concrete specimens of the SB group belonged to the same batch, characterized by $f_{cm} = 13$ MPa. For each $f_{cm}$ value, specimens were grouped in smaller test sets depending on the layout of the SRP strip, in particular:

- the steel tape density identified as: low (L), low-medium (LM), medium (M) and High (H);
- the length of the bonded strip, $l_b$. (ranging between 100 and 350 mm);
- the width of the bonded strip $b_f$ ($b_f = 100$ mm, except for some SB specimens for which $b_f = 20$ and 40 mm were also investigated).

Table 1 collects the main information about geometry and mechanical properties of the dry steel fabric investigated in the experimental program, i.e.: the mentioned tape density (ρ), the cord spacing ($c_s$), the mass density (γ), the equivalent design thickness ($t_f = A_{cord} \cdot \rho$); the tensile strength ($f_{f,u}$), the Young's modulus ($E_f$), and the ultimate strain ($\varepsilon_{f,u}$).

The main mechanical properties of the thixotropic epoxy mineral adhesive used for steel cords impregnation can be found in technical sheets provided by the manufacturer (Kerakoll 2020).

The experimental results – omitted here for the sake of brevity – provided the following main conclusions:

1) like carbon or glass FRP-concrete joints, the dominant failure mode was cohesive debonding within the concrete substrate whatever the concrete surface finish and the steel fiber density. Conversely, depending on the surface finish, the thickness of the separated concrete layer varied between 0.5 and 1.0 mm for the GR finish while ranged between 2.0 and 3.0 mm for the BH and SB ones;
2) the load-slip response exhibited a brittle or seemingly ductile behavior based on the used bonded lengths $l_b$ and tape densities ρ, irrespective of the concrete surface finish;
3) the L density strips assure a better exploitation of the strengthening system than the stiffer strips. Indeed, since failure is basically governed by the concrete strength, an increase of the equivalent thickness of the steel layer does not yield to a proportionally higher failure load;
4) the concrete grinding treatment – which produces the lowest surface roughness among those considered – provided the highest scatter of the debonding loads within each test set. On average, the maximum loads achieved during the tests were lower than those obtained for SB and BH surfaces, whatever the concrete strength and the steel tape density.

## 3 SRP-Concrete Debonding: A Simple Mechanical Model

In a recently published paper, Ascione et al. (2021) developed a simple mechanical model describing the problem of the SRP-concrete debonding, which is governed by the following second order differential equation in the interfacial slip *s(x)*:

$$E_f t_f \frac{d^2 s(x)}{dx^2} = \tau[s(x)] \tag{1}$$

By integrating Eq. 1 and adopting an appropriate constitutive law $\tau(s)$, it is possible to evaluate the unknown function $s(x)$ which describes the trend of the slip, s, along the steel tape.

A closed form solution of Eq. 1 could be obtained only if:

- a constitutive law $\tau[s(x)]$ suitable to be integrated in closed form is assumed;
- the SRP system, which is subjected to a traction force F, is perfectly bonded to the concrete substrate (i.e., when the developed length $l_d$ is shorter than, or equal to, the bonded length $l_b$).

If the last condition is verified, the abscissa $x$ should be measured from a point to a distance equal to $l_d$ from the system loaded end (see Fig. 2); of course, from that point up to the system free end the slip $s$ will be equal to 0.

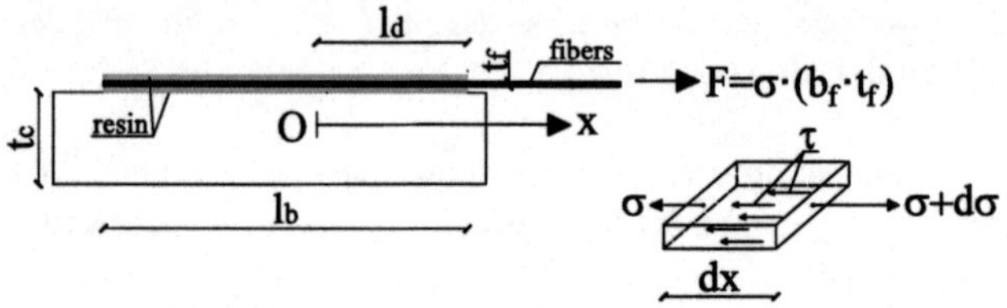


**Fig. 2.** Scheme of the DSLS test used in the modelling.

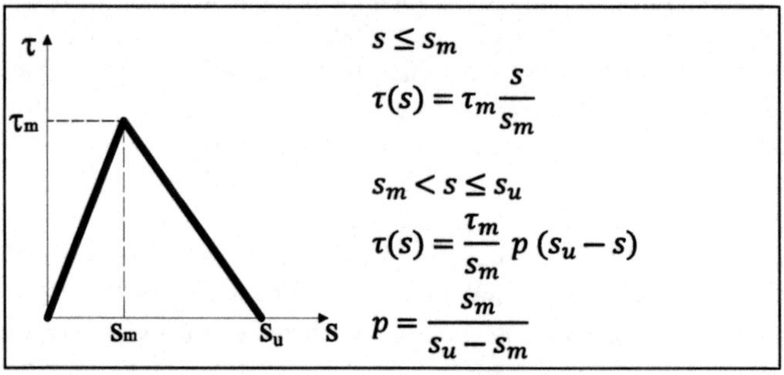


**Fig. 3.** Bilinear τ-s law used in the analysis.

Finally, once $s(x)$ has been determined, the normal and shear stress distribution along the fiber $-\sigma(x)$ and $\tau(x)$ - can be derived as follows:

$$\sigma(x) = E_f \varepsilon_f(x) \cong E_f \frac{ds(x)}{dx}$$
$$\tau(x) = E_f t_f \frac{d^2 s(x)}{dx^2} \cong t_f \frac{d\sigma}{dx} \quad (2)$$

by neglecting the concrete deformation in tension.

### 3.1 Analytical Solution of the Problem: Implementation of a Bilinear Shear Stress –Slip Law

In Ascione et al. (2021), the authors provided the solution of the problem by considering three constitutive bond-slip, $\tau(s)$, laws which were defined by a different number of characterizing parameters. In this paper, the analytical solution found only in the case of the bilinear 3-parameter $\tau(s)$ model ($\tau_m$, $s_m$, $s_u$), illustrated in Fig. 3, is taken under consideration. Therefore, by implementing the describing equation for $\tau(s)$ in Eq. 1, specified for the two branches, i.e., for $s \leq s_m$ and $s_m < s \leq s_u$, integrating such equation and assuming two boundary conditions for each branch, the solutions for the function $s(x)$ and, consequently, for $\sigma(x)$ and $\tau(x)$ (from Eq. 2) can be obtained, as summarized in Fig. 4.

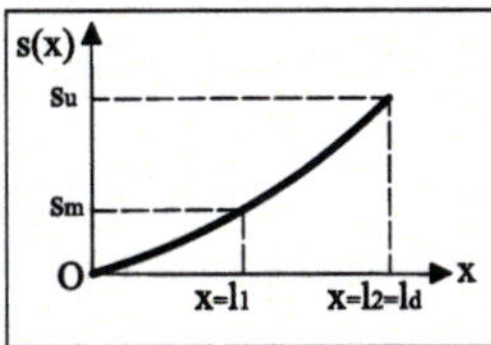


for $x \le l_1$ (i.e., $s \le s_m$)

$$\frac{s(x)}{s_m} = \frac{\sinh\left(\sqrt{\beta_2}\,x\right)}{\sinh\left(\sqrt{\beta_2} l_1\right)}$$

for $l_1 < x \le l_d$ (i.e., $s_m < s \le s_u$)

$$\frac{s(x)}{s_m} = \frac{1}{p}\{1 + p - \cos[\omega(x - l_1)] + \sqrt{p}\sin[\omega(x - l_1)]\}$$

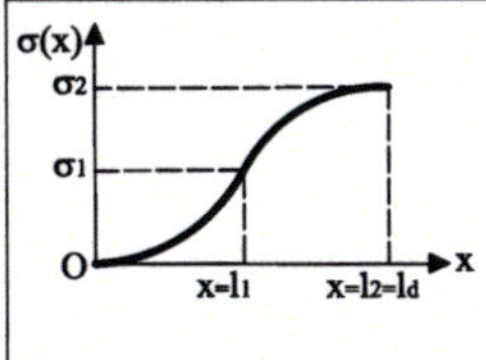


for $x \le l_1$ (i.e., $s \le s_m$)

$$\frac{\sigma(x)}{\sigma_1} = \frac{\cosh\left(\sqrt{\beta_2}\,x\right)}{\sinh\left(\sqrt{\beta_2} l_1\right)} = \coth\left(\sqrt{\beta_2}\,x\right)\frac{s(x)}{s_m};\ \text{when } x = l_1 \rightarrow \frac{\sigma(x)}{\sigma_1} = \chi \coth\left(\sqrt{\beta_2} l_1\right)$$

for $l_1 < x \le l_d$ (i.e., $s_m < s \le s_u$)

$$\frac{\sigma(x)}{\sigma_1} = \sqrt{\frac{1}{p}}\{\sin[\omega(x - l_1)] + \sqrt{p}\cos[\omega(x - l_1)]\}$$

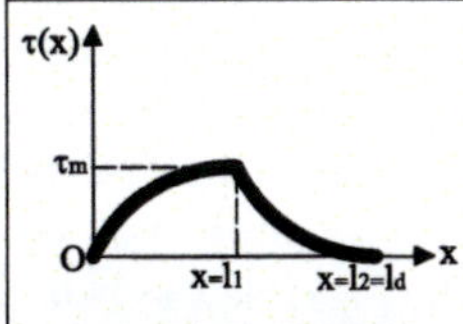


for $x \le l_1$ (i.e., $s \le s_m$)

$$\frac{\tau(x)}{\tau_m} = \frac{\sinh\left(\sqrt{\beta_2}\,x\right)}{\sinh\left(\sqrt{\beta_2} l_1\right)} = \frac{s(x)}{s_m}$$

for $l_1 < x \le l_d$ (i.e., $s_m < s \le s_u$)

$$\frac{\tau(x)}{\tau_m} = \cos[\omega(x - l_1)] + \sqrt{p}\sin[\omega(x - l_1)]$$

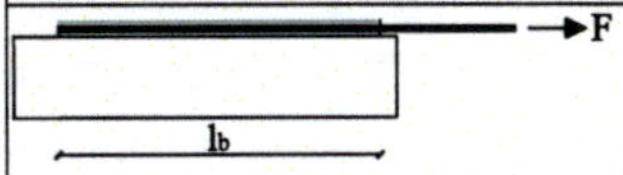


$$\sigma_1 = \sqrt{\frac{E_f}{t_f}\tau_m s_m} \qquad \sigma_2 = \sqrt{\frac{E_f}{t_f}\tau_m s_u}$$

$$l_1 = \frac{t_f}{\tau_m}\sigma_1 \,\text{arccoth}\left(\frac{1}{\chi}\right) \qquad l_2 = l_d = \frac{t_f}{\tau_m}\sqrt{\frac{s_m}{s_u}}\sigma_2\left[\text{arccoth}\left(\frac{1}{\chi}\right) + \frac{1}{\sqrt{p}}\text{arccot}(\sqrt{p})\right]$$

where:

$$\beta_2 = \frac{1}{E_f t_f}\frac{\tau_m}{s_m} \qquad \omega = \sqrt{\frac{\tau_m}{E_f t_f}\frac{1}{s_u - s_m}}$$

$95\% \le \chi < 100\%$; $\chi = 97\%$ is assumed (Ascione et al. 2021)

**Fig. 4.** Solution of the differential Equation in terms of slip s(x), axial stress σ(x) and shear stress τ(x): summary.

# 4 Use of the Analytical Solution

The analytical solution presented in the previous section (see Fig. 4) can be effectively used once the parameters ($\tau_m$, $s_m$, $s_u$) of the bilinear $\tau(s)$ law are known. To this purpose, Ascione et al. (2019, 2020) selected some popular shear stress-slip constitutive models proposed for the FRPs and extended them to the case of the SRPs. Particularly, the authors recalibrated the key parameters of these models by using the experimental data of DSLS tests performed on concrete substrates with GR, BH and SB surface.

The set of unknown parameters of the considered interface laws were determined by applying the Mean Squared Error (MSE) minimization technique between the theoretical and the experimental shear stress, both normalized with respect to the square root of $f_{cm}$. In particular, the experimental values were evaluated by using the strain readings provided by the gauges bonded to the SRP strip and by applying the procedure described in Ascione et al. (2019).

By focusing on the bilinear model, Table 2 provides, for the three concrete surface finishes (SB, BH, GR), the set of parameters found through the mentioned MSE minimization process. Therefore, once the value of these parameter is known, the analytical procedure can be directly applied to yield the numerical solutions to the equations summarized in Fig. 4, some of which plotted in Ascione et al. (2021).

**Table 2.** Calibrated values of the set of parameters featuring the bilinear τ-s law.

| Surface finish | Set of parameters $\{\overline{\tau}_m; s_m; s_u\}$ | | |
|---|---|---|---|
| | $\overline{\tau}_m = {\tau_m}/{\sqrt{f_{cm}}}$ | $s_m$ [mm] | $s_u$ [mm] |
| GR finish | 0.475 | 0.024 | 0.444 |
| BH finish | 0.616 | 0.035 | 0.494 |
| SB finish | 0.669 | 0.032 | 0.566 |

Note: $f_{cm}$ in MPa

Conversely, another interesting application of the proposed modeling consists of indirectly identifying the relevant parameters of the local bond-slip interface laws through the so-called "back-analysis" which makes use of the experimental load-slip or axial stress-slip curves available from the experimental investigation. The application of this procedure requires the analytical derivation of the axial stress – slip [$\sigma(s)$] law as better detailed in the next section.

### 4.1 Identification of the Local Shear Stress – Slip Law: Back Analysis

In the back analysis, the application of the procedure requires the availability of a reliable experimental database of DSLS tests and, specifically, of a significant number of axial load-loaded end slip curves (*F*-*s*) or, equivalently, $\sigma$-$s$ curves in which the axial load is divided by the cross-section area of the steel tape ($A_f = t_f \cdot b_f$). Then, the identification process of the parameters $\overline{\tau}_m$, $s_m$, $s_u$ can be obtained by minimizing the scatter between the experimental and theoretical results by varying the parameters upon which the constitutive bond-slip law is dependent. In particular, by starting from a tentative triad of values for the parameters $\{\overline{\tau}_m; s_m; s_u\}^I$, for each experimental value of the slip $s_i$ measured at the loaded end of the strip at the i-th step of loading – provided that $s_i$ is lower than the tentative value $s_u^I$ – the corresponding theoretical value of the shear stress $\tau^{th}(s = s_i)$ can be calculated based on the expressions reported in Fig. 3, specified for the cases of $s_i \leq s_m$ and $s_m < s_i \leq s_u$.

Hence, the function $\sigma^{th}(s)$ can be derived based on the following relationship, which represents an energy-based formulation (Cosenza et al. 1997, Pecce et al. 2001):

$$A_\tau(s) = t_f A_\sigma(\sigma) \tag{3}$$

in which $A_\tau$ and $A_\sigma$ represent, respectively, the area underneath the τ-s curve and the area underneath the σ-ε curve (see Fig. 5).

By assuming for the steel fabric a linear elastic σ-ε constitutive law, $A_\sigma(\sigma)$ is given by:

$$A_\sigma(s) = \frac{1}{2}\frac{\sigma^2}{E_f} \tag{4a}$$

whereas $A_\tau(s)$ specifies in the following expressions, suitable for $s \leq s_m$ and $s_m < s \leq s_u$:

$$s \leq s_m$$

$$A_\tau(s) = \frac{1}{2}\frac{\tau_m}{s_m}s^2 \quad (4b)$$

$$s_m < s \leq s_u$$

$$A_\tau(s) = \frac{\tau_m}{2}\left\{s_m + (s - s_m)\left[1 + \frac{p}{s_m}(s_u - s)\right]\right\} \quad (4c)$$

By replacing Eqs. 4a–4c in Eq. 3, it is possible to define the function $\sigma(s)$:

$$s \leq s_m$$

$$\sigma^{th}(s) = \sqrt{\frac{E_f}{t_f}\frac{\tau_m}{s_m}s} \quad (5a)$$

$$s_m < s \leq s_u$$

$$\sigma^{th}(s) = \sqrt{\frac{E_f}{t_f}\tau_m\left\{s_m + (s - s_m)\left[1 + \frac{p}{s_m}(s_u - s)\right]\right\}} \quad (5b)$$

Therefore, by plugging $s = s_i$ in Eq. 5a, 5b, the current value of the error between the theoretical $\sigma(s_i)^{th}$ and the experimental $\sigma(s_i)^{exp}$ value of the axial stress can be estimated as:

$$E_i = \left[\sigma(s_i)^{th} - \sigma(s_i)^{exp}\right]^2 \quad (6)$$

Finally, the optimal set of values for the parameters $\{\overline{\tau}_m; s_m; s_u\}$ is obtained by minimizing the Root Mean Square Error (RMSE), calculated as:

$$RMSE = min\sqrt{\frac{\sum_{n=1,N_{test}}\sum_{i=1,k}E_i}{N_{meas}}} \quad (7)$$

where k represents the number of steps of loading considered for each test, $N_{test}$ is the number of tests considered together in the error minimization process and $N_{meas}$ is the total number of considered measures.

In the calibration process, the first tentative set of values for the parameters $\{\overline{\tau}_m; s_m; s_u\}^I$ always coincided with the triad of values found in Ascione et al. (2020) thrgh the direct calibration process and specified for the concrete surface type (see Table 2).

However, in the use of the experimental data related to a given test, it is important to check that - for each tentative set of parameters - the development length $l_d$ needed to achieve the theoretical maximum axial stress $\sigma_2$ be lower than the bonded length $l_b$ (the analytical expression for the estimate of $l_d$ and $\sigma_2$ are provided in the summary of Fig. 4). If this is the case, the $\tau(s)$ law in Fig. 3 is completely developed up to $s = s_u$

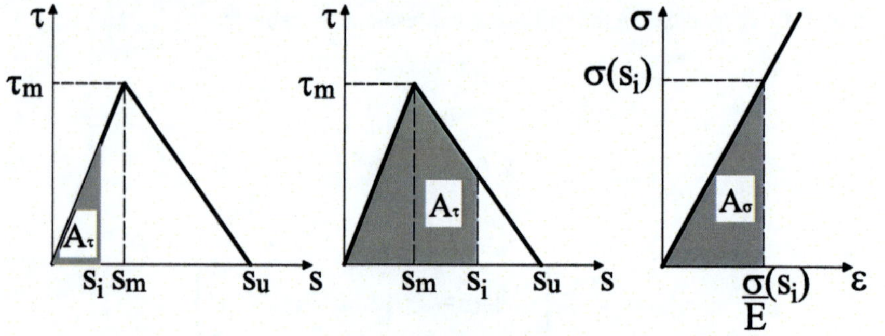


**Fig. 5.** Areas under τ-s (bilinear law) and σ-s laws.

and, consequently, all the experimental slip values up to $s_u$ have to be included in the minimization process, while greater experimental values are not taken into account in the RMSE estimate. Conversely, if $l_d \geq l_b$, the $\tau(s)$ law is developed up to $s(x) = s(x = l_b)$, meaning that the theoretical maximum axial stress is lower than $\sigma_2$ and estimated by plugging $x = l_b$ in one of the two equations $\sigma(x)$ reported in Fig. 4 (suitable for $x = l_b \leq l_1$ and $x = l_b > l_1$, respectively). Figure 6 better clarifies the two mentioned analysis cases.

In the theoretical model, when $l_b = l_d$, the axial stress-slip curve just stops at $s = s_u$, i.e., at the achievement of the maximum stress at debonding initiation $\sigma_2$. Conversely, when $l_b > l_d$, the maximum stress $\sigma_2$ is kept constant under values of the slip higher than $s_u$, since the interface is capable of redistributing the released stresses upon initiation of debonding; the more or less long plateau depends on how much $l_b$ is greater than $l_d$ to allow for the gradual propagation of debonding toward the free end of the SRP strip. (see *Case 1* in Fig. 6a).

Of course, the brittle response is always expected when $l_b < l_d$ and debonding initiation occurs for stress values lower than the maximum achievable ones (see *Case 2* in Figs. 6b, c).

### 4.2 Application of the Back Analysis and Results

The procedure described in Sect. 4.1 was applied to the whole experimental database, in which the three different concrete surface finishes (SB, BH, GR) were treated separately. Furthermore, within each surface finish, tests were considered together or grouped in smaller sets based on concrete strength value, $f_{cm}$; as a result, several local bond stress-slip laws were identified through the back analysis, suitable either for any concrete strength value or for a specific $f_{cm}$ value.

For the sake of brevity, this paper presents only some results of the back analysis performed by considering the experimental DSLS tests with SB finish only; results related to the other surfaces can be found in Ascione et al. (2022). In particular, the DSLS tests under consideration are 21 over 29, in which the concrete blocks were all characterized by $f_{cm} = 13$ MPa and by the same sandblasting typology (differences in the type of sandblasting treatment are described in Ascione et al. 2020), while the width of the SRP strip was always equal to $b_f = 100$ mm.

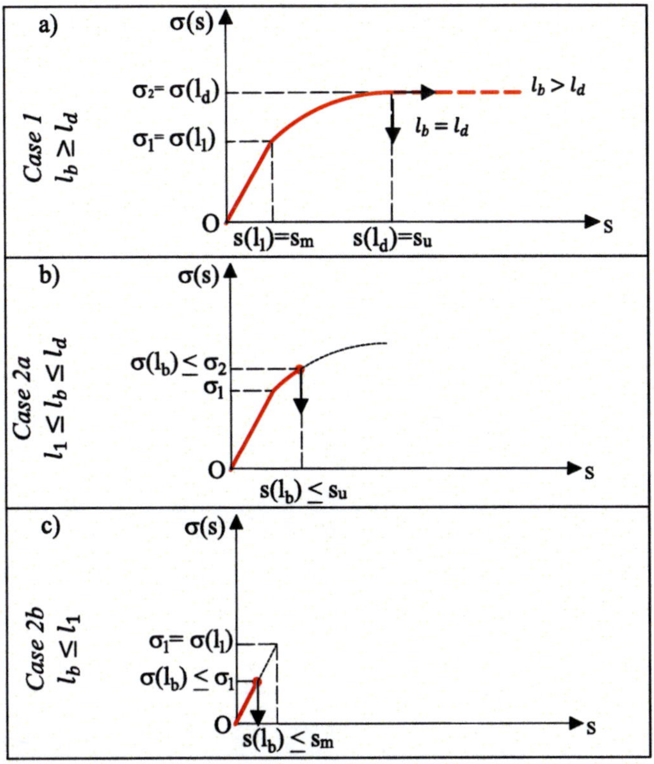


**Fig. 6.** Theoretical axial stress-slip curves: Case 1 (a); Case 2a (b); Case 2b (c).

Table 3 reports the main details of these tests, i.e.: the tape density and the bonded length; N identifies the number of identical tests with same tape density and $l_b$. In the minimization process, these tests were considered together ($N_{test} = 21$ in Eq. 7) thus neglecting the influence of the tape density in the identification of the interface ($\tau$-s) law.

In Fig. 7a, the shear stress-slip law calibrated via back analysis (let's say indirect method) is compared with that previously obtained, for the same surface treatment, through the direct calibration process (the values of the key parameters are those in Table 2).

As observed, the two interface laws are significantly different, with the indirect method yielding a peak stress which is about the 70% of the value found per direct way (1.69 vs 2.41 MPa). Conversely, the interfacial slip at peak $s_m$ is equal to 0.228 mm (against 0.032, Table 2), resulting in a first branch significantly less stiffer than the response obtained through the direct method; in terms of ultimate slip $s_u$, instead, the two methods do not differ so much ($s_u = 0.683$ mm vs 0.576 mm, Table 2). Finally, in terms of fracture energy parameter ($\Gamma_F$), defined by the area underneath the ($\tau$-s) law, the value provided by the direct method is 17% higher than $\Gamma_F$ estimated through the back analysis, that implying a higher maximum stress calculated at debonding.

In Fig. 7b, instead, some experimental-numerical comparisons in terms of force-slip (F-s) curves are reported; the different colours used in the plots refer to the variation of the tape density from low (L, green curve), low-medium (LM, red curve), medium (M, blue curve) to high (H, black curve).

**Table 3.** Details of the DSLS tests with SB finish used in the back analysis.

| Tape density | $l_b$ [mm] | N |
|---|---|---|
| L | 150 | 2 |
| | 250 | 3 |
| LM | 150 | 3 |
| | 250 | 4 |
| M | 150 | 3 |
| | 250 | 3 |
| H | 300 | 3 |

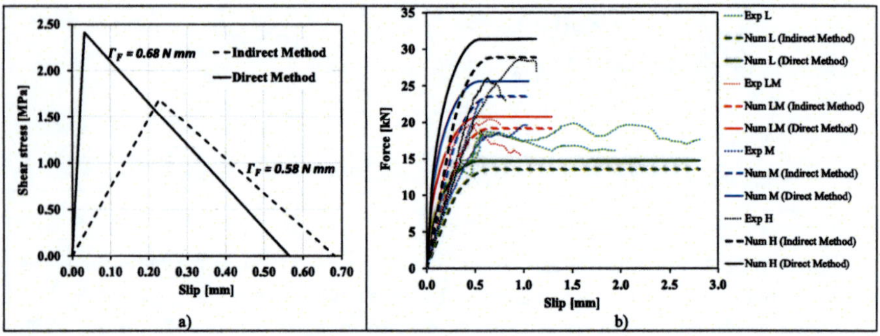


**Fig. 7.** Shear stress-slip (τ-s) laws: comparison between indirect and direct method (a); axial force-slip curves: experimental-numerical comparisons (b).

For each density, the couple of representative experimental curves is compared with two numerical simulations, one making use of the (τ-s) law calibrated through the direct method (solid line curve) and the other one employing the interface law obtained from the back analyses (dotted line curve).

Both numerical simulations are obtained by imposing increasing values of the slip up to the largest ultimate value achieved in each set experimental tests and calculating the corresponding stress value through Eq. 5a, 5b; then, the force F is obtained by multiplying the stress value by the cross-section area of the dry steel fabric ($= b_f \cdot t_f$). Of course, for slip values greater than the values of $s_u$ calibrated in the found (τ-s) laws, the corresponding axial stress is always equal to the maximum one $\sigma_2$, provided that the condition $l_b > l_d$ is satisfied.

The comparisons in Fig. 7b show a general good agreement between the experimental and the numerical curves with a better matching when using the (τ-s) law calibrated via indirect method, mainly in terms of stiffness in the pre-peak phase. As expected, the (τ-s) law obtained by direct method yields higher values of the debonding force (the fracture energy $\Gamma_F$ is higher). Whatever the interface law, the numerical simulations obtained for L density steel tape always provides conservative predictions meaning that further investigation is needed on the role of the greater or lesser cord spacing on the local (τ-s) response.

Figures 8a, b, d, e show the experimental-numerical comparisons in terms of both (F-s) and (σ-s) curves with reference to sets of tests related to LM and M density steel tapes; the main data on the performed simulations are provided in Figs. 8c, f.

The comparisons allowed, again, for verifying the accuracy of the simulations in terms of both stiffness and axial stress mainly when using the (τ-s) law calibrated through the back analysis. Particularly, a better agreement is observed in the modelling the set of tests involving LM density strips.

In order to better investigate the accuracy of the numerical simulations, the bar chart in Fig. 9 reports the per cent mean errors between the experimental axial stress values and the corresponding numerical ones estimated on both the single dataset of homogenous DSLS tests and all the 21 tests together. In particular, the datasets under consideration are four and are identified based on the density of the steel strip, i.e.: 5 tests for L density, 7 tests for LM density, 6 and 3 tests for M and H density, respectively.

In Fig. 9, the model errors are estimated according to the Mean Absolute Percentage Error (MAPE) which is expressed by:

$$MAPE = \frac{\sum_{n=1,N_{test}} \sum_{i=1,k} \left| \frac{\sigma(s_i)^{exp} - \sigma(s_i)^{th}}{\sigma(s_i)^{exp}} \right| \cdot 100}{N_{meas}} \tag{8}$$

where the symbols are the same used in Eq. 7.

For each experimental test, the error was calculated up to a slip value equal to $s_u$ of the considered (τ-s) laws only if the bonded length $l_b$ the DSLS test was at least equal to the development length $l_d$ theoretically estimated through the analytical procedure (see Fig. 4). Conversely, the error count was stopped to a slip value equal to $s(l_b)$ – to which corresponds $\sigma(l_b)$ – if the bonded length $l_b$ was lower than $l_d$; again, $s(l_b)$ and $\sigma(l_b)$ were estimated according to the relationships reported in Fig. 4.

Figure 9 shows that the use of the new (τ-s) model provides very accurate simulations in the case of LM density and less good for the other densities; however, it generally provides MAPE errors significantly lower than the values obtained by changing the interface law, while comparable errors among the two interface models are observed only in the case of L density.

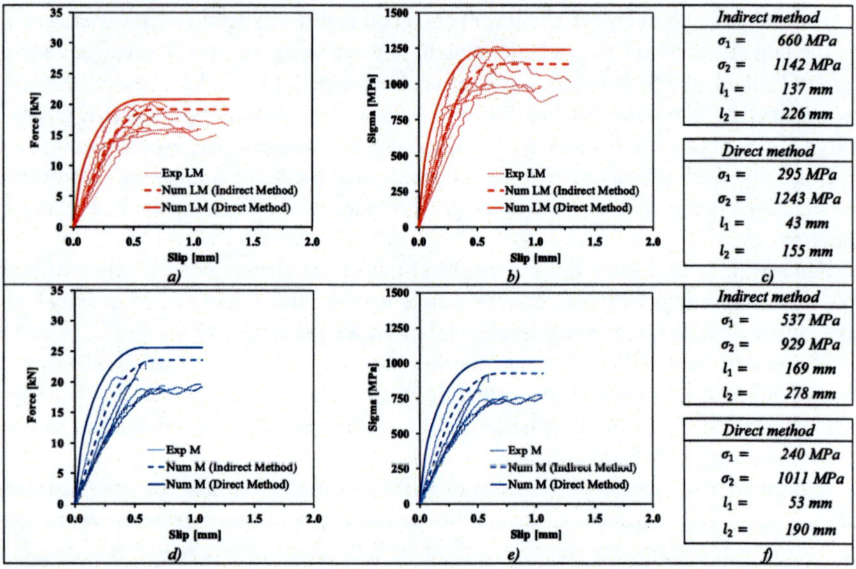


**Fig. 8.** Axial force-slip (a, d) and axial stress-slip (b, e) curves: experimental-numerical comparisons in the case of DSLS tests with LM and M density strip; main data on the performed simulations (c, f).

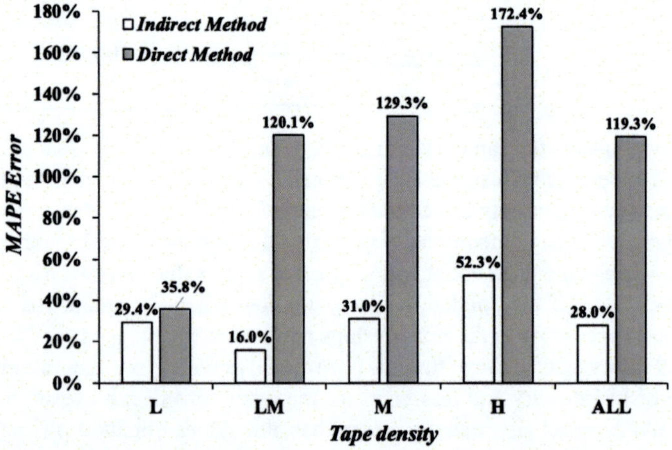


**Fig. 9.** Simulation of experimental curves: comparison between direct and indirect method in terms of model errors.

## 5 Conclusions

An analytical procedure has been presented for an indirect identification of the concrete-SRP interface law. To this purpose, experimental results from direct single-lap shear (DSLS) tests on SRP systems bonded to concrete blocks were taken into account. In particular, by focusing on DSLS tests performed on concrete with sandblasted surface finish, the new calibrated shear stress-slip law was compared to that previously found by the authors through a direct calibration method based on the use of experimental strain measures. For a better comparison, the two stress-slip models were, then, implemented within the proposed analytical procedure for reproducing the global load-slip curves of the experimental DSLS tests. The first results highlighted a good experimental-numerical agreement especially when using the interface model calibrated by back analysis. However, further research is undergoing to better investigate the SRP-concrete interface modeling.

**Acknowledgements.** The financial support by ReLUIS (Network of the Italian University Laboratories for Seismic Engineering - Italian Department of Civil Protection) is gratefully acknowledged (Executive Project 2022-24 - WP14).

## References

Ascione, F., Lamberti, M., Napoli, A., Razaqpur, A.G., Realfonzo, R.: Modeling SRP-concrete interfacial bond behavior and strength. Eng. Struct. **187**, 220–230 (2019)

Ascione, F., Napoli, A., Realfonzo, R.: Experimental and analytical investigation on the bond of SRP systems to concrete. Compos. Struct. **242**, 112090 (2020)

Ascione, F., Napoli, A., Realfonzo, R.: Interface bond between FRP systems and substrate: analytical modeling. Compos. Struct. **257**, 112942 (2021)

Ascione, F., Lamberti, M., Napoli, A., Realfonzo, R.: Bond-slip law for SRP-concrete interface: Back Analysis. In: Proceedings of the 6th edition of the Workshop "The New Boundaries of Structural Concrete", NBSC 2022, Lecce, Italy (2022)

Cosenza, E., Manfredi, G., Realfonzo, R.: Behaviour and Modeling of bond FRP rebars to concrete. J. Compos. Constr. **1**(2), 40–51 (1997)

De Santis, S., de Felice, G., Napoli, A., Realfonzo, R.: Strengthening of structures with steel reinforced polymers: a state-of-the-art review. Compos. B Eng. **104**, 87–110 (2016)

Kerakoll Spa. www.kerakoll.com. Accessed May 2020

Pecce, M., Manfredi, G., Realfonzo, R., Cosenza, E.: Experimental and analytical evaluation of bond properties of GFRP bars. J. Mater. Civ. Eng. **13**(4), 282–290 (2001)

Prota, A., Manfredi, G., Nanni, A., Cosenza, E., Pecce, M.: Flexural strengthening of R/C beams using emerging materials: ultimate behavior. In: Proceedings of the 2nd International Conference on FRP Composites in Civil Engineering, CICE 2004, Adelaide, Australia, pp. 163–170 (2004)

Wobbe, E., et al.: Flexural capacity of R/C beams externally bonded with SRP and SRG. In: Proceedings of Society for the Advancement of Material and Process Engineering Symposium, Long Beach, Ca, USA, pp. 20–27 (2004)

# Limits of Current Design Approaches in the Analysis of Wind Turbine Foundations

Matteo Colombo[1], Paolo Martinelli[1(✉)], Bruno Dal Lago[2], and Marco di Prisco[1]

[1] Department of Civil and Environmental Engineering, Politecnico di Milano, Milan, Italy
paolo.martinelli@polimi.it

[2] Department of Theoretical and Applied Sciences, Università degli Studi dell'Insubria, Varese, Italy

**Abstract.** The design of massive structures such as the shallow foundations of wind turbines is not a trivial task also in light of the fact that this structural typology is not specifically covered in technical Standards. The structural modelling of onshore wind turbine shallow foundations deals with combined complex phenomena, such as material nonlinearities, soil-structure interaction and local stress concentration in both reinforced concrete foundation and soil. The design practice basically adopts two methods: (i) an elastic approach where wind tower foundations are usually designed by modelling the foundation as a slab with variable thickness using shell finite elements and by introducing very simplified assumptions for the soil behaviour (i.e. Winkler model), and (ii) an approach where wind tower foundations are designed by using limit analysis. The paper aims to highlight the limits and the safety margins of the current design practice by comparing the aforementioned methods (i) and (ii) with nonlinear numerical simulations fully accounting for soil-structure interaction based on three-dimensional elements for both concrete foundation and soil, contact surfaces for their interface, and beam elements for reinforcing bars.

**Keywords:** onshore wind turbine design · shallow foundations · reinforced concrete · soil-structure interaction (SSI) · nonlinear FE analyses

## 1 Introduction

The design of wind tower foundation deals with several aspects ranging from the geotechnical consideration, structural design and soil-structure interaction. Wind turbine structures with steel mast are conventionally founded on what may be classified as large shallow footings. Shallow foundations, specific for wind turbines, are typically axisymmetric truncated-cone shaped with large thickness to breadth ratios reinforced both circumferentially and radially in order to withstand the large actions associated to the overturning moment originated from extreme wind conditions. The foundation also relies on its self-weight and soil over-burden to provide sufficient resistance to overturning, and on the soil shear strength and compressibility characteristics for vertical resistance. The design methods for such foundations are also universally accepted and well understood, as well


M. A. Aiello and A. Bilotta (Eds.): ICC 2022, LNCE 435, pp. 174–189, 2024.
https://doi.org/10.1007/978-3-031-43102-9_15

as this foundation concept being applicable to a wide range of subgrade strengths, making this the simplest and most common form of foundation for wind turbines (Morgan and Ntambakwa 2008). Shallow footings for wind turbine structures were pioneered in the United Kingdom, where early wind farm developments took place in the highlands on competent soils and rock (Bonnett 2005). Octagonal and circular footings have proved to be more suitable than rectangular ones because at least four reinforcement layers can be provided in the bottom layers, as opposed to the two orthogonal directions provided by a rectangular footing (Maunu 2008). This aids the moment carrying capacity of the foundation as well as construction efficiency and time. Furthermore, circular footings, approximated by an octagon or hexa-decagon, are most economical for the support of large towers where the direction of overturning moment is not fixed (Bowles 1996).

## 2 Building Description

### 2.1 Soil Bearing Capacity

The concept of bearing capacity was first explored by Pauker in 1850 and later by Bell and Prandtl in the early 1900s by exploring the behaviour of a metal cone passing through a metal sheet (Fellenius 1999). It was Terzaghi who extended this research and coined the first formal bearing capacity relationship in 1943 through the analysis of rigid continuous strip footings. This was an empirical solution based on the notion that the soil would undergo a perfectly plastic failure in shear. Fundamental to Terzaghi's solution was the implementation of the Mohr-Coulomb failure criterion concerning the shear strength of the material. This is one of the most common models within the field of soil mechanics extensively used in engineering practice.

Because the foundation is subjected to eccentric loading, the ultimate bearing capacity can be calculated using the ultimate bearing capacity equation proposed by Vesic (1975) with effective dimensions proposed by Meyerhof (1963). The Vesic's general bearing capacity formula is:

$$q_u = c' N_c s_c d_c i_c b_c g_c + q' N_q s_q d_q i_q b_q g_q + \frac{1}{2}\gamma' B N_\gamma s_\gamma d_\gamma i_\gamma b_\gamma g_\gamma \quad (1)$$

where $N_c$, $N_q$ and $N_\gamma$ are the bearing capacity factors, $s_c$, $s_q$ and $s_\gamma$ are the shape factors, $d_c$, $d_q$ and $d_\gamma$ are the depth factors, $i_c$, $i_q$ and $i_\gamma$ are the load inclination factors, $b_c$, $b_q$ and $b_\gamma$ are the base inclination factors, $g_c$, $g_q$ and $g_\gamma$ are the ground inclination factors, $B$ is the width of foundation and $q'$ and $\gamma'$ are the effective overburden pressure and effective unit weight, respectively. Several approaches can be used to compute the previous coefficients (see for example Meyerhof 1963; Brinch Hansen 1970; Vesic 1975); in the present paper, the Vesic's formulation is used.

### 2.2 Effective Area Approach

The bearing capacity expression (Eq. (1)) is applicable only for cases where the load passes through the centroid of the foundation. Wind tower foundations are subjected to axial load, lateral load and bending moment. Therefore, Eq. (1) must be modified to account for the moment and the eccentricity. In this study, the effective area method proposed by Meyerhof (1963) is used to consider the eccentricity in loading. In this method, the effective dimensions of the foundation are calculated and used in the bearing capacity equation. In order to determine the effective dimensions of a circular raft, the process described in DNV/Risø (2002, 2010) is used. According to this method, the effective area of the circular foundation with radius *R* is represented by an ellipse as shown in Fig. 1.

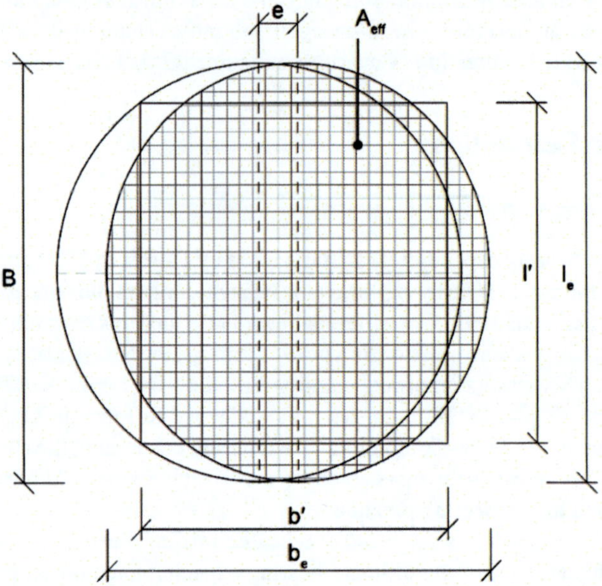


**Fig. 1.** Circular footing with effective area marked (adapted from DNV/Risø 2002)

The area of the ellipse and its axes (major axis, $l_e$ and minor axis, $b_e$) are calculated using the following equation:

$$\begin{cases} A_{eff} = 2\left[R^2\cos^{-1}(e/R) - e\sqrt{R^2 - e^2}\right] \\ b_e = 2(R - e) \\ l_e = 4A_{eff}/\pi b_e \end{cases} \tag{2}$$

where the eccentricity $e$ is given by $e = M/Q$ being $M$ the bending moment applied and $Q$ the total axial force and self-weight of the foundation. This effective area allows to compute also the distribution of bearing pressure under the foundation according to a linear elastic distribution as shown in Fig. 2.

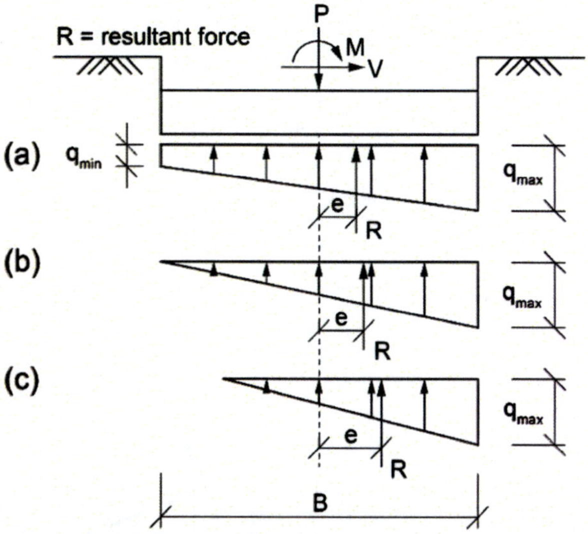


**Fig. 2.** Distribution of bearing pressure under foundation: (a) e < B/6, (b) e = B/6 and (c) e > B/6 (adapted from DNV/Risø 2002)

### 2.3 Cracking Moment of the Foundation

In order to predict the first cracking of concrete, equilibrium equations are adopted to compute the internal bending moment ($M_{int}$) in the critical cross-section as shown in Fig. 3.

The situation corresponding to an internal bending moment equal to the cracking bending moment of the cross-section ($M_{int} = M_{cr}$) is computed. The on-set of cracking in bending for the cross-section is here defined by means of a simplified plane section analysis assuming elastic materials both in tension and in compression and considering the situation in which the tensile stress of concrete at the bottom edge of the section reaches the average value of the tensile strength ($f_{ctm}$). It is interesting to observe that, due to the fixed value of $P$ corresponding to the axial force applied to the tubular shaft, the force $P$ can move at the border of the foundation collar, respecting the maximum compressive strength of the concrete under the steel ring.

Each further increase of the bending moment is limited by the tensile force exerted by the vertical bolt. This means that the part of the foundation lifted by the tensile force applied at the collar by the stretched bolts, can be equilibrated at the rotation only by the stretching force applied to the radial and circumferential reinforcements at the onset of cracking and at the vertical direction by the shear transmitted along the critical section. The eccentricity ($e$-$R_i$) can be positive only if the tensile force applied to the bolt is equilibrated by the shear force applied along the crack.

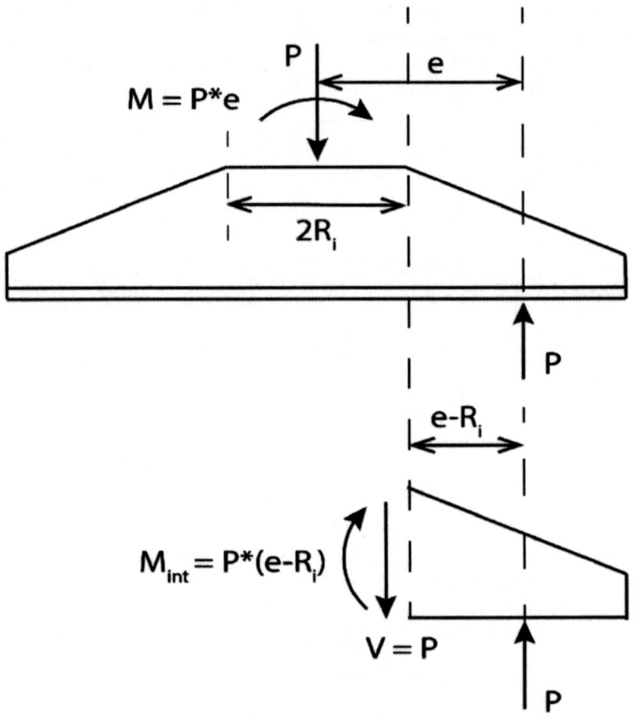


**Fig. 3.** Equilibrium scheme to define the internal bending moment at the critical cross-section

## 2.4 Ultimate Bearing Capacity of the Foundation

The ultimate bearing capacity of the foundation, due to the squat geometry, can be assessed by means of a strut-and-tie approach, neglecting the failure of the soil. An ad hoc strut-and-tie model is set in accordance to the static theorem of limit analysis to consider the three-dimensional stress distribution inside the relatively squat foundation. The scheme of this model is represented in Fig. 4.

The yielding of the radial rebars as well the yielding of the circumferential rebars is considered. The resultant force of the circumferential rebars ($Z_c$) is defined by considering the most external position of the resultant of the soil reaction force $P$ and consequently only the corresponding circular rebars involved, according to the static theorem of limit analysis.

Both tensile resultant forces of the radial rebars ($Z_r$) and compressive inclined struts (C1, C2, C3) are defined dividing the circular sector interested by the mechanism in three slices, characterized by a circular sector corresponding to the arch length able to produce a local vertical stress under the steel ring ($t$ = 20 mm) equal to the uniaxial cylindrical compressive strength ($f_c = 45$ MPa): for $P = 88.1$ kN this corresponds to a global circular sector characterized by an angle of 108°, that corresponds to three sectors of 36° each one.

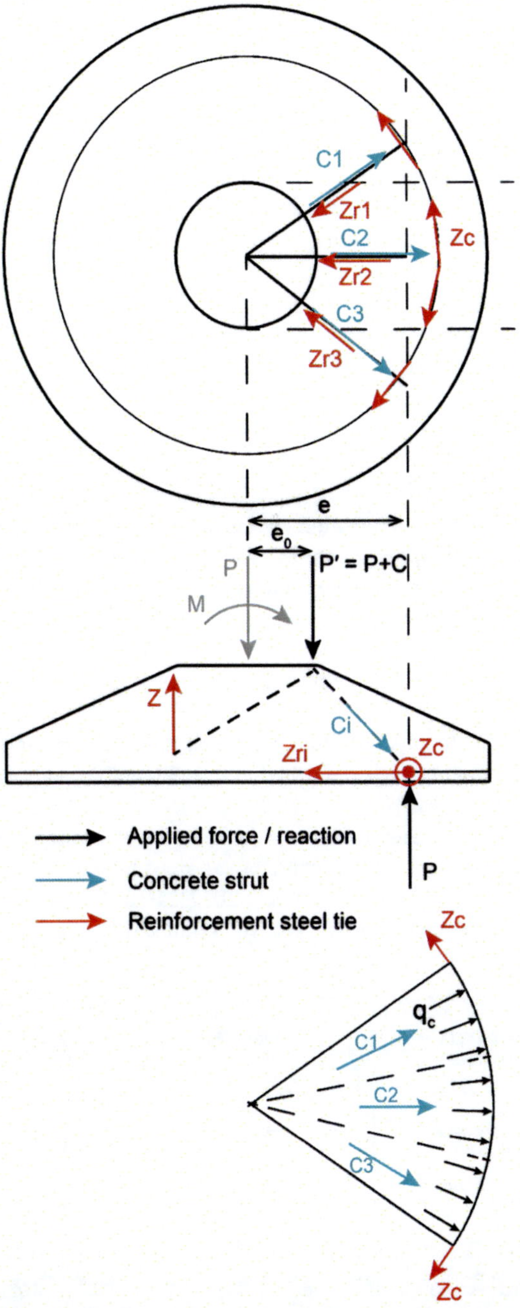


**Fig. 4.** Scheme of the strut-and-tie model for the foundation

The maximum eccentricity parameter *e*, taking into account the geometry of the strut-and-tie configuration, can be determined knowing the ultimate vertical load *P* by equilibrium of the whole circular sector and assuming the yielding of all the rebars involved by the specific *e* selected. Knowing the load *P*, the total bending moment *M* can be computed in relation to the number of circular rebars considered, as the addition of the contribution of the three slices.

The maximum eccentricity parameter *e*, taken as failure condition, is the one that leads to the yielding of the minimum number of circular reinforcements (4 rebars) able to equilibrate the total vertical soil reaction *P*, that corresponds to the applied axial force. The prediction of the ultimate bearing capacity of the foundation is performed considering, for the yielding strength of the steel, the average value obtained from the test of the rebars ($f_{ym} = 537.7$ MPa).

## 2.5 Fatigue Behaviour

The context in which the wind towers is placed implies a very high number of loading cycles. During the whole life span wind towers are constantly subjected to strong wind actions and, during energy production, to cyclic loads from the rotor blades. This can lead to fatigue problems in the structure, including the foundation, and thus the ultimate limit state is not the only aspect that needs to be considered in the design process.

According to Thun (2006) fatigue is commonly divided into different categories, low-cycle and high-cycle fatigue, with the limit between the two at roughly $10^3$ cycles. In recent years, the concept of super-high cycle fatigue has been lifted for structures subjected to more than $10^7$ cycles. This is however a quite uncommon situation (Thun 2006). The fatigue categories are described in Fig. 5 together with structures typically exposed to these kinds of loads.

According to Fig. 5, wind towers fall within the high-cycle fatigue category. However, according to Gasch and Twele (2012) wind towers are subjected to more than $10^9$ cycles during a 20-year life span (Gasch and Twele 2012). Hau (2013) places the number of cycles for a wind turbine between $10^7$ and $10^8$ (Hau 2013). Therefore, the foundation

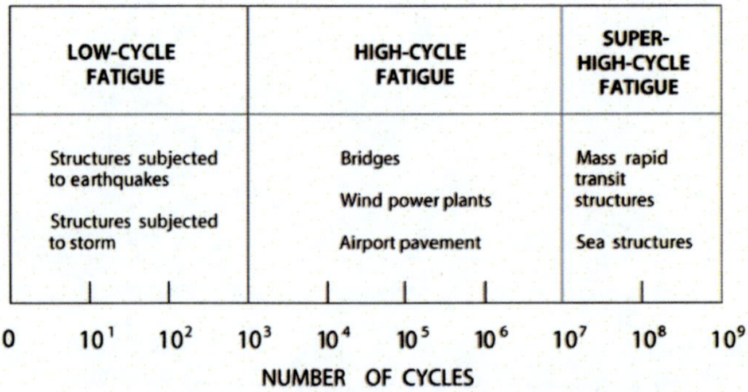


**Fig. 5.** Spectra of fatigue loading in different types of structures (Source: http://www.structuremag.org/?p=10735)

should probably be seen as a super-high cycle structure. Unfortunately, there are no fully-reliable methods for accounting for so many cycles today. A failure of a structure caused by a load lower than the design ultimate load is characterized as a fatigue failure. The failure is caused by a continuous decomposition of the material due to a high number of cycles. The progress of the decomposition is dependent on factors like the mean stress level, the amount of cycles and the stress range, which is the difference between maximum and minimum stress (Heffernan 1997).

## 3 Soil-Foundation-Tower Scaled Prototype

Some of the simplified approaches presented in the previous section are applied to the interpretation of the test results from a scaled (1:15) soil-foundation-tower prototype tested as part of a larger experimental/numerical research programme at the LPMSC laboratory of Politecnico di Milano. A general overview of the research project including both 1:15 and 1:4 scaled models is here omitted, but can be found in di Prisco et al. (2022). The experimental tests conducted on the 1:15 prototype together with the respective results are presented and discussed in Dal Lago et al. (2022). A sketch and a picture of the test set-up of the scaled soil-foundation-tower prototype are shown in Fig. 6. Reinforcement details (shear reinforcement is omitted) of the scaled foundation prototype are shown in Fig. 7. Full details on the reinforcement are given in Dal Lago et al. (2022).

Nonlinear finite element (FE) simulations aimed at investigating the role of nonlinearities in both soil and concrete foundation, emphasizing the parameters and the nonlinear constitutive laws that significantly affect the foundation response, are presented in Martinelli et al. (2022).

## 4 Numerical FE Models

Two finite element models are used in the analysis of the behaviour of the prototype foundation. The first FE model, labelled as "num. 2D" hereafter, reflects the most common and simple modelling technique employed in practice. It was built and processed by using the software Midas GEN (2019). The foundation was modelled by using shell finite elements as a slab with variable thickness, while the soil was modelled with independent elastic compression-only Winkler springs ($k_w = 123 \times 103$ kN/m$^3$).

The second FE model, labelled as "num. 3D (N2)" hereafter, is a nonlinear model fully accounting for soil-structure interaction and is based on three-dimensional elements for both concrete foundation and soil (Fig. 8). It adopts compression-only contact surfaces for concrete-soil interface, and beam elements for reinforcing bars. The model was built and processed by using the software Abaqus 6.14-5 (2016). The inclusion of the steel tower in the FE model allows to perform displacement-controlled analyses during the application of a top horizontal load. The use of this second model in this context is to highlight the limitations/advantages of the simplified models previously discussed with an eye to the safety limits as well.

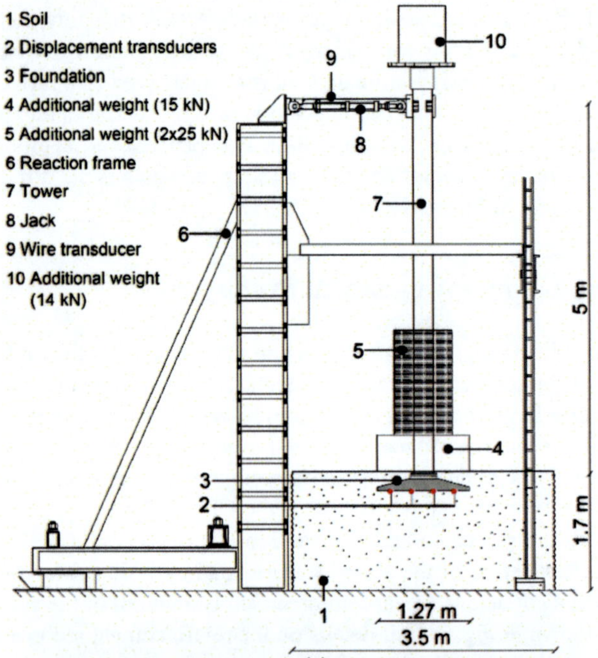


(a)

(b)

**Fig. 6.** Test set-up of the scaled soil-foundation-tower prototype: (a) sketch (adapted from Martinelli et al. 2022), (b) picture

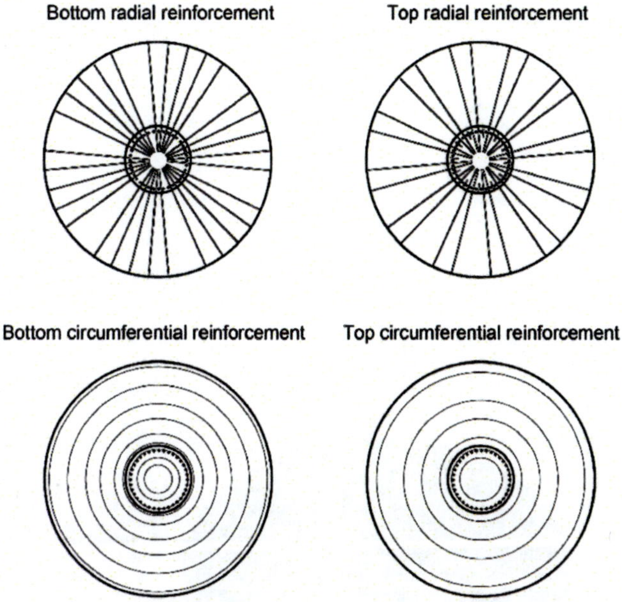


**Fig. 7.** Sketch of the reinforcement details of the scaled foundation prototype

Concrete inelastic behaviour was simulated by means of the Concrete Damaged Plasticity (CDP) model. An elasto-J2 plastic material model was assumed in the analyses for steel rebars. The mechanical behaviour of the soil was simulated by means of a non-associated elasto-perfectly plastic constitutive law with a Mohr-Coulomb failure criterion. A detailed description of geometrical and material modelling of both FE models is here omitted, but can be found in Martinelli et al. (2022).

## 5 Results and Discussion

The reliability of the simplified approaches described in Sect. 2 is verified against the experimental data of the 1:15 soil-foundation-tower prototype. For a better understanding of the phenomena, the response of both 2D linear and refined 3D nonlinear FE models are also analysed. Numerical and experimental results are presented in terms of moment at the base ($M$) vs. rotation at the base ($\theta$), while the simplified approaches taken into consideration provide only information on bending moments and do not allow to make any consideration on the deformability of the system.

The initial stiffness of 2D FE model is significantly smaller than the ones obtained experimentally and with the 3D FE model (Fig. 9). As highlighted in Martinelli et al. (2022), the constant $k_w$ affecting the service behaviour of the turbine tower strongly depends on the assumptions related to the estimation of the simplified Winkler coefficient. In particular, the assumption adopted in the simplified model is that the soil condition is not virgin, and the Winkler coefficient was calculated following the approach proposed by Terzaghi (1955), based on the results of a series of ad hoc local plate

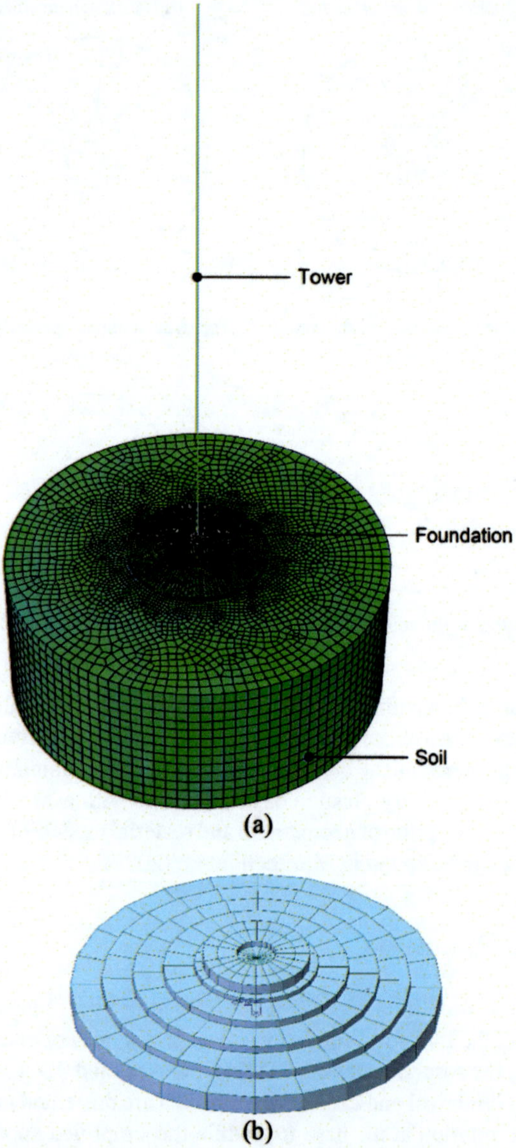


**Fig. 8.** FE tower-foundation-soil system: (a) brick model of foundation and soil generated in Abaqus (adapted from Martinelli et al. 2022), (b) shell model of foundation resting on unilaterally elastic soil generated in Midas GEN

tests carried out on undisturbed sand after deposition and prior to the execution of the first test series, assuming the reload stiffness, as it better reflects the state of the soil before the execution of the monotonic test, where it experienced various cyclic load applications.

Always dealing with soil behaviour, Eq. (1) is first applied to compute the ultimate resistance of the soil adopted in the experiments. In the case at study, a friction angle of 40°, a specific weight of 16.5 kN/m$^3$ and cohesion equal to 0 are assumed providing a limit capacity $q_u = 640$ kPa.

The application of the effective area approach with a linear distribution of the soil pressure (as presented in Sect. 2.2) allows to define the stress distribution for a given eccentricity of the applied vertical force (equal to 88.1 kN in the case under investigation) and therefore allows to compute the bending moment that can be applied to the foundation to reach a maximum pressure on the soil ($q_{max}$ in Fig. 2) corresponding to the limit value $q_u$. The results of this computation allow detecting that the point at which the maximum pressure is reached at the border of the foundation corresponds to a bending moment of 25.4 kNm (dashed cyan line in Fig. 9), that, for the case under investigation, is very close to the loose of linearity of the experimental curve.

Figure 9 shows as both 3D and 2D FE models are characterised by a marked nonlinearity in the response which is attributed to the geometric nonlinearity associated with the partial detachment of the foundation from the soil. The value of the 1$^{st}$ order moment associated with the detachment of the foundation from the soil ($M_{det}$) can be calculated as:

$$M_{det} = N \cdot u = 88.1 \cdot 0.159 \cong 14.0\,\text{kNm} \tag{3}$$

where $N$ is the axial load applied and $u$ is the size of the central core of inertia for a circular cross-section.

The moment associated to the first detachment of the foundation is plotted in Fig. 9 as a dashed green line. It is worth noting that the moment in the foundation associated with the ultimate capacity of the soil, imagining a Winkler soil, (even if the soil under the extreme portion of the foundation is not confined!) is 1.8 times the moment associated to the first detachment of the foundation.

The simplified methods described in Sect. 2.1 and 2.2 for calculating the moment in the foundation associated with the ultimate capacity of the soil provide a conservative estimate of the moment for the case under consideration. This is mainly due to the fact that the simplified approach does not consider any plastic redistribution of stresses, but simply defined the point at which just one point of the soil reaches for the first time the limit capacity.

Considering the behaviour of the slab foundation, also cracking can play an important role especially referring to Serviceability Limit State (SLS) and durability. The method described in Sect. 2.3 for calculating the first cracking moment in the foundation (dashed cyan line in Fig. 10) is compared with the experimental and numerical responses in Fig. 10. The load and rotation levels for which the first flexural crack is activated in the 3D FE model is also identified in Fig. 10 by a red full circle.

The first flexural crack in the 3D FE model has to be intended as a fully developed crack along a chord. It is worth to note that the significant difference between the values provided by the simplified approach and by the nonlinear FE approach (FEA) can be related to the definition of this value because the simplified approach refers to a sectional analysis and assumes that first cracking arise when the bottom edge of the cross-section reaches a stress value equal to the average tensile strength while, in the case of nonlinear FEA, the points correspond to the end of the crack propagation through the whole chord of the base circle. Therefore, the value proposed for the FE analysis considers all the sectional stress redistribution that is not considered in the simplified approach. Finally, the comparison between the results of the ultimate load prediction calculated following the method described in Sect. 2.4 and the experimental and numerical results of the 1:15 foundation model is shown in Fig. 11. The simplified method provides a very good estimate of the ultimate condition.

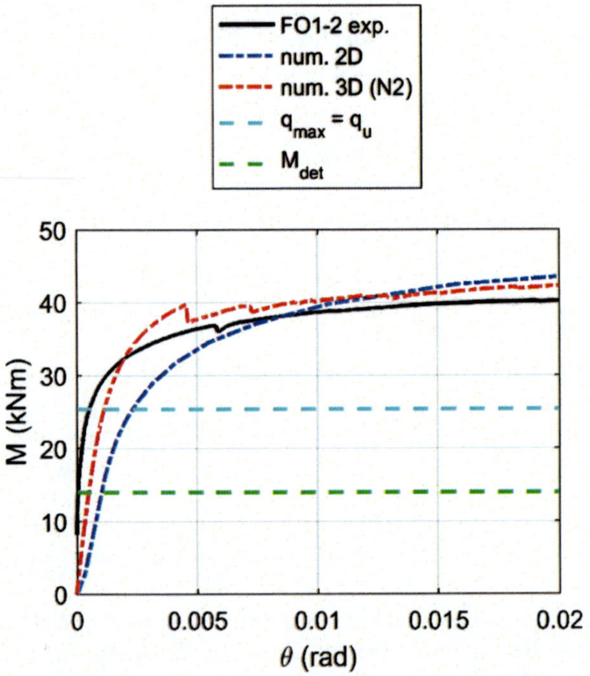


**Fig. 9.** Prediction of the condition $q_{max} = q_u$ for the foundation

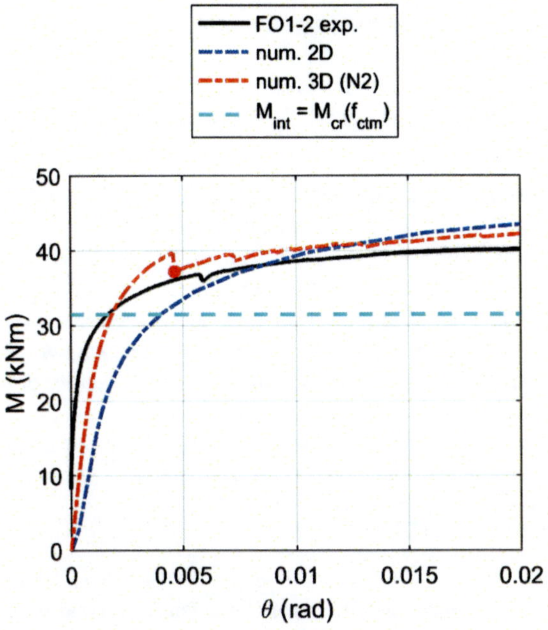


**Fig. 10.** Prediction of the first cracking of concrete for the foundation

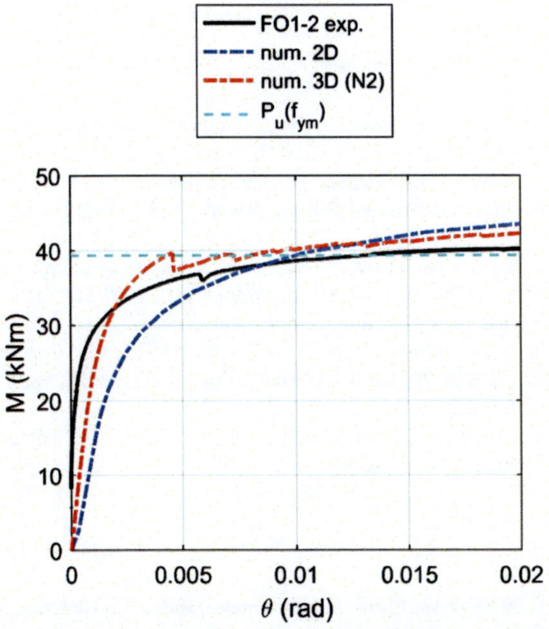


**Fig. 11.** Prediction of the ultimate load capacity of the foundation

## 6 Conclusions

The paper has presented the comparison of simplified approaches and numerical analyses in predicting the behaviour of circular wind tower foundations. The results allow to draw the following conclusions:

- the effective area approach, together with the soil ultimate capacity estimation proposed by Vesic (1975), allow to provide a reliable estimation of the bearing capacity of the soil even if this cannot be considered as an ultimate value of the bearing capacity, because it does not consider any plastic redistribution within the soil;
- the simplified estimation of first cracking bending moment can be regarded as a safe side estimation that allows to detect the arise of first cracking within the foundation that can be considered an important parameter in the durability design of such kind of infrastructures;
- the 3D strut-and-tie model developed and here applied to the scaled foundation seems to be very effective in the prediction on the ultimate bearing capacity of this kind of infrastructure.

**Acknowledgements.** This research has been supported by Enel Green Power Spa in the framework of a joint research contract between Enel Green Power Spa and several departments of Politecnico di Milano.

## References

Bonnett, D.: Wind turbine foundations – loading, dynamics and design. Struct. Eng. **83**(3), 41–45 (2005)

Brinch Hansen, J.: A Revised and Extended Formula for Bearing Capacity. Copenhagen (1970)

Bowles, J.E.: Foundation Analysis and Design, 5th edn. McGraw-Hill Companies Inc, Peoria (1996)

Dal Lago, B., et al.: Experimental tests on shallow foundations of onshore wind turbine towers. Struct. Concr. **23**(5), 2986–3006 (2022). https://doi.org/10.1002/SUCO.202100655

di Prisco, M., et al.: Wind tower FRC foundations: research and design. In: Serna, P., Llano-Torre, A., Martí-Vargas, J.R., Navarro-Gregori, J. (eds.) Fibre Reinforced Concrete: Improvements and Innovations II. BEFIB 2021. RILEM Bookseries, pp. 831–842 (2022). https://doi.org/10.1007/978-3-030-83719-8_71

DNV/Risø 2002: Guidelines for Design of Wind Turbines, 2nd edn. Det Norske Veritas (DNV) & Risø National Laboratory, Copenhagen

DNV/Risø 2010: Design of Offshore Wind Turbine Structures. Det Norske Veritas (DNV) & Risø National Laboratory, Copenhagen

Fellenius, B.H.: Bearing Capacity of Footings and Piles — A Delusion? Dearborn, Michigan (1999)

Gasch, R., Twele, J.: Wind Power Plants: Fundamentals, Design, Construction and Operation, 2nd edn. Springer, Heidelberg (2012)

Hau, E.: Wind Turbines: Fundamentals, Technologies, Application, Economics, 2nd edn. Springer, Heidelberg (2013)

Heffernan, P.: Fatigue behaviour of reinforced concrete beams with CFRP laminates (Doctoral dissertation, Royal Military College of Canada) (1997)

Martinelli, P., Flessati, L., Dal Lago, B., Fraraccio, G., di Prisco, C., di Prisco, M.: Role of numerical modelling choices on the structural response of onshore wind turbine shallow foundations. Structure **37**, 442–458 (2022). https://doi.org/10.1016/j.istruc.2022.01.002

Maunu, P.: Design of Wind Turbine Foundation Slabs. Lulea University (2008)

Meyerhof, G.G.: Some recent research on the bearing capacity of foundations. Can. Geotech. J. **1**, 16–26 (1963). https://doi.org/10.1139/t63-003

Midas GEN: Midas GEN 2019 (V2.2) Analysis Manual for Midas GEN (2019)

Morgan, K., Ntambakwa, E.: Wind turbine foundation behavior and design considerations. In: AWEA Wind Power Conference. AWEA, Houston, pp. 1–14 (2008)

Systèmes, D.: Abaqus analysis user's manual - version 6.14. Dassault Systèmes, Providence (2016)

Terzaghi, K.: Evalution of coefficients of subgrade reaction. Géotechnique **5**(4), 297–326 (1955). https://doi.org/10.1680/geot.1955.5.4.297

Thun, H.: Assessment of fatigue resistance and strength in existing concrete structures (Doctoral dissertation, Luleå University of Technology) (2006)

Vesic, A.S.: Bearing Capacity of Shallow Foundations. In: Winterkorn, H.F., Fang (eds.) Foundation Engineering Handbook. Vam Nostrtand, New York (1975)

# Use of Plastic Waste for the Development of Green Lightweight Structural Concrete

Alice Sirico[1(✉)], Patrizia Bernardi[1,3], Beatrice Belletti[1,3], and Alessio Malcevschi[2,3]

[1] Department of Engineering and Architecture, University of Parma, Parma, Italy
alice.sirico@unipr.it
[2] Department of Chemistry, Life Sciences and Environmental Sustainability, University of Parma, Parma, Italy
[3] Centre for Energy and Environment-CEE/CIDEA, University of Parma, Parma, Italy

**Abstract.** With a production of more than 30 billion tons each year, concrete is the building material most widely used in the world, which means a high environmental impact, especially in terms of carbon dioxide emissions and consumption of raw materials. Moreover, some of the industrial wastes are unavoidable, hence there is an increasing interest in finding new solutions for their recycling. To this end, this work aims at the development of innovative concretes able to meet the new trends related to sustainability, from a circular economy perspective. In particular, two different kinds of waste are investigated to be inserted in the concrete mix: plastic, which derives from industrial processing waste but is mixed in composition, so making it difficult to reuse or recycle, as well as biochar, which is the solid carbonaceous by-product resulting from wood-waste pyro-gasification. The developed concretes are characterized in terms of physical-mechanical performance, in order to assess the feasibility of using these industrial wastes as secondary raw materials for the building industry.

**Keywords:** Recycled plastic · biochar · sustainable concrete · mechanical characterization

## 1 Introduction

According to the estimations (European Parliament - News 2018), during the past 70 years, the amount of plastic produced worldwide has always grown, except for a short pause in the first months of 2020 due to Covid-19 pandemic. The production of plastic has indeed passed from 1.5 million tons in 1950 to 359 million tons in 2018. In Europe, only 32.5% is recycled, 24.9% is still disposed of in landfills, whereas 42.6% is incinerated for energy recovery. Since most plastics are not biodegradable, if dumped into landfills, they can resist in the environment for decades with the risk of releasing toxic compounds. However, also plastic incineration is not ideal, since, although representing an energy source, it causes the release into the atmosphere of more than 850 million tons of greenhouse gases worldwide, with the additional risks of producing toxic compounds.


M. A. Aiello and A. Bilotta (Eds.): ICC 2022, LNCE 435, pp. 190–203, 2024.
https://doi.org/10.1007/978-3-031-43102-9_16

The major part of plastic production (about 40%) is devoted to packaging. On one hand, one of the main objectives should be prevention, by promoting awareness and improving the design of the products to avoid plastic use, but often this industrial waste is unavoidable. Nowadays, Italy produces each year about 20 kg/ per person of plastic packaging waste, and the recycling rate ranges from 40 to 50%; hence, there is still a void to reach the ambitious targets that the EU set toward sustainability. These goals, which are devoted to lowering the environmental impacts and promoting a transition toward circular economy, were set within European Green Deal and include that 55% of plastic packaging waste should be recycled by 2030 (European Parliament - News 2018).

The need for recycling is not only related to the management of the enormous quantities of waste produced every day, but also to minimize the environmental footprint in terms of saving natural resources.

Within this context, the use of plastic waste in the building industry could represent an ideal way of recycling, due to the enormous consumption of raw materials used each year worldwide. Among the construction materials, concrete is the most widely produced, with quantities that reach over 30 billion tons and a trend still increasing, especially in developing countries, mainly due to the rapid growth of industrialization, urbanization, and the rising necessity of transport facilities (Monteiro et al. 2017, Lehne and Preston 2018). This also implies that more than 25 billion tons of aggregates are used for concrete production every year, which means the largest volume of solid material extracted globally, so negatively changing the riverbeds, costs, and land. Moreover, the concrete industry contributes for around 8% of annual global $CO_2$ emissions, so efforts to lower its environmental impact is expected to play a vital role to reach the EU targets and mitigate climate change.

For this reason, in the last decades, research has made many attempts to develop cementitious materials that incorporate waste, especially industrial waste considered unavoidable and difficult to reuse. One research goal was to find ways for recycling, so as to develop innovative concretes with added economic value, possibly to be used also for structural applications.

Some waste, such as slags, fly ashes, silica fumes, and rice husk ash, were studied to partially replace clinker or Portland cement, so helping in reducing the green gases related to cement production (Nwankwo et al. 2020, Chandra 1996), as well as in saving energy and natural resources. Other waste materials, such as marble dust, ceramic powder, biochar, were instead used as filler or additive to modify concrete properties with the aim of enhancing the performance in both fresh and hardened states. In this case, the environmental benefits related to the increase of the recycling rate of the waste meet also the market goals of developing high added-value concretes, characterized by improved microstructure, rheology, mechanical properties, and durability. Finally, other wastes were used to partially replace natural aggregates, such as plastics, glass, and construction and demolition waste (Tam et al. 2018, Saikia and De Brito 2012, Shi and Zheng 2007).

Extensive research on the use of plastic waste in concrete demonstrated good improvement of non-structural performance, such as thermal and acoustic insulation (Belmokaddem et al. 2020) as well as reduction in density; whereas, a significant worsening in the mechanical behavior can be recognized, especially in terms of compressive strength (Sharma and Bansal 2016, Gu and Ozbakkaloglu 2016, Siddique et al. 2008). However, the presence of plastic aggregates seemed to improve some other material properties, i.e., impact resistance, and energy absorption capacity (Saxena et al. 2018). Thanks to these features, but most of all to the undoubted environmental, economic, and social impacts related to the recycling of plastic waste in concrete, the research for the development of green structural concrete with plastic aggregates is still noteworthy to be investigated.

This work lies in this line of research, since it analyses structural concretes that incorporate industrial plastic waste up to a replacement equal to 20% by volume of natural aggregates. The main material properties are determined and compared to that of conventional concrete without waste, so to highlight the influence of plastic aggregates in terms of workability, compressive strength, flexural strength, and fracture energy. Also the bond between a reinforcing bar (with 6 or 10 mm diameter) and the developed concretes is investigated by performing pull-out tests, so to obtain the load-slip curves and assess the role of plastic aggregates both in case of pull-out and splitting failure. Moreover, to overcome the reduction of mechanical properties due to plastic waste addition, the combined use of another lightweight waste is proposed: biochar. It is defined as the solid by-product derived from the thermo-chemical decomposition of biomasses during pyrolysis or gasification processes, aimed at heat and energy production. Biochar was widely studied and applied in the agriculture field as soil amendment; however, some recent research has highlighted its promising use as filler in cementitious materials to improve physical and mechanical properties (Danish et al. 2021, Maljaee et al. 2021). Moreover, biochar particles are able to sequester a high volume of stable carbon in their structure, thus locking it, indirectly, for decades in civil buildings, which can be seen as long-term carbon sinks (Sirico et al. 2021). Hence, bearing in mind the high environmental benefits related to the use of both plastic waste and biochar, in this work, the combination of plastic and biochar is explored to develop green lightweight structural concretes, with the aim of taking advantage of the specific properties of each waste.

## 2 Experimental Program, Mix Design and Preparation of Specimens

In order to assess the influence of plastic waste on concrete performance, three different types of concrete were investigated, which were characterized by a replacement (in volume) of 0% (P0), 13.33% (P13), and 20% (P20) of natural aggregates with recycled plastic grains.

The three mixes were prepared by following the proportion reported in Table 1, where the amounts refer to the saturated-surface-dry (SSD) condition for the aggregates.

For all the concrete mixes, cement Type II A-LL 42.5 R was used, whereas calcareous sand (d/D = 0/4 mm) and siliceous gravel (d/D = 2/8 mm) were used as natural aggregates. Their grain size distributions, reported in Table 2, were properly combined so to achieve an optimal formulation, as suggested by Bolomey grading curve.

Recycled plastic grains derived from regranulation of industrial processing waste mainly devoted to packaging, and were here used as partial replacement for natural

**Table 1.** Mix design in kg/m$^3$.

| Mix | Cement | Cement | Sand | Gravel | Plastic | SP |
|---|---|---|---|---|---|---|
| P0 | 408 | 204 | 1126 | 562 | – | 3.88 |
| P13 | 408 | 204 | 900 | 526 | 74 | 1.92 |
| P20 | 408 | 204 | 900 | 449 | 111 | 1.47 |

* SP: Superplasticizer Mapei Dynamon Xtend W202R

**Table 2.** Aggregate size distribution of sand and gravel in terms of cumulated sieve passing vs. sieve mesh size.

| | sand | gravel |
|---|---|---|
| mesh size | cumulated passing | cumulated passing |
| mm | % | % |
| 10.00 | 100.0 | 100 |
| 8.000 | 100.0 | 83.9 |
| 5.600 | 100.0 | 19.2 |
| 4.000 | 98.9 | 2.8 |
| 2.000 | 81.8 | 1.2 |
| 1.000 | 63.4 | 1.0 |
| 0.500 | 40.7 | 0.4 |
| 0.250 | 14.5 | 0.2 |
| 0.125 | 3.3 | 0.1 |
| 0.036 | 0.9 | 0 |

aggregates. They were characterized by a variable composition of low-density polyethylene (LDPE) and polyamide (PA), with the latter ranging from 25 to 75 wt%. This plastic waste is generally disposed of in landfills, since its not homogenous composition makes it difficult to reuse or recycle. The plastic grains were characterized by a roughly lentil shape, with a base of 3–4 mm and a height of 1–2 mm.

The materials were mixed in a drum-type mixer, by following the standard recommendation and mixing sequence for concrete without waste. For P13 and P20 mixes, the plastic grains were added in the initial stage of mixing, together with the natural aggregates. All the mixes showed a high homogeneity, without any problem of segregation or bleeding.

In order to analyze compressive strength and tensile behavior, in terms of both flexural strength and fracture energy, for each concrete mix, three cubes (150 × 150 × 150 mm) and four prismatic specimens (100 × 100 × 400 mm) were cast and tested after 28 days of water curing; three additional cubes were cast to determine the hardened density.

Finally, reinforced specimens were cast to perform pull-out tests. These cylindrical specimens, with diameter and height equal to 60 mm, were cast by placing a single reinforcing bar in the center of the mold. 6 mm and 10 mm diameter deformed B450C steel bars were chosen, so to assess the influence of the plastic grains in case of different concrete cover / bar diameter ratios $C_c/d_b$. This ratio was equal to 4.5 and 2.5 for 6 mm and 10 mm bar, respectively, so leading to different failure modes, i.e. pull-out vs. splitting failure.

By following RILEM-RC6 recommendations, a bonded length of 5 bar diameters (i.e. 30 mm) was set for 6 mm rebar. Then it was unbonded for other 5 diameters by using a polyethylene sheath, so to guarantee an almost uniform state of stress in specimens along the bonded length and avoid the yielding of the steel bar during testing. The same dimensions, i.e. 30 mm bonded and 30 mm unbonded, were chosen also for the samples reinforced with 10 mm bar. Before casting, all the rebars were painted with epoxy resin for all the length except for the bonded one, to avoid corrosion during the subsequent 28 days of water curing of concrete samples. Then the specimens were subjected, for one year, to wet and dry (w/d) cycles, which consisted of 4 days of immersion in a saturated $Ca(OH)_2$ solution and 3 days of drying. The use of $Ca(OH)_2$ prevented concrete carbonation, so allowing to assess the role of plastic grains on bond behavior, by limiting the parameters involved.

## 3 Results and Discussion

### 3.1 Density

Density measurements, performed according to EN 12390–7 and reported in Fig. 1, highlight the beneficial effect offered by recycled plastic use.

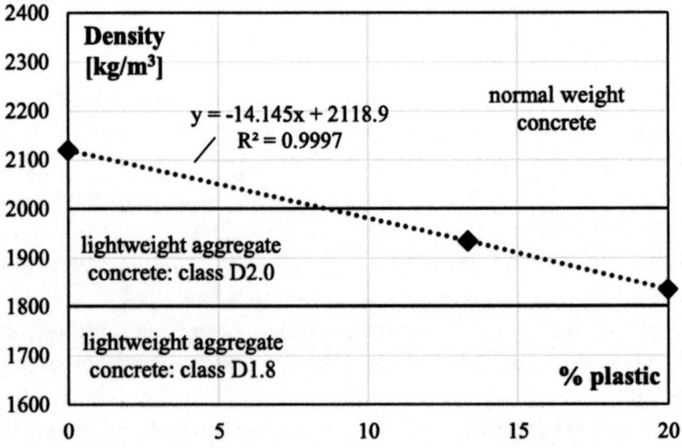


**Fig. 1.** Correlation between density and percentage of plastic substitution with respect to the total volume of aggregates.

Plastic grains present indeed low specific gravity (i.e. 0.868 g/cm$^3$) with respect to that of natural aggregates (i.e. 2.646 and 2.640 g/cm$^3$ for the sand and the gravel, respectively), which results in a linear decrease of density as the percentage of substitution of natural aggregates increases, by passing from 0% for reference concrete (P0), to 13% and 20% for P13 and P20, respectively. This allows classifying both P13 and P20 as lightweight aggregate concrete (density class D2.0), according to MC2010 definition.

## 3.2 Workability

Workability was evaluated with slump tests according to EN 12350–2:2019, in terms of superplasticizer required to maintain about the same slump value as reference one (P0), as reported in Fig. 2. P13 and P20 required a much smaller amount of superplasticizer with respect to control. This means that the substitution of part of natural aggregates with the plastic grain enables to enhance concrete workability, leading to reduced cost, or increased strength and durability, when water contents are reduced instead of reducing plasticizer.

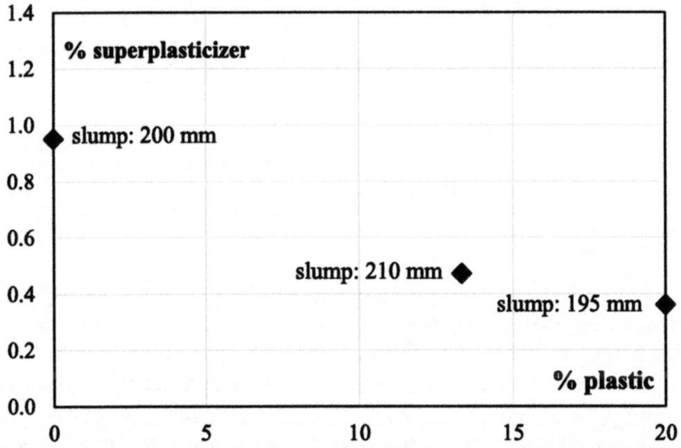


**Fig. 2.** Percentage (by weight of cement) of superplasticizer to maintain the same slump class as a function of percentage of plastic substitution with respect to the total volume of aggregates.

## 3.3 *Compressive Strength*

Compressive strength was evaluated at 28 days, by means of a Universal Testing Machine METROCOM PV P30, by following EN 12390–3 recommendation.

By analyzing the average compressive strength reported in Fig. 3 for each mix, an almost linear relationship can be observed between this latter and the percentage of substitution of plastic. Hence, for the tested mixes, a maximum substitution up to 13% by volume is possible to obtain a structural concrete.

The lack of physical-chemical compatibility between the plastic grains and the cement matrix causes a low bonding strength and weaker interfacial transition zone

(ITZ), so worsening the compressive behavior of concrete. The hydrophobic nature of plastic, which interacts badly with the substantially hydrophilic cement matrix, causes indeed the presence of not absorbed water, so an empty interface between the two materials. This condition, together with the less hardness of plastic grains compared to natural aggregates, considerably reduces concrete compressive strength, as already observed in the literature (Saikia and De Brito 2012, Bahij et al. 2020).

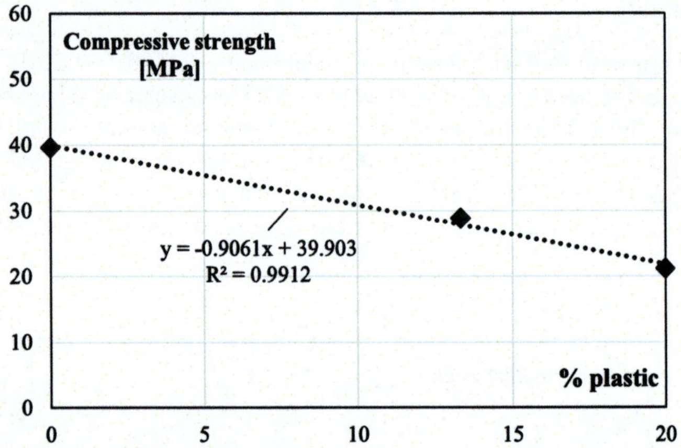


**Fig. 3.** Correlation between average compressive strength and percentage of plastic substitution with respect to the total volume of aggregates.

### 3.4 Tensile and Fracture Behavior

The tensile and fracture behavior of concrete was investigated in terms of flexural strength and fracture energy, at 28 days of curing. To this aim, three-point bending (TPB) tests were performed under Crack Mouth Opening Displacement (CMOD) control, by means of an Instron 8862 Universal Test Machine. The specimens were tested over a net span of 300 mm, and prior to testing were notched at their mid-length, with a notch width equal to 2 mm and a depth equal to 30 mm, i.e. 0.3 times specimen depth, by following JCI-S-001–2003 recommendations. The average flexural strength was evaluated for each concrete as the mean of the modules of rupture of the four notched specimens tested; whereas the average fracture energy was computed according to JCI-S-001–2003 as the ratio between the area under the load-CMOD curve and the area of the nominal ligament, by also taking into account the work done by the deadweight of specimen and loading equipment.

As can be seen in Figs. 4, tensile behavior in terms of both strength and ductility, appears more positively influenced by the use of recycled plastic grains than compressive strength.

P13 mix shows a comparable flexural strength with respect to control (P0), whereas a higher amount of plastic (i.e. 20% by volume) reduces flexural strength by 35%.

However, P20 mix displays a considerable increase of fracture energy (i.e. 95%) with respect to P0; hence, the reduction of flexural strength is largely counterbalanced by the more ductile behavior. Also P13 shows a significant increase in fracture energy, by 30% with respect to control. The capacity of absorbing a higher amount of fracture energy can be attributed to the elastic and non-brittle nature of recycled plastic grains, which enable to slow down the propagation of micro-cracks, so providing concrete characterized by larger displacement at failure and higher residual loads in the post-peak branch of P-CMOD curve, as can be seen from the results reported in Table 3. The average CMOD at failure corresponds to a residual load of 0.1 kN, whereas the average residual flexural tensile strengths $f_R$ were computed on the basis of the residual loads registered for CMOD equal to 0.5, 1.5, and 2.5 mm.

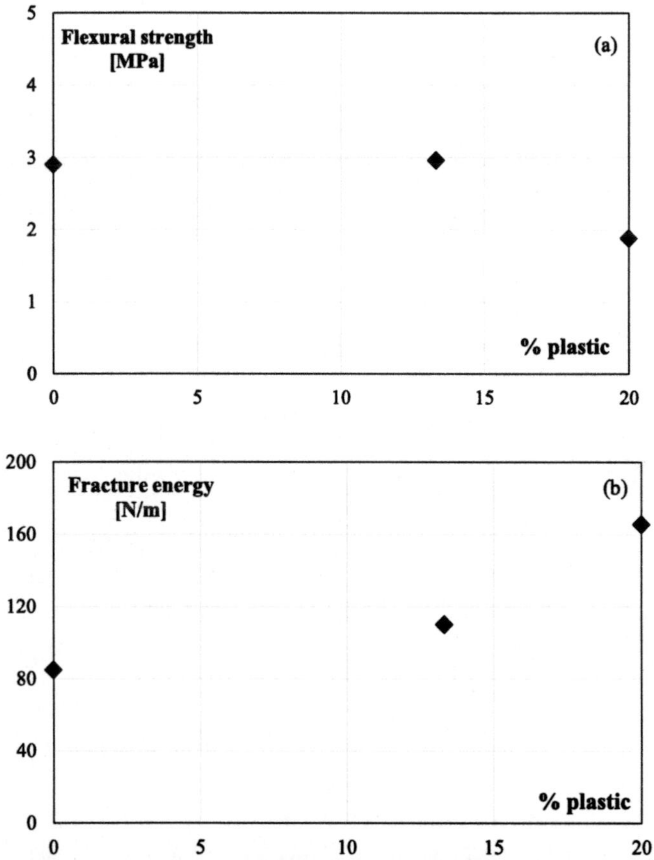


**Fig. 4.** Correlation between percentage of plastic substitution with respect to the total volume of aggregates and: (a) average flexural strength; (b) average fracture energy

A remarkable beneficial effect on post-cracking can be observed when plastic is used as partial replacement of natural aggregates, so making these concretes with recycled

**Table 3.** Average CMOD value at failure and residual strengths for different CMOD values.

| Mix | CMOD at failure | $f_{R,CMOD=0.5}$ | $f_{R,CMOD=1.5}$ | $f_{R,CMOD=2.5}$ |
|---|---|---|---|---|
| | mm | MPa | MPa | MPa |
| P0 | 1.00 | 0.331 | – | – |
| P13 | 1.79 | 0.466 | 0.119 | – |
| P20 | 3.07 | 0.887 | 0.263 | 0.131 |

plastic grains particularly suitable for several in situ applications where ductility is a key factor. One example is improving the earthquake resistance of buildings, as suggested by Saikia and De Brito (2012).

### 3.5 Bond Behavior

To assess the bond behavior between reinforcing bars and concrete containing recycled plastic grains, pull-out tests were carried, by using a Instron 8862 Universal Test Machine and adopting the test setup as shown in Fig. 5, according to RILEM RC-6 recommendations.

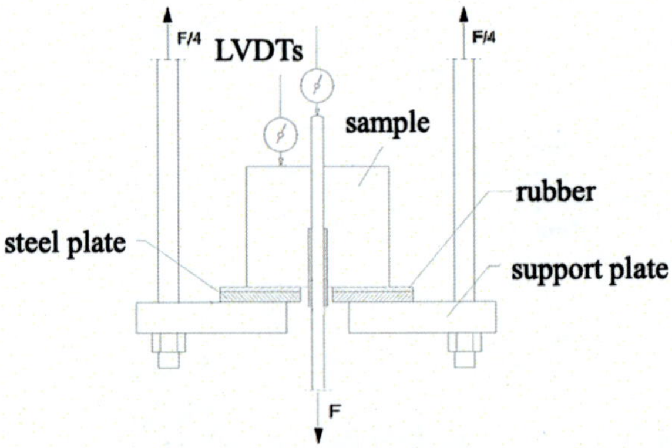


**Fig. 5.** Setup of pull-out tests.

The relationship between the bond stress τ and the slip *s* of the reinforcing bar with respect to the surrounding concrete was obtained by assuming that the bond stress was constant along the bonded length $l_b$, equal to 30 mm, as follows:

$$\tau = \frac{P}{\pi d_b l_b} \tag{1}$$

where $P$ is the pull-out force and $d_b$ is the bar diameter, equal to 6 mm or 10 mm.

As can be seen from Fig. 6, in case of 6 mm bar ($C_c/d_b = 4.5$), all the specimens experienced pull-out failure, whereas when 10 mm bars ($C_c/d_b = 2.5$) were used the cracks propagate in an unstable manner in the concrete cover, so leading to the brittle splitting failure.

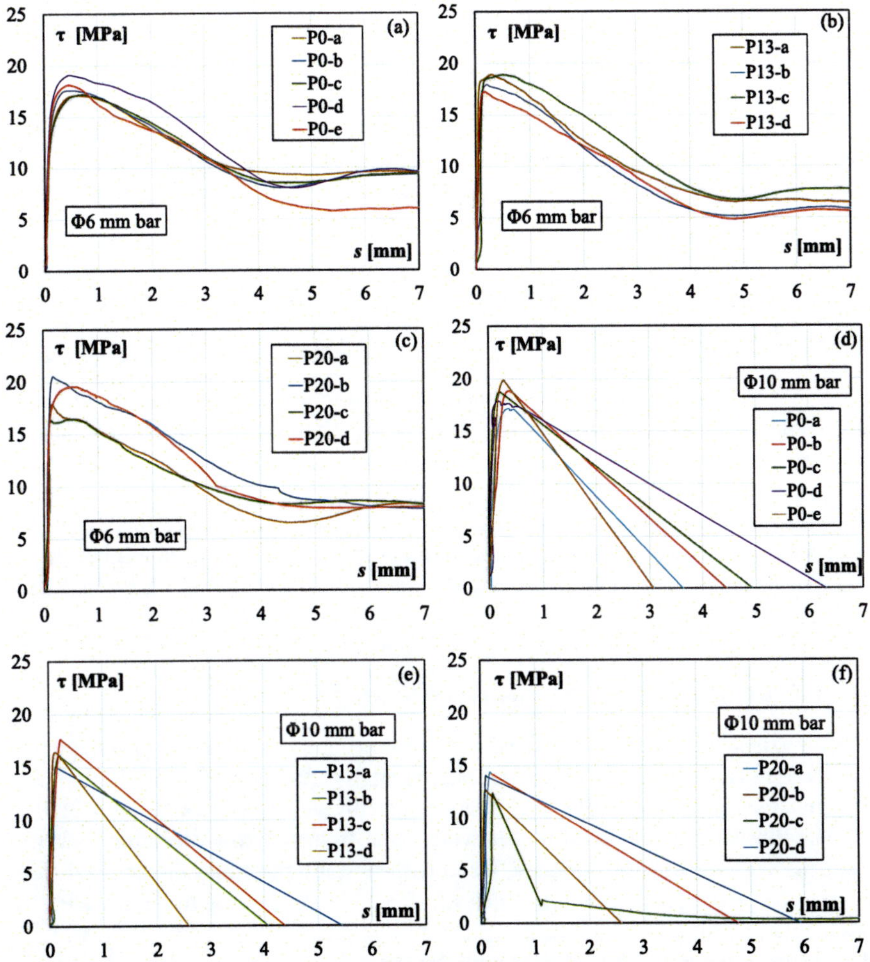


**Fig. 6.** Bond-slip curves for pull-out specimens characterized by different amount of plastic (0%, 13% and 20% by volume of total aggregates), reinforced with 6 mm diameter bar: (a) P0, (b) P13, (c) P20; and 10 mm diameter bar: (d) P0, (e) P13, (f) P20.

In case of $C_c/d_b = 4.5$, the replacement up to 20% by volume of natural aggregates with recycled plastic grains does not change the failure mode, and moreover, a negligible influence on the maximum bond strength $\tau_{max}$ is observed. This happens despite the lower compressive strength that P13 and P20 present with respect to control (P0) and it is probably related to the ability of plastic grains to slow down the propagation of microcracks.

On the other hand, in case of spitting failure ($C_c/d_b = 2.5$), concretes that contain the recycled plastic grains experience a reduction of maximum bond strength $\tau_{max}$. This can be attributed to their lower compressive strength, which in turn derives from the weaker ITZ characterized by more porosity with respect to concrete without waste, so reducing the amount of confinement that the surrounding concrete is able to provide to avoid spitting failure.

Considering both the cases of splitting and pull-out failure, it can be stated that, for the tested mixes, the use of plastic maintains bond between rebar and concrete within acceptable limits, especially in case of replacement of natural aggregates up to 13%. Since, in this case, the reduction of $\tau_{max}$ is registered only for spitting failure and the value decreases only by 12%.

# 4 Influence of Biochar Addition on Physical-Mechanical Behavior of Plastic concrete

On the basis of the results of this experimental campaign, and the findings in the literature (e.g. Saikia and De Brito 2012, Bahij et al. 2020), the most negative effect related to the use of recycled plastic in concrete concerns compression strength. To this aim, a concrete mix with both biochar and plastic grains was analyzed. Biochar was chosen since it is a lightweight carbon-negative waste, which was successfully applied as filler in concrete (Sirico et al. 2021). The purpose is to try to get the maximum benefits of both plastic and biochar, while reducing their individual negative effects through their combined use.

## 4.1 Mix Design, Workability and Preparation of Specimens

The investigated concrete mix, named in the following P13B5, was characterized by a replacement (in volume) of 13.33% of natural aggregates with recycled plastic grains and contained an additional 5% biochar (by weight of cement) with the function of filler.

The biochar particles used in this work came from a pyro-gasification industrial plant, which uses, as input biomass, wood chips resulting from locally sourced wood waste. The power plant, which is a downdraft gasifier, is devoted to energy and heat production, and it produces biochar as solid by-product, which represents a waste to dispose of in landfills. The biochar particles were used as received from the gasifier without any pre-treatment, such as sieving or grinding, to make the recycling process more sustainable.

The mix proportion was the same reported in Table 1 for P13, with the addition of 5% biochar and a different superplasticizer (SP) dosage (i.e. 4.69 kg/m$^3$). In order to maintain the same slump value, for B5P13 it was indeed necessary to use more than double the amount of superplasticizer required for the same mix without plastic (P13), since biochar particles absorb a high amount of water, as known in the literature (Gupta and Kua 2018, Sirico et al. 2021). Nevertheless, the amount of superplasticizer needed for B5P13, which is equal to 1.15% by weight of cement, was only 20% more than that needed for concrete without waste, so still lying within an acceptable range for practical uses.

For the same tendency of biochar particles of retaining part of the mixing water for gradually releasing it during the curing stage, the mixing sequence was slightly modified by pre-soaking biochar in about 25% of the water computed for the mix design, for about 48h before mixing. Then biochar particles, together with the 25% of water were added at the first stage of the mixing sequence, together with the natural aggregates and the recycled plastic grain.

Then the specimens (cubes and prisms) were cast and cured in water for 28 days until testing.

### 4.2 *Compressive Strength*

As can be seen in Table 4, the addition of 5% biochar partially counterbalances the reduction of compressive strength due to the 13% replacement of natural aggregates with plastic grains. Average compressive strength ($R_c$) of B5P13 mix is indeed about 20% more than that of P13, while providing a reduction of only 13% with respect to concrete without any waste (P0). The improvement in compressive strength offered by biochar can be attributed to two mechanisms. The first is the filler effect: biochar particles fill the voids of the cement matrix, which increases when plastic is used. The second is linked to biochar ability to absorb water: the water not absorbed due to the hydrophobic nature of plastic grains is soaked up by biochar particles, so resulting in an improvement of compressive strength.

**Table 4.** Average values of mechanical properties of B5P13 mix with respect to P0 and P13.

| Mix | $R_c$ | $f_{ct,fl}$ | $G_f$ |
|---|---|---|---|
| | MPa | MPa | MPa |
| P0 | 39.58 | 2.90 | 84.73 |
| P13 | 28.81 | 2.96 | 109.89 |
| B5P13 | 34.28 | 2.29 | 210.62 |

### 4.3 *Tensile and Fracture Behavior*

As can be seen in Table 5, the addition of 5% biochar leads to a reduction of about 20% with respect to the average flexural strength ($f_{ct,fl}$) of both P0 and P13. This effect can be probably related to a not optimal dispersion of biochar particles in the cement matrix and an increased formation of air voids in the tensile plane, as also suggested by Sirico et al. (2021). However, this slight reduction in flexural strength is largely compensated by a considerable enhancement of fracture energy ($G_f$), with an increase of about 200% and 110%, with respect to P0 and P13 mix, respectively. Also ductility is greatly improved since B5P13 presents a CMOD at failure equal to 2.88 mm, meaning an increase of about 190% and 60%, with respect to the value P0 and P13 mix, reported in Table 3.

Even if the ability of biochar to increase the toughness and ductility is known in the literature (Restuccia and Ferro 2016, Suarez-Riera et al. 2020) for cement pastes and mortars, its effect in case of concretes seems not so significant (Sirico et al. 2021). This means that the combined use of biochar and plastic grains is able to provide a greater beneficial effect as regards the post-cracking behavior of concrete, with respect to their individual use.

## 5 Conclusions

In this work, the physical-mechanical behavior of concretes that incorporate recycled plastic grains is experimentally investigated. On the basis of the obtained results the following conclusions can be drawn:

- the substitution of 13% and 20% of natural aggregates by volume with plastic enables reducing the density, so obtaining lightweight concrete;
- the use of 13% and 20% plastic enables strongly enhancing workability, so requiring almost half or third amount of superplasticizer needed to obtain the same slump class as concrete without waste;
- compressive strength decreases linearly for increasing plastic amounts, due to the poor physical-chemical compatibility between the plastic grains and the cement matrix;
- 13% plastic enhances tensile behavior of concrete, especially in terms of post-cracking performance, by increasing fracture energy by 30%, residual strength by 40%, and CMOD at failure by 80%;
- as regards the tested pull-out specimens, plastic grains do not significantly influence both the maximum and the residual bond strength value in case of pull-out failure, whereas in case of splitting failure bond strength decreases, but the reduction is low (i.e. equal to 12% for 13% plastic);
- the addition of biochar allows a general improvement of the mechanical performance of concrete with recycled plastic grains, and a strong increase is registered for post-cracking behavior, in terms of both fracture energy and ductility.

**Acknowledgements.** This work was financially supported by Emilia Romagna Region (Italy) in the framework of 2014/20 POR-FESR project "IMPReSA-betonplast: Development of innovative construction materials including non-recyclable plastic particles acting as reinforcement", grant number PG/2018/632099. Authors also gratefully acknowledge Ecoplast Srl, Cementirossi SpA, Pizzarotti & C. SpA, and Mapei SpA for providing the materials needed for the preparation of specimens.

## References

Bahij, S., Omary, S., Feugeas, F., Faqiri, A.: Fresh and hardened properties of concrete containing different forms of plastic waste–a review. Waste Manag. **113**, 157–175 (2020)

Belmokaddem, M., Mahi, A., Senhadji, Y., Pekmezci, B.Y.: Mechanical and physical properties and morphology of concrete containing plastic waste as aggregate. Constr. Build. Mater. **257**, 119559 (2020)

Chandra, S.: Waste Materials Used in Concrete Manufacturing. Elsevier Science, Amsterdam (1996)

Danish, A., Mosaberpanah, M.A., Salim, M.U., Ahmad, N., Ahmad, F., Ahmad, A.: Reusing biochar as a filler or cement replacement material in cementitious composites: a review. Constr. Build. Mater. **300**, 124295 (2021)

EN 12350-2:2019, Testing fresh concrete - Part 2: Slump test

EN 12390-3:2019, Testing hardened concrete - Part 3: Compressive strength of test specimens

EN 12390-7:2019, Testing hardened concrete - Part 7: Density of hardened concrete

Gu, L., Ozbakkaloglu, T.: Use of recycled plastics in concrete: a critical review. Waste Manag. **51**, 19–42 (2016)

https://www.europarl.europa.eu/news/en/headlines/society/20181212STO21610/plastic-waste-and-recycling-in-the-eu-facts-and-figures

Gupta, S., Kua, H.W.: Effect of water entrainment by pre-soaked biochar particles on strength and permeability of cement mortar. Constr. Build. Mater. **159**, 107–125 (2018)

JCI-S-001-2003. Method of test for fracture energy of concrete by use of notched beam. Japan Concrete Institute Standard

Lehne, J., Preston, F.: Making Concrete change - Innovation in low-carbon cement and concrete. Chatham House Report, The Royal Institute of International Affairs, London (2018)

Model Code. Final draft, fib Bulletin, pp. 65-66 (2010)

Monteiro, P.J., Miller, S.A., Horvath, A.: Towards sustainable concrete. Nat. Mater. **16**, 698–699 (2017)

RILEM-RC6. Bond test for reinforcement steel. 2. Pull-out test, in RILEM Technical Recommendations for the testing and use of construction materials, pp. 218–220 (1983)

Tam, V.W., Soomro, M., Evangelista, A.C.J.: A review of recycled aggregate in concrete applications (2000–2017). Constr. Build. Mater. **172**, 272–292 (2018)

Maljaee, H., Madadi, R., Paiva, H., Tarelho, L., Ferreira, V.M.: Incorporation of biochar in cementitious materials: a roadmap of biochar selection. Constr. Build. Mater. **283**, 122757 (2021)

Nwankwo, C.O., Bamigboye, G.O., Davies, I.E., Michaels, T.A.: High volume Portland cement replacement: a review. Constr. Build. Mater. **260**, 120445 (2020)

Restuccia, L., Ferro, G.A.: Promising low cost carbon-based materials to improve strength and toughness in cement composites. Constr. Build. Mater. **126**, 1034–1043 (2016)

Saikia, N., De Brito, J.: Use of plastic waste as aggregate in cement mortar and concrete preparation: a review. Constr. Build. Mater. **34**, 385–401 (2012)

Saxena, R., Siddique, S., Gupta, T., Sharma, R.K., Chaudhary, S.: Impact resistance and energy absorption capacity of concrete containing plastic waste. Constr. Build. Mater. **176**, 415–421 (2018)

Sharma, R., Bansal, P.P.: Use of different forms of waste plastic in concrete–a review. J. Clean. Prod. **112**, 473–482 (2016)

Shi, C., Zheng, K.: A review on the use of waste glasses in the production of cement and concrete. Resour. Conserv. Recycl. **52**, 234–247 (2007)

Siddique, R., Khatib, J., Kaur, I.: Use of recycled plastic in concrete: a review. Waste Manag. **28**, 1835–1852 (2008)

Sirico, A., et al.: Biochar from wood waste as additive for structural concrete. Constr. Build. Mater. **303**, 124500 (2021)

Suarez-Riera, D., Restuccia, L., Ferro, G.A.: The use of Biochar to reduce the carbon footprint of cement-based materials. Procedia Struct. Integr. **26**, 199–210 (2020)

# Infrastructures: Development and Upgrading

# Comparison of Different Approaches to Derive Global Safety Factors for Non-linear Analyses of Slender RC Members

Diego Gino and Paolo Castaldo(✉)

Department of Structural, Geotechnical and Buildings Engineering, Politecnico di Torino, Turin, Italy
{diego.gino,paolo.castaldo}@polito.it

**Abstract.** The present study relates to comparison between different approaches for definition of global safety factors for non-linear analysis of slender RC members with reference to new or existing structures. Firstly, a benchmark set of 40 experimental results on reinforced concrete columns is presented. After the description of the main features of the benchmark test sets the related non-linear numerical models have been realized using fiber-modelling as solution strategy. Then, appropriate assumptions concerning aleatoric and epistemic uncertainties have been performed with the aim to run probabilistic analysis of global resistance for each one of the 40 columns. The results of the probabilistic analysis are useful to define global safety factors in line to the global resistance method. Finally, the comparison between different approaches to derive global safety factors is presented and discussed.

**Keywords:** reinforced concrete · global safety factors · non linear analysis

## 1 Introduction

The non-linear analysis (NLAs) is nowadays widely used to perform safety verifications of structures and infrastructures (Castaldo et al. 2017, Vecchi & Belletti 2021, Gino et al. 2021, Castaldo & Amendola 2021, Castaldo et al. 2020a, Castaldo et al. 2022, Troisi 2022, Troisi & Alfano 2022a, Troisi & Alfano 2022b). In this framework, different safety formats conceived for NLAs have been developed and investigated over the last decade (*fib* Model Code 2010, Castaldo et al. 2019). As a general concept, the mentioned above safety formats are based on the global resistance format introduced by *fib* Model Code 2010 and allow to estimate the design structural resistance $R_d$ of a reinforced concrete (RC) system as:

$$R_d = \frac{R_{NLA,m}}{\gamma_R^{GL}} \tag{1}$$

where: $R_{NLA,m}$ is the mean value of global structural resistance estimated with NLA; $\gamma_R{}^{GL}$ is the global safety factor that includes the influence of aleatoric uncertainties (i.e., materials and geometric properties) and epistemic uncertainty (i.e., model uncertainty) (Castaldo et al. 2020b, Castaldo et al. 2020c, Engen et al. 2017).


M. A. Aiello and A. Bilotta (Eds.): ICC 2022, LNCE 435, pp. 207–220, 2024.
https://doi.org/10.1007/978-3-031-43102-9_17

This work is focused on the comparison of different approaches to derive the global safety factor $\gamma_R{}^{GL}$ concerning the case of slender RC columns. As first, a benchmark set of 40 experimental results on reinforced concrete columns is presented according to Gino et al. (2021). The experimental tests have been reproduced by non-linear numerical (NLN) models using a validated solution strategy adopting the software OpenSees (McKenna et al., 2000). Then, probabilistic analysis of structural resistance has been performed using the Latin Hypercube Sampling (LHS) method to account for aleatoric and epistemic uncertainties (Gino et al. 2021, JCSS 2001). The results of the probabilistic investigation permit to compare the efficiency of three approaches (denoted as *Approach 1*, *2* and *3*) useful to derive the global safety factor $\gamma_R{}^{GL}$. These outcomes can be useful to discuss about the best approach to define global safety factors in next generation of design codes (e.g., EN1992-1-1 2014, *fib* Model Code 2020).

### 1.1 Approaches to Derive the Global Safety Factor

The global safety factor $\gamma_R{}^{GL}$ can be derived with:

- *Approach 1:* this approach is the one adopted by *fib* Model Code 2010 and separates the partial safety factors accounting for *aleatoric* uncertainties $\gamma_R$ and epistemic ones (i.e. model) $\gamma_{Rd.}$ The $\gamma_R{}^{GL}$ *can be evaluated as the product:*

$$\gamma_R^{GL} = \gamma_{Rd}\gamma_R \tag{2}$$

with:

$$\gamma_R = \exp(\alpha_R\beta_t V_R) \tag{3}$$

$$\gamma_{Rd} = \frac{\exp(\alpha'_R\beta_t V_\vartheta)}{\mu_\vartheta} \tag{4}$$

where $V_R$ is the coefficient of variation of global resistance from the results of probabilistic analysis of global resistance including sampling of aleatoric uncertainties only; $\mu_\vartheta$ is the mean value of model uncertainty; $V_\vartheta$ is the coefficient of variation of model uncertainty; $\alpha_R$ and $\alpha'_R$ are the FORM factors for dominant and non-dominant variables, fixed, respectively to 0.8 and 0.32 (Hasofer & Lind 1974) while $\beta_t$ is the target reliability index (*fib* Model Code 2010).

In line to *fib* Model Code 2010 $\gamma_R$ and $\gamma_{Rd}$ are determined assuming a lognormal probabilistic distribution.

- *Approach 2:* this approach consists in the evaluation of $\gamma_R{}^{GL}$ in line to the assumption of lognormal probabilistic distribution for structural resistance as:

$$\gamma_R^{GL} = \frac{\exp(\alpha_R\beta_t V_R^{GL})}{\mu_\vartheta} \tag{5}$$

where the coefficient of variation of global resistance $V_R{}^{GL}$ is inclusive of aleatoric and epistemic uncertainties and can be obtained in simplified manner as (Yu et al. 2021):

$$V_R^{GL} = \sqrt{(V_R)^2 + (V_\vartheta)^2} \tag{6}$$

- *Approach 3:* this approach is the more general and estimates the ${\gamma_R}^{GL}$ in line to Eq. (5) but, as a difference, it determine ${V_R}^{GL}$ from the results of probabilistic analysis of global resistance including sampling of both aleatoric and epistemic uncertainties. This approach can be adopted as the reference one.

Finally, with reference to the mentioned above approaches, the application of Eq. (1) can be performed using as for $R_{NLA,m}$ the outcome from NLAs of the 40 RC columns performed using experimental values (i.e. mean) of main properties according to Allaix et al. (2013). Equations (3)–(6) are valid until the values of coefficient of variation of main variables are lower or equal than 0.3 (with an error lower than 5%).

## 2 Benchmark Results and Modelling

In this study the benchmark experiments summarized in Gino et al. (2021) are adopted.

The database collects results of tests in compression of 40 RC columns (Mehmel et al. 1969, Saenz & Martin, 1963, Foster & Attard 1997, Pancholi 1977, Dracos 1982, Iwai et al. 1986, Chuang & Kong 1997, Barrera et al. 2011, Baumann 1935) characterized by wide range of slenderness, materials properties and reinforcement configuration and fulfils the limits of applicability of EN1992-1-1 2014 and *fib* Model Code 2010.

Figure 1 shows tests sets configurations and also related assumptions for numerical modelling according to Gino et al. (2021). The test sets configuration presents both compression tests with incremental axial load (type A, D), compression tests with eccentric incremental load (type B), test with incremental horizontal force with constant axial load (type C).

The NLN models of the 40 RC columns have been realized using force-based fiber elements implemented in Opensees. The solution strategy herein adopted correspond to the optimum one that minimizes the model uncertainty in line to the "*Model 8*" as described by Gino et al. (2021) and is inclusive of both geometric and materials non-linearities. The NLN models consider the effect of confinement provided by stirrups (Saatcioglu et al. 1992) and of instability of reinforcement bars loaded in compression (Dhakal & Maekawa 2022). The ultimate axial load reached within NLA is established as the one associated to the last load-step able to fulfil the convergence criteria (Gino et al. 2021). The loading process have been defined according to the procedure of execution of the experimental tests including the influence of self-weight.

Table 1 reports general information associated to the tests sets and the results in terms of comparison between NLAs performed using the experimental values of both materials and geometrical properties $R_{NLA,m}$ and experimental evidence $R_{exp}$. Further and detailed information about the NLN models can be acknowledged in Gino et al. (2021).

## 3 Probabilistic Analysis of Structural Resistance

The NLN models determined according to previous section has been used to perform probabilistic analysis of structural resistance of the 40 RC columns.

**Table 1.** Benchmark results collected by Gino et al., (2021) and results from numerical modelling using Opensees.

| Exp. Test | Spec. | Slenderness λ | $R_{NLA,m}$ | $R_{exp}$ |
|---|---|---|---|---|
| | | [–] | [kN] | [kN] |
| 2L20-30 | B | 15 | 694.3 | 750.0 |
| 2L20-60 | | | 736.4 | 700.0 |
| 2L8-120R | | | 1152.7 | 1092.0 |
| 4L8-30 | | | 1032.9 | 1100.0 |
| 4L20-120 | | | 830.7 | 900.0 |
| 4L8-120R | | | 1319.5 | 1247.0 |
| C000 | A | 17 | 560.6 | 559.6 |
| C020 | B | | 328.5 | 327.3 |
| B020 | | 52 | 263.7 | 271.5 |
| RL300 | | 56 | 423.3 | 474.3 |
| A-17-0.25 | B | 48 | 1367.4 | 1181.4 |
| C-31.7-0.25 | | 94 | 280.1 | 333.4 |
| 3.3 | | 59 | 856.4 | 782.6 |
| 5.1 | | | 810.8 | 735.5 |
| 4.1 | | 88 | 391.7 | 367.7 |
| N30-10.5-C0-3-30 | C | 68 | 16.6 | 16.6 (280)* |
| H60-10.5-C0-1-30 | | | 17.9 | 17.2 (412)* |
| III | A | 74 | 347.3 | 343.2 |
| Va | | | 680.7 | 684.5 |
| 2 | | 83 | 236.8 | 235.4 |
| I | | 104 | 258.0 | 264.8 |
| VI | | 106 | 363.2 | 392.3 |
| 15 | | 136 | 560.3 | 549.2 |
| 3 | | 137 | 563.4 | 666.9 |
| 8 | B | 83 | 236.8 | 235.4 |
| 9 | | 135 | 205.9 | 205.9 |
| 12 | | | 112.2 | 112.8 |
| 6 | | 137 | 227.6 | 225.6 |
| 24D-2 | D | 104 | 192.8 | 198.4 |
| 15E-2 | A | 139 | 129.3 | 161.0 |

(*continued*)

**Table 1.** (*continued*)

| Exp. Test | Spec. | Slenderness λ | $R_{NLA,m}$ | $R_{exp}$ |
|---|---|---|---|---|
| | | [–] | [kN] | [kN] |
| S28 | B | 167 | 49.9 | 44.0 |
| S30 | | | 53.4 | 48.0 |
| S25 | | 200 | 42.3 | 36.0 |
| 5 | A | 208 | 78.7 | 72.7 |
| 6 | | | 82.3 | 72.2 |
| 17A | | 225 | 37.1 | 31.9 |
| 20 | | 243 | 39.8 | 37.9 |
| 18 | | | 39.8 | 33.9 |
| 8 | | 274 | 31.0 | 31.9 |
| 20 | | 15 | 32.3 | 29.9 |

* Value of the axial load applied to the RC column during the laboratory test.

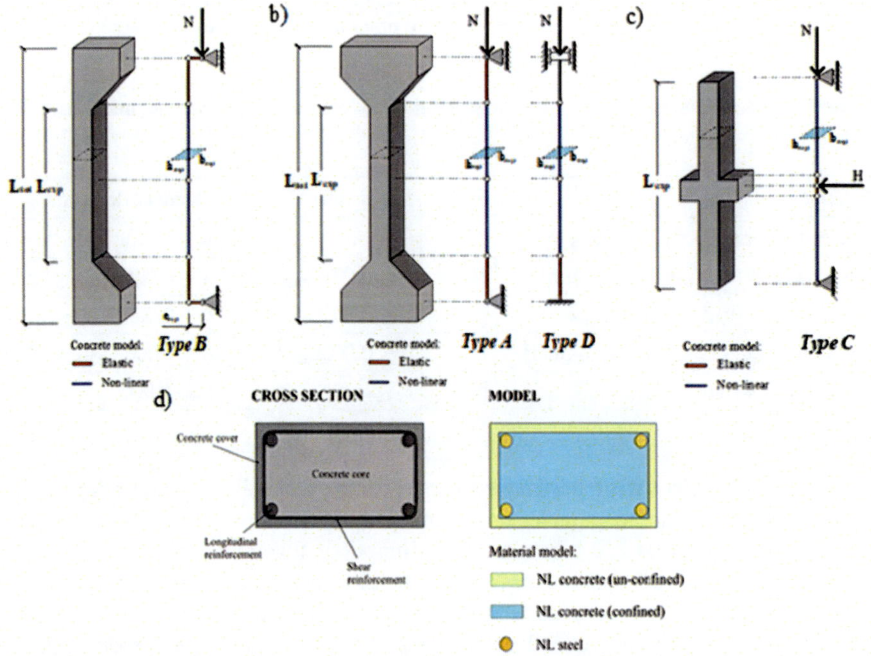


**Fig. 1.** Test sets configuration and strategy for numerical modelling of the 40 RC columns (Gino et al. 2021).

**Table 2.** Probabilistic model for material properties random variables (aleatoric) in line to JCSS 2001.

| Random variable | Distr. | Mean value | Coefficient of variation [-] | Statistical correlation* |
|---|---|---|---|---|
| Concrete cylinder compressive strength $f_c$ [MPa] | LN | $f_{c,exp}$ | 0.15 | – |
| Reinforcements yielding strength $f_y$ [MPa] | LN | $f_{y,exp}$ | 0.05 | $f_u$ (0.75), $\varepsilon_u$(−0.45) |
| Reinforcements yielding strength fu [MPa] | LN | $f_{u,exp}$ | 0.05 | $f_y$ (0.75), $\varepsilon_u$(−0.60) |
| Reinforcements Young's modulus $E_s$ [MPa] | LN | 210000 | 0.03 | – |
| Reinforcements ultimate strain $\varepsilon_u$ [-] | LN | 0.075 | 0.09 | $f_y$ (−0.45), $f_u$ (−0.60) |

* (–) coefficient of linear correlation with respect to other material random variables.

**Table 3.** Probabilistic model for geometric properties random variables (aleatoric) in line to JCSS 2001.

| Random variable | Distr. | Mean value | Standard deviation |
|---|---|---|---|
| Concrete cover (C) $Yc = C\text{-}C_{exp}$ [mm] | N | 0 | 5 |
| Cross section (b) $Y_b = b\text{-}b_{exp}$ [mm] | N | $0 \leq 0.003b_{exp} \leq 3$ | $4 + 0.006\ b_{exp} \leq 10$ |
| Cross section (h) $Y_h = h\text{-}h_{exp}$ [mm] | N | $0 \leq 0.003h_{exp} \leq 3$ | $4 + 0.006\ h_{exp} \leq 10$ |
| Column length (L) $Y_L = L\text{-}L_{exp}$[mm] | N | $0 \leq 0.003L_{exp} \leq 3$ | $4 + 0.006\ L_{exp} \leq 10$ |
| Eccentricity e [mm] | N | $e_{exp}$ | $L_{exp}/1000$ |

In Tables 2, 3 and 4 it is reported the summary of the probabilistic model adopted for the main involved random variables in according to JCSS 2001 and Gino et al. (2021).

In particular, Tables 2, 3 illustrates the assumptions for probabilistic modelling of material and geometric uncertainties (i.e., aleatoric) whereas Table 4 reports the probabilistic idealization for model uncertainty according to Gino et al. (2021) (i.e., epistemic). The data of Table 4 relates to the herein adopted solution strategy and are determined assuming limited influence of experimental uncertainty (Gino et al. 2021).

The Latin Hypercube Sampling method (Mckey et al. 1979, Castaldo et al. 2019) have been adopted to realize 100 samples for each NLN numerical model related to the RC columns of Table 1.

Instead, the *set II* reports the characterization of the resistance random variable ${R_{NLA}}^{GL}$ as a function of both aleatoric and epistemic (i.e., model) uncertainties with reference to Table 2, 3 and 4. The Eq. (8) is consistent with the multiplicative approach of JCSS (2001) for determination of model uncertainty. The outcomes from the probabilistic analysis are reported in Fig. 2. In particular, the ratio δ between mean value μ of structural resistance derived from probabilistic investigation (i.e., $\mu_R$ for *set I* and ${\mu_R}^{GL}$ for *set II*) and the $R_{NLA,m}$ determined using experimental values (i.e., mean values) of geometric and material properties is illustrated in Fig. 2(a). It can be observed that these ratios, in average, are below the unit. This result is a confirmation that the first order approximation, which consists in the equality between $R_{NLA,m}$ and the mean value derived from probabilistic analysis, is a safe approximation (Allaix et al. 2013, Castaldo et al. 2019). Figure 2(b) reports the summary of the coefficient of variation of structural resistance estimated by *set I* ($V_R$) and *set II* LHS samplings. In both cases, the coefficient of variation of structural resistance is strongly influenced by geometric uncertainty when the slenderness grows. The contribution of model uncertainty to overall variability is more significant for slenderness values lower that 150 with respect to slenderness higher than 150. This is due to the assumption related to the model uncertainty random variable (Table 4) and to the increasing influence of geometric uncertainty for high values of slenderness.

The statistical parameters related to mean values $\mu$ and coefficients of variation $V$ reported in Fig. 2 are the maximum likelihood estimators (MLE) (Faber 2012) assuming that the structural resistance pertains to lognormal probabilistic distribution.

**Table 4.** Probabilistic model for model uncertainty (epistemic) in line to Gino et al 2021.

| Random variable | Distr. | Mean value | Standard deviation |
|---|---|---|---|
| Model uncertainty $\vartheta$ $\lambda \leq 150$ | LN | 1.01 | 0.07 |
| Model uncertainty $\vartheta$ $\lambda > 150$ | LN | 0.91 | 0.03 |

In particular, two sets of probabilistic analyses have been carried out investigating the following resistance random variables:

- *Set I*:

$$R_{NLA} = R_{NLA}\left(f_c, f_y, f_u, E_s, \varepsilon_u, Y_C, Y_b, Y_h, e\right) \tag{7}$$

- *Set II:*

$$R_{NLA}^{GL} = \vartheta \cdot R_{NLA}\left(f_c, f_y, f_u, E_s, \varepsilon_u, Y_C, Y_b, Y_h, e\right) \tag{8}$$

The *set I* of proposes the evaluation of the resistance random variable $R_{NLA}$ as a function of the main aleatoric random variables according to Table 2 and 3.

The assumption of lognormal probabilistic model for structural resistance have been verified according to Anderson-Darling statistical test with a level of significance 5% (Anderson & Darling 1953).

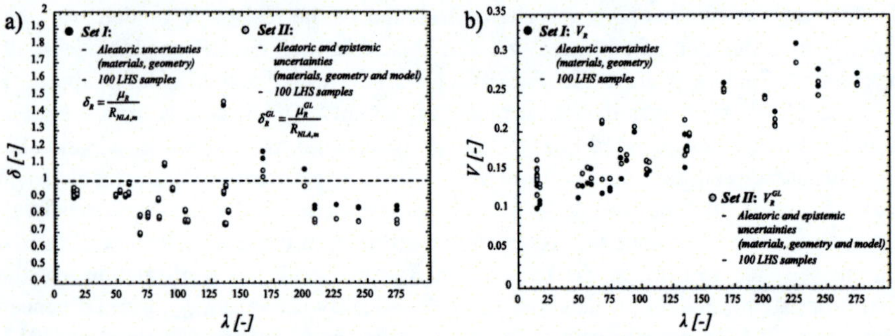


**Fig. 2.** Results from probabilistic analysis of structural resistance of the 40 RC column: δ ratio a); coefficient of variation of structural resistance V b) for both *set I* and *set II*.

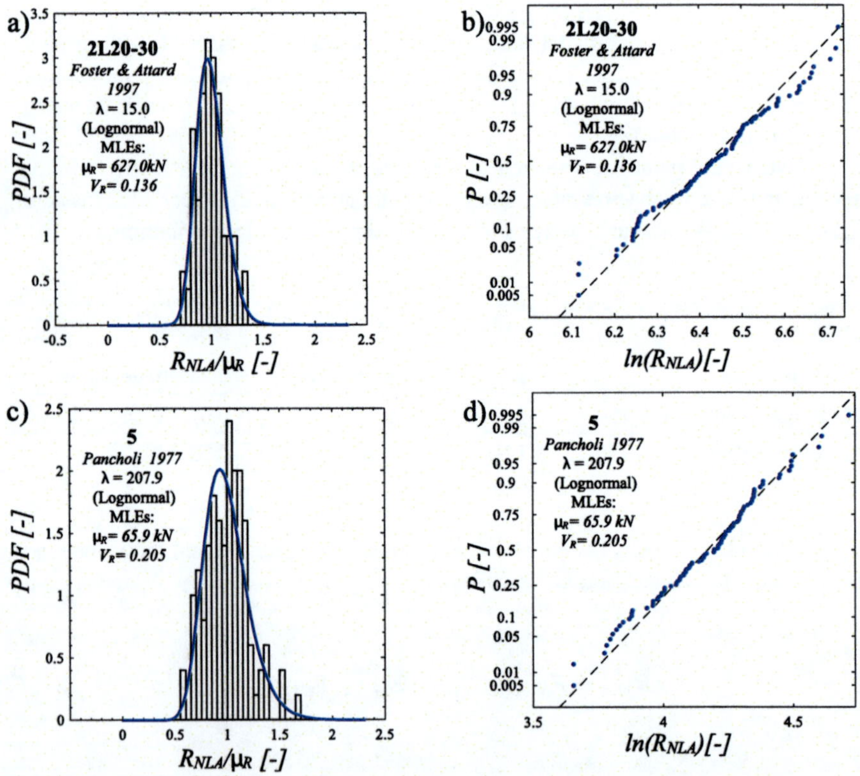


**Fig. 3.** Frequency histograms, lognormal fitting distributions a), c) and probabilistic plot b), d) for 2 columns (Gino et al. 2021) considering the LHS – *set I* (only aleatoric uncertainty).

Figures 3 and 4 illustrate the frequency histograms, lognormal fits and probabilistic plots (lognormal) related to two columns of the 40 considered for probabilistic investigation.

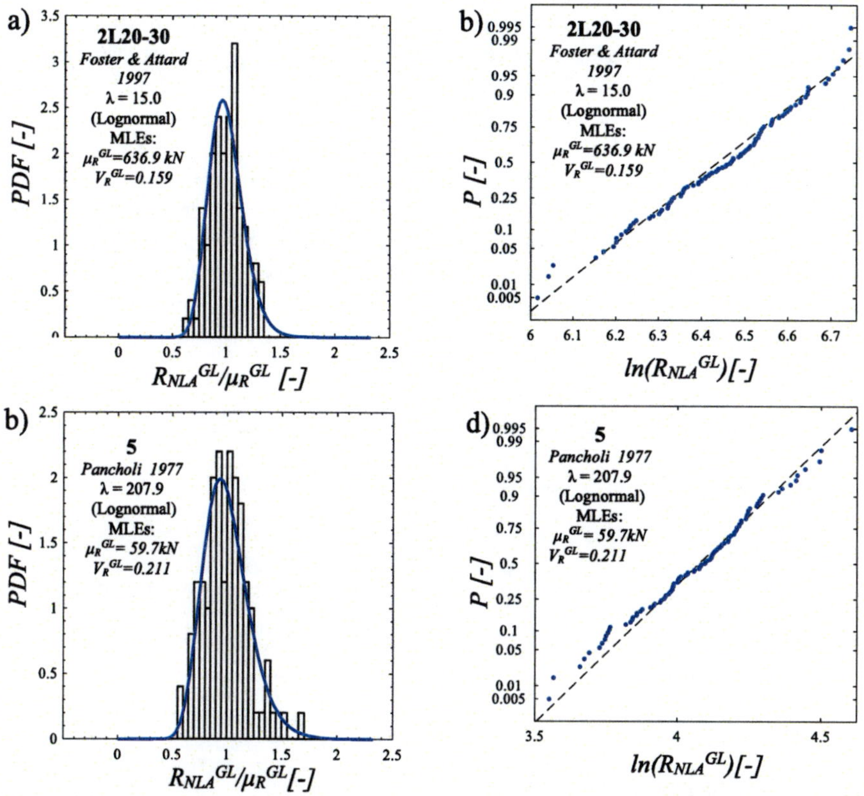


**Fig. 4.** Frequency histograms, lognormal fitting distributions a),c) and probabilistic plot b),d) for 2 columns (Gino et al. 2021) considering the LHS – *set II* (aleatoric and epistemic uncertainty).

The results of the probabilistic analysis can be used to estimate the global safety factor $\gamma_R{}^{GL}$ according to description of Sect. 1.1.

## 4 Comparison Between Global Safety Factors

In the next, the comparison between the Approaches 1, 2 and 3 to derive the global safety factor $\gamma_R{}^{GL}$ is proposed. The comparison is performed between global safety factors derived for the same target level of reliability.

In detail, concerning reinforced concrete structures of new realization and reference period set equal to 50 years (*fib* Model Code 2010, *fib* Bulletin 80 2016 and ISO 2394 2015), the target value of reliability index can be set equal to $\beta_t = 3.8$. In case of existing structures, reference to *fib* Bulletin 80 can be made with aim to determine the most appropriate value of the target reliability index.

With reference to the *Approach 1*, the derivation of the global resistance factor $\gamma_R$ according to Eq. (3) have been carried out using the results in terms of coefficient of variation $V_R$ which derives from probabilistic results of LHS sampling *Set I*.

The model uncertainty safety factor $\gamma_{Rd}$ have been determined through the Eq. (4) using the statistical properties reported din Table 4. In this way, the global safety factor $\gamma_R{}^{GL}$ can be obtained as the product between $\gamma_R$ and $\gamma_{Rd}$ (Eq. (2)). Both Eq. (3) and Eq. (4) grounds to the assumption of lognormal probabilistic distribution. As introduced in Sect. 1.1, this approach separates the aleatoric uncertainties from the epistemic ones and adopts fixed values of FORM factors (Hasofer & Lind 1974) able to cover the major part of loading and structural configurations related to practical cases. Within the Approach 1, as commented in *fib* Model Code 2010, the aleatoric uncertainties are, in general, considered as dominant variables while the epistemic ones as non-dominant. This assumption is clearly a simplification and can turns out to be not proper in cases where the model uncertainty mostly contributes to probability of structural failure with respect to aleatoric ones.

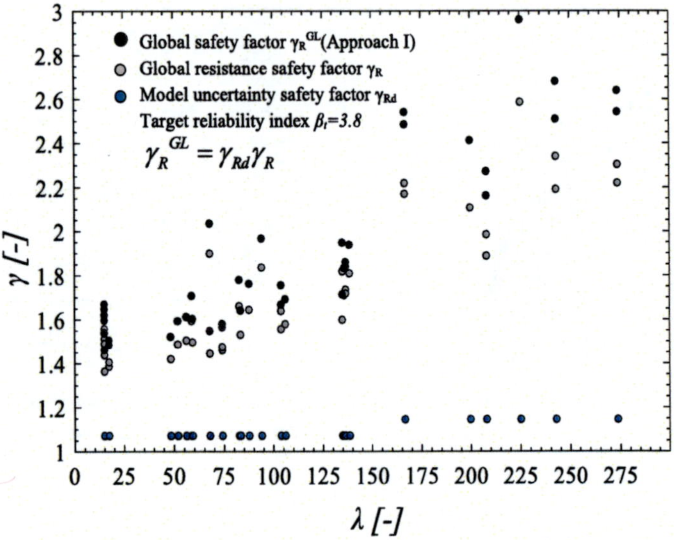


**Fig. 5.** Results from the determination of global safety factor $\gamma_R{}^{GL}$ according to the *Approach 1.*

This situation verifies in cases where, as an example, the coefficient of variation describing the probabilistic distribution of model uncertainty is higher than the ones associated to probabilistic distribution of materials and geometric random variables (*fib* Bulletin 80). This approach is the one that have been implemented within the *fib* Model Code 2010. Figure 5 illustrates the results of the application of *Approach 1* to the estimation of $\gamma_R{}^{GL}$. Due to the relevant influence of geometric uncertainties for high values of slenderness, the maximum value of $\gamma_R{}^{GL}$ is close to 3.0 whereas the minimum one is close to 1.45 for slenderness values around to 15.

The *Approach 2* consist in the determination of $\gamma_R{}^{GL}$ considering together sources of uncertainty of both aleatoric (i.e., material and geometric) and epistemic (i.e., model) nature. This approach leads to the evaluation of the $\gamma_R{}^{GL}$ global safety factor according to Eq. (5) in line to assumption of lognormal probabilistic distribution. The Eq. (5)

can be used adopting statistical parameters for model uncertainty according to Table 4 and the outcomes from probabilistic analysis related to *Set I* of LHS. In particular, the Eq. (6) allow to estimate the coefficient of variation ${V_R}^{GL}$ inclusive of both aleatoric and epistemic uncertainty with a simplified approach depending from the coefficient of variation $V_R$ and $V_\vartheta$. Note that with the *Approaches 1* and *2*, as an alternative and in absence of probabilistic analysis of structural resistance, the term $V_R$ can be estimated in simplified manner in line to estimation of coefficient of variation methods as described by Cervenka (2013) and Novak & Novak (2021).

In analogy with the *Approach 2*, the *Approach 3* is able to include directly in ${\gamma_R}^{GL}$ the random variability of both aleatoric and epistemic uncertainties. The global safety factor ${\gamma_R}^{GL}$ can be determined always using the Eq. (5) but, this time, the outcomes from the probabilistic analysis related to *Set II* LHS are used to determine the coefficient of variation ${V_R}^{GL}$. This is the most general approach but, in practice, it is not of simple application as it requires necessarily a probabilistic evaluation of structural resistance including both epistemic and aleatoric uncertainties according to Eq. (8). In the case of the present investigation, the *Approach 3* is considered as the reference one as descends from refined probabilistic analysis of structural resistance. The comparison between the *Approaches 1*, *2* and *3* is reported in Fig. 6.

As first, it can be recognized that the *Approach 1* is the most conservative and leads to the determination of global safety factors ${\gamma_R}^{GL}$ in average 5% higher with respect to the one of the *Approach 3* and *3%* with respect to the *Approach 2*.

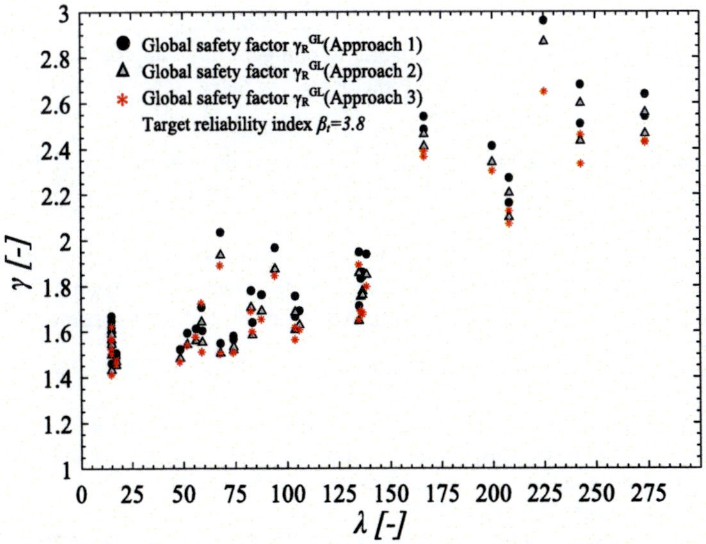


**Fig. 6.** Comparison of the results in terms of ${\gamma_R}^{GL}$ achieved using the *Approach 1*, *2* and *3*.

The adoption of fixed values for FORM factors $\alpha_R$ and $\alpha'_R$ relating to dominant and non-dominant variables (respectively 0.8 and 0.32 in line to Hasofer & Lind (1974) lead to achieve extra safety that justifies the choice performed by *fib* Model Code 2010.

However, too much extra safety may become inconvenient in design processes where the economic aspect is of crucial relevance.

The *Approach 2*, in comparison to *Approach 1*, requires the same input data and allow to take into account each random variable affecting structural resistance with its actual variability without differentiating between dominant or non-dominant variables. The *Approach 2* leads to global safety factors $\gamma_R{}^{GL}$ that are, in average, higher than 2% with respect to the ones obtained with the *Approach 3*. This highlight that the *Approach 2* keep a certain level of extra-safety with respect to pure probabilistic approach but that is less that the one associated to the *Approach 1*. The results of the investigation may suggest that the *Approach 2* can be considered as the most consistent to be adopted for implementation in next generation design codes, changing the road traced, for example, by *fib* Model Code 2010.

## 5 Conclusions

The present study relates to comparison between different approaches for definition of global safety factors for non-linear analysis of slender RC columns with reference to new or existing structures. A set of 40 benchmark test on RC columns has been considered in line to current design codes. The solution strategy for NLN modelling have been selected with the aim to minimize the epistemic uncertainty related to the definition of numerical model. The numerical models have been useful to perform two different sets of probabilistic analysis considering aleatoric (Set I) and both aleatoric and epistemic uncertainty (Set II) using the Latin Hypercube Sampling. The outcomes of the probabilistic investigation have been useful to compare three approaches (*Approach 1*, *2* and *3*) to determine the global safety factor $\gamma_R{}^{GL}$ related to the 40 RC columns. The following conclusions can be drawn:

- the *Approach 1* is the safer one as it ground on the assumption of fixed values for FORM factors differentiating between dominant (aleatoric) and non-dominant (epistemic) variables;
- the *Approach 2* proposes intermediate results with respect to *Approaches 1* and *3* having the same input variable as required by *Approach 1*. The *Approach 2* seems to be the most convenient to be implemented in drafting of next generation of design codes.

**Acknowledgements.** This work is part of the collaborative activity developed by the authors within the framework of the Commission 3 – Task Group 3.1: "Reliability and safety evaluation: full-probabilistic and semi-probabilistic methods for existing structures" of the International Federation for Structural Concrete (*fib*).

This work is also part of the collaborative activity developed by the authors within the framework of the "PNRR": SPOKE 7 "CCAM, Connected Networks and Smart Infrastructure" - WP4.

This work is also part of the collaborative activity developed by the authors within the framework of the WP 11 – Task 11.4 – ReLUIS.

## References


Allaix, D.L., Carbone, V.I., Mancini, G.: Global safety format for non-linear analysis of reinforced concrete structures. Struct. Concr. **14**(1), 29–42 (2013)

Anderson, T.W., Darling, D.A.: Asymptotic theory of certain "goodness-of-fit" criteria based on stochastic processes. Ann. Math. Stat. **23**, 193–212 (1952). https://doi.org/10.1214/aoms/1177729437

Barrera, A.C., Bonet, J.L., Romero, M.L., Miguel, P.F.: Experimental tests of slender reinforced concrete columns under combined axial load and lateral force. Eng. Struct. **33**(12), 3676–3689 (2011)

Baumann, O.: Die Knickung der Eisenbeton-Säulen. PhD thesis, ETH Zurich (1935)

Castaldo, P., Amendola, G.: Optimal DCFP bearing properties and seismic performance assessment in nondimensional form for isolated bridges. Earthquake Eng. Struct. Dynam. **50**(9), 2442–2461 (2021). https://doi.org/10.1002/eqe.3454

Castaldo, P., Amendola, G., Gino, D., Miceli, E.: Seismic performance of bridges isolated with DCFP devices. In: Proceedings of the International Conference on Structural Dynamic, EURODYN, pp. 1704–1721 (2020a)

Castaldo, P., Gino, D., Mancini, G.: Safety formats for non-linear analysis of reinforced concrete structures: discussion, comparison and proposals. Eng. Struct. **193**, 136–153 (2019). https://doi.org/10.1016/j.engstruct.2018.09.041

Castaldo, P., Gino, D., Amendola, G., Miceli, E.: Model uncertainties in NLFEAs of RC systems under cyclic loads. In: Proceedings of the International Conference on Structural Dynamic, EURODYN, pp. 3527–3547 (2020c)

Castaldo, P., Gino, D., Bertagnoli, G., Mancini, G.: Resistance model uncertainty in non-linear finite element analyses of cyclically loaded reinforced concrete systems. Eng. Struct. **211**(2020), 110496 (2020). https://doi.org/10.1016/j.engstruct.2020.110496

Castaldo, P., Nastri, E., Piluso, V.: FEM simulations and rotation capacity evaluation for RHS temper T4 aluminium alloy beams. Compos. B Eng. **115**, 124–137 (2017)

Castaldo, P., Gino, D., Marano, G.C., Mancini, G.: Aleatory uncertainties with global resistance safety factors for non-linear analyses of slender reinforced concrete columns. Eng. Struct. **255**, 113920 (2022)

Cervenka, V.: Global safety formats in fib model code 2010 for design of concrete structures. In: Proceedings of the 11th International Probabilistic Workshop, Brno (2013)

Chuang, P.H., Kong, F.K.: Large-scale tests on slender, reinforced concrete columns. Struct. Eng. **75**(23), 410–416 (1997)

Dhakal, R.P., Maekawa, K.: Modeling for post-yield buckled of reinforcement. J. Struct. Eng. **128**(9), 1139–1147 (2002)

Dracos, A.: Long slender reinforced concrete columns. PhD thesis, University of Bradford (1982)

EN 1992-1-1. Eurocode 2 – Design of concrete structures. Part 1–1: general rules and rules for buildings. CEN, Brussels (2014)

Engen, M., Hendriks, M.A.N., Köhler, J., Øverli, J.A., Åldtstedt, E.: A quantification of modelling uncertainty for non-linear finite element analysis of large concrete structures. Struct. Saf. **2017**(64), 1–8 (2017)

Faber, M.H.: Statistics and Probability Theory. Springer, Cham (2012)

Fib Bulletin N°80. Partial factor methods for existing concrete structures, Lausanne, Switzerland (2016)

Fib Model Code for Concrete Structures 2010. Fib, Lausanne (2013)

Fib Model Code for Concrete Structures 2020. In: Drafting

Foster, S.J., Attard, M.: Experimental tests on eccentrically loaded high strength concrete columns. Struct. J. **94**(3), 295–303 (1997)

Gino, D., Castaldo, P., Giordano, L., Mancini, G.: Model uncertainty in non-linear numerical analyses of slender reinforced concrete members. Struct. Concr. **22**(2), 845–870 (2021)
Hasofer, A.M., Lind, N.C.: An exact and invariant first-order reliability format. J. Eng. Mech. **100**, 111–112 (1974)
ISO 2394. General principles on reliability for structures. Genéve (2015)
Iwai, S., Koichi, M., Minoru, W.: Stability of slender reinforced concrete columns subjected to biaxially eccentric loads. Bull. Disaster Prev. Res. Inst. **36**(3–4), 137–157 (1986)
JCSS. JCSS Probabilistic Model Code (2001)
Lukas, N., Novak, D.: Estimation of coefficient of variation for structural analysis: the correlation interval approach. Struct. Saf. **92**(4), 102101 (2021). https://doi.org/10.1016/j.strusafe.2021.102101
McKenna, F., Fenves, G.L., Scott, M.H.: Open system for earthquake engineering simulation. University of California, Berkeley, CA, USA (2000)
Mckey, M.D., Conover, W.J., Beckman, R.J.: A comparison of three methods for selecting values of input variables in the analysis from a computer code. Technometrics **21**, 239–245 (1979)
Mehmel, A., Schwarz, H., Karperek, K.H., Makovi, J.: Tragverhalten ausmittig Beanspruchter Stahlbetondruckglieder. Institut für baustatik, EHT, deutscherausschuss für stahlbeton, Heft 204 (1969)
Pancholi, V.R.: The instability of slender reinforced concrete columns. A buckling study of very slender reinforced concrete columns between the slenderness ratios of 30 and 79 Including essential creep investigations, and leading to design recommendations. PhD thesis, University of Bradford (1977)
Saatcioglu, M., Razvi, S.R.: Strength and ductility of confined concrete. J. Struct. Eng. **118**(6), 1590–1607 (1992)
Saenz, L.P., Martin, I.: Test of reinforced concrete columns with high slenderness ratios. J. Proc. **60**, 589–616 (1963)
Troisi, R.: Illegal land use by Italian firms: an empirical analysis through the lens of isomorphism. Land Use Policy **121**, 106321 (2022). https://doi.org/10.1016/j.landusepol.2022.106321
Troisi, R., Alfano, G.: Proximity and inter-firm corruption: a transaction cost approach. Small Bus. Econ. (2022). https://doi.org/10.1007/s11187-022-00649-y
Troisi, R., Alfano, G.: The re-election of corrupt mayors: context, relational leadership and level of corruption. Local Gov. Stud. (2022). https://doi.org/10.1080/03003930.2022.2087060
Vecchi, F., Belletti, B.: Capacity assessment of existing RC columns. Buildings **11**(4), 161 (2021). https://doi.org/10.3390/buildings11040161
Yu, Q., Valeri, P., Ruiz, M.F., Muttoni, A.: A consistent safety format and design approach for brittle systems and application to textile reinforced concrete structures. Eng. Struct. **249**(2021), 113306 (2021). https://doi.org/10.1016/j.engstruct.2021.113306

# North-West Ring Road of Merano - 2nd Lot: Cut&Cover Tunnel

Massimiliano Donelli[1], Matteo Moja[1], Enrico Maria Pizzarotti[1(✉)], Filippo Prati[1], Pierfrancesco Readaelli[2], Luigi Regondi[1], and Johannes Strimmer[3]

[1] Pro Iter Srl - Progetto Infrastutture e Territorio, Milan, Italy
enrico.pizzarotti@proiter.it
[2] Consorzio San Benedetto, Merano, Italy
pierfrancesco.redaelli@pacspa.it
[3] Provincia Autonoma Di Bolzano, Bolzano, Italy
johannes.strimmer@provincia.bz.it

**Abstract.** Following a brief description of the context and of the project as a whole, the report presents the solution adopted for the excavation of the Cut&Cover Tunnel planned at the beginning of the 2nd lot, before the excavation of the mining method tunnel. The Cut&Cover Tunnel, 136 m long, will house the main axis, characterized by a platform with a total width of 8.5 m (two lanes of 3.5 m plus two docks of 0.75 m) and two ramps, one for the entrance and one for the exit, with lane widths of 5.2 m and 5.5 m respectively. The geometric complexity and the resulting size of the work, as well as the boundary conditions typical of a highly anthropized environment, led to a top-down excavation method, up to a depth of approximately 20 m, with large diameter (1.2 m) perimetral piles, 1.5 m spacing and intermediate 1,0 m piles, the casting of the roof slab in cast in situ reinforced concrete and temporary tie rods during the lowering phases of the excavation, before the construction of the inner shell in cast in situ reinforced concrete.

**Keywords:** Road tunnels · Top-down · Urban environment · BIM

## 1 Introduction

The Merano North-West Ring Road, Lot 2, is a project of the Autonomous Province of Bolzano (APB) aimed at completing the access and exit routes from the city center and connecting the existing Bolzano-Merano "Mebo" freeway to the West to the Val Passiria valley to the East.

The Detailed Design was made by the design team consisting of Dr. Ing. A. Gretzer, Dr. Ing. K. Bergmeister, Dr. Ing. M. Ebner and Dr. Ing. W. Weis.

The contract for the construction of the works, for an amount of about 100 M €, was awarded to the consortium San Benedetto - Merano Scarl, made up of the mandatary Carron Bau S.r.l., Mair Josef & Co S.a.s., Di Vincenzo Dino & Co S.p.A. and PAC S.p.A., which has entrusted the consultancy for the For Construction Design to the engineering company Pro Iter S.r.l.

In this article, after a brief description of the entire project, the choices made for the construction of the Cut&Cover Tunnel at the beginning of Lot 2 are described.


M. A. Aiello and A. Bilotta (Eds.): ICC 2022, LNCE 435, pp. 221–233, 2024.
https://doi.org/10.1007/978-3-031-43102-9_18

## 2 Description of the Project

The road layout of Lot 2 starts in the North-West part of Merano, near the railway station, in continuity with Lot 1, which has already been completed and is currently in operation, and runs eastwards under the city and then connects to the existing road system of Passiria Valley (Fig. 1).

With the exception of the open-air sections to the East, consisting of three roads, that branch off from the "Monte Zeno" traffic circle, located close to the northern entrance of the tunnel, to connect the "Maia Alta" district (L = 380 m), the "Monte Zeno" artisanal area (L = 140 m) and the Passiria Valley (SS44), Lot 2 mainly consists of a single tunnel with a total length of about 2200 m, which in turn is divided into two parts; an initial section of the Cut&Cover Tunnel (L = 136 m), connected to Lot 1 and the surface roads, and a subsequent section of natural tunnel (L = 2064 m).

The latter, for the first km, crosses the city of Merano and is characterized by the presence of loose soil and low overburdens; moving eastward, vice versa, the tunnel crosses the rock formations of Mount San Benedetto with overburdens that increase up to a maximum of about 100 m (Fig. 1).

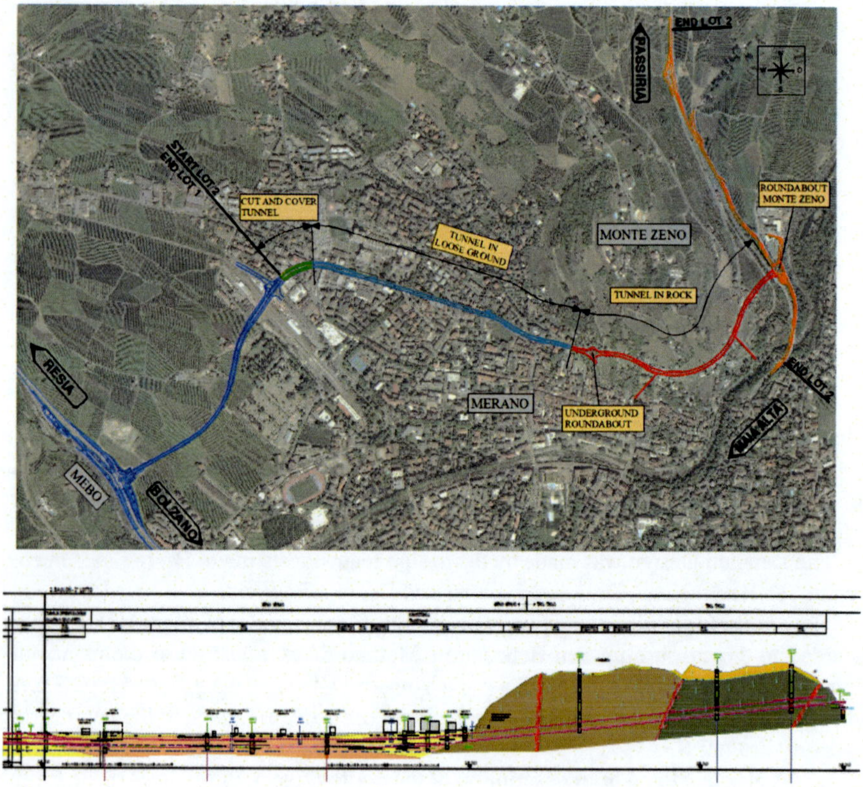


**Fig. 1.** General plan and Lot 2 longitudinal profile

The standard section adopted is Q7A of the APB, with a single two-way carriageway, characterized by 2 lanes of 3.5 m and two 0.75 m platforms, for an overall width of 8.5 m. The road platform widens in correspondence to the 4 lay-bys provided along the route (2 in the rocky section and 2 in the section in loose soil), in the section approaching the "Monte Zeno" traffic circle, to accommodate the entry and exit lanes, and, for the same reason, in the first section of the natural tunnel in continuity with the previous lot.

The Cut&Cover Tunnel planned at the western entrance, as better described below, hosts two ramps, one entrance and one exit, whose acceleration and deceleration lanes involve the first stretch of natural tunnel, for a development of about 160 m, implying a significant increase in the width of the road platform (up to 15.5 m) and consequently the excavation area of the section, which increases from about 150 m$^2$ (L $\approx$ 15 m x h $\approx$ 11 m) to about 240 m$^2$ (L $\approx$ 21 m x h $\approx$ 13 m).

Along the development of the rock tunnel, at the height of the "Tappeiner" promenade, an underground traffic circle with an external diameter of about 42 m is planned, supported by a central rock pillar with a diameter of 8 m, which will allow access to the underground parking lot of Mount San Benedetto, to be built in the future.

## 3 Cut&Cover Tunnel

The first part of the project to be realized, as well as the subject of study of this article, is the Cut&Cover Tunnel. This latter has a length of 136 m, is located to the North-West of Merano, in the area delimited by Cantiere Road, Goethe Road and Laurin Road, and begins to the West as a continuation of the works of the previous lot and ends to the East with the connection of the natural section (Fig. 2).

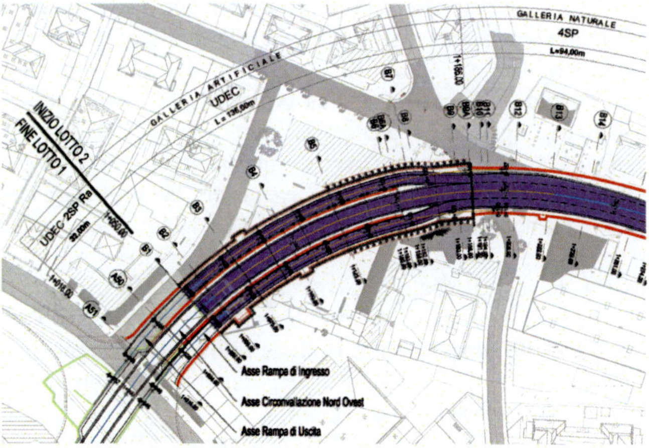


**Fig. 2.** Plan of the Cut&Cover Tunnel

The Cut&Cover Tunnel has a two-lane, two-way roadway on the main axis and two ramps on the sides that house the entrance and exit lanes respectively (Fig. 3).

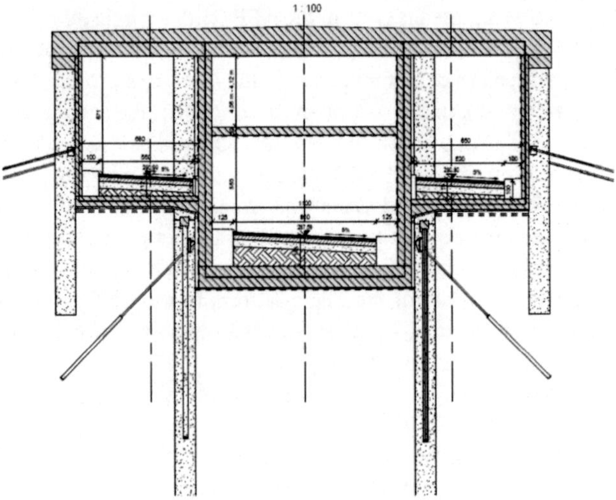

**Fig. 3.** Typical section

The road surface of the ramps is at a higher level than that of the main axis with a maximum height difference of slightly more than 6 m at the beginning of the tunnel. These differences in height are gradually reduced as one proceeds towards West until they are eliminated at the junction of the ramps with the main axis, just before the beginning of the natural excavations. In this last section the Cut&Cover Tunnel has a four-lane carriageway.

A further complication from a construction point of view is the presence of a ventilation channel, with a minimum area of 15 $m^2$, located, in the initial section in continuity with Lot 1, below the main axis, which involves a lowering of the bottom excavation level of about 3.5 m.

This channel, approximately 20 m after the beginning of the work, passes from below to above the main axis by means of a lateral conduit as shown in Fig. 4.

Considering the finished structure, the coverings are of modest entity, varying from a little more than one meter to a maximum of five meters, the width reaches about 30 m, while the depth of the bottom of the excavation reaches, at the lowest point, about -22 m from ground level.

As far as the geological context is concerned, the Cut&Cover Tunnel develops entirely within alluvial deposits of the Adige, the Rio Nova and the Passiria Rivers. These loose materials are composed of silty sands with modest cohesion in the shallowest portion (approximately 2 $\div$ 4 m) and underlying gravelly sands without cohesion, with an estimated shear resistance angle of 38°.

The hydrogeological study for the Detailed Design (2013) showed the presence of an aquifer at a depth of about 20 m from ground level, which interferes with the lower part of the Cut&Cover Tunnel structures. This interference has been the subject of an in-depth study during all the design phases, also considering the results of piezometric measurements carried out in the last decade, which have highlighted strong fluctuations of the level of the water table, with an increased risk of affecting the excavation.

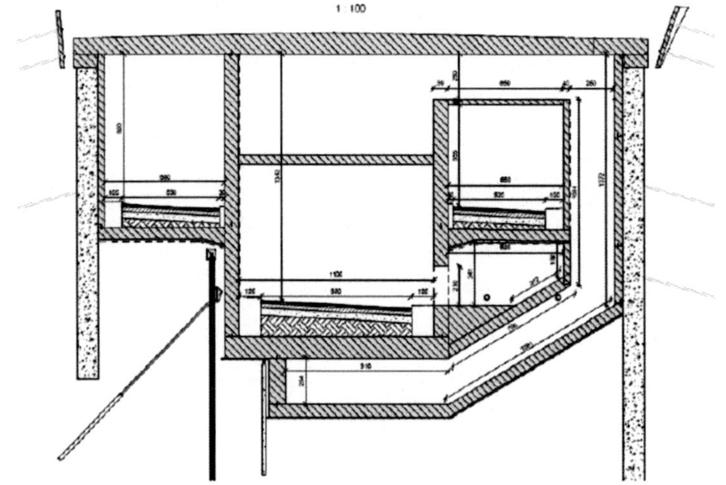

**Fig. 4.** Section in correspondence of the ventilation channel

## 4 Detailed Design of the Cut&Cover Tunnel (UTAG)

The Detailed Design envisaged the construction of the Cut&Cover Tunnel (named UTAG) for a total length of approximately 130 m.

In extreme synthesis, the excavation for the realization of the tunnel was planned by means of sub-vertical excavation supported by soil-nailing, as shown in Fig. 5. Once the bottom of the excavation would have been reached, the realization of a box in reinforced concrete was planned to house both the main road axis and the two entrance and exit ramps (Fig. 6).

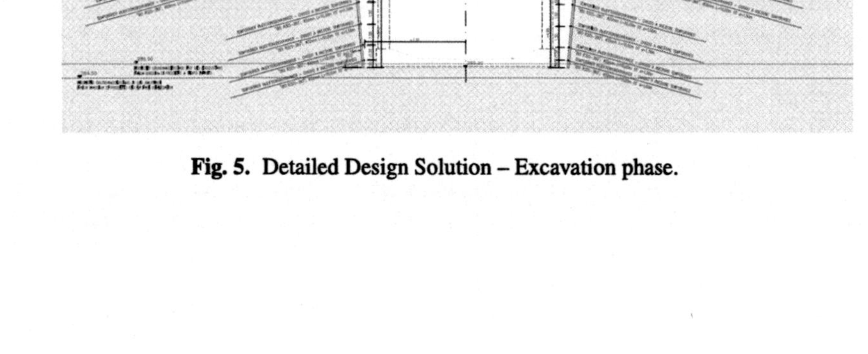

**Fig. 5.** Detailed Design Solution – Excavation phase.

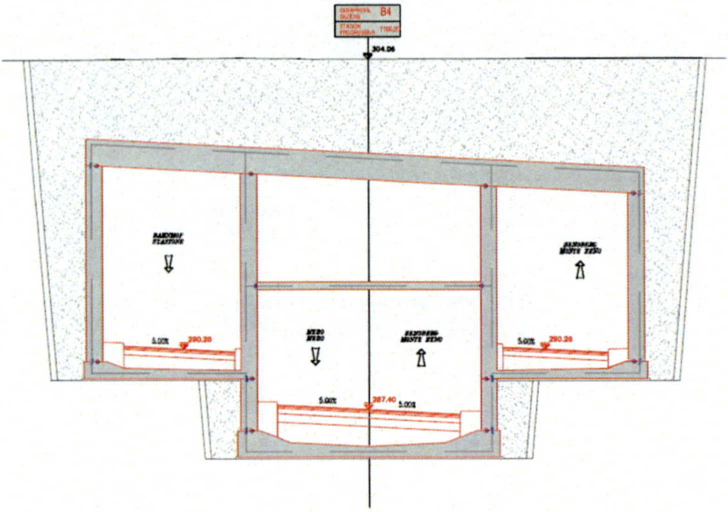

**Fig. 6.** Detailed Design Solution – Final lining

## 5 For Construction Design of the Cut&Cover Tunnel (UDEC)

The solution proposed in the For Construction Design provides, on the one hand, a minimum optimization of the road project in order to anticipate the achievement of the coplanarity of the main axis with the ramps, thus allowing the natural excavations to be attached with a smaller excavation Sect. (21 m wide) than provided for in the Detailed Design (24 m) and, on the other, a substantial modification of the construction method of the Cut&Cover Tunnel, passing from a solution with down-top construction (UTAG) to a top-down method (called UDEC).

The changes to the road project can be summarized as follows:

1. reduction of the longitudinal slope of the main axis;
2. slight increase of the longitudinal slope of the inbound ramp;
3. reduction of the height connection radius of the inbound ramp.

At the same time as the route modifications, the Cut&Cover Tunnel was lengthened by 6 m, thus achieving the objective of reducing the size of the connection section of the natural tunnel. This advantage is of considerable importance, especially in view of the loose nature of the soil, in which the excavation of the 240 $m^2$ tunnel section must be carried out, and of the relative boundary conditions (low cover and presence of buildings).

As far as the realization of the Cut&Cover Tunnel is concerned, after an approach excavation supported by soil nailing with a maximum slope of 5/1 and a height never higher than 7 m, it consists in the execution of two lateral retaining walls of large diameter piles $\phi$1200/1.5 m, the realization of the reinforced concrete cover slab and the subsequent lowering excavation under the slab (top-down method), before the realization of the inner shell and finishing works (Fig. 7).

During the excavation, the stability of the retaining walls is guaranteed not only by the constraint provided by the roof slab, but also by an order of temporary active tie-rods; during the final phase, the contribution of the tie-rods is neglected and the function of contrast to the support works is entrusted to the reinforced concrete slabs of the inner structure.

To optimize the sizing of the roof slab, characterized by large spans (about 30 m), during the excavation phase the realization of temporary intermediate discrete piles is foreseen to break the length of the beam into 3 parts. The same static scheme of the beam on 4 supports is also guaranteed in the final phase, replacing the temporary intermediate piles, which will be removed, with the internal walls of the reinforced concrete structure.

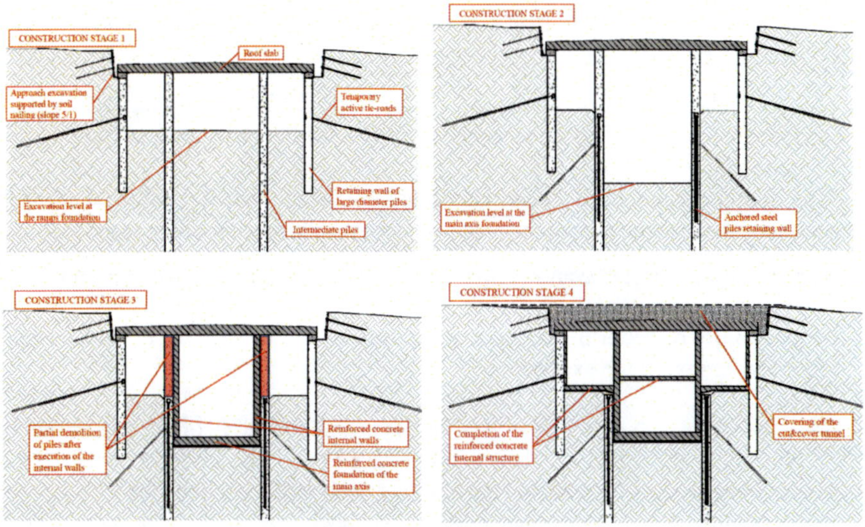


**Fig. 7.** Construction stages

Only in correspondence of the last segments, before the connection of the natural tunnel, where it is not possible to realize the intermediate supports because of the configuration of the road platform, the static scheme of the roof slab is that of a beam on two supports. In this section, to cope with the demanding load conditions (28.5 m width with an overburden of about 2.5 m), it was necessary to use a double-T concrete slab of a significant height, 3.65 m (Fig. 8), compared to the 1.5 m thick reinforced concrete slab used along the rest of the tunnel. Moreover, despite the slab height, it was necessary to install 4 layers of reinforcements $\phi$30/15 cm in the lower slab.

The use of cast-in-place reinforced concrete slabs, unusual for such large spans, was preferred to prestressed prefabricated beams, which were evaluated in a first design phase, mainly for logistic reasons, in consideration of the high spans and the consequent difficulties of transport and mobilization within the urban context of Merano, as well as following a cost/benefit analysis and other structural and durability aspects.

Finally, in the initial section of the Cut&Cover Tunnel, where the road surface of the main axis is significantly deeper than that of the ramps, after an initial excavation phase

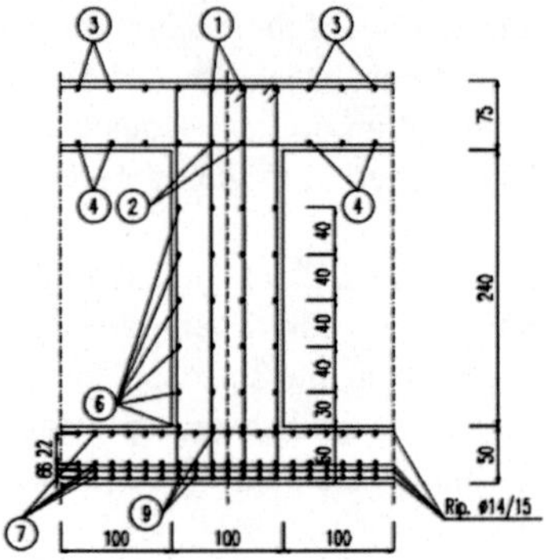


Fig. 8. Double-T reinforced concrete slab

up to the foundation level of the ramps, a temporary steel piles retaining wall will be built, contrasted by means of active tie-rods, to reach the bottom of the excavation level of the main axis.

The solution proposed in the For Construction Design made it possible to achieve, substantially at the same cost of the Detailed Design solution, the following goals:

- greater safety during excavation of the Cut&Cover Tunnel, guaranteed by the presence of large-diameter piles in place of the soil-nailing;
- the possibility of better managing any fluctuations in the water table, which could affect the bottom of the excavation;
- the reduction of the impact of the surrounding environnment, in terms of dust and noise, working undercover instead of in the open, as well as of the site traffic on IV Novembre Road and Cantiere Road, thanks to the possibility of access to the excavation area directly from the traffic roundabout of the previous lot;
- the possibility of immediately backfilling the last segments of the Cut&Cover Tunnel, reducing interference with Goethe Road. The Cut&Cover Tunnel top-down solution adopted was completed approximately in one year (1 shift/day, 5 days/week), despite all the constraints to which the job site was subjected, in comparison to a completion time of the down-top Detailed Design solution reasonably 40% greater.
- the increase in excavation safety even in the first section of the natural tunnel following the reduction in the size of the section of the natural excavation.

## 6 BIM Model in for Construction Design

The infrastructure foreseen, both for final and temporary works, is particularly articulated. This allows the opportunity to build a complete informative model, to allow both to understand in the best way, already in the planning phase, the insertion in the context of the work in geometric terms, and to catalogue and enrich with 'information' all the planning parts, to make easy the consultation from the model itself, extrapolating both the graphic elaborations and the quantity schedules in an automatic and analytical way. The BIM process followed has involved the management of coordinated models for:

- the topographic component and the insertion in the territory;

- the road component with the route optimization;
- the geotechnical component in relation to the modes of support of excavations and phasing;
- the structural component of the definitive and temporary works both for the carpentry and for the reinforcement cages.

From the acquisition of data of the territory, integrated with detailed surveys of the areas previously approached in relation to the lot already executed and the preparation of site areas, it was possible to reproduce the as built state. Subsequently it has been analysed and optimized the project road layout defining the standard sections and applying them to the layout through the modeler. In the next step the modelling of the road has been done in all its main components, having the possibility to keep under control the covering degrees and the respect of the road geometry and of the marginal elements (Fig. 9).

An accurate check of the end sections was carried out, on the one hand to guarantee the connection with the already built part of the previous lot, and on the other hand for the congruence with the entrance of the natural tunnel that develops in continuity with the cut&cover part. Once the general structure of the work had been determined, each situation was analysed, proceeding to the modelling of the excavation geometries in compliance with the territorial constraints, to simplify and minimise operations on site (Fig. 10). The level of detail, compatible with what is necessary to the preparation of a detailed project and in compliance with the provisions of UNI 11337, has been attested to the LOD D, but preparing the model for the possible evolution to LOD E.

**Fig. 9.** 3D model of the Cut&Cover Tunnel in the topographical context

Once defined the excavation geometries and introduced the temporary works, where the limitation of the encumbrances and the guarantee of safety required it, it has been proceeded to transpose in the model the construction details related to the fundamental components (micropiles and piles, tie-rods, nailed walls) as well as all the protection layers (slopes screeds, waterproofing, non-woven fabric, protection screeds, etc.). In the

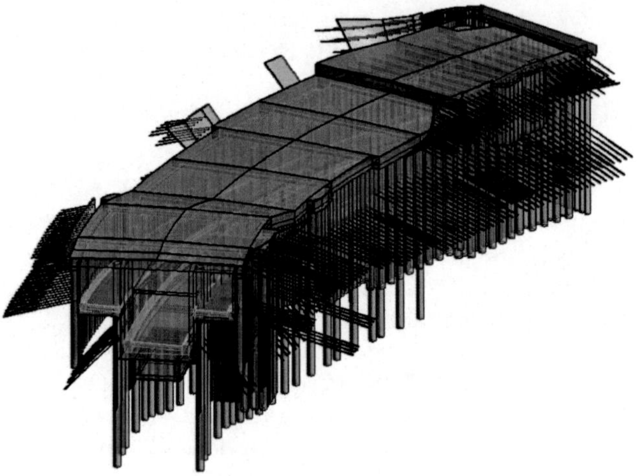

**Fig. 10.** 3D model of the Cut&Cover Tunnel combined with the road platform model

next step, following calculations, sizing and structural checks, the reinforcement cages were modelled in their entirety, in order to obtain both the graphic drawings and the detailed quantity tables.

An ACDat platform has been prepared and structured for the sharing and the publication of models and drawings, making available to stakeholders the consultation of the model and tables directly from a web browser. The coherence of the elaborates and of the model is automatically guaranteed just because within the followed process and exploiting the chosen digital platform, the elaborates 'photograph' the model views without adding or removing contents compared to what has been modelled (Fig. 11 and Fig. 12).

The work done already in the early stages of modelling has allowed to simplify and ensure the necessary level of coordination between models (LC1 - Single model coordination; LC2 - Coordination between several single models; LC3 - Coordination between data/info/content between models and processed). Through an automated process, it was also possible to perform the checks of level V1 and V2 (V1 - internal verification of correctness on individual models, guaranteed by the manager and coordinator; V2 - verification on individual and aggregated models, readability, traceability and consistency).

In all the phases of realization and control of the models, all the actors involved in the process, everyone for his own role, have been involved and have participated to the collaborative phase (the Specialist for the project and the modelling, the Coordinator for the analysis and the verifications of the contents and the Manager for the respect of the prefixed standards). To forehead of a greater engagement in the phase of modelling and in the generation of the elaborates, the unquestionable tied up benefit to the process is the guarantee of the coherence of the plan, the possibility to derive successive details regarding those chosen for the representation, the immediate visualization of the more articulated parts (as an example extracting more sections or using specific points of visualization), the opportunity to proceed to the revisions without losing trace of the previous

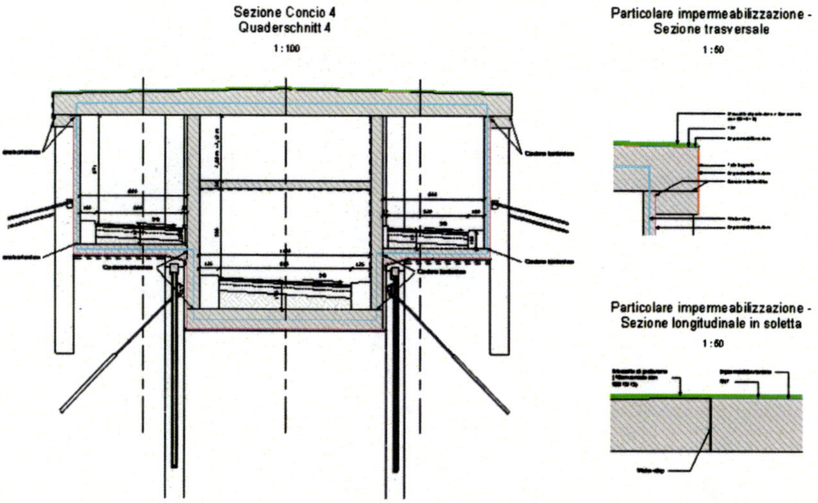


**Fig. 11.** Waterproofing detail section

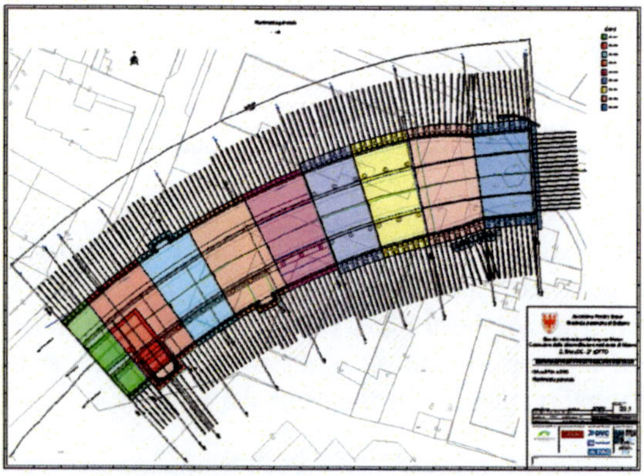

**Fig. 12.** Extrapolation of the table for the identification of the segments (visualization according to the identification parameters of the model)

versions of the plan, the certainty of the correspondence of the project documentation to the more modernized versions of the model.

## 7 Progress of Work

To date, for the realization of the Cut&Cover Tunnel, the approach excavation (Fig. 13) until reaching the plan for the realization of the piles was completed with a maximum 5/1 slope supported by soil nailing; the large diameter piles were made (Fig. 14), both the lateral ones that constitute the retaining wall, and the intermediate ones needed to support the roof slab; the roof slab was cast (Fig. 14) and then the excavation was carried out with top-down methodology until reaching the foundation level of the ramps (Fig. 15) with the simultaneous realization of the contrast tie-rods.

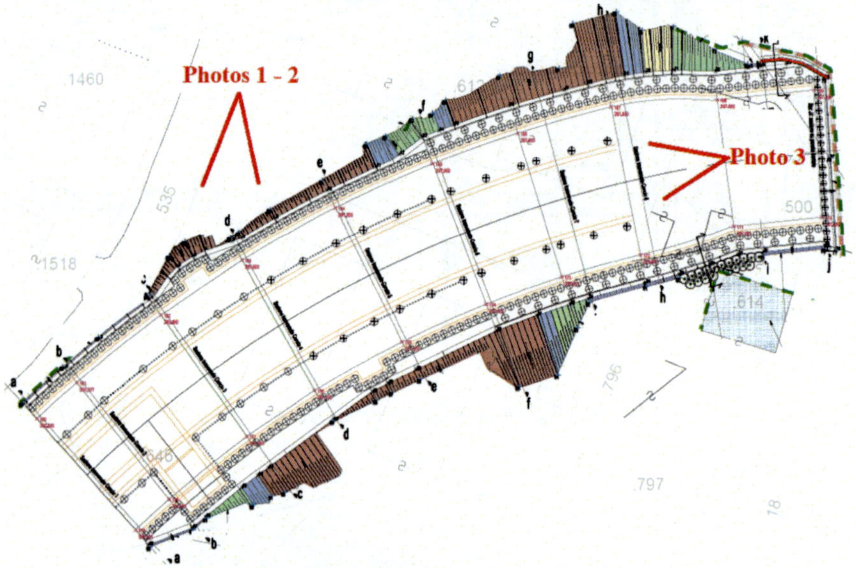


**Fig. 13.** Excavation plan with identification of the photos 1, 2 and 3 points of view

In April, the excavation until the foundation level of the main axis near the frontal retaining wall was completed and the excavation of the natural tunnel began.

The internal structures of the artificial tunnel will be built in phases at the same time as the excavation of the natural tunnel, using alternately the three carriageways (the main axis and the two lateral ramps) for the transit of site vehicles.

**Fig. 14.** Photo 1 – Realization of the large diameter piles (on the left); Photo 2 - Casting of the roof slab (on the right)

**Fig. 15.** Photo 3 – Excavation with top-down method until the level of foundations of the ramps

**Acknowledgements.** The authors would like to thank, for their contribution to the realization of the project and this work, the designers of the Detailed Design, Eng. Aribo Gretzer, Eng. Manfred Ebner, Eng. Konrad Bergmeister, Eng. Walter Weis as well as the Construction Management: Ingenieurteam Bergmeister S.r.l., EUT Engineering Srl, Valdemarin S.r.l., Plan Team S.r.l., Kauer Seehauser Engineering, Pfeifer Planung Srl, Ingenieurgemeinschaft Eng. Aribo Gretzer & Partner GMK, Sint Ingegneria S.r.l., Eng. Manfred Ebner and Ferro Studio Ingegneria S.r.l. as well as the Project Manager of Carron Bau S.r.l., consortium San Benedetto leader, Eng. Massimo Dal Sasso.

# Analysis of Failure Mechanisms of Gerber Half-Joint Specimens Through Digital Image Correlation Technique

Filippo Molaioni[1](✉), Diego Alejandro Talledo[2], Manuel Bartoli[1], and Fabio Di Carlo[1]

[1] Department of Civil Engineering and Computer Science Engineering (DICII), University of Rome Tor Vergata, Rome, Italy
filippo.molaioni@students.uniroma2.eu, manuel.bartoli@uniroma2.it, di.carlo@ing.uniroma2.it

[2] Department of Architecture, Construction and Conservation, University IUAV of Venice, Campus Terese, Dorsoduro 2206, 30123 Venezia, Italy
diego.talledo@iuav.it

**Abstract.** Reinforced concrete Gerber half-joints are characterized by D-Regions in which the de Saint Venant theory is not valid. Therefore, in the capacity assessment of existing Gerber half-joints, the correct assumptions about the stress fields and the strut-and-tie models play a crucial role. Digital Image Correlation (DIC) is an optical technique for the determination of displacements and strains of specimens subjected to mechanical actions. In this work, the DIC technique is applied within an experimental campaign aimed at identifying the nonlinear behavior of Gerber half-joints according to different construction details. The results obtained using the DIC technique allow to identify the shape of the compression fields (struts) and to predict the development of the crack patterns. Therefore, the outcomes are compared with the expected shapes of the strut-and-tie models and interpreted to analyze the different failure mechanisms oh the half-joints.

**Keywords:** Gerber half-joints · digital image correlation · reinforced concrete · experimental campaign · existing bridges · strut-and-tie models

## 1 Introduction

Reinforced concrete (RC) half-joints are a connection system between deck's components, widely adopted during the construction of the Italian road network thanks to their construction advantages. Most of the Gerber bridges were built between 1930 and 1980 (Giannetti et al. 2021); therefore, given their age, some are exposed to a high deterioration risk. Furthermore, considering that the half-joints are usually cramped, they are characterized by inspection and maintenance issues, which could increase the structural deterioration (e.g. rebars corrosion), leading the structure unsafe (Di Prisco et al. 2018). In this context, the recent *"Guidelines on risk classification and management, safety assessment and monitoring of existing bridges"* (C.S.LL.PP. 2020) classify it as "critical," leading to the need for a detailed safety evaluation (Santarsiero et al. 2021).


M. A. Aiello and A. Bilotta (Eds.): ICC 2022, LNCE 435, pp. 234–244, 2024.
https://doi.org/10.1007/978-3-031-43102-9_19

From these reasons arises the urgent need to develop reliable analytical and numerical tools based on experimental evidence. An experimental campaign on the structural behavior of Gerber half-joints, even in the presence of corrosion phenomena, was carried out at the "*Laboratorio di strutture e prove materiali*" of the University of Rome "Tor Vergata" by some of the authors, with the aim to study their structural behavior and provide an experimental benchmark for analytical and numerical validation (Di Carlo et al. 2023). Gerber half-joints are characterized by D-regions, where the compression fields develop and are balanced by steel ties (Desnerck et al. 2018, fib Bulletin 100 2021). Then for these members, the direction and the width of compression struts, and therefore the crack pattern evolution, is an essential aspect in evaluating the structural behavior.

Digital Image Correlation (DIC) is a non-contact optical technique for evaluating displacements and strains of specimens subjected to mechanic loads. It is based on the comparison of the image of the specimen captured before and during the test (Pan et al 2009). Given its accuracy, it is widely used in experimental tests on materials (Lecompte et al. 2007, Caminero et al. 2013, Khechai et al. 2018, Belloni et al 2019). Furthermore, given the technique's potential, many applications on RC elements in real scale have recently been performed (Blikharskyy & Kopiika 2020, Słoński & Tekieli 2020). Although the optimal parameters for applications on this specimen's scale (i.e., pattern, camera distance, acquisition) are still being defined, this research has shown satisfactory results as regards: flexural crack assessment on RC members (Hamrat et al. 2016); shear crack opening examination (Tambusay et al. 2018, Li et al. 2020); evolution of the crack pattern in bridges slabs (Christensen et al. 2021, Mousa et al. 2021).

It is therefore believed that DIC can be a suitable tool to support the interpretation of the structural behavior of Gerber half-joints mapping the crack pattern until failure; moreover, it can provide a benchmark of displacement and strain fields for the validation of numerical finite element 2D models.

In this paper, an open-source 2D-DIC software named *Ncorr* (Blaber et al. 2015), is applied to support the study of failure mechanisms of four half-joints specimens. First, the main characteristics of the specimens and of the experimental campaign are presented, then the acquisition system, the pattern characteristics, and the DIC software adopted are briefly described. Finally, the results are reported in terms of: (1) vertical displacement of the half-joint; and (2) shear strains of the half-joints. Furthermore, a comparison between the DIC results and the experimental results (i.e. traditional sensors) is made.

## 2 Experimental Tests

The analysis presented in this paper refers to the experimental tests given in Di Carlo et al. (2023). The experimental campaign aims to analyze the structural behavior of Gerber half-joints considering different: (a) reinforcement layout; (b) reinforcement amount; (c) deterioration condition due to corrosion.

The DIC technique is used for evaluating the nonlinear mechanism of the un-corroded specimens. As they govern the mechanism, the reinforcement amount and layout of the analyzed specimens are presented in this section, while the geometry and materials of the specimen are omitted in this paper for the sake of brevity. More information about the experimental survey can be found in Di Carlo et al. (2023).

Specimens are designed according to the STMs " proposed by EN1992–1-1, named here "Model A" and "Model B", and illustrated in Fig. 1. Two different reinforcement layouts are considered. The first refers to model "A" (i.e., longitudinal reinforcement in the nib, stirrups at the back of the nib, and longitudinal tension reinforcement in the beam). The second is made by both model "A" and model "B" (i.e., diagonal reinforcement in the half joint are added). Gerber half-joints non-linear mechanism is influenced by the weakest element of the strut and tie model. Therefore, to analyze several mechanisms of the half-joints, the specimens were designed according to two reinforcement amounts: (a) "low" reinforcement amount to have a mechanism governed by steel fracture, (b) "high" reinforcement amount to have a mechanism governed by failure of struts in compression. A summary of the main information of the specimens is reported in Table 1, with the number of rebars that characterize the main ties (i.e., longitudinal tie T1, stirrups tie T2, and diagonal tie TD).

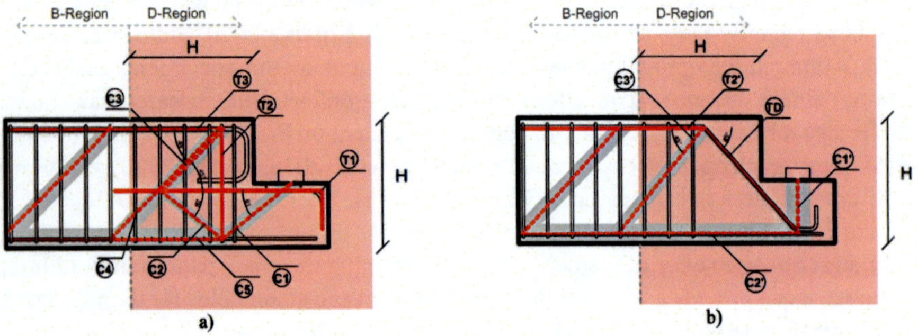


**Fig. 1.** Strut and Tie Models (STMs) proposed in EN 1992–1-1: a) Model A and b) Model B.

**Table 1.** Reinforcement pattern of the tested samples.

| specimen | STM | Amount | T1 | | T2 | | TD | |
|---|---|---|---|---|---|---|---|---|
| | | | Φ | n° | Φ | n° | Φ | n° |
| | | | mm | | mm | | mm | |
| G1 | A + B | LOW | 12 | 2 | 10 | 2 | 12 | 3 |
| G2 | A + B | HIGH | 12 | 4 | 10 | 3 | 12 | 3 |
| G3 | A | LOW | 12 | 2 | 10 | 2 | 12 | - |
| G4 | A | HIGH | 12 | 4 | 10 | 3 | 12 | - |

## 3 Image Acquisition Setup

The camera used during the test is a "LUMIX DMC-TZ70". It is equipped with a MOS sensor 1/2.3 and a LEICA DC VARIO-ELMAR. For practical reasons, a 25 fps Full High Definition (FHD) video was recorded during the test, then the images for the subsequent DIC analyses were selected from the video relatively to the main stages of the test (i.e., first crack, first yielding, maximum load, collapse). The resolution of the obtained frames is 3840x2160 px. Both exposure and aperture were automatically calibrated before each test to obtain the best focus according to the lighting conditions. To ensure a homogeneous brightness on the specimen's surface, this was illuminated with a spotlight placed at 1m. The acquisition setup is shown in Fig. 2; the camera was arranged through a laser level so that the sensor was parallel to the specimen surface. The sensor was one meter from the surface of the specimen and was placed in the center of the Region Of Interest (ROI) of the DIC analysis (depicted in Fig. 3). It is worth noting that the loading speed was about 1 kN/s, and the corresponding displacement velocity was about 0.05 mm/s during the linear behavior of the specimens.

**Fig. 2.** Image acquisition setup.

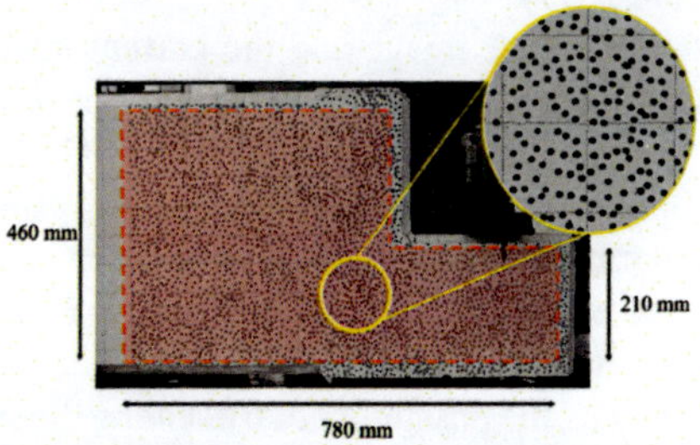


**Fig. 3.** Region of Interest (ROI) and adopted speckle pattern.

## 4 Pattern

Optimal pattern characteristics in terms of specimen surface coverage and speckles dimensions are described in Reu (2015) and Su et al. (2022). These state that an optimal coverage of the pattern might be around 50% and suggest a range of values for the speckles dimension (the diameter, in the case of circular speckle) of 3 to 5 pixels. The pixel's size depends on the camera setup, and particularly on: (1) the distance of the sensor from the specimen; (2) the image resolution; (3) and the size of the ROI.

In the present experimental setup, the actual dimension of a pixel of the image is equal to 0.27 mm; therefore, the "suggested" circular speckles might have a diameter between 0.8 and 1.3 mm. However, this work aims to implement a straightforward and useful DIC application on full-scale structural elements (i.e., using an ordinary camera with a hand-made pattern). For example, a 1 mm diameter speckles pattern would require significant effort to cover 50% of the specimen's surface, which is 325′000 mm$^2$. Therefore, a hand-made pattern using a 4 mm tip pen was chosen, and medium coverage of the specimen's surface of about 20%, evaluated as reported in (Su et al. 2022), is thus obtained. The pattern is shown in Fig. 3, it covers the entire ROI set for the analysis; it has the shape of the half-joint, and it covers all the specimen's height and extends up to 500 mm from the nib-beam connection.

## 5 DIC Software

This paper evaluates the displacement and strain of Gerber half-joint specimens through the open-source 2D DIC package *Ncorr* (Blaber et al. 2015). It represents a useful and flexible Matlab software based on the Reliability Guided DIC method (Pan 2009). It is based on the subset method (i.e. local DIC), correlating independent subsets of the image, and therefore does not guarantee the compatibility between adjacent subsets. The tool has been widely used in several experimental studies, obtaining satisfying results in material characterization tests (Hariral and Ramji 2014) and RC structural members experiments (Tambusay et al. 2018). The intuitive and easy-to-use graphical user interface (GUI) allows the analysis to be carried out through the flow presented below (Blaber & Antoniou 2017).

First, the input of the reference image and the current images are required. Then, a region of interest (ROI) must be selected; this will be the area where displacements and deformations will be evaluated. Subsequently, the DIC input parameters must be defined; these concern options for subsets (i.e., radius and spacing) and options for the iterative solver (i.e., the norm of the difference vector and number of iterations). In this phase, it is also possible to enable the high strain analysis (i.e., the reference image is updated during the analysis to facilitate the solution of the solver for large strains) and the discontinuous analysis (i.e., the possibility of truncating the subsets when intersected by a crack tip). Furthermore, taking advantage of the fact that with the subset method, each subset is solved independently, it is possible to choose to partition the ROI allowing the multithreaded analysis, significantly speeding up the analysis process.

Once the analysis has been carried out, it is possible to inspect the displacement field evaluated for the ROI in actual length units, once the size of an image pixel is provided by the user. Also, options for excluding "bad" results through a correlation-coefficient cutoff and accounting lens distortion are available.

The strain radius (i.e., radius of a circle that includes displacement values to fit plane to) is required to compute the strain analysis. Even in this phase, it is possible to enable the subset truncation for the discontinuous analysis.

## 6 Results

The results of the 2-D DIC analysis with *Ncorr* for the measure of displacement and strains of Gerber half-joint specimens are described below.

For the sake of clarity, the nonlinear behavior observed for the analyzed specimens is briefly summarized. Specimens G1 and G3, characterized by low reinforcement amounts, show a combined flexure shear failure at the nib-beam corner governed by the rupture of the steel reinforcement and a maximum load of 294.8 kN and 140 kN respectively. For specimen G1 also spalling of concrete under the nib was observed. Specimens with high reinforcement amount (i.e. G2 and G4) exhibited a failure governed by nib struts compression with yielded reinforcement. Maximum observed load was 331 kN and 220 kN, respectively. For more information about the structural behavior of the specimens, reference is made to Di Carlo et al. (2023).

Used DIC analysis parameters are described in the following, and it is remarked that some sensitivity analyses were needed in order to identify the optimal output of the results. The subset radius was selected equal to 80 pixels, and the subset spacing was set equal to 5 pixels. For the iterative solver options, 1e-06 is set as the difference norm vector threshold, and 50 is selected as the maximum number of iterations. The subset truncation discontinuous analysis is enabled since, as specimens are tested until failure, high displacement and, therefore, crack width are expected. Therefore, to investigate the behavior until the collapse, high strain step analysis is enabled with the option "seed propagation" in which the reference image is updated based on the correlation coefficient and the number of iterations to the convergence of the seeds. To better evaluate the whole mechanism of the specimen, no correlation coefficient cutoff is introduced. Strains calculations are made using a 5-pixels strain radius and enabling the subset truncation.

The video acquisition was synchronized with the purchase of the sensors to compare the DIC results with the experimental ones (i.e. sensors). Therefore, for each analyzed image, the corresponding load and displacement of the specimen are known.

### 6.1 Displacements

Vertical displacements of the half-joints were evaluated for the four specimens. As these are usually measured to describe the non-linear behavior of the specimens, two wire encoders were placed at the load application section on both sides of the specimen. 3)The comparison of the vertical displacement at load axis obtained with DIC analysis with those measured by wire encoders at the same point is useful to test the goodness of the DIC analysis. Comparing the aforementioned quantities for the entire loading phase, costant errors (Mean Absolute Error) of 2.06 mm, 0.93 mm, 1.03 mm, and 0.99 mm (4.2%, 1.3%, 2.0%, 1.7% of the maximum displacement experienced by the specimens), were identified for the loading point's displacement, respectively, for G1, G2, G3, and G4.

For the sake of example, only vertical displacements of specimen G1, captured in the main stages of the test, are shown in Fig. 4.

### 6.2 Strains

The output of the specimen's strains measurement is available in terms of the two-dimensional strain tensor components, relatively to the reference system shown in Figs. 4, 5, 6, 7, and 8. In this work, the shear strain $\varepsilon_{xy}$ is used to better interpret the crack evolution as the geometry of the half-joints involves the formation of compressed diagonal concrete struts. The results for the four specimens, G1, G2, G3, and G4, are shown in Figs. 5, 6, 7, and 8, respectively.

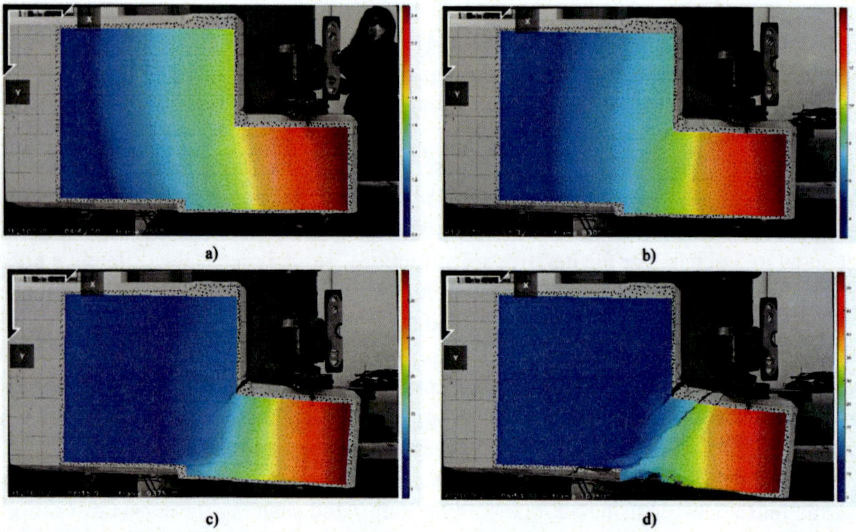

**Fig. 4.** Vertical displacement for specimen G1 during different phases: a) first crack; b) yielding; c) maximum load; d) collapse.

Observing the results, it is worth noting that, despite some noise in the results in the elastic regime, the first crack is well captured by the DIC analysis for all the four specimens. In the following stages, DIC identifies the cracks and shows the mechanism evolution; in general, the compression struts' shapes assumed by the STMs are confirmed. In addition, the following consideration on nonlinear behavior could be made: the flexure-shear crack is the first to appear and it develops with the increasing load, while the compression strut in the nib is the first compression crack to appear. Cracks in the beam zone highlight the involvement of compression fields in the beam, which, however, seem interested by multiple cracks, highlighting stresses redistribution, unlike the single crack in the nib. Failure happens in the nib zone for each of the four specimens,

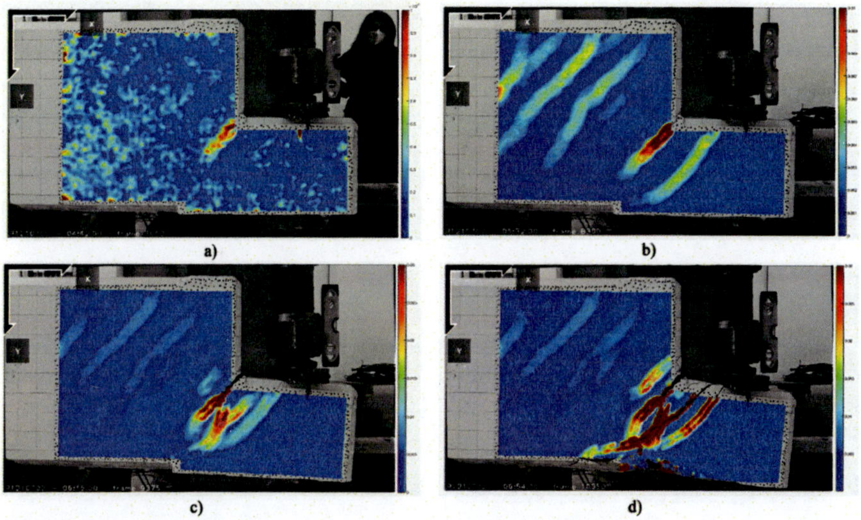

**Fig. 5.** Shear strain for specimen G1 during different phases: a) first crack; b) yielding; c) maximum load; d) collapse.

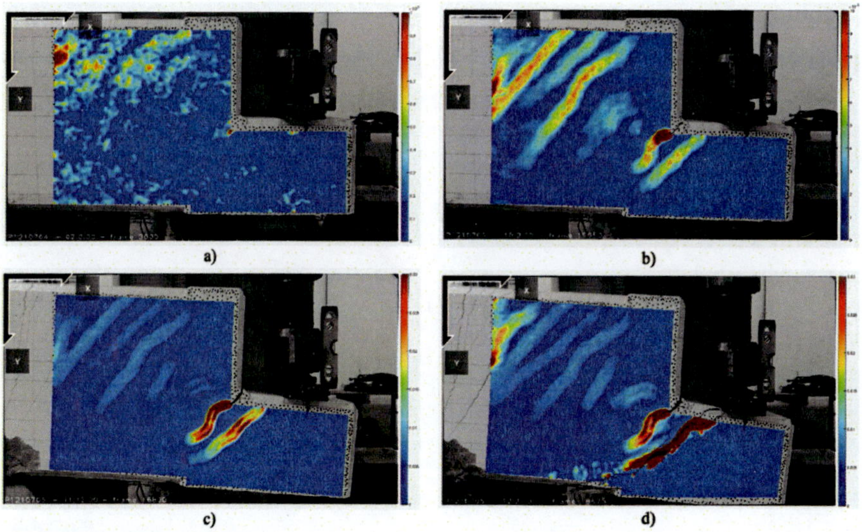

**Fig. 6.** Shear strain for specimen G2 during different phases: a) first crack; b) yielding; c) maximum load; d) collapse.

and it is possible to appreciate the significant involvement of the concrete (cracks) in the specimens characterized by a high amount of reinforcement (G2 and G4).

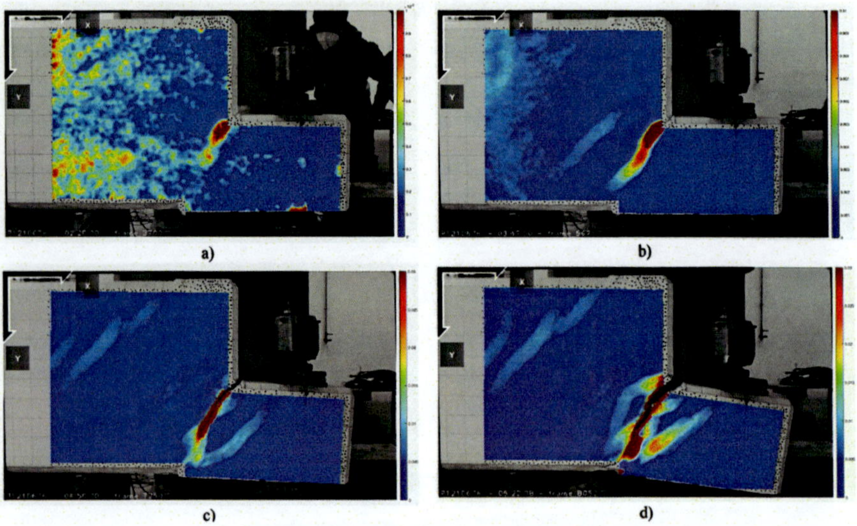


**Fig. 7.** Shear strain for specimen G3 during different phases: a) first crack; b) yielding; c) maximum load; d) collapse.

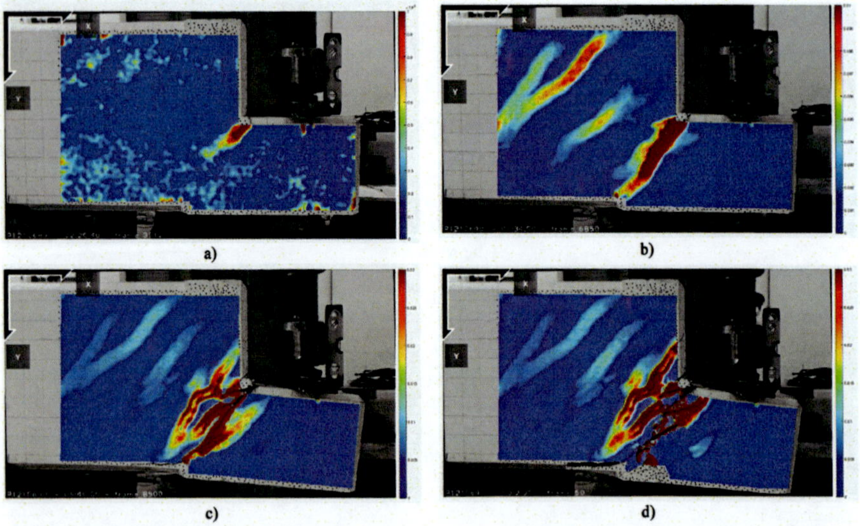


**Fig. 8.** Shear strain for specimen G4 during different phases: a) first crack; b) yielding; c) maximum load; d) collapse.

## 7 Conclusion

In this paper, the first outcomes of Digital Image Correlation (DIC) application on RC full-scale specimens, carried out through the "Laboratorio di strutture e prove materiali" of "University of Rome Tor Vergata" are presented. In particular, the DIC technique is applied on four Gerber half-joints specimens, designed and cast according to the Strut and Tie models proposed by EN1992, and to different reinforcement amounts (i.e. "low" and "high").

In this work, a useful and easy-to-use solution has been adopted to apply the DIC on RC structural elements scale. This consists of a coarser pattern than those recommended for small specimens, which can be easily hand-made using a pen, and by an acquisition system that involves the use of a digital camera capable of recording movies in full high definition (FHD) placed parallel to the plane of the specimen.

The key frames from the different records were extracted and then analyzed with an open-source software based on subset methods (local DIC) called *Ncorr*, available on the Matlab platform. The results in terms of displacements measured by DIC compared with the experimental measurements were satisfactory, with mean absolute errors around 1 mm. The study of the deformation fields with DIC allows to better understand the different failure mechanisms that are established in the specimens.

**Acknowledgements.** The paper presents some of the results obtained in the framework of the ReLUIS project (accordo attuativo DM 578/2020).

## References

Belloni, V., Ravanelli, R., Nascetti, A., Rita, M.D., Mattei, D., Crespi, M.: Py2dic: A new free and open source software for displacement and strain measurements in the field of experimental mechanics. Sensors (Switzerland) **19**(18), 1–19 (2019). https://doi.org/10.3390/s19183832

Blaber, J., Adair, B., Antoniou, A.: Ncorr: open-source 2D digital image correlation MatLab software. Exp. Mech. **55**(6), 1105–1122 (2015). https://doi.org/10.1007/s11340-015-0009-1

Blaber J., Antoniou, A.: Ncorr. Instruction manual (2017)

Blikharskyy, Y.Z., Kopiika, N.: Digital image correlation method for analysis of reinforced concrete structures. Bull. Odessa State Acad. Civil Eng. Archit., 27–33 (2020). https://doi.org/10.31650/2415-377X-2020-78-27-33

Caminero, M.A., López-Pedrosa, M., Pinna, C., Soutis, C.: Damage assessment of composite structures using digital image correlation. Appl. Compos. Mater. **21**, 91–106 (2014)

Christensen, C.O., Schmidt, J.W., Halding, P.S., Kapoor, M., Goltermann, P.: Digital image correlation for evaluation of cracks in reinforced concrete bridge slabs. Infrastructures **6**(7), 99 (2021). https://doi.org/10.3390/infrastructures6070099

C.S.LL.p.. Linee guida per la classificazione e gestione del rischio, la valutazione della sicurezza ed il monitoraggio dei ponti esistenti (2020)

Desnerck, P., Lees, J.M., Morley, C.T.: Strut-and-tie models for deteriorated reinforced concrete half-joints. Eng. Struct. **161**(February), 41–54 (2018). https://doi.org/10.1016/j.engstruct.2018.01.013

Di Prisco, M., Colombo, M., Martinelli, D., Coronelli, D.: The technical causes of the collapse of Annone overpass on SS.36. In: Italian Concrete Days, October 2016, pp. 1–16 (2018)

EN 1992-1-1: Eurocode 2: Design of Concrete Structures (2004)

Fib Bulletin 100. Design and assessment with strut-and-tie models and stress fields: from simple calculations to detailed numerical analysis. State of the art report (2021)

Giannetti, I., Mornati, S., Coccia, S., Di Carlo, F., Rinaldi, Z.: Safety assessment of existing post-war reinforced concrete bridges. the case study of 'Gerber Girders' bridges in Italy, pp. 1–12 (2021). https://doi.org/10.23967/sahc.2021.217

Hamrat, M., Boulekbache, B., Chemrouk, M., Amziane, S.: Flexural cracking behavior of normal strength, high strength and high strength fiber concrete beams, using digital image correlation technique. Constr. Build. Mater. **106**, 678–692 (2016). https://doi.org/10.1016/j.conbuildmat.2015.12.166

Harilal, R., Ramji, M.: Adaptation of open source 2D DIC software Ncorr for solid mechanics applications. In: 9th International Symposium on Advanced Science and Technology in Experimental Mechanics, pp. 1–6 (2014)

Khechai, A., Tati, A., Guerira, B., Guettala, A., Mohite, P.M.: Strength degradation and stress analysis of composite plates with circular, square and rectangular notches using digital image correlation. Compos. Struct. **185**(November 2017), 699–715 (2018). https://doi.org/10.1016/j.compstruct.2017.11.060

Lecompte, D., Smits, A., Sol, H., Vantomme, J., Van Hemelrijck, D.: Mixed numerical – experimental technique for orthotropic parameter identification using biaxial tensile tests on cruciform specimens. Int. J. Solids Struct. **44**, 1643–1656 (2007). https://doi.org/10.1016/j.ijsolstr.2006.06.050

Li, W., Huang, Y., Jiang, Y., Shi, T., Xing, F.: Application of DIC technology to shear crack measurement of concrete beam. In: Kountchev, R., Patnaik, S., Shi, J., Favorskaya, M.N. (eds.) Advances in 3D Image and Graphics Representation, Analysis, Computing and Information Technology. SIST, vol. 180, pp. 339–346. Springer, Singapore (2020). https://doi.org/10.1007/978-981-15-3867-4_39

Di Carlo, F., Meda, A., Molaioni, F., Rinaldi, Z.: Experimental evaluation of the corrosion influence on the structural response of Gerber half-joints. Eng. Struct. (2023)

Mousa, M.A., et al.: Application of digital image correlation in structural health monitoring of bridge infrastructures: a review. Infrastructures **6**, 176 (2021). https://doi.org/10.3390/infrastructures6120176

Pan, B., Qian, K., Xie, H., Asundi, A.: Two-dimensional digital image correlation for in-plane displacement and strain measurement : a review (2009). https://doi.org/10.1088/0957-0233/20/6/062001

Pan, B.: Reliability-guided digital image correlation for image deformation measurement. Appl. Opt. **48**(8), 1535–1542 (2009). https://doi.org/10.1364/AO.48.001535

Reu, P.: All about speckles: speckle density. Exp. Tech. **39**(3), 1–2 (2015). https://doi.org/10.1111/ext.12161

Santarsiero, G., Masi, A., Picciano, V.: Durability of Gerber saddles in RC bridges: analyses and applications (Musmeci Bridge, Italy). Infrastructures **6**(2), 1–23 (2021). https://doi.org/10.3390/infrastructures6020025

Słoński, M., Tekieli, M.: 2D digital image correlation and region-based convolutional neural network in monitoring and evaluation of surface cracks in concrete structural elements. Materials **2020**(13), 3527 (2020). https://doi.org/10.3390/ma13163527

Su, Y., Zhang, Q.: Glare : a free and open-source software for generation and assessment of digital speckle pattern. Opt. Lasers Eng. 148(August 2021), 106766 (2022). https://doi.org/10.1016/j.optlaseng.2021.106766

Tambusay, A., Suryanto, B., Suprobo, P.: Visualization of shear cracks in a reinforced concrete beam using the digital image correlation. Int. J. Adv. Sci., Eng. Inf. Technol. **8**(2), 573–578 (2018). https://doi.org/10.18517/ijaseit.8.2.4847

# Quality Control of Prestressed Concrete Girder Decks in Existing Bridges: From Diagnostics to Numerical Analysis

Dario De Domenico(✉), Davide Messina, and Antonino Recupero

Department of Engineering, University of Messina, Contrada Di Dio, 98166 Sant'Agata, Messina, Italy
dario.dedomenico@unime.it

**Abstract.** The inherent aging of existing bridges together with adverse surrounding environmental conditions that have accelerated material degradation phenomena emphasize the need of a comprehensive quality control and management system of the Italian roadway bridge stock. In this context, this paper presents a procedure oriented to a comprehensive quality control and structural safety assessment of prestressed concrete (PC) girder decks in existing bridges. The proposed procedure combines experimental material characterization, field tests in serviceability conditions (static load tests and free vibration tests) and numerical simulation through a finite element model of the PC girder deck (linear and nonlinear analysis at ultimate limit states). The procedure, which is here illustrated in the context of a case study represented by the Zappulla viaduct (Sicily, Italy) built in the first 1970s, accounts for both serviceability and ultimate loading conditions, thus representing a reliable tool for the quality control of existing bridge structures.

**Keywords:** bridges · structural safety · quality control · prestressed concrete girder decks

## 1 Introduction

More than 80% of existing bridges in the Italian bridge stock were designed and built before the 1980s, following design regulations that are different from those in force today (Cosenza and Losanno 2021). Most of them (around 90%) were realized with a girder scheme with simply supported prestressed concrete (PC) beams and transversal diaphragms and an overlying slab made in reinforced concrete (RC) (Miluccio et al. 2021). These structures are quite vulnerable to material deterioration phenomena that might have taken place over their structural lifetime, especially corrosion of steel (Colajanni et al. 2016). In many cases, durability aspects were not carefully considered at the design stage and not regarded as a crucial performance requirement.

As a matter of fact, a scarce maintenance activity was performed over the years, and in most cases, there has been little attention paid to periodical management plans for safety checks and local repair/retrofitting actions. The occurrence of recent, unacceptably


M. A. Aiello and A. Bilotta (Eds.): ICC 2022, LNCE 435, pp. 245–259, 2024.
https://doi.org/10.1007/978-3-031-43102-9_20

frequent cases of collapse of existing bridges in Italy (Di Prisco et al. 2018; Morgese et al. 2020; Nuti et al. 2020) has emphasized the presence of an outdated infrastructure system in need of a rational, widespread, and reliable quality control and management system.

This need is even more important in view of the large number of bridges in Italy, with approximately one bridge every two kilometers of the infrastructure network. It is of extreme importance to systematically assess the structural health conditions of existing bridges through standardized, and comprehensive procedures that simultaneously consider serviceability and ultimate limit states (De Domenico et al. 2021).

The increasing importance of safety assessment of existing bridges has motivated the development of a document entitled "Guidelines for classification and risk management, safety assessment, and structural health monitoring of existing bridges", based on which an extensive classification of the entire existing bridge stock is being performed, based on a simplified risk analysis combined with a multi-hazard methodology.

Inspired by these guidelines, this contribution presents a procedure oriented to a comprehensive quality control and structural safety assessment of PC girder decks in existing bridges. The proposed procedure combines experimental material characterization, field tests in serviceability conditions (static load tests and free vibration tests) and numerical simulation through a finite element model of the PC girder deck (linear and nonlinear analysis at ultimate limit states). The procedure is here illustrated in the context of a case study represented by the Zappulla viaduct (Sicily, Italy) built in the first 1970s, and accounts for both serviceability and ultimate loading conditions, thus representing a reliable tool for the quality control of existing bridge structures (De Domenico et al. 2022).

## 2 The Zappulla Viaduct

### 2.1 Design Drawings and Structural Typology

The Zappulla viaduct was constructed around fifty years ago (first 1970s) in the municipality of Rocca di Capri Leone, approximately 2 km far from the Tyrrhenian Sea. Belonging to the A20 highway connecting Messina to Palermo, the multi-span viaduct is composed of two separate carriageways and includes 18 spans (of equal length 45.2 m and structural scheme).

The structural typology of the deck of the Zappulla multi-span viaduct is quite peculiar in the class of viaducts made of prestressed concrete. Indeed, the deck is formed by two box-shaped girders (U-shaped PC girders closed on the upper part by the overlying RC slab) mutually collaborating in the transverse direction through five PC transverse diaphragms per span, three of which along the span. Photographs and transverse section of the bridge deck are illustrated in Fig. 1. The presence of the (sufficiently rigid) transverse diaphragms near the supports and along the span of the bridge alters the structural response and increases the mutual collaboration of the two girders.

Each PC girder is equipped with 12 prestressing tendons (6 + 6 on the two sides of the section); each tendon is realized with 42 prestressing steel wires (ultimate tensile strength ≥1650 MPa) having a parabolic configuration, as shown in Fig. 2. The girders are simply supported on four neoprene bearings.

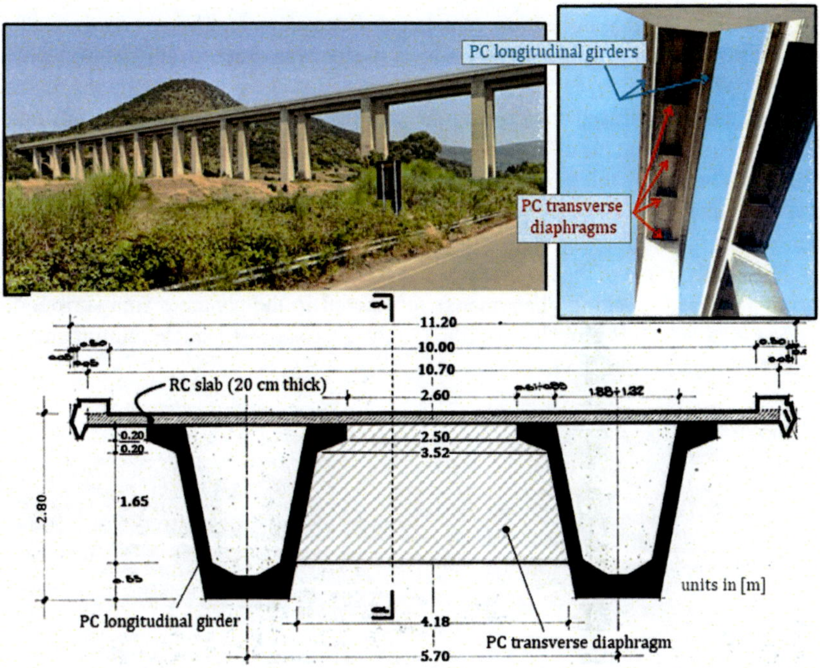


Fig. 1. Photographs (top) and transverse section of the deck of the Zappulla viaduct (bottom).

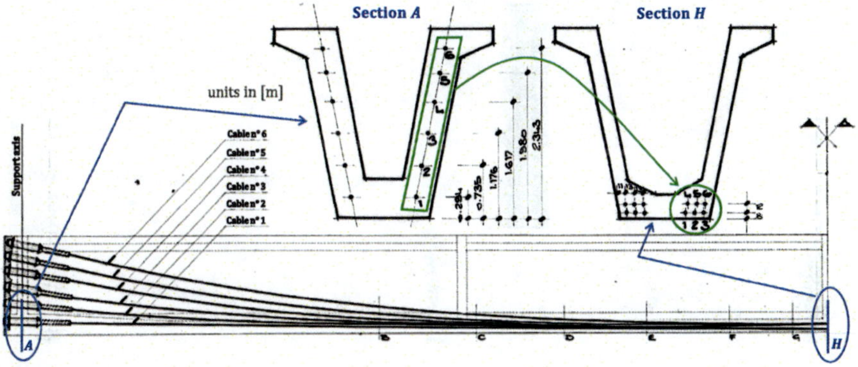


Fig. 2. Longitudinal profile of prestressing tendons and cross sections near the support and at mid span.

## 2.2 Visual Inspection and Defect Identification

The quality control of the bridge deck was first conducted through a series of inspections aimed at locating possible damage spots, cracking phenomena or material degradation signs that are macroscopically visible.

To this end, a bridge inspection vehicle cantilevered from different portions of the span (and from different spans of the viaduct) was used in order to take photographs on girders, diaphragms, slab, supports and abutments. Based on the photographs reported in Fig. 3, a state of diffused cracking on the external faces of the PC girders was noted, which may be symptomatic of degradation phenomena triggered by the chloride-induced corrosion. A relatively thick part of the concrete cover was removed by the aggressive action of the corrosive agents. Moreover, a set of damp stains were observed on the PC girders and at the soffit of the slab.

Additionally, some of the neoprene bearings were also characterized by an excessive deformability, and cracking phenomena were noted in the concrete area surrounding these bearings, which is probably ascribed to the absence of adequate reinforcement to transfer the concentrated load.

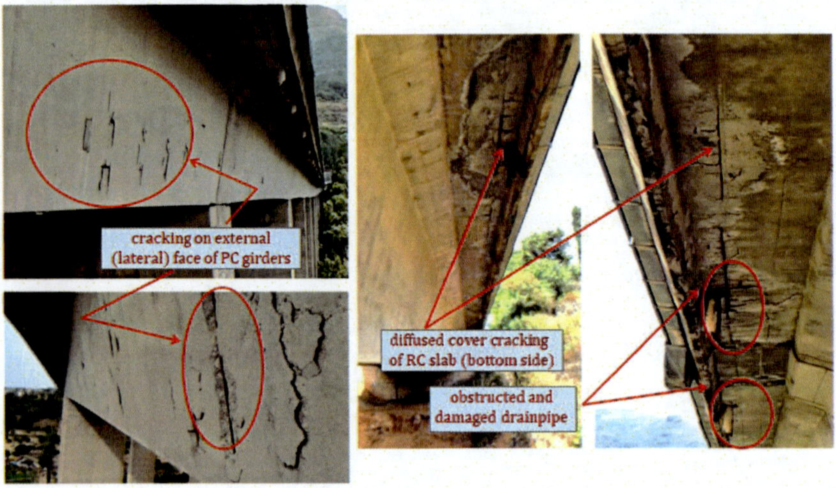


**Fig. 3.** Photographs of recurrent defects observed in the Zappulla viaduct.

## 3 Experimental Campaign for Material Characterization

A comprehensive in-situ testing campaign was planned and executed on the Zappulla viaduct. The experimental tests included: 1) 8 concrete cores, extracted from the RC slab and tested in laboratory to determine the compressive strength of concrete; 2) SonReb tests (sclerometric tests for the determination of the rebound index, combined with ultrasonic pulse velocity measurements) was performed in 16 different locations of the PC girders, which were used to determine, through suitable calibration equations, indirect information on the concrete compressive strength; 3) pachometer tests (in 16 locations), used to check that the location of the reinforcement bars is consistent with that indicated in the design drawings; 4) georadar tests (in 8 locations) to check the location of the prestressing tendons.

More specifically, cores having unitary height-to-diameter ratio ($h/d = 1$) were extracted from different portions of the RC slab (different spans) and tested in laboratory. The core strength, ranging from 32.3 MPa to 65.3 MPa, was corrected to account for the disturbance induced during the extraction process. The resulting concrete cubic strength ranged from 34.3 MPa to 65.3 MPa, with a mean value of 40 MPa and a coefficient of variation (CoV) of 10.7%.

SonReb tests were executed by measuring the rebound index $I_r$ and the ultrasonic pulse velocity $V_{us}$, which are then used within literature calibration formulae to estimate the concrete compressive strength. In particular, the average strength calculated from three popular power-law expressions was considered, in order to minimize the uncertainties inherent to available literature formulae in predicting strength data without a proper calibration of their empirical coefficients. The three expressions considered in this study are:

$$R_c = 9.27 \cdot 10^{-11} \cdot V_{us}^{2.6} \cdot I_r^{1.4} \quad \text{RILEM NTD4 (1993)} \tag{1}$$

$$R_c \quad 8.06 \cdot 10^{-8} \cdot V_{us}^{1.85} \cdot I_r^{1.246} \quad \text{Gašparik (1992)} \tag{2}$$

$$R_c = 1.2 \cdot 10^{-9} \cdot V_{us}^{2.446} \cdot I_r^{1.058} \quad \text{Di Leo \& Pascale (1994)} \tag{3}$$

The values of the concrete compressive strength obtained by Eqs. (1)–(3) are plotted in Fig. 4: it can be seen that three formulations provide quite comparable results. For this reason, the so-called average SonReb (SR) strength can be assumed as a realistic estimate of the concrete compressive strength.

Based on the results of concrete strength detected from compressive tests on concrete cores and from SonReb tests, a lognormal distribution is fitted to determine statistically meaningful parameters. The mean compressive strength obtained from the fitting lognormal distribution is $R_{cm} = 46.53$ MPa, and the CoV is 17.7%.

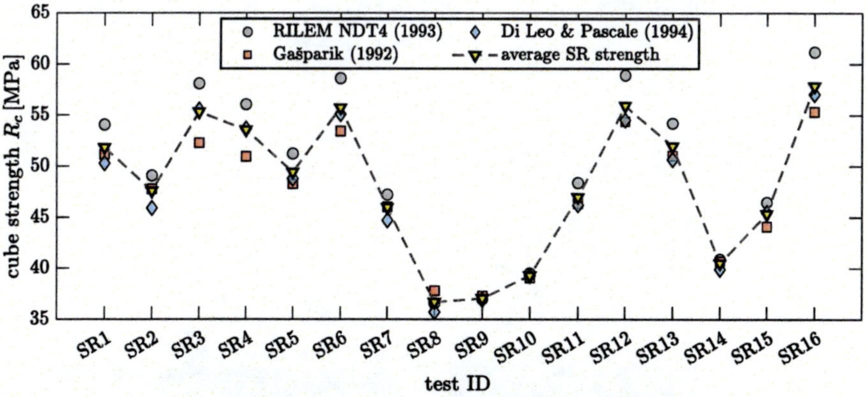


**Fig. 4.** Concrete compressive strength determined from combined SonReb tests.

Pachometer tests were conducted to examine areas located on the soffit and lateral face of the PC girders and to inspect the reinforcement arrangement. The bar distribution was consistent with that indicated in the design drawings. The concrete cover identified from the tests ranged from 20 to 45 mm, with values equal to 30–35 mm in most cases. These values are in line with the minimum cover for durability requirements of steel in accordance with Eurocode 2 (2005) provisions (Table 4.4N) for structural class S4 and exposure class XD1/XS1, consistently with the environmental conditions of the Zappulla viaduct.

Finally, georadar tests were conducted to study the profile of prestressing tendons. In this case, the tests were conducted on both lateral and bottom faces of the PC girders, and the results were obtained in the form of grayscale georadar two-dimensional sections (radargrams), from which heterogeneities and irregularities ascribed to a contrast of impedance were identified. Figure 5 shows some representative georadar sections. Not only the presence of six prestressing tendons of each inclined face of the PC girder

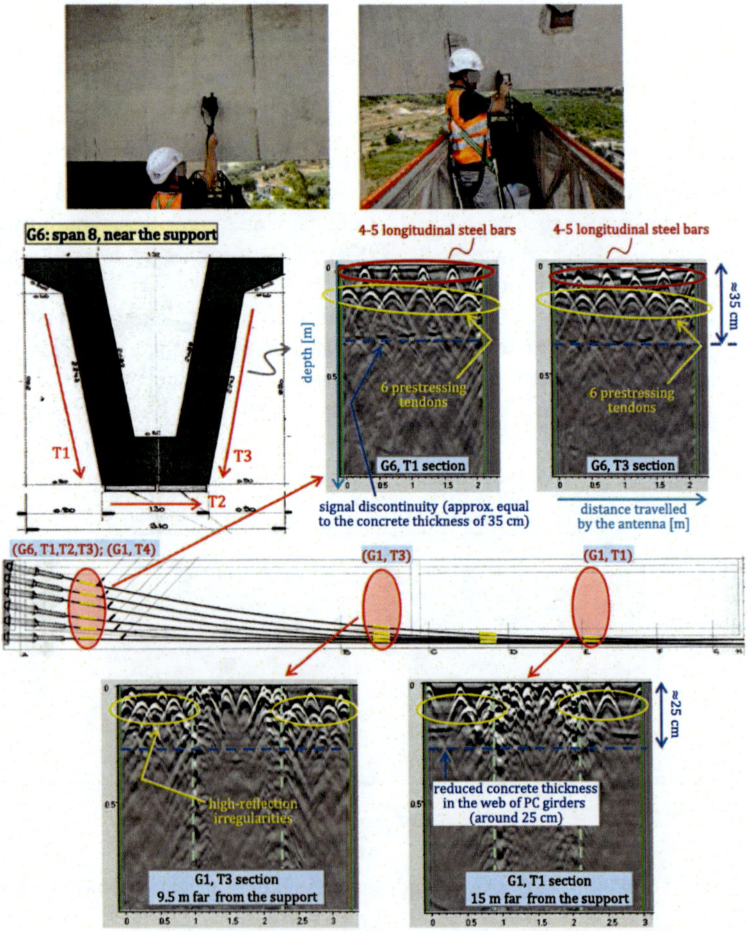


**Fig. 5.** Georadar tests for identification of prestressing tendons profile.

(highlighted within a yellow ellipse), but also of some longitudinal steel bars in the shallower layers (skin reinforcement) was identified. In general, georadar tests turn out to be helpful in those cases in which one has a prior knowledge, such as drawings, to properly interpret the results. These tests are also useful to locate possible imperfections at higher depths, e.g., defects occurring during concrete casting, like segregation and internal voids near the tendon ducts, or concrete delamination.

## 4 Field Tests in Serviceability Conditions

The global structural health conditions of the Zappulla viaduct were investigated through field tests, namely free vibration tests and static load tests. Free vibration tests were performed on eight spans of the multi-span viaduct for the dynamic identification of natural frequencies and mode shapes. Four vertical-axis accelerometers were deployed along the span of the viaduct (three along one side of the deck, and one along the opposite side), and free vibrations were triggered by an artificial excitation consisting in the transit of a heavy truck (around 490 kN gross weight) passing through an artificial step (Fig. 6).

The acceleration time histories were filtered and processed through the frequency domain decomposition technique, and natural frequencies were identified: the first frequency of vibration was identified in the range [2.90–3.10] Hz, depending on the considered span; the second frequency was slightly higher, falling in the range [5.20–5.40] Hz; the third frequency was identified in the range [9.90–10.00] Hz. Such values of natural frequencies are consistent with those found in a preliminary numerical model of

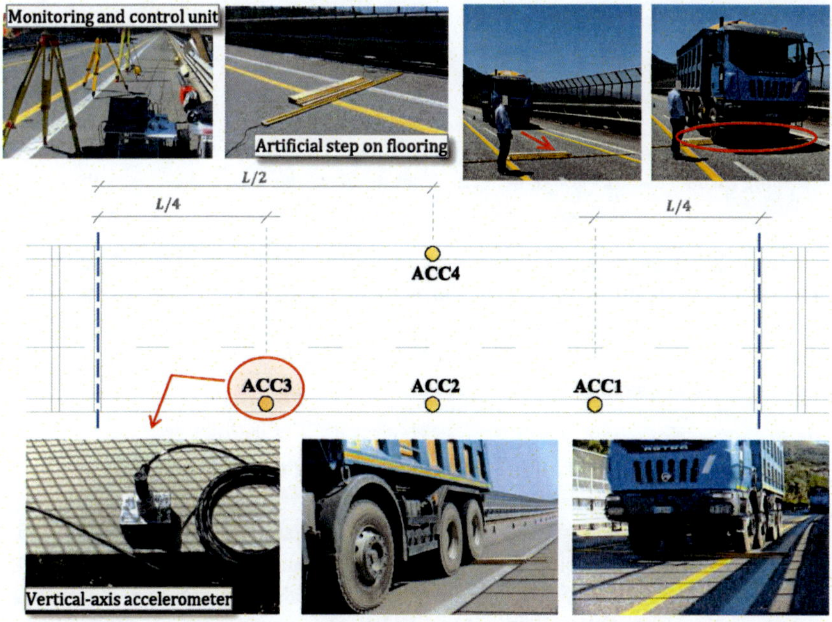


**Fig. 6.** Free vibration tests conducted through the transit of a heavy truck along an artificial step.

the Zappulla bridge deck, assuming uncracked sections. This result is symptomatic of a global undamaged state of the viaduct under the service loads.

Additionally, static load tests (Cai and Shahawy 2003; Hernandez and Myers 2019) were performed on eight spans of the multi-span viaduct to investigate the deflection response under the maximum serviceability loads prescribed by the Italian Technical Code (2018), termed NTC2018 hereinafter. Traffic load models of the NTC2018 involve distributed and tandem (concentrated) loads that are conventional loads. In reality, in the static load tests the loads were applied by means of four-axle heavy trucks filled with coarse gravel so as to reach a gross weight of 490 kN per each truck. Position of the trucks was designed to produce comparable stress state to that produced by the code compliant traffic load models. The loads were introduced in two loading phases (1 and 2) to carefully check the occurrence of any cracking sign or unexpected deflection response (see Fig. 7). The deflections were monitored in each loading phase through six digital levelling rods located in the bridge deck (points P1-P6 in Fig. 7).

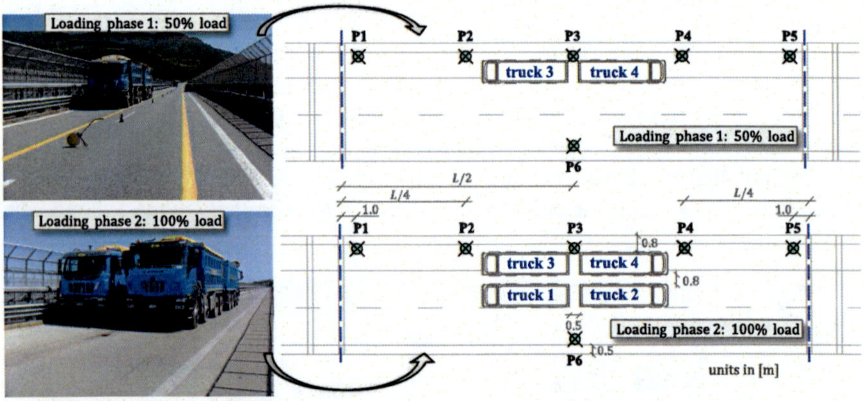


**Fig. 7.** Static load tests conducted through a particular arrangement of four-axle heavy trucks in two loading phases, and location of deflection measurement points.

After loading phase 2, the loads were removed, and it was checked that residual deflection in any point of the bridge did not exceed the 15% threshold of the maximum values as recommended by the NTC2018 provisions. The maximum deflection (at mid span) was around 18 mm, in line with the expectations and previously performed numerical simulations.

## 5 Numerical Simulation: Linear Finite Element Analysis

Both dynamic and static tests were numerically simulated through a simplified finite element model (FEM) developed in SAP2000 (Computers and Structures Inc. 2016), using beam elements (with 6 degrees of freedom per node) and adopting consistent sectional properties. In particular, both girders and diaphragms incorporated a collaborating portion of RC slab for shear lag effects (Fig. 8).

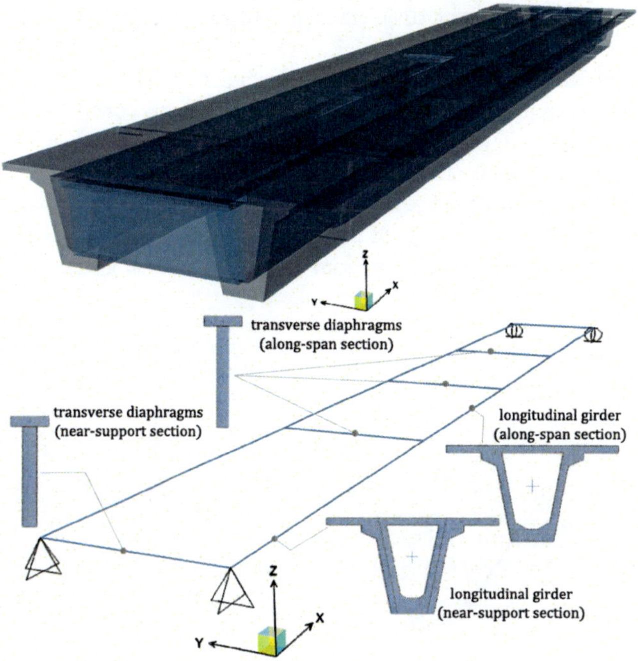


**Fig. 8.** Finite element model of the Zappulla viaduct deck using beam elements.

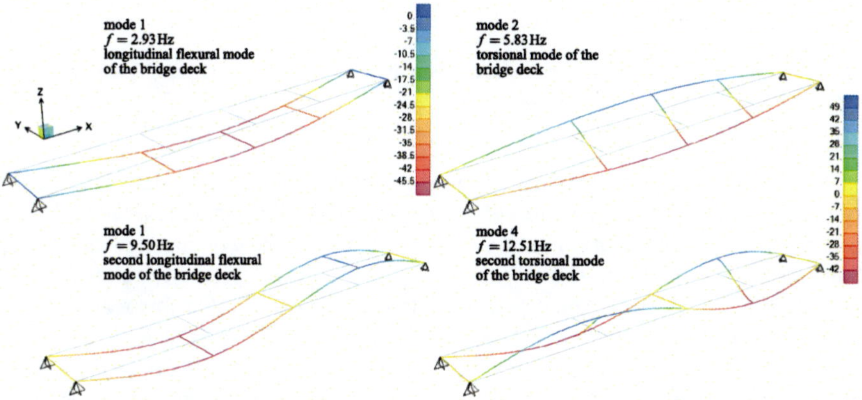


**Fig. 9.** Mode shapes and natural frequencies obtained from modal analysis on the FEM.

The FEM was validated by comparison between numerical natural frequencies (calculated from modal analysis and reported in Fig. 9 together with the mode shapes) and those obtained from free vibration tests. In general, the FEM provided a satisfactory prediction of the first three modes of vibration, and the discrepancies between the natural frequencies were in the order of 10%, which is a relatively good result considering the simplicity of the adopted structural model.

Similarly, deflection response induced by the static load tests was simulated through a linear static analysis in the FEM, by using load arrangements consistent with the heavy truck loads in the field tests (in terms of spacing and magnitude of equivalent truck loads). To calculate equivalent loads on the longitudinal girders generated by the trucks located on the bridge deck, a simplified scheme for a reasonable transversal distribution was adopted. In particular, the reactions of the supports from the deck analysis are applied as concentrated loads (in opposite direction) to the longitudinal girders of the FEM.

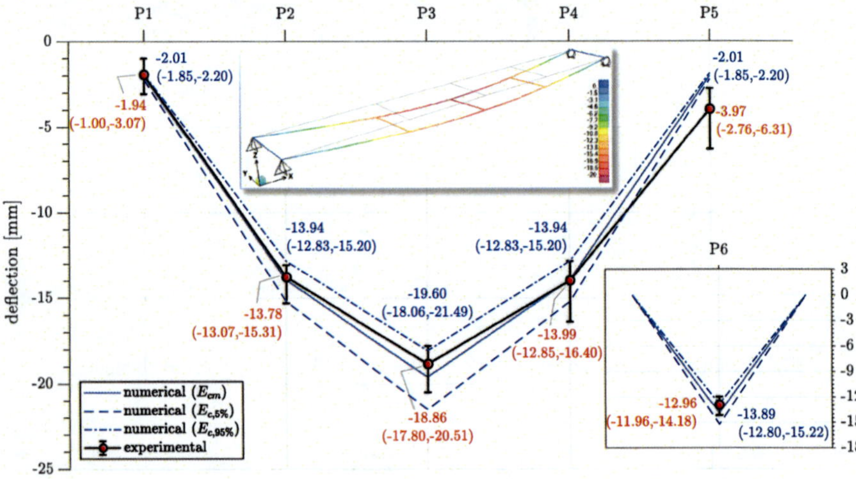


**Fig. 10.** Comparison between experimental (static field tests) and numerical (FEM) deflection response.

In the comparison between experimental and numerical deflections shown in Fig. 10, attention was paid to the deflections of those nodes of the FEM that are close to the six locations of the measurement points P1-P6 considered in the experimental campaign. The deflection response under serviceability loads is highly dependent upon the elastic properties of the deck, i.e., the elastic modulus of concrete $E_c$. For this reason, three numerical simulations were carried out by considering three values of $E_c$ to cover the inherent variability of the material properties among the different spans of the multi-span viaduct, namely the mean elastic modulus $E_{cm}$, and the lower and upper 5% fractile of the distribution, $E_{c,5\%}$ and $E_{c,95\%}$, respectively, calculated by processing the probability density function of the concrete compressive strength $p_{Rc}$ (i.e., the previously determined fitting lognormal distribution) and assuming an empirical relationship between $R_c$ and $E_c$ as per the Eurocode 2 (2005) provisions. From the comparison depicted in Fig. 10, it can be observed that a reasonable agreement was obtained between the numerical simulation (with $E_{cm}$) and the experimental findings, with relative errors less than 10%. Moreover, the range of variability of the deflection response obtained with the two limit values of the elastic modulus ($E_{c,5\%}$ and $E_{c,95\%}$) seems to cover, with reasonable accuracy, the interval of variation of the experimental deflection response observed during the static load tests. These results confirmed the validity of the developed FEM.

## 6 Numerical Simulation: Nonlinear Finite Element Analysis

After developing and validating the FEM against field test results under service load conditions, the numerical model was used to investigate the ultimate limit state behavior of the bridge deck. In particular, starting from the load configuration of the loading phase 2 of the static load tests (maximum serviceability loads), the loads were increased monotonically until reaching the collapse of the deck. This nonlinear static analysis is called in the literature "pushdown analysis", as the incremental load is vertical and directed downwards (similar to the pushover analysis utilized in the seismic context). A concentrated plasticity approach is adopted (see Fig. 11) by incorporating flexural plastic hinges along the PC girders to capture the variation of the flexural resistance and deformation capacity depending on the parabolic configuration of the prestressing tendons. A plastic hinge length equal to the depth of the cross section was assumed, and prestressing losses in the prestressing tendons were calculated and considered in the hinge calibration procedure.

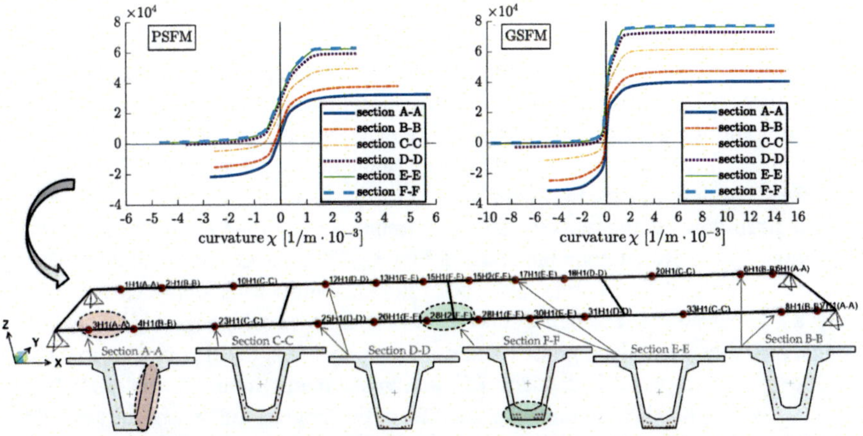


**Fig. 11.** Nonlinear FEM with concentrated plasticity approach and moment-curvature relationships calibrated for different sections of the PC girders.

For any step $s$ of the incremental analysis, the vertical load can be expressed as $F(s) = \lambda(s) \cdot F_0$, where $F_0$ is the initial load and $\lambda(s)$ denotes a dimensionless (positive) load multiplier at the examined step. Only the variable (traffic) loads were increased while keeping the permanent loads equal to their design value $\gamma_g\ G_k$. In this manner, it is possible to quantify the "extra" safety margin of the girder deck with respect to code-conforming traffic loads, i.e., the traffic loads computed at the end of the loading phase 2 of the field tests, in compliance with the Eurocode 0 (2002). Safety checks were performed by comparing the collapse load multiplier for variable loads obtained at the end of the nonlinear static analysis $\lambda_{qu}$ (corresponding to the ultimate load $F_u$) with the code-prescribed partial safety factor $\gamma_q$ (corresponding to the design load $F_d$).

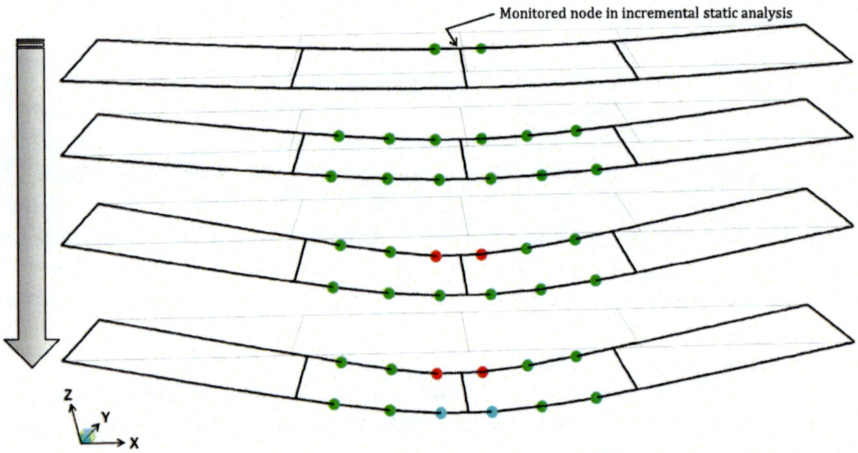


**Fig. 12.** Plastic hinge development for increasing load intensities in the push-down analysis.

Displacement-controlled nonlinear static analysis was performed in the FEM, and the "vertical push-down curve" was extracted in order to identify the ultimate (collapse) load and the safety margin of the bridge deck with respect to the maximum serviceability loads (starting point of the analysis). As reasonably expected, the development of the plastic hinges under monotonically increasing loads indicates that the most critical sections were those located near the mid span (Fig. 12).

The push-down curve plots the total maximum vertical load $F$ versus the mid-span deflection $\Delta$ (monitored node indicated in Fig. 12). The results are reported in Fig. 13 in terms not only of $F$-$\Delta$ curves, but also of dimensionless load multiplier for variable loads $\lambda_q$ versus normalized deflection $\Delta/L$, where $L$ is the bridge span (45.2 m). It can be observed that the deck response remains in a nearly linear elastic range up to the load level corresponding to the Eurocodes (EC) permanent loads amplified with safety factors (i.e., $\gamma_g\, G_k \approx 11800$ kN). A marked reduction of the tangent stiffness of the push-down curve is observed shortly after the attainment of the EC design load $F_d$ (including both permanent and variable loads amplified by partial safety factors). The load multiplier for variable loads $\lambda_q$ (bottom part of Fig. 13) quantifies the extra safety margin with respect to code-conforming traffic loads. In particular, it is noted that the collapse load multiplier for variable loads $\lambda_{qu} = 2.90$ largely exceeded the partial safety coefficient $\gamma_q = 1.35$ assumed in the Eurocodes, with a ratio $\lambda_{qu}/\gamma_q$ equal to 2.15. This result demonstrates that the bridge deck analyzed in this study exhibits an ample margin of safety with respect to code-conforming traffic loads when considering the design resistances given by the Eurocode semi-probabilistic approach.

No visible signs of corrosion of tendons were noted during the visual inspection, and corrosion of reinforcement bars was observed only in limited parts along the external face of some PC girders; moreover, carbonation tests performed on concrete cores

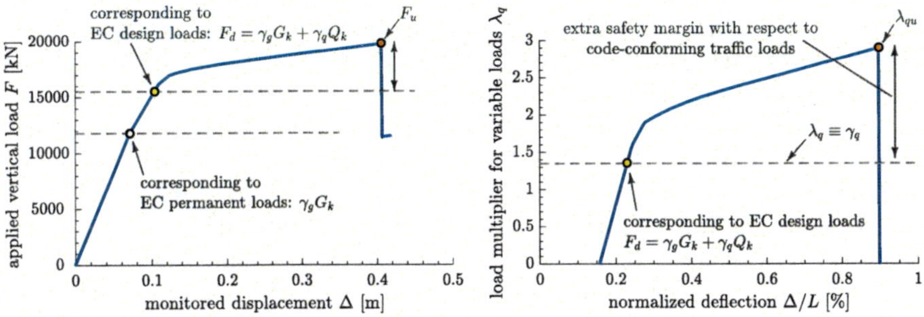


**Fig. 13.** Push-down curves of the bridge deck obtained by nonlinear static analysis, and quantification of safety margin with respect to Eurocodes (EC) design loads.

extracted from the RC slab revealed no carbonation depth. These findings indicate that the corrosion phenomena, broadly present in other bridges belonging to the same highway A20 Messina-Palermo, e.g., the deck of the Longano viaduct (De Domenico et al. 2021), might not represent a critical problem for the structural safety assessment of the deck of the Zappulla viaduct analyzed in this work. For future research, an integration of the testing protocol is envisaged through non-invasive corrosion potential mapping to estimate the actual level of corrosion in girders, diaphragms, and slabs, accompanied by further numerical analyses that incorporate a reduction of material properties (e.g., corrosion of tendons) via empirical models. A life-cycle assessment taking into account possible repairing actions to implement is also object of ongoing research.

## 7 Conclusions

A systematic approach for the quality control and safety assessment of existing PC bridge decks through numerical analysis assisted by field test results has been presented. The proposed methodology has been illustrated in the context of a case study represented by the Zappulla multi-span viaduct, built in the first 1970s and whose deck is characterized by a static scheme recurrently used for medium-span bridges, i.e., simply supported prestressed concrete girders with overlying reinforced concrete slab. The procedure has combined in-situ experimental investigations, field-test results (static and dynamic loading tests) and numerical FE simulation under both service and ultimate loading scenarios, taking into account design code requirements. In-situ experimental tests have been performed for the determination of concrete compressive strength, and for the identification of location and profiles of steel reinforcement and prestressing tendons. Moreover, field tests have been executed and properly interpreted to acquire knowledge on the bridge global structural health conditions. These field tests have included free vibration tests for dynamic identification of natural frequencies and mode shapes of the bridge deck, which are useful indicators of the potential ongoing damage of the structure, and static load tests to compute the deflection response of the bridge deck under the maximum allowed (code-conforming) serviceability loads. Apart from giving valuable information on the bridge safety conditions under serviceability loading, these tests have been used to calibrate and validate a numerical finite element model. This model has been adopted for the structural safety assessment of the bridge deck at ultimate limit states. Based on experimental outcomes and numerical simulation results, it has been concluded that the considered bridge deck exhibits an ample margin of safety with respect to code-conforming traffic loads (provided by the traffic load models of the current Eurocodes). These results are by no means trivial because the bridge was designed based on old (1962) standards and the corresponding traffic load models considered at the design stage were quite different from those prescribed by current regulations and considered in the numerical analyses. The proposed procedure can rapidly identify critical portions of a large infrastructure network in order to give the bridge management companies a preliminary outlook of the bridge structural health conditions and some general insights into the number, extent, and type of repairing action to implement, if necessary.

**Acknowledgements.** The authors express their gratitude to the Italian High Council of Public Works (CC.SS.LL.PP.) and the Network of University Laboratories of Seismic Engineering (RELUIS). The results of this study were achieved in the national agreement for implementing the agreement pursuant to art. 15 law 7 August 1990, No. 241 between the Superior Council of Public Works and RELUIS.

## References

Cai, C.S., Shahawy, M.: Understanding capacity rating of bridges from load tests. Pract. Period. Struct. Des. Constr. **8**(4), 209–216 (2003)

Colajanni, P., Recupero, A., Ricciardi, G., Spinella, N.: Failure by corrosion in PC bridges: a case history of a viaduct in Italy. Int. J. Struct. Integr. (2016). https://doi.org/10.1108/IJSI-09-2014-0046

Computers and Structures Inc.: CSi Analysis Reference Manual for SAP2000®, Berkeley, CA, USA (2016)

Cosenza, E., Losanno, D.: Assessment of existing reinforced-concrete bridges under road-traffic loads according to the new Italian guidelines. Struct. Concr. **22**(5), 2868–2881 (2021)

De Domenico, D., Messina, D., Recupero, A.: A combined experimental-numerical framework for assessing the load-bearing capacity of existing PC bridge decks accounting for corrosion of prestressing strands. Materials **14**(17), 4914 (2021)

De Domenico, D., Messina, D., Recupero, A.: Quality control and safety assessment of prestressed concrete bridge decks through combined field tests and numerical simulation. Structures **39**, 1135–1157 (2022)

Di Leo, A., Pascale, G.: Prove non distruttive sulle costruzioni in cemento armato. Il Giornale delle Prove non Distruttive Monitoraggio Diagnostica 4 (1994)

di Prisco, M., Colombo, M., Martinelli, P., Coronelli, D.: The technical causes of the collapse of Annone overpass on SS.36. In: Italian Concrete Days, pp. 1–16 (2018)

European Committee for Standardization: Eurocode 0: Basis of structural design - EN 1990:2002 +A1:2005, Brussels, Belgium (2002)

European Committee for Standardization 2005. Eurocode 2: Design of Concrete Structures - Part 1–1: General Rules and Rules for Buildings (EN 1992-1-1:2004), Brussels, Belgium

Gašparik, J.: Prove Non Distruttive nell'edilizia. Quaderno didattico Associazione Italiana Prove non Distruttive (AIPnD), Brescia (1992). (in Italian)

Hernandez, E.S., Myers, J.J.: Strength evaluation of prestressed concrete bridges by load testing. In: Life-Cycle Analysis and Assessment in Civil Engineering: Towards an Integrated Vision - Proceedings of the 6th International Symposium on Life-Cycle Civil Engineering, IALCCE 2018 (2019)

Italian High Council of Public Works: Circolare CSLLPP 14/02/1962, n. 384. Norme relative ai carichi per il calcolo dei ponti stradali, Rome, Italy (1962). (in Italian)

Italian Ministry of Infrastructures and Transportations: NTC2018. Aggiornamento delle «Norme tecniche per le costruzioni». Decreto 17 Gennaio 2018, Supplemento ordinario alla Gazzetta ufficiale n. 42, 20 February 2018 (2018). (In Italian)

Miluccio, G., Losanno, D., Parisi, F., Cosenza, E.: Traffic-load fragility models for prestressed concrete girder decks of existing Italian highway bridges. Eng. Struct. **249**, 113367 (2021)

Morgese, M., Ansari, F., Domaneschi, M., Cimellaro, G.P.: Post-collapse analysis of Morandi's Polcevera viaduct in Genoa Italy. J. Civ. Struct. Heal. Monit. **10**(1), 69–85 (2020)

Nuti, C., Briseghella, B., Chen, A., Lavorato, D., Iori, T., Vanzi, I.: Relevant outcomes from the history of Polcevera Viaduct in Genova, from design to nowadays failure. J. Civ. Struct. Health Monit. **10**(1), 87–107 (2020)

RILEM NDT 4: Recommendation for in Situ Concrete Strength Determination by Methods, Combined Non-destructive, Compendium of RILEM Technical Recommendations, London, UK (1993)

# Recent Developments of an Optimisation Procedure for Seismic Retrofit of RC Frames

Francesco Nigro, Roberto Falcone, and Enzo Martinelli(✉)

Department of Civil Engineering, University of Salerno, Salerno, Italy
{fnigro,rfalcone,e.martinelli}@unisa.it

**Abstract.** The spirit of the ecological transition applied to seismically deficient structures should inspire the development of simple methods capable of identify the "optimal" retrofitting solution (with reference to some predefined criteria). In order to support the intervention choice, an "objective" approach could be implemented making use of recently-invented Artificial Intelligence (AI) procedures. Specifically, the application of Genetic Algorithms (GAs) is usually thought as a suitable way in civil engineering optimization problems. In the present paper, a parametric study on a RC structure based on a GA procedure is reported. Consequently, the design of retrofit interventions results to be based on one objective criterion related to their cost-effectiveness, rather than on the highly subjective "engineering judgement". Moreover, the adoption of GAs enables to set more than one optimization criteria, in order to achieve not only strictly economic objectives, but also relevant targets intended to make seismic retroftting or upgrading interventions more sustainable.

**Keywords:** RC buildings · structural optimization · seismic retrofitting · Artificial Intelligence · Genetic Algorithm

## 1 Introduction

Nowadays, the retrofit of seismically-deficient reinforced concrete (RC) structures is one of the most relevant issues, since - as reported by the European Commission (2019) - the vast majority of European reinforced concrete (RC) structures were built in areas classified as non-seismic at the time of construction and before seismic codes came into force.

Faella et al. (2008) have clearly shown that combining member- and structure-level techniques could be a rational strategy to achieve the desired levels of seismic protection for existing RC structures.

However, since a huge number of combinations among those techniques can be theoretically figured out, finding out the "fittest intervention" is generally a challenging hard task. In principle, the search for the "fittest" seismic retrofitting intervention should be approached as a constrained intervention optimization problem.

In this regard, the application of Artificial Intelligence (AI) techniques to civil engineering problems could be an option when it comes to "select" the "fittest" interventions


M. A. Aiello and A. Bilotta (Eds.): ICC 2022, LNCE 435, pp. 260–273, 2024.
https://doi.org/10.1007/978-3-031-43102-9_21

out of the set of technically feasible ones. As a matter of fact, in the latest decades the use of AI-based techniques to solve engineering and applied science problems (including civil engineering ones) has been gaining wide and wide consensus. More specifically, Genetic Algorithms (GAs) have been usually preferred for structural optimization problems, since they are able to manage many discrete variables. Falcone et al. (2020) realized a literature review concerning the most relevant applications of Soft-Computing techniques to civil engineering problems.

Moreover, Falcone et al. (2019) have recently proposed a Soft-Computing approach (SC), formulating a Genetic Algorithm (GA) procedure for an "optimal" retrofit of existing Reinforced Concrete (RC) structures. The procedure may help to find the "fittest" retrofit solution, with regard to one predefined optimization criterion (e.g. the initial cost of intervention).

At each step of the procedure, a "population" of possible solutions (namely, individuals) is defined, in order to explore the features of different retrofit solutions that combine member-level and structure-level techniques. At the end of each iteration, the "fitness" of the "individuals" (with regard to the predefined objective function) is evaluated and the main Genetic Operators (selection, crossover and mutation) are handled in order to generate a new "generation" characterised by individuals with enhanced "fitness" to the purpose. A more detailed description of the procedure workflow (in its first formulation) has been given by Falcone (2017).

After a brief explanation of the main aspects of the procedure (Sect. 2), the present paper proposes a parametric study intended at highlighting how the "fittest" retrofitting/upgrading solution obtained by means of a GA procedure similar to the one proposed by Falcone et al. (2020) can be influenced by some relevant "engineering parameters". Specifically, Sect. 3 describes the main feature of the structure under consideration and Sect. 4 summarises the obtained results.

## 2 Overview of the Procedure

The procedure starts from the definition of the first population composed by $N_{ind}$ individuals (that represent different potential retrofit solutions).

### 2.1 Definition of the Individuals

Each individual is basically represented by a string (a row vector), that includes both its member-level (namely, the number of FRP layers $n_L$ used for the confinement of single columns) and its structure-level interventions (mainly, the section ID, $ID_{sec}$, employed for the realization of a concentric steel bracing system). Figure 1 shows how the generic individual vector $\boldsymbol{x}_i$ is structured, as it is subdivided in a first substring related to the member-level ("local") techniques and a second substring describing the ("global") structure-level ones.

More specifically, the "local" part of the individual defines the number of FRP layers $n_{L,hk}$ associated to the h-th column of the k-th storey (of each candidate retrofit solution or individual) so that the corresponding "Concrete01" material is modified by implementing the Kent and Park (1971) constitutive law for confined concrete in the

OpenSEES workspace (Mazzoni et al. 2006). The effects of confinement are taken into account both in terms of peak stress and ultimate strain, according to the Italian code provisions (CNR-DT 200/2013).

Moreover, the "global" part of the individual defines all the bays characterized by the presence of a concentric steel bracing system, which at the first storey is realized by means of the $N_b$-th section ID profile. Usually, the elements of the vector $\boldsymbol{x_i}$ are also called "individual genotype".

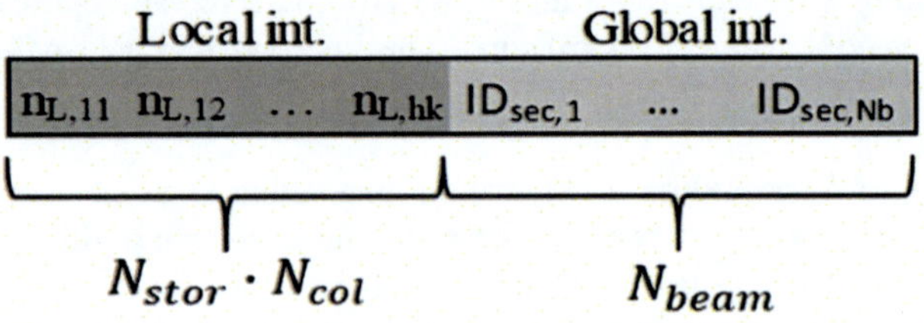


**Fig. 1.** Representation of the generic individual "$\boldsymbol{x_i}$"

The first version of the procedure was characterized by a binary encoding of the interventions although this aspect usually limited the number of the possible steel profile to be used as bracings. In order to avoid this limitation, in the latest version of the procedure a decimal encoding is adopted.

Equation 1 resumes the number of design variables that characterize each individual:

$$N_{\text{var}} = N_{stor} \cdot N_{col} + N_{beam} \tag{1}$$

where $N_{stor}$ = number of storeys; $N_{col}$ = number of columns of the single storey; $N_{beam}$ = number of beams (of the single storey). It is worth mentioning that, in order to reduce the number of design variables, the description of the bracing system depends only on the profile (bracing area) at the bottom floor: bracing area at the other (superior) levels are proportional to it.

## 2.2 Definition of the Objective Function and the Solution Constraints

The fundamental aspect of any optimization procedure consists in defining the objective function $f(x)$, which represents the criterion regulating the application of the genetic operators. The objective function also helps measure the "fitness" (quality) of the individuals composing each population.

In the present work the initial cost of intervention is chosen as the objective function. It includes not only the "direct" costs related to the local and global interventions (including eventual costs of demolition and restorations), but also the cost of the possible upgrading of foundations, which may be required especially where the braces are realized. Such costs have been defined with reference to the unit costs typical of the considered region (Prezzario Campania LL. PP. 2016) and it is worth stating that their meaning stays not in their absolute values, but on their relative values.

Since the optimization procedure is applied to the civil engineering problem concerning the retrofit of RC structures, it is also necessary to define one or more "constraints" for the objective function. The constraints impose some requirements that the optimal solutions have to fulfil.

Consequently, in order to improve the effectiveness of the procedure, it is worth to add to the aforementioned costs a "penalty cost" for those individuals that do not fulfil the constraints requirements. Equation 2 shows the definition of the objective function:

$$f(x) = C_{tot}(x) = C_{loc}(x) + C_{glob}(x) + C_{found}(x) + \Phi_{pen}(x) \tag{2}$$

where $f(x) = C_{loc}$ = objective function or total intervention cost (to be minimized), $C_{loc}$ = cost of local intervention, $C_{glob}$ = cost of global intervention, $C_{found}$ = cost of intervention at the foundation level, and $\Phi_{pen}$ = penalty cost function whose meaning and purpose will be explained later on.

Equation 3 represents the main goal of the present optimization procedure, which consists in the minimization of the objective function defined into the search space $S$, considering only the technically-admissible solutions, belonging to the feasible region $\Omega_f$:

$$\mathrm{x_{opt}} = \mathrm{argmin}[f(\mathrm{x})], \mathit{with} : \mathrm{x_{opt}} \in \Omega_\mathrm{f} \subseteq \mathrm{S} \tag{3}$$

where $x_{opt}$ is the optimal individual, $f(x)$ is the objective function, $S$ is the search space, $\Omega_f$ is the feasible region. Generally speaking, if a certain individual belongs to the feasible region $\Omega_f$, its penalty cost function $\Phi_{pen}$ is equal to zero.

The first constraint is represented by Eq. 4, which should be fulfilled to achieve the retrofit for each relevant Limit State (LS). Such constraint has been applied to all the analyses reported herein.

$$\mathrm{g_{LS,i}(x) = C_{LS,i} - D_{LS,i} \geq 0, \forall_i = 1, \ldots, n_{LS}} \tag{4}$$

where $g_{LS}$ = difference between the Capacity and the Demand for a certain LS; $C_{LS}$ = structural Capacity for a certain LS; $D_{LS}$ = structural Demand for a certain LS; $n_{LS}$ = number of the relevant LSs.

In Sect. 3.3 a "symmetry constraint" has been applied, in order to achieve symmetrical solutions. Such constraint is expressed by Eqs. 5–6:

$$\left| \min_{\mathrm{i}} \left[ \frac{\mathrm{C_{LS,i}(x)}}{\mathrm{D_{LS,i}(x)}} \right]_{\mathrm{dirX+}} - \min_{\mathrm{i}} \left[ \frac{\mathrm{C_{LS,i}(x)}}{\mathrm{D_{LS,i}(x)}} \right]_{\mathrm{dirX-}} \right| = 0, \forall_\mathrm{i} = 1, \ldots, \mathrm{n_{LS}} \tag{5}$$

$$\left| \min_{\mathrm{i}} \left[ \frac{\mathrm{C_{LS,i}(x)}}{\mathrm{D_{LS,i}(x)}} \right]_{\mathrm{dirY+}} - \min_{\mathrm{i}} \left[ \frac{\mathrm{C_{LS,i}(x)}}{\mathrm{D_{LS,i}(x)}} \right]_{\mathrm{dirY-}} \right| = 0, \forall_\mathrm{i} = 1, \ldots, \mathrm{n_{LS}} \tag{6}$$

where $C_{LS}$ = structural Capacity for a certain LS; $D_{LS}$ = structural Demand for a certain LS; $n_{LS}$ = number of the relevant LSs.

Finally, it is worth clarifying the meaning of the penalty function $\Phi_{pen}$: if the constraint expressed in Eq. 4 is imposed to the analyses, a penalty cost is added to the cost of those individuals (candidate retrofit solutions) that do not fulfil the Eq. 4 requirement. Such cost increase causes a huge reduction of the fitness of those individuals that do not fulfil the constraint avoiding them in the following populations. Equation 7 expresses such condition:

$$\Phi_{pen}(x) = \begin{cases} 0; & if \min_i [g_{LS,i}(x)] \geq 0 \\ 10^6 \cdot \mathrm{C}_{\mathrm{tot}}(\mathrm{x}); & if \min_i [g_{LS,i}(x)] < 0 \end{cases} \quad (7)$$

### 2.3 Overview of the Genetic Operators

In order to obtain the desired result, the proposed GA relies on the application (at each iteration of the procedure) of three main Genetic Operators, which mimic the drive mechanisms of the evolution of the species. The Genetic Operators are applied, at each iteration of the procedure, after the evaluation of the structural performance and the total cost (objective function) of all the individuals of the current population.

The aim of the Genetic Operators is to improve the fitness (quality) of the individuals of each population during the analysis. A critical discussion concerning the main features of Genetic Operators has been given by Falcone (2017). For the sake of brevity, in the following only the main innovations (aimed at improving the effectiveness of the GA) to such operators are reported:

- concerning the *selection operator*, the rate of application of the Random Pairing strategy has been increased passing from 50% to 85%;
- in the case of the *crossover operator*, only one binary "mask" has been applied;
- the number of times in which the *mutation operator* is applied has been reduced by means of an additional binary random vector.

### 2.4 Seismic Analysis

In order to guess the seismic behaviour of the structure in the present work Non-Linear Static (Pushover) Analyses have been utilized. The pushover analyses are run under displacement control, imposing a target displacement equal to 0.50 m, that corresponds to the 3.5% of the structure total height.

#### 2.4.1 Definition of the FE Model

The "as built" model of the RC structure is defined using the OpenSEES distributed-plasticity "nonlinearBeamColumn" element (Mazzoni et al. 2006) for both beams and columns, having previously defined the fibre cross sections that define such elements. The integration is conduced defining 5 Lobatto points for each element.

Starting from the "as built" model, a specific subroutine associated to all the individuals of the current population a nonlinear mechanical model, decoding their genotype, as defined in Fig. 1, Sect. 2.1.

Moreover, the bracing systems are implemented similarly to the RC members by means of the "nonlinearBeamColumn" element discretizing their cross section in fibers, as well. The modelling of buckling is gained imposing to each brace a geometric imperfection (according to EN 1993-1:2005) without modifying the steel material properties.

#### 2.4.2 Seismic Demand Determination

The displacement demand for the SDOF equivalent to the given MDOF structure can be easily derived suing the N2 method, proposed by Fajfar (1999).

#### 2.4.3 Relevant Performance Parameters

The structural behaviour is studied considering two relevant Limit States for the Italian Code (DM 20/02/2018): SLD (also mentioned as Damage Limitation LS) and SLV (also referred as Life Safety LS).

The determination of the capacity step for each relevant Limit State relies on Eq. 8 which takes into account the biaxial response of members representing an "interaction domain" of the interstorey drifts "δ" (along both X and Y directions):

$$C_{LS}(x) : \left[ \left( \frac{\delta_D}{\delta_{C,LS}} \right)_X^{\alpha} + \left( \frac{\delta_D}{\delta_{C,LS}} \right)_Y^{\alpha} \right] = 1 \tag{8}$$

where $C_{LS}$ = capacity of the structure in a certain LS; $\delta_D$ = drift demand at a certain analysis step; $\delta_{C,LS}$ = drift capacity at a certain analysis step in a certain LS; $\alpha$ = exponent, assumed equal to 1 to be on safety side. It is worth to notice that the drift capacity $\delta_{C,LS}$ at the SLV is computed according to "Equation A.1" of EN 1998-3:2005, taking into account its variation due to the variation of axial force, whereas the capacity in the SLD is assumed constantly equal to 0.5%, as fixed by Italian Code (2018).

## 3 Parametric Analyses

### 3.1 "As Built" Structure Presentation

In the present study a 4 storey RC structure regular in plan and elevation has been considered. It is composed by 5 bays in the X direction and 3 bays in the Y direction. The structure is characterized by uniform span length of 5 m and uniform storey height of 3.50 m. It is ideally drawn from a Gravitational Load Designed (GLD) residential building.

The typical column cross section consists in a square with a side of 0.35 m, while beams are typically characterized by a $0.35 \times 0.60\ \text{m}^2$ rectangular cross section. All the members are reinforced with 12 mm reinforcement bars; the stirrup spacing is assumed to be equal to 0.25 m, so that the confinement effect can be neglected. Figure 2 represents schematically the 3D model of the structure, while Fig. 3 offers a plan view of the typical storey of the structure. Finally, Fig. 4 represents the pushover curve of the structure in its "as built" configuration.

### 3.2 "Optimal" Retrofit Solutions Obtained for Different Hazard Scenarios

The first application proposed herein consists in the evaluation of the GA effectiveness in finding the "optimal" retrofit solution, considering three different seismic hazard scenarios, associated to three Italian cities as possible locations of the structure. The hazard is, indeed, gradually reduced passing from the site of Sant'Angelo dei Lombardi (AV) to Acerno (SA).

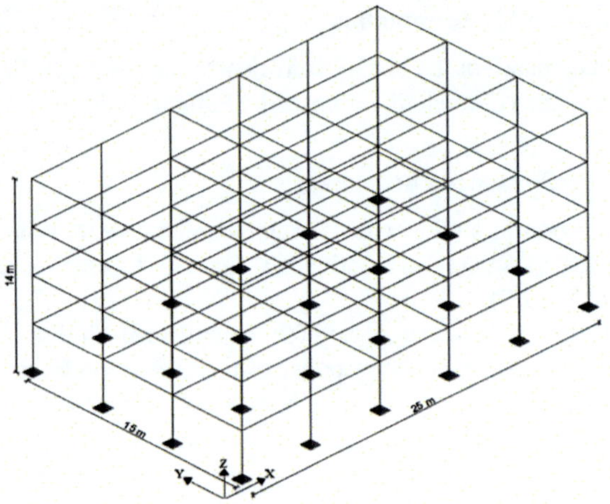


**Fig. 2.** 3D scheme of the structure under consideration.

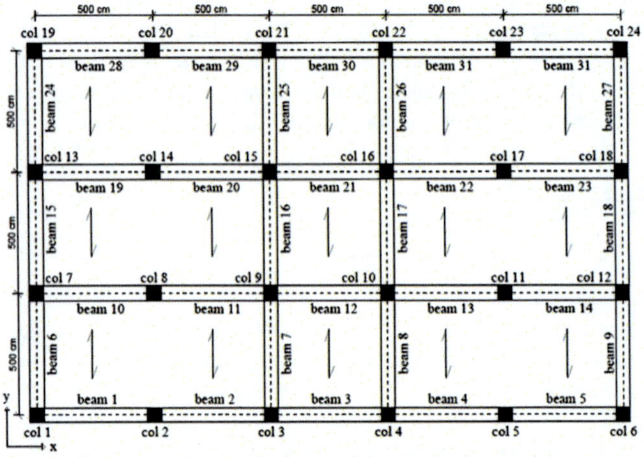


**Fig. 3.** Plan of the typical storey.

### 3.2.1 Sites Considered in the Analyses

Table 1 summarizes the PSHA for the considered sites, considering only the two relevant LS for the structure.

The "optimal" solutions given by the GA are presented by:

- Figures 5, 6 and 7 for the site of Sant'Angelo dei Lombardi (AV);
- Figures 8, 9 and 10 for the site of Avellino (AV);
- Figures 11, 12, Table 2 for the site of Acerno (SA).

Such figures not only represent the convergence of the solution –in other words, the objective function of the best individuals of each population - but also the plan distribution of the bracings and the pushover curves of the retrofitted structures. The maximum values

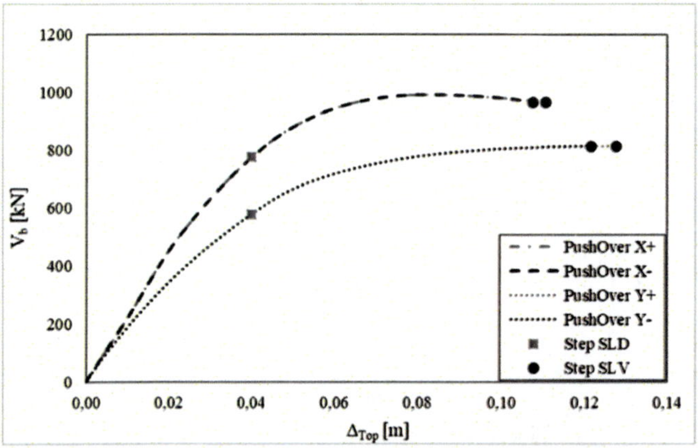


**Fig. 4.** Pushover curves of the structure in the "as built" state.

**Table 1.** PSHA of the sites considered in the analyses, soil class "C"

| Site | Sant'Angelo dei Lombardi | | Avellino | | Acerno | |
|---|---|---|---|---|---|---|
| Limit State | SLD | SLV | SLD | SLV | SLD | SLV |
| $^{*}a_g/g$ [-] | 0.08 | 0.27 | 0.07 | 0.20 | 0.06 | 0.15 |
| $^{**}F_0$ [-] | 2.33 | 2.28 | 2.32 | 2.37 | 2.43 | 2.48 |
| $^{***}S$ [-] | 1.50 | 1.36 | 1.50 | 1.42 | 1.50 | 1.48 |

$^{*}a_g/g$ is the PGA normalized with respect to gravitational acceleration g;
**$F_0$ is a factor that quantifies the maximum spectral acceleration;
***S is the soil factor

of the "demand-to-capacity" ratio $(D/C)_{max}$ in terms of top displacement of the structure are represented, as well, in the same figures concerning the convergence.

### 3.3 "Optimal" Retrofit Solutions Obtained Considering the Symmetry Constraint

The third application aims at assessing the GA-based procedure in finding the "optimal" retrofit solution, ideally locating the structure in Sant'Angelo dei Lombardi (AV), considering also the symmetry constraint. Consequently, the "optimal" solution has to fulfil not only the Eq. 4 constraint, but also the symmetry constraint, expressed by Eqs. 5–6.

Figure 13 summarizes the convergence procedures considering the symmetric constraint (Symm.) and without imposing such constraint (Asymm.).

The description of the asymmetric solution has been already given by Figs. 5–7, while Fig. 14 express the main features of the symmetric solution, which might represent the absolute minimum of the objective function.

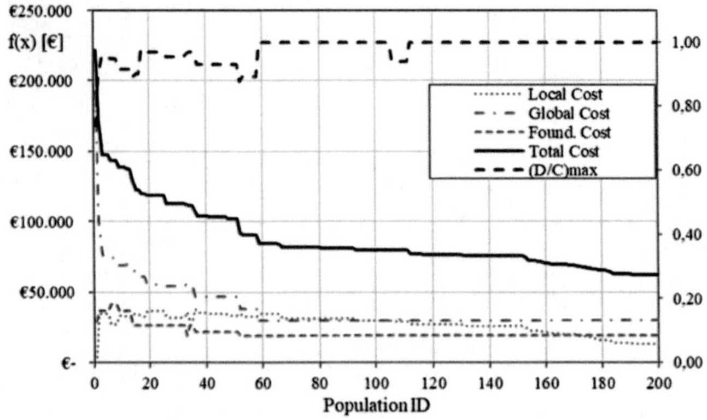


**Fig. 5.** Optimization (convergence) of the best individual in the case of Sant'Angelo dei Lombardi.

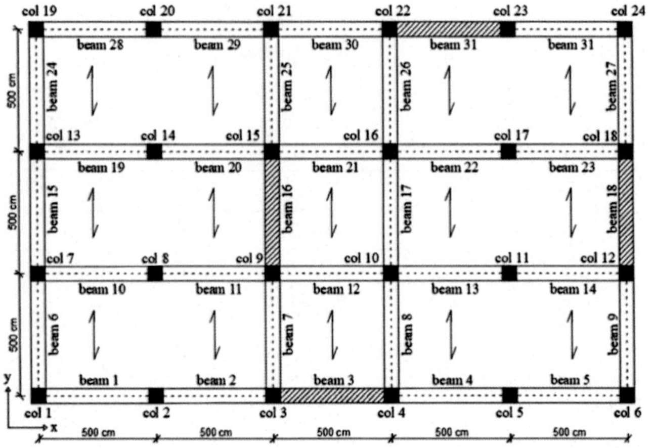


**Fig. 6.** Plan distribution of bracings of the best individual in the case of Sant'Angelo dei Lombardi.

## 3.4 Result Discussion

In this section the results obtained for the parametric analyses previously presented are discussed.

From the parameter analysis reported in Sect. 3.2, concerning the design of retrofit interventions, varying the hazard of the construction site, it is clear that:

- the proposed GA is able to account for the variation of the seismic hazard, leading to more economic individuals (retrofit solutions) as soon as the hazard is reduced;
- when the difference between the structural capacity in the "as built" configuration is huge, the GA leads to individuals (retrofit solutions) characterized by a combination of local and global interventions, since an increase of strength and ductility are both required in these cases; whereas, if such difference is not excessive the "optimal" solution is characterized by local interventions only, since an increase of structural ductility is sufficient to meet the required seismic performance;

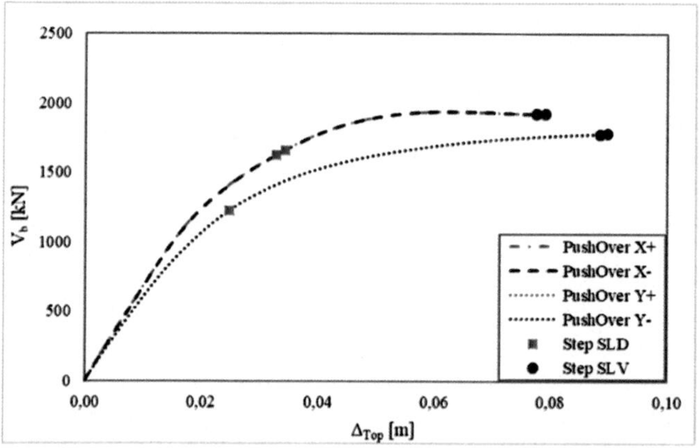


**Fig. 7.** Pushover curves of the retrofitted structure in the case of Sant'Angelo dei Lombardi.

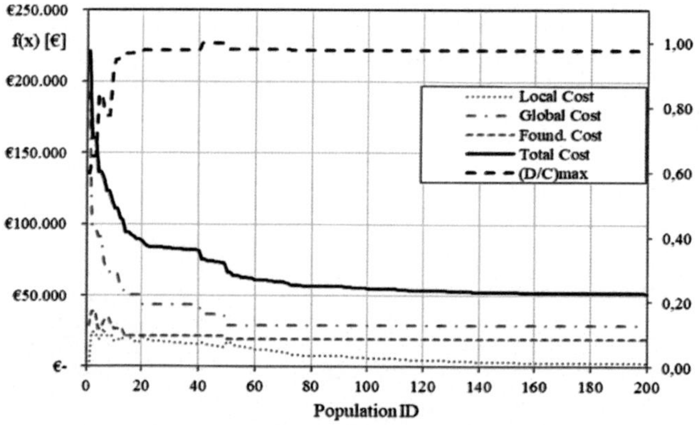


**Fig. 8.** Optimization (convergence) of the best individual in the case of Avellino.

- when applying only local interventions (as in the case of Acerno) the GA tends to prefer the confinement at the first level, recognizing that the columns of the first storey may evidence a lack of ductility, if compared to the ones of the upper storeys;
- in general, at the end of the analyses, the maximum displacement "demand-to-capacity" ratio $(D/C)_{max}$ of the best individuals (according to the GA) results to be close to the unit value. This may be a consequence of the definition of the objective function: since the fitness of the individuals is measured only in terms of initial cost of intervention, values of $(D/C)_{max}$ minor than 1 do not influence the individual fitness.

From the parameter analysis exposed in Sect. 3.3, concerning the influence of different constraints to the design of retrofit interventions, locating the structure in the site of Sant'Angelo dei Lombardi, it might be observed that:

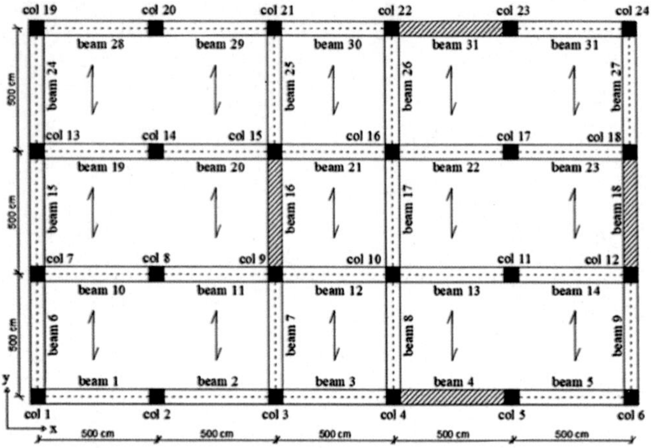


**Fig. 9.** Plan distribution of bracings of the best individual in the case of Avellino.

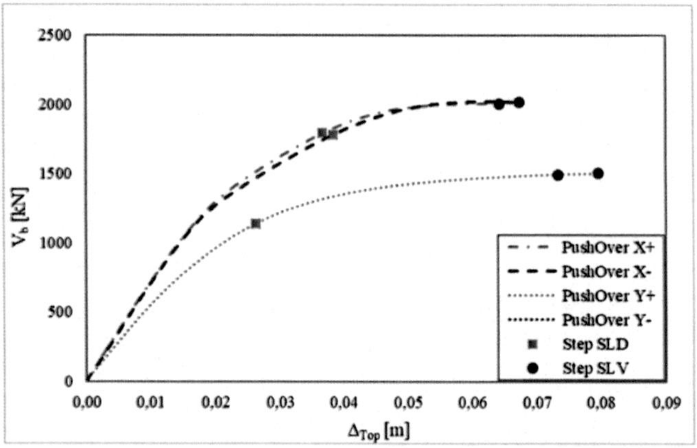


**Fig. 10.** Pushover curves of the retrofitted structure in the case of Avellino.

- the symmetry constraint is able to speed up the convergence process, reducing the actual number of design variables in the "global part" of each individual passing from $N_{beam} = 32$ (total number of bays of a single storey) to 10 (number of groups of symmetrical bays);
- if the hazard of the construction site is huge, the symmetry constraint causes the exclusion of the presence of local interventions;
- the symmetry constraint may lead to the absolute minimum of the objective function (Fig. 14) if the search space is explored properly.

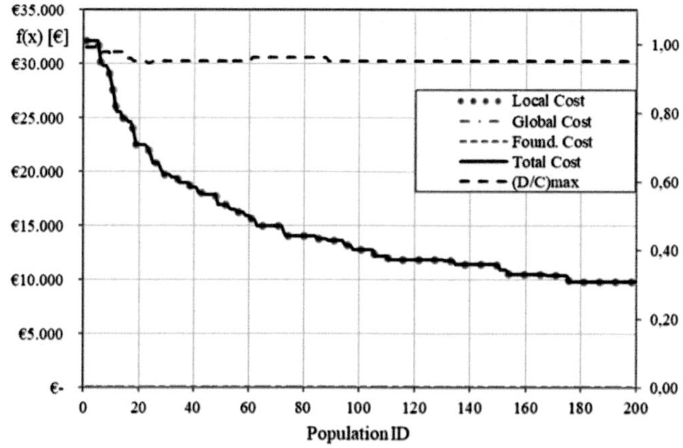


**Fig. 11.** Optimization (convergence) of the best individual in the case of Acerno.

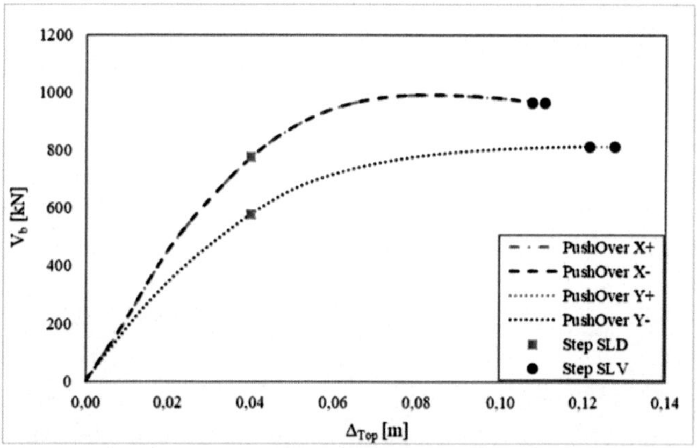


**Fig. 12.** Pushover curves of the retrofitted structure in the case of Acerno.

**Table 2.** Confinement layers for each storey in the case of Acerno.

| Storey | Number of confined columns | Total number of FRP layers |
|---|---|---|
| 1st | 19 | 20 |
| 2nd | 0 | 0 |
| 3rd | 1 | 2 |
| 4th | 2 | 3 |

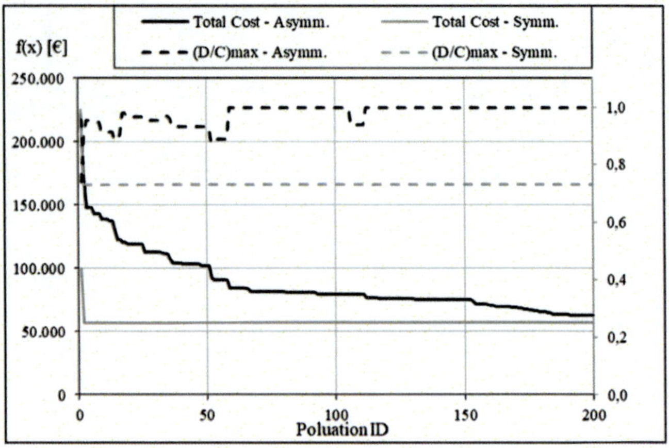


**Fig. 13.** Optimization (convergence) of the best individual imposing the symmetry constraint or not.

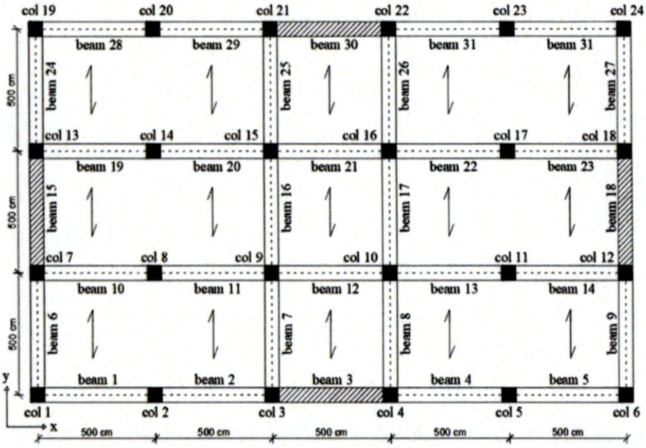


**Fig. 14.** Plan distribution of bracings of the best individual imposing the symmetric constraint.

# 4 Conclusions

The goal of the present paper is to highlight the potentialities of a Soft-Computing procedure based on a Genetic Algorithm, firstly formulated by Falcone (2017) and recently revised and updated by the Authors. Some parametric analyses have been carried out in order to get a better understanding of the GA effectiveness in this kind of engineering problem. Making use of the aforementioned results, it should be observed that:

- the outcomes of the GA are usually in good agreement with the imposed constraints, although the presence of only one constraint (as in Sect. 3.2) may lead to solutions that seem quite distant from the" intuitive engineering judgment" of the problem, with regard to the location of the additional bracing systems;

- one of the possible way to improve the GA effectiveness stays in modifying the method of analysis employed herein, which is not able to catch the presence of eventual asymmetries in the distribution of the bracing systems;
- as usual for numerical procedures, the quality of the solutions (namely, best individuals) can be considerably affected by the setting of algorithmic parameters;
- the GA could be employed also modifying the objective function passing to a multi-objective one or adapting the objective function to the needs of the user.

Nevertheless, the above-discussed examination of the analyses results gives the possibility to notice (and correct) the critical aspects of this Artificial Intelligence technique, in order to enhance its efficiency. To this aim, it is relevant to state that the main objectives of the future works will consist in improving the ability of the GA to explore the search space, in order to achieve an optimization solution which is as nearer as possible to the absolute minimum of the objective function, being at the same time consistent from an engineering point of view.

## References


CNR-DT 200 R1/2013: Istruzioni per la progettazione, l'esecuzione ed il controllo di interventi di consolidamento statico mediante l'utilizzo di compositi fibrorinforzati (2013)

DM 17/01/2018: Italian Technical Code of Constructions; Ministerial Decree, Rome, Italy (2018)

DM 65, 07/03/2017, Allegato A: Linee guida per la classificazione del rischio sismico delle costruzioni. Ministerial Decree, Rome, Italy (2017)

EN 1993-1:2005. Design of steel structures, Part 1-1: General rules and rules for buildings. European Committee for Standardization, Bruxelles (2005)

EN 1998-3:2005. 2005 Design of structures for earthquake resistance. Part 3: Assessment and retrofitting of buildings. European Committee for Standardization, Bruxelles (2005)

European Commission: EU Buildings Factsheets (2013). https://ec.europa.eu/energy/eu-buildings-factsheets_en

Faella, C., Martinelli, E., Nigro, E.: A rational strategy for seismic retrofitting of RC existing buildings. In: Proceedings of the 14th World Conference on Earthquake Engineering, Beijing, China, 12–17 October 2008 (2008)

Fajfar, P.: Capacity spectrum method based on inelastic demand spectra. Earthq. Eng. Struct. Dyn. **28**(9), 979–993 (1999)

Falcone, R.: Optimal seismic retrofitting of existing RC frames through soft-computing approaches. Ph.D. Course on risk and sustainability in civil, architectural and environmental engineering systems, XXX Cycle (2017)

Falcone, R., Carrabs, F., Cerulli, R., Lima, C., Martinelli, E.: Seismic retrofitting of existing RC buildings: a rational selection procedure based on genetic algorithms. Structures **22**, 310–326 (2019)

Falcone, R., Lima, C., Martinelli, E.: Soft computing techniques in structural and earthquake engineering: a literature review. Eng. Struct. **207**, 110269 (2020)

Kent, D.C., Park, R.: Flexural members with confined concrete. J. Struct. Div. **97**(7), 1969–1990 (1971)

Mazzoni, S., McKenna, F., Scott, M.H., Fenves, G.L., et al.: Open system for earthquake engineering simulation user command-language manual (2006)

Prezzario Campania LL.PP.: Prezzario Regionale dei Lavori Pubblici Anno 2016. Delibera della Giunta Regionale n. 359 del 13/07/2016 (2016)

# Degradation and Rehabilitation of Gerber Saddles of Concrete Bridges

Michele Fabio Granata[1(✉)], Lidia La Mendola[1], Davide Messina[2], and Antonino Recupero[2]

[1] Department of Engineering, Università di Palermo, Palermo, Italy
michelefabio.granata@unipa.it
[2] Department of Engineering, Università di Messina, Messina, Italy

**Abstract.** In recent decades, many concrete bridges, built in the 60s and 70s of the 20th Century with a Gerber-girder scheme, show severe damage of the saddles due to concrete degradation and reinforcement corrosion. On the basis of the recent Ministerial Guidelines for the assessment of existing bridges, it is necessary to evaluate the state of the Gerber saddles in a large number of bridges on both main and secondary roads. To do this, it is necessary to apply flexible assessment procedures that can be adapted to several situations and that allow a speedy assessment of the intervention priorities. The methodologies of intervention represent the crucial point for a rapid rehabilitation of these infrastructures. In this paper, on the basis of some case-studies, assessment and strengthening methodologies are proposed that can be applied to different real cases, in consideration of the budget available, with the aim of drafting an operative list of priorities and improving as many bridges as possible, currently in service.

**Keywords:** bridges · concrete degradation · corrosion · Gerber saddle

## 1 Introduction

The Italian Ministerial Guidelines on the assessment of bridges (Italian Ministry of Infrastructures 2020) establish that the damaged saddles of Gerber-girder bridges must be always considered with the highest degree of attention, leading to classify the bridge with the highest vulnerability class, regardless of the actual degree of damage and the intensity or extension of the defect. Although this is justified from collapses and local failures occurred in the last years that involved this typology of bridges, this kind of evaluation implies that each Gerber bridge, regardless of the configuration of the saddle, becomes part of the special supervised bridges. This is also due to the lack of robustness, intrinsic of the isostatic structural scheme of Gerber girders, leading to the safety assessment of level 4 foreseen by the Guidelines, that is an attentive and complete analysis of the bridge behaviour, based on in situ tests and investigations and on advanced structural models and analyses, up to the evaluation of the transit ability, scheduled for structures that present a high class of dangerousness and vulnerability in service life.


M. A. Aiello and A. Bilotta (Eds.): ICC 2022, LNCE 435, pp. 274–287, 2024.
https://doi.org/10.1007/978-3-031-43102-9_22

The huge variety of configurations of the so-called half-joints, as the Gerber saddles are named too, involves a variety of structural behaviours and therefore of strength mechanisms. The presence of crossbeams or heads of prestressing anchors often completely changes the behaviour respect to that of the classic two dapped-end beams with the bearing on the girder web. This means that for each bridge, the state of degradation of concrete and the corrosion of the saddle reinforcements can have very different consequences, based on the case examined.

In this paper, after a discussion on the Guidelines provisions, two different configurations of Gerber saddles are considered, in absence and in the presence of prestressing, based on two case-studies. The safety assessment of these bridges is presented illustrating how several saddles show a greater robustness than expected. Aspects relating to strengthening through global retrofitting interventions with static scheme change and external prestressing are discussed also for the case-studies presented.

## 2 Gerber-Girder Bridges

In Gerber-girder bridges, the half-joints have a wide variability of configurations, depending on the typology of the deck and the age of design. The configuration of the Gerber saddles is different for grillage decks composed of main girders and transverse beams, for double-beam decks, for box girders and for slab bridges and it also depends on the presence or absence of prestressing tendons in the beam webs.

The saddles on the girder webs can be single without crossbeams, with transverse beams on the corbels only, with rigid diaphragms on the entire saddle, or on box sections, with or without prestressing anchors inside (Fig. 1).

The presence of prestressing is generally beneficial, also because the internal tendons have the anchor head right in correspondence with the saddles, entering those areas with a favourable inclination for the shear behaviour, through the vertical component of inclined tendons that is in the opposite direction to that of the external load. Moreover, the prestressing provides a further benefit through the presence of the axial force (Fig. 2).

The crossbeam configuration is very important because it can allow or prevent the redistribution of the reactions among the saddles of adjacent girders near failure, although its presence in correspondence with the corbels only or with the entire saddle could cause a failure in shear of the section of the girder before the saddle. In these cases, the saddle is not the most dangerous element and failure shifts in the section of the girder with maximum shear just before the saddle. Furthermore, the lower slab of cantilever sections has a beneficial effect for lower corbels as well as the upper slab supplies a contribution to strength for upper corbels: the presence of slabs and crossbeams increase the strength of the saddle making critical the section where the main girder attaches to the crossbeam.

Several Gerber-girder bridges present the signs of concrete degradation and reinforcement corrosion, with decrease of safety factors; hence, it is necessary to evaluate what is the actual safety factor against failure of these half-joints. The large number of bridges which is subject to this type of degradation and the lack of indications in the Ministerial Guidelines about the visual evaluation of the severity of the defect during in situ inspections, leads to the need for speedy procedures that allow engineers to give a

**Fig. 1.** Gerber saddle configurations on bridges

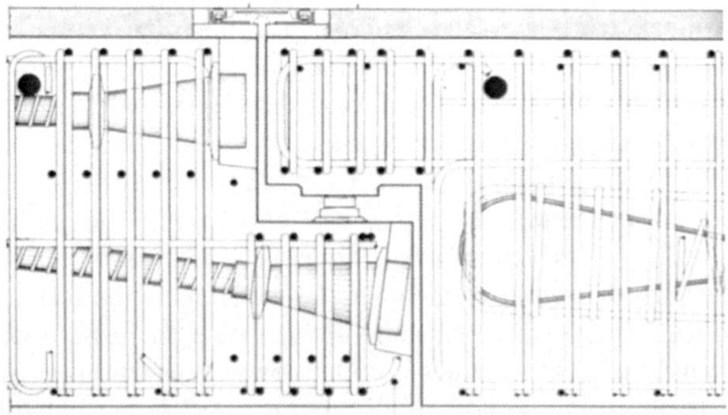

**Fig. 2.** Gerber saddles with and without prestressing anchor heads

reliable safety assessment of these bridges, which often present problems concentrated on the saddles or in any case in the points of water percolation from the deck.

In this study, non-linear analyses are presented in comparison to classic strut-and-tie methods, with different hypotheses of reducing strength properties due to corrosion, showing how in many cases, even with very penalizing hypotheses, these bridges keep acceptable safety levels, thanks to the configuration of the saddle, to the amount of reinforcement provided in the original design and in some cases for the presence of prestressing.

However, due to the static scheme, the overall vulnerability of these bridges remains, and it is convenient to address the deterioration of saddle areas through global retrofitting interventions that introduce a greater degree of structural robustness rather than local strengthening of the half-joint through classical solutions of corbels plating, FRP wrapping or local reinforcement addition. In the following, this global strategy is proposed for two case-studies with different saddle configurations.

## 3 The Evaluation of Defects on Gerber Saddles in the Italian Guidelines for Bridge Assessment

Generally, the main damage found through visual inspections in the most common cases of deteriorated saddles is linked to longitudinal bar and stirrup corrosion together with concrete cover detaching and spalling.

Existing Reinforced Concrete (RC) bridges show common shortcomings that are the consequence of durability not considered in the initial design and poor checks performed during construction. The most common defects related to structural elements are:

- Poor concrete quality and consequent acceleration of carbonation phenomena over time.
- Inadequate design of the reinforcement due to insufficient standards that do not take into account the effects of dynamic amplification, the possible increase in axle moving loads, shrinkage, thermal effects, etc.
- Insufficient concrete cover compromising the protection against carbonation accelerating the reinforcement corrosion.
- Detail weakness: it is closely related to the lack of durability rules in the original design. Often, details such as Gerber saddles are not well arranged, with the overlapping bars not properly anchored associated with ineffective solutions adopted for casting during construction without guaranteeing the minimum required concrete cover.

All these shortcomings are widespread in Italian infrastructures and saddles show often all these aspects, becoming particularly delicate elements, especially for the lower corbel that is particularly prone to receiving aggressive agents from the upper platform because of draining waters inside expansion joints. Furthermore, in bridges the state of Gerber saddles is generally made worse by permanent cracking due to tensile stresses in the corbels and in the cantilever areas; in this connection, tensile stresses occur with crack openings, which are preferential ways of water penetration, making corrosion faster. This is magnified by the opening of wider and deeper cracks when moving loads act on the deck, reducing the durability of the structural elements, especially with the increasing traffic loads.

From the inspection point of view, Gerber saddles are difficult to investigate, due to their reduced accessibility and consequently classical techniques are not always able to provide an accurate estimation of the level of degradation. In these cases, endoscopic inspection of the expansion joint allows engineers to estimate the level of degradation in points of the saddle not otherwise accessible. In addition, non-destructive testing (NDT) techniques can supply contribution to the knowledge of the saddle degradation such as measurement of corrosion potential together with reinforcement resistivity and ground penetrating radar (GPR) tests at high frequencies. These techniques provide information on reinforcements or hidden defects such as fractures and voids as well as on the risk of corrosion. They are very useful also for saddles with prestressing tendon anchorages, to know the actual configuration of tendons. However, in general, the identification of a quantitative measure of damage levels is really complex.

For the reasons considered above, the Italian Guidelines for Assessment and Maintenance of Bridges consider the defects in the Gerber saddles among the most dangerous ones, paying particular attention to them. The level of the defect and its extent is defined in the defect sheets attached to the main text of the Guidelines, reporting the most common damages on bridges. The defect sheet especially provided for Gerber saddles fix the "weight of the defect", i.e. the dangerousness parameter, always to the maximum value ($G = 5$), regardless of the extent and intensity of the actual damage (Fig. 3). This choice, although it is reasonable in light of the risk that a brittle and sudden break of the saddles can entail losses in terms of human life or in any case the attainment of the Ultimate Limit State, immediately leads, through the procedure supplied by the Guidelines, to a maximum threshold of attention for the bridge that has such a defect (Granata et al. 2020).

In fact, the flow chart of the procedure entails that to each defect identified during the inspection, a numerical weight is attributed and, through the total grade of defects found, the *defect class* is assigned; afterwards this combines with the *vulnerability class* and finally the global *attention class* is obtained, providing an overall assessment of the state of the bridge investigated. In the case of Gerber saddles, every defect found, whether of minor or major importance, immediately leads to the definition of maximum alert with a high attention class. The immediate consequence of this evaluation is that the bridge must be moved to a higher level of evaluation (level 4), in which the tests and investigations in situ, the analytical and numerical evaluations on the structural model and the definition of retrofitting interventions are the essential steps for the resolution of the problem. Awaiting these interventions and depending on the deficit found in the level 4 analysis, a transit ability assessment of the bridge is also necessary with loads and strength partial coefficients are referred to 5-year reference times, applying reduced loads compared to the current ones of the Italian standards (or Eurocode 1, which is the same).

Actually, Gerber saddles can be built with different arrangements, as seen above and they can host prestressing tendon anchorages. The presence of prestressing and crossbeams which allow redistribution of forces among adjacent girders near failure modify the behaviour of the classical dapped-end beam, i.e. that of the saddle on the web of the girder only in which the contribution of mild reinforcement and concrete strength are the fundamental features of the half-joint capacity.

In this last case, the likelihood of brittle failure with catastrophic events is very high and the situation has to be considered extremely dangerous, while in the other ones the consequences of damage are less important, because corrosion is generally concentrated

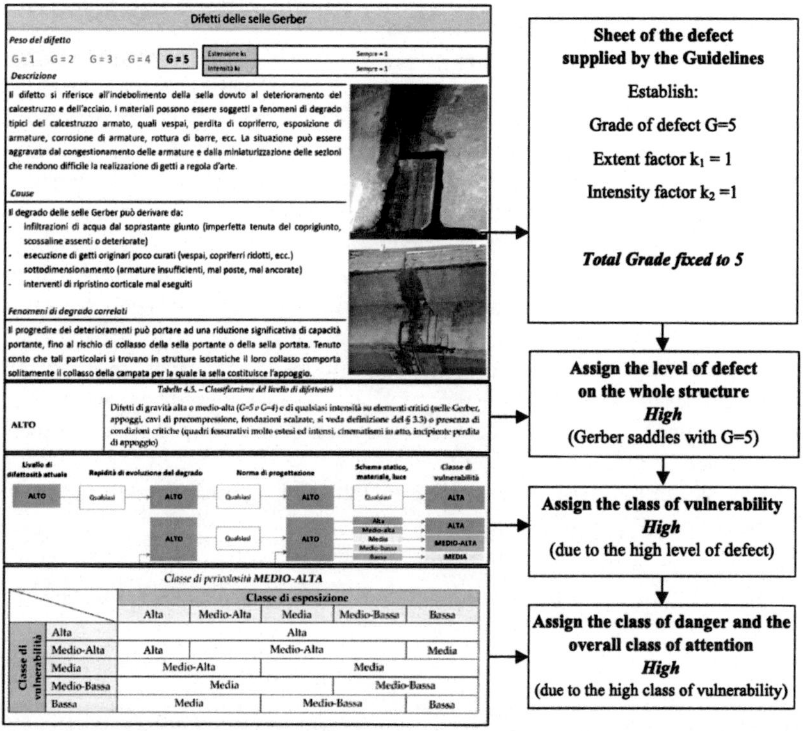


**Fig. 3.** Defect sheet of Ministerial Guidelines and flow-chart for defining vulnerability and attention classes.

in the outermost elements where water drains or enters the joints but the contact between corbels occur on the entire deck width, thanks to crossbeams.

These considerations suggest differentiating the weights of the defect found in Gerber saddles considering two aspects:

- the consequences of the defect due to the actual static scheme and to the arrangement of the Gerber saddle, which can have different effects on the structural behaviour and on the probability of achieving the ULS;
- the intensity and extent of damage due to corrosion, concrete spalling, and reduced contact area of the structural members.

For this reason, it would be appropriate the introduction of two weights in the defect sheet of Gerber saddle, as is done for other defects (extent $k_1$ and intensity $k_2$) to take into account different situations, graduating different levels of risk. In the present form of the Guidelines, they are set always to 1, bringing the total degree of defect to 5 regardless of these aspects. Hence, the inspector cannot graduate the level of damage and cannot give any judgement of the incidence of the defect on the structural behaviour. It could be appropriate, instead, to distinguish Gerber saddles on main beams with limited contact between structural members from the case of saddles along transverse beams or in slab bridges with wide areas of contact, differentiating the values of $k_1$ and $k_2$. In this way, when corrosion is advanced in reduced corbels the maximum level of risk can be assessed otherwise the defect has less importance.

A graduation of the defect in the phase of visual inspection (level 2) might seem at first sight contrary to a criterion of prudence in terms of safety. In reality, it must be considered that the Guidelines are created to carry out a quick classification of the vulnerability of bridges, considering the defects detected in the visual inspections. Afterwards, with the assignment of a class of attention, each Road Authority can draft a priority list of level 4 analysis for the major risks detected. When a single defect immediately leads to the maximum attention, without the possibility of grading its actual importance, the consequence is that a significant number of structures would immediately be classified as being at maximum risk due to a defect that is not necessarily globally dangerous for the structure. A large number of these cases could lead to an excess of work for the Technical Offices of Owners making the procedure too onerous, although it was initially thought to be quick and easy. This circumstance could slow down the actual execution of maintenance work.

A proposal for defect grading is given in Table 1 for possible modification of the Guidelines.

**Table 1.** Proposal for defect grading for Gerber saddles with different configurations

| Configuration | Grade G | Extent $k_1$ | Intensity $k_2$ |
|---|---|---|---|
| Saddle with open cracks, regardless of the girder typology | 5 | 1 | 1 |
| Saddle on the girder web only, without crossbeams and prestressing | 5 | $0{,}7 \div 1$ | $0{,}7 \div 1$ |
| Saddle on the girder web only, without crossbeams and with prestressing | 4 | $0{,}7 \div 1$ | $0{,}7 \div 1$ |
| Saddle with crossbeams in the corbels only, with or without prestressing | 4 | $0{,}7 \div 1$ | $0{,}7 \div 1$ |
| Saddle on box girder webs without diaphragms, with prestressing | 5 | $0{,}7 \div 1$ | $0{,}7 \div 1$ |
| Saddle on box girder webs with diaphragms and prestressing | 4 | $0{,}5 \div 1$ | $0{,}5 \div 1$ |
| Saddle with rigid crossbeams on the corbels and on the girders, with or without prestressing | 3 | $0{,}5 \div 1$ | $0{,}5 \div 1$ |
| Saddle transversely continuous on slab bridges, with or without prestressing | 3 | $0{,}5 \div 1$ | $0{,}5 \div 1$ |

With the aim of helping Road Authorities by reducing the workload, it could be appropriate to differentiate the defect weights on Gerber saddles and to give useful guidelines on local or global interventions targeting the problem, without the necessity of reaching level 4 of study associated with a high computational burden and time to spend before intervening on site. A more detailed classification of defects in Gerber saddles could allow Owners to plan fast evaluations and organize retrofitting interventions, reducing times and costs and upgrading the safety level, especially when the damage is concentrated mainly in these structural elements.

In the following section two case-studies are considered in which the visual inspections, onsite investigations and level 4 assessment were conducted.

# 4 Case-Studies

## 4.1 Bridge Over the Salso River

The bridge over the Salso river in Licata (Sicily, Southern Italy) was designed by Riccardo Morandi. It is a Niagara-type cantilever girder bridge with side spans of 33.10 m and a central span with a total length of 49.60 m, the central beam between Gerber saddle being 32 m long (Fig. 4). The total length is 115.80 m while the deck width is 19.00 m, composed of 8 double T-beams with crossbeams and upper slab 20 cm thick; in cantilever areas a bottom slab is present with variable width.

**Fig. 4.** Views of the bridge over the Salso river

The central beams are prestressed through Morandi's patent M5 type tendons. Concrete strength of side spans and cantilever is $f_{ck} = 25$ MPa, while for prestressed girder $f_{ck} = 35$ MPa; reinforcement steel is of the ALE-TOR type, yielding strength $f_{yk} = 440$ MPa; prestressing steel has tensile failure strength $f_{ptk} = 1700$ MPa.

Figure 5a shows the geometry of the saddle and the arrangement of the reinforcements. The lower corbel of the saddle is considered for safety checks because the upper corbel is prestressed showing higher strength.

The particular geometry of the saddle designed by Morandi, which presents a soft connection between the girder cross-section and the corbel, instead of the classic dapped-end beam, leads to the need of conceiving a specific strut-and-tie scheme for this case

(Fig. 5b); in fact, the schemes suggested by Eurocode 2 can be applied to dapped-end beams and they not agree with the arrangement of reinforcement because an existing bridge differs a lot from the usual models that are mainly conceived for design and not for checks of existing cases.

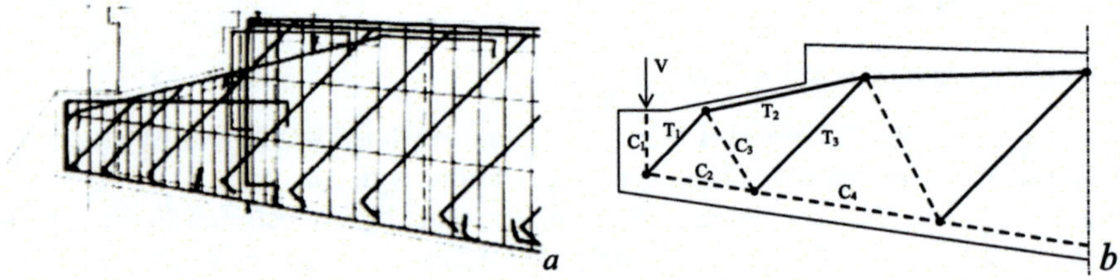


**Fig. 5.** Reinforcement arrangement of the cantilever (*a*) and strut-and-tie model (*b*)

Results showed that the weakest element is tie $T_2$ with capacity (in terms of maximum allowable reaction) of 1939 kN and safety coefficient 1.63. The check was repeated taking all the partial safety coefficients to 1, using the yielding strength of the reinforcements and obtaining an estimated capacity of 2230 kN with an actual safety coefficient of the saddle equal to 2.55.

The above evaluations are made in the intact design conditions. The subsequent assessment of the degradation conditions of the saddle due to the percolation of the water and therefore to corrosion of the most exposed reinforcements, leads to the evaluation of the decreased safety factor due to deterioration.

In this connection, for ties it was considered an effective area of reinforcement equal to about 85% of the initial one and a reduction in the concrete strength of about 20%, according to the studies of Desnerck et al. (2017). The steel yielding strength was not changed, as the degradation does not significantly influence its value (Desnerck et al. 2018). With these assumptions, the weakest element is again corroded tie $T_2$ and a check on the saddle supplies a minimum safety factor of 2.04 with capacity of 1784 kN, fully acceptable. This confirms the robustness of the saddle designed by Morandi, considering that the redistribution offered by the transverse beam, has been disregarded in the strut-and-tie model although it represents an additional contribution to safety (the eight beams cannot be contemporarily stressed at the failure level for traffic loads, the external one being the most stressed one for eccentric loads).

A nonlinear FE model of the entire cantilever shows that the saddle is not the most critical section of the girder. In fact, in the original condition a slight insufficiency was found in the cantilever sections out of the crossbeam, between the pier and the saddle (Granata et al. 2022); this result, found through the application of *M*-*V* interaction diagrams, is confirmed by the nonlinear FE models for which, the failure is governed by a shear-flexure mechanism in the area with the maximum values of negative bending moments and shear. Figure 6 shows the crack pattern found through the FE model for the case of original design.

It is worth noting that the behavior at failure changes clearly in the presence of the degradation of the saddle, with the corroded reinforcements of the saddle area and the

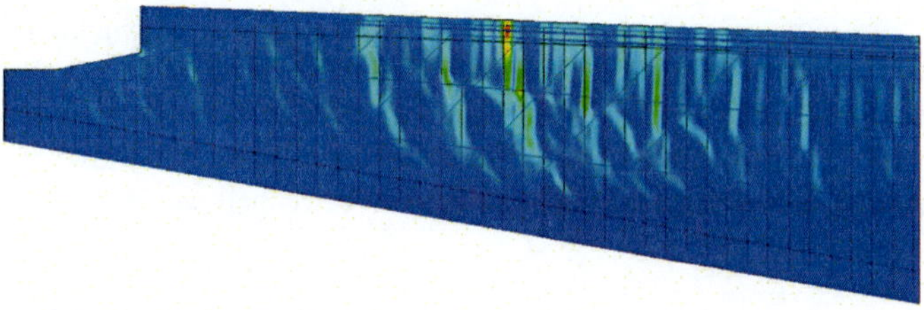

**Fig. 6.** Nonlinear FE model, strain pattern at failure of original behaviour

reduced strength of concrete. In fact, it is obtained from the analysis that the critical area becomes that of the saddle and not the cantilever, leading to the failure of the lower corbel with a wide crack in this area, in accordance with the strut-and-tie model, in which the critical element is that of inclined reinforcement (Fig. 7). Furthermore, the failure value coincides with the one provided by the strut-and-tie model with excellent approximation, that is capacity of 1880 kN for the FE model and 1774 kN for the strut-and tie model. The slight difference of about 5% is in agreement with the fact that the ST model is always a lower-bound solution and that crossbeam cannot be taken into account (even though it is present only on the corbel).

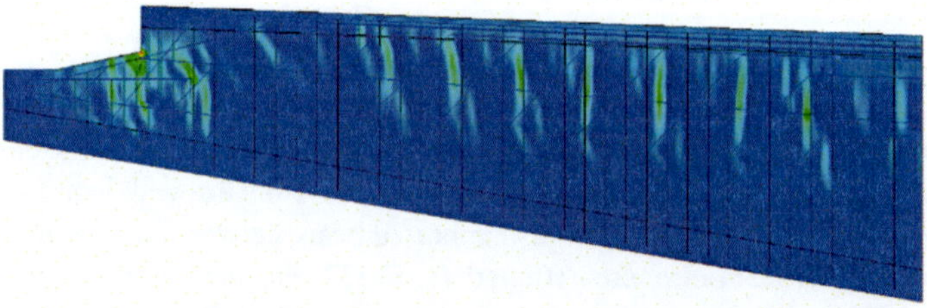

**Fig. 7.** Nonlinear FE model, strain pattern at failure with degradation

This result implies that the geometry and amount of reinforcement in the saddle supply a certain degree of local robustness and that the global behavior of the isostatic cantilever shows instead a significant level of vulnerability. The saddle degradation worsens the initial behavior, shifting the critical area towards the saddle and modifying the initial behavior. However, it remains a sufficient local safety coefficient, although the deterioration hypotheses of the saddle are quite penalizing. This leads to thinking that a possible retrofitting intervention of the bridge should not be concentrated on a local saddle strengthening, which is generally the most commonly adopted solution in similar cases, but to modify the global behavior by providing a greater degree of global robustness to the structure.

This is possible by closing the saddles, changing the static scheme from the isostatic one to the continuous hyperstatic girder and adopting external prestressing for enhancing the global behaviour.

In this case it is possible to exploit the natural camber of the bridge, inserting a straight tendon that is almost centred in the saddle area and rises in the cantilever area remaining below the cross-section centroids of the central span, thus simplifying the introduction of the tendon into the girders and optimizing the eccentricity due to prestressing (Fig. 8), without the need of deviators. The presence of crossbeams, spaced 8 m, is beneficial for hosting sliding guides, maintaining the tendon straight along the bridge and avoiding intermediate deflections and accidental deviations (Petrangeli and Fieno 2021).

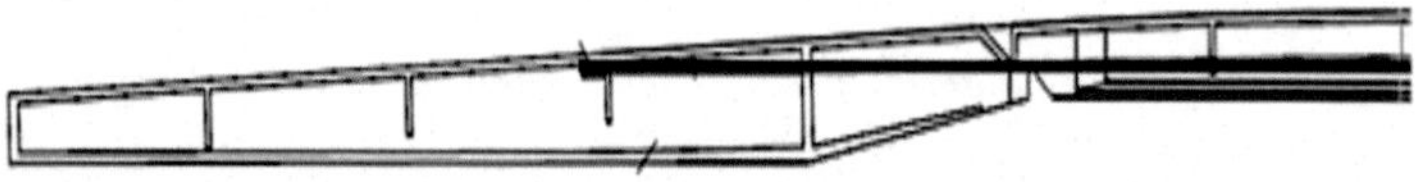

**Fig. 8.** Hypothesis of strengthening with external prestressing on the half bridge (straight tendon)

Prestressing permits to nullify the tensile stress due to the load combination with traffic at the lower edge of the saddle for the positive bending moments induced by moving loads in the continuous girder. Tensile stresses at the upper edge are maintained below the first cracking (tensile stress $f_{ct}$) of concrete, with the introduction of new reinforcements in the upper slab when the saddle is closed through concrete infill. For this case-study, 2 tendons 7T15 (7 strands 0.6″ ) were considered for the outermost beams and 2 tendons 4T15 (4 strands 0.6″) for the other beams, introducing a maximum prestressing force of 2400 kN.

### 4.2 Motorway Overpass

The second case-study is a motorway overpass with a classical Niagara-type scheme (Fig. 9). The deck is composed of 4 prestressed double-T beams, with total length 89 m, central span 44 m long while the simply supported beam between saddles is 31 m long.

Material properties are: concrete strength $f_{ck} = 33$ MPa, reinforcement yield strength $f_{yk} = 440$ MPa and prestressing steel strength $f_{ptk} = 1700$ MPa.

The safety check of the saddle was carried out through the schemes recommended by Eurocode 2, because in this case the two dapped end beams agree with the schemes of the Code. In order to take into account the prestressing tendons inside the saddle Gerber, the component in vertical direction of axial force is considered as a reduction of shear acting to the corbel (Fig. 10).

The capacity was evaluated at ULS, using partial safety coefficients of loads and materials provided by Eurocode; the value of minimum shear force was 2016 kN and the weakest elements are always the vertical and inclined ties of the Code schemes. The safety coefficient at ULS is 1.30, considering the fundamental contribution of prestressing. The evaluation with all partial safety coefficients fixed to 1 gives instead an estimated capacity of 2542 kN with demand 1133 kN, and the actual safety coefficient is equal to 2.24. This result is satisfactory, but it is found in the hypotheses of no degradation.

From onsite investigations, the loss of reinforcement area due to corrosion could be set, in this case, to 20%. Prestressing loss of 30% was considered, due to the loss at anchorages, subjected to deterioration, and to the loss of the tendon along the girder, attributable to an inadequate sheath grouting (local debonding of prestressing and reduced effective area). The bond of mild reinforcement is guaranteed by hooks; hence,

**Fig. 9.** Motorway overpass and details of Gerber saddles with crossbeams

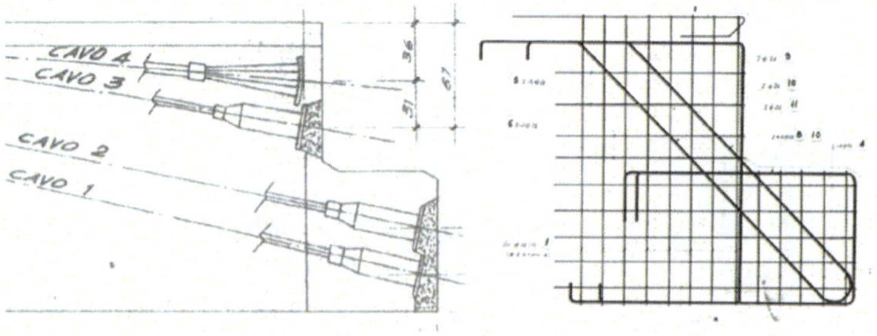


**Fig. 10.** Arrangement of prestressing tendons and reinforcements in the lower corbel of the saddle

no decrease of bond was considered for reinforcements but only reduction of concrete strength and reinforcement area.

No redistribution is allowed between adjacent girders for this case because cross-beams are placed in the rear of corbels, not affecting the saddle in the bearing area.

The assumptions made above are very penalizing for the saddle, bringing the total capacity to 1969 kN and the demand to 1133 kN, with the safety coefficient reduced to 1.74, respect to the previous case of intact saddle, but still satisfactory. The weakest elements of the ST model are always the ties while the struts show greater capacities. With corroded reinforcements and concrete damage, the check at ULS shows a capacity/demand ratio near to 1: this implies that the damaged saddle satisfies the safety check at its minimum value.

For this reason, a global intervention of strengthening was considered for this case-study too, analogously to the previous case but with important differences of the adopted solution. In fact, three conditions have been considered in this case:

1) the presence of the original prestressing that is still effective along the entire deck allows to set the intervention in limited areas around the saddle.
2) The configuration of crossbeams just before the corbels on the two sides of the saddle is ideal to make adequate contrast to prestressing forces.
3) The cantilever sections as well as central beams do not require further strengthening (i.e. additional prestressing), even with the change of the static scheme from the Gerber-girder to the continuous girder and for the different distribution of bending moments due to moving loads.

Hence, the solution adopted for this case-study is to fill the joint of the saddle with concrete and to introduce Dywidag bars to tighten the joint (Fig. 11), with an intervention that is easy to implement, at a limited cost.

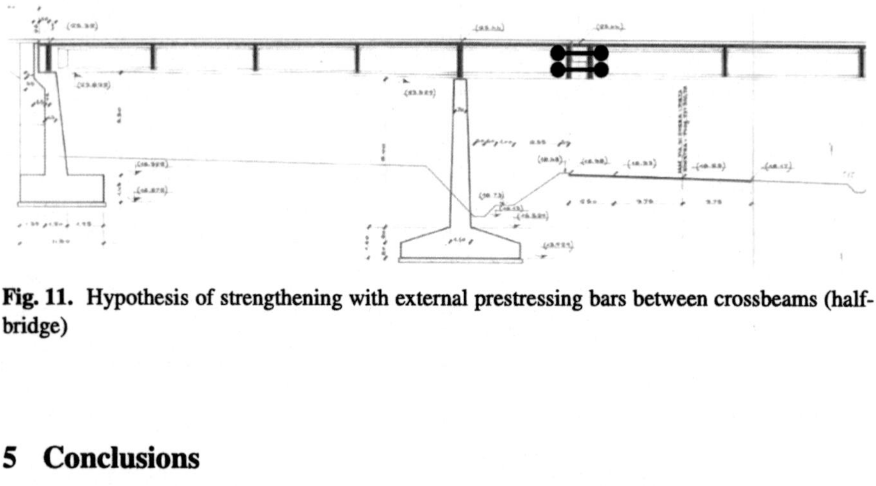

**Fig. 11.** Hypothesis of strengthening with external prestressing bars between crossbeams (half-bridge)

## 5 Conclusions

A careful analysis of the configurations of Gerber saddles in the existing bridges and their condition of degradation allows engineers to evaluate the real risk conditions associated with the failure of these delicate elements in the isostatic structural schemes of the cantilever bridges, which lack of structural robustness.

In particular, the application of the Italian Ministerial Guidelines does not allow, in the current version, a grading of the damage, bringing to a univocal assessment of maximum danger even of very different situations with the consequence of classifying situations of major or minor risk in the same way, partially frustrating the objective of drawing up a list of intervention priorities for the bridges with greatest risk. A modification proposal is reported in this paper, in the attempt of providing a contribution to the work of the Road Authorities.

Furthermore, through the in-depth study of two real situations of cantilever bridges with Gerber saddles on girders with or without prestressing, it can be seen how, in the presence of careful design of the reinforcements, these elements present an unexpected structural robustness, in some cases shifting the problem in the cantilever areas rather than in the saddle area. Of course, the deterioration and the consequent damage attributable to the reinforcement corrosion and to the degradation of concrete due to the effect of percolating waters from the joint, modify the original situation, reducing the safety coefficients of the deteriorated saddles bringing the problem of the fragility of half-joints in the foreground. It follows that although the saddles can show reassuring safety coefficients, even under the critical conditions of degradation, it is good to operate a retrofitting intervention that provides local or global robustness to the bridge at low costs. A solution was preferred here, alternative to those of local strengthening commonly used, that increases the global robustness of the bridge through the modification of the static scheme from Gerber to continuous beam with the total closure of the saddle joint and the introduction of prestressing through bars or tendons, providing the appropriate strengthening along the deck in its new configuration and in the most critical areas.

## References

Desnerck, P., Lees, J.M., Morley, C.T.: The effect of local reinforcing bar reductions and anchorage zone cracking on the load capacity of RC half-joints. Eng. Struct. **152**, 865–877 (2017)

Desnerck, P., Lees, J.M., Morley, C.T.: Strut-and-tie models for deteriorated reinforced concrete half-joints. Eng. Struct. **161**, 41–54 (2018)

Granata, M.F., La Mendola, L., Messina, D., Recupero, A.: Case-studies of corroded reinforced concrete bridges in Southern Italy. In: Proceedings of CACRCS DAYS 2020, Capacity Assessment of Corroded Reinforced Concrete Structures. Fib, CTE, AICAP (2020)

Granata, M.F., Messina, D., Colajanni, P., La Mendola, L., Lo, G.E.: Performance of a historical cantilever reinforced concrete bridge with half-joint degradation. Structures **37**, 561–575 (2022)

Italian Ministry of Infrastructures: Linee guida per la classificazione e gestione del rischio, la valutazione della sicurezza ed il monitoraggio dei ponti esistenti. Edited by the Consiglio Superiore dei Lavori Pubblici (2020). (in Italian)

Petrangeli, M., Fieno, L.: L'impiego della precompressione esterna nella riparazione e nell'adeguamento statico dei ponti. Ingenio (2021). ISSN 2307-8928. (in Italian)

# Experimental Tests on Post-tensioned PC Girders with Grouting Defects Under Different Prestressing Levels: Preliminary Results

Daniele Losanno(✉), Simone Galano, Fulvio Parisi, Maria Rosaria Pecce, and Edoardo Cosenza

Department of Structures for Engineering and Architecture, University of Naples Federico II, Naples, Italy
daniele.losanno@unina.it

**Abstract.** During last decades, prestressed reinforced concrete (PC) was widely used in the construction of bridge decks. Recent assessments of existing bridges demonstrated how degradation phenomena and construction defects can significantly reduce performance levels of PC elements. In the case of post-tensioned tendons, insufficient grouting of ducts is one of the most common types of defects in existing girders. This paper presents preliminary results of an experimental testing program on 1/5 scale PC bridge girders. The specimens were casted in different configurations including local voids and lacking grout within ducts, while varying the prestress level in the tendons. Experimental outcomes provide useful information for load-bearing capacity assessment of PC girders with defective grout.

**Keywords:** prestressed reinforced concrete · experimental testing · post-tensioned tendons · ducts · defective grout

## 1 Introduction

In last decades, the worldwide heritage of prestressed reinforced concrete (PC) bridges and the importance of checking their safety and durability has significantly increased, promoting accurate inspections on tendons of post-tensioned (PT) structures. The PT technique was developed since the 1950s, reaching a widespread use especially between 1960 and 1970. Therefore, a huge amount of PT bridges are older than 50 years.

Several investigations evidenced that PT girders are often affected by defective grouting of the ducts, highlighting voids or even long ungrouted segments (CEB-FIB 2001). Grout defects do not always reduce durability because, in line of principle, corrosion cannot occur in ducts due to low amount of oxygen. Nevertheless, the majority of corrosion problems in tendons was caused by chloride-contaminated water infiltration once steel ducts have been damaged by corrosion. In most inspections, end anchorages of external tendons located at the highest point of the tendon profile were found to be partially grouted. In this respect, VSL (2002) reported on the probable identification of defects


M. A. Aiello and A. Bilotta (Eds.): ICC 2022, LNCE 435, pp. 288–301, 2024.
https://doi.org/10.1007/978-3-031-43102-9_23

and the number of surveys that are required to perform a condition-based assessment of bridges.

Defects in grouted ducts have another important role in the structural performance because they influence the bond between tendons and concrete. Bond allows a significant increase in the prestressing force in a cracked section after decompression, also permitting steel tendons to reach yielding or even ultimate strength. This has significant effects on the flexural behavior of the PC girder cross sections, crack pattern, and deformability after cracking (Wang et al. 2014). Another major benefit from bond is that local defects in tendons (e.g. reduced effective cross section due to corrosion) produce local effects, i.e. the tendon force is not affected over its entire length but only over a part of it, which is needed to restore anchorage (Asp et al. 2020).

Due to an increasing interest in the performance of existing PC bridges that in Italy are numerous and recently showed significant defects and corrosion damage, the authors of this paper designed a wide experimental campaign on 15 post-tensioned PC girders. The design of the experimental program was developed in order to study the effects of defects, including higher loss of prestress, grout voids in ducts (resulting in unbonded tendons), and reduction of tendons section due to corrosion. Specifically, the influence of such parameters on flexural response can be detected through global parameters (e.g. midspan deflection, load-bearing capacity), local curvature measurements and analysis of observed crack patterns. In this paper, preliminary test results on 4 specimens with 2 different levels of pre-stressing and different grout conditions (i.e. fully grouted and ungrouted ducts) are presented.

## 2 Experimental Program

The experimental campaign was designed so that three different types of defects could be investigated, as follows:

- higher loss of prestress in tendons;
- lack of bond due to defective grouting of ducts;
- reduction of prestressing steel sectional area.

Fifteen equal specimens consisting of 6.6-m-long PC girders (according to a 1:5 reduced scale size) were designed and were characterized by one of the abovementioned defects.

This paper focuses on 4 specimens with intact tendons, including a reference beam T1 (with fully grouted ducts and target prestress force of 150 kN) and 3 specimens with different prestressing level (75 kN) and/or lacking grout (Table 1). The highest prestress level (corresponding to approximately 50% of ultimate tendon load) is deemed representative of the residual prestress level in existing PC bridge girders, unless the occurrence of a prestress loss higher than expected.

In addition to different prestress levels, the remaining specimens to be tested in future stages of the experimental program will include partially grouted ducts and damage to tendons (in the form of either single or multiple cuts), in the latter case to simulate the effects of corrosion at different locations. Shear tests are also expected to be carried out on some specimens.

**Table 1.** Characteristics of specimens in terms of duct grouting and initial prestress levels.

| Specimen | | Duct | Prestress level |
|---|---|---|---|
| T1 | B-HP | Fully grouted (B) | High (HP) |
| T2 | U-HP | Ungrouted (U) | High (HP) |
| T3 | B-LP | Fully grouted (B) | Low (LP) |
| T4 | U-LP | Ungrouted (U) | Low (LP) |

## 2.1 Description of the Specimens

Laboratory facilities required a 1:5 scaled geometry of the PC girders. Each specimen had a T-shaped cross section with 60-mm-thick slab, web and top flange width of 150 and 480 mm, respectively, and total depth of 440 mm. The total length (L) of each specimen was equal to 6600 mm (Fig. 1a). The girder web included equally spaced reinforcing bars with nominal diameter of 8 mm and concrete cover of 30 mm. A welded wire mesh with diameter and mesh size equal to 8 mm and 200 × 200 $mm^2$, respectively, was installed in the top flange. Shear reinforcement consisted of stirrups with nominal diameter of 8 mm and 100 mm longitudinal spacing.

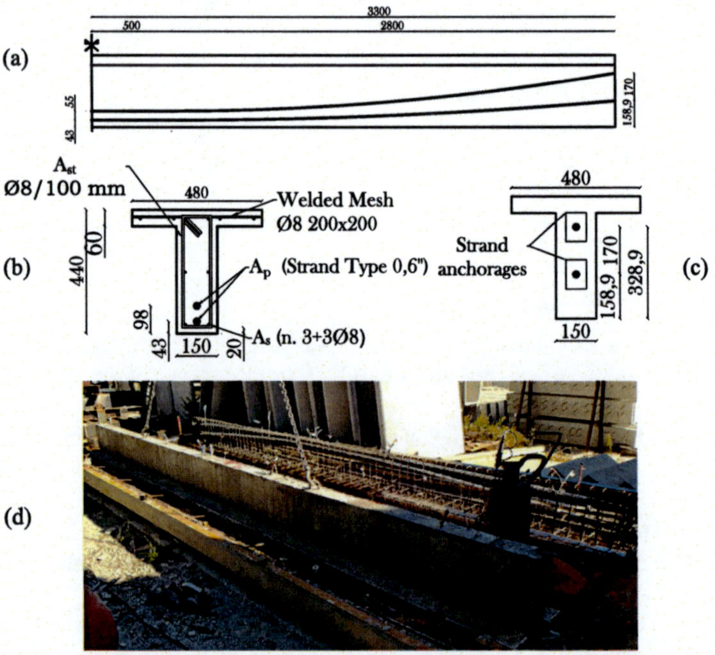


**Fig. 1.** PC girder half-span geometry with steel reinforcement and tendons: a) tendons profile, b) mid-span cross section, c) end cross section, d) picture of two specimens during construction (dimensions in mm).

The specimen was prestressed via two seven-wire PT tendons (Fig. 1a) with parabolic profile and equivalent area of 150 mm$^2$ per strand (corresponding to nominal strand diameter equal to 0.6″). In the mid-span cross section, the clear depth of the upper and lower tendons (i.e. their clear distance from the top flange) was respectively equal to 342 mm and 407 mm (Fig. 1b). To ensure no bending moment at supports, the clear depth of the upper and lower tendons reduced to 111 mm and 281 mm (Fig. 1c). Two single-strand anchorages were used at girder end sections to restrain each tendon (Fig. 1d).

Dealing with material properties, the mean value of concrete compressive strength evaluated on 6 cubic samples after 28 days of curing was 34 MPa. Tensile tests on strands outlined the following properties: Young's modulus $E_p = 203,400$ MPa; conventional yield stress (i.e. stress corresponding to 1% of total strain) $f_{py} = 1782$ MPa; ultimate tensile strength $f_{pu} = 1969$ MPa.

The prestress level at one end of each specimen was monitored through either one load cell (installed on lower tendon of T2 and T4 specimens) or two load cells (installed on both upper and lower tendons of T1 and T3 specimens). Table 2 highlights the difference between initial prestressing force (i.e. jacking force $P_j$), prestressing force after instantaneous stress losses due to anchorage devices ($P_i$) and residual prestressing force on the day of testing ($P_t$, , i.e. prestressing force after both instantaneous and time-dependent losses including creep, shrinkage and relaxation) for the monitored tendons.

**Table 2.** Prestressing force levels measured at different stages on each specimen

| Specimen | Tendon | $P_j$ [kN] | $P_i$ [kN] | $P_t$ [kN] |
|---|---|---|---|---|
| T1-B-HP | Upper | 150 | 123 | 104 |
| | Lower | | 118 | 111 |
| T2-U-HP | Lower | 150 | 125 | 115 |
| T3-B-LP | Upper | 75 | – | – |
| | Lower | | 48 | 36.3 |
| T4-U-LP | Upper | 75 | – | – |
| | Lower | | 50 | 38.3 |

## 2.2 Test Procedure

Each specimen was quasi-statically tested under four-point bending to evaluate the non-linear flexural response of PC girders, according to a symmetric loading scheme. The couple of point loads had spacing equal to 850 mm, resulting in a shear span length $L_v = 2800$ mm. Each specimen rested on a couple of rubber bearings, allowing any rotation at supports. Lateral restraints were installed only to prevent parasite out-of-plane rotations and buckling, particularly in ultimate conditions.

Figure 2 shows the experimental set-up of the tests carried out at the Department of Structures for Engineering and Architecture of the University of Naples Federico II. The testing machine consists of a rigid steel basement ($1.0 \times 4.0$ m$^2$ in plan), four columns fixed at the basement and a moving crossbar on top, which can slide in the vertical direction. Based on a vertical servo-controlled hydraulic jack (actuator) placed on the crossbar, specimens can be tested with either displacement (maximum stroke of 150 mm) or force control (compressive load capacity of 3000 kN, tensile load capacity of 2400 kN).

The girder deflection at mid-span cross section $d$ was measured using a vertical linear variable differential transducer (LVDT), as shown in Fig. 2. A load cell was used to measure the applied load $F$ (see set-up front view in Fig. 2). In order to evaluate the PC girder behavior under different load levels in both cyclic and monotonic loading conditions, a two-phase loading protocol was imposed. In the first phase (P1, Fig. 3), the specimen was tested under 6 stair-stepped loading cycles (20 min duration per cycle) with three different increasing amplitudes (two loading cycles for each amplitude, i.e. $F_1 = 33.1$ kN, $F_2 = 48.3$ kN, $F_3 = 72.4$ kN) in load-controlled mode as per ACI 437.1R-07 [ACI 2007]. During cyclic loading, the different levels of peak amplitude were aimed at representing different performance levels of the bridge girder. According to Italian code provisions for bridge design under traffic loads (IMIT 2018), the following threshold values were calculated: serviceability level ($F_1$), design level ($F_2$) and 1.5 times design level ($F_3$). During testing, slightly different values from target values of $F_1$, $F_2$, $F_3$ were achieved due to the control system of the universal testing machine.

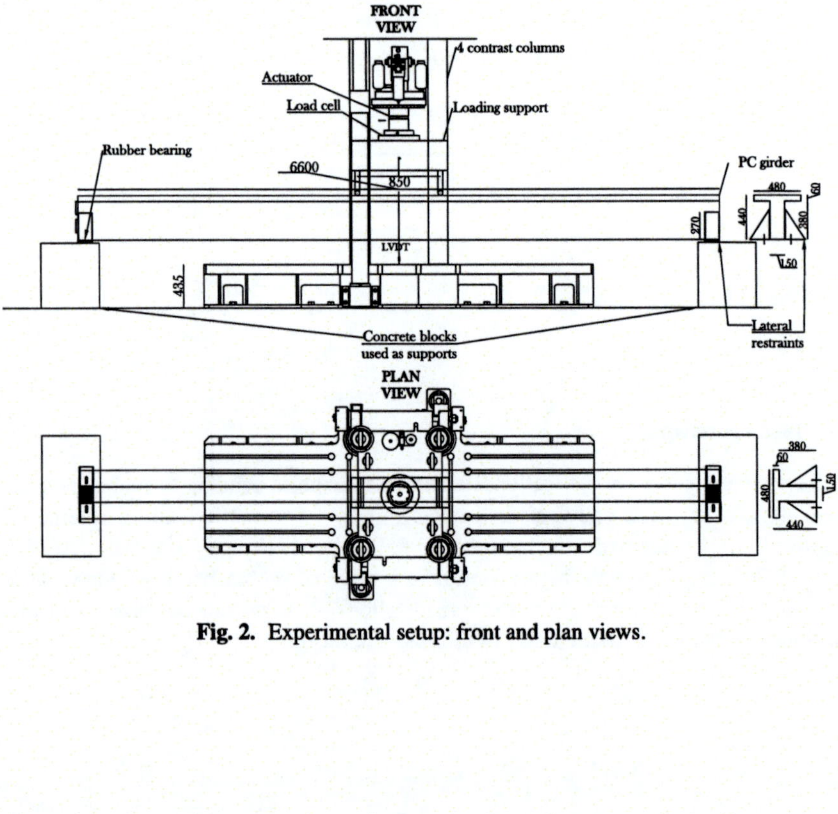


**Fig. 2.** Experimental setup: front and plan views.

In the second loading phase (P2), the specimen was monotonically tested up to failure with displacement control, according to a displacement rate of 0.05 mm/s. In order to increase displacement capacity despite the actuator stroke limit of 150 mm, a temporary restraint system based on external belts was installed on T3 and T4 specimens during testing. While unloading the specimen by the actuator and transferring external load to supplementary belts (transfer force around 80 kN), the moving head was translated downward, thus resuming a full stroke of 150 mm. That procedure allowed the specimens to reach near-collapse conditions.

In this paper, the cyclic response and the complete monotonic envelope response of specimens under the two-phase P1-P2 protocol is discussed before and after cracking up to maximum load $F_{max}$.

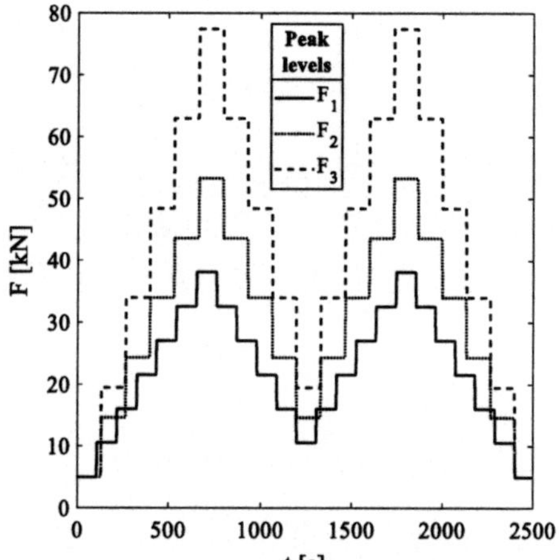


**Fig. 3.** Cyclic loading protocol P1.

# 3 Experimental Results

## 3.1 Cyclic Response

The cyclic force–displacement curves of the four specimens are shown in Fig. 4, where the combined action of prestressing force and self-weight of the girder is not explicitly considered. In each plot, peak loads ($F_1, F_2, F_3$) and final residual deflection are marked for the sake of comparison.

In Fig. 5, peak deflections ($d_{max}$) are associated with each force level of loading protocol P1.

Under $F_1$ (i.e. service conditions), only T1 and T2 specimens with high prestress level were able to prevent cracking. As a further advantage, the high prestress level ensured linear response up to $F_2$ and no damage increase under $F_3$. The influence of grouting becomes clear once cracking of concrete is achieved. In both cases of HP and

LP, ungrouted ducts provided higher flexibility. The main difference between T2 and T4 specimens (i.e. those with unbonded tendons) is that the former developed a higher recentering capability due to higher prestress level.

Just after unloading of each specimen, the permanent deflection was found to be 1.69 mm on T1 specimen, 1.61 mm on T2 specimen, 3.50 mm on T3 specimen, and 26.5 mm on T4 specimen, showing an increasing trend under decreasing prestress level and lacking grout. Those values of deflections tend to reduce over time due to elastic recoverable deformation in the tendons, which did not yield. Before starting protocol P2, no residual deflection and open cracks were detected on each specimen.

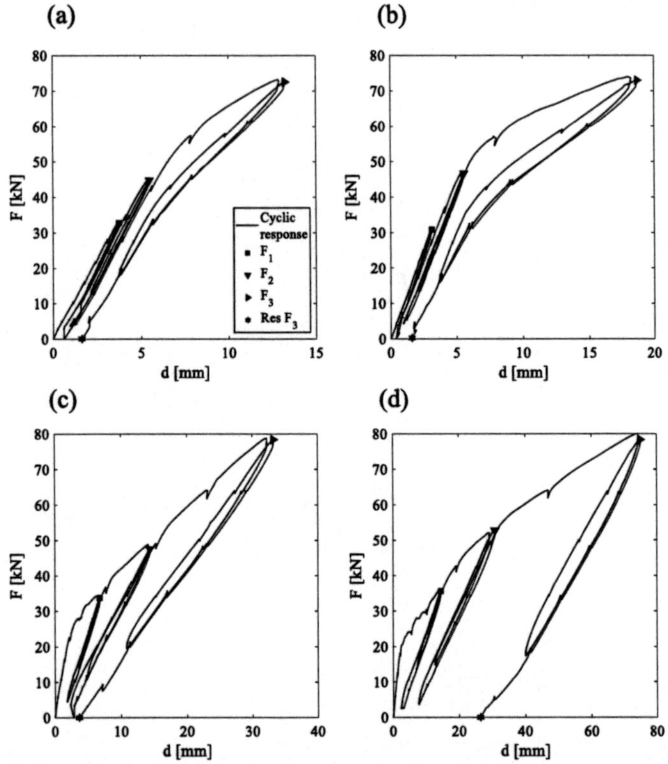


**Fig. 4.** Experimental curves corresponding to cyclic loading protocol P1: a) T1-B-HP, b) T2-U-HP, c) T3-B-LP, d) T4-U-LP.

### 3.2 Monotonic Response

After loading protocol P1, displacement-controlled protocol P2 was imposed up to maximum load. The monotonic envelope force–displacement curves of the four PC girders over both P1 and P2 loading phases are shown in Fig. 6.

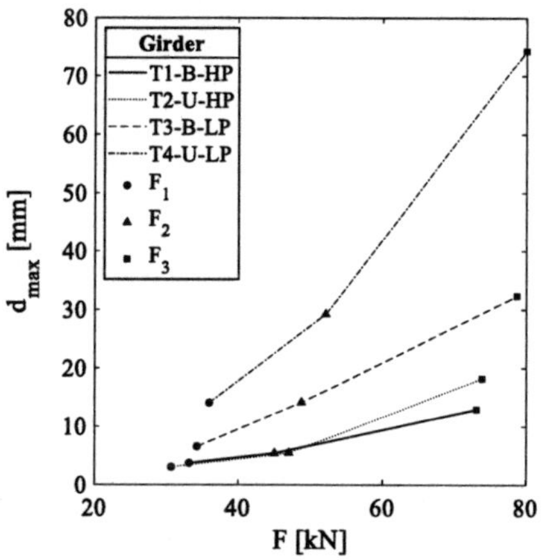


**Fig. 5.** Peak deflection versus load levels under loading protocol P1.

The four PC girders share the same initial vertical stiffness $K_1$, i.e. up to cracking onset of concrete. $K_1$ depends mostly on the girder geometry, while appearing to be independent on both the initial prestress level and grout conditions.

The level of prestress had a major influence on the attainment of cracking, which occurred at approximately the same load levels in specimen pairs (T1,T2) and (T3,T4). The slightly different values of $F_{cr}$ in Table 3 for a given level of target prestress (i.e. T1 vs T2 and T3 vs T4) are due to the effective prestress level at the time of testing $P_t$ (see Table 2).

Beyond the initially uncracked linear behavior, the girders' response was significantly affected by both grout conditions (B/U) and prestress level (HP/LP). Under the same level of initial prestressing (T1 and T2, T3 and T4), larger deflections were attained in specimens with ungrouted ducts under the same load, demonstrating the different effect of tension stiffening and its lower impact on global response of ungrouted specimens compared to their grouted counterparts. Moreover, under the same grout conditions (T1 and T3, T2 and T4), low prestress levels produced larger deflections. After cracking, both lower prestress and lack of grout resulted in higher levels of girder flexibility.

As outlined in Fig. 6, softening phase of non-linear response was not investigated. The ultimate test condition of each PC girder was limited by the maximum stroke of the actuator. T3 and T4 specimens reached peak deflections that were larger than those experienced by T1 and T2 specimens, due to temporary restrainers installed during protocol P2 as described in Sect. 2.2. For this reason, T3 and T4 specimens returned increasing force values up to displacement values of 160 mm and 180 mm, respectively (Fig. 6). Regarding ultimate conditions, the effect of initial prestress in bonded tendons became negligible, i.e. both T1 and T3 specimens were able to sustain the same peak load $F_{max}$ due to yielding of tendons. In case of unbounded tendons, peak force values $F_{max}$ are 15% to 20% lower than their bounded counterparts due to lower tensile strain in the tendons. In case of specimens with unbonded tendons, a limited increase in tensile force beyond cracking took place due to second-order effects provided by mid-span deflection, which resulted into tendon elongation.

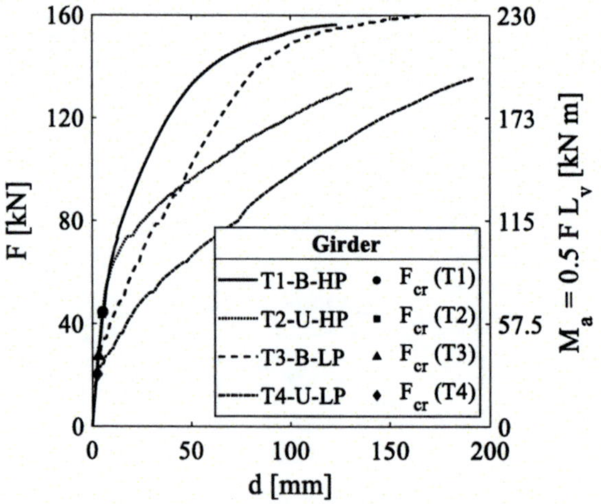


**Fig. 6.** Monotonic (envelope) force–displacement response curves.

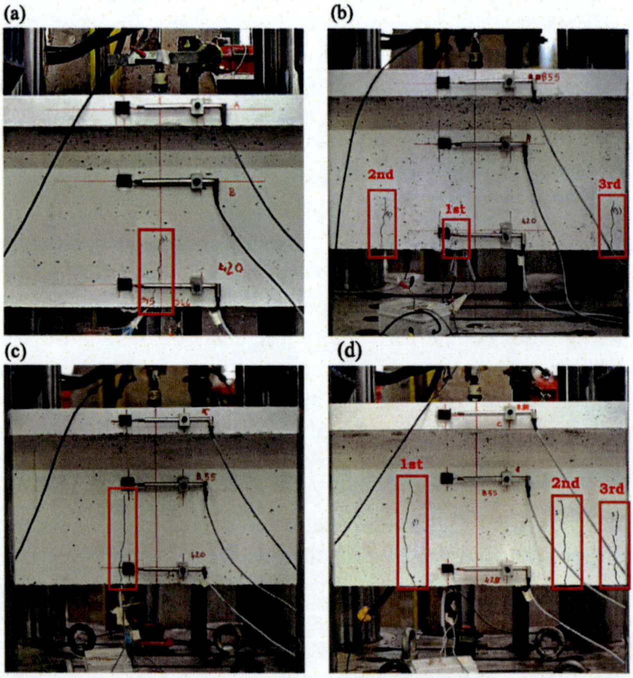


**Fig. 7.** First cracks detected by visual inspection: a) T1-B-HP, b) T2-U-HP, c) T3-B-LP, d) T4-U-LP.

### 3.3 Cracking Pattern

The first crack formed around midspan under cyclic protocol P1, specifically as a result of a force level achieved in the first phase ($F < F_1$) and third phase ($F < F_3$) in case of LP and HP specimens, respectively.

Figure 7 shows the onset of cracking detected on each girder during testing. For specimens with bonded tendons, a single crack appeared at the bottom of the girder (due to tensile stress around the mid-span section) as shown in Figs. 7a and 7c. In girders with unbonded tendons, three cracks were detected almost simultaneously under the same load. This could be the result of lower tension stiffening contribution in case of specimens with unbonded tendons due to lacking transfer of tensile stresses from tendons to cracked concrete.

Figure 8 shows the cracking pattern at peak load $F_{max}$. In case of high prestress, cracks were more spaced from each other, reaching smaller widths in case of T1 specimen compared to T2 specimen. Specimens with unbonded tendons suffered larger cracks. In case of high prestress, the bonded specimens showed a lower number of cracks than their unbonded counterparts. The opposite outcome was found in case of low prestress, even if cracks with significantly larger width (i.e. more than 10 mm) were observed. In most cases, crack spacing was around 100 mm (i.e. equal to stirrups spacing), reaching a maximum value of approximately 200 mm.

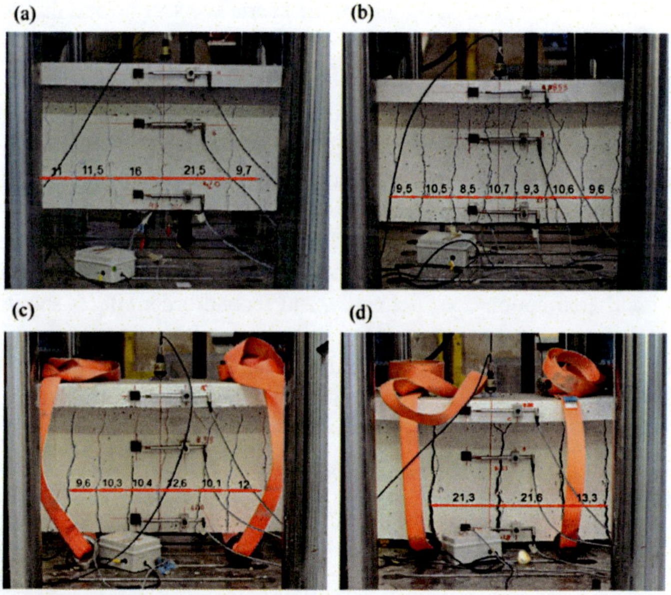


**Fig. 8.** Cracking pattern at peak load: a) T1-B-HP, b) T2-U-HP, c) T3-B-LP, d) T4-U-LP (external belts installed on T3 and T4 specimens only).

# 4 Theoretical Interpretation

## 4.1 Analytical Formulation

In order to predict the experimental response, relevant girder properties were calculated in terms of cracking bending moment, ultimate bending moment and vertical initial stiffness at mid-span.

Cracking bending moment of the mid cross section $M_{e,cr}$ was calculated as follows:

$$M_{e,cr} = \left(\frac{P_t}{A_n} + f_{ct}\right)\frac{I_1}{y_{G,i}} + M_p - M_G \quad (1)$$

where:

- $P_t$ is the prestressing force at the time of testing.
- $A_n$ is the total equivalent area of the reinforced concrete (RC) cross section (unbonded tendons are not taken into account).
- $f_{ct}$ is the concrete tensile strength (positive).
- $I_1$ is the second-order moment of inertia of the uncracked section (unbonded tendons are not taken into account).
- $y_{G,i}$ is the distance between the centroid of the uncracked section and bottom concrete fiber.
- $M_p = F_p(d_p - y_{G,s})$ is the prestressing bending moment with respect to the centroid of the uncracked section ($d_p$ is the distance of the tendon from the top concrete fiber; $y_{G,s}$ is the distance between the center of the uncracked section and the top concrete fiber).
- $M_G$ bending moment due to self-weight of the girder.

Ultimate bending moment $M_u$ was calculated by assuming perfect steel-concrete bond according to classical theory of RC section in bending. For concrete in compression, a stress-block constitutive law was assumed while neglecting concrete in tension. Prestressing steel and mild steel were modeled using an elastic-plastic with hardening stress-strain behavior and elastic-plastic constitutive law, respectively.

In case of unbonded tendons, a proper formulation is worth to be developed because the perfect bond assumption does not apply, and second-order effects should be taken into account.

The PC girder deflection was analyzed according to the static scheme shown in Fig. 9. Before cracking of concrete, the mid-span deflection can be obtained using the following equation.

$$d = F/K_1 \quad (2)$$

where:

$$K_1 = \frac{48EI_1}{f(L_v)} \quad (3)$$

being $K_1$ the vertical stiffness of the uncracked cross section due to external loading, $E_c$ the instantaneous compressive modulus of concrete (30,000 MPa) and $f(L_v)$ a function of the shear-span length. For $L_v = L/2$ the value of the function is $L^3$ and Eq. (3) returns the well-known solution for a simply supported beam subjected to a point load at mid-span.

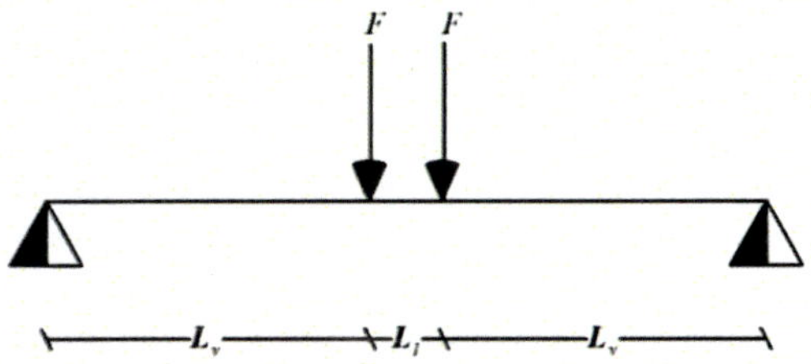

**Fig. 9.** Structural model of girder specimens.

## 4.2 Analytical-Experimental Comparison

The experimental values of the initial stiffness are shown in Table 3 and were obtained as $K_1 = F_{cr}/d_{cr}$, being $F_{cr} = 2M_{a,cr}/L_{cr}$ the cracking load and $d_{cr}$ the corresponding cracking deflection. The analytical value of $K_1$ was equal to 9.87 kN/mm, resulting very close to the average experimental estimate of stiffness.

The cracking load $F_{cr}$ depends mostly on the initial prestress level. Moving from high to low prestress level, $F_{cr}$ reduced by 62% and 58% in case of specimens with bonded and unbonded tendons, respectively.

The bonding of tendons appears to play a minor role on the uncracked behavior: different values between T1–T2 and T3–T4 specimen pairs depends on the effective prestress $F_p$ (see Table 2) rather than grout conditions.

Values of $F_{cr}$ are plotted in Fig. 6 with different markers for each girder. It can be noted that cracking pattern in Fig. 7 was attained for $F > F_{cr}$ due to formation of cracks beyond the attainment of concrete tensile strength.

**Table 3.** Deflection, load, bending moment and stiffness at cracking of each specimen.

| Specimen | $d_{cr}$[mm] | $F_{cr}$[kN] | $M_{e,cr}$[kNm] | $K_1$ [kN/mm] |
|---|---|---|---|---|
| T1-B-HP | 5.68 | 49.0 | 68.6 | 8.62 |
| T2-U-HP | 5.00 | 50.0 | 70.0 | 10.0 |
| T3-B-LP | 1.99 | 17.4 | 24.3 | 8.74 |
| T4-U-LP | 2.20 | 20.0 | 28.0 | 9.09 |

Table 4 shows the comparison between experimental and analytical values of ultimate bending moment. While values of maximum force and corresponding displacement represent test values, bending moment $M_{\max}^{\exp}$ is compared to $M_{\max}^{an}$ also combining girder self-weight to external action (i.e. $M_{\max}^{\exp} = M_{a,max} + M_G$).

In case of specimens with bonded tendons, experimental and analytical values closely match with each other with approximately 5% error. Specimens with unbonded tendons (i.e. T2 and T4) showed that maximum experimental values are 15% to 20% lower than those obtained in case of T1 and T3 specimens, even if such values do not correspond to ultimate conditions for which larger deflections should be imposed (see hardening response in Fig. 6). For both T1 and T3 specimens, the ultimate condition was caused by concrete crushing (Fig. 10).

The ultimate response of the girders does not depend on the initial prestress level. This is confirmed by both experimental and analytical results (see T1 vs T3 in Table 4). Furthermore, grout conditions may affect the ultimate response, playing a major role on the corresponding deformation levels to be attained.

**Table 4.** Peak deflection, load, and experimental and analytical bending moments.

| Specimen | $d_{max}$ [mm] | $F_{max}$ [kN] | $M_{max}^{exp.}$ [kNm] | $M_{max}^{an.}$ [kNm] |
|---|---|---|---|---|
| T1-B-HP | 124 | 156 | 230 | 228 |
| T2-U-HP | 130 | 132 | 196 | - |
| T3-B-LP | 165 | 161 | 236 | 228 |
| T4-U-LP | 202 | 138 | 204 | - |

**Fig. 10.** Deformed configuration of T1 specimen at ultimate conditions.

## 5 Conclusions

This paper investigated the flexural response of four, reduced-scale, post-tensioned PC bridge girders. The specimens were designed according to a combination of two different initial prestressing force and grout conditions: high (150 kN) and low (75 kN) levels of initial prestress along with fully grouted ducts (resulting in bonded tendons) or ungrouted ducts (resulting in unbonded tendons). The girders were subjected to both cyclic and monotonic quasi-static loading protocols. The cyclic and monotonic envelope responses were investigated.

The experimental findings confirmed that: i) different initial prestress levels and grout conditions do not affect the uncracked behavior of the girders, mostly depending on gross sectional geometry; ii) the cracking load mainly depends on the initial prestress level rather than bond condition of tendons; and iii) unbonded tendons provide significantly lower stiffness after cracking due to lower tension stiffening contribution.

An analytical-experimental comparison was than carried out based on existing formulations from the literature. A satisfactory agreement was found between experimental outcomes and analytical prediction unless ultimate bending moment in unbonded tendons is considered, demanding for a proper model to be developed.

A future study will address the role of tension stiffening in PC girders with different grout conditions. Further investigation will be carried out considering the effect of partial grouting of tendons and local damage on girder's flexural response.

**Acknowledgements.** This study was developed in the framework of PON INSIST (Sistema di monitoraggio INtelligente per la Sicurezza delle infraSTrutture urbane) research project, which was funded by the Italian Ministry for Education, University and Research (Programma Operativo Nazionale "Ricerca e Innovazione 2014–2020", Grant No. ARS01_00913). The support by Eng. Elia Acconcia (University of Naples Federico II), Eng. Davide Esposito (TECNO IN s.p.a.), Eng. Simone Barile (MAPEI s.p.a.) and laboratory staff (particularly Eng. Emanuele Scaiella and Eng. Gennaro Maddaloni) is gratefully acknowledged.

## References

American Concrete Institute (ACI): Load Tests of Concrete Structures: Methods, Magnitude, Protocols, and Acceptance Criteria. Standard (2007)

Asp, O., Tulonen, J., Kuusisto, L., Laaksonen, A.: Bond and re-anchoring tests of post-tensioned steel tendon in case of strand failure inside cement grouting with voids. Struct. Concr. **2021**(22), 2373–2390 (2020)

CEB-FIB: Durability of post-tensioning tendons. In: Proceedings of a Workshop Held at Ghent University on 15–16 November 2001. Technical report (2001)

DM 17/01/2018: Aggiornamento delle «Norme tecniche per le costruzioni». Italian Ministry of Infrastructures and Transportation, Rome, Italy (2018). (in Italian)

Eurocode 2: Design of Concrete Structures: British Standard. London: BSi (2008). Print

US Department of Transportation Federal Highway Administration: Guidelines for Sampling, Assessing, and Restoring Defective Grout in Prestressed Concrete Bridge Post – Tensioning Ducts. Publication No. FHWA-HRT-13-028 (2013)

VSL International LTD: Grouting of Post-tensioning Tendons. VSL Report Series (2002)

Wang, L., Zhang, W., Zhang, J., Ma, Y., Xiang, Y., Liu, Y.: Effect of insufficient grouting and strand corrosion on flexural behavior of PC beams. Constr. Build. Mater. **53**(2014), 213–224 (2014)

# Dapped-End Beams: Experimental Tests and Capacity Models in the Literature

Danilo D'Angela[1], Chiara Di Salvatore[1](✉), Massimo Acanfora[2], Edoardo Cosenza[1,3], and Gennaro Magliulo[1,3]

[1] Department of Structures for Engineering and Architecture, University of Naples Federico II, Via Claudio 21, 80125 Naples, Italy
{danilo.dangela,chiara.disalvatore,edoardo.cosenza, gmagliul}@unina.it

[2] Strutture and Servizi s.r.l, Viale Augusto 140, 80125 Naples, Italy
massimo.acanfora@libero.it

[3] Construction Technology Institute, National Research Council, Via Claudio 21, 80125 Naples, Italy

**Abstract.** Half joints are very common construction details, often used for precast beams connections or bridge deck supports, since they allow the creation of an effective joint without increasing the height of the nodal region. If not adequately designed, dapped-end beams can show premature shear failure, due to the abrupt variation in the cross-section dimension. Given the very low shear span, preventing the application of the Bernoulli's beam theory, strut and tie models need to be taken into account. In this context, experimental tests provide a powerful instrument to detect the stress flows and define the associated capacity models. The presented work supplies a wide literature review of experimental tests and numerical formulations, aiming to characterize the shear behavior of dapped-end beams. It can be a useful tool for design and assessment procedures of such beams.

**Keywords:** dapped-end beams · experimental shear tests · strut and tie models

## 1 Introduction

Half joints are often realized in precast structures in order to allow a convenient support condition for beams, for example on a corbel or on another beam, containing the overall height. If a proper design for such joints is not performed, a potential vulnerability is generated against shear loads. Indeed, the peculiar configuration of dapped-ends entails a squat element behavior characterized by high shear stresses. For this reason, empirical strut and tie models are usually referred to for design and assessment of such joints (Schlaich and Schafer 1991). It appears clear that experimental campaigns investigating the shear response of dapped-end beams represent the most valid tool to understand the resistant mechanisms of half joints, to detect the stress flows and to develop increasingly more accurate, yet simplified, computational approaches (Nanni and Huang 2002; Moreno-Martínez and Meli 2014; Desnerk et al. 2016).


M. A. Aiello and A. Bilotta (Eds.): ICC 2022, LNCE 435, pp. 302–312, 2024.
https://doi.org/10.1007/978-3-031-43102-9_24

In this paper, a wide review of both experimental and numerical research outcomes in the field of dapped-end beams evaluation is reported. Different strut and tie calculation approaches are illustrated and applied to the experimental results in order to verify their accuracy in predicting the shear capacity of dapped-end beams. The comparison between experimental and numerical outcomes can provide useful information and insights for design and assessment procedures to be applied to half joints in precast elements assembling.

## 2 Experimental Campaigns

Four experimental works from the literature are considered in this paper, in chronological order: Mattock and Chan (1979), Lu et al. (2003), Nagrodzka-Godycka and Piotrkowski (2012), and Mata-Falcón et al. (2019). Authors tested several dapped-end beams, accounting for different reinforcement layouts and load conditions. Cracked pattern evolution and failure mode are always detected and analyzed in detail, providing a clear and accurate description of the specimens' behavior under shear loads.

### 2.1 Mattock and Chan (1979)

Eight Gerber's joints, representing the extremities (A and B) of four beams (from 1 to 4), were tested at the laboratory of structural engineering of the University of Washington. All the specimens had the same geometry, for both the full depth beam and the nib. In particular, beams were characterized by a rectangular cross-section with dimensions 127 $\times$ 610 mm$^2$ and a length of 3.050 m, whilst dapped-ends presented a depth equal to half the beam's depth (305 mm) and same width, with a length of 0.203 m. Reinforcement details, instead, changed for each beam, i.e. for each couple of Gerber's joints. Three groups of steel bars were deemed to participate in the shear resistance mechanism: the close stirrups near the interface between the full depth beam and the nib ($A_{vh}$), the main horizontal reinforcement of the nib ($A_s$) and the horizontal ties along the nib's height ($A_h$). It is worth highlighting that no diagonal reinforcement was provided for the specimens. In order to investigate the influence of an acting horizontal force, four specimens (extremity A) were subjected to vertical load only, whilst the remaining four (extremity B) were tested under an inclined force, i.e. a combination of a vertical and a horizontal force, both applied on the top of the beam. However, for all the cases, the first crack appeared at the interface between the full depth beam and the nib, more specifically at the re-entrant corner with an inclination of about 45° with respect the longitudinal axis of the beam; as the load increased, more cracks developed, and the existing ones became increasingly longer and wider. The influence of the loading position was investigated too, so as to consider, for the first six specimens, the failure mechanisms only in the nib, and for the last two specimens (4A/4B), the variation of behavior induced by the crack propagated from the bottom of the full depth beam.

For all the tests, it was detected that the nib's main reinforcement yielded before the attainment of the maximum shear load; from the definition of the vertical plane in which flexural occurred, a consistent real shear span ($l_v$) for the Gerber's joint can be determined. It resulted that $l_v$ was always equal to the distance between the centre of the support under the nib and the centre of the hanger reinforcement, i.e. the line of action of the stirrups resultant. Specimens 1A and 1B were the reference specimens, which showed very poor shear performance with respect to both the ultimate and the service conditions. The parameter characterizing the ultimate limit states was assumed to be the shear resistance of the dapped-end beam, whilst for the service limit state, the crack width was measured. Specimens 2A/2B and 3A/3B investigated the possibility of reducing and increasing respectively the amount of hanger reinforcement, demonstrating worsen and better performance, respectively, with respect the reference specimen at both the ultimate and the service limit conditions. Finally, specimens 4A/4B, which were identical to 3A/3B and investigated the influence of a different loading position, showed a good response, even better than specimens 3A/3B, at the service limit state and slightly worse than 3A/3B at the ultimate states.

## 2.2 Lu et al. (2003)

Authors tested twelve dapped-end beams with the aim of assessing the influence of concrete strength (high-strength concrete was also considered), main reinforcement amount and shear span-to-depth ratio on the shear capacity of the joints. All the beams had the same dimensions, with a 200 × 600 mm$^2$ cross section and a length of 3 m. Beams' ends were equally shaped too, presenting both depth and length equal to 300 mm. Steel reinforcement of Gerber's joint was constituted by: main horizontal reinforcement in the nib, horizontal hoops in the nib and hanger reinforcement close to the sudden depth variation. No diagonal reinforcement was provided. As usual, beams were tested at one end (one Gerber's joint) first; then, since damage was limited to the loaded nodal region, beams were turned, and the other end tested. A static vertical load was applied to the specimens, observing the crack pattern and measuring the strain in the steel bars. It was detected that concrete diagonal compression and main reinforcement yielding always occurred before shear failure, for all the tested joints. Force-displacement curves for each specimen are provided in the paper, highlighting their poorly ductile response. As expected, the shear capacity of dapped-end beams increased increasing both concrete strength and main reinforcement amount. On the contrary, shear resistance decreased with the increase of the shear span-to-depth ratio.

## 2.3 Nagrodzka-Godycka and Piotrkowski (2012)

Experimental program conducted by the Authors included six beams, tested at both ends. Specimens had all the same geometry: the full depth part presented a 400 × 200 mm$^2$ cross-section and a length of 2 m, whilst both nibs, spanning 0.2 m at the ends, were characterized by a depth equal to half the full depth (200 mm) and by the same width of the beam (200 mm). Specimens were named with number from 1 to 6. The right-side nib was referred to as P, whereas the left-side nib was indicated with L. The experimental work primarily aimed to investigate the effect of inclined forces on dapped-end beams; for this reason, three specimens (1, 3 and 5) were loaded with a combination of horizontal and vertical forces, and outcomes were compared to the ones from corresponding identical specimens (2, 4 and 6) loaded with vertical force only. Specimens 1–2 presented main horizontal reinforcement in the nib with a geometrical ratio $\rho_{sH} = 0.76\%$ and hanger reinforcement (4 ϕ 10) in the section depth variation zone; no diagonal reinforcement was provided. Specimens 3–4 had doubled reinforcement, i.e. $\rho_{sH} = 1.52\%$ and 8 ϕ 10 as hanger bars; again, no diagonal reinforcement was considered. For the first four specimens, reinforcement was assumed equal to both the nibs of the beam. Finally, specimens 5–6 presented a lower amount of main and hanger bars with respect the previous specimens, but they were provided with diagonal bars, different at the two ends; in particular, right end has doubled diagonal bars with respect to the left end (2 ϕ 16 vs 1 ϕ 16).

Loads were monotonically applied with 5 kN increments in vertical force, keeping the horizontal load (if present) always equal to half the vertical load, i.e. $H/F_V = 0.5$. For each loading step, strains in both reinforcement bars and concrete were measured, and the crack pattern evolution was detected. The first visible difference between the cases of inclined and vertical loads could be recorded in the final crack pattern: in the former case, cracks mostly propagated in the vicinity of the recessed, corner showing a subvertical trend, whereas, in the latter case, cracks concentrated in the nib region with a low inclination with respect to the longitudinal axis of the beam. As concern the shear capacity, it was demonstrated that, in the presence of horizontal components, the resistance of the dapped and beam was about 25% lower than the case of vertical load only, regardless the amount of reinforcement. In the case of diagonal reinforcement, the reduction was even more significant. Furthermore, increasing the amount of both main reinforcement and hanger bars (i.e. from 1–2 specimens to 3–4 specimens) led to an increase in the shear resistance of about 50%. Failure mode changed too, since for the specimens with low amount of reinforcement only the main horizontal bars yielded before the maximum load was reached, whilst for specimens with large amount of reinforcement both main and hanger bars yielded before the shear collapse of the nib. The highest value of shear strength is recorded when the highest amount of diagonal bars were provided, in the case of vertical load only (6-P specimen). The enhancement in the capacity, given by the increase in the diagonal reinforcement amount, is more important in the case of vertical load only, reaching about the 35% against the 16% of the inclined force case.

## 2.4 Mata-Falcón et al. (2019)

Authors investigated the shear response of 15 beams with dapped-ends, for a total of 30 experimental tests. Unfortunately, in two cases, the load applied for the first end joint induced non negligible damages to the other end, affecting the experimental outcomes; for this reason, results from pre-damaged specimens were discarded and the number of experimental tests was reduced to 28. Same geometrical features were chosen for all the specimens: beams presented a (250 × 600) mm$^2$ cross-section and a length of 3 m, whilst dapped-ends were characterized by half the depth of the beam (300 mm) and the same width, with a length of 0.35 m. Reinforcement amount and layout changed for each beam, i.e. for each couple of specimens. Two sets of specimens were considered: set 1 and set 2. The former had no diagonal reinforcement and took into account three different values of the ratio between main reinforcement in the nib ($A_{sH}$) and hanger reinforcement near the critical section ($A_{sV}$); the reference layout provided a ratio $A_{sH}/A_{sV}$ almost equal to 1.5, whereas the other two investigated the cases of poor horizontal reinforcement

($A_{sH}/A_{sV} \approx 0.9$) and poor hanger reinforcement ($A_{sH}/A_{sV} = 3.9$), respectively. The influence of the number of stirrups near the critical section was also evaluated in the first two layouts, noting that the case of one single stirrup was the most favourable for the concrete spalling condition in the upper side of the beam. The latter set of specimens was provided with 47° inclined diagonal bars in three different configurations such that, according to a strut and tie design approach, they collaborated to the total shear strength for about 40%, 60% and 80%, respectively. Load was applied through a three points non-symmetric bending, in displacement control with a rate of 0.25 mm/min.

The crack pattern observation allowed to detect the evolution of damage as the load increased; a diagonal crack always occurred, starting from the top of the support and propagating up to the top of hanger reinforcement, where concrete cover detachment was recorded at peak load. In some cases, after spalling phenomenon took place, a brittle failure was observed. A three-rule criterion was set by the Authors to identify the spalling failure taking into account the crack propagation and the load-displacement response of each specimen: i) concrete cover detachment was recorded in correspondence of the peak load, ii) load-displacement response curve showed no horizontal plateau before the peak load, and iii) after the peak load was reached, a drop 20% drop in resistance was detected for a 20% increase in displacement. As concerns the reinforcement strain level, it was observed that, when a single layer of vertical stirrups was provided, horizontal and vertical reinforcement yielded more or less at the same load value; when more than one level of stirrups constituted the hanger reinforcement, the first steel bar to yield was the closest stirrup to the critical section, then yielding propagated to the other vertical bars up to the furthest with respect to the critical section; horizontal reinforcement reached the elastic limit always before the last stirrup. This happened for both the first and the second set of specimens. When diagonal bars were provided, they were always the first ones to yield. However, the large plastic redistribution capacity allowed both the inclined and the orthogonal mechanism to deploy expound the maximum resistance upon the peak load.

## 3 Strut and Tie Models

Four strut and tie models from codes and scientific literature are considered and evaluated through their application to the previously presented experimental tests. The following studies/methods were considered as a reference: Cinuzzi and Gaudiano (1992), Lu et al. (2003), Eurocode 2 (2004), and Wang et al. (2005).

### 3.1 Cinuzzi and Gaudiano (1992)

Cinuzzi and Gaudiano (1992) propose two different shear resistant mechanisms for dapped-end beams, named "A" and "B", depending on the reinforcement layouts (Fig. 1). In both the models, the compression and the tension zones of the full depth beam contribute to the global shear resistance. In model A, two 45° inclined diagonal struts are generated: the former starting from the top of the support and the latter from the inferior node of the beam. Authors also provide the experimentally derived maximum width that both the struts can assume; this value is a function of the nib's effective height, d, and it is equal to 0.2·d and 0.3·d for the strut above the support and the strut in the inferior part of the beam, respectively. Moreover, two ties are considered, consisting in the nib main reinforcement and in the hanger bars near the critical section. In model B, instead, inclined reinforcement is provided, activating a diagonal mechanism together with a vertical strut upon the support.

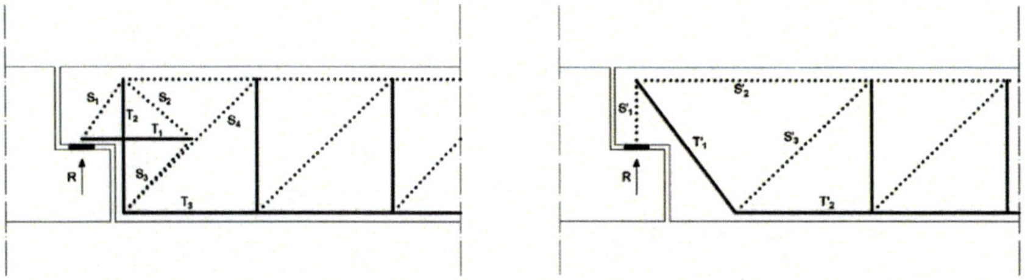


**Fig. 1.** Strut and tie mechanisms according to Cinuzzi and Gaudiano (1992) model: a) mechanism A and b) mechanism B.

Each mechanism ensures a certain shear capacity, corresponding to the minimum resistance among the ones offered by each of the element constituting the considered mechanism, i.e. each strut and each tie. Such resistance values are determined through simple equilibrium equations, as usual for strut and tie models, which derive from the lower bound theorem of the structural limit analysis. Then, shear capacity of model A and shear capacity of model B have to be summed in order to obtain the overall dapped-end beam capacity. Clearly, in the cases in which no diagonal reinforcement is provided, model B is not applicable, and the global shear resistance coincides with the capacity of mechanism A. It is worth specifying that, in the present study, the maximum width is assumed for both the diagonal struts in model A, thus optimizing their shear resistance. This choice derives from the attempt to maximize the lower bound capacity, provided by the application of a strut and tie method, due to its own nature.

### 3.2 Lu et al. (2003)

The strut and tie model from Lu et al. (2003) requires the use of two mechanisms: the horizontal and the diagonal mechanism (Fig. 2). The former consists in a diagonal compression strut (D), inclined of an angle θ with respect the beam's longitudinal axis; the width of the strut depends on its boundary conditions, dictated by the compression zone between the nib and the full depth beam. The latter, instead, is made up of one horizontal tie ($F_h$), generated by horizontal hoops in the nib, and two flat struts. The shear capacity of the nib ($V_{dv}$) can be calculated through Eq. 1:

$$V_{dv} = -D \cdot sen\theta + F_h \cdot tan\theta \tag{1}$$

The inclination of the diagonal strut conditions the prevailing mechanism among the possible ones. In particular, if $\theta \geq \tan^{-1}(2)$, the applied shear is totally absorbed by the horizontal mechanisms, whilst if $\theta \leq \tan^{-1}(1/2)$, the applied shear is carried by the diagonal mechanism only. For values of θ ranging between the limits, both the mechanisms contribute to the shear capacity of the nib. It is worth noticing that the model by Lu et al. (2003) is not able to take into account the diagonal mechanism provided by diagonal reinforcement. Such strut and tie resistance is compared to both the flexural capacity of the nib and the capacity provided by the yielding of hanger bars: the minimum capacity value among the ones considered represents the shear capacity of the dapped-end beam.

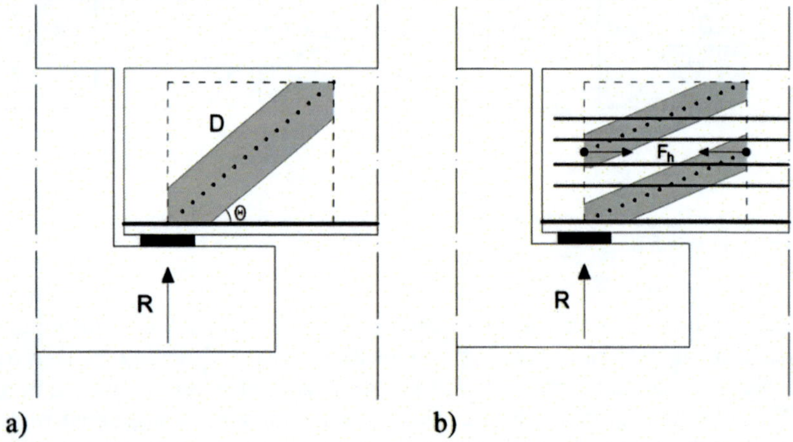


**Fig. 2.** Strut and tie mechanisms for Lu et al. (2003) model: a) diagonal mechanism and b) horizontal mechanism.

### 3.3 Eurocode 2 (2004)

European codes suggest the use of two different mechanisms, referred to as model "A" and model "B". The reference mechanisms are almost in line with the one presented by Cinuzzi and Gaudiano (1992), while the procedure applied to obtain the overall shear capacity is the exactly the same (for each mechanism the joint resistance is equal to the lowest resistance offered by the element participating to the mechanism). The only difference concerns the diagonal struts in mechanism A. First, the inclination angle of the strut above the support, according to Eurocode 2 (2004), is not fixed and equal to 45° but depends on both the internal level arm in the nib and the position of the hanger reinforcement resultant. Thus, for the same nib's internal level arm, increasing the stirrups number and/or spacing in the critical region leads to a decrease in the strut's inclination; on the contrary, decrease the hanger reinforcement number and/or spacing involves an increase in the strut's inclination. Clearly, this variation affects the shear capacity provided by the strut, which increases with the increasing of the inclination angle. Another significant change introduced by the Eurocode 2 (2004) concerns the width of the struts, which can be evaluated in a more accurate manner. In particular, the dimension depends on the support width, the length of horizontal bars extending beyond the strut, the distance between the longitudinal bar axis and the concrete edge and the spacing between two or more horizontal reinforcement layers. In the present case, since no information is provided in the experimental campaigns about the length of main reinforcement protruding beyond the strut, the simplest case is considered, where the width of the strut only depends on the support dimension.

### 3.4 Wang et al. (2005)

Wang et al. (2005) calibrated their theoretical strut and tie model on the results of an experimental campaign on 22 dapped-end beams. The considered schemes are analogous to the ones in Fig. 1, but the proposed formulation for the shear strength calculation is now additive. In particular, three addends are considered: the concrete contribution ($V_c$), provided by both the uncracked material and the aggregate interlock between the crack edges, the vertical stirrups contribution ($V_s$) and the diagonal reinforcement contribution ($V_b$), if present. The overall shear strength ($V_N$) can be evaluated as:

$$V_N = V_c + V_s + V_b \tag{2}$$

Each contribution is firstly evaluated through simple equilibrium considerations and then corrected through the insertion of a different regression factor ($\beta_1$, $\beta_2$, and $\beta_3$ respectively), derived from a statistical analysis on the tested specimens.

Flexural strength of the nib, provided by its main reinforcement, and yielding of hanger reinforcement are also evaluated. If flexural strength is lower than the value provided by Eq. 2, the related mechanism could affect the global shear strength. Indeed, global shear capacity is the minimum between the shear resistance provided by the strut and tie model (Eq. 2) and the shear resistance corresponding to the flexural failure.

## 4 Application and Comparison

In Fig. 3, the application of the abovementioned strut and tie models with regard to the considered experimental tests is shown. A different plot is depicted for each of the experimental campaign, characterized on the horizontal axis by the experimentally derived shear resistance of the specimens and, on the vertical axis, by the ratio between the numerical value and the experimental value of the shear strength.

A different plot is depicted for each of the experimental campaign, characterized on the horizontal axis by the experimentally derived shear resistance of the specimens and, on the vertical axis, by the ratio between the numerical value and the experimental value of the shear strength.

As concerns Mattock and Chan (1979), it can be noted that Cinuzzi and Gaudiano (1992) model is the one that approximates better the shear resistance of the specimens, being always at the safe side. However, the dapped-end beams failure mode only in the first two cases (1A/1B specimens) coincides with the experimentally detected one, which is main reinforcement yielding, whilst for the other specimens, the model detects a failure due to the yielding of hanger reinforcement. Eurocode 2 (2004) provides similar results with respect to Cinuzzi and Gaudiano (1992), even if more cautionary. This is

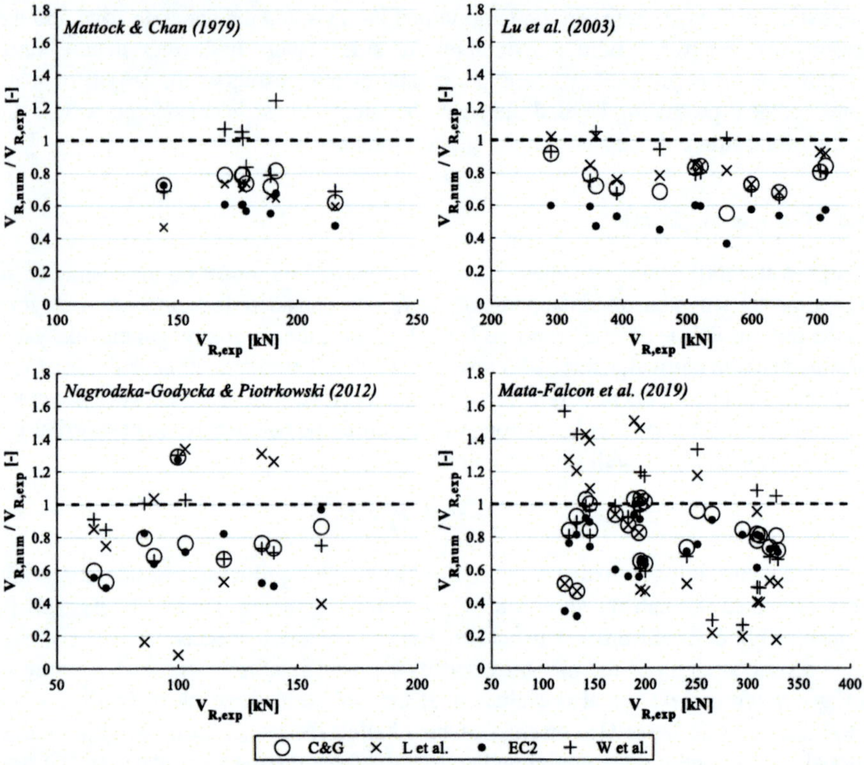


**Fig. 3.** Numerical strut and tie models (C&G = Cinuzzi and Gaudiano (1992), L et al. = Lu et al. (2003), EC2 = Eurocode 2 (2004), W et al. = Wang et al. (2005)) applied to the considered experimental tests.

not surprising since the assessment approaches are very similar for both the models. The evaluated failure modes are consistent too; the only difference is detected for specimen 1B, for which the failure is given by the diagonal concrete strut originated from the inferior corner of the full depth beam. This can be explained considering that Eurocode 2 (2004) suggests the use of a concrete compressive strength reduction factor ($\nu$) in order to account for the real stress state of concrete struts, which are loaded not only in their axial direction. The strut and tie model by Lu et al. (2003) proves to be always on the safety side. However, it fails to capture the experimental failure modes for all the specimens; indeed, according to such model, since it is always $\theta \geq \tan^{-1}(2)$, the shear mechanism is governed by the horizontal hoops in the nib. Such collapse is detected for all the specimens, except specimen 2A, for which hanger reinforcement provides the lowest strength. Finally, Wang et al. (2005) additive model can dangerously overestimate the nib resistance against shear loads, especially when, in the reality, a contribution is strongly prevalent on the others. This occurs only when a shear failure is detected in the nib, i.e. for specimens loaded with an inclined force; for the remaining specimens, loaded with a vertical force only, the model provides flexural failures, correctly predicting the dapped-end beam behavior.

Similar comments can be drawn from the comparison between the considered numerical models and the experimental outcomes by Lu et al. (2003). Cinuzzi and Gaudiano (1992) proves to be always on the safety side, even if failure modes are not consistent with the ones indicated by the Authors of the experimental work. Eurocode 2 (2004) model seems to excessively underestimate the shear resistance of the specimens, providing the lowest predictions. In most cases, the failure is associated to the attainment of the concrete maximum strength. It is worth specifying that, since no information can be found in Lu et al. (2003) about the support dimensions, a 100 mm support width is assumed; as already stated, this parameter can affect the concrete struts shear resistance. Thus, a level of uncertainty inevitably undermines the reliability of the results. The model proposed by the Authors themselves (Lu et al. (2003)) proved to be almost always at the safety side; in two cases only the predicted shear resistance slightly exceeds the experimental one. Ultimately, Wang et al. (2005) model detects flexural failures for all the specimens except the first, for which the strut and tie additive formulation provides the minimum strength.

Experimental outcomes from Nagrodzka-Godycka and Piotrkiwski (2012) are well interpreted by Cinuzzi and Gaudiano (1992) in terms of failure mode (yielding of horizontal/diagonal bars) and shear resistance. Eurocode 2 (2004) offers a good prediction of the nib's strength too, but several failures are associated with the attainment of concrete struts ultimate resistance, due to the reduction factor $\nu$. Lu et al. (2003) always provides flexural collapses in accordance with the experimental evidence. However, since the model is not able to take into account the contribution of diagonal bars, it significantly underestimates the shear strength when inclined reinforcement is considered. Wang et al. (2005) is able to detect the flexural failure of all the specimens; indeed, the given flexural strength is lower than the resistance offered by the strut and tie scheme.

The case of Mata-Falcón et al. (2019) is of more complex interpretation since the experimental failure is often associated to concrete spalling phenomenon in correspondence of the node upon the hanger reinforcement, on the upper face of the beam. This changes the layout of the strut and tie mechanisms within the dapped-end beams. However, the same qualitative considerations of the previous experimental campaign (Nagrodzka-Godycka and Piotrkowki (2012)) can be done. In this case, Cinuzzi and Gaudiano (1992) provide the most accurate model with the lowest dispersion.

Further and detailed comments can be found in Di Salvatore et al. (in prep.).

## 5 Conclusions

The paper selects from the literature the most significant strut and tie models for dapped-end beams, and provides their evaluation in terms of shear capacities, considering several experimental shear tests as a reference. The work offers a valuable aid for the capacity assessment of such elements, mostly spread in precast structures and bridges. As a result, the following comments can be drawn.

- Cinuzzi and Gaudiano (1992) model provides shear strength predictions always at the safe side, even if it is not always able to correctly identify the collapse mode;

- Lu et al. (2003) model is not able to take into account the contribution of diagonal bars; thus, its application should be avoided in the cases in which such reinforcement is implemented. Furthermore, since the model is calibrated on dapped-end beams made by high strength concrete, the results might be unreliable for ordinary concrete elements;
- Eurocode 2 (2004) model excessively underestimates the shear resistance, often due to the concrete compressive strength reduction factor, introduced by the code. In the context of capacity assessment, this might not be the preferable solution, even though it might be the most conservative;
- Wang et al. (2005) model can overestimate the shear strength of dapped-end beams, due to its additive formulation, which sums the contributions of concrete, vertical stirrups, and diagonal bars without considering the single component failure as a limiting condition. This might also lead to a misconception in the failure mode of the whole element, since no distinction is made between the different component of the truss system; moreover, the flexural capacity can easily become lower than the one provided by the strut and tie mechanism.

**Acknowledgements.** This research study is developed in the framework of the national project ReLUIS project prot. n°31 of 23/06/2021 - WP2-WP4 T4.4. The contribution by Eng. Ester Calce for the computation is gratefully acknowledged.

## References

CEN. Eurocode 2: Design of concrete structures. EN 1992 Brussels, Belgium (2004)

Cinuzzi, A., Gaudiano, S.: Tecniche di progettazione per strutture di edifici in cemento armato. Masson (1992)

Desnerk, P., Lees, J.M., Morley, C.T.: Impact of the reinforcement layout on the load capacity of reinforced concrete half-joints. Eng. Struct. **127**, 227–239 (2016)

Di Salvatore, C., D'Angela, D., Cosenza, E., Magliulo, G.: Capacity assessment of dapped-end beams considering multiple strut and tie models and parametric analysis. In prep.

Lu, W., Lin, I., Hwang, S., Lin, Y.: Shear strength of high-strength concrete of dapped-end beams. J. Chin. Inst. Eng. **26**(5), 671–680 (2003)

Mata-Falcón, J., Pallarés, L., Miguel, P.F.: Proposal and experimental validation of simplified strut-and-tie models on dapped-end beams. Eng. Struct. **183**, 594–609 (2019)

Mattock, A.H., Chan, T.C.: Design and behavior of dapped-end beams. PCI J. **24**(6), 28–45 (1979)

Moreno-Martínez, J.Y., Meli, R.: Experimental study on the structural behavior of concrete dapped-end beams. Eng. Struct. **75**(152), 163 (2014)

Nagrodzka-Godycka, K., Piotrkowski, P.: Experimental study of dapped-end beams subjected to inclined loads. ACI Struct. J. **109**(1), 11 (2012)

Nanni, A., Huang, P.: Validation of an alternative reinforcing detail for the dapped ends of prestressed double tees. PCI J. (2002)

Schlaich, J., Schafer, K.: Design and detailing of structural concrete using strut-and-tie models. Struct. Eng. **69**(6), 113–125 (1991)

Wang, Q., Guo, Z., Hoogenboom, P.C.J.: Experimental investigation on the shear capacity of RC dapped end beams and design recommendations. Struct. Eng. Mech. **21**(2), 221 (2005)

# The Role of Capacity and Flexibility of Floor Diaphragms in the Seismic Retrofit of Existing RC Buildings

Chiara Passoni, Elena Casprini(✉), Alessandra Marini, and Andrea Belleri

Department of Engineering and Applied Sciences, University of Bergamo, Dalmine, BG, Italy
elena.casprini@unibg.it

**Abstract.** The renovation of existing buildings to improve the safety and resilience of our building stock is now recognized as a priority, and many retrofit techniques have been recently proposed, especially carried out from the outside to avoid inhabitants' relocation. The ability of the existing floors to act as seismic diaphragms is however rarely addressed, if not in terms of diaphragm flexibility. Analyzing data from numerical and experimental studies carried out on a real beam and block floor system, some considerations about the role of existing diaphragms on the structural response of RC buildings are made with particular emphasis on in-plane capacity, deformability, and collapse mechanisms. Practical implications for the design of seismic retrofit interventions are defined.

**Keywords:** Seismic retrofit · RC buildings · beam and block floor systems

## 1 Introduction

The construction sector and the built environment are responsible of a great share of environmental impacts and carbon emissions. In addition, 40% of existing buildings has already exhausted their nominal service life and were designed in a lack of seismic standards, thus also requiring urgent structural upgrading. The need to renovate the existing building stock to foster sustainability, safety, and resilience is thus now recognized as a priority (Marini et al. 2017; Passoni et al. 2021).

Focusing on existing RC buildings, many techniques were recently proposed for the seismic upgrading of the building stock; however, the ability of the existing floors to act as seismic diaphragms is rarely addressed. When considering the seismic response of a structure against earthquake, the major functions of a diaphragm are, among others: guaranteeing the integrity of the structure by connecting the vertical elements of the Lateral Force Resisting System (LFRS) and enabling their activation, transferring the floor inertial loads to the vertical elements, providing load paths for lateral forces, and providing lateral constraints to the vertical elements against buckling and second order forces (Moehle et al. 2010).

Although this is usually not critical for existing structures, this may become an issue when buildings are reinforced by means of additional walls or shell systems, which do


M. A. Aiello and A. Bilotta (Eds.): ICC 2022, LNCE 435, pp. 313–324, 2024.
https://doi.org/10.1007/978-3-031-43102-9_25

not just increase the global lateral stiffness of the retrofitted building and, consequently, the seismic inertial forces, but also lump this additional stiffness in a few elements of the LFRS. This may be an issue especially when the retrofit is applied from the outside of the building (Passoni et al. 2020), without relocating the inhabitants, since the floor cannot be strengthened to increase its in-plane capacity, e.g. by casting an additional topping overlay to the floor.

In the present paper, the capacity of beam-and-clay block floor systems, which is common in Italian and Mediterranean existing RC buildings, is investigated. A tied-arched model is proposed to represent the resistant mechanism of these floors (Marini et al. 2022). Focusing on the sole inertia forces, possible in-plane load paths and collapse mechanism connected to different retrofit solutions are discussed. Data from numerical and experimental studies are presented to validate the model, and practical implications for the design of seismic retrofit interventions are defined.

## 2 Capacity and Flexibility of Beam-and-Clay Block Floor Systems

### 2.1 The Tied-Arch Analytical Model to Estimate the Ultimate In-Plane Floor Capacity

Beam-and-clay block floor systems consist in a series of parallel joists, usually made of cast-in-place RC, precast RC, or mixed masonry and RC beams, distanced by rows of hollow clay blocks, which may have different sections and dimensions. The system may be completed with a 2 to 5 cm thickness RC topping. Being these floors made of a series of different units, especially in the lack of the RC topping, possible inertial load paths should involve the single units and their interfaces, and their potential brittle behaviour may influence the global in-plane capacity of the diaphragm (Fleischman et al. 2005).

As for the load path, in case of regular floor plan and in absence of large openings, the inertial forces may be represented as distributed forces along the floor extrados $f_f$, which are transferred through the diaphragm to the LFRS elements by means of arches balanced by a bottom chord (Bull 2004) (Fig. 1).

According to this model, the ultimate capacity of beam-and-block floor systems may be associated to a tied-arched resistant mechanism, which depends on the tied-arch net span ($L$) and rise ($z_d$).

Considering the case of floors without the topping layer, the ultimate capacity of the diaphragm may be associated to one of the 3 failure mechanisms (Fig. 2) (Marini et al. 2022): A) compression failure of the blocks at the arch key, which are compressed orthogonally to their holes; B) combined compression and shear failure at the supports, usually connected to either the shear failure of the block or the failure of the block/joist interface; C) tensile failure at the bottom chord. In case of RC topping, the loads are distributed between the topping and the beam-and-block system as a function of their relative stiffness, up to the brittle failure of one of the layers.

According to Marini et al. (2022), the maximum capacity of a beam-and-block floor diaphragms may be preliminary estimated as the minimum of the floor shear reactions $V_f$ associated with the onset of these three mechanisms (Eqs. 1–4).

$$V_{fd} = \min\{V_{fA}; V_{fB}; V_{fC}\} \quad (1)$$

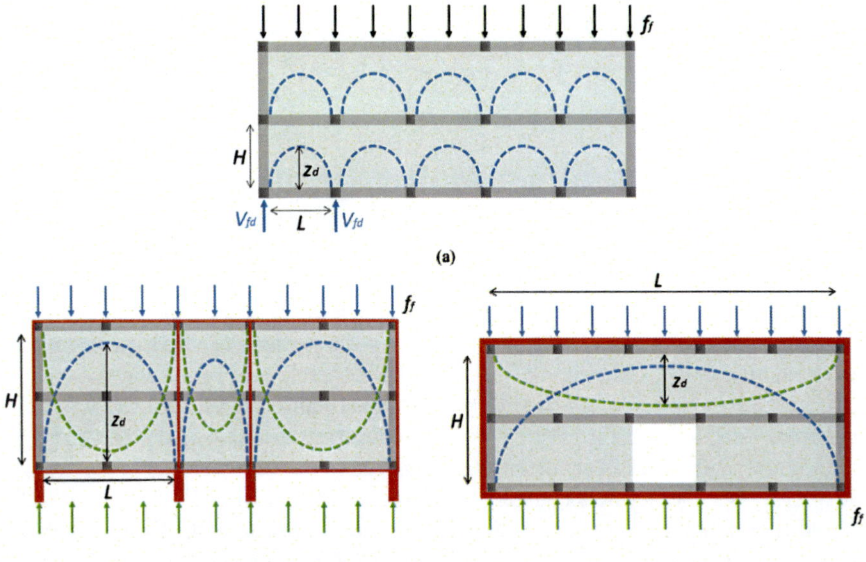


**Fig. 1.** In plane inertial load paths in case of building in as-is condition (a) and building retrofitted either with walls (b) or with a shell exoskeleton (c) (adapted from Marini et al. 2022)

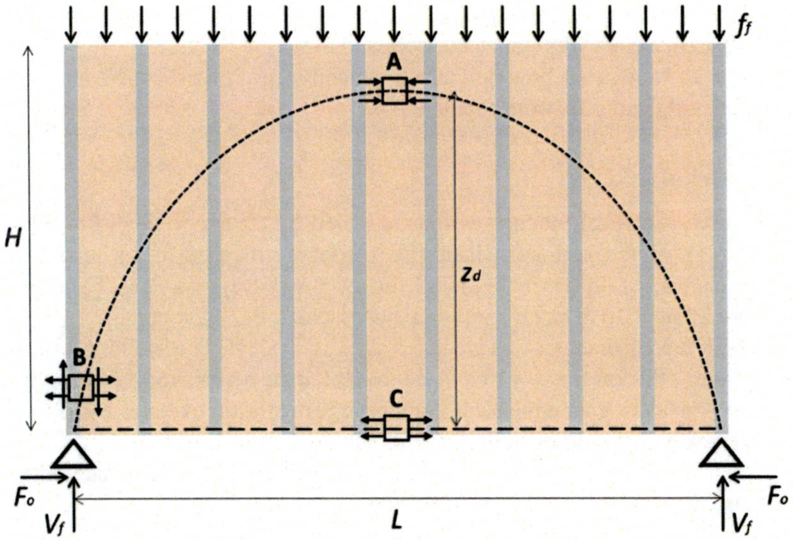


**Fig. 2.** In-plane tied arch mechanism and possible failure mechanisms (adapted from Marini et al. 2022)

$$V_{fA} = 4\left(\frac{z_d}{L}\right)\min\left[f_{c,b\perp}\left(A_b + \frac{E_c}{E_b}A_c\right); f_c\left(A_b + \frac{E_b}{E_c}A_b\right)\right] \tag{2}$$

$$V_{fB} = \min\left[\frac{\tau_{ud}H}{\beta}\left(t_{eq} + \frac{G_c}{G_b}t_c\right); \frac{\tau_{cd}H}{\beta}\left(t_c + \frac{G_b}{G_c}t_{eq}\right)\right] \tag{3}$$

$$V_{fC} = 4\left(\frac{z_d}{L}\right)\max\left[f_t\left(\alpha_1 A_c' + \frac{E_s}{E_c}\alpha_2 A_s + \frac{E_s}{E_c}A_{Tie}\right); f_{yd}\alpha_2 A_s + f_{ydTie}A_{Tie}\right] \tag{4}$$

where: $f_{c,b\perp}$and $\tau_{ud}$ are the design ultimate compressive strength in the block weakest direction and shear strength; $E_b$ and $G_b$ are the elastic and shear modulus of the blocks; $A_b = t_{eq}b^*$ is the resisting area of the blocks considering a homogenous equivalent brick slab with thickness $t_{eq}$ and width $b^*$ of the arch. $f_c$, $f_t$, and $\tau_{cd}$ are the design ultimate compressive, tensile and shear strength of the concrete; $E_c$ and $G_c$ are the elastic and shear modulus of the concrete; $A_c = t_c b^*$ is the resisting area of the concrete topping slab with thickness $t_c$ and width $b^*$ and $A_c'$ is the concrete area of the edge beam. $F_{yd}$ and $f_{ydTie}$ are the steel design tensile strength of the reinforcement and of an eventual additional steel tie; $E_s$ is the elastic modulus of the steel; $A_s$ and $A_{Tie}$ are the area of the continuous longitudinal rebars in the edge beam and of the tie. Finally, $H$ is the depth of the floor; $\beta$ is a coefficient accounting for the shear stress distribution ($\beta = 1.3 - 2.0$); $\alpha_1$ and $\alpha_2$ are reduction factors, which, for instance, allow to consider the combined effects of the seismic tensile axial force and the bending moment due to gravity loads.

It should be noted that, differently from other studies which investigate the role of diaphragm flexibility (Pecce et al. 2019), this model considers for the diaphragm a stiff behaviour up to a brittle failure. When structures in the as-is condition are analysed, the model may thus be applied just in case of regular plan without great openings; however, when a LFRS is added, it may be adopted to define the position of the additional elements of the system in order to regularize the floor and avoid the collapse.

In order to validate such model, numerical and in situ experimental analyses were carried out, and the results were compared.

## 2.2 Numerical Analyses

The proposed analytical model was first validated by means of simplified numerical analyses. A 2D nonlinear finite element (FE) model of a diaphragm of a building before and after seismic retrofit was realized with the software Abaqus (v.6.11). The considered floor was a 25 m × 10 m beam-and-block floor lacking the RC topping.

In the FE model, the floor was modelled as an equivalent plate adopting 4-node shell elements with a thickness $t_{eq} = 35$ mm, equivalent to the net thickness of the clay block webs, and considering an equivalent homogeneous inelastic material implementing a smeared crack model to describe the fracture. Material properties are reported in Table 1. In particular, the tensile strength of the plate was calibrated to represent the shear failure of the system at the arch supports (Mechanism B) and was calibrated by means of laboratory tests on a floor subassembly ($\tau = 1.7$ MPa, Marini et al. 2022). Values of $f_{t,e} = 0.5 - 1.0 - 2.0$ MPa correspond to shear strength $\tau = 0.5 - 0.9 - 1.7$ MPa. The equivalent compressive strength represents instead the compressive strength of the block in the weakest direction (Mechanism A).

The edge beam was modelled with 2 nodes truss elements, considering an area $A_c = 0.6$ m × 0.3 m and an amount of continuous steel reinforcement able to develop the tie action equal to 0.2%$A_c$.

As for the LFRS, a 5 m × 5 m column grid, modelled as springs with stiffness equal to 4.5 kN/mm, has been considered in the as-is condition; while for the retrofitted configuration, two springs with stiffness equal to 162 kN/mm were placed at the bottom ends of the floor, thus simulating an additional shell system or 2 walls aligned with the short façades of the building. More about the model may be found in Marini et al. (2022).

**Table 1.** FE model material properties

| Equivalent plate elements | |
|---|---|
| Elastic modulus, $E_{c,e}$ | 23000 MPa |
| Poisson coefficient, $\nu_{c,e}$ | 0.2 |
| Compressive strength, $f_{c,e}$ | 8 MPa |
| Tensile strength, $f_{t,e}$ | 0.5–1.0–2.0 MPa |
| Concrete damage plasticity | |
| Dilation angle, $\varphi$ | 31 |
| $f_{bo}/f_{co}$ | 1.0 |
| $K$ | 0.667 |
| Viscosity parameter | 0.0004 |
| RC edge beam | |
| Elastic modulus, $E_c$ | 30000 MPa |
| Poisson coefficient, $\nu_c$ | 0.2 |
| Compressive strength, $f_c$ | 20 MPa |
| Tensile strength, $f_t$ | 0.75 MPa |

In Fig. 3, the plots of the compression principal stresses for the building in the as-is situation and after the retrofit are reported. The development of the tied-arch mechanism between the element of the LFRS may be noted (red arrows in figure).

From this figure, it becomes clear how the introduction of the retrofitting LFRS may be critical for the floor diaphragms. In particular, the addition of elements with high stiffness (about 2 to 3 times the initial stiffness of the building, according to Labò et al. 2020, among others) increases the inertial forces on the floor and, contextually, reduces the capacity of the diaphragm by increasing the arch span $L$ (Fig. 2).

A parametric analysis was then carried out in order to evaluate the influence of significant parameters for the retrofitted building. In particular, the tensile strength of the plate (i.e. its shear capacity) was varied according to Table 1; the influence of the columns was checked by considering or disregarding their contribution; finally, an additional steel tie was introduced ($A_{Tie} = 15$ cm$^2$) to evaluate its efficiency in increasing the ultimate in-plane capacity of the diaphragm in the retrofitted building configuration (Fig. 4).

From the analyses, it may be observed that, in general, Mechanism A is rarely activated, while Mechanisms B and C are more critical. However, while Mechanism C may be easily avoided by introducing an additional steel tie (as modelled in Fig. 4, b),

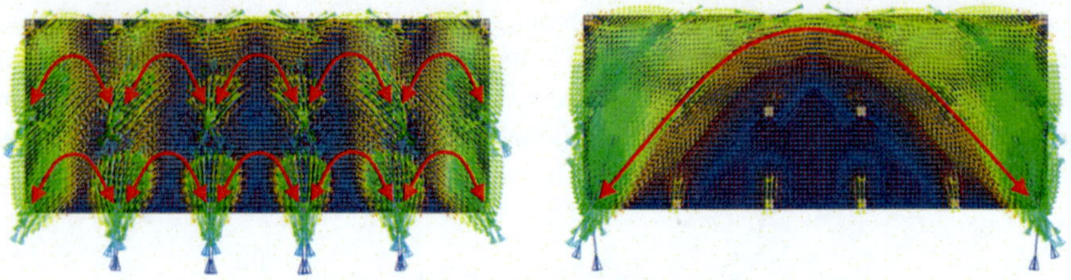

(a) (b)

**Fig. 3.** Flow of the compression principal stresses in the as-is (a) and retrofitted (b) configurations (adapted from Marini et al. 2022)

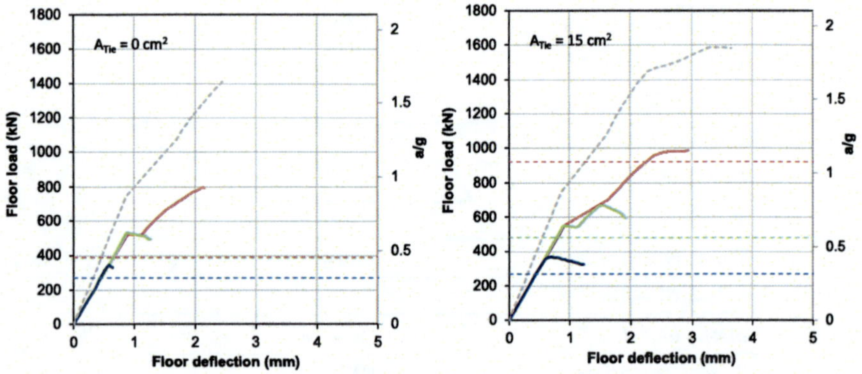


(a) (b)

**Fig. 4.** Floor load and mid-span net deflection for $\tau = 0.5 - 0.9 - 1.7$ MPa without columns (blue, green, red lines) and for $\tau = 1.7$ MPa with columns (grey line) considering $A_{Tie} = 0$ (a) and $= 15\ \text{cm}^2$ (b). Horizontal dashed lines are the ultimate capacity calculated with Eq. 1–4 (adapted from Marini et al. 2022)

Mechanism B, which depends on the shear capacity $\tau$ of the floor at the supports, may not be solved without strengthening the entire floor. Mechanism B thus represent the ultimate capacity of a beam and block diaphragm (as shown by Table 2 and by the blue curves of Fig. 4, which correspond to $\tau = 0.5$ MPa, and which do not change even when the additional steel tie is introduced).

**Table 2.** Collapse mechanism for different properties of the diaphragm

| | $\tau = 0.5$ MPa | $\tau = 0.9$ MPa | $\tau = 1.7$ MPa |
|---|---|---|---|
| $A_s = 0\ \text{cm}^2$ | B | C | C |
| $A_s = 5\ \text{cm}^2$ | B | B | C |
| $A_s = 25\ \text{cm}^2$ | B | B | B |
| $A_s = 20\ \text{cm}^2$ | B | B | B |

The comparison between the numerical curves and the estimation of the ultimate capacity with the analytical model by adopting Eq. 1–4 (horizontal dashed lines in Fig. 4) are in good agreement. Considering these preliminary results, it may thus be stated that calculating the diaphragm capacity with Eq. 1–4 provides an estimation of the capacity which is on the safe side.

As for the flexibility of the beam-and block floor systems, the maximum floor net deflection at the diaphragm's collapse was also evaluated. For all the investigated cases, deflections to span ratios smaller than 1/6000 were obtained, thus validating the hypothesis of stiff behaviour for the system.

### 2.3 Experimental Analyses

In order to validate the adoption of the tied-arched model for the estimation of the in-plane capacity of diaphragms, an in-field experimental campaign on an existing beam-and-block floor was also carried out. The aim of the research was to evaluate the actual capacity of the floor and its capability to redistribute an in-plane load, such as the seismic reaction force in correspondence of a new shear wall, by developing a tied-arch mechanism (Fig. 5).

The in-field test was carried out on a portion of a floor of an abandoned building built in 1960. The floor was composed by RC joists with base 12 cm and height 20 cm and clay blocks 44 cm large, having 5 and 4 horizontal clay webs with 10 mm thickness in the lateral and central sections, respectively ($t_{eq}$ = 50 mm and 40 mm). The floor section was completed with an unreinforced concrete topping with average thickness $t_c$ = 15 mm and a good-quality plaster layer on the bottom with thickness of 20 mm. The properties of the materials are reported in Table 3.

**Table 3.** Material properties of the experimental specimen

| Properties | Concrete | Clay | Plaster |
|---|---|---|---|
| E [MPa] | 36000 | 15000* | 8000 |
| ν [-] | 0.2* | 0.1* | 0.24* |
| G [MPa] | 15000 | 6818 | 3226 |
| t [mm] | 200 | 40–50 | 20 |
| $\sigma_{lim}$ [MPa] | 35 | 18* | 7 |
| $\tau_{lim}$ [MPa] | 3.5 | 1.6–1.9 | 0.7 |

*Reference values from literature

A strip composed by 3 joists and the interposed 2 rows of blocks were isolated, considering 4 blocks in each row. An in-plane point load F was then applied at the centroid axis of the central joist with a hydraulic jack, and the displacements at the beginning and at the end of the central joists (D1, D2) and the strains in the central portion of the joist (E1 to E4) were measured by means of LVDTs and strain gauges, respectively (Fig. 6).

Two tests were carried out on the floor: with and without topping. In the present paper, the sole results of the floor without topping are briefly presented. More may be found in Casprini et al. (2020 and 2023).

In Fig. 6, the axial strains $\varepsilon$ in the central joist are plotted in relation to the applied point load $F$. It may be noted that, in correspondence to the same applied load, the level of strain decreased from segment 1 (the closest to the load application point) to segment 4 (the farthest from the application point), and that an almost regular spacing is detected between the four measurements. This entails that the joist is able to distribute almost uniformly the load among all the rows of blocks.

The same may be stated by observing the crack pattern at collapse in Fig. 7. Various inclined struts developed, spanning from the central to the side joists, and the shear component of the inclined forces led then to the failure of the clay blocks at the interface between the lateral and central sections of the blocks, where the change of the number of the horizontal webs occurred, as shown by the propagation of the main crack in the right-hand panel.

This experimental test allowed to confirm the ability of beam-and-block floor systems to develop a tied-arch resistant mechanism, even when lacking the RC topping. Considering the displacements recorded at collapse, the test validated also the stiff-brittle behaviour of the system up to collapse. Finally, the test was also able to confirm that the most vulnerable failure of this floor system is the shear failure of clay blocks (Mechanism B), which happened for an average shear stress $\tau = 1.6 - 1.9$ MPa, confirming the value obtained with the laboratory test ($\tau = 1.7$ MPa, Marini et al. 2022) also adopted in the numerical analyses.

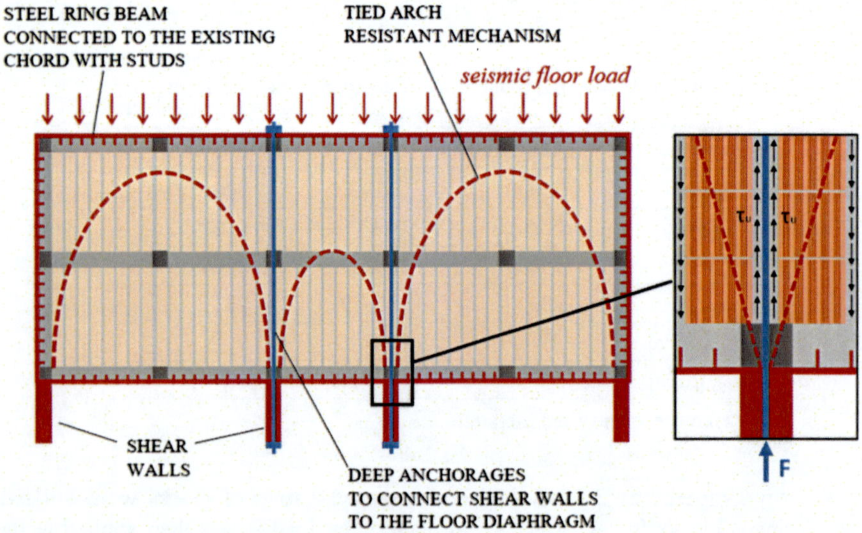


**Fig. 5.** Tied-arched resistant mechanism in case of introduction of 4 shear walls and reaction force diffusion in a beam-and-block floor system (adapted from Casprini et al. 2023)

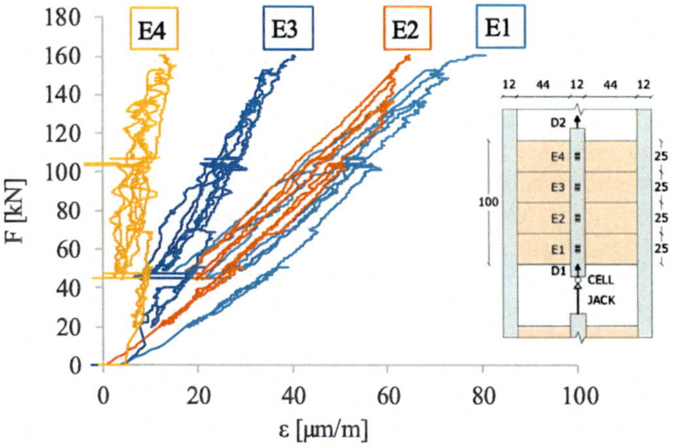


**Fig. 6.** Plot of the axial strains ε in the central joist in relation to the applied point load F and test instrument layout (adapted from Casprini et al. 2023)

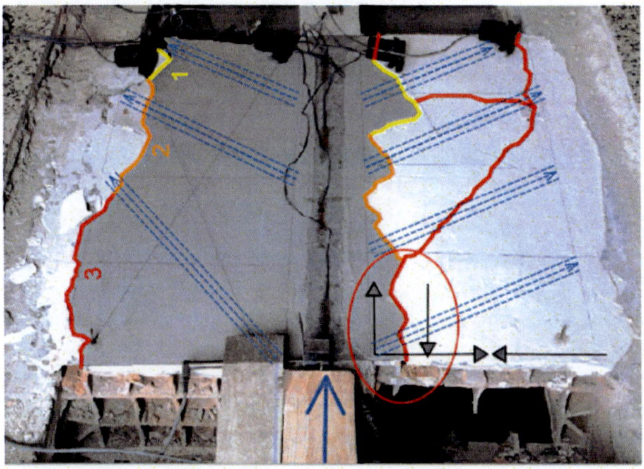

**Fig. 7.** Crack pattern at collapse and hypothesis of formation of struts in the single blocks (adapted from Casprini et al. 2023)

## 3 Design Implications

In case of earthquake, the role of seismic diaphragms is essential to guarantee the integrity of the structure and a good behaviour of the LFRS. The evaluation of the in-plane capacity of diaphragms should thus be evaluated both for the seismic vulnerability assessment of a building and for the conceptual design of the retrofit intervention. However, as previously discussed, the in-plane capacity of diaphragm is usually not an issue for the building in the as-is condition, which usually collapse for the activation of more vulnerable mechanisms. The evaluation of the ultimate capacity of existing floors is instead critical when stiff retrofitting elements are added to the structure.

The in-plane capacity of the existing beam-and-block floor systems may thus influence the design of the retrofit intervention in two different ways. When the position of the elements of the additional LFRS are already determined, the evaluation of the diaphragm capacity may determine whether the floors need to be strengthened, depending on the floor geometry and on the seismic action. Possible strengthening techniques for the diaphragms, even alternative to the traditional cast-in-place RC slab, which is not compatible with an intervention from outside, are presented in Fig. 8 (Marini et al. 2022). In addition, considering a tied-arched mechanism, the possible need of additional steel ties along the floor perimeter or deep anchorages to either act as a tie for the arches or connect shear walls to the diaphragms (as in Fig. 5) may be highlighted.

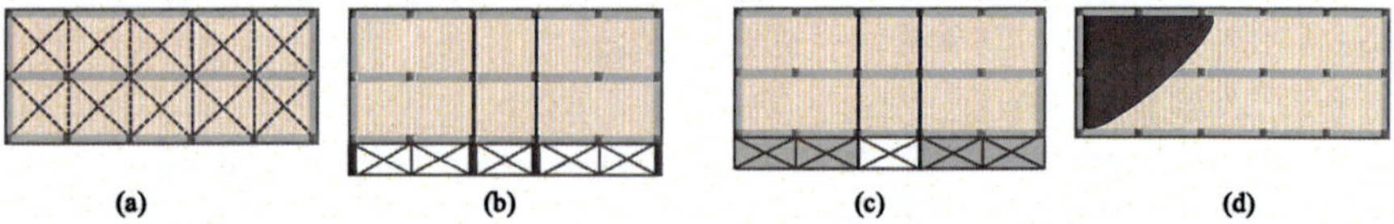


**Fig. 8.** Possible floor strengthening techniques: (a) intrados steel trusswork (concealable with false ceilings), (b) external diaphragms between adjacent walls, (c) external diaphragms by reengineering existing balconies, (d) RC overlay (adapted from Marini et al. 2022)

When the position of the elements of the additional LFRS are not yet given, the in-plane capacity of the diaphragms may enable determining the maximum spaces between adjacent elements of the new system ($L_{sw,max}$) (Eq. 5) with the aim of avoiding the floor strengthening. Indeed, by limiting the span of the arches, the ultimate floor capacity increases.

$$L_{sw,\max} = \frac{2V_{fd}}{f_{f,d}} \tag{5}$$

### 3.1 Example

In order to clarify the design implication of in-plane floor capacity, an example is here proposed with reference to an existing RC building built in 1976 and having a very irregular in-plane shape (Fig. 9).

In the as-is situation, the building presents major vulnerabilities that lead to its collapse, and the floors do not represent an issue. However, in this case, the tied-arched model cannot be applied due to the irregularity of the floors, and the flexibility of the diaphragm needs to be considered in the FE model.

On the other hand, when the additional LFRS of Fig. 9, consisting in a discrete number of shear walls in each direction, is applied to the structure, the maximum in-plane capacity of the floor should be estimated.

In this case, the building floor can be divided in a subassembly of different regular units (numbered in figure), each one having stiff seismic resistant walls on each side. With such a configuration, the tied-arch model may be applied to each unit. The floors may be modelled as rigid diaphragm in the FE model and may be verified ex-post by

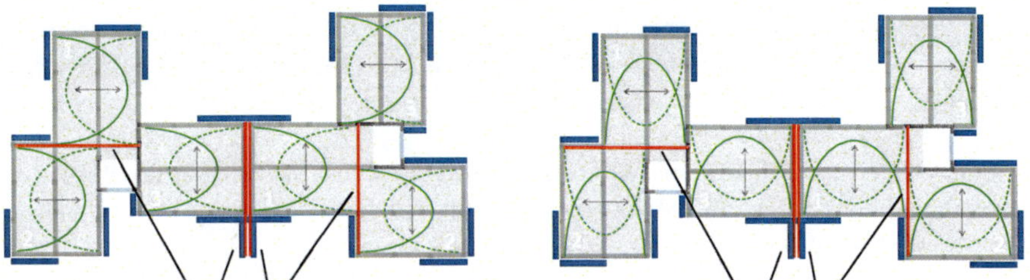


**Fig. 9.** In plane layout of the tied-arch mechanisms after the introduction of a new LFRS in an existing RC building considering an earthquake in X (a) and Y (b) direction

comparing the reaction in the walls at each floor and the ultimate capacity $V_f$ of the floor, (Passoni et al. 2020; Labò et al. 2020). The ultimate capacity may be calculated as the minimum force triggering mechanisms A, B, and C, by means of Eq. 1 to 4. In addition, the need to introduce additional ties (in red in Fig. 9) to equilibrate the arches in both earthquake directions and verses may be immediately defined from the observation of the in-plane layout of the tied-arched mechanisms.

## 4 Concluding Remarks

The renovation of existing RC buildings to ensure sustainability, safety and resilience is a priority. When seismic retrofit interventions are applied to existing structures, the diaphragms play a major role in the response of the building against earthquakes. Even though the ultimate capacity of existing floors does not usually represent a great vulnerability for existing building, it may become so for those buildings retrofitted by introducing a new stiff LFRS.

In this paper, the main findings of a major research investigating the in-plane capacity of beam-and-clay block floor systems are presented. In particular, the possibility of adopting an analytical tied-arch model to estimate the ultimate capacity of existing floors is introduced and validated by means of numerical and experimental analyses. Such analyses validated:

- the development of a tied-arch resisting mechanism for floors with regular geometry and without great openings, even when lacking RC topping;
- a stiff behaviour of the diaphragm up to a brittle failure in correspondence of very limited lateral deflections;
- the vulnerability of Mechanism B (combined compression and shear failure at the supports, usually connected to the shear failure of the blocks).

The implications of floor capacity in the design of seismic retrofit interventions were finally outlined, even with reference to a practical example.

Future development of this research implies the study of the connections between the existing floors and the elements of the additional LFRS, and the role of the 'transfer forces' (i.e. internal actions that develop in the floors consequently to displacement incompatibilities or discontinuities in elevation of the elements of the LFRS) in the estimation of the internal load path and of the diaphragm capacity.

**Acknowledgements.** This research was partly carried out within the Italian national research project ReLUISDPC 2019–2021. Authors acknowledge Prof. Ezio Giuriani and Prof. Paolo Riva for their contribution in the study.

## References

Bull, D.K.: Understanding the complexities of designing diaphragms in buildings for earthquakes. Bull. N. Z. Soc. Earthq. Eng. **37**(2), 70–88 (2004)

Casprini, E., Passoni, C., Marini, A., Belleri, A., Giuriani, E.: In-plane capacity of beam and block floor systems: an in-field experimental study. In: 17th International Brick and Block Masonry Conference (IB2MaC) from historical to sustainable masonry, 5–8 July 2020, Krakow, Poland (2020)

Casprini, E., Passoni, C., Belleri, A., Marini, A.: In plane capacity of RC beam-and-clay-block floor systems: in-situ experimental tests. Bull. Earthq. Eng. (2023). Phase of submission to

Fleischman, R.B., Naito, C.J., Restrepo, J., Sause, R., Ghosh, S.K.: Seismic design methodology for precast concrete diaphragms Part 1: design framework. PCI J. (2005). https://doi.org/10.15554/pcij.09012005.68.83

Labò, S., Passoni, C., Marini, A., Belleri, A.: Design of diagrid exoskeletons for the retrofit of existing RC buildings. Eng. Struct. **220**, 110899 (2020). https://doi.org/10.1016/j.EngStruct.2020.110899

Marini, A., et al.: Combining seismic retrofit with energy refurbishment for the sustainable renovation of RC buildings: a proof of concept. Eur. J. Environ. Civ. Eng. (2017). https://doi.org/10.1080/19648189.2017.1363665

Marini, A., Belleri, A., Passoni, C., Feroldi, F., Giuriani, E.: In-plane capacity of existing post-WWII beam-and-clay block floor systems. Bull. Earthq. Eng. **20**(3), 1655–1683 (2022). https://doi.org/10.1007/s10518-021-01301-y

Moehle, J.P., Hooper, J.D., Kelly, D.J., Meyer, T.R.: Seismic design of cast-in-place concrete diaphragms, chords, and collectors: a guide for practicing engineers. NEHRP Seismic design technical brief no. 3, produced by the NEHRP Consultants Joint Venture, a partnership of the Applied Technology Council and the Consortium of Universities for Research in Earthquake Engineering, for the National Institute of Standards and Technology, Gaithersburg, MD, NIST GCR 10–917–4 (2010)

Passoni, C., Guo, J., Christopoulos, C., Marini, A., Riva, P.: Design of dissipative and elastic high-strength exoskeleton solutions for sustainable seismic upgrades of existing RC buildings. Eng. Struct. **221**, 111057 (2020). https://doi.org/10.1016/j.engstruct.2020.111057

Passoni, C., Marini, A., Belleri, A., Menna, C.: Redefining the concept of sustainable renovation of buildings: state of the art and an LCT-based design framework. Sustain. Cities Soc. **64**, 102519 (2021). https://doi.org/10.1016/j.scs.2020.102519

Pecce, M.R., Ceroni, F., Maddaloni, G.: In-plane deformability of RC floors: assessment of the main parameters and influence on dynamic behavior. Bull. Earthq. Eng. **17**, 297–311 (2019)

# The Tower of Cable Stayed Railway Bridge Over Anji River

Mario Paolo Petrangeli[1](✉), Roberto Di Bianco[2], and Andrea Polastri[1]

[1] Mario Petrangeli and Ass. Srl, Rome, Italy
info@mpaing.it
[2] ITALFERR SpA Gruppo FS Italiane, Rome, Italy

**Abstract.** In the high seismic hazard region of Kashmire, close to the Himalayan mountains, an important cable stayed railway bridge, with a main span of 290 m crossing the Anji valley 190 m above the river bed, is under construction. Due to the particular orography of the valley, the bridge shall have an asymmetric structural scheme, with a 200 m high single tower and 20 m diameter well foundation. The main features of the already completed tower will be illustrated more specifically with reference to heavily stressed post-tensioned transition pieces..

**Keywords:** well foundation · self compacting concrete · post-tensioning · transition pieces · stresses spreading · stay cables

## 1 Introduction

Northern Railways of India is managing the construction of an approximately 300 km long railway line in the J&K State, India. This line has some segments with only one track and shall link Udhampur-Srinagar-Baramulla.

The very difficult morphology of the area implies to bore many tunnels with a total length of about 90 km, and the construction of important bridges. One of these crosses the Anji River 190 m over the bottom of the valley.

A steel truss arch, spanning 260 m, was the initial structural solution; some piers of the approach viaduct were also erected.

Anyway, the construction was stopped immediately after the beginning because of the difficulties faced to cut the sub-vertical slope on the left side of the valley. The extremely fractured and sheared rock was difficult to stabilize also with expensive works.

For this reason, IRCON International Limited (a Government of India Undertaking under the Ministry of Railways) called for an international bid to prepare a new design and to supervise the construction works.

The tender was assigned to the Italian Company ITALFERR with LoA dated April 2016, while the subsequent tender for Proof Checking was assigned to the Company COWI (UK).


M. A. Aiello and A. Bilotta (Eds.): ICC 2022, LNCE 435, pp. 325–342, 2024.
https://doi.org/10.1007/978-3-031-43102-9_26

A new tender for construction was called on October 2016 based on a new Preliminary Design developed by Italferr, then assigned to the Indian Contractor HCC.

The Construction works started on February 2017 and are currently in progress supervised by Konkan Railway Corporation Limited (Fig. 1).

**Fig. 1.** View from Reasì side (right side) of the construction of the Anjikhad Bridge

## 2 Bridge's Main Features

### 2.1 Key Considerations

The situation of the site concerning both orography and geological/geotechnical aspects can be summarized by the following points:

- The first layers of the rock are highly fractured (with many "shear zones") on both sides of the valley;
- Any cut in these layers can cause land slide, as the service road built in the past shows; expensive preliminary works, lasting long time, should be requested to avoid the mentioned slides;
- The deep slope on the left (Katra) side is very inclined from the bottom up to the end of the future tunnel T2 (now completed), while on the right (Reasi) side, there is a flat area, about 350 m long, between the slope of the river and the beginning of the already built tunnel T3.

As a consequence of these points, a solution that avoids any cut (or able to minimize the cuts) on the Katra side slope was preferred to any other.

The consolidation of very fractured rock is always possible but, cost apart, it expresses has has incertitude that forces to foresee long time scheduled works.

On the basis of the existing orography and geotechnical characteristics of the site, all the workshops, batching plant and so on, had to be located at Reasi side, no room being available on Katra side.

The bridge has been subdivided in three WBS:

1. a 130 m long approach viaduct (called "Ancillary") on Reasi side;
2. a 473 m long cable-stayed bridge, crossing the deep valley, with a main span of 290 m;
3. a central embankment, located between the main bridge and the approach (ancillary) viaduct.

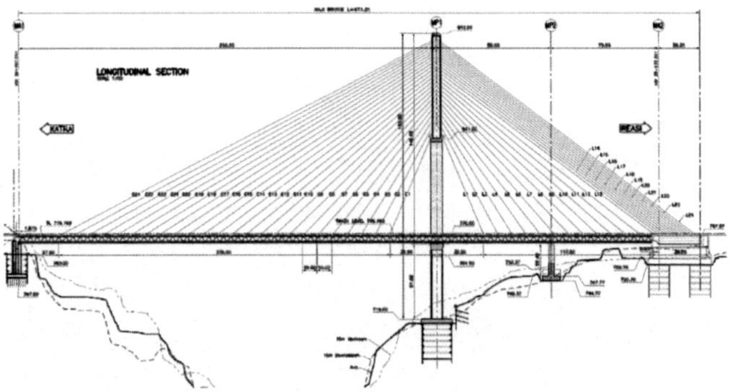

**Fig. 2.** Longitudinal profile of the cable-stayed bridge

In the final configuration, the central embankment will result in being wider than the bridge deck in order to have the necessary room for auxiliary equipment or vehicles road devoted to maintenance and emergency conditions.

It shall be also used as an access road to the railway line, coming from the already built service road from Reasi.

Hereafter the presentation of a short description of the main bridge, both the ancillary viaduct and the embankment being of ordinary features.

## 2.2 The Cable-Stayed Bridge

Due to the above considerations, the use of an asymmetric scheme of the bridge was compulsory. Different solutions were compared before the develop of the Preliminary design and finally a cable stayed bridge, with only one tower placed on Reasi side, in a position where the disturb to the slope is reduced, has been adopted (see Fig. 2).

In order to limit as much as possible the cutting, a well foundation has been adopted for the tower. So doing, it has been possible to reach the sound strata of the rock without disturbing the slopes. Moreover, it has been possible to check "de visu" the depth necessary to get this target; with a 20 m in diameter, the well excavation, temporary sustained by $\phi$ 300 mm micropiles and concrete ribs spaced @ 4 m before the final concrete pouring, has reached 20 m in deepness (Fig. 3).

**Fig. 3.** Tower well foundation during the excavation

Two steel trusses of 5 m constant height, connected by cross girders that support a concrete slab, compose the deck (see Fig. 4). The choice of a composite bridge section (steel and concrete) is considered convenient: the deck of the bridge in reinforced concrete assures a high resistance to the environmental actions (wind, rain, ice) reducing maintenance interventions; the steel trusses guarantee light dead load compared to a high level of resistance. The concrete slab collaborates with the steel elements, the continuity being assured by steel connectors, which are very common and tested all around the world. Besides, the global section is considered a box girder, so it has a very good torsional stiffness.

The bridge deck is 15 m wide; it is designed to carry a single railway track and a 3.75 m wide vehicles road (for maintenance and emergency conditions).

**Fig. 4.** The steel truss girder

## 2.3 Design Criteria

Design has been based on Indian Codes ref. [1] integrated by Eurocodes, where necessary ref. [2–6].

The design speed of the line is 100 km/h, limit that does not pose problems for the train-structure interaction.

The area is classified as seismic with a PGA of 0,17 g for the Service Limit State analysis, and 0,27 g for the Ultimate Limit State.

Because of the high flexibility of this type of bridges, the seismic analysis was carried out with the elastic spectrum defined by the Indian Code, without any reduction, as prescribed by the Eurocode 8.

The first flexural frequency is 0,40 Hz; the first torsional is 1,05 Hz.

The following design assumptions guarantee the "robustness" of the bridge:

- Two consecutive stays missing: the bridge remains in service for transit of trains at a limited speed (30 km/h) and a reduced weight (50%);
- Three consecutive stays missing: no collapse of the bridge under the permanent loads;

- Explosion of 40 kg (TNT equivalent) on the deck: no collapse of the bridge under the permanent loads and possible quick repair with limited cost.

## 2.4 Construction Method

The steel trusses, subdivided in ten meters long segments, are carried in a workshop located on the central embankment; here a segment of the full deck, without the slab, is assembled adding the cross beams and horizontal bracings.

Nowadays the entire tower and the first phase of incremental deck launching of the rear span plus the first 50 m of main span, with a total length of 196 m (see Fig. 4 & 5), have been completed including the concreting of the slab.

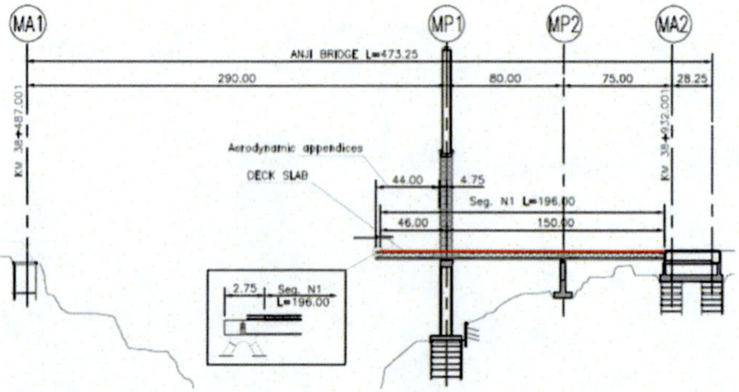


**Fig. 5.** Up to date execution

After the installing of a launching gantry on the bridge tip (see Fig. 6), the remaining 240 m of the main span construction will go on by cantilevering, launching 10 m long segments and installing the stay cables until landing on the MA1 Katra abutment.

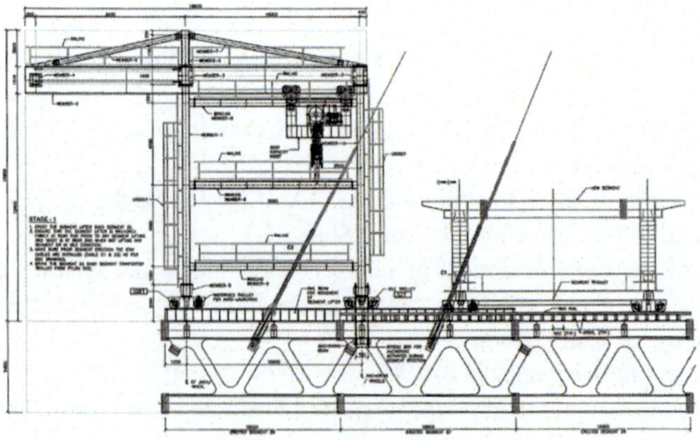

**Fig. 6.** Launching gantry

The casting of the slab, with a total length of 437 m, will be executed in 3 macro-segments; the first one has been just completed (as above mentioned). The other two segments, with a length of 100 and 141 m, will be casted during specific phases of the construction in order to optimize the favorable compressive stresses produced by the stay cables tensioning and retensioning.

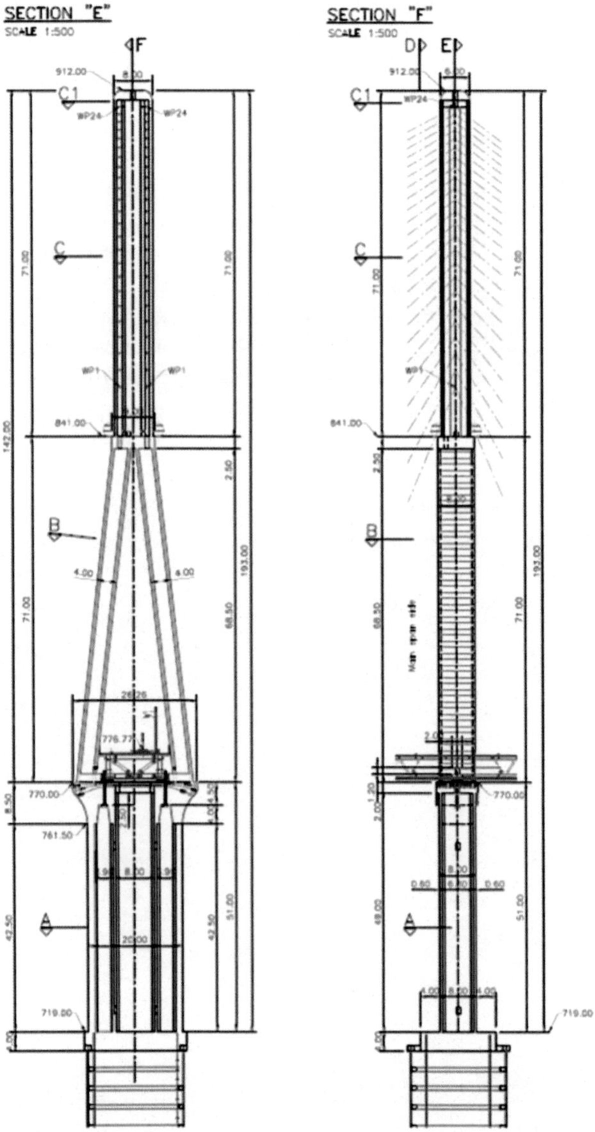


**Fig. 7.** The main tower – cross and longitudinal sections

## 3 The Main Tower

The main tower, the so called MP1, with a total high of 193 m, has been completed few months ago (Fig. 7).

The tower can be subdivided in five sub-elements that are, the lower part, the pier cap, the inclined legs, the upper transition piece, and the upper tower, the so called "antenna", where the $24 \times 4 = 96$ stay cables will be anchored; all the tower elevation, with the only exception of the inclined legs, has been executed using self-compacting concrete.

### 3.1 The Lower Part of the Tower

The lower part of the tower, from the well cap level at 719 m up to the bottom of the pier cap, has a total height of 42.5 m and has a reinforced concrete three cellular box section. The section is constant along the height with external dimensions that are $20 \times 8$ m along the cross and longitudinal direction, respectively (Fig. 8).

The pier walls are generally 60 cm in thickness with the only exception for the external longitudinal walls that are enlarged up to 150 cm due to compressive strength requirements, especially in the upper part right below the pier cap (Fig. 9).

### 3.2 The Pier Cap

The first transition piece, the pier cap, is located at level 770 m; the dimension is a little more than 26 m along the transversal direction and 9 m longitudinally.

The total depth of the transition element is 8.5 m with 2 m minimum thickness of the top slab in the central part (see Fig. 10).

**Fig. 8.** Execution of the main tower and deck launching

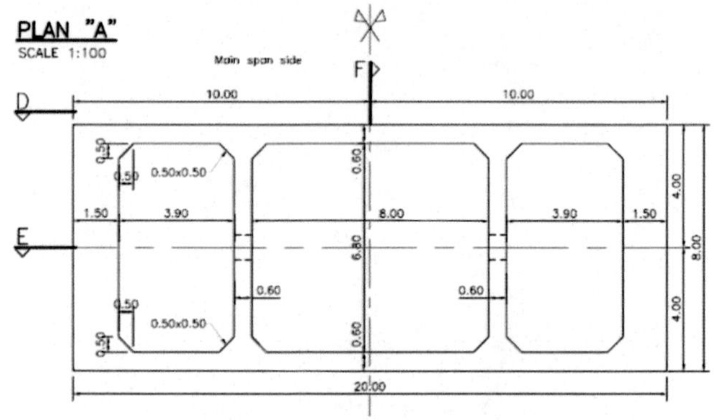


Fig. 9. Horizontal section of the tower lower part

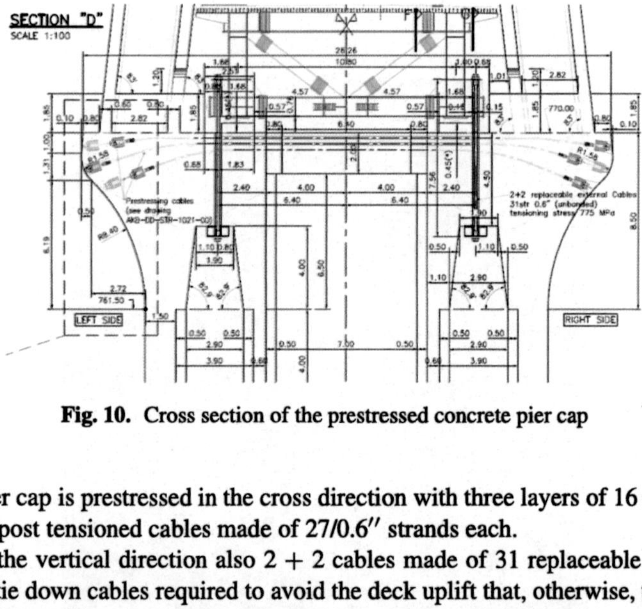


Fig. 10. Cross section of the prestressed concrete pier cap

The pier cap is prestressed in the cross direction with three layers of 16 cables for a total of 48 post tensioned cables made of 27/0.6″ strands each.

Along the vertical direction also 2 + 2 cables made of 31 replaceable strands are present as tie down cables required to avoid the deck uplift that, otherwise, the extreme seismic actions of the site would produce.

The vertical compression forces spread from the inclined legs, through the pier cap, toward the lower part of the tower producing high levels of compressive tensions and transversal horizontal tension forces that are absorbed by the post-tensioning cables above described.

Analysis for checking the maximum compressive tensions and for dimensioning the post tensioning system has been carried out using both a strut and tie approach (see Fig. 11) and a 3D solid finite elements model (see Fig. 12).

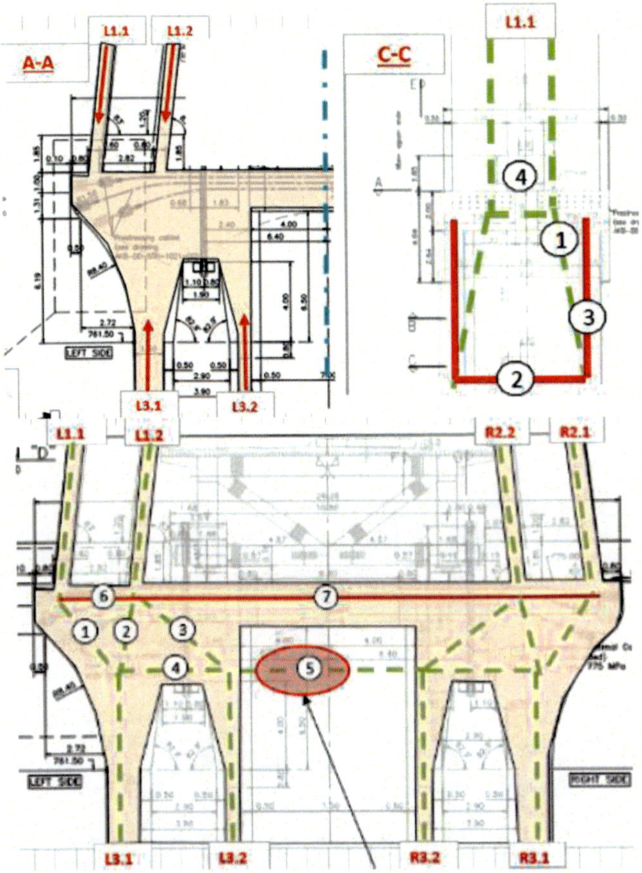


**Fig. 11.** Strut and tie model

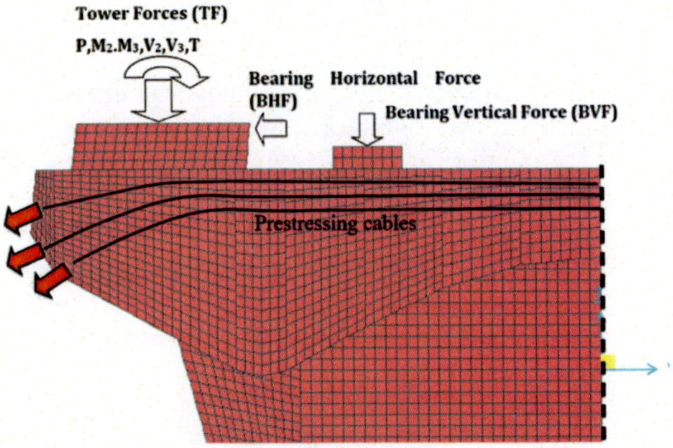


**Fig. 12.** 3D FEM

### 3.3 The Inclined Legs of the Inverted Y

Starting from pier cap level at 770 m, where the two inclined legs are spaced 16.60 m to allow the deck of passing through the tower, they converge after 68.50 m, right below the upper transition piece, at 838.5 m level.

The inclined legs are made of reinforced concrete with a 8.00 × 4.00 m box section and 60 cm thick walls.

### 3.4 The Upper Transition Piece

The transition of internal forces from the upper three cellular box section of the tower "antenna" and the two inclined legs occurs throughout a prestressed concrete 2.5 m deep slab (Fig. 13).

**Fig. 13.** Upper transition piece

This transition slab is prestressed along the longitudinal direction with 14 cables with 27/06″ strands each and with 12 cables of the same size along the cross direction (see Fig. 14).

Also in this case, due to the high level of stress, the analysis for checking the maximum compressive tension and for dimensioning the post tensioning system has been carried out using both a strut and tie approach and with a local 3D solid finite elements model (see Fig. 15): this figure represents the transition element, the prestressing system and a significant length of the upper antenna and the lower inclined legs of the tower.

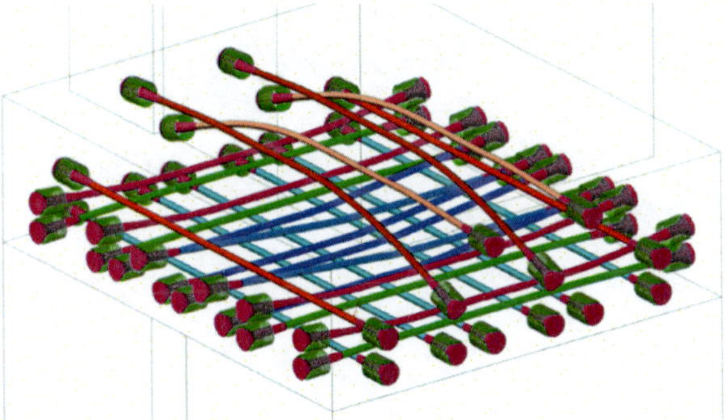

**Fig. 14.** 3D drawing of post-tensioning system of upper transition piece

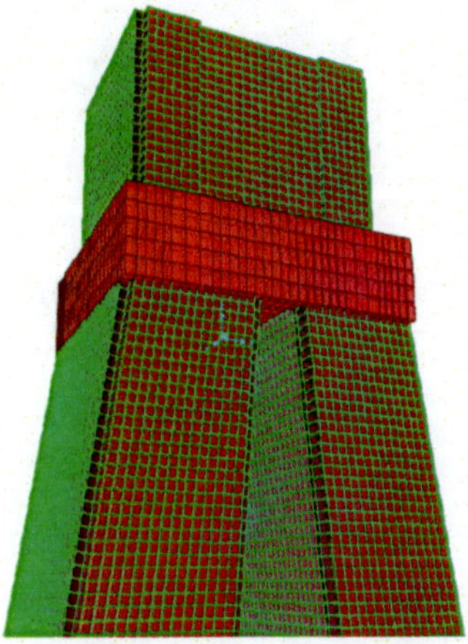

**Fig. 15.** 3D model of upper transition piece

### 3.5 The Upper Tower

The last section of the tower, the so called "antenna", with a total length of 71 m, starts at level 841 m and reaches level 912 m, at the top of the tower.

The section is a three cellular box with an outer dimension that is 6 × 8 m, along the longitudinal and transversal direction, respectively.

The stay cables will be tensioned from the top bearing so that, the tower anchor chambers require of having a sufficient dimension for operating with the tensioning equipment.

Two external accesses are present at the bottom and the top of the two chambers in order facilitate the ventilation and the tensioning operations (Fig. 16).

The central chamber, with a net dimension of 5 × 3.4 m, will be equipped with an elevator that will be used for the future inspection and maintenance of the bridge.

The two outer chambers, where the stay cables will be anchored, are made of a composed steel-concrete section; the steel plates of the box work as the link of a chain

**Fig. 16.** View of the tower from the top

that resist to the horizontal traction force **T** produced by forces the opposed stays (**$S_C$** and **$S_L$**) coming from the Central (the Main span) and the Lateral span (see next Fig. 17).

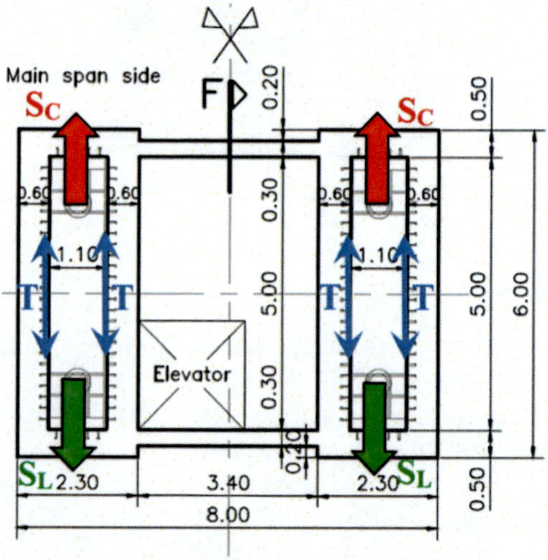


**Fig. 17.** Horizontal section of the upper part of the tower

For each relevant class of stay anchor the evaluation of the stresses for ULS and fatigue resistance check, as well as at SLS, has been executed by means of local 3D FEMs model, as showed in next Fig. 18.

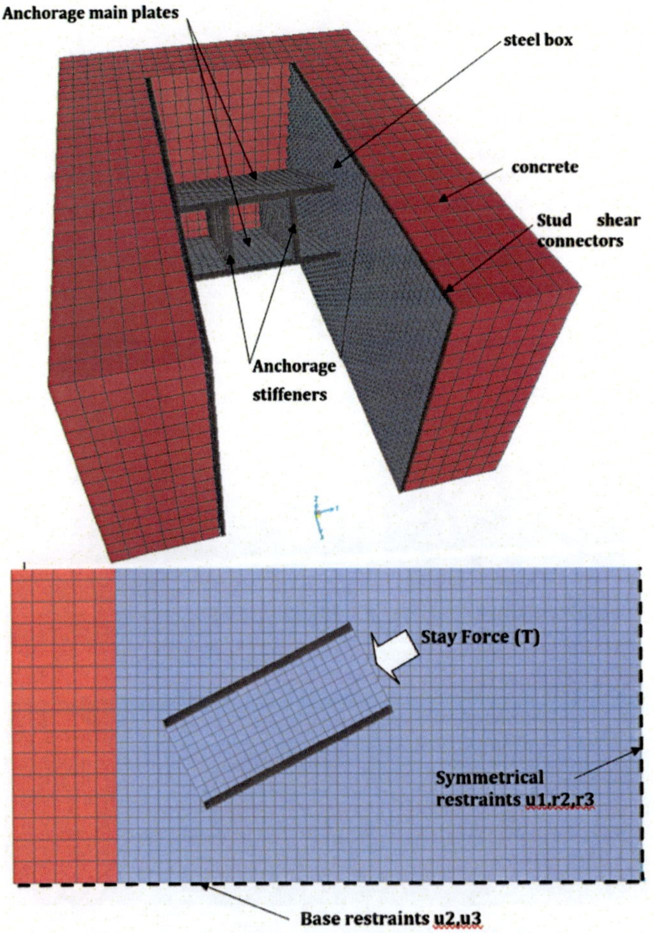


**Fig. 18.** 3D local FEM of composite steel-concrete box section and stay anchor

All the elements of the stay cables anchorages have been dimensioned allowing the possibility of a future 5% increasing of the stay forces (Fig. 19).

## 4 The Stay Cables and the to Be Executed Construction Stages

Each one of the two stay planes of the bridge will contain 24 stays for the Central span and 24 for the Lateral one, with a total number of 96 stays; the stay cable system is provided and installed by VSL. Three stay classes will be adopted with 31, 36 and 43 strands, as a maximum dimension. The stays length varies from a minimum of 83 m for the stay N.1 near the tower up to a maximum of 295 m for the stay N. 24.

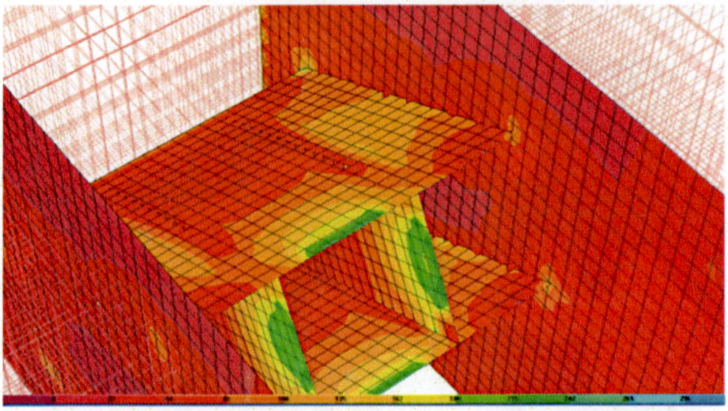

**Fig. 19.** Von Mises stresses under ULS condition for the 43 strands stay anchorage

**Fig. 20.** View of one of the stay guide pipe on deck side

Having nowadays completed the incremental launching of the first 200 m macro segment of the bridge and the slab casting (see Fig. 20), the next step will be the installing and tensioning of stays 1C/1L and subsequently the number 2C/2L, operating simultaneously from the Lateral and Central span.

After the installing of a launching gantry on the bridge tip, the construction will go on with a cantilever incremental launching of 10 m long stayed segments; during this construction phase stays from N.3 up to N.24 will be progressively installed and tensioned operating simultaneously from the central and lateral span. Taking into consideration that two planes of stays are present, for each newly launched segment, four stays will be installed and simultaneously tensioned (Fig. 21).

**Fig. 21.** View of the completed lateral span

After the cantilever incremental construction of the second macro segment of the main span and slab casting (100 m long) a first retensioning, involving stays from 3C/L to 12C/L on each side (40 stays, overall) will be executed. Right after the landing of the deck on abutment MA1 (Katra side), having completed the incremental cantilever construction of the last 141 m macro segment, a general retensioning of all stay cables will be executed; after that the third slab macro segment of the slab will be casted.

With all the structures of the bridge completed, a final recheck of the tension of all the stay cables will be executed using a threaded tube tensioning equipment instead of the mono-strand jack allowed during the construction phases. A recheck of cable tensioning by means of natural frequencies measuring will be also executed in order to correlate the results obtained by the direct measurement and indirect frequencies method. After the finishing works, with the bridge loaded with all the non-structural dead loads, a further final check of stay cable tensioning, using the indirect frequencies method, will be executed.

## 5 Conclusions

The Anji bridge is a typical example of a modern structure where steel and concrete have the same importance, each material being chosen according to its best performance. This fact, together with the possibility of inspecting and replacing the most important parts, make these bridges more and more similar to a "machine".

**Acknowledgements.** Authors of the article would like to thank:
Sh Sandeep Gupta (Northern Railway Chief Engineer).
Sh B. K. Sharma (Northern Railway Deputy Chief Engineer).
Sh R K Hegde (KRCL Executive Director).
Sh Gopal Raju, (KRCL Chief Engineer).

## References


1. Indian standards:- IRS Bridge Rules (Third Reprint-2014 duly incorporating correction slip up to 46 dated 02.05.2014) - Indian Railway Bridge Manual (Latest available update). - IRS Fabrication Specification B1-2001 (Reprint-2008 duly incorporating correction slip up to 4 & further inserted correction slips 5 to 10 dated 22.03.2016).- IRS Steel Bridge Code (Reprint-2003 duly incorporating correction slip up to 17 & further inserted correction slips 18 to 20 dated 16.01.2015). - IRS Bridge Substructure & Foundation Code (Second Revision-2013 & further inserted correction slip 1 dated 17.04.2014). - IRS Concrete Bridge Code (Reprint Sep-2014 duly incorporating correction slip up to 13 & further inserted correction slips 1 to 3 dated 20.01.2015). - IRC:6-2014 Sec II Loads and Stresses (for roadway loads definition). - IITK-RDSO GUIDELINES ON SEISMIC DESIGN OF RAILWAY BRIDGES – 2010
2. BS-EUROCODES 2 - Design of Concrete Structures & British National Annex
3. BS-EUROCODE 3- Design of Steel Structures & British National Annex
4. BS-EUROCODE 4 - Design of composite steel and concrete structures & British National Annex
5. BS-EUROCODE 7 - Geotechnical design & British National Annex
6. BS-EUROCODE 8 - Design of structures for earthquake resistance & British National Annex

# Numerical Simulation of Prestressed Concrete Girders Through Different Modelling Approaches

Simone Galano(✉), Giacomo Miluccio, Daniele Losanno, Fulvio Parisi, and Maria Rosaria Pecce

Department of Structures for Engineering and Architecture, University of Naples Federico II, Naples, Italy
simone.galano@unina.it

**Abstract.** The structural safety assessment of existing infrastructures is a central issue in civil engineering, especially after recent disasters that have hit different Italian bridges. Several data collected showed how existing Italian bridges are mainly composed by simply supported, beam-type, prestressed concrete decks. In this work, the response of a scaled prestressed concrete bridge girder was investigated through experimental test, different types of numerical models and analytical methods. Nonlinear models were developed according to the finite element method and the applied element method. The two numerical approaches were considered: the Finite Element Method and the Applied Element Method, computing the accuracy and the computational efficiency of each one. The results of the analytical prediction and the numerical simulations are then compared with the experimental test outcomes, showing good agreement on both local and global response parameters.

## 1 Introduction

During the last decades an increasing number of Italian bridges reached their design service life, with a number of collapse due to aging, corrosion or lack of maintenance. Thus, the attention on the motorways infrastructures is growing up due to potential risk of floods, collisions, lack of maintenance and overloads failures (Cook et al. 2013, Deng et al. 2016, Di Prisco 2019, Nuti et al. 2020).

Literature studies highlighted how most of the Italian existing bridges have simply supported, beam type, prestressed concrete (PC) deck (Borzi et al. 2015). Furthermore, an almost equal distribution of pre-tensioned and post-tensioned PC bridges in the existing Italian PC bridges stock was found from literature reviews (Galano et al. 2022).

In a recent study (Miluccio et al. 2021) the vulnerability of a class of Italian PC bridge decks was studied against the traffic load (i.e. vertical loads) models proposed by the Italian building code (IMIT 2018).

The flexural response of existing PC bridge girders depends on several factors, such as the levels of initial prestressing (Lehký et al. 2019, Galano et al. 2023). Due to long-term losses, the residual prestressing force in a single tendon may greatly reduce over


M. A. Aiello and A. Bilotta (Eds.): ICC 2022, LNCE 435, pp. 343–355, 2024.
https://doi.org/10.1007/978-3-031-43102-9_27

time (Weichen et al. 2018, Kralovanec et al. 2022). Thus, the influence of the prestressing level on the significant flexural response parameters of the PC bridge girder needs to be investigated.

With this aim in mind, the authors designed an experimental test campaign This paper studies the flexural behavior of a scaled simply supported post-tensioned PC bridge girder (T1) through three main approaches: (i) experimental, (ii) analytical, and (iii) numerical. Two different numerical methods were used, namely the Finite Element Method (FEM) and the Applied Element Method (AEM).

## 2 Experimental Test

In this study, the results of the experimental test on prototype T1 are presented. The following sections describe the sample's geometry and construction details as well as the test protocol.

### 2.1 Description of the Specimen

Laboratory facilities required a 1:5 scaled geometry of the PC girder. The specimen has a T-shaped cross section with a 60 mm thick slab, web and top flange width of 150 and 480 mm, respectively, and a total height of 440 mm; the total length (L) is equal to 6600 mm (Fig. 1a). The girder web presents equally spaced reinforcing bars with nominal diameter of 8 mm and concrete cover of 30 mm; an 8 mm 200 × 200 welded wire mesh was used in the top flange. The shear reinforcement is provided by vertical stirrups with nominal diameter of 8 mm and 100 mm longitudinal spacing.

The specimen is prestressed with two seven-wire single strand tendons (Fig. 1a) with a parabolic eccentricity and an equivalent area of 150 $mm^2$ each (nominal diameter 0,6"). The clear distance from the top flange at mid-span is equal to 342 mm for the upper tendon and 407 mm for the lower tendon (Fig. 1b); at the supports these distances reduce to 111 mm and 281 mm for the two tendons, respectively (Fig. 1c). Two single strand anchorages were used at girder end cross sections to restrain each tendon (Fig. 1c).

The value of the concrete compressive strength obtained from cubic samples has been found equal to 30–35 MPa. The strands were tested showing a ultimate tensile strength of 1969 MPa, a conventional yielding stress (i.e. stress at 1% of total strain) of 1782 MPa and an elastic modulus of 203,400 MPa.

Each strand has been initially tensioned up to a target stress of 1000 MPa (i.e., a prestressing force equal to 150 kN for each tendon), that is approximately half of their ultimate capacity. The prestressing force has been monitored through a load cell at support, and a reduction up to 111 kN and 104 kN was measured for lower and upped tendon, respectively (i.e., approximately 700 MPa), due to immediate and time dependent losses (i.e. creep, shrinkage and relaxation). Due to prestressing action, the girder upward displacement was found equal to –2.5 mm.

### 2.2 Test Procedure

The specimen has been tested under quasi-static loading protocol in a four-point bending configuration to evaluate the girder flexural response. The distance between the two load

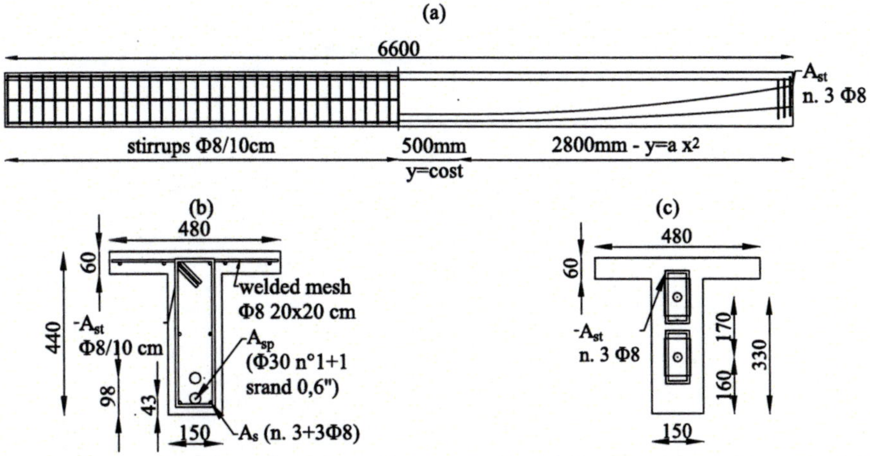


**Fig. 1.** PC girder geometry with steel reinforcement and tendons: a) longitudinal view, b) Mid-span cross section, c) End cross section.

points is equal to 850 mm resulting in a shear span length $L_v$ of 2800 mm. Given the applied total vertical force $F_e$, the maximum bending moment is $M_e = 0.5F \cdot L_v$.

Figure 2 shows the layout of the setup for the experimental test carried out at the Department of Structures for Engineering and Architecture of the University of Naples Federico II. The test frame consists of a rigid steel base ($1.0 \times 4.0\,\mathrm{m}^2$), four columns fixed at the base and a moving crossbar at the top which can slide in vertical direction. Through a vertical actuator placed on the crossbar, both displacement (max stroke 150 mm) and force control (max compression load 3000 kN, max tensile load 2400 kN) protocols can be imposed.

The girder deflection at mid-span cross section was measured using a vertical Linear Variable Differential Transducer (LVDT) sensor (Fig. 2); a load cell sensor is used to evaluate the vertical applied load (see set-up front view of Fig. 2). The specimen has been tested using a displacement-control test procedure with a constant rate of 0.05 mm/s up to failure.

### 2.3 Experimental Test Results

Figure 3 shows the experimental force – displacement curve of the T1 specimen. The first crack was detected on the bottom part (tensile region) of the mid-span cross section (Fig. 4a), when the specimen was subjected to an external vertical load of 49 kN, corresponding to a vertical displacement of about 5.54 mm. With the shear-span length given in Sect. 2.2, the resulting cracking bending moment $M_{cr}$ resulted equal to 68.6 kNm.

As the imposed displacement increases, an increasing number of cracks were observed in the tensile region across the mid-span cross section (Fig. 4b). The extension of the cracked zone leads to a noticeable reduction of the vertical stiffness, as can be seen from Fig. 3 beyond cracking.

The maximum applied vertical displacement was about 120 mm under a force 156 kN, corresponding to a maximum bending moment of 218 kNm. The maximum stroke

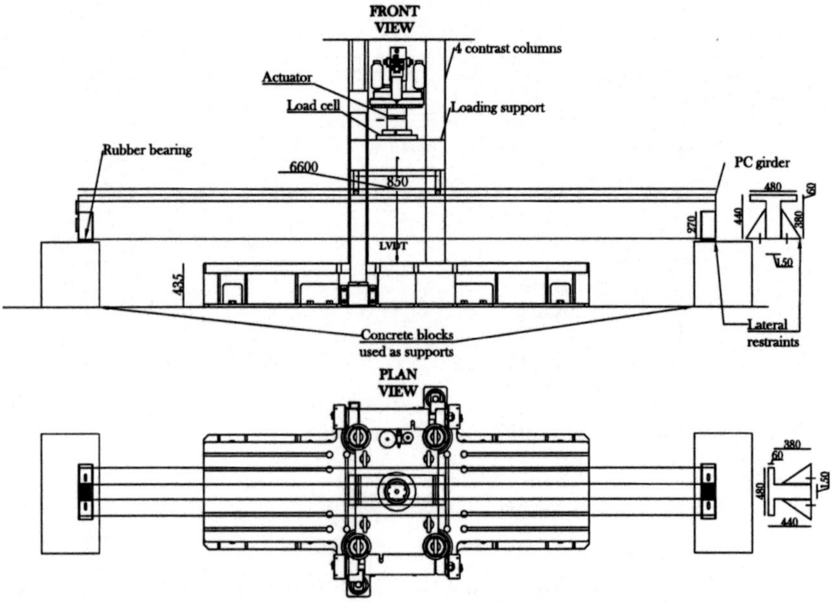


**Fig. 2.** Experimental setup: front and plan view.

of the actuator was reached around peak vertical force of the girder. Thus, no post-peak response (i.e. softening and collapse) was investigated in this paper.

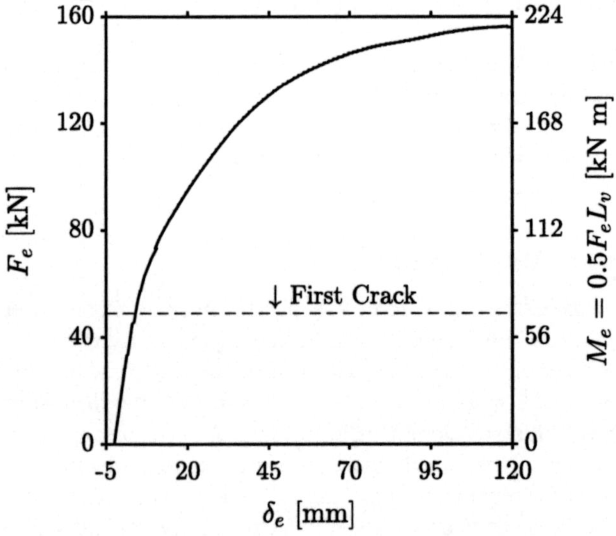


**Fig. 3.** Experimental (vertical) force – displacement curve.

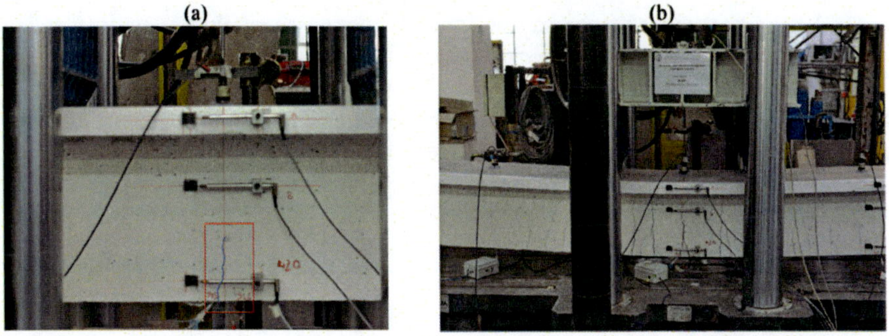


**Fig. 4.** Specimen under test: a) first detected crack, b) crack pattern at ultimate load condition of midspan.

## 3 Analytical Approach

An analytical prediction of the PC girder flexural behaviour has been carried out for mid-span cross section. Threshold bending moment and corresponding curvature values have been calculated including the following condition: initial prestress, decompression of concrete, cracking of concrete, yielding of the bottom rebar, yielding of the bottom tendon, concrete compressive failure. These values and the corresponding vertical force in a four-point bending test are shown in Table 1.

The cross-section geometric details are shown in Fig. 1. The following material features have been used in the analytical approach (IMIT 2018):

- Parabola-rectangular uniaxial concrete compressive stress-strain behaviour, with a compressive strength $\sigma_{cc}$ equal to 30 MPa and an elastic modulus $E_c$ of 30,000 MPa. Conventional yielding strain $\varepsilon_{cy}$ and ultimate strain $\varepsilon_{cu}$ are assumed equal to 0.002 and 0.0035, respectively.
- Linear uniaxial pre-cracking concrete tensile stress-strain behaviour, with a tensile strength $\sigma_{ct}$ equal to 2.9 MPa (IMIT 2018).
- No tensile stress for cracked concrete.
- Elastic-plastic with hardening uniaxial stress-strain response for mild steel, with a yielding stress $\sigma_{sy}$ and an ultimate stress $\sigma_{st}$ equal to 450 MPa and 540 MPa, respectively. The elastic modulus $E_s$ was set equal to 210,000 MPa. Accordingly yield strain is $\varepsilon_{sy} = \sigma_{sy}/E_s = 0.0021$ and ultimate strain $\varepsilon_{su}$ was set equal to 0.0675 (IMIT 2018).
- Elastic-plastic with hardening stress-strain constitutive law for prestressing steel, using a yield stress $\sigma_{py}$ and an ultimate stress $\sigma_{pt}$ of 1782 MPa and 1969 MPa, respectively. Assuming an elastic modulus $E_p$ of 203,400 MPa, the yielding strain $\varepsilon_{py}$ and the ultimate strain $\varepsilon_{pu}$ are equal to 0.0088 and 0.060, respectively.

The decompression load is reached when a null value of the compressive stress is attained at the external concrete fibre at mid-span: this point corresponds to a decompression bending moment $M_{dec}$ of 57.5 kNm. The first cracking (i.e., attainment of tensile strength at the external concrete fibre) is reached with a moment $M_{cr}$ equal to 64.9 kNm. The mild steel yielding point was obtained for a bending moment $M_{sy}$ equal to 130 kNm. The lower tendon and top concrete yielding correspond to a bending moment of 206 kNm and 211 kNm, respectively, whereas the concrete compressive failure is attained at 219 kNm (i.e. 156 kN total vertical force).

In Table 2 a comparison between analytical prediction and experimental values at cracking and ultimate condition is shown. In this table, $M_{cr}$ and $M_u$ stand for cracking and ultimate bending moment, respectively. A good agreement is found, with a maximum error of 6% between analytical prediction and experimental results.

**Table 1.** Mid-span cross-section noticeable bending moment and applied force values.

| *Threshold* | $M_e$ *[kN m]* | $F_e$ *[kN]* |
|---|---|---|
| Decompression of concrete | 57.5 | 41.1 |
| Cracking of concrete | 64.9 | 45.9 |
| Yielding of rebar | 130 | 92.9 |
| Yielding of tendon | 206 | 147 |
| Yielding of concrete | 211 | 151 |
| Concrete failure | 219 | 156 |

**Table 2.** Experimental-analytical comparison between external forces at cracking and ultimate condition.

| | *Analytical* | *Experimental* | *An/Exp* |
|---|---|---|---|
| $M_{cr}$ [kNm] | 64.9 | 68.6 | 1.06 |
| $M_u$ [kNm] | 219 | 218 | 1.00 |

# 4 Numerical Models

Numerical models of T1 girder were developed using two different approaches, i.e. FEM and AEM, whose details are presented in the following sections.

## 4.1 Material Modeling

FEM analysis has been carried out using Abaqus FEA (Smith 2009), while the commercial software Extreme Loading for Structures (ELS) (ASI, 2020) has been considered for the AEM numerical model. In both software the non-linear behavior of the elements can be considered through specific material modelling and the same values of the material properties have been used for the sake of comparison. According with materials specifications given in Sect. 2.1, the following parameters were used:

- Concrete peak compressive strength $\sigma_{cc} = 30$ MPa and Young's modulus $E_c = 30,000$ MPa.

- Tendons yielding stress $\sigma_{py}$ and failure strength $\sigma_{pt}$ of 1782 and 1969 MPa, respectively; Young's modulus $E_p$ equal to 203,400 MPa.
- Mild steel with yielding stress $\sigma_{sy} = 450$ MPa and Young's modulus $E_s = 210,000$ MPa.

The concrete compressive part (Fig. 5) is defined by the following equation (Fib 2013):

$$\sigma_c = f_{cm} \cdot \frac{k\eta - \eta^2}{1 + (k-2)\eta} \tag{1}$$

The tensile response is divided into a linear stress – strain relation for the uncracked concrete (Fig. 5a):

$$\sigma_{ct} = \varepsilon_{ct} \cdot E_c \tag{2}$$

while a bilinear equation for the stress – crack opening is used (Fig. 5b):

$$\begin{aligned} \sigma_{ct} &= f_{ctm} \cdot \left(1 - 0.8\frac{w}{w_1}\right) \quad \text{for } w \leq w_1 \\ \sigma_{ct} &= f_{ctm} \cdot \left(0.25 - 0.05 \cdot \frac{w}{w_1}\right) \quad \text{for } w_1 < w \leq w_c \end{aligned} \tag{3}$$

All the parameters of these models are given in (Fib 2013). In both software, the prestressed-steel has been modeled using an elastic – plastic with hardening stress-strain behavior, while for the ordinary steel an elastic – plastic constitutive law has been adopted.

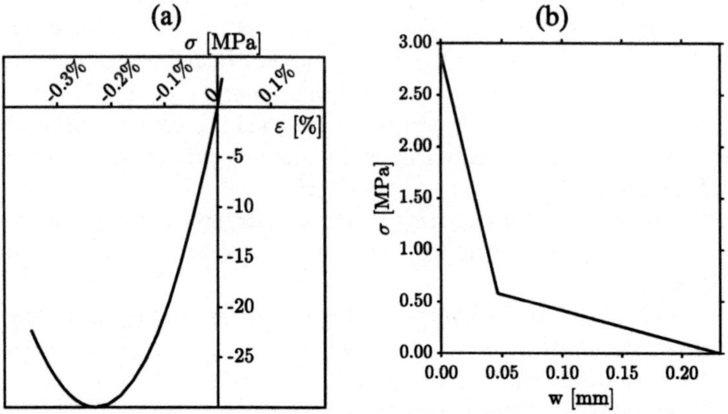


**Fig. 5.** Uniaxial response of concrete: a) compression (negative) and pre-cracking tension (positive), b) post-cracking uniaxial tension.

### 4.2 Models' Specification

Concrete material has been modeled in Abaqus using Concrete Damage Plasticity Model (CPDM), a continuum, plasticity – based, damage model (Smith 2009). In CPDM, two main failure mechanisms are assumed: tensile cracking and concrete crushing. The evolution of the yield (or failure) surface is controlled by two hardening variables, namely the tensile equivalent plastic strain $\overline{\varepsilon}_t^{pl}$ and the compressive equivalent plastic strain $\overline{\varepsilon}_c^{pl}$. Detailed information on this method can be found in (Lee 1998, Lubliner 1989).

The eight-node first-order element with one reduced integration rule (element C3D8R in Abaqus library) has been used for concrete elements.

The embedded interaction technique has been used in Abaqus for rebars (i.e. longitudinal bars and stirrups) and tendon steel, simulating the perfect bonding condition with the surrounding concrete. Two-node truss elements (element T3D2 in Abaqus library) have been used to model both ordinary and prestressed steel.

A temperature approach is used in Abaqus to apply the initial prestress to tendons. For one-dimensional rod element, the effective tendon initial prestress $\sigma_{p,i}$ could be obtained using the following equation:

$$\sigma_{p,i} = E_p \alpha_t \Delta T \tag{4}$$

being $E_p$ the elastic modulus, $\alpha_t$ the thermal expansion coefficient of the prestressing steel and $\Delta T$ the applied thermal variation.

The AEM (Meguro 2000, Tagel-Din 2000, Meguro 2001, Tagel-Din 2008, Christy 2017) takes into account non-linear behavior starting from the elastic stage up to collapse of each element. The entire girder is discretized into a number of cubic elements; thus, a predefined number of normal and shear non-linear springs are generated on the elements' surfaces. The stress-strain constitutive models of Fig. 5 were then assigned to the axial springs, while the default Mohr-Coulomb behavior was used for the shear springs. The strain level of each spring defines the damage reached in that part of the element. Six degrees of freedom are assigned to each element centroid (i.e., in a 3D problem) and the behavior of the defined springs depends on the relative distance on the same surface of the element. Moreover, stresses and deformations of each spring are related to the volume of the element material they are connected.

The stiffness matrix of each spring is defined starting from the element centroid displacements. Then, the global stiffness matrix is obtained by combining different stiffness contributions.

The reinforcing bars and strands were modelled as specific springs. Both tendons and rebars were defined using the default procedure of ELS for reinforcements, i.e. axial springs embedded within the main element.

The AEM approach allows to take into account the failure of each spring based on stress- or strain-criteria. Moreover, when stress and/or strain of the single spring reach their capacity, cracks are formed (i.e., defined as "open cracks" in ELS software). Further or wider cracks develop under increasing load, whereas under cyclic loading (i.e., loading/unloading pattern) these cracks may close. When maximum strain attains the ultimate deformation capacity (i.e., "separation strain" in ELS) the springs became inactive, and elements are disconnected. Thus, a permanent crack is formed, and the elements are separated.

In this software the prestress has been applied as force along the strands instead of thermal deformation.

As known, using a damaged material model with softening branch, induces mesh-dependent results. The optimal mesh size of the concrete element has been defined under a mesh-sensitivity analysis. Changing the side of a single element, the convergence of the models was tested up to ultimate displacement and a mesh size of 50 mm was set.

Figures 6a and 6b show a lateral view of the 3D models of the T1 girder in Abaqus and ELS software, respectively.

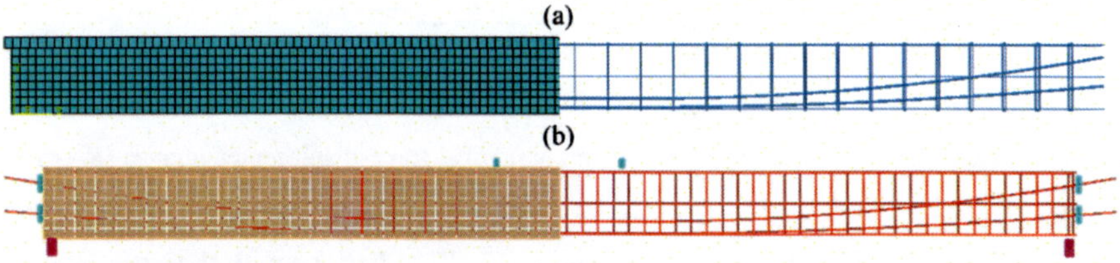


**Fig. 6.** Numerical models with mesh specification in (a) Abaqus and (b) ELS.

Both in Abaqus and ELS the analysis is divided into two different stages: (i) prestressing, and (ii) vertical loading.

In the first stage, the actual prestress (i.e., 717 MPa, see Sect. 2.1) is applied to both tendons, whereas the gravity loads are automatically considered from the software. Then, increasing vertical displacement is applied to the control points according with the experimental layout.

### 4.3 Experimental – Numerical Comparison

Figure 7 shows the comparison between numerical and experimental force-displacement curves. This plot shows a good matching between experimental test and both numerical models, as the non-linear behavior of T1 girder is accurately traced up to maximum displacement. An almost perfect match between experimental and numerical curves can be identified in the linear response part, i.e., before the first crack forms (see Table 1). In both numerical cases the girder counter deflection due to initial prestressing is accurately predicted, i.e.–2.93 mm in Abaqus and –2.76 mm in ELS. Figure 7 shows how past the cracking threshold (approximately 50 kN), the response of the PC girder is strongly non-linear due to concrete cracking.

The ordinary steel yielding was attained by an experimental force of about 92.5 kN (Table 1), similar to the numerical value of 91.5 kN.

As for the ultimate flexural stage of the girder, the maximum experimental test force is equal to 156 kN, whereas the numerical analyses return a value of 159 kN for Abaqus ($F_{max}^{Test}/F_{max}^{Abaqus} = 0.981$) and 156 kN for ELS ($F_{max}^{Test}/F_{max}^{ELS} = 1.00$). The global $F - \delta$ curves and corresponding cracking, yielding and ultimate points comparison all confirm

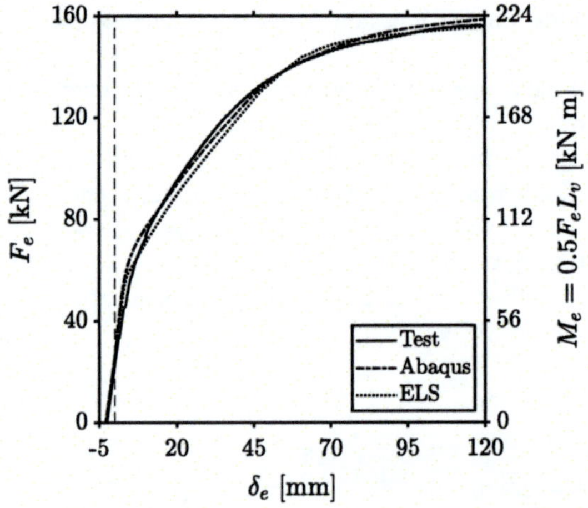


**Fig. 7.** Comparison between experimental and numerical force-displacement curves ($F - \delta$).

an accurate match between analytical and numerical predictions with the experimental response of T1 girder.

Ultimate response obtained from experimental tests and numerical models is shown in Fig. 8. In this figure, the crack pattern at the beam central region can be easily tracked. The first crack was observed at mid-span cross-section in both experimental test and numerical simulations (see also Fig. 4a) as per load condition symmetry. Then, cracking propagation occurred and the bottom parts of the girder (i.e. the tensile zones) gradually showed increasing number of wider cracks. The crack pattern (cf. Figure 8a) is provided in terms of tension damage contour plots in Abaqus (damage index $\leq 1$, cf. Figure 88), while in ELS software the cracks can be estimated in terms of tensile strains (calculated including crack width, cf. Figure 8c.

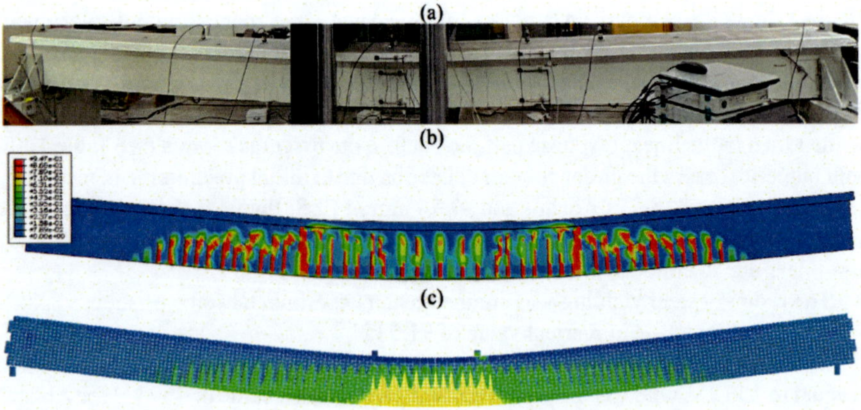


**Fig. 8.** Ultimate response comparison in terms of cracking pattern: (a) experimetal, (b) Abaqus and (c) ELS.

### 4.4 Brief Comparison Between Numerical Methods

Under both numerical methods, the PC girder response was predicted with good accuracy. The CDPM used in Abaqus provides detailed response in terms of concrete incremental damage. In ELS, damage is implicitly considered through the failure of the connection springs between adjacent elements.

In contrast to good accuracy of both models in reproducing the overall force-displacement response, significantly different computational burden is related to considered modelling strategies. Using the same mesh size (i.e. brick side equal to 50 mm), it takes more than one hour to complete a FEM analysis, while only 3 min are required for an AEM analysis (i.e. less than 5% of computational time). Analysis time difference is due to reduced number of degrees of freedom considered for each element in the AEM approach, since the equilibrium equations are written at the centroid of each element and then correlated to the adjacent springs. Moreover, the PARADISO direct sparse solver (Schenk & Gartner, 2004) was used in ELS software that allows to have a more efficient equation's solver.

This suggests that detailed FEM can be adopted for analysis of local phenomenon in the girder (e.g. potential local damage or defect), whereas the AEM would be a very convenient option for faster analyses of the global response (e.g. girder failure or progressive collapse of deck etc.) without local phenomenon to be accurately investigated (e.g. lack of injection or damage to tendons.

## 5 Conclusions

This study presented the results of a four-point bending test on one scaled post-tensioned PC girder. The test was run under displacement-control protocol up to potential flexural failure of the girder. The main thresholds of bending moment values at mid-span section were predicted by analytical method and compared with experimental ones. Two different numerical methods were adopted to predict the flexural behavior of the PC girder: Finite Element Method (FEM) and Applied Element Method (AEM) implemented in Abaqus FEA and Extreme Loading for Structures software, respectively.

The results obtained showed how numerical force-displacement curves accurately predict the experimental ones. Crack patterns are also shown, and a direct comparison between experimental test and numerical prediction is provided. Thus, both numerical methods (e.g., FEM and AEM) were able to describe the PC girder flexural behavior with good accuracy, even with different computational efforts.

Results of this paper demonstrated the efficiency of different numerical methods (e.g., FEM or AEM) in predicting the flexural response of post-tensioned concrete bridge girders, laying the basis for further development of the models, such as study the shear response or the residual prestressing force due to long-term losses.

**Acknowledgements.** This study was developed in the framework of PON INSIST (Sistema di monitoraggio INtelligente per la Sicurezza delle infraSTrutture urbane) research project, which was funded by the Italian Ministry for Education, University and Research (Programma Operativo Nazionale "Ricerca e Innovazione 2014–2020", Grant No. ARS01_00913).

## References

Borzi, B., Ceresa, P., Franchin, P., Noto, F., Calvi, G.M., Pinto, P.E.: Seismic vulnerability of the Italian roadway bridge stock. Earthq. Spectra **31**, 2137–2161 (2015)

Christy, L., Pillai, M., Nagarajan, P.: Analysis of concrete beams using applied element method. IOP Confer. Ser. Mater. Sci. Eng. **330**(1), 012117 (2017)

Cook, W., Barr, P.J., Halling, M.W.: Bridge failure rate. J. Perform Construct Fac. **29**(3), 04014080 (2013)

Deng, L., Wang, W., Yu, Y.: State-of-the-art review on the causes and mechanisms of bridge collapse. J. Perform Construct Fac. **30**(2), 04015005 (2016)

Di Prisco, M.: Critical infrastructures in Italy: state of the art, case studies, rational approaches to select the intervention priorities. In: Proceedings of fib Symposium 2019 Concrete: Innovations in Materials, Design and Structures, Krakow, Poland, 27–29 May 2019 (2019)

DM 17/01/2018. Aggiornamento delle Norme tecniche per le costruzioni. Italian Ministry of Infrastructures and Transportation, Rome, Italy (2018). (in Italian)

EN1992–1–1. Eurocode 2: Design of concrete structures – Part 1–1: General rules and rules for buildings. Comité Européen de Normalisation, Brussels, Belgium (2004)

Fédération internationale du béton. Fib model code for concrete structures 2010. Lausanne, Switzerland: Ernst & Sohn, a Wiley brand, (2013)

Galano, S., Losanno, D., Miluccio, G., Parisi, F.: Multidimensional nonlinear numerical simulation of post-tensioned concrete girders with different prestressing levels. Struct. Concr. (2023). https://doi.org/10.1002/suco.202300272

Galano, S., Losanno, D., Pecce, M.R.: Prestressed reinforced concrete bridges: state of art of the Italian existing stock. In: Italian Concrete Conference. Naples, 14th–15th October (2022)

Kralovanec, J., Bahleda, F., Moravcik, M.: State of prestressing analysis of 62-year-old bridge. Materials **15**, 3583 (2022). https://doi.org/10.3390/ma15103583

Lee, J., Fenves, G.L.: Plastic–damage model for cyclic loading of concrete structures. J. Eng. Mech. **124**(8), 892–900 (1998)

Lehký, D., Pan, L., Novák, D., Cao, M., Šomodíková, M., Slowik, O.: A comparison of sensitivity analyses for selected prestressed concrete structures. Struct. Concr. **20**, 38–51 (2019). https://doi.org/10.1002/suco.201700291

Lubliner, J., Oliver, J., Oller, S., Onate, E.: A plastic–damage model for concrete. Int. J. Solids Struct. **25**(3), 299–326 (1989)

Maekawa, K., Okamura, H.: The deformational behavior and constitutive equation of concrete using the elasto-plastic and fracture model. J. Fac. Eng. Univ. Tokyo (B), **37**(2), 253–328 (1983)

Meguro, K., Tagel-Din, H.: Applied element method for structural analysis: theory and application for linear materials. Struct. Eng./Earthq. Eng. **17**(1) (2000)

Meguro, K.: Applied element method: a new efficient tool for design of structure considering its failure behavior. ICUS/INCEDE, 13–30 (2001)

Miluccio, G., Losanno, D., Parisi, F., Cosenza, E.: Traffic-load fragility models for prestressed concrete girder decks of existing Italian highway bridges. Eng. Struct. **249**, 113367 (2021)

Nuti, C., Briseghella, B., Chen, A., Lavorato, D., Iori, T., Vanzi, I.: Relevant outcomes from the history of Polcevera Viaduct in Genova, from design to nowadays failure. J. Civil Struct. Health Monitor. **10**, 87–107 (2020)

Okamura, H., Maekawa, K.: Nonlinear Analysis and Constitutive Models of Reinforced Concrete. Gihodo, Tokyo (1991)

Schenk, O., Gärtner, K.: Solving unsymmetric sparse systems of linear equations with PARDISO. J. Future Gen. Comput. Syst. **20**(3), 475–487 (2004)

Smith, M.: ABAQUS/Standard User's Manual, Version 6.9. Dassault Systèmes Simulia Corp. United States (2009)

Tagel-Din, H.: Applied Element Method as a Practical Tool for Progressive Collapse Analysis of Structures. Nafems, 98 (2008)
Tagel-Din, H., Meguro, K.: Analysis of a small-scale RC building subjected to shaking table tests using applied element method, 12WCEE, 1–7 (2000)
Weichen X., Min D., Chi, H., Jie, L.: Long-term behavior of prestressed composite beams at service loads for one year. J. Struct. Eng. **134**(6), 930–937 (2008)

# Towards Quantitative Prioritization Schemes for Bridge Portfolios in Italy

Andres Abarca, Ricardo Monteiro(✉), and Gerard O'Reilly

Scuola Universitaria Superiore IUSS Pavia, Pavia, Italy
{andres.abarca,ricardo.monteiro,gerard.oreilly}@iusspavia.it

**Abstract.** The Italian government has recently instated guidelines on risk classification, management, safety assessment and monitoring of existing bridges. Such guidelines can also be used as a rapid prioritisation method that, based on the limited information available about assets in the inventory, allows the identification of bridges requiring special attention in the form of inspection, detailed analysis, monitoring and possible retrofitting. However, these guidelines are based on qualitative indicators that can produce overly conservative results. In this paper, the recent Italian guidelines are explored and a quantitative partial modification is proposed, based on average annual loss prioritization results from detailed risk assessment. The proposed modification is evaluated on a case study of 617 reinforced concrete bridges, using seismic hazard to demonstrate its potential large-scale implementation. The results show a notable improvement in the identification of high-risk assets in the portfolio, encouraging the adoption of similar strategies to other hazards and bridge typologies, with a view to multi-hazard and data-driven quantitative prioritization schemes for the Italian territory.

**Keywords:** bridges · prioritization · transportation infrastructure · regional risk

## 1 Introduction

The bridge inventory of developed countries can reach thousands of assets that have been built over several decades by different administrations (Calvi, et al. 2019), creating a challenge for the institutions currently managing these large portfolios of bridges for which there is incomplete information about their current structural condition and limited resources available to upgrade or maintain them. In the case of Italy, a great portion of its current infrastructure was built during a construction surge of freeways that happened all over Europe in the 1960s (Calvi, et al. 2019). The longevity of the current inventory, aided by the difficulties of management agencies in providing proper maintenance, has led to a generalized problem of deterioration that increases the vulnerability of these structures, a condition that has become evident by the number of bridge collapses in recent years.

For example, a non-exhaustive list of collapses collected from reports in the media is presented in Table 1, where it can be observed that several months or even years can pass for a bridge to be reopened following its full or partial collapse.


M. A. Aiello and A. Bilotta (Eds.): ICC 2022, LNCE 435, pp. 356–373, 2024.
https://doi.org/10.1007/978-3-031-43102-9_28

Considering the situation described above, there is a need for bridge management institutions to determine rapid prioritization methods that, based on limited information available about assets in the inventory, allow the identification of the assets requiring special attention in the form of inspection, detailed analysis, monitoring and possible retrofitting. Such prioritization methodologies have been the source of multiple research efforts worldwide. A summary is available in a recent technical report by the United States Department of Transportation (Chase, Adu-Gyamfi, Aktan, & Minaie 2016). It documents the evolution and application of different bridge health indices used by bridge management agencies interested in preserving the condition of bridge structures or prioritizing the maintenance or replacement projects within their bridge inventory. Recent Italian examples include the simplified index-based methods developed by Pellegrino et al. (2011) and D'Apuzzo et al. (2019), both of which are based on detailed inspection-level information to assess the deterioration status of the bridges and combine it with the importance of each asset to the overall network by incorporating an additional index based on road typology and traffic flows. More recently, the Italian Superior Council of Public Works, within the Ministry of Infrastructure and Transport (MIT), issued a technical report with guidelines on risk classification and management, safety assessment and the monitoring of existing bridges (Consiglio Superiore dei Lavori Publici 2020), which has already become part of the mandatory legislation for bridge management institutions and concessionaries in Italy (Ministero delle Intrastrutture e dei Trasporti 2020). This document, which will be referred to from this point forward as the 2020 MIT Guidelines, intends to standardize the procedure with which existing bridges in Italy are assessed at a large scale by a multi-level and multi-component approach that classifies bridges in risk categories via a combination of qualitative metrics.

**Table 1.** Recent bridge collapses in Italy

| Province | Bridge Name/Location | Length (m) | Collapse Date |
|---|---|---|---|
| Pordenone | Viadotto del Chiavalir | 25.0 | Dec-04 |
| Genova | Carasco | 258.0 | Oct-13 |
| Nuoro | Oliena-Dorgali | 130.0 | Nov-13 |
| Agrigento | Lauricella-Petrulla | 476.0 | Jul-14 |
| Lecco | Annone | 56.0 | Oct-16 |
| Ancona | Ancona | 45.0 | Mar-17 |
| Genova | Viadotto Polcevera | 1182.0 | Aug-18 |
| Savona | Madonna del Monte | 30.0 | Nov-19 |
| Massa-Carrara | Albiano Magra | 290.0 | Apr-20 |
| Novara | Romagnano Sesia | 156.0 | Oct-20 |

When looking for an established metric that allows the consideration of the entire scope of the problem in a single value, average annual loss (AAL) is a risk metric that has seen growing use within the structural engineering community (O'Reilly &

Calvi 2019) (Shahnazaryan & O'Reilly 2021), even being proposed as a target metric to be used in new methods for structural design and assessment (Calvi, O'Reilly, & Andreotti 2021). AAL, also referred to in some sources as expected annual loss (EAL), is a product of risk assessment that represents long-term expected economic losses per year, averaged over many years, that are produced by specific hazards of varying intensities and their respective annual exceedance rates, or return periods. In this paper, a seismic risk methodology is applied to a case study of 617 bridges in the Italian province of Salerno to determine prioritization of assets based on AAL, which is then used for two main purposes: as a benchmark to compare with the results obtained using the recent 2020 MIT Guidelines and as a possible guiding parameter to determine the relative importance of each factor affecting the determination of priority, with a view to moving towards a more optimized but still simple prioritization approach.

## 2 Case Study Bridge Inventory

### 2.1 Database Description

A bridge database comprising 308 bridges from the National Autonomous Roads Corporation ANAS (*Azienda Nazionale Autonoma delle Strade)* inventory, collected and managed by the Eucentre Foundation, was considered to create the case study for this research. These bridges form a part of the Italian road network, and their actual geographic location is scattered along the primary highway grid of Italy, as shown in Fig. 1.

**Fig. 1.** Location of the 308 case-study assets in the ANAS bridge inventory.

The information available in the database represents a complete account of geometrical and structural properties of the bridges, allowing detailed structural numerical models of each asset to be created. Each asset in the database is a reinforced concrete (RC) bridge with two or more spans, a predominant configuration in the Italian road network (Zelaschi, Monteiro, & Pinho 2016). In terms of general dimensions, the overall number of spans ranges from 2 to 36, which translates to an overall bridge length range of 50 m to 1250 m. The height of piers ranges between 5 m and 45 m in the overall inventory and it is typical to observe large variation of the pier height within the same asset, leading sometimes to irregular dynamic configurations within straight bridges. In terms of static configuration, most of the case-study assets have spans that are simply supported upon the piers with thin elastomeric pads, and only a small percentage has continuous deck and bearings that can be either elastomeric or isolators.

The construction year was available for all assets, ranging between 1953 and 2000, with most of them built during the 1960 s and 1970 s. As is common for regular Italian bridges of those decades, none of them are expected to have been specifically designed to meet appropriate seismic requirements, especially considering that the first national seismic regulation in Italy that addressed the entire national territory was instated in 2003 (Consiglio dei Ministri 2003). In general, the reinforcement percentages in the piers, both in longitudinal ($A_{sl}/A_c$) and transverse ($A_{st}/A_c$) directions, are low in comparison to current design standards and are quite similar across the different pier sections. This is atypical under current design practices, however, both the reinforcement ratios and the properties of the materials used for construction are in line with the age of construction of the inventory. In terms of dynamic properties, a structural model was created for each asset to determine the modal periods in both orthogonal horizonal directions. Since, for the case of bridges, the first mode does not typically account for a significative percentage of the total modal mass, an appropriate number of modes were evaluated for each asset to include 85% of the modal mass in each direction. The distributions for the first modal period ($T_1$) and the modal period at which 85% of the modal mass is obtained ($T_{85\%}$) as shown in Fig. 2.

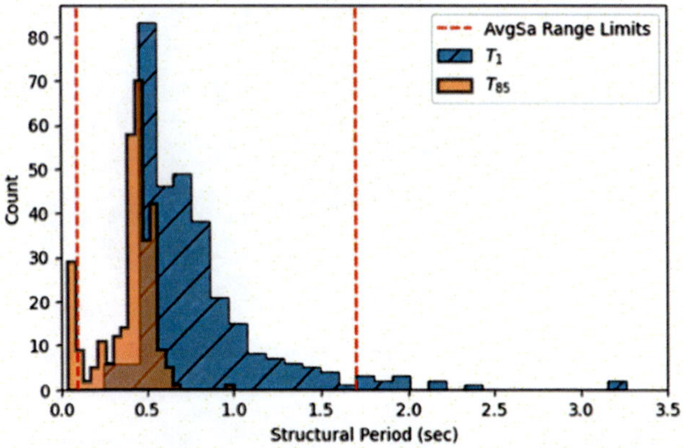


**Fig. 2.** Results for modal structural periods of the entire inventory and definition of AvgSa range

The intensity measure chosen to perform hazard and fragility calculations was average spectral acceleration (AvgSa), for which the collective results of $T_1$ and $T_{85\%}$ were used to define the period range. As shown in Fig. 2, the selected range was 0.1 s to 1.7 s, which was defined as per O'Reilly (2021) as 1.5 times the 84th percentile to account for period elongation of the first mode and 0.5 times the 16th percentile to account for higher mode contributions of the $T_1$ and $T_{85\%}$ periods, respectively, for the entire inventory.

## 2.2 Case Study Definition

As shown in Fig. 1, the bridges in the ANAS database are scattered geographically all over the Italian territory and not directly connected, therefore, their real location is not ideal to define a case study, since the consideration of the collective and individual role of each asset in the road network would be an unfeasible exercise. Ideally, if a case study of bridges closely connected within the same territory were available, it could be explored and fully analysed to represent a benchmark with which to evaluate the performance of simplified prioritisation frameworks. For this reason, and taking advantage of the fact that even in locations with different seismic hazard demands, bridge design practices did not vary considerably among the Italian territory for the construction period of the bridges in the database (Borzi, et al. 2015), a synthetic case study was created. To do so, the road network of a region for which the location of bridges and road properties was known was taken, with a bridge from the 308-asset database being assigned to each location.

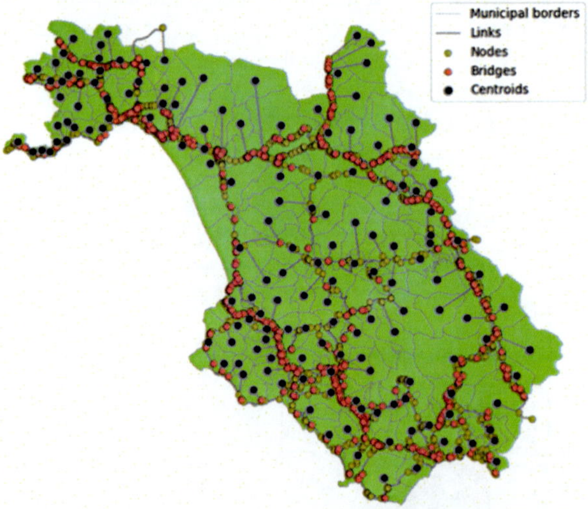


**Fig. 3.** Road network model for the case study region of Salerno built on AequilibraE (www.aequilibrae.com) based on OpenStreetMap data

The Salerno province was selected for having a transportation network that relies heavily on the vehicular road system, a relatively low number of bridges and a varying seismicity level. Information about the road network of Salerno was taken from the OpenStreetMap database (OpenStreetMap contributors 2020), which comprises all roads within the highway, primary and secondary systems, including 2929 nodes and 3086 links, of which 617 represent bridges. The centroid locations of the 158 municipalities in the Salerno province were used as traffic attraction zones (centroids) from which all trips were assumed to occur to and from. The 308 bridges in the database were therefore randomly assigned to the 617 possible locations of bridges in the Salerno network using a sampling with replacement scheme. Once the final distribution of assets in the case study was defined, a transportation network model was created using the software AequilibraE (www.aequilibrae.com), an open-source Python and QGIS package to perform transportation network analysis, to determine the baseline traffic conditions that are fundamental to assess the importance of each bridge in the network. A graphical representation of the network model is shown in Fig. 3.

A database containing travel pattern information for work and study purposes performed in 2011 was taken from the Italian Institute of Statistics (ISTAT 2014) and used to define origin-destination demands between the different municipalities of Salerno. Furthermore, in order to account for congestion in the network, previous research regarding Italian road characteristics (Maratini 2008) was used to obtain the volume-delay function modelling parameters according to the commonly used BPR model (Bureau of Public Roads 1964) for the different road types in the network. Free flow speed was taken as the speed limit reported for each road in the OpenStreetMap database. A trip distribution based on the minimization of travel time of each user was carried out using a bi-conjugate Frank-Wolfe algorithm (Mitradjieva & Lindberg 2013) to determine the baseline traffic conditions of the fully operational road network, as shown in Fig. 4.

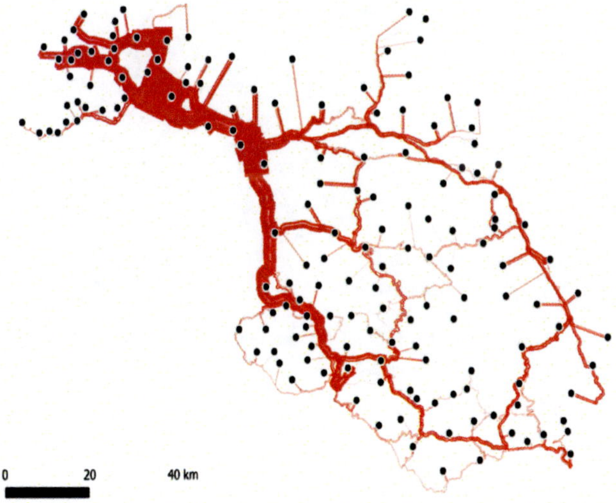


**Fig. 4.** Baseline traffic flows for case study region (line thickness is proportional to traffic flow)

## 3 Seismic Risk Analysis

### 3.1 Seismic Hazar

In terms of hazard curves, the SHARE hazard model (Woessner, et al. 2015), implemented in the OpenQuake Engine (Silva, Crowley, Pagani, Monelli, & Pinho 2014), was used to determine the probability of exceedance of different levels of AvgSa for an investigation period of 50 years at each bridge site. In terms of ground motion record selection, a conditional spectrum scheme (Lin, Haselton, & Baker 2013) was adopted using a modification that allows the conditioning of the spectra for AvgSa (Kohrangi, Bazzurro, Vamvatsikos, & Spillatura 2017). Given the large number of bridge locations, and to minimize the computational burden of performing disaggregation at each location, all assets were assigned to four hazard zones and two soil classes (i.e., soft and stiff soil differentiated by a $V_{s,30}$ threshold of 360 m/s) as illustrated in Fig. 5. Following this, a complete hazard disaggregation analysis was carried out for the eight possible zone-soil combinations. For each combination, sets of 30 bidirectional ground motion records were selected from the NGA West-2 Strong-motion Database (Ancheta et al. 2014) for nine return periods ranging from 98 years to 9975 years and were used for NLTHA.

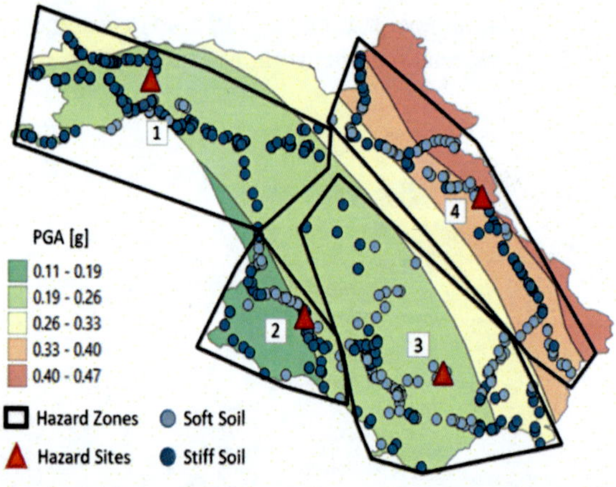


**Fig. 5.** Hazard zones and soil sites defined for the case study region (PGA values for a return period of 475 years are shown for reference).

### 3.2 Seismic Risk

#### 3.2.1 Fragility Assessment

In this study, an element-based approach implemented by Borzi et al. (2015) was adopted to evaluate the seismic fragility of each bridge in the case study portfolio. Since the focus of this study was not on the derivation of fragility curves for bridges, the specificities of the numerical models created and the overall analysis procedure is not explained herein. The collapse limit state was focused on since it is the limit state directly related to the complete loss of the bridge connectivity, rendering a straightforward evaluation of indirect losses possible. The results obtained for the fragility curves of each element in the inventory are shown in Fig. 6, where the mean fragility curve is shown for reference.

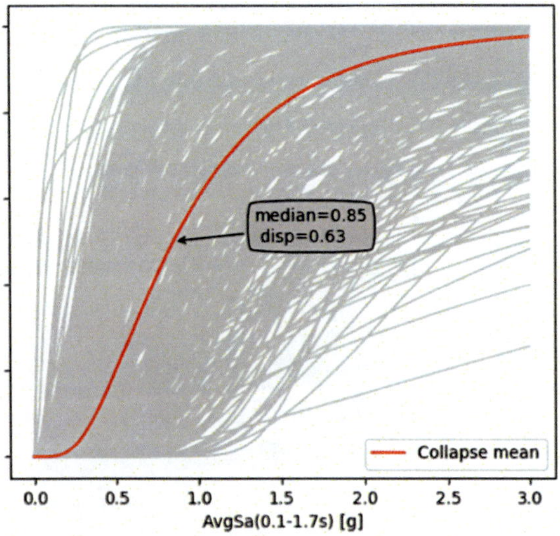


**Fig. 6.** Fragility curves for collapse limit state obtained for the 308 bridges in the database

### 3.2.2 Direct Loss Assessment

The calculation of direct losses associated with the collapse limit state was carried out using the basic formulation from the Pacific Earthquake Engineering Research Center's Performance-Based Earthquake Engineering (PEER PBEE) framework (Porter 2003). A straightforward implementation of the formulation is possible by including only the collapse limit state, where the product of the annual probability of exceedance of the limit state and the direct replacement cost will result in the direct collapse-based AAL, as per Eq. 1.

$$\mathrm{AAL} = p(LS_C) \cdot €L|LS_C p(LS_C) \cdot €L|LS_C = APE_C \cdot €RCAPE_C \cdot €RC \tag{1}$$

where:

$LS_C$: Collapse Limit State.

$p(LS_C)p(LS_C)$: probability of occurrence of $LS_C$

$€L|LS_C€L|LS_C$: direct economic losses associated to $LSC$

$APE, LS_CAPE, LS_C$: annual probability of exceedance of $LS_C$

€RC: bridge replacement cost

The annual probability of exceedance (APE) for the limit state was obtained by combining the fragility and hazard curves obtained for each bridge in the case study, evaluating the probability of exceedance in terms of the IML and the respective annual probability of exceeding that IML. The integration over the entire IML range results in the APE for each asset. The replacement cost for each bridge was taken as proportional to the deck area, considering a generic cost per square meter of €930, taken from the mean replacement cost per area obtained by Perdomo et al. (Perdomo, Abarca, & Monteiro 2020) for a similar Italian bridge inventory. The results for direct collapse-based AAL are show in Fig. 7, where it can be observed that higher values of loss are concentrated in the areas with higher seismic hazard.

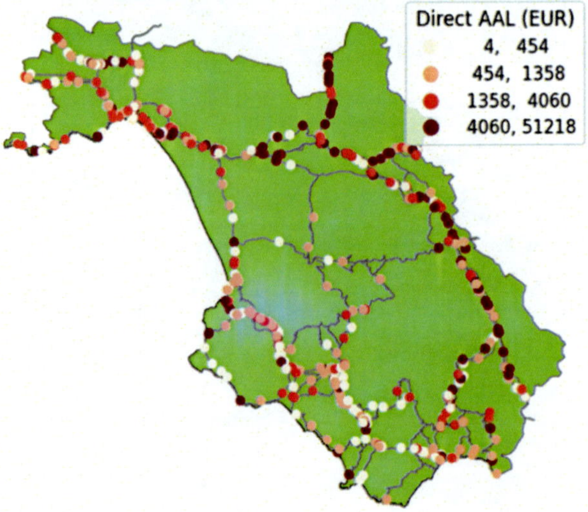


**Fig. 7.** Direct collapse-based average annual losses in Euros

### 3.2.3 Indirect Loss Assessment

In general, the same underlying concept and formulation presented previously to calculate direct losses can be used to determine the corresponding indirect ones; however, the difficulty lies in determining the indirect replacement cost associated to the collapse of the bridge. To determine the indirect replacement cost in this study, the previously described road network model, used to determine the baseline conditions of the network when all bridges are operational, was explored. Two main metrics were obtained from the model: the vehicle hours travelled (VHT), and the vehicle distance travelled (VDT), corresponding to the total amount of time and distance, respectively, that all the users in the network experience daily during their respective travels. Both metrics were then combined with median costs for automobile fuel efficiency, fuel prices and hourly salary rates appropriate for the Salerno province (ISTAT 2020). This allowed the calculation of a baseline daily cost (BDC) of operation of the road network in its current configuration.

Subsequently, the road network was modified by assuming the collapse of each bridge in the network, removing the associated link in the model and rerunning the daily operation cost with the modified network configuration to determine a Modified Daily Cost (MDC) associated with the collapse of each bridge. The total indirect cost of each bridge was then calculated as the difference between the BDC and the MDC multiplied by the repair time in days assumed for each bridge. The computation of the repair time to use for each calculation represented another challenge. In this study, the data from the 10 recent collapses in Italy shown in Table 1 was used to fit the lognormal distribution shown in Fig. 8 and the median value of 710 days was found and used as a deterministic value for all elements in the case study.

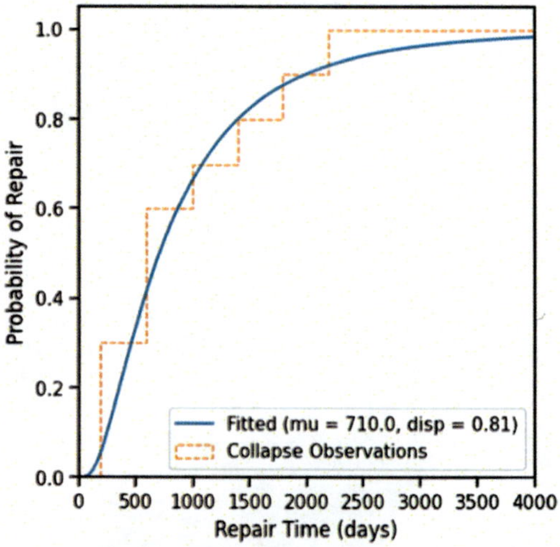


**Fig. 8.** Cumulative histogram and log-normal fit for repair time observations based on recent collapses in Italy.

The results for indirect AALs are shown in Fig. 9, where it can be seen that the indirect losses were concentrated near the coast of Salerno where the traffic is generally higher, even though the seismic hazard in this area was relatively low. Once both direct and indirect loss components were determined, the total collapse-based AALs were aggregated for each bridge, resulting in the distribution shown in Fig. 10. Analyzing the overall results, it is seen that the indirect losses represent 78% of the total losses and that the overall losses have a very similar spatial distribution to the one found for the indirect losses alone, which is expected given that these are much greater than the direct loss component.

## 4 Machine Learning Prediction of Aal-Based Ranking

A supervised machine learning model was evaluated using the case-study AAL results to assess the feasibility of predicting losses based on limited data, and to gain insights on the effect and relative importance of simple bridge parameters on the prioritization, defined by sorting bridges based on their individual AAL results. For this case study, the

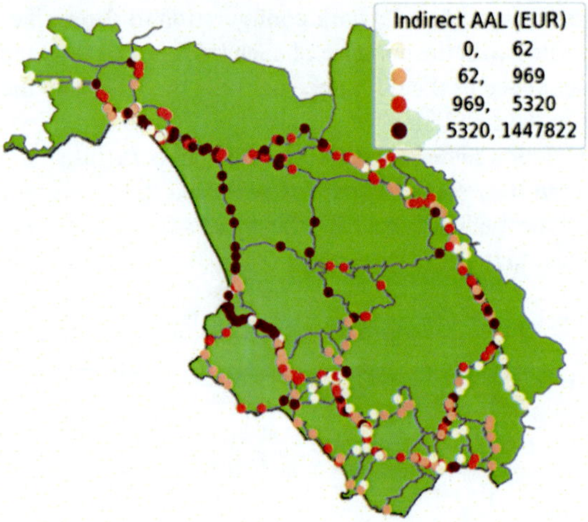


**Fig. 9.** Indirect average annual losses.

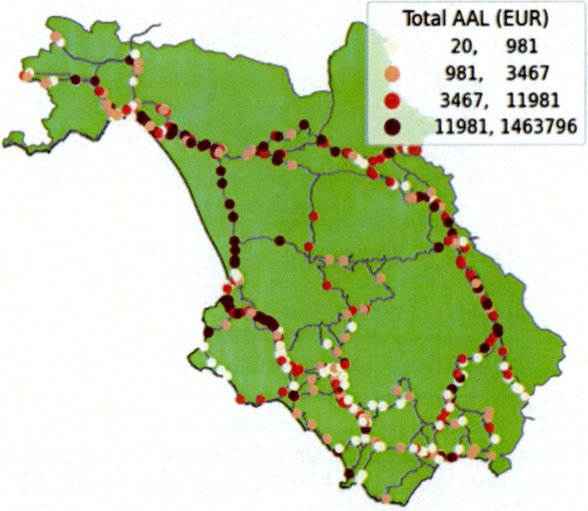


**Fig. 10.** Total average annual losses.

intention of the machine learning modelling process is not to create a model to be used on bridges outside of the current case study, but to take advantage of the capabilities of this method to infer relationships between independent features (simple bridge parameters and reference hazard values in this case) and their impact on target values of interest (AAL estimates). It is envisaged that these can be later used to guide improvement proposals for the 2020 MIT guidelines.

A random forest regression model was chosen given its recently demonstrated good performance when compared to other machine learning algorithms for similar applications (Mangalathu, Hwang, Choi, & Jeon 2019), and the ability of this algorithm to evaluate the relative importance of each independent variable. A database was assembled using the AAL results for each bridge in the case study to train the random forest model. For this purpose, the AAL representing the dependent variable (target) and a vector of independent variables (or features) was retrieved for each bridge structure. A set of six features were used for each bridge: maximum span length, maximum pier height, daily traffic flow, seismic intensity measure level for a return period of 475 years, number of spans and total replacement cost.

A set of useful regression performance metrics is presented in Table 2, and the relative feature importance is shown in Fig. 11. In general, the model does not have an ideal prediction performance, which is to be expected given the small amount of data points and features used to attempt to predict a complex value such as AAL, which depends on multiple variables that cannot be included in this type of model in a straightforward manner.

**Table 2.** Performance metrics for the machine learning model on the entire dataset

| Parameter | Value |
|---|---|
| Root-mean-squared error (RMSD) | € 52,279.8 |
| Mean absolute error (MAE) | € 10,888.2 |
| Median absolute error (MedAE) | € 3,398.4 |
| Coefficient of determination (R2) | 0.542 |
| Total $AAL_{pred}$ / $AAL_{calc}$ | 0.962 |

Overall, in terms of model performance, daily traffic flow has the highest relative importance over all the evaluated features, which is a consequence of the fact that the indirect losses represent the majority of the losses calculated and are directly related to the daily traffic. Moreover, maximum pier height was found to be the second most relevant feature when trying to predict AAL, which is a parameter that is not currently accounted for in the 2020 MIT Guidelines and has been shown to have a correlation with the dynamic properties of bridges in previous studies (Zelaschi, Monteiro, & Pinho 2016). The maximum span length, which has a great impact in the risk classification of the 2020 MIT Guidelines, as will be shown in the following section, has the lowest relative importance as per the machine learning model exercise implemented.

## 5 Italian Guidelines for Bridge Portfolio Assessment

The 2020 MIT Guidelines propose a multi-level and multi-component approach that classifies bridges in risk categories through the processing of qualitative metrics, specific to each of the considered hazards: a) structural/foundational, including eventual

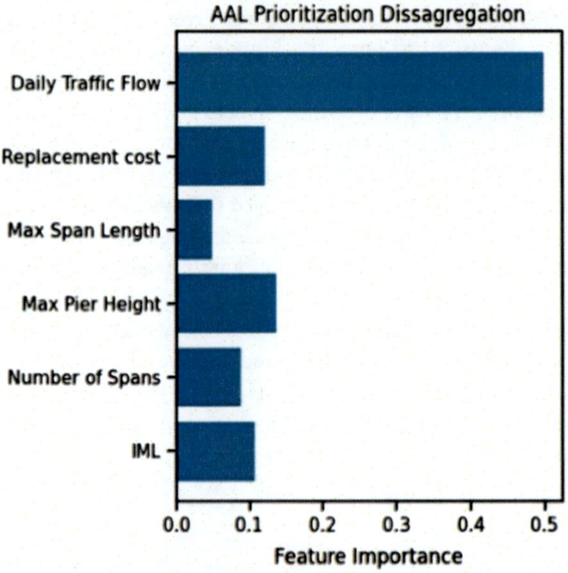


**Fig. 11.** Feature importance found from machine learning model.

degradation; b) seismic; and c) flood/landslide. These guidelines have been recently analyzed and evaluated by Santarsiero et al. (2021), where a thorough summary of the entire classification methodology is presented. In such study, the simple application of the seismic and degradation components of the guidelines to an inventory of 48 bridges concluded that the obtained classification leads to conservative results. In the study presented herein, the focus will be only on the treatment of the seismic risk classification of bridges, since it is the only component for which the benchmark AAL calculations performed in the previous sections is applicable for comparison. For what concerns seismic risk, as with the other considered risk types, the procedure is divided in the three well-known main components: a) hazard; b) exposure; and c) vulnerability, each of which being assigned one of five possible attention levels that range from low to high. This is done by processing qualitative characteristics of each bridge using a specific set of tabular values, as described in the following paragraphs. After each risk component is processed and a classification is made, all components are convoluted into an overall seismic risk attention class. Once each component has been characterised, they are combined to determine an overall seismic risk class, as per the indications shown graphically in Fig. 12. As noted by Santarsiero et al. (2021), the overall classification is very much affected by the vulnerability component; for example, if this component is high, then the seismic risk class will be assigned the highest category, almost regardless of the other components.

The methodology foreseen by the guidelines was applied to the case study inventory, providing the results shown in Fig. 13. It can be seen that both the hazard and vulnerability components are mostly classified in the highest possible option, leading to an overall seismic risk class with mostly the high category. This is attributed to the fact that the

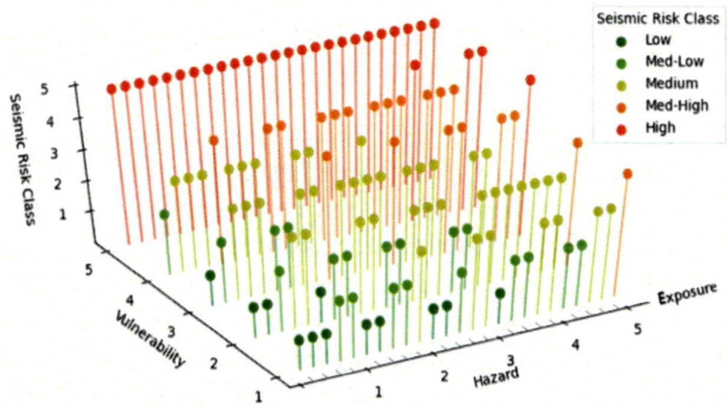


**Fig. 12.** Determination of seismic risk class based on the partial classification of hazard, exposure and vulnerability, adapted from Santarsiero et al. (2021)

vulnerability component dominates for simply supported bridges with spans longer than 20m that have not been seismically designed, which correspond to the predominant characteristics in the case study and to a large portion of the Italian bridge stock. The fact that there are only two resulting categories and the predominance of the high class creates a problem for the effective implementation of these guidelines as a tool for efficient decision-making and resource prioritization. As per the 2020 MIT Guidelines, 498 bridges from the 531 in the inventory that were classified into the high category would require the immediate development of detailed structural analysis, implementation of periodic inspections and the installation of monitoring systems. This would clearly require a great number of resources to comply with and be, in some respects, not fulfilling the need of being able to prioritize effectively.

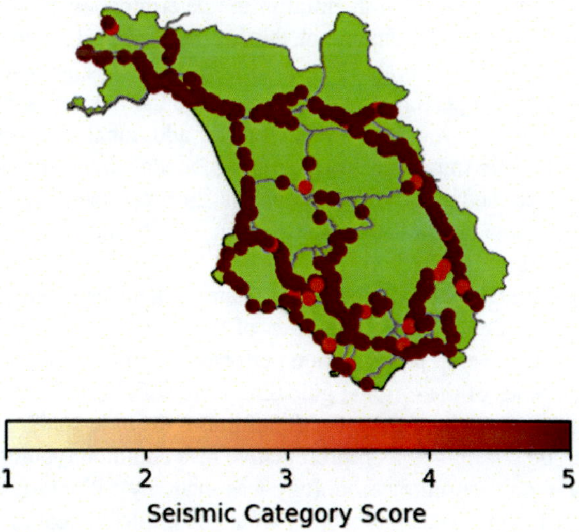


**Fig. 13.** Results for application of 2020 MIT Guidelines to case study inventory.

# 6 Directions for Improvement of Prioritization Scheme

Using the insights gained by the application of the seismic risk quantification to the case study, along with the influential features found via machine learning techniques, a possibly improved methodology to perform bridge prioritization, based on the 2020 MIT Guidelines and their observed performance, is outlined and discussed here. In general, the 2020 MIT Guidelines constitute a robust and well-structured methodology for bridge management. The shortcomings that were observed during its implementation are specifically related to the thresholds used to characterize each of its components in a simple and schematic manner, as well as the high relative importance that the vulnerability component has on the overall risk class. While this conservatism in the vulnerability component was likely a conscious decision made to prioritize bridge safety, it has the downside of classifying a large number of bridges, even those with low associated losses, in the categories of highest priority, which is not in agreement with the findings from a complete quantitative exercise based solely on economic losses. Furthermore, the definition of only five risk classes creates an additional limitation since it can be restrictive when a large, thus more diverse, inventory is considered.

To potentially improve the results obtained by the application of the guidelines, the definition of fixed risk classes could be, for instance, changed to an approach based on a point system per component without establishing a limit. The overall seismic risk score would then be composed of the sum of the scores of each component with the available number of points per component being defined as proportional to the findings from the machine learning model, by giving a higher importance to the exposure component and the daily traffic flows, to further stress the importance of the indirect losses. Regarding the exposure component, the thresholds for span lengths could be modified to reduce the impact of this parameter on the overall results. Also, traffic flows would be reduced to increase its sensitivity, given that this parameter was observed during the machine learning exercise to be the most influential in the determination of annual losses. For what concerns the vulnerability component, the threshold values for number of spans and maximum span length were calibrated by iterating on different values and observing their effect in the classification performance with respect to the AAL ranking. Furthermore, the maximum pier height would be included as an additional parameter since it was recognised as a relatively important feature during the machine learning experiment.

Adopting the described modification proposals, the proposed modified methodology was applied to the same case study, leading to the results shown in Fig. 14. It can be observed that there is a higher resolution of results for each of the components (i.e., no saturation with the high limit), which also translates in a wider range of risk scores for the overall inventory. The spatial distribution of the scores is more in agreement with the loss results and the overall prioritization performance appears greatly improved with respect to the outcomes of the original guideline's methodology.

It is important to note that, while the definition of the case study and its properties were designed to be considered as representative of a common typology of the bridge network of Italy, the proposed methodology was made by calibrating values from the available database therefore its applicability would be limited to real case databases that would be created following the same methodology as the one used herein, particularly in terms of road network modelling.

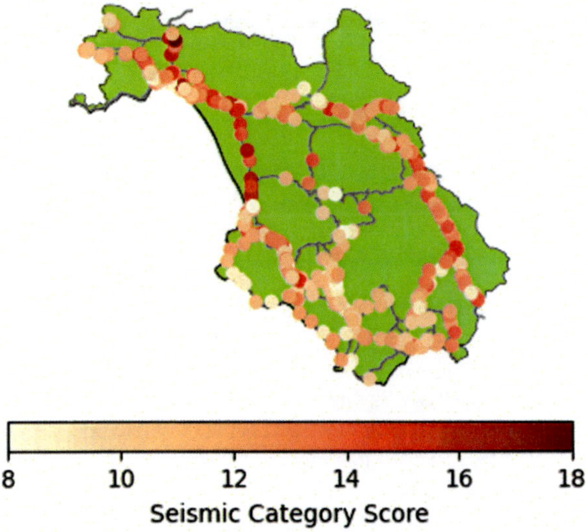


**Fig. 14.** Proposed modified seismic risk classification's prioritisation.

## 7 Conclusions

In general, it can be concluded that, when possible, AAL can be considered an alternative or complementary metric by which assets within a bridge portfolio can be prioritized. This also leads to an amplification of the indirect loss component in prioritization (highly correlated to daily traffic flow data), since it has a much higher economic loss contribution than direct losses. Finally, the application of the current version of the 2020 MIT Guidelines leads to large portions of the inventory classified to the highest-risk available category, creating a challenge to use it efficiently to classify bridge priorities and resource allocation. To address this, a proposal for modification was proposed herein, which performed better in comparison to the current guidelines on the case study evaluated.

**Acknowledgements.** The work presented in this paper has been developed within the framework of the projects INFRA-NAT, co-funded by the European Commission ECHO – Humanitarian Aid and Civil Protection (project reference: 783298 – INFRA-NAT – UCPM-2017-PP-AG), FIRMITAS, co-funded by the Italian Ministry of University and Research (Grant No. 2020P5572N) and "ReLUIS-CSLLPP", funded by the Italian Department of Civil Protection.

## References

Abarca, A., Monteiro, R., O'Reilly, G.J., Zuccolo, E., Borzi, B.: Evaluation of intensity measure performance in the regional assessment of reinforced concrete bridge inventories. Struct. Infrastruct.(2021). https://doi.org/10.1080/15732479.2021.1979599

Borzi, B., Ceresa, P., Franchin, P., Noto, F., Calvi, G.M., Pinto, P.E.: Seismic vulnerability of the Italian roadway bridge stock. Earthq. Spectra **31**(4), 2137–2161 (2015). https://doi.org/10.1193/070413eqs190m. Accessed 18 Jan 2019

Bozorgnia, Y., et al.: NGA-West2 research project. Earthq. Spectra **30**(3), 973–987 (2014). https://doi.org/10.1193/072113eqs209m. 21 Sept 2021

Bureau of Public Roads: Traffic Assignment Manual. U.S. Department of Commerce, Urban Planning Division, Washington, D.C. (1964)

Calvi, G.M., et al.: Once upon a time in Italy: the tale of the morandi bridge. Struct. Eng. Int. **29**(2), 198–217 (2019). https://doi.org/10.1080/10168664.2018.1558033. Accessed 20 Sept 2021

Calvi, G., O'Reilly, G., Andreotti, G.: Towards a practical loss-based design approach and procedure. Earthq. Eng. Struct. Dynam. **50**, 1–13 (2021). https://doi.org/10.1002/eqe.3530

Chase, S., Adu-Gyamfi, Y., Aktan, A., Minaie, E.: Synthesis of National and International Methodologies Used for Bridge Health Indices. McLean (VA): Federal Highway Administration (2016). https://doi.org/10.13140/RG.2.1.1558.1683

Consiglio dei Ministri. Primi elementi in materia di criteri generali per la classificazione sismica del territorio nazionale e di normative tecniche per le costruzioni in zona sismica. Roma: G.U. n. 105 del 8 maggio 2003 - S.o. n.72 (2003)

Consiglio Superiore dei Lavori Publici. Linee Guida per la Classificazione e Gestione del Rischio, Valutazione della Sicurezza ed il Monitoraggio dei Ponti Esistenti. Roma: Ministero delle Infrastrutture e dei Trasporti (2020)

D'Apuzzo, M., Evangelisti, A., Nicolosi, V., Rasulo, A., Santilli, D., Zullo, M.: A simplified approach for the prioritization of bridge stock seismic retrofitting. In: 29th European Safety and Reliability Conference (ESREL). Hannover: Research Publishing Services, pp. 3277–3284 (2019). https://doi.org/10.3850/978-981-11-2724-3_0592-cd

Iatsko, O., Nowak, A.: Revisited live loads for single-span bridges. J. Bridge Eng. (2021).https://doi.org/10.1061/(ASCE)BE.1943-5592.0001647

ISTAT. Matrici del Pendolarismo (2014). https://www.istat.it/it/archivio/139381

Kohrangi, M., Bazzurro, P., Vamvatsikos, D., Spillatura, A.: Conditional spectrum-based ground motion record selection using average spectral acceleration. Earthq. Eng. Struct. Dyn. (2017)

Mangalathu, S., Hwang, S.-H., Choi, E., Jeon, J.-S.: Rapid seismic damage evaluation of bridge portfolios using machine learning techniques. Eng. Struct. **201**, 109785 (2019). https://sciencedirect.com/science/article/pii/s0141029619328068. Accessed 14 Sept 2021

Maratini, R.: Strumenti per l'analisi dei sistemi di trasporto alla scala regionale:modelli di simulazione e sistemi informativi. PhD Dissertation, Università degli Studi di Trieste (2008). http://hdl.handle.net/10077/2752

Ministero delle Intrastrutture e dei Trasporti. Decreto M 578 del 17–12–2020. Roma: Consiglio Superiore dei Lavori Pubblici (2020)

O'Reilly, G., Calvi, G.: Conceptual seismic design in performance-based earthquake engineering. Earthq. Eng. Struct. Dyn. 389–411 (2019).https://doi.org/10.1002/eqe.3141

OpenStreetMap contributors. Planet dump (2020). https://planet.osm.org: https://www.openstreetmap.org

O'Reilly, G.: Seismic intensity measures for risk assessment of bridges. Bull. Earthq. Eng. 3671–3699 (2021).https://doi.org/10.1007/s10518-021-01114-z

Pellegrino, C., Pipinato, A., Modena, C.: A simplified management procedure for bridge network maintenance. Struct. Infrastruct. Eng. **7**(5), 341–351 (2011). https://tandfonline.com/doi/abs/https://doi.org/10.1080/15732470802659084. Accessed 20 Sept 2021

Perdomo, C., Abarca, A., Monteiro, R.: Estimation of seismic expected annual losses for multi-span continuous RC bridge portfolios using a component-level approach. J. Earthq. Eng. (2020). https://doi.org/10.1080/13632469.2020.1781710

Porter, K.A.: An overview of PEER's performance-based earthquake engineering methodology. In: Ninth International Conference on Applications of Statistics and Probability in Engineering. San Francisco, California (2003)

Santarsiero, G., Masi, A., Picciano, V., Digrisolo, A.: The Italian guidelines on risk classification and management of bridges: applications and remarks on large scale risk assessments. Infrastructures **6**(8), 111 (2021). https://doi.org/10.3390/infrastructures6080111

Shahnazaryan, D., O'Reilly, G.: Integrating expected loss and collapse risk in performance-based seismic design of structures. Bull. Earthq. Eng. 987–1025 (2021).https://doi.org/10.1007/s10518-020-01003-x

Silva, V., Crowley, H., Pagani, M., Monelli, D., Pinho, R.: Development of the OpenQuake engine, the Global Earthquake Model's open-source software for seismic risk assessment. Nat. Hazards, **72**(3), 1409–1427 (2014). https://doi.org/10.1007/s11069-013-0618-x. Accessed 23 July 2020

United Nations Office for Disaster Risk Reduction. Infrastructure and Disaster. Third U.N. World Conference on Disaster Risk Reduction. Sendai (2015). https://www.preventionweb.net/files/40429_infrastructure.pdf

Woessner, J., et al.: The 2013 European seismic hazard model: key components and results. Bull. Earthq. Eng. **13**(12), 3553–3596 (2015). https://doi.org/10.1007/s10518-015-9795-1. Accessed 22 Sept 2021

Zelaschi, C., Monteiro, R., Pihno, R.: Parametric characterization of RC Bridges for seismic assessment purposes. Structures (7), 14–24 (2016)

Zelaschi, C., Monteiro, R., Pinho, R.: Simplified period estimation of Italian RC bridges for large-scale seismic assessment. ECCOMAS Conference 2016. Crete, Greece (2016). https://doi.org/10.7712/100016.2237.16427

Zilske, M., Neumann, A., Nagel, K.: OpenStreetMap for traffic simulation (2011). https://depositonce.tu-berlin.de/bitstream/11303/4976/2/zilske_neumann_nagel.pdf. Accessed 21 Oct 2021

# Pier Foundation of Railway Bridge with Long Raked or Vertical Piles

Paolo Stellati[1(✉)], Stefano Palumbo[1], Alfredo D'Angiò[1], and Emanuele Mastrangelo[2]

[1] TEAM Engineering S.P.A., Rome, Italy
{pstellati,spalumbo,adangio}@teamengineering.it
[2] Aya Engineering, Accra, Ghana
e.mastrangelo@ayaengineering.it

**Abstract.** This paper intends to discuss the design aspects of the foundations of bridge piers with long piles, with particular reference to the case of the railway bridge over the Volta River in Ghana, designed for both conditions and finally built with four/six long raked piles for the four piers in water. Due to the significant length of the foundation piles, raked piles are more effective and efficient than vertical piles to take the high seismic and train horizontal loads and to satisfy the requirements of the bridge-track interaction analysis. The design with vertical piles requires a significant increase of foundation quantities. The lateral load test between piers with raked piles is a very effective method to check the compressive/tensile piles' capacity and the actual foundation horizontal displacement and stiffness, in order to confirm the bridge-track interaction analysis.

**Keywords:** raked piles · vertical piles · bridge-track interaction analysis · structural optimization

## 1 Introduction

This paper intends to discuss the design aspects of the foundations of bridge piers with long piles, with particular consideration of their resistance to horizontal loads induced by railway actions and/or seismic events, such as the case of the Volta Bridge, recently built in Ghana.

### 1.1 Vertical or Raked Piles?

The bridge foundation with vertical piles is always preferable, whenever possible, because their construction procedure is simpler than that required by the raked piles. Raked piles become more convenient than vertical piles when the following conditions are encountered:

- High depth of the sound foundation soil, for the presence of water (i.e. off shore structures) and/or swampy top soil;
- High horizontal loads induced by railway actions and seismic events;


M. A. Aiello and A. Bilotta (Eds.): ICC 2022, LNCE 435, pp. 374–387, 2024.
https://doi.org/10.1007/978-3-031-43102-9_29

In order to understand the best design solution, two designs have to be developed, one with vertical piles and one with raked piles.

Raked piles are more structurally effective and efficient than vertical piles, because the horizontal loads are mainly taken by the pile axial load.

The comparison will be carried out taking into account the following parameters:

- Quantities of construction materials
- Equipment / workmanship;
- Construction time;
- Construction risks;

Each of the above mentioned parameters has to be evaluated with a corresponding cost and their sum determines the cost of each of the two design options; the cheapest design is the winning one. The Construction Contractor has to finally carry out this cost comparison, taking into consideration the available equipment and the costs/risks of their adoption. The evaluation of the construction risks is always difficult and is based on confidence, experience and know-how acquired from similar works. The collaboration with a specialist external company has to be considered in this phase, because the cooperation with a competent and reliable supplier is always a key factor for the success of any project.

## 2 Vertical or Raked Piles for the Volta Bridge?

The comparison between vertical and raked foundation piles made for the design of the Volta Bridge is presented in the next paragraph.

The design was initially made with vertical piles, but the cost of this solution is very significant; as a consequence, a second design with raked piles has been studied. The pier foundation design requires 10 or 8 vertical piles of 2.0 m diameter, instead of 4 or 6 raked piles of 1.6m diameter, with a consequent saving of foundation quantities.

In fact, the raked piles design was finally chosen by the Construction Contractor (Afcons), and the Consultant (TEAM) has approved their proposal, with some engineering fear given the perceived risks, also due to the fact that the length of these raked piles is a world record.

The Volta Bridge has five spans of 60 m, for a total length of 300 m; the deck is made up of two steel girders and a top slab in reinforced concrete.

This railway bridge, for single standard gauge track, crosses the Volta River and is located about half way between the Akosombo and the Akuse dams in Ghana. The four piers of this bridge are founded on piles embedded in the rock for a minimum depth of 10.0 m, and with layers of dense and soft sand on top of the rock, with pile lengths ranging between the 40 m of pier P1 to the 77 m of pier P3.

The bridge layout with raked piles is shown in the Fig. 1.

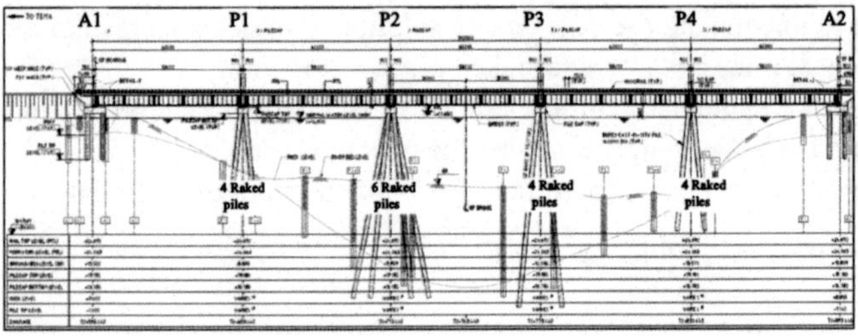


**Fig. 1.** Bridge side view with five spans of 60.0 m

## 2.1 Design Loads of the Volta Bridge

The design standards used for the structural check of the bridge are the Ghana Building Code Ed. 2018 and the Eurocodes.

The total structural and superimposed weight for each pier is approximately 10,000 kN, and the maximum calculated vertical train load for each pier is approximately 5,300 kN.

The longitudinal forces are due to train traction and braking; these loads act at the top of the rails. The longitudinal braking load for each span is 1,200 kN, that is slightly more onerous than the maximum traction force of 1,000 kN.

The water pressure on the piles is also calculated and assessed, but it is not critical for the piles' design.

The wind actions on structures are less critical than seismic actions.

The seismic loads are considered in accordance with the Ghana Building Code (Ed. 2018) that classifies the area of Ghana into four zones (0, 1, 2 & 3). The bridge in subject falls in zone 3, with a horizontal design peak ground acceleration ($a_g$) equal to 0.35 g.

The soil is profile "A" (S1 for the Ghana Code) for the normalized elastic spectra; in fact, the structure is well anchored in the rock (ground type A), but the presence of top layers of sand puts it in the nominal ground type of class D or other. The analysis has been carried out assuming ground type A and considering the structural interaction with the sandy soil.

The overall horizontal seismic action is determined by two orthogonal components, assumed as being independent and characterized by the same response spectrum. In order to consider the seismic kinematic and inertial soil-structure interaction, a general procedure for the estimation of the piles' forces is adopted (Carbonari 2011), in accordance with the bridge stratigraphic profile that is different for each pier; in particular, no sand is encountered at pier P1, a small layer of loose sand at pier P4 and dense and loose sand for the long piles of piers P2 and P3.

The spectrum contribution of the sandy soil layers is automatically considered in the bridge finite element model (SAP 2000), because the lateral stiffness and mass contributions of the sandy layers are calculated and included in the bridge analysis. The contribution of the water is also considered, increasing the mass of the piles in the horizontal directions; the total effective mass is equal to the sum of the actual pile mass (without allowance for buoyancy) and the water mass corresponding to the volume of the immersed pile; in accordance with Annex F of the Eurocode EN 1998–2.7.

The kinematic loads induced by the soil acceleration are also considered. The "free field" acceleration is defined as the soil acceleration without the interaction of the structure. The kinematic forces are induced by the difference between the acceleration of the structure, that is engaged in the bottom rock layer, and the "free field" acceleration of the sandy layers. The analysis is performed using Winkler-type springs for the soil, and the relative calculated soil interaction forces can be determined from the spring support forces.

## 2.2 Method of Analysis

The structural check is made by performing a linear analysis with the two following analysis methods:

1. Pseudo-Static Analysis;
2. Dynamic Modal Response Spectrum Analysis.

### 2.2.1 Pseudo-Static Analysis

The design of the piles can be carried out with this simplified static analysis, based on conventional engineering concepts. This is a first important step for the bridge analysis.

In the pseudo-static analysis, the peak ground acceleration is converted into pseudo-static inertia forces applied to the structural members of the bridge proportionally to their masses. The design seismic inertia forces (FH) are applied in the horizontal directions X and Y, respectively. The Mononobe-Okabe approach is adopted for the determination of seismic earth pressures acting on the piles, similarly to the retaining walls, but doubling the load for the presence of soil in both sides of the pile.

The piles' response with the simplified pseudo-static approach is evaluated, taking into account the pressure, stiffness and strength of the surrounding soil, in accordance with the pile schematic layout shown in the Fig. 2.

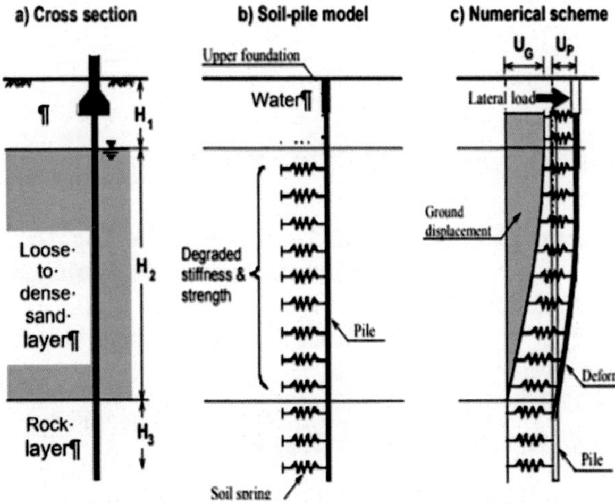


**Fig. 2.** Typical FEM beam-spring model for pseudo-static analysis of piles

### 2.2.2 Dynamic Modal Response Spectrum Analysis

The structural check is made in accordance with an elastic structural analysis based on a "design spectrum", without consideration of the ductility of the foundation piles; it is noted that the eventual formation of flexural plastic hinges in the piles will provide an additional resource of strength, that is conservatively not considered in the analysis. As a consequence, the behaviour factor (K), reflecting the pile ductility capacity, is assumed to be equal to 1.0.

The method used for determining the seismic effects is the modal response spectrum analysis, using a linear-elastic model of the structure. The response of all modes of vibration contributing significantly to the global response is taken into account, and said responses are properly combined with each other under the design seismic excitation, using the CQC method (Complete Quadratic Combination).

The evaluated structural response for each of the two horizontal components (x and y axes of the structures) of seismic action is then combined.

The structural analysis, for different soil types, shall be carried out with the linear dynamic analysis and response spectrum method (item 3.2.2.2 of BS-EN-1998), where the elastic response spectrum Se(T), with T the vibration period, is outlined by the following Fig. 3.

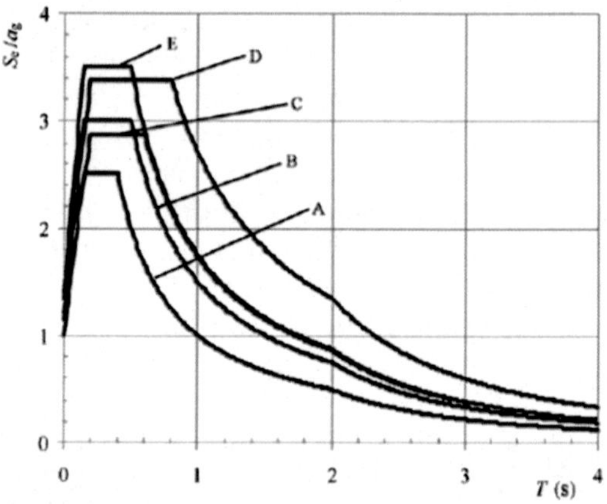


**Fig. 3.** Graph for Elastic Response Spectra from "BS-EN-1998"

The analysis results are quite different for the derived bending moments and shear forces in the piles, because the spectrum analysis captures modal shapes that cannot be determined by the static analysis.

It is noted that the spectrum analysis leads to more onerous stresses in the piles than the pseudo static analysis.

## 3 Piers of Volta Bridge with Vertical Piles

Bored cast-in-situ piles of 2.0 mm diameter were designed with a 3.0 m thick pile cap. The spacing of the piles is 6.0 m in both directions.

The number of vertical piles is no. 8 for piers P1, P3 & P4 and no.10 for pier P2, with lengths ranging between the 38m of pier P1 to the 75 m of pier P2) see Figs. 4 and 5).

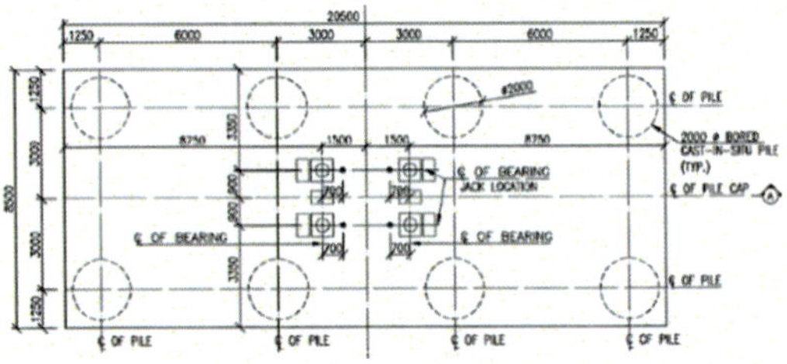

**Fig. 4.** Foundation plan of piers P1, P3 & P4 with 8 vertical piles of 2.0 m diameter

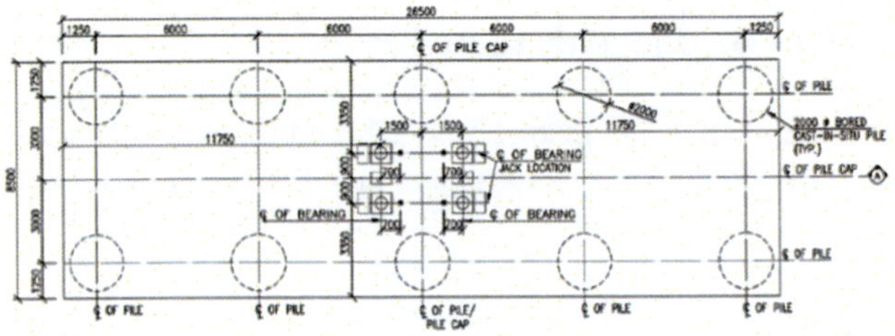

**Fig. 5.** Foundation plan of pier P2 with 10 piles of 2.0 m diameter

In the following Table 1 the construction foundation quantities of concrete and steel are summarized:

**Table 1.** Quantities of construction materials for the vertical piles design

| PIER | Cap. Vol ($m^3$) | Piles Vol. ($m^3$) | Cap Steel (ton) | Piles Steel. (ton) |
|---|---|---|---|---|
| P1 | 523 | 913 | 55 | 201 |
| P2 | 676 | 2565 | 68 | 564 |
| P3 | 523 | 1851 | 55 | 407 |
| P4 | 523 | 1176 | 55 | 259 |

The steel incidence is 100 kg/m$^3$ for pile cap and 220 kg/m$^3$ for piles.

The seismic analysis of the piers has been carried out using the local models, taking into account the impedance between the rock and the loose sand layer.

The FE models of the bridge pier with eight and ten vertical piles are shown in the Fig. 6 below.

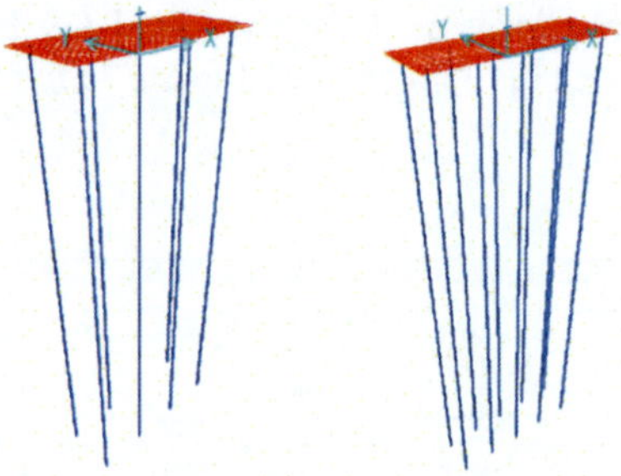

**Fig. 6.** Pier FE model with 8 vertical piles (P1, P3 & P4) and 10 piles (P2)

The piles are subject to significant shear and bending moment for the effects of the longitudinal and transversal seismic and train loads.

## 4 Piers of Volta Bridge with Raked Piles

Bored cast-in-situ piles of 1.6 m diameter have been proposed with a 2.4 m thick pile cap. The number of raked piles is 4 for piers P1, P3 & P4 and 6 for pier P2 with the lengths ranging between the 40 m of pier P1 to the 77 m of pier P3 (see Figs. 7 and 8). The pile cap dimensions are significantly reduced because the pile spacing of three diameters is not requested for raked piles at their top.

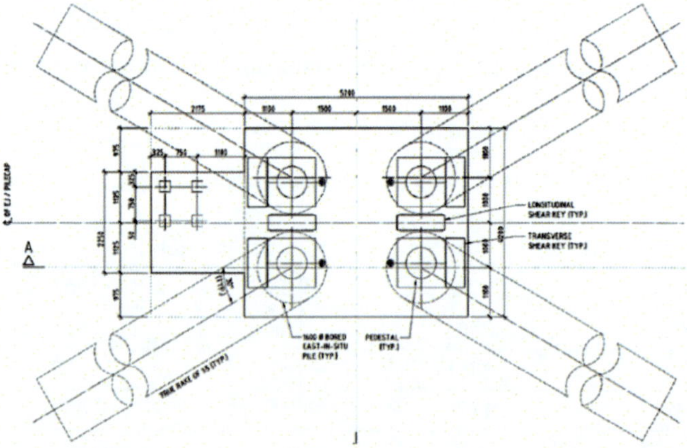

**Fig. 7.** Foundation plan of pier P3 (P1 & P4) with 4 piles of 1.6 m diameter

The four corner piles are inclined of 30° in respect to the longitudinal axis and with a true rake of 1:5 in the vertical direction.

The Table 2 shows the construction foundation quantities of concrete and steel (Tables 3 and 4):

The incidences are 100 kg/m$^3$ for pile cap and 290 kg/m$^3$ for piles.

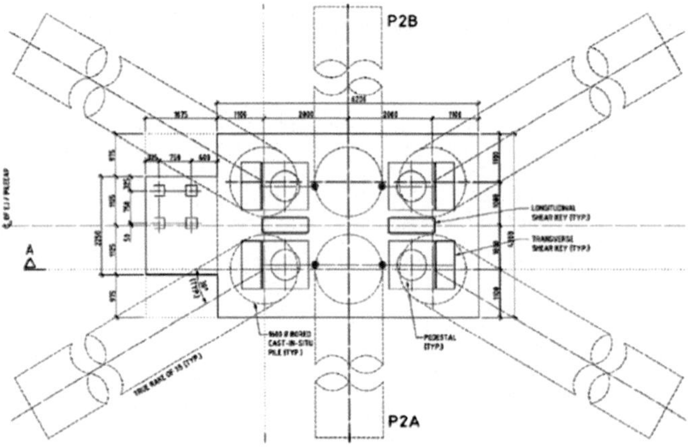


**Fig. 8.** Foundation plan of pier P2 with 6 piles of 1.6 m diameter

**Table 2.** Quantities of construction materials for raked piles solution/Quantità dei materiali di costruzione per soluzione conpali inclinati

| PIER | Cap. Vol ($m^3$) | Piles Vol. ($m^3$) | CapSteel (ton) | PilesSteel. (ton) |
|---|---|---|---|---|
| P1 | 52.4 | 320 | 5.0 | 93 |
| P2 | 62.5 | 880 | 6.5 | 255 |
| P3 | 52.4 | 620 | 5.0 | 180 |
| P4 | 52.4 | 410 | 5.0 | 119 |

**Table 3.** Difference and ratio of concrete volumes between the design with raked and vertical piles

| PIER | Cap Vol. Diff. ($m^3$) | Piles Vol. Dif. ($m^3$) | Cap Vol. Raked/Vertical. (%) | Piles Vol. Raked/Vertical. (%) |
|---|---|---|---|---|
| P1 | 471 | 593 | **10.0** | **35.0** |
| P2 | 614 | 1685 | **9.2** | **34.3** |
| P3 | 471 | 1231 | **10.0** | **33.5** |
| P4 | 471 | 766 | **10.0** | **34.9** |

The FE models of the bridge pier with four and six raked piles are shown in the Fig. 9.

The piles are subject to significant axial load, but also shear and bending moment.

**Table 4.** Difference and ratio of steel quantities between the designs with raked and vertical piles

| PIER | Cap Steel diff. ($m^3$) | Piles Steel dif. ($m^3$) | Cap Steel Raked/Vertical. (%) | Piles Steel Raked/Vertical. (%) |
|---|---|---|---|---|
| P1 | 50 | 108 | **9.1** | **46.2** |
| P2 | 62 | 309 | **9.6** | **45.3** |
| P3 | 50 | 227 | **9.1** | **44.2** |
| P4 | 50 | 140 | **9.1** | **45.9** |

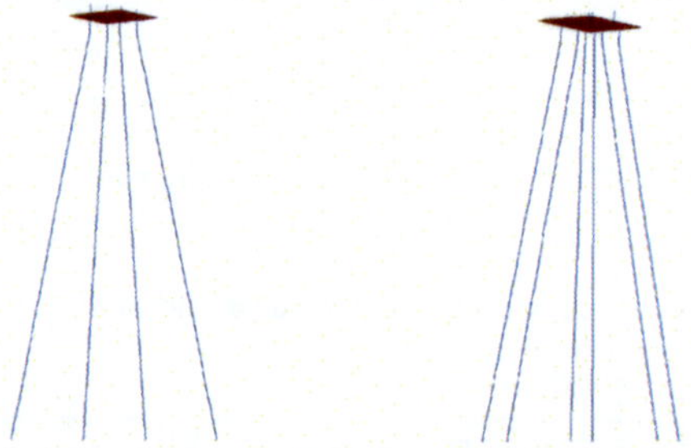

**Fig. 9.** Pier FE model with 4 raked piles (P1, P3 & P4) and 6 raked piles (P2)

## 5 Comparison Between Vertical and Raked Piles

Raked piles provide higher horizontal stiffness than vertical piles for the contribution of the pile axial capacity; in fact, raked piles sustain the horizontal loads principally with this push-pull mechanism. Vertical piles fully rely on the lateral soil displacements demand and its soil-structure interaction behaviour to fully utilise the benefit of this solution.

As a consequence, the use of raked piles leads to a significant savings of reinforced concrete for the piles and for the foundation base, as demonstrated in the tables below:

In conclusion, the saving of reinforced concrete is of approximately 90% for the pile cap and of 35% and 45% for the piles' concrete and reinforcement respectively.

## 6 Construction Equipment for Raked Piles

To allow the construction of the raked piles it is necessary to install a temporary platform (founded on circular hollow steel piles) composed of a horizontal base frame and a pair of guides fixed 4 m apart from each other, with the purpose of driving the liners and ensuring that the drilling has the same design slope of the raked pile (1:5 in the Volta Bridge case). The construction of the platform and the subsequent drilling operations for the piles in water require the support of a barge, previously manufactured in place, where the machineries and the laborers can safely operate (see Fig. 10).

Given the significant depth to be bored, R.C.D. (Reverse Circulation Drilling) technology is used.

**Fig. 10.** Equipment adopted for the construction of the Volta Bridge raked piles

The system consists of a tool fitted with roller cutter bits, which penetrates continuously into the soil or rock by rotation, thrust and weight.

After the completion of the boring, the pre-assembled reinforcement cages are lowered into the liners and the longitudinal bars are joined with mechanical couplers.

## 7 Tests of the Piles

Notwithstanding the specialized workmanship and supervision during all the construction stages, large diameter raked bored piles involve challenges and risks that needs to be taken into account during the development of the method statement, in which the type and number of tests to be carried out on the piles are decided.

Indeed, in order to assess the concrete quality and to check the pile integrity, *Ultrasonic Crosshole Testing* were carried out on all the piles of piers and abutments, measuring the propagation time and relative energy of an ultrasonic pulse between parallel access ducts (4 nos. Installed in each pile).

Furthermore, for the purpose of testing the four groups of piles, two *Lateral Load Tests* were conducted between the piers P1-P2 and P3-P4.

The three main objectives of the test were:

1. to carry out the compressive and tensile load tests of the piles;
2. to assess the extent of depth in rock where the pile stresses were still significant;
3. to measure the actual lateral pile cap displacement under a sequential load increment, to be compared with the theoretical values, in order to determine/confirm the correct pier stiffness to be used in the bridge-track interaction analysis.

The *Lateral Load Test* was carried out with a pulling system composed of 12 strands (6 on each side of the pile caps) of nominal diameter of 15.2mm, fixed to the flanges of two floating steel girders placed over temporary platforms anchored to the pile caps, as shown in the Figs. 11 and 12:

**Fig. 11.** Details of the test arrangement

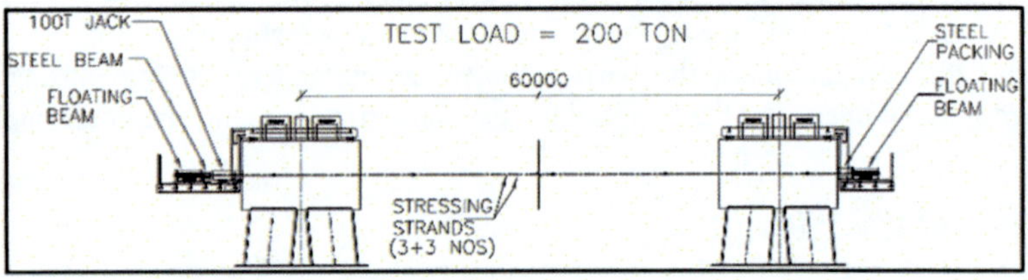


**Fig. 12.** Lateral static load test between adjacent piers

The test load of 2000 kN was applied by 4 hydraulic jacks installed one of the piers, in 9 gradual increments, followed by the same number of unloading steps.

All the piles were instrumented with multiple strain gauges (vibrating wire type) previously fixed to the reinforcement cages at different levels (4 m above and 3 m below the liner termination level and 3 m above the pile tip) and embedded in the poured concrete (ref. Fig. 13).

The Fig. 14 shows the measurements, step-by-step, of 3 + 3 strain gauges placed along two rebars of two piles of the pier P4 (at the aforementioned levels 1, 2 and 3). It demonstrates that the adopted socketing length in rock of 6 pile diameters (≈10 m) was sufficient to ensure the complete transmission of the shaft stresses to the rock, being the microstrains almost null at levels 3.

The displacements of the pile caps were measured also step-by-step, with two different total stations fixed on the riverbank.

The measured force-displacement graphs were then compared to the theoretical diagrams predicted with the FE model, to confirm the stiffness values of the system soil-bridge assumed in the analysis (Fig. 15).

Such values, derived with the help of literature correlations, are by nature affected by uncertainties and hence they needed to be confirmed by direct measurements on site.

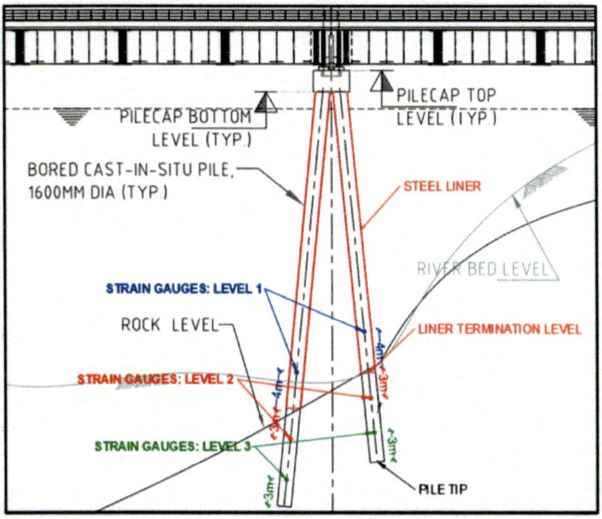


**Fig. 13.** Scheme of the pile test instrumentation

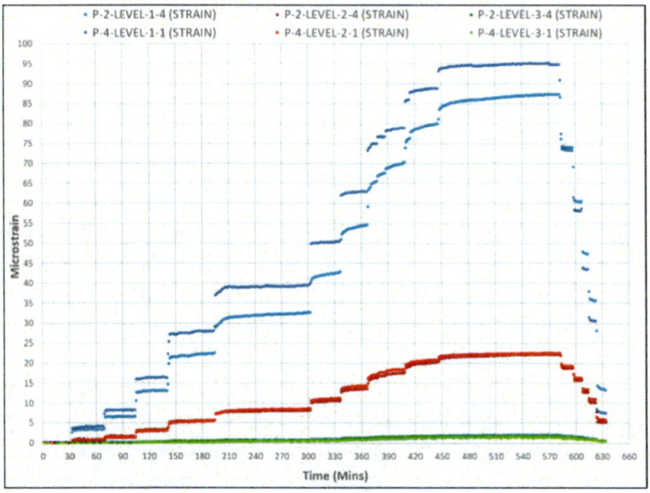


**Fig. 14.** Results of six strain gauges in two piles of Pier P4

It can be noticed, from the above figure, that the slope (stiffness) of the predicted F-u curve is very similar to the average slope of the measured diagrams, with a deviation of the maximum displacement of the pier cap of only 4 mm.

Such confirmation is extremely important to validate the assumptions on the soil-structure interaction and to confirm the correspondence between the behavior of the actual structure and the one modelled in the FE analysis.

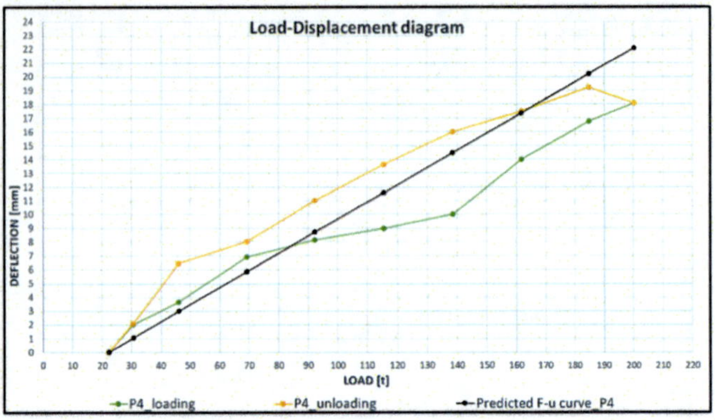


**Fig. 15.** Measured and predicted Load-Displacement curves of Pier

# 8 Conclusions

The foundation layout with raked piles is more effective and efficient than that with vertical piles in the case of foundation caps in water and/or in soils with poor mechanical properties, as demonstrated for the Volta Bridge case.

The main advantages of the raked piles compared with the vertical piles are the following:

1. Reduced quantities of the pier cap reinforced concrete due to the more compact shape of the basement, being the pile spacing of three diameters not requested for raked piles at their top; this reduction is of approximately **90%** in the case of the Volta Bridge.
2. Reduced quantities of the piles reinforced concrete as a result of the reduced number of piles and of their diameter; this reduction, in the case of the Volta Bridge, is of approximately **35%** for the concrete and **45%** for the reinforcement.
3. Reduced environmental impact and pollution during construction, for the compact shape of the foundation basement.
4. Preferable landscaping of the bridge with a foundation base of reduced dimensions.
5. Reduced construction time, as a consequence of the reduced number of piles.
6. Load test of the piles carried out with a simpler temporary system (pulling strands, as shown in the previous paragraph), instead of an expensive static load test with counterweight beams required for the vertical piles.

The main disadvantages of the raked piles compared to the vertical piles are the following:

1. The equipment required for the construction of the raked piles is more expensive and requires specialist workmanship.
2. The construction of raked piles with a big diameter may require the engagement of a specialist company, with consequent time and risks derived from the definition of tender requirements and contract details.
3. The construction risks can be evaluated with less accuracy given the less work experience of general contractors.

The design choice must come from the critical analysis of all advantages and disadvantages, but two designs, one with raked and one with vertical piles, are necessary in order to define the solution that is best thought out for each project.

**Acknowledgements.** The Authors of this article wish to give special thanks to the technical staff of: Afcons Ltd (Construction Contractor); Structurae (Designer of the Volta Bridge), TEAM Engineering SpA (Consultant) and Aya Engineering (Supervisor of the works).

## References

Ghana Building Code GhBC GS1207:2018: "Building and Construction (2018)

Eurocode EC-8. Design of structures for earthquake resistance. Part 5: foundations, retaining structures and geotechnical as-pects. Brussels: European Committee for Standardization 1998

Bieniawski, Z.T.: Engineering Rock Mass Classifications: A Complete Manual for Engineers and Geologists in Mining, Civil, and Petroleum Engineering. John Wiley & Sons (1989)

Joseph, E., Bowles, R.E., S.E.: Foundation Analysis and Design. The McGraw-Hill Companies, Inc. (1997)

Cairo, R., Dente, G.: Kinematic interaction analysis of piles in layered soils, ISSMGE-ERTC 12 Workshop Geotechnical Aspects of EC8. Pàtron Editore, Bologna, Madrid (2007)

Carbonari, S., Dezi, F., Leoni, G.: Seismic soil-structure interaction in multi-span bridges: application to a railway bridge. Earthquake Engineering and Structural Dynamics (2011)

Stellati, P., Palumbo, S., D'Angiò, A., Mastrangelo, E.: Seismic soil structural interaction for foundation piles embedded in rock with top layers of dense and loose sands. Italian concrete days - Giornate AICAP 2020 Congresso Aicap CTE (2021)

# Risk Classification and Preliminary Safety Evaluation for a Network of Existing RC Bridges: An Application of the Italian Guidelines 2020

Andrea Miano(✉), Antimo Fiorillo, Annalisa Mele, and Andrea Prota

Department of Structures for Engineering and Architecture, University of Naples Federico II, Naples, Italy
{andrea.miano,a.fiorillo,annalisa.mele,aprota}@unina.it

**Abstract.** Existing national road and highway bridges require periodical inspections and, in some cases, urgent structural measures. The Italian Guidelines for risk classification and management, safety assessment and monitoring of existing bridges (LG20) outline a procedure for estimating approximately the risk associated to bridges, by defining for each of them an Attention Class, on the basis of hazard, vulnerability and exposure parameters. For each Attention Class, consequential actions in terms of surveys/monitoring/verifications are required, in order to prevent inadequate levels of damage. A set of 4 existing RC bridges are considered herein as case study. The static scheme is simply supported beam. Initially, the Attention Classes are obtained. Then, the preliminary safety evaluation is provided. Based on the outcomes, a priority classification of the bridges to be further investigated is estimated. Potentials and criticisms of the LG20 are also discussed in the work.

**Keywords:** reinforced concrete bridges · bridges guidelines 2020 · bridges structural risk · bridges seismic risk

## 1 Introduction

In the recent years, the preliminary safety evaluation of bridges has become a central theme in Italy, both because of the tragic collapse of many bridges and because of the age of the majority of the existing Italian bridges, that is preparing to reach or has already passed their nominal project life. The heritage of Italian bridges is very vast and varied because of the geographical and orographic conformation of the territory. With the emission of the Italian Guidelines for risk classification and management, safety assessment and monitoring of existing bridges (LG20, Consiglio Superiore dei Lavori Pubblici 2020), it has been introduced a methodology for the risk classification and preliminary safety evaluation of all the existing typologies of bridges on the national territory. In particular, existing national road and highway bridges are subjected to periodical inspections and, in some cases, urgent structural measures.


M. A. Aiello and A. Bilotta (Eds.): ICC 2022, LNCE 435, pp. 388–398, 2024.
https://doi.org/10.1007/978-3-031-43102-9_30

Some analysis of the LG20 have been reported in Salvatore et al. (2020), Baratono et al. (2021) and Buratti (2021). The LG20 outline a procedure for estimating approximately the risk associated to bridges, by defining for each of them an Attention Class on the basis of hazard, vulnerability and exposure parameters. Based on the Attention Class defined for the bridge under consideration, consequential actions are required, as surveys, prevision of different types of monitoring, execution of preliminary or accurate structural verifications, aimed to prevent dangerous levels of damage. In particular, the theme of the structural monitoring is very important (FIB Task Group 2002), since the observation of significant parameters through traditional monitoring systems (e.g., composed by accelerometers, inclinometers, LVDT, Zhang et al. 2017 and Rainieri et al. 2020) or most innovative ones (e.g., satellite remote sensing, Di Carlo et al. 2021, Talledo et al. 2022, Mele et al. 2022 and 2023, and Miano et al. 2022a and b) can highlight problematic performance of the structure in a timely manner.

In this work, an application of the LG20 methodology is reported with reference to a set of 4 existing reinforced concrete (RC) bridges. In all the cases, the static scheme is simply supported beam.

Initially, once collected all the information related to the necessary parameters, the Attention Classes are defined for each bridge. Then, based on the Attention Classes results, the subsequent interventions undertaken for each individual viaduct are illustrated. In this case, for all the set of bridges the performing of the preliminary assessments is necessary. Based on the outcomes of the preliminary assessment, a priority classification of the bridges, to be further investigated, is evaluated.

In detail, Sect. 2 presents a synthesis of the procedure to determine the Attention Class of a bridge (Sect. 2.1) and to perform the preliminary evaluations (Sect. 2.2). The case study application id illustrated in Sect. 3, with the description of the case study bridges (Sect. 3.1), the determination of the Attention Classes for the 4 bridges (Sect. 3.2) and the presentation of the results of the preliminary verifications (Sect. 3.3). Finally, potentials and criticisms of the LG20 are also discussed in Sect. 3.4.

## 2 Methodology

### 2.1 Attention Classes

This paragraph presents a discussion about the determination of the Attention Class. The attention class is based on 4 sub-classes for 4 risk categories: i) Structural and Foundational Attention Class; ii) Seismic Attention Class; iii) Landslides Attention Class; iv) Hydraulic Attention Class. Five possible results can be obtained for each sub-Attention class: High, Medium-High, Medium, Medium-Low and Low. For each of them, the definition of the Attention Class is inspired by the well-known scheme of risk definition, which is the result of the combination of three main factors: hazard, vulnerability and exposure (Fig. 1).

At the end, the combination of the results of the 4 sub-Attention classes gives the overall Attention Class. Table 1 and Table 2 summarize the parameters and their explanation for the definition of the two classes more related to structural and seismic engineering: Structural and Foundational Attention Class and Seismic Class. Based on the intersection of the results of these parameters can be obtained the hazard, the vulnerability and

**Fig. 1.** Combination of hazard, vulnerability and exposure to evaluate the Attention Class of a bridge.

the exposure and then the overall Structural and Foundational and Seismic Attention Classes.

**Table 1.** Synthesis of the parameters needed for the definition of the Structural and Foundational Attention Class.

| Parameter | Explanation |
|---|---|
| **Hazard-5 classes (from Low to High) based on the intersection of the sub-parameters** | |
| Road class | 5 classes based on the maximum admittible mass on the road |
| Frequency of vehicles | 3 classes based on frequency of passage of commercial vehicles |
| **Vulnerability-5 classes (from Low to High) based on the intersection of the sub-parameters** | |
| Level of defectivity | 5 classes based on the current state of conservation of the structure |
| Speed of the degradation evolution | 3 classes based on the period of construction or the last significant maintenance |
| Design Regulation | 3 classes based on the design year |
| Static scheme, material, span length | 5 classes based on the intersection of these parameters |
| **Exposition-5 classes (from Low to High) based on the intersection of the sub-parameters** | |
| Average daily traffic and span length | 5 classes based on the intersection of these two parameters |
| Presence of alternative roads | 2 classes based on presence of alternative roads |
| Type of bypassing entity | 3 classes based on the consequences of the bridge collapse on the entity |

Similarly, the other two classes (Landslides and Hydraulic Attention Classes) are defined.

As mentioned, at the end of the process, with a multi-risk analysis, intersecting the results for the 4 risk categories, the overall Attention Class is defined. Based on the

**Table 2.** Synthesis of the parameters needed for the definition of the Seismic Attention Class.

| *Parameter* | *Explanation* |
|---|---|
| **Hazard-5 classes (from Low to High) based on the intersection of the sub-parameters** | |
| Peak acceleration | 5 classes based on the peak ground acceleration ($a_g$), with probability of exceedance over 50 years of 10%, on rigid soils |
| Topographic class | 5 classes based on the features of the topographic surface |
| Subsoil category | 2 classes based on the subsoil typology |
| **Vulnerability-5 classes (from Low to High) based on the intersection of the sub-parameters** | |
| Static scheme, material, span length | 5 classes based on the intersection of these parameters |
| Design Criteria | 2 classes based on the design criteria |
| Level of defectivity | 5 classes based on the current state of conservation of the structure |
| Static scheme, material, span length | 5 classes based on the intersection of these parameters |
| **Exposition-5 classes (from Low to High) based on the intersection of the sub-parameters** | |
| Structural Attention Class | 5 classes based on the estimation of the Structural and Foundational Attention Class Exposition |
| Strategic importance Level | 2 classes based on the strategic importance of the bridge |

Attention Classes results, the interventions to be undertaken for each individual viaduct, are defined. The measures for each of the 5 classes are herein summarised:

- High Class: it is appropriate to immediately have an accurate assessment, both in terms of safety assessments and in-depth analysis of geotechnical and/or structural characteristics, where necessary. For High Attention Class bridges, regular inspections and, if necessary, extraordinary periodic inspections, and the installation of periodic or continuous monitoring systems, are required.
- Medium-High Class: it is foreseen the use of preliminary evaluations and the execution of routine periodic inspections and, if necessary, of extraordinary periodic inspections, and the installation of periodic or continuous monitoring systems. The management team can verify, case-by-case, the need to carry out accurate assessment, based on the type and quality of defects found and on the results of the preliminary evaluations.
- Medium Class: it is necessary to carry out preliminary evaluations and regular periodic inspections. If, on the basis of the inspections, rapidly changing degradation phenomena have been detected, special periodic inspections should also be organized. The management team can verify, case-by-case, on the basis of the preliminary evaluations, whether it is necessary to install periodic or continuous monitoring systems (reclassifying the bridge in the Medium-High Class) and/or whether to carry out accurate safety assessments.

- Medium-Low and Low Classes: there are no evaluations or analyses other than those already carried out but the execution of frequent periodic inspections.

### 2.2 Preliminary Evaluations

The preliminary evaluations are carried out where the attention class is found to be medium or medium-high. These evaluations aim to assess the quality and type of defects detected at the periodic inspections and to estimate, in advance, the resources of the viaduct according to the design age regulations. It is assumed that the viaduct design has been drawn up in accordance with the regulatory guidelines in force at the time of its construction unless there are clear indications to the contrary that show macroscopic design errors, and that it has been optimized to withstand to the corresponding traffic loads. Then, the ratio between the demand induced on the various viaduct components (slabs, traverses, beams and/or main structures, piles, shoulders, retaining devices and foundations) by the traffic loads required by the rules at the time construction and the demand obtained using the traffic models provided for by the current rules is calculated. In this case, for all the elements considered critical for the viaduct, the maximum bending/shear/axial load calculated with the regulation in force at the time of construction are compared with the maximum bending/shear/axial load calculated according to NTC 2018. Then, as mentioned before, the management team can assess, case-by-case, the need to carry out accurate assessments, based on the type and quality of defects found through the different types of inspection levels provided (assessing whether, for example, they may have been caused by vertical traffic loads) and on the preliminary evaluations results. Additional information from the monitoring systems, where applicable, may assist in the preliminary evaluations.

## 3 Application

The LG procedure is applied to a set of 4 existing RC bridges. In Sect. 3.1, a brief description of the RC bridges is reported. In Sect. 3.2, the determination of the Attention Class for each of them is illustrated. Finally, in Sect. 3.3, the results of preliminary verification are shown, where required.

### 3.1 Description of the Case Study Bridges

The first bridge presents a single span, with a static scheme of a supported beam in prestressed RC. The deck is made up of 8 beams placed side by side, 1.20 m high, connected by a slab 0.20 m thick and by two end traverses. The theoretical span light is 20.00 m. The deck rests directly on the foundations, consisting of a block of c.a. above the poles. A scheme of the transversal cross section of the bridge n.1 is reported in Fig. 2a.

The second bridge has 4 bays with a static scheme of supported beams in prestressed RC. The deck is made up of 8 beams placed at a distance of 1.65 m, 1.20 m high, connected by a slab 0.20 m thick and by two end traverses and two central traverses. The theoretical length of the span is 16.90 m. The structure in elevation of the viaduct

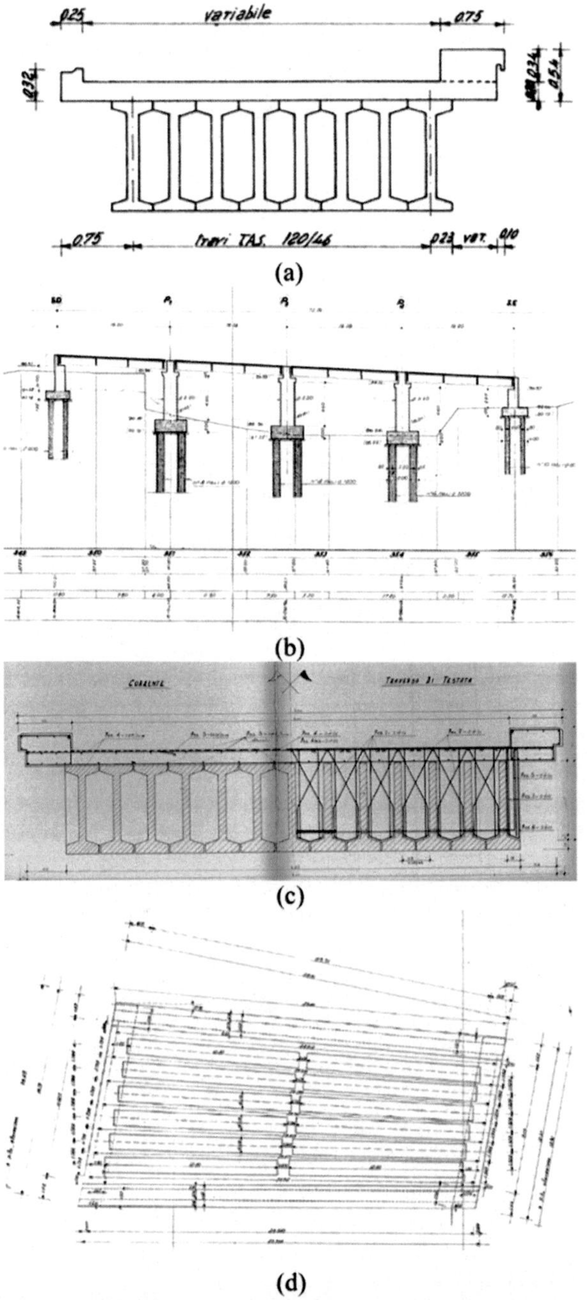


**Fig. 2.** (a) Transversal cross section of the bridge n.1; (b) longitudinal section of the bridge n.2.; (c) transversal cross section of the bridge n.3; (d) plan view of one of the two decks of the bridge n.4.

consists of c.a. stacks with circular cross-section, pulvinus on which rests the deck, and shoulders. The foundations consist of plinths on poles. A scheme of the longitudinal section of the bridge n.2 is reported in Fig. 2b.

The third bridge has a single span, with a static scheme of supported beam in prestressed RC. The deck is made up of 13 adjacent beams, 1.20 m high, connected by a slab 0.20 m thick and by 2 end traverses. The theoretical span length is 22.05 m. This bridge has a shoulder in common with another adjacent bridge. The foundations are on poles. A scheme of the transversal cross section of the bridge n.3 is reported in Fig. 2c.

The fourth bridge has 1 span schematized as a supported beam. The deck consists of a lightened insole in prestressed RC, consisting of an upper and a lower slab in c.a. and ribs in c.a.p. placed at a variable distance between 1.35 m and about 2 m. The overall height of the section is 1,53 m. The insoles have a theoretical length of about 28.5 m. There are two end traverses and one traverse in the middle of the decks. The decks have a width variable between 12.51 m and 14.79 m for the left sole and 12.78 and 15.97 m for the right sole. The deck rests on the two lateral shoulders, while the foundations are on poles. A scheme of the plan view of the one of the two decks of the bridge n.4 is reported in Fig. 2d.

### 3.2 Definition of the Attention Classes

The procedure outlined by the LG20 for estimating approximately the risk associated to bridges has been applied to the 4 bridges described in the previous section. For each of them, the relative Attention Class has been defined on the basis of hazard, vulnerability and exposure parameters of the 4 sub-Attention Classes. The evaluation of the Structural and Foundational Attention Classes is reported in Table 3. The first three blocks report the information needed to evaluate the hazard, the vulnerability and the exposition, and in the last row the Structural and Foundational Attention Classes that derive are reported.

**Table 3.** Evaluation of the Structural and Foundational Attention Class for the 4 case-study bridges.

| | Bridge 1 | Bridge 2 | Bridge 3 | Bridge 4 |
|---|---|---|---|---|
| Road class | Class A | Class A | Class A | Class A |
| Frequency of vehicles | Low | Low | Low | Low |
| **Hazard** | **Medium-High** | **Medium-High** | **Medium-High** | **Medium-High** |
| Level of defectivity | Medium-High | Medium-Low | Medium-High | Medium-Low |
| Speed of the degradation evolution | 1945–80 | 1945–80 | 1945–80 | 1945–80 |
| Design Regulation | Class B | Class B | Class B | Class B |
| Static scheme, material, span length | Medium | Medium-High | Medium | Medium-High |

(*continued*)

**Table 3.** (*continued*)

| | Bridge 1 | Bridge 2 | Bridge 3 | Bridge 4 |
|---|---|---|---|---|
| **Vulnerability** | **Medium-High** | **Medium** | **Medium-High** | **Medium** |
| Average daily traffic and span length | Medium | Medium | Medium-Low | Medium-High |
| Presence of alternative roads | Yes | Yes | Yes | Yes |
| Type of bypassing entity | Low | Low | High | High |
| **Exposition** | **Medium-Low** | **Medium-Low** | **Medium** | **High** |
| ***Structural and Foundational Attention Class*** | ***Medium-High*** | ***Medium*** | ***Medium-High*** | ***Medium-High*** |

The evaluation of the Seismic Attention Classes is reported in Table 4. The first three blocks report the information needed to evaluate the hazard, the vulnerability and the exposition, and in the last row the Seismic Attention Classes that derive are reported. It is important to specify that all the presented bridges were built without following seismic criteria.

**Table 4.** Evaluation of the Seismic Attention Class.

| | Bridge 1 | Bridge 2 | Bridge 3 | Bridge 4 |
|---|---|---|---|---|
| Peak acceleration | Medium-High | Medium-High | Medium-High | Medium-High |
| Topographic class | T1 | T3 | T1 | T1 |
| Subsoil category | C | C | C | C |
| **Hazard** | **High** | **High** | **High** | **High** |
| Static scheme, material, span length | Medium-High | High | Medium-High | High |
| Design Criteria | Not seismic | Not seismic | Not seismic | Not seismic |
| Level of defectivity | High | Medium | High | Medium |
| **Vulnerability** | **High** | **Medium-High** | **High** | **Medium- High** |
| Structural Attention Class | Medium-Low | Medium-Low | Medium | High |
| Strategic importance Level | Strategic | Strategic | Strategic | Strategic |
| **Exposition** | **Medium** | **Medium** | **Medium-High** | **High** |
| ***Seismic Attention Class*** | ***High*** | ***Medium-High*** | ***High*** | ***High*** |

A specific discussion can be introduced about the level of defectivity (both structural and seismic). This parameter is crucial in the assignment of the overall class and should be obtained based on the defectivity grades proposed in the. Clearly, the role of the inspection is fundamental and should be as much as possible objective through an entire system of bridges in order to reduce the dependency on the subjective opinion of the single inspectors.

With respect to the level of defectivity, it is to highlight the importance of the definition of the critical elements, for the evaluation of both the static and the seismic vulnerability. In fact, a significant defect level on those critical elements can led to an High level of defectivity, and consequently to an High Attention Class (at least with respect to the Structural and Foundational Attention Class).

Another key aspect is linked to the RC bridges with post-tense cables, for which it is necessary to acquire additional information, as mentioned in the paragraph of the LG20 focused on the "special inspections". The special inspections are mandatory for the definition of the Attention Class of bridges with such structural typology. For this reason, they should have absolute priority with respect to the other bridges of the set, in order to accelerate the inspections process.

The Hydraulic and Landslides Attention Classes are reported in Table 5. The results were provided by experts.

**Table 5.** Hydraulic and Landslides Attention Class.

| | Bridge 1 | Bridge 2 | Bridge 3 | Bridge 4 |
|---|---|---|---|---|
| *Hydraulic and Landslides Attention Class* | ***Medium*** | ***Low*** | ***Low*** | ***Low*** |

Finally, the overall Attention Classes are reported in Table 6. Three over four bridges are characterized by a Medium-High Overall Attention Class, one of them by a Medium Overall Attention Class.

**Table 6.** Overall Attention Class.

| | Bridge 1 | Bridge 2 | Bridge 3 | Bridge 4 |
|---|---|---|---|---|
| *Overall Attention Class* | ***Medium-High*** | ***Medium*** | ***Medium-High*** | ***Medium-High*** |

### 3.3 Results of the Preliminary Verifications

The Overall Attention Classes for the four bridges are Medium and Medium-Hight, thus it is necessary to perform the preliminary evaluations. The verifications consist in the assessment of the relationship between the demand induced on the critical structural elements of the bridge by the traffic loads provided for by the regulations of the design epoch and the demand obtained using the traffic models provided by the rules currently in force. The four bridges have different structural configurations, so the critical elements considered in the following verifications differ from a bridge to another one. Herein, a summary of the results of the preliminary evaluations is reported:

- For the bridge n.1, the critical elements are beams, slab, and supports. The most critical condition regards the beams, with a maximum flexional ratio of 0.62;
- For the bridge n.2, the critical elements are beams, slab, traverses, pulvinus, supports, piles and shoulders. The most critical condition regards the slab, with a maximum flexional ratio of 0.51;
- For the bridge n.3, the critical elements are beams, supports and shoulders. The most critical condition regards the shoulder, with a load ratio of 0.48;
- Finally, for the bridge n.4, the critical elements are deck and shoulders. The most critical condition regards the shoulder, with a load ratio of 0.76.

## 4 Conclusions

In this work, an application of the LG20 methodology has been presented with reference to a set of 4 existing RC bridges. Initially, once collected all the information related to the necessary parameters, the Attention Classes have been defined for each bridge. Then, based on the Attention Classes results, the subsequent interventions undertaken for each individual viaduct have been illustrated. The critical issues in the assignment of the Attention Classes have been discussed. In this case, for all the set of bridges the performing of the preliminary evaluations has been necessary. Based on the outcomes of these preliminary evaluations, a priority classification of the bridges, to be further investigated, can be evaluated. In particular, for each given Attention Class, the bridges can be put in a priority order of future interventions, based on the lowest coefficient of the preliminary evaluations. This gives to the management teams an important methodology to plan the future operations on big systems of infrastructures.

## References

Baratono, P., Cosentino, A., Puggelli, S., Renzi, E., Salvatore, W.: Ingenio, Le nuove Linee Guida per la classificazione e gestione del rischio, la valutazione della sicurezza dei ponti ed il monitoraggio dei ponti esistenti approvate dal Consup (2021)

Buratti, G., et al.: The new guidelines of Italian Ministry of Infrastructures for the structural risk classification of existing bridges: genesis, examples of application and practical considerations. In: Pellegrino, C., Faleschini, F., Zanini, M.A., Matos, J.C., Casas, J.R., Strauss, A. (eds.) EUROSTRUCT 2021. LNCE, vol. 200, pp. 835–844. Springer, Cham (2022). https://doi.org/10.1007/978-3-030-91877-4_95

Consiglio Superiore dei Lavori Pubblici, Ministero delle Infrastrutture e dei Trasporti. Linee guida per la classificazione e gestione del rischio, la valutazione della sicurezza ed il monitoraggio dei ponti esistenti (2020). (in italian)

Di Carlo, F., et al.: On the integration of multi-temporal synthetic aperture radar interferometry products and historical surveys data for buildings structural monitoring. J. Civ. Struct. Heal. Monit. **11**, 1–19 (2021)

FIB Task Group 5.1. Monitoring and Safety Evaluation of Existing Concrete Structures, State-of-the-Art Report, Final draft (2002)

Mele, A., et al.: On the joint exploitation of satellite DInSAR measurements and DBSCAN-based techniques for preliminary identification and ranking of critical constructions in a built environment. Remote Sens. **14**(8), 1872 (2022)

Mele, A., Crosetto, M., Miano, A., Prota, A.: ADAfinder tool applied to EGMS data for the structural health monitoring of urban settlements. Remote Sens. **15**(2), 324 (2023)

Miano, A., Mele, A., Prota, A.: Fragility curves for different classes of existing RC buildings under ground differential settlements. Eng. Struct. **257**, 114077 (2022)

Miano, A., et al.: GIS Integration of DInSAR measurements, geological investigation and historical surveys for the structural monitoring of buildings and infra-structures: an application to the Valco San Paolo urban area of Rome. Infrastructures **7**(7), 89 (2022)

Rainieri, C., Notarangelo, M.A., Fabbrocino, G.: Experiences of dynamic identification and monitoring of bridges in service ability conditions and after hazardous events. Infrastructures **5**(10), 86 (2020)

Salvatore, W., Renzi, E., Baratono, P., Puggeli, S., Cosentini, A.: Ingenio, Analisi delle Linee Guida per la classificazione e gestione del rischio, la valutazione della sicurezza dei ponti (2020)

Talledo, D.A., et al.: Satellite radar interferometry: Potential and limitations for structural assessment and monitoring. J. Build. Eng. **46**, 103756 (2022)

Zhang, W., Sun, L.M., Sun, S.W.: Bridge-deflection estimation through inclinometer data considering structural damages. J. Bridg. Eng. **22**(2), 04016117 (2017)

# Semi-automated Transit Authorization of Exceptional Transport Vehicles: Case-Study Application on a Prestressed Concrete Bridge

Antonio Grella[1(✉)], Giusiana Testa[1], Carmine Lauro[2], Georgios Baltzopoulos[1], Alessio Lupoi[2,3], and Iunio Iervolino[1]

[1] Department of Structures for Engineering and Architecture, University of Naples Federico II, Naples, Italy
antonio.grella@unina.it
[2] Studio SPERI Società di Ingegneria, Rome, Italy
[3] Department of Structural and Geotechnical Engineering, University of Rome La Sapienza, Rome, Italy

**Abstract.** A possible methodology for streamlining the structural safety verifications, needed for authorizing the transit of exceptional transportation vehicles (ETV) on road bridges, is discussed. ETVs, and mainly those classified as such due to total weight, entail a legal obligation of supporting the granting or not of transit permits, by structural calculations that verify adequate load-bearing capacity of the highway bridges the ETV will go over. This method treats vehicle weight and number of axles of a generic fictitious ETV as parameters and is based on progressively increasing the value of both. Cross-sectional forces on bridges due to thusly defined ETVs are expediently obtained using influence lines, then inserted in code-mandated load combinations and compared to structural capacity that is calculated using material properties considered in the original design, deduced from in-situ testing data, or evaluated from simulated design based on the practice at time of construction. Although virtually applicable to any structural typology, a case-study application to an existing underpass with prestressed concrete deck is considered here.

**Keywords:** exceptional transportation vehicles · influence lines · safety verifications

## 1 Introduction

The transit of exceptional transport vehicles (ETV) over road bridges typically requires an authorization issued by the transportation infrastructure operator, that should be supported by structural safety analysis. This is also the case in Italy, where road traffic regulations (Nuovo Codice della Strada, CdS; Ministero delle Infrastrutture e dei Trasporti 1992) define what constitutes an ETV, according to vehicle mass, geometry and payload. In principle, any ETV transit authorization would entail a series of vehicle-specific structural analyses covering all bridges along the intended route of the vehicle. As an alternative, a case study for a semi-automated method of ETV transit authorization applied


M. A. Aiello and A. Bilotta (Eds.): ICC 2022, LNCE 435, pp. 399–412, 2024.
https://doi.org/10.1007/978-3-031-43102-9_31

to a prestressed concrete (PC) highway underpass, tailored to be applicable without an a priori specified ETV mass and geometry, is presented herein. This simplified methodology calculates the structural demand, needed for safety checking, of just the primary load-bearing elements of bridge decks, by considering a set of fictitious ETVs whose axle loads and geometry are used as parameters. The two parameters considered are the total number of axles per ETV and the load carried by any axle, which is modeled as a force acting on the bridge deck. While axle spacing is held constant at 1.25 m, which is considered a minimum value for tandem axles, the total number of axles and the tandem axle load are progressively increased.

Using influence lines obtained from a simple plane-grid modelled using the finite element method (FEM; Cook et al. 2001), the structural demand, *D*, is calculated, in terms of sectional forces, for various values of the two parameters, according to the load combinations imposed by the relevant Italian building codes (Norme Tecniche per le Costruzioni, Consiglio Superiore dei Lavori Pubblici 2018; Linee Guida per la Classificazione e Gestione del Rischio, la Valutazione della Sicurezza ed il Monitoraggio dei Ponti Esistenti, Consiglio Superiore dei Lavori Pubblici 2020; hereafter NTC18 and LL.GG.2020, respectively). The results of these analyses are then expressed in terms of capacity-over-demand ratios, *C/D*, where structural capacity, *C*, at ultimate limit state (ULS) is calculated from cross section geometry and mechanical properties of the materials. This information is obtained either from in-situ and laboratory testing, if available, or is deduced either from design documentation or a simulated design based on engineering practice at the time of construction.

The point of this parametric approach is that, due to the practically minimum axle spacing considered, any real ETV with the same total weight and length as one of the parametric vehicles satisfying code-based structural safety checks along the intended route, could be authorised to transit. The premise is that its effect on the road infrastructure in terms of *C/D* will not exceed that of its parametric counterpart, with the additional implicit assumption that structural safety is governed by the behaviour of the primary elements of the deck. Other elements, such as bearings, piers, and foundations, are considered to have higher safety margins.

This article presents a case-study application of this method on a highway underpass with a PC deck. After an overview of the methodology, the case-study bridge is briefly described along with its numerical model.

## 2 Semi-automated Method

The structural analyses in support of ETV transit authorization over existing bridges, outlined above, can be carried out repeatedly while varying the ETV axle-load and axle-number parameters, in a semi-automatic manner. In this case-study, structural verifications are carried out for two code-mandated conditions (see LL.GG.2020), where a bridge may be declared *immediately transitable* (type one condition or T1) or *transitable under heavy traffic* (type two condition or T2). Note that condition T1 entails application of load model 1 of NTC18 and ostensibly supersedes T2, which is a case where a maximum weight limitation has been imposed on the vehicles that may transit the transportation infrastructure. Both loading conditions are described in the following and it should be noted that, in this case, variable structural actions such as wind, snow, temperature, and braking loads were neglected.

## 2.1 Method Description

According to the adopted methodology, structural analysis is repeatedly performed for a series of conventional ETVs of increasing total weight and length. Given a constant tandem axle load, a fictitious ETV with two axles is first considered and the number of axles is thereafter gradually increased by one, up to a maximum number of *n* axles, that is either the largest number of axles passing the safety check or the number of axles that can fit in the deck span, as shown in Fig. 1, where the constant spacing between tandem axles of 1.25 m is also indicated.

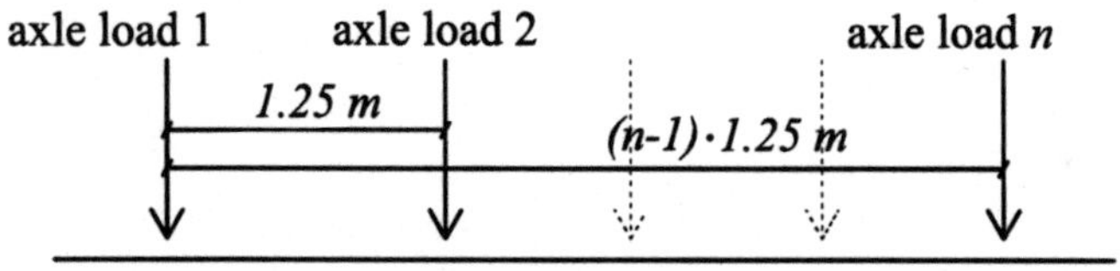


**Fig. 1.** Conventional configuration of exceptional transportation vehicle, ETV, in longitudinal direction.

The resulting tandem-load convoy representing the ETV is intended to be positioned on the structure, in the longitudinal sense, to establish the most unfavourable situation in terms of sectional forces at the critical cross-sections of the bridge deck's primary load bearing elements, all the while maintaining a constant position in the transverse sense. The method is semi-automated, as it involves iterations described by the flow-chart shown in Fig. 2; for each tandem axle load and *transit condition* (to follow) considered, each step entails increasing the number of axles by one, to be repeated until the first cross-section has a *C/D* ratio reduced to unity or until the total length of the conventional ETV exceeds that of the bridge. The procedure is then repeated, considering a higher tandem axle load. The procedure is summarized here.

- Four values of axle load are considered, from 120 to 150 kN / axle with increments of 10 kN.
- For each value of axle load, a progressively increasing number of axles is considered to constitute the ETV, always maintaining a constant axle spacing of one meter and twenty-five centimetres, assumed to represent a minimum distance.
- At each iteration, safety verifications at select critical cross-sections of the primary load-bearing elements of the deck are performed.
- The procedure is repeated until the first safety check is no longer satisfied.

The preceding steps are repeated for T1 and T2 level safety verifications and for four different transit conditions: regular transit, reduced speed transit, exclusive transit, and exclusive transit with reduced speed. These different transit conditions simulate how ETV can transit the road infrastructure, combining it with other standard traffic loads or not, and considering the possibility of a reduction in its transit speed for more unfavourable conditions.

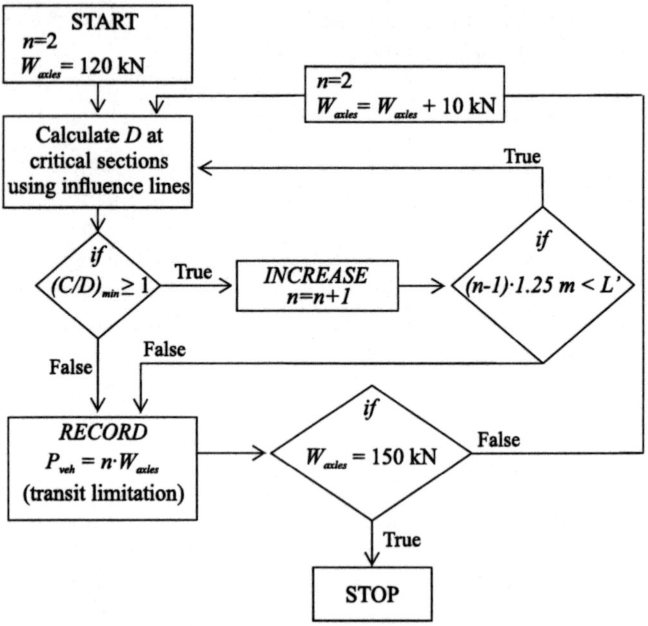


**Fig. 2.** Flow-chart of the semi-automated method.

## 3 Case-Study Bridge

The case-study bridge used to illustrate this methodology is an underpass located on the A3 Naples-Salerno highway in southern Italy. The single-span underpass is composed of two almost identical PC decks separated by a longitudinal joint, one per carriageway, one for traffic directed north (towards Naples) and one going south (towards Salerno).

In this example, safety checks are performed for the northbound carriageway, whose deck is composed by eleven precast pretensioned concrete girders with a double-T cross-section, joined together via four cast-in-situ transverse beams and deck slab, shown in Fig. 3.

The support lines of the girders are skewed with respect to the highway axis at an angle of $\vartheta \approx 14°$ and the span length, $L'$, measured between support points along the main girders is equal to 14.20 m. Total deck width, $B$, for the northbound carriageway is 17.30 m, as can be seen in Fig. 3a. According to the plan view in the figure, two of the transverse beams are at the supports, while the other two are further along the span.

Main girder spacing, measured normal to the beam axis is 1.60 m, and the precast beams have a height of 0.80 m, with a wider lower flange and variable web thickness which increases towards the supports to accommodate shear stresses (Fig. 3b). The cast-in-situ reinforced concrete deck slab has a thickness of 0.21 m.

## 4 Simplified Numerical Model

A simplified linear-elastic plane beam grid FEM model that accounts for the effect of staged construction in the distribution of internal forces, in considered. The model was used for calculating internal forces (flexural moments and shear forces) at a few critical sections of the main girders alone.

The model, which is shown in Fig. 4a, is a set of simply supported beam elements representing the main longitudinal load-bearing girders, complemented by a series of transverse beam elements that represent the stiffness contribution of the deck slab and transverse beams. Pinned supports on the abutments are assumed. The structural analyses were based on influence lines obtained from the FEM model; as an example, Fig. 4b shows one such influence line corresponding to the bending moment of one of the main girders at midspan.

The loads considered for the bridge can be divided into permanent loads and traffic loads. In the case of the former, the structural and non-structural permanent loads such as road pavement, guardrails and protective barriers were considered. Apart from the conventional parametric ETVs described above, other traffic loads considered were the so-called load model 1 (LM1, using the designation of EN-1991–2) defined in NTC18 for condition T1, while for T2 the specific load model defined by LL.GG.2020 for heavy traffic conditions was used. Single-lane examples for both load models are provided in Fig. 5. Figure 5a shows the primary conventional lane load of NTC18's LM1, consisting of tandem axle loads represented by concentrated forces with a resultant of 600 kN, and a uniformly distribution load of 9 kN / m$^2$. On the other hand, Fig. 5b shows the typical conventional lane load from LL.GG.2020 for heavy traffic allowance: a five-axle truck with a total weight of 440 kN and length of 11 m, interrupting the accompanying uniformly distributed 9 kN / m$^2$ (note that LM1 allows the superposition of tandem and distributed loads instead).

It should also be noted that a dynamic amplification coefficient, $\phi$, is applied to the loads corresponding to the ETVs, to account for inertial effects during transit, in accordance with the provisions of LL.GG.2020. This coefficient is a function of a characteristic length of the structure, $L$, given by Eq. 1. In the case of simply supported beams, $L$, coincides with the main girders' theoretical span $L'$. In this application, it is assumed that $\phi$ can be allowed to vary according to the ETV transit speed considered, which is coherent with the stipulations of EN-1991–2. More specifically, it is considered that $\phi$ assumes the full value given by Eq. 1 for a transit speed of 75 km / h and can be reduced to unit value for a speed of 5 km / h, with values for intermediate speeds obtainable via linear interpolation. In this case, two transit speed scenarios of 75 km / h and 20 km / h were considered, and the corresponding dynamic amplification coefficient values are reported in Table 1.

$$\phi = \begin{cases} 1.40, \ L \leq 10m \\ 1.40 - \dfrac{L-10}{150}, \ 10m < L \leq 70m \\ 1.00, \ L \geq 70m \end{cases} \quad (1)$$

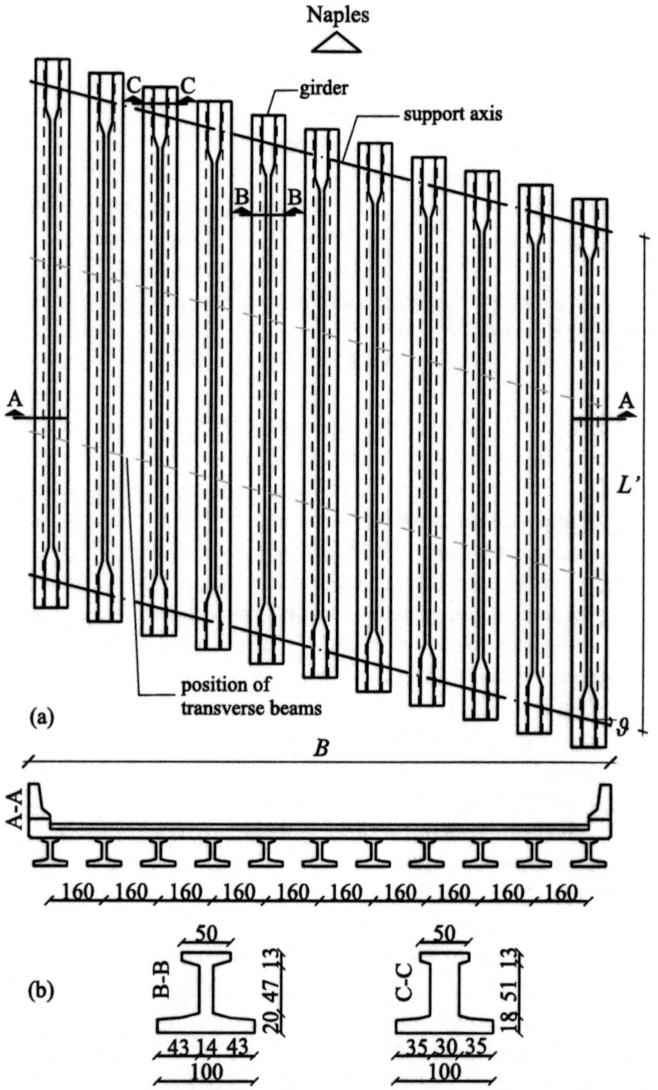


**Fig. 3.** (a) Plan view and cross-section of skew underpass deck; (b) precast beam cross-sections. Dimensions in cm.

As anticipated, the application of traffic loads on the structural model entails the division of the usable deck cross-section into three-meter-wide conventional lanes, as defined by NTC18 and LL.GG.2020 and as shown in Fig. 6. As shown in the figure, the least eccentric lane, with respect to the deck's centre of rotation, is considered reserved for the ETV transit.

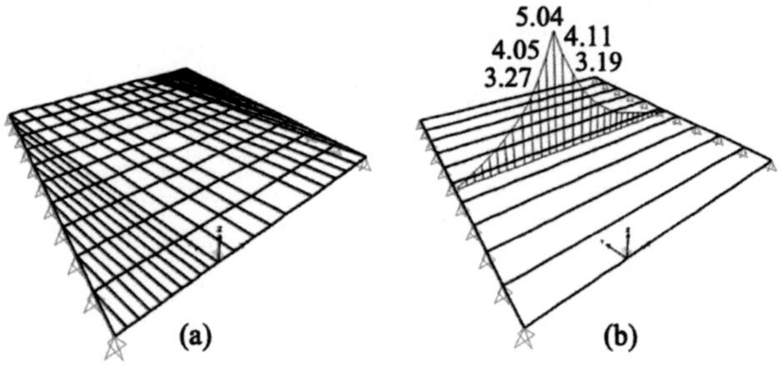


**Fig. 4.** (a) Finite element numerical model of the underpass; (b) influence line of bending moment for an example section.

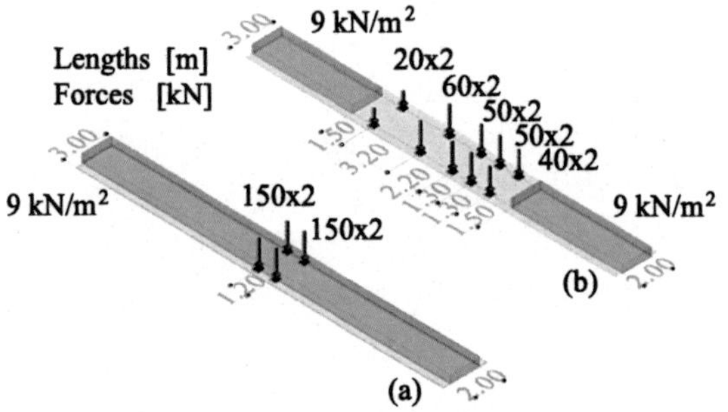


**Fig. 5.** (a) Principal lane load of the load model 1 of NTC18 with total weight of 600 kN; (b) possible distribution of load corresponding to the vehicles of 440 kN.

**Table 1.** Dynamic amplification coefficient for different transit scenarios considered.

| Transit speed | $\phi$ |
|---|---|
| [km / h] | [–] |
| 5 | 1.00 |
| 20 | 1.19 |
| 75 | 1.51 |

The fundamental load combination used for safety verification at ULS, that is for the demand *D*, is given by Eq. 2:

$$D = \gamma_{G1} \cdot G_1 + \gamma_{G2} \cdot G_2 + \gamma_{ETV} \cdot \phi \cdot Q_{ETV} + \gamma_{Qk} \cdot (Q_k + q_k), \quad (2)$$

where $G_1$ represents actions (sectional forces) due to permanent structural load, $G_2$ due to permanent non-structural load, $Q_{ETV}$ due to ETV traffic load, $Q_k$ due to axle traffic loads (vertical concentrated forces other than the ETV's), and $q_k$ due to distributed traffic loads. On the other hand, $\gamma_{G1}$, $\gamma_{G2}$, $\gamma_{ETV}$ and $\gamma_{Qk}$ are the code-mandated partial safety factors for ETV transit safety checks, also reported in Table 2 below.

**Table 2.** Partial safety factors.

| Safety factors | T1 | T2 |
|---|---|---|
| | [-] | [-] |
| $\gamma_{G1}$ | 1.26 | 1.26 |
| $\gamma_{G2}$ | 1.26 | 1.26 |
| $\gamma_{ETV}$ | 1.10 | 1.10 |
| $\gamma_{Qk}$ | 1.20 | 1.60 |

As mentioned previously, four distinct transit scenarios are considered. While in this application the ETV is always constrained to the single conventional lane of minimum eccentricity (a plausible alternative could be to consider the ETV transit on the slow, that is, the outermost, lane). The remaining lanes may or may not be loaded with the T1 and T2 situation load models, depending on whether ETV transit can be exclusive or otherwise. The normal- and slow-speed transit scenarios result from corresponding application of $\phi$ as previously defined.

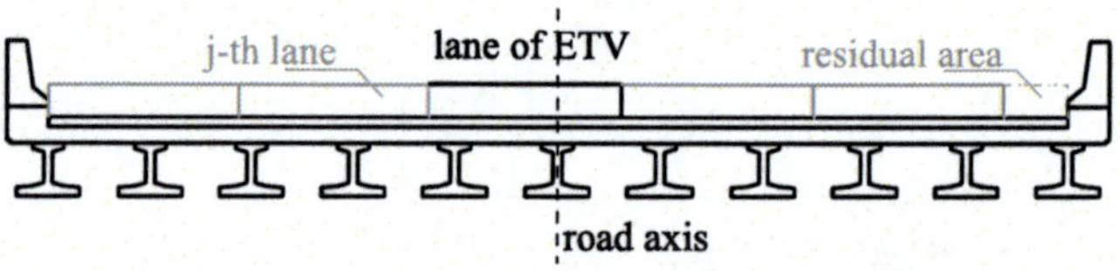


**Fig. 6.** Definition of conventional three-meter-wide lanes and lane of ETV transit.

Apart from the calculation of structural demand *D,* evaluation of the corresponding structural capacity is also required. For this calculation the use of so-called confidence factors (*FC*), was adopted in accordance with Italian standards (§C8.5.4 of Circolare n.7, Consiglio Superiore dei Lavori Pubblici 2019); these correspond to the *knowledge factors* postulated by EN-1998–3 and are coefficients intended to reflect the level of knowledge reached about the structure from inspections, in-situ and lab tests and from available documentation.

For the case-study underpass, the calculation of sectional capacity is based on material properties reported in the technical documentation produced during design and construction, dating from 2001. More specifically, the superstructure's precast girders were constructed using high-strength concrete, with a characteristic cubic compressive strength, $R_{ck}$, of 55 N / mm$^2$ while the prestressing steel used had a characteristic tensile strength, $f_{ptk}$, of 1900 N / mm$^2$ and characteristic nominal yield strength, $f_{p01k}$, of 1700 N / mm$^2$. For the cast-in-situ deck elements, such as the slab and transverse beams, concrete with $R_{ck}$ of 40 N / mm$^2$ and steel reinforcement with a characteristic yield strength of 440 N / mm$^2$ were used.

These values were used to determine each material's design strength, $f_d$, according to Eq. 3,

$$f_d = \frac{f_k}{\gamma_M \cdot FC} \tag{3}$$

where $f_k$ is the characteristic strength value (cylinder compressive strength for concrete, assumed equal to $0.83 \cdot R_{ck}$ or characteristic yield strength for steel) and $\gamma_M$ is the safety factor for each material.

It should be noted that Eq. 3 is intentionally more conservative than what is stipulated in LL.GG.2020, according to which $f_k$ could have been substituted by mean strength. The reason for this choice is that, in this case, no in-situ or laboratory testing was available to determine the mechanical properties of the materials, which had to be taken from design documentation dating from the time of construction. In Table 3 every parameter for the strength design computation is reported, distinguishing between materials and on the quality of information about them by choosing different confidence factors.

These values were in turn used to calculate the capacities of a typical girder's critical sections, that is ultimate moment resistance, $M_{rd}$, and ultimate shear resistance, $V_{rd}$, which are reported in Table 4.

**Table 3.** Partial safety factors, confidence factors and design strengths.

| Materials | $\gamma_M$ | $FC$ | $f_k$ | $f_d$ |
|---|---|---|---|---|
| | [-] | [-] | [N / mm$^2$] | [N / mm$^2$] |
| precast concrete | 1.50 | 1.20 | 45.65 | 21.56 |
| in situ-concrete | 1.50 | 1.35 | 33.20 | 15.68 |
| prestressing steel | 1.10 | 1.20 | 1700.00 | 1287.90 |
| reinforcement steel | 1.15 | 1.20 | 440.00 | 296.30 |

**Table 4.** Structural capacity.

| Critical cross-section | $M_{rd}$ | $V_{rd}$ |
|---|---|---|
| of the beams | [kN m] | [kN] |
| Section at midspan | 2958.0 | 734.2 |
| Section at support | 2958.0 | 1535.9 |
| Reduced section | 2958.0 | 734.2 |

## 5 Structural Verifications: Results and Discussion

The results of the structural analysis are expressed in terms of minimum capacity over demand ratios, $(C/D)_{min}$, and reported in the Tables 5, 6 (minimum over all examined sections of the bridge deck); the critical sections considered, given the static scheme of the specific underpass, are the sections at midspan and at the support of the most stressed girder per transit scenario, where flexural and shear safety checks were performed, respectively. Another considered cross-section was the outermost section with reduced web thickness, which was also considered at ULS against shear forces. Tables 5, 6 are organized according to the transit scenarios described above, depending on assumed ETV transit speed. For each line representing the fictitious exceptional transportation vehicle, the axle weight, $W_{axles}$, the number of axles, *n*, the total weight of the ETV, $P_{veh}$, and the minimum ratio $(C/D)_{min}$ are given. The line highlighted in grey represents, for each transit mode and only for $(C/D)_{min}$ greater than one, the limit configuration of ETV that can transit on the underpass, obtained by the successive increments in axle number and weight, until the stopping criterion of the method is reached.

According to Table 5, the semi-automated method suggests that any ETV up to 13.75 m long weighing 1440 to 1800 kN could be authorized for exclusive transit at a speed of 75 km / h. In this case, the most onerous flexural verification, for $P_{veh}$ of 1800 kN, results in a $(C/D)_{min}$ ratio equal to 1.88, indicating that the method stopped only because ETV length surpassed the underpass span length, rather than due to unit capacity-over-demand being reached. In the case of ETV transit in combination with LM1, the outcome of the structural analysis is governed by the shear verification at the reduced section. For this scenario, it can be observed that the verification criterion is defined by reaching a nominal shear failure condition in the most stressed reduced section, already for values of total ETV weight of 300 kN. This is because shear demand at the cross-section of interest is dominated by the tandem axle loads of LM1. On the other hand, for the combination of the ETVs with the load model that includes a 440 kN vehicle, safety checks are once again determined by underpass span length, albeit with lower $(C/D)_{min}$ ratios than those obtained for ETV in exclusive transit. This is to be expected, given the presence of the additional traffic load considered during ETV transit.

The same safety checks were carried out under the assumption of reduced transit speed for the ETV. The results for these cases, in terms of $(C/D)_{min}$ ratios, are given in Table 6, and follow the same patterns observed for the 75 km / h transit speed, albeit with ratios being higher due to application of a lower coefficient $\phi$. Another noteworthy aspect concerns the results of the shear verification at the reduced section for the scenario with the low-speed transiting ETV, in combination with LM1. In this case, the $(C/D)_{min}$ ratios are almost the same as in Table 5, despite the reduction in the transit speed of the ETV, which can be explained by the increase in the total weight of the exceptional vehicle. This increase is due to the reduction of the maximum demand that allowed further iteration in the definition of the fictitious configuration of ETV, with the addition of another pair of axles prior to stopping the method's iterations.

**Table 5.** Results of safety verifications for different scenarios and transit speed 75 km / h.

| ETV in exclusive transit, $v$=75 km / h | $W_{axles}$ [kN] | $n$ [-] | $P_{veh}$ [kN] | $(C/D)_{min}$ [-] |
|---|---|---|---|---|
| Flexural verifications at girder midspan | 120 | 12 | 1440 | 2.11 |
| | 130 | 12 | 1560 | 2.03 |
| | 140 | 12 | 1680 | 1.95 |
| | 150 | 12 | 1800 | 1.88 |
| Shear verifications at girder support section | 120 | 12 | 1440 | 2.42 |
| | 130 | 12 | 1560 | 2.29 |
| | 140 | 12 | 1680 | 2.17 |
| | 150 | 12 | 1800 | 2.06 |
| Shear verifications at reduced section near supports | 120 | 12 | 1440 | 2.75 |
| | 130 | 12 | 1560 | 2.64 |
| | 140 | 12 | 1680 | 2.53 |
| | 150 | 12 | 1800 | 2.44 |
| **ETV and NTC18; $v$=75 km / h** | **$W_{axles}$ [kN]** | **$n$ [-]** | **$P_{veh}$ [kN]** | **$(C/D)_{min}$ [-]** |
| Flexural Verifications at girder midspan | 120 | 12 | 1440 | 1.34 |
| | 130 | 12 | 1560 | 1.32 |
| | 140 | 12 | 1680 | 1.29 |
| | 150 | 12 | 1800 | 1.27 |
| Shear verifications at girder support section | 120 | 12 | 1440 | 1.65 |
| | 130 | 12 | 1560 | 1.64 |
| | 140 | 12 | 1680 | 1.62 |
| | 150 | 12 | 1800 | 1.61 |
| Shear verifications at reduced section near supports | 120 | 3 | 360 | 0.99 |
| | 130 | 2 | 260 | 1.00 |
| | 140 | 2 | 280 | 1.00 |
| | 150 | 2 | 300 | 1.00 |
| **ETV and 440 kN; $v$=75 km / h** | **$W_{axles}$ [kN]** | **$n$ [-]** | **$P_{veh}$ [kN]** | **$(C/D)_{min}$ [-]** |
| Flexural verifications at girder midspan | 120 | 12 | 1440 | 1.22 |
| | 130 | 12 | 1560 | 1.20 |
| | 140 | 12 | 1680 | 1.17 |
| | 150 | 12 | 1800 | 1.15 |
| Shear verifications at girder support section | 120 | 12 | 1440 | 1.83 |
| | 130 | 12 | 1560 | 1.76 |
| | 140 | 12 | 1680 | 1.69 |
| | 150 | 12 | 1800 | 1.62 |
| Shear verifications at reduced section near supports | 120 | 12 | 1440 | 1.18 |
| | 130 | 12 | 1560 | 1.17 |
| | 140 | 12 | 1680 | 1.17 |
| | 150 | 12 | 1800 | 1.17 |

**Table 6.** Results of safety verifications for different scenarios and transit speed 20 km / h.

| ETV in exclusive transit, $v$=20 km / h | $W_{axles}$ [kN] | $n$ [-] | $P_{veh}$ [kN] | $(C/D)_{min}$ [-] |
|---|---|---|---|---|
| Flexural verifications at girder midspan | 120 | 12 | 1440 | 2.37 |
| | 130 | 12 | 1560 | 2.28 |
| | 140 | 12 | 1680 | 2.20 |
| | 150 | 12 | 1800 | 2.13 |
| Shear verifications at girder support section | 120 | 12 | 1440 | 2.84 |
| | 130 | 12 | 1560 | 2.69 |
| | 140 | 12 | 1680 | 2.56 |
| | 150 | 12 | 1800 | 2.45 |
| Shear verifications at reduced section near support | 120 | 12 | 1440 | 3.08 |
| | 130 | 12 | 1560 | 2.97 |
| | 140 | 12 | 1680 | 2.86 |
| | 150 | 12 | 1800 | 2.77 |
| **ETV and NTC18; $v$=20 km / h** | $W_{axles}$ [kN] | $n$ [-] | $P_{veh}$ [kN] | $(C/D)_{min}$ [-] |
| Flexural verifications at girder midspan | 120 | 12 | 1440 | 1.41 |
| | 130 | 12 | 1560 | 1.39 |
| | 140 | 12 | 1680 | 1.37 |
| | 150 | 12 | 1800 | 1.35 |
| Shear verifications at girder support section | 120 | 12 | 1440 | 1.69 |
| | 130 | 12 | 1560 | 1.68 |
| | 140 | 12 | 1680 | 1.67 |
| | 150 | 12 | 1800 | 1.65 |
| Shear verifications at reduced section near support | 120 | 3 | 360 | 1.00 |
| | 130 | 3 | 390 | 1.00 |
| | 140 | 3 | 420 | 0.99 |
| | 150 | 3 | 450 | 0.99 |
| **ETV and 440 kN; $v$=20 km / h** | $W_{axles}$ [kN] | $n$ [-] | $P_{veh}$ [kN] | $(C/D)_{min}$ [-] |
| Flexural verifications at girder midspan | 120 | 12 | 1440 | 1.29 |
| | 130 | 12 | 1560 | 1.27 |
| | 140 | 12 | 1680 | 1.25 |
| | 150 | 12 | 1800 | 1.23 |
| Shear verifications at girders support section | 120 | 12 | 1440 | 2.06 |
| | 130 | 12 | 1560 | 1.99 |
| | 140 | 12 | 1680 | 1.92 |
| | 150 | 12 | 1800 | 1.85 |
| Shear verifications at reduced section near support | 120 | 12 | 1440 | 1.19 |
| | 130 | 12 | 1560 | 1.19 |
| | 140 | 12 | 1680 | 1.18 |
| | 150 | 12 | 1800 | 1.18 |

## 6 Final Remarks

The present study describes the code-conforming structural analysis carried out on an existing prestressed concrete underpass of the A3 Naples-Salerno highway to authorise the transit of exceptional transportation vehicles, or ETVs. The results of this analysis were obtained through a semi-automated method that treats ETV weight and geometry as parameters, modelled as a series of closely-spaced axles (concentrated vertical forces), whose number and weight are progressively increased for as long as the bridge safety checks continue to be satisfied. The analysis method refers to simplified elastic-linear finite element structural calculation models and takes also into account the effect of staged construction in the distribution of internal forces. The structural elements of the numerical model were considered one-dimensional and used to derive the internal forces of the main longitudinal beams; this choice of modeling is the result of the desire to only investigate the effects of the main elements of the deck, neglecting the safety verification of structural elements considered secondary, such as the transversal beams and the slab. For this reason, for each scenario of transit and ETV configuration, the maximum internal forces were obtained at the critical sections identified: for this structural model of simply supported deck, the shear at support sections and at outermost reduced-web-thickness sections of the girders were considered, as well as bending moment at midspan. The result of applying the method is therefore the determination of the ratio between the corresponding structural capacity, *C*, and the maximum internal force demand during ETV transit, *D*. The former is established as a function of the maximum number of axles that can transit, for each of the four alternative axle weights considered. In other words, having fixed an axle load, it was possible to establish the maximum number of axles for an ETV that can transit the bridge in conventional safety. It was shown that, for this case-study underpass, the structural verifications for the authorization of the defined exceptional vehicles transit are always satisfied to the largest possible configuration of the ETV, with $(C/D)_{min}$ ratios greater than unity for the scenarios in which ETV passes on the bridge exclusively and in combination with the vehicle of 440 kN, with and without transit speed limitation. With regard to the combination scenarios of the ETV with the Load Model 1 of NTC18, the safety checks are governed by shear at the reduced web thickness sections, for which the maximum total weight of the ETV is 300 kN for the transit speed of 75 km / h and 390 kN for 20 km / h.

**Acknowledgements.** This study was developed within the activities of the Convenzione Tangenziale di Napoli – DIST 2020–2022.

## References

CEN. Eurocode 1: Actions on structures – Part 2: Traffic loads on bridges (2003)
CEN. Eurocode 8: Design of structures for earthquake resistance – Part 3: Assessment and retrofitting of buildings (2005)
Consiglio Superiore dei Lavori Pubblici. Norme Tecniche per le Costruzioni. Gazzetta Ufficiale della Repubblica Italiana (2018)
Consiglio Superiore dei Lavori Pubblici. Circolare n.7. Gazzetta Ufficiale della Repubblica Italiana (2019)

Consiglio Superiore dei Lavori Pubblici 2020. Linee Guida per la Classificazione e Gestione del Rischio, la Valutazione della Sicurezza ed il Monitoraggio dei Ponti Esistenti. Gazzetta Ufficiale della Repubblica Italiana (2020)

Cook, R.D., Malkus, D.S., Plesha, M.E.: Concept and Applications of Finite Element Analysis. John Wiley & Sons Inc (2001)

Franciosi, V.: I principi di reciprocità e le linee di influenza. Scienza delle Costruzioni Vol. III (Teoria delle Strutture). Napoli: Liguori editore (1979)

Ministero delle Infrastrutture e dei Trasporti. Nuovo Codice della Strada. Gazzetta Ufficiale della Repubblica Italiana (1992)

# Structural Health Monitoring of a Prestressed Concrete Bridge Deck

Gabriele Bertagnoli(✉), Emiliano Ciccone, and Mario Ferrara

Dipartimento di Ingegneria Strutturale, Edile e Geotecnica (DIESG), Politecnico di Torino, Torino, Italy
gabriele.bertagnoli@polito.it

**Abstract.** In recent years it has become clear that many western countries infrastructures require scrupulous and continuous monitoring. The main purpose of Structural Health Monitoring (SHM) is the evaluation of the safety of existing structures and the guide for maintenance interventions, where necessary. One of the techniques used in structural monitoring is the measurement of angles using clinometers Nevertheless, many issues on the reliability and the correct use of measures done with clinometers have to be addressed to achieve a trustworthy SHM. In this paper the most relevant issues related to the f.e.m. modelling of a prestressed concrete bridge deck related to the use of clinometers for SHM are presented. The study presents a test-case deck that has been under continuous monitoring for many months. A brief explanation of the data-cleaning process and the interpretation of the clinometers outputs is also given, stressing out the limitations of the technology and possible outcomes.

**Keywords:** Structural Health Monitoring · bridges · finite elements · prestressed concrete

## 1 Introduction

Most Western World countries built their backbone infrastructures after WWII between 1950 and 1980. This heritage of roads, railways and highways bridges and tunnels is nowadays becoming old and it is often suffering serious deterioration problems (Bertagnoli et al., 2019, 2020, Gino et al. 2020, Castaldo et al. 2017, 2020*a*,*b*).

During the last ten years, the evolution of low cost sensors derived from TLC industry, the development of broadband internet communication, the rise of cloud based services and the growth of big data platforms, have changed the scenario of Structural Health Monitoring (SHM) that can now be deployed on large scale to infrastructures as a standard option and not only when specific pathologies are found (Boller et al., 2009).

The SHM system presented in this work is made of high precision clinometers. It was installed before the beginning of repair and strengthening works on an old prestressed concrete girder deck. It has been active on the bridge deck for several months during repair operations. Traffic has been kept partially open on the deck during repair by means of carriageways reduction.


M. A. Aiello and A. Bilotta (Eds.): ICC 2022, LNCE 435, pp. 413–426, 2024.
https://doi.org/10.1007/978-3-031-43102-9_32

The initial test and calibration of the SHM system was done applying a static load in absence of traffic: four lorries were placed on one side of the deck to maximize the torque effect and therefore the bending action on the most external longitudinal beam.

A finite element model (f.e.m.) is developed to be used during the continuous monitoring phase to provide mechanical interpretation to the data coming from the monitoring system.

The geometry of the deck is taken from the original blueprints and the material properties are known from a widespread test campaign on several specimens taken from the deck.

The f.e.m is calibrated in order to fit the deformed shape of the deck measured by the clinometers during the load test done before the beginning of the monitoring activity.

The f.e.m. model is calibrated with a refinement procedure in order to understand which structural mechanisms and characteristics are playing a major role in determining the deformation of the deck under service variable loads.

The study starts with a simple f.e.m. model and step by step increases the complexity until a good accordance with measured data is achieved.

During the monitoring period, a process of cleaning of the clinometers' outputs was carried out.

The measured angles were related to the temperatures in order to find the possible correlations between rotations and the temperatures themselves.

## 2 Bridge Deck Description

The monitored bridge is made of several isostatic decks each one with a span of 45m and a transverse width of 19.1m. A throughout description of the deck is provided by (Bertagnoli et al., 2021).

Each deck is made of six longitudinal beams and four transverse beams as can be seen in Fig. 1. The top slab is 20cm thick. The transverse cross section is saddlebacked as both carriageways are supported by the same deck; the two central beams are at the same level, whereas each beam is 20cm lower moving towards the edge of the deck.

Each longitudinal beam of the deck is prestressed using 94 prestressing strands with 93mm$^2$ cross section each; 70 strands are straight running at the bottom of the beam and 24 are deviated upwards at the extremities of the beams to reduce prestressing moment on the supports.

One post tensioning bonded tendon made of 32φ7 wires with a total cross section area of 1232 mm$^2$ is also present with a parabolic curve from the bearing to 15m and then a linear layout at the bottom of the beam until midspan.

A throughout testing campaign on the materials was done taking from the deck: twelve cylindrical specimens φ = 94 h = 94 mm from the beams and six specimens with the same dimensions from the slab, nine ordinary reinforcements bars segments (3φ8, 3ϕ10 and 3ϕ16), five pre-stressing strands segment and five wires from the post-tensioning tendon.

The deck was instrumented with eight self-compensated mems clinometers placed at the extremities of the internal beams (beams 2 to 5): four on the north side (N) and four on the south side (S), as can be seen in Fig. 1.

Tiltmeters are fixed on 2m long aluminum bars in order to measure the mean rotation of the first 2 m of each beam.

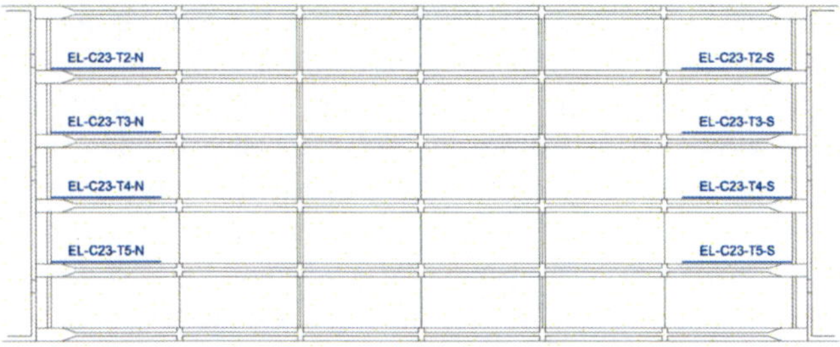


**Fig. 1.** Bridge deck and arrangement of the sensors

## 3 Load Test Description

After the installation of the monitoring system the deck has been statically tested using four lorries weighting 34t each. The position of the lorries on the deck, their shape and their weight on each tire is given in Fig. 2. The trucks were positioned with the greatest possible eccentricity. This induces maximum twisting effect and maximum bending actions on the midspan of the external beam. During the test the deck was closed to traffic and it was therefore possible to set the "zero" condition for all sensors, with the bridge unloaded and load test ambient temperature.

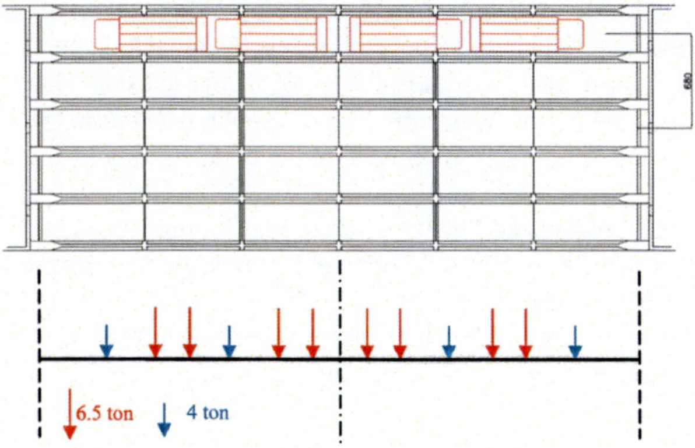


**Fig. 2.** Load test

## 4 Finite Elements Models

The deck was modelled using simple 3D beams elements. The collaborating width of the slab is calculated according to paragraph 5.3.2.1 of EN 1992–1-1 for all beams *(CEN,* 2004*).*

Two different cross sections for longitudinal beams were used. The first one, assigned to external (edge) beams, includes 2.75m of top slab. The slab is eccentric with respect to the web vertical axis because the width of the kerb is smaller than the interaxis between longitudinal beams. The second one, assigned to internal beams, includes 3.4m of centered top slab that is exactly the interaxis between longitudinal beams.

Two different sections for transverse beams were also used. The first section is used for the first and the fourth transverse beams (the ones connecting bearings). It includes only 1.4m of top slab because the effective span of these elements is only the interaxis of 3.4m between longitudinal beams. The second transverse beam cross section is assigned to the two central transverse beams and includes 6.33m of top slab. The effective span of these beams is indeed taken equal to 17m that is the transverse distance between the two external longitudinal beams. This assumption is very important as the stiffness of these transverse beams is highly dependent on the amount of collaborating slab.

The remaining part of the slab, which is not considered as the top flange of the transverse beams is then divided into stripes of about one meter of width and modelled as transverse slab stripes as can be seen in Fig. 3 (a).

Plain concrete properties are calculated taking into account the difference between slab and beams concrete grade, whereas homogenized properties are calculated taking into account the presence of reinforcement and prestressing. Homogenization is done with respect to the Young modulus of the concrete of the beams.

According to design blueprints, the north side of the deck is provided with fixed bearings in longitudinal direction whereas the south side has free rollers realized by means of double pendulum bearings. No specification is given by blueprints on the bearings behaviour in transverse direction. In the model on the north side all bearings are fixed in longitudinal direction and only one is considered to be fixed in transverse direction. On south side all bearings are free in longitudinal direction and only one is fixed in transverse one. Beam nr.3 has been chosen by the authors to get the transverse fixed bearing both north and south. Results do not change substantially if another beam is chosen. During load test no horizontal load are applied, therefore the only horizontal forces that occur in the bearings are related to the compatibility of the deformation of the deck.

The role of the bearings and the type of the reaction they can transfer plays a fundamental role in the behaviour of the deck under serviceability conditions, like the one represented by the load test.

Several different f.e.m. models have been developed to test the level of accuracy needed to obtain displacement results comparable to the ones measured during the load test.

The rotations obtained in each model, loaded with the load pattern shown in Fig. 2, are compared to real one measured on the bridge during the load test.

Model 1, pictured in Fig. 3 (a), is perfectly flat. It does not take into consideration the vertical eccentricities between beams, slab, and bearings. This kind of model was commonly used by designer in the past because of its simplicity.

The results obtained by this model overestimate the deformability of the deck of about 26% on the most deformed instrumented beam.

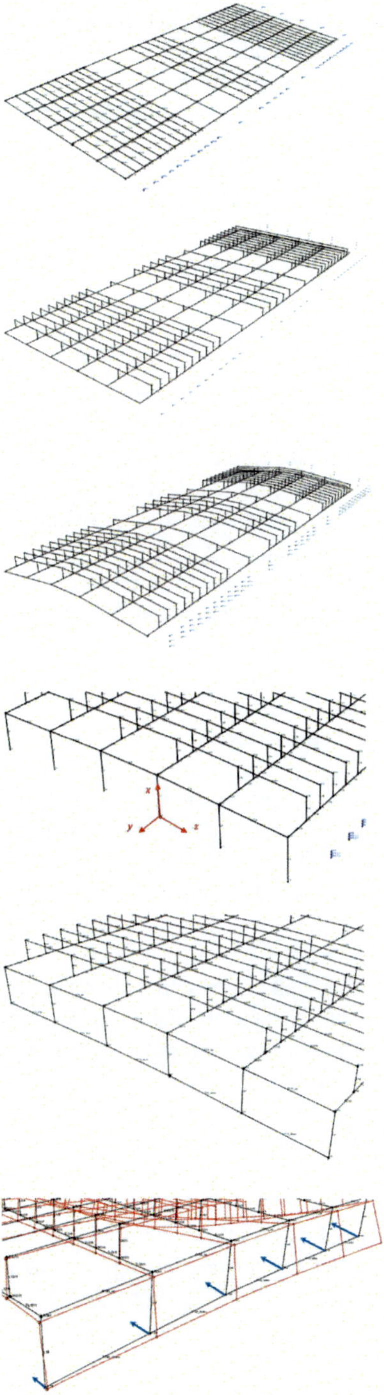

(a)

(b)

(c)

(d)

(e)

(f)

**Fig. 3.** f.e.m. models: (a) model 1, (b) model 2, (c) model 4, (d) model 5, (e) model 6, (f) model 7

In the second model the vertical offsets between the centroids of longitudinal and transverse beam and top slab are introduced. Results are similar to the ones of the first model, registering an overestimation of deformability of 24%.

Model 3 is based on the geometry of the second one but cross sectional properties of beams and slab are calculated taking into account homogenization due to effective reinforcement bars and prestressing strands and tendons layout.

The results are stiffer than the ones obtained with model 2, reducing the maximum difference between calculated and measured displacements to about 16%.

Model 4 is derived from the third by introducing the transverse saddlebacked slope, appreciable in Fig. 3 (c). No significative difference is appreciated between the results obtained using model 3 and model 4.

Model 5 introduces in model 4 the vertical offsets of the bearings positions, as shown in Fig. 3(d). The vertical elements that are connecting the bearings to the centroids of the longitudinal beams are 1.76m long and are modelled with a cross section of 0.75 by 1.0m trying to represent the transverse bending deformability of the web of the longitudinal beams that are 0.75m thick at the beginning of the deck. The difference between measured displacements and calculated ones obtained with model 5 is about 14%. Better than model 4 but not good enough.

Model 6 is an evolution of the fifth one. In order to model the transverse restraint offered by transverse beams also at the level of the bearings, the 1st and 4th transverse beam are split into two elements: one at the level of the bearings and one at the level of the real centroid of the transverse beams. The sum of the area and the inertia of the two elements is equal to the one of the real transverse beam to keep the global stiffness unchanged. This modification increases the overall stiffness reducing the maximum error to $9 \div 12\%$.

As the results of model 6 are still quite far from the measured ones, the authors decide to introduce a non-linear behaviour within the bearing devices to take into account the friction in free bearings. The north side of the deck is fixed in longitudinal direction, therefore no friction effect is considered, but the south side has free longitudinal bearings for all six beams. In model 7 a friction coefficient in longitudinal direction equal to 5% is introduced in all south bearings.

The bearing behaviour is rigid (until maximum horizontal force allowable by friction is reached) then perfectly plastic. This means that the static friction force is maintained during sliding, ignoring the reduction between static and sliding friction coefficient. An example of friction forces is shown in Fig. 3(f).

The horizontal resultant due to friction is about 200 kN that is about 30% of the vertical reaction due to the lorries weight. This horizontal force reduces the deformation of the deck leading the differences between numerical and experimental values to less than 5%.

It can be therefore concluded that friction within free bearings plays a fundamental role in monitoring of bridge deck and cannot be neglected as it is commonly done in design.

# 5 Data Analysis

## 5.1 Data Acquisition

Data acquisition began on January 2019, and measurements are available until June 2019. During this time, different sampling frequencies were used, which were kept constant for a certain period of time. In particular, sampling frequencies equal to 5, 15, 60, 30 and 20 min were used. Only the data relating to the period with constant sampling frequency equal to 20 min, are considered valid for the analysis to be carried out in this paper.

## 5.2 Preliminary Data Cleaning

Temperatures were also recorded in addition to the rotations.

The chosen positive direction of rotation is shown in Fig. 4, having assumed negative rotations that involve a downward inflection of the deck.

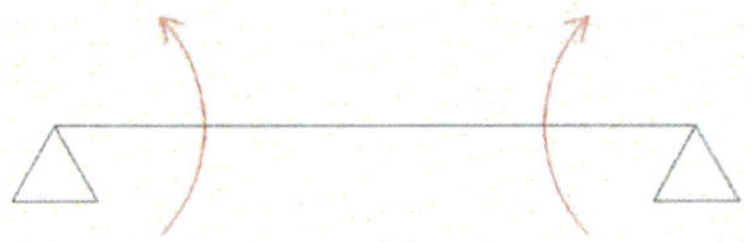

**Fig. 4.** Positive direction of rotation

A first correction of the data is made by translating it with respect to a new zero value obtained during the loading tests, when the bridge was unloaded and traffic was closed. All the data acquired before the load test was then aligned on the new zero reference.

This operation is described in Eq. (1). The series of raw rotations ('raw') is the starting data-set on which all the signal processing operations will be carried out.

$$\theta_{raw}(t) = \theta(t) - \theta(t_0) \tag{1}$$

where $\theta_{raw}(t)$ is the 'raw' rotation at instant t, $\theta(t)$ is the rotation measured by the sensor at instant t and $\theta(t_0)$ is the rotation measured with unloaded deck.

Some statistical operations are carried out on the 'raw' series of rotations, such as the calculation of the average value of the series ($\overline{x}$), the standard deviation ($\sigma_x$), the cantered moving average calculated over time windows of 2 h and 20 min (MR7), 3 h (MR9) and 3 h and 40 min (MR11) (the abbreviations derive from the fact that considering a sampling time of 20 min, these are respectively relative to 7, 9 and 11 measures).

Similarly, starting from the series of measured temperatures, moving averages were calculated over 60 (MT3), 80 (MT4), 100 (MT5) and 120 min (MT6) prior to the instant considered (respectively relating to 3, 4, 5 and 6 measurements for a sampling time of 20 min). The central point of the moving average over 2 preceding hours (6 measures), therefore, is one hour behind the considered time t.

Moving averages are useful for cleaning the instantaneous effects of traffic on the deck and instantaneous temperature changes, which have a prevalent effect on the instrumentation and not on the deck (*Cigada et al.* 2021 *a,b*).

Moving averages of temperatures over previous 120 min (MT6) were then calculated with a delay from 1 to 23 h. It should be noted that with a delay of 23 h, the central value of the average has a delay of 24 h with respect to the instant considered. The moving averages calculated with delay are indicated by an abbreviation of the type MT6_r1, where T stands for Temperature, 6 is the number of measures on which the moving average is calculated, r stands for delay and, 1 is the delay value expressed in hours. For example, the abbreviation of a moving average on temperatures made over a time window of 120 min and a delay of 7 h will be MT6_r7.

This operation is done to take into account the fact that the response of the bridge may not be immediately following a temperature change because of the thermal inertia of the deck.

### 5.3 Rotation – Temperature Correlation

After the calculation of the previously indicated statistical parameters, the correlation values between the 'raw' rotations and temperatures are calculated. Correlation was evaluated through Pearson's (Pearson, 1895) correlation coefficient ($\rho_{xy}$). Considering two data series X and Y both having length equal to N, it is defined as (2).

$$\rho_{xy} = \frac{\sigma_{xy}}{\sigma_x \cdot \sigma_y} \tag{2}$$

where $\sigma_{xy}$ is the covariance between the two data series and $\sigma_x$, $\sigma_y$ are standard deviations of the two data series respectively. In particular, the correlation between the instantaneous and averaged rotations and the instantaneous and averaged temperatures is calculated. By way of example, the results relating to the inclinometer placed on beam 2, north side, are reported in Table 1, but the calculation has been made for all sensors and the response is similar.

**Table 1.** Rotation-temperature correlation factors

| ROT. | TEMPER. | | | | |
|---|---|---|---|---|---|
| | instant. Temp | MT3 | MT4 | MT5 | MT6 |
| Inst. Rot | −0.630 | −0.605 | −0.597 | −0.589 | −0.581 |
| MR7 | −0.743 | −0.714 | −0.706 | −0.697 | −0.688 |
| MR9 | −0.747 | −0.719 | −0.711 | −0.703 | −0.694 |
| MR11 | −0.750 | −0.722 | −0.714 | −0.706 | −0.697 |

It is observed that the instantaneous rotations have a high correlation, in absolute value, with temperatures, but the averaged rotations, regardless of the width of the interval over which the average occurs, have a even higher correlation with temperatures.

Nevertheless, if the average on temperatures is extended over a longer period, all the correlation values drop in absolute value, contrary to what one might think following a real action on the structure.

The highest correlation remains on the instantaneous temperatures, suggesting that it is a response of the instrumentation and not of the structure to the thermal effect.

### 5.4 Instantaneous Thermal Drift

The series of 'raw' rotations is then corrected for the instantaneous temperatures via Eq. (3).

$$\theta_k(t) = \theta_{raw}(t) - D * [T(t) - T(t_0)] \tag{3}$$

where $\theta_K(t)$ is the series of corrected rotations, $\theta_{raw}(t)$ is the series of 'raw' rotations, $D$ is the instantaneous thermal drift, $T(t)$ is the series of instantaneous temperatures and $T(t_0)$ is the temperature relating to the unloaded bridge measurement.

The value for the instantaneous thermal drift D is calculated by minimizing the correlation between the corrected rotations and the instantaneous temperature, using the bisection method.

For sensor EL-C23-T2-N (span 23, beam 2, north side) the instantaneous thermal drift obtained is equal to -0.0161 mrad/°C. The same order of magnitude values are found for all other sensors. It can be noticed that the thermal drift of the instruments (declared by the manufacturer) should be less than 0.020 mrad/°C, therefore drifts close to or less than this value are fully compatible with the manufacturing tolerances.

Rotations corrected or compensated with instantaneous drift are called "k rotations" and are no longer correlated with instantaneous temperatures.

### 5.5 Corrected Rotations – Temperature Correlation

The correlation between corrected rotations ('k') and the temperatures is calculated.

In this case, the delayed moving averages on temperature are also considered.

Moving average over 9 rotations readings (180 min) (MR9) and moving average over 6 temperature readings (120 min) with a delay of 'n' hours (MT6_rn) are considered.

Results are summarized in Fig. 5, in which the abscissa axis represents the delay that has the central value of the average of the temperatures with respect to the instant considered, while the ordinates represent the relative correlation coefficient.

The shape of the curve is similar for all sensors, and has a sinusoidal trend, as it is correct to expect from a physical phenomenon that has a periodicity of 24h.

Three groups of curves can be distinguished:

- T3-N sensor (green curve): maximum with delay of −6 h;
- T4-S and T5-S sensors (brown and gray curves): maximum with delay of 0 h approximately;
- T2-N, T2-S, T3-S, T4-N, T5-N sensors (other curves): maximum with delay of −4.5 h approximately.

Practically speaking, if the air temperature rises at instant t, the deck seems to inflect with a certain delay (the most probable is 4.5h), compatibly with its thermal inertia.

Therefore, it is assumed that the 'raw' data is a measurement affected by both a thermal effect on the sensor and a thermal effect on the deck.

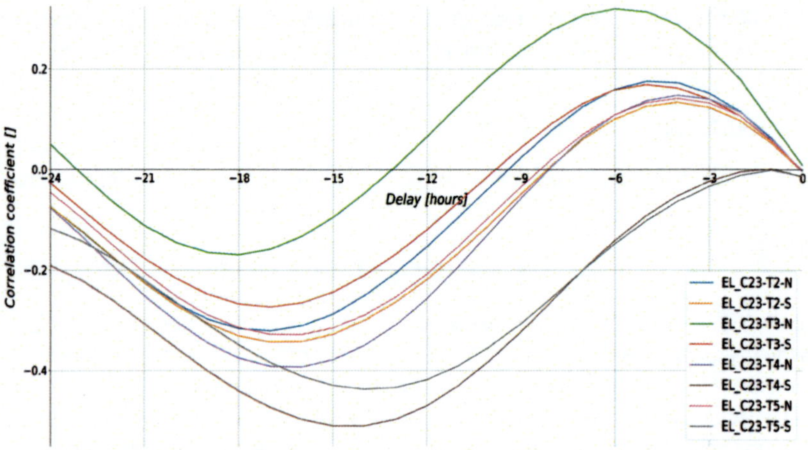


**Fig. 5.** Correlation coefficients between corrected rotations averaged over 180 min and temperature averaged over 120 min with different delays (from 0h to -24h)

### 5.6 Deferred Thermal Drift

Equation (4) aims to correct the series of compensated rotations ('k') so that the doubly corrected series of rotations ('kk') is not correlated with the moving average of temperatures calculated with the delay relative to the maximum of correlation index.

$$\theta_{kk}(t) = \theta_k(t) - DD * [T(t) - T(t_0)] \tag{4}$$

where $\theta_{KK}(t)$ is the series of doubly corrected rotations, $\theta_K(t)$ is the series of corrected rotations, *DD* is the deferred thermal drift, *T(t)* is the series of moving averages of temperatures calculated with the delay relative to the maximum of the correlation index and $T(t_0)$ is temperature relating to the unloaded bridge measurement.

To obtain the optimal value of the deferred thermal drift, the same method mentioned above is used, minimizing the correlation value between double compensated rotations and the series of moving average of the temperature calculated with the delay relative to the maximum of the correlation index (*T(t)*).

## 6 Results

Average and standard deviation for 'raw', corrected 'k' and doubly corrected 'kk' rotations are shown in Table 2.

Moving from 'raw' values to corrected 'k' values the average of the signals falls from -0.072 mrad to -0.028 mrad, thus approaching the zero value (of unloaded bridge) and the sensitivity of the instruments. The standard deviation falls in proportion much less than the average values from 0.094 mrad to 0.071 mrad. As a result, the 'k' values are closer to zero, but noisier than the 'raw' values.

Moving from corrected 'k' values to double corrected 'kk' values the average of the signals moves away from zero (unloaded bridge), passing from –0.028 mrad to -0.034

mrad. Standard deviation remains practically constant. Consequently, the signal noise is slightly reduced. The difference between 'k' and 'kk' values is much smaller than that between raw and 'k' values.

This is a very important issue as it shows that the thermal movements of the bridge are at least one order of magnitude smaller than the thermal noise on the sensor. Therefore the biggest part of the thermal troubles related to the use of clinometers is due to the effect of temperature on the instruments and not to the effect of temperature on the bridge itself (*Balagesa et al.,* 2006, Sohn et al. 2004).

**Table 2.** Average and standard deviation for 'raw', 'k', and 'kk' rotation

| Sensor | AVG 'RAW' [mrad] | ST. DEV. 'RAW' [mrad] | AVG. K [mrad] | ST. DEV. K [mrad] | AVG. KK [mrad] | ST. DEV. KK [mrad] |
|---|---|---|---|---|---|---|
| T2-N | −0.072 | 0.093 | −0.017 | 0.070 | −0.026 | 0.069 |
| T2-S | − .058 | 0.072 | −0.024 | 0.062 | −0.030 | 0.062 |
| T3-N | 0.006 | 0.102 | −0.037 | 0.090 | −0.054 | 0.088 |
| T3-S | −0.072 | 0.086 | −0.019 | 0.063 | −0.026 | 0.062 |
| T4-N | −0.048 | 0.066 | −0.019 | 0.057 | −0.025 | 0.057 |
| T4-S | −0.065 | 0.086 | −0.031 | 0.077 | −0.031 | 0.077 |
| T5-N | −0.058 | 0.077 | −0.01 | 0.062 | −0.016 | 0.061 |
| T5-S | −0.205 | 0.169 | −0.067 | 0.087 | −0.067 | 0.087 |
| AVG | −0,072 | 0,094 | −0,028 | 0,071 | −0,034 | 0,070 |

Furthermore, even in the doubly corrected series, a downward trend was noted during the data acquisition period, as shown in Fig. 6.

Rotations become more and more negative moving from end of February to beginning of June. It probably indicates that deck's rotations are still affected by seasonal temperature variation; it cannot be excluded, however, that this phenomenon is due to a drift of the instrumentation.

It could be possible to understand which of the previous two hypothesis is the correct one, only by processing the data of a complete year. In this way, the effects of seasonal thermal variations could be eliminated from the measurements, and the real cause of the observed trend can be found.

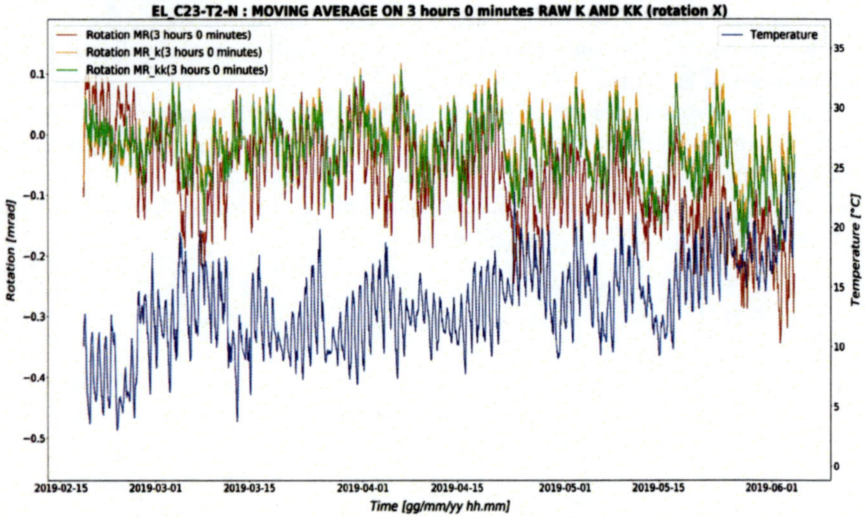


**Fig. 6.** Moving average on 9 values of the 'raw', 'k' and 'kk' rotations

## 7 Conclusions

The model proposed for the bridge deck studied in this paper has given the following results.

The correct evaluation of the collaboration width of the flange of transverse beams is a key parameter. The effective span to of the transverse beams to be used in this procedure is not the interaxis between the longitudinal beams, but it may be close to the transverse width of the whole deck. Longitudinal beams provide, in fact, a kind of vertical elastic restraint to transverse ones, which is by far less stiff than a rigid one.

The model of the first and last transverse beam, that are connecting the bearings should be different from the one used for the transverse placed along the span, to take into consideration the correct level of restraint.

The vertical eccentricity of the bearings positions with respect to the longitudinal beams axis plays a fundamental role and friction in the free bearings should be considered.

Friction in free bearings can play a determinant role in modifying the deformability of the structure. Being friction a non-linear behaviour it may complicate remarkably the model, shifting it from elastic linear to non-linear and therefore avoiding to use the simple superposition of effect law that is very useful in ordinary design.

Rigid perfectly plastic friction model was used in this work, but a simplification of this behaviour using linear spring is under study.

The deck is prestressed both longitudinally and transversely, therefore linear elastic behaviour under service load is a verified hypothesis. Modelling not prestressed decks may become more difficult if cracking issues can affect the deformation in a non-linear way.

Despite the fact that the sensors are declared self-compensated by the manufacturer, the analysis of the data shows a strong dependence of the measures on the instantaneous temperature.

After instantaneous temperature cleaning, a correlation with a sinusoidal trend is observed between rotations and temperatures. The sinusoid shows the maximum correlation between rotations and temperatures recorded with a delay of about 4.5 h. This can be due to the thermal inertia of the deck, which consequently shows a delayed response to daily temperature variations.

The data were also cleaned from delayed thermal drift, coming from structural response. Despite this, a further trend has been observed which may be due to seasonal variations in temperature, that cannot be fully appreciated because the monitoring window covers only late winter and spring.

**Acknowledgements.** This study was developed in the framework of PON INSIST (Sistema di monitoraggio INtelligente per la SIcurezza delle infraSTrutture urbane) research project, which was funded by the Italian Ministry for Education, University and Research (Programma Operativo Nazionale "Ricerca e Innovazione 2014–2020", Grant No. ARS01_00913).

## References

Balageas D., Fritzen C.-P., Güemes A.: Structural Health Monitoring. Wiley, Hoboken (2006). ISBN: 9781905209019

Bertagnoli G., Anerdi, C., Ferrara, M.: Structural health monitoring issues using inclinometers on prestressed concrete girder bridge decks. In: IOP Conference Series Materials Science and Engineering, vol. 1203, no. 3, p. 032101 (2021)

Bertagnoli, G., Malavisi, M., Mancini, G.: Large scale monitoring system for existing structures and infrastructures. In: IOP Conference Series: Materials science and Engineering, vol. 603, no. 5, p. 052042 (2019)

Bertagnoli, G., Lucà, F., Malavisi, M., Melpignano D., Cigada, A.: A large scale SHM system: a case study on pre-stressed bridge and cloud architecture. In: Dynamic of Civil Structures Vol. 2 – Proceedings of the 37th IMAC, A Conference and Exposition on Structural Dynamics, pp. 75–83 (2020)

Boller, C., Chang, F.-K., Fujino, Y.: Encyclopedia of Structural Health Monitoring. Wiley, Hoboken (2009). ISBN: 978–0–470–05822–0

Castaldo, P., Amendola, G.: Optimal DCFP bearing properties and seismic performance assessment in nondimensional form for isolated bridges. Earthq. Eng. Struct. Dyn. **50**(9), 2442–2461 (2021)

Castaldo, P., Amendola, G.: Optimal sliding friction coefficients for isolated viaducts and bridges: a comparison study. Struct. Control Health Monit. **28**(12), e2838 (2021)

Castaldo, P., Palazzo, B., Ferrentino, T., Petrone, G.: Influence of the strength reduction factor on the seismic reliability of structures with FPS considering intermediate PGA/PGV ratios. Compos. B Eng. **115**, 308–315 (2017)

CEN, Comité Européen de Normalisation. EN 1992–1–1 Eurocode 2: Design of concrete structures - Part 1–1: General rules and rules for buildings (2004)

Cigada, A., Lucà, F., Malavisi, M., Mancini, G.: A damage detection strategy on bridge external tendons through long-time monitoring. In: Pakzad, S. (ed.) Dynamics of Civil Structures, Volume 2. CPSEMS, pp. 159–168. Springer, Cham (2021). https://doi.org/10.1007/978-3-030-47634-2_18

Cigada, A., Lucà, F., Malavisi, M., Mancini, G.: Structural health monitoring of a damaged operating bridge: a supervised learning case study. In: Pakzad, S. (ed.) Dynamics of Civil Structures, Volume 2. CPSEMS, pp. 169–177. Springer, Cham (2021). https://doi.org/10.1007/978-3-030-47634-2_19

Gino, D., Castaldo, P., Bertagnoli, G., Giordano, L., Mancini, G.: Partial factor methods for existing structures according to fib bulletin 80: assessment of an existing prestressed concrete bridge. Struct. Concr. **21**(1), 15–31 (2020)

Pearson, K.: Notes on regression and inheritance in the case of two parents. Proc. R. Soc. Lond. **58**, 240–242 (1895)

Sohn, H., et al.: A review of structural health monitoring literature: 1996–2001. Los Alamos National Laboratories, Los Alamos, NM (2004)

# Numerical Modeling of the Monotonic and Cyclic Behavior of Exterior RC Beam-Column Joints

Ernesto Grande[1](✉), Maura Imbimbo[2], Annalisa Napoli[3], Riccardo Nitiffi[2], and Roberto Realfonzo[3]

[1] Department of Engineering Science, University Guglielmo Marconi, Roma, Italy
e.grande@unimarconi.it

[2] Department of Civil and Mechanical Engineering, University of Cassino and Southern Lazio, Cassino, FR, Italy

[3] Department of Civil Engineering, University of Salerno, Fisciano, SA, Italy

**Abstract.** Past and recent seismic events have highlighted undesirable brittle failure in RC frame structures, indicating that a substantial damage may result from beam-column joints in non-seismically designed buildings. This paper presents a numerical investigation on exterior RC joints under seismic loading based on a macro-modelling approach, which employs the "scissors model" to schematize the shear behavior of the joint and the bond-slip of the longitudinal steel rebars at the beam-joint interface. A set of literature experimental tests have been used to develop a model for the estimate of the maximum shear strength and a new multi-linear shear stress-strain law of the joint element. Cyclic analyses have been also performed by implementing the "pinching4" material parameters available in OpenSees, governing the hysteresis rules and pinching effect.

**Keywords:** RC beam-column joints · nonlinear numerical analysis · shear capacity

## 1 Introduction

In the seismic assessment of existing RC buildings, great attention must be deserved to the beam-column joints, by correctly evaluating their local strength and deformability capacity and, consequently, their contribution to the seismic response of the whole structure. To this purpose, relevant contributions can be found in the literature, dealing with experimental investigations of exterior RC joints which are the most vulnerable ones under seismic loading (Hassan et al. 2018, De Risi et al. 2016).

Other studies focus on the numerical modeling of the joints' behavior. Among these, several modeling approaches were developed, i.e.,: lumped–plasticity approaches, multi–spring macro–models, nonlinear finite element (FE) modeling (De Risi et al. 2016).


M. A. Aiello and A. Bilotta (Eds.): ICC 2022, LNCE 435, pp. 427–441, 2024.
https://doi.org/10.1007/978-3-031-43102-9_33

This paper provides a contribution on the modeling aspects of exterior RC joints, by presenting a new multilinear backbone joint shear stress-strain law, properly defined through its key parameters.

Specifically, by following the numerical outcomes presented in Grande et al. (2021a) a new formula for the estimate of the shear strength of exterior joints is proposed on the basis of a multivariate linear regression analysis calibrated on a set of 15 experimental tests collected from the literature and described in detail in a recently published paper (Grande et al. 2021b). Moreover, the strain parameters were calibrated by means of a sensitivity analysis, from which a number of monotonic load-drift numerical curves were derived and compared with the experimental ones by properly varying the shear strain values.

Finally, cyclic analyses have been performed by accurately reproducing the experimental response of the 15 tests collected in the database. To this aim, the "Pinching4" material available in OpenSees (McKenna et al. 2010) was used to set the parameters governing the hysteresis rules and the pinching effect. In particular, some literature proposals were considered to set proper values of these parameters. Then, to identify the model which best approximate the cyclic behavior of the tests, an assessment was carried out on the basis of the difference between the numerical and the experimental response in terms of dissipated energy and stiffness degradation at each cycle. The results of the assessment procedure, summarized in Grande et al. (2021b), are herein better discussed with reference to 3 representative tests included in the experimental database and allow to draw relevant information on the cyclic modelling.

## 2 Proposed Model for Exterior Joints

The macro-modeling approach adopted to simulate the seismic behavior of exterior RC joints, recently published in Grande et al. (2021a), consists of modelling beams and columns through fiber discretization of the cross section, while the "scissors model" is used for the joint element. In the latter, two nonlinear rotational springs are set in series to simulate the two main mechanisms governing the overall behavior of exterior joints: the shear deformation of the panel and the "fixed-end-rotation" of the beam due to the debonding of the longitudinal steel rebars at the beam-joint interface.

The numerical analyses presented in this paper are carried out through the open-source finite element platform OpenSees (McKenna et al. 2010). The behavior of the two aforementioned rotational springs were modeled through the "Pinching4 Uniaxial Material", which is based on a multilinear moment-rotation (M-θ) backbone law directly derived from multilinear shear stress-strain (τ-γ) law and from bond-slip (τ-s) constitutive law, respectively.

Since the collected experimental database mostly concerns joints which experienced the shear failure of the panel zone, the present study mainly focuses on the definition of the parameters characterizing the multilinear τ-γ constitutive law assigned to the joint shear spring. More details on both the modeling of the bond-slip spring and the collected experimental database can be found in Grande et al. (2021a).

The shear behavior of exterior beam-column joints can be modeled by means of a multilinear shear stress-strain (τ-γ) constitutive law (Fig. 1), that is defined by four characteristic points in terms of stress ($\tau_1$, $\tau_2$, $\tau_3$, $\tau_4$) and strains ($\gamma_1$, $\gamma_2$, $\gamma_3$, $\gamma_4$).

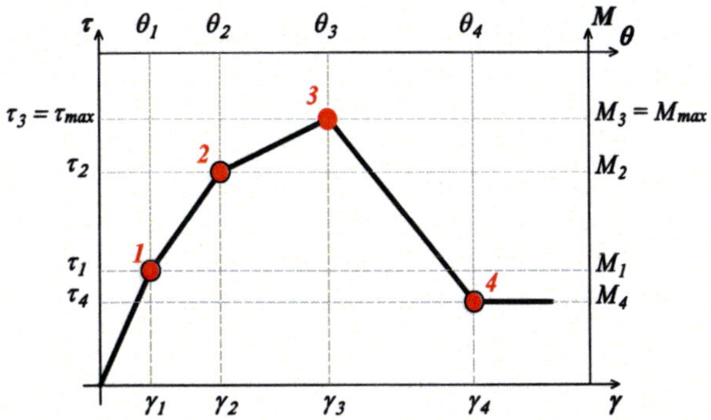


**Fig. 1.** Multilinear shear stress-strain (or moment-rotation) law assigned to the shear spring.

For the estimate of the shear strength $\tau_3$ (indicated as $\tau_{max}$), the following literature models have been selected: model 1 by Kim & LaFave (2009), model 2 by Vollum & Newman (1999), model 3 by Ortiz (1993), model 4 by Hwang & Lee (1999), model 5 by Jeon et al. (2015). For the shear stress at cracking ($\tau_1$), the formula proposed by Uzumeri (1977) was adopted, while for the other stress parameters ($\tau_2$, $\tau_4$) and the four strains ($\gamma_1$, $\gamma_2$, $\gamma_3$, $\gamma_4$) the following models have been considered: model A proposed by De Risi et al. (2016), models B1 and B2 by Celik and Ellingwood (2008), models C1 and C2 by Shin and LaFave (2004) and model D by Sharma et al. (2011).

The above models are combined together in order to reproduce the shear response ($\tau$-$\gamma$) of the panel zone. In particular, the shear strength $\tau_3$ given by each of the models 1, 2, 3, 4, 5 is combined with the other parameters given by models A, B1, B2, C1, C2, D. As a result, a total of 30 $\tau$-$\gamma$ relationships are obtained and implemented in the numerical analyses.

In Grande et al. (2021a) preliminary numerical analyses were performed by considering the mentioned experimental database which includes a set of fifteen cyclic tests performed on specimens representative of typical exterior "2D" beam-column joints. In each case, the monotonic behavior was studied by implementing the 30 $\tau$-$\gamma$ constitutive laws in the numerical model. The numerical curves in terms of force-drift were compared with the envelope of the experimental hysteretic cycles from the collected tests. The performed analyses allowed to identify the combinations of the models that better describe the behavior of the specimens during the tests. To this purpose, for a given drift, the scatter in terms of force between the numerical and the experimental curves of each specimen was estimated according to the mean absolute percentage error (MAPE), which is expressed by the following formula:

$$E_{rr} = \frac{\sum_{i=1}^{n} \frac{|F_{exp,i} - F_{num,i}|}{F_{exp,i}} \cdot 100}{n} \quad (1)$$

where: $F_{exp,i}$ and $F_{num,i}$ are, respectively, the i-th experimental force and the corresponding numerical one; n is the total number of measures considered in the analysis performed on each test specimen.

The bar chart in Fig. 2 shows the best six model combinations for which the MAPE values calculated on the whole monotonic curve for all the 15 experimental joints were below the 25% threshold. In particular, the plot shows that three combinations include model 5 (Jeon et al. 2015) for evaluating the shear strength); similarly, two combinations entail model 1 (Kim and La Fave 2009). It is important to underline that these two models were developed by considering the same parameters (i.e., the beam reinforcement index "BI" and the concrete compressive strength "$f_c$"), which are employed in this study in the calibration procedure for the estimate of the shear strength.

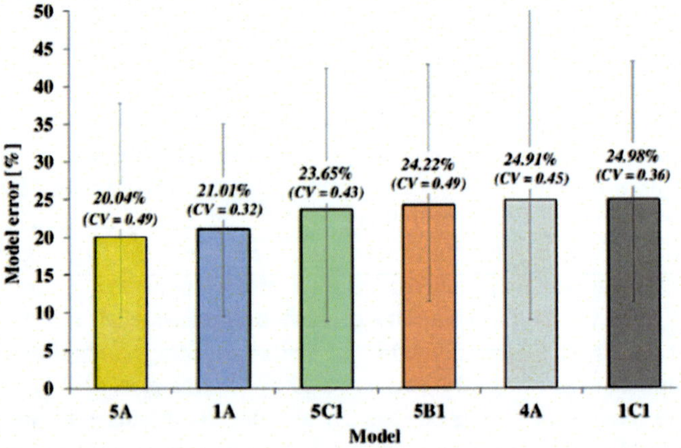


**Fig. 2.** Ranking of the best models yielding a mean absolute percentage error (MAPE) calculated on all the monotonic envelopes.

# 3 Calibration of a Constitutive Shear Stress-Strain Law

The calibration procedure implemented to define the shear stress-strain constitutive model regarded, on one side, the estimate of the shear strength $\tau_{max}$, and on the other side, the identification of the remaining stress and strain parameters of the multilinear $\tau$–$\gamma$ relationship (Fig. 1). A detailed description of the procedure can be found in Grande et al. (2021b).

## 3.1 Joint Shear Strength Model Calibration

The assessment of the literature models in predicting the shear strength of the experimental database, described in Grande et al. (2021a), showed that the models proposed by Kim and LaFave (2009) and by Jeon et al. (2015) provided the best accurate predictions. These two models only differ for the numerical coefficients calibrated by the authors on their own accounted set of experimental cases.

In this study, considering an experimental database containing 2D-joint configurations only, a multivariate linear regression analysis is performed by taking into account the same predictor variables (BI and $f_c$) and the same structure of the formula used by model 1 and model 5 for deriving the shear strength:

$$\tau_{max} = a \cdot (BI)^b \cdot (f_c)^c \tag{2}$$

where *a, b* and *c* are the numerical coefficients to be opportunely calibrated.

The use of a multivariate linear regression analysis allows to study the relationship between a dependent variable (in this case, the shear strength $\tau_{max}$) and more than one independent variables (in this case, BI and $f_c$), through the following functional equation expressed in a logarithmic scale:

$$ln(Y) = \beta_0 + \beta_1 \ln(X_1) + \beta_2 \ln(X_2) + \cdots + \beta_n \ln(X_n) \tag{3}$$

where: $Y$ is the response (dependent) variable; $X_j$ ($j = 1,2...n$) are several predictor (independent) variables; $\beta_j$ ($j = 1,2...n$) are the coefficients of the regression analysis.

Therefore, Eq. 3 can be equivalently rewritten in the multiplicative form like Eq. 2.

In order to examine the accuracy of the two predictor variables BI and $f_c$ on the response variable $\tau_{max}$, an analysis of variance (ANOVA) was performed in a log-transformed space. In the ANOVA analysis, two different statistic tests have proved the accuracy of the regression method adopted: the F statistic test and t statistic test. In particular, the F statistic test checks the hypothesis of no significant relationship between the predictive independent variables and the single dependent variable (null hypothesis). On the other hand, the t statistic test is used to check the hypothesis of no significant relationship between a single independent variable and the dependent variable (null hypothesis). Both tests must meet the condition of significance at the 95% level, or equivalently, a p-value less than the 5%, so that the null hypothesis can be rejected.

The regression analysis, in conjunction with the ANOVA test, was performed by considering both the positive and the negative values of the shear strength $\tau_{max}$ available for each experimental test collected in the database, except for two tests, for which the two negative values of $\tau_{max}$ were excluded from the analysis since they are associated to the occurrence of bond failure.

Finally, the results of the calibration procedure, deeply described in Grande et al. (2021b), allowed to identify the following formulation for the estimate of the joint shear strength:

$$\tau_{max} = 0.569 \cdot (BI)^{0.445} \cdot (f_c)^{0.783} \tag{4}$$

The comparison between the proposed strength model with the literature formulas by Kim and LaFave (2009) and by Jeon et al. (2015) is provided in Table 1, which shows the mean errors in terms of $\tau_{max}$ calculated according to MAPE for all the experimental tests by considering the following three case: a) $\tau_{max}$ in the push ("+") loading direction only; b) $\tau_{max}$ in the pull ("−") loading direction only, and c) $\tau_{max}$ in both loading directions. As observed, the application of proposed model yields an overall error reduction of about 13–15% and 20–30% with respect to model 1 and model 5, respectively.

**Table 1.** Comparison between the new shear strength proposal and the models 1 & 5 in terms of MAPE.

| Model | Model Error | | |
|---|---|---|---|
| | Push direction | Pull direction | Both directions |
| 1 - Kim & La Fave | 12.60% | 13.69% | 13.11% |
| 5 – Jeon et al | 13.23% | 15.72% | 14.39% |
| Proposal | 11.06% | 11.89% | 11.43% |

### 3.2 Calibration of the Joint Shear Stress and Strain Parameters

**Cracking Shear Stress $\tau_1$**

Based on the examined literature studies, the cracking shear stress $\tau_1$ can be reasonably estimated according to the formula proposed by Uzumeri (1977).

**Pre-peak Stress $\tau_2$**

The pre-peak strength $\tau_2$ can be assumed equal to $0.85\tau_{max}$, according to the model proposed by De Risi et al. (2016), which showed the best agreement in predicting the pre-peak branch of the monotonic experimental envelopes (Grande et al. 2021a).

**Residual Stress $\tau_4$**

The residual strength point $\tau_4$ of the backbone law is assumed equal to $0.3\tau_{max}$ which is the value approximately suggested by the most of considered literature models (Celik & Ellingwood 2008, Sharma et al. 2011, Shin & LaFave 2004).

**Strain Parameters ($\gamma_1$, $\gamma_2$, $\gamma_3$, $\gamma_4$)**

The assessment of the monotonic analyses in terms of load-drift published in Grande et al. (2021a) provided preliminary indications on the range of strain values which best meet the experimental results. Therefore, a sensitivity analysis was carried out on all the 15 experimental tests with the aim to identify a set of strain parameters to propose in the $\tau$–$\gamma$ constitutive model for exterior joints. The monotonic analyses previously performed by implementing the 30 mentioned model combinations showed that a larger variation of this parameter does not lead to a significant scatter in terms of errors between experimental and numerical envelopes in the first loading branch. Thus, regarding the strains corresponding to the cracking strength and the pre-peak strength, only three values were accounted in the proposed range ($\gamma_1$ = 0.0004–0.0008–0.0013; $\gamma_2$ = 0.0017–0.002–0.004).

Within the ranges related to the parameters $\gamma_3$ and $\gamma_4$, a thickening of the values was considered, since higher uncertainty was identified in the numerical analyses ($\gamma_3$ = 0.004–0.006–0.007–0.009–0.010–0.012–0.014–0.016–0.018–0.020; $\gamma_4$ = 0.025–0.030–0.035–0.0441–0.050–0.060–0.070–0.080).

In order to assess the influence of each $\gamma_j$ parameter on the global response in terms of force–drift results, the sensitivity analysis was performed by varying every single value of the strain parameter within the predetermined range, thus requiring a hard computation effort (a total of 21600 monotonic analyses were carried out). Each of the monotonic curves resulting from the sensitivity analysis was compared to the experimental envelope of each test and the MAPE value was calculated according to Eq. 1.

For each experimental test, the combination of ($\gamma_1, \gamma_2, \gamma_3, \gamma_4$) minimizing the MAPE value on the monotonic envelope was, then, identified. The set of $\gamma_j$, obtained by averaging the values resulting from all the 15 experimental tests, provided a mean error of 10.87%. Such set of $\gamma_j$ values is that proposed by the authors in conjunction with the stress parameters $\tau_j$ to define a new backbone joint shear stress-strain law to implement and validate in the numerical analyses.

For better clarity, the four couples of ($\tau_j$-$\gamma_j$) parameters of the proposed law are summarized in Table 2.

Then, a validation of the proposed model was carried out by performing new monotonic analyses in the positive and the negative direction of the loading and the MAPE was again calculated between the experimental and the numerical curve.

In Fig. 3, the mean errors related to the implementation of the proposed constitutive law (namely "proposed model") are compared with those obtained by considering the best three literature $\tau$–$\gamma$ models identified in Grande et al. (2021a) (i.e. model 1A, model 5A and model 5C1). Moreover, the errors resulting from considering the optimal set of $\gamma_j$ parameters - found test by test in the minimization error procedure – are reported for a better comparison ("optimal solution" in Fig. 3).

Figure 4 shows the resulting force-drift monotonic curves plotted for three of the fifteen tests: specimen T01 by Hadi & Tran (2016), specimen O1 by Tsonos (2002) and BS-L by Wong (2005). In each graph both red and green curves represent the monotonic envelope numerically obtained by implementing the $\tau$–$\gamma$ law identified as "optimal solution" and "proposed model", respectively; the black dotted curve, instead, is the experimental monotonic envelope.

**Table 2.** Shear stress and strain parameters of the proposed constitutive law.

| backbone point | Shear stress, $\tau_j$ | Shear strain, $\gamma_j$ |
|---|---|---|
| 1 | $\tau_1 = 0.29\sqrt{f_c}\sqrt{1 + 0.29\sigma_j}$ | $\gamma_1 = 0.001087$ |
| 2 | $\tau_2 = 0.85 \cdot \tau_{max}$ | $\gamma_2 = 0.003273$ |
| 3 | $\tau_{max} = 0.56 \cdot 9(BI)^{0.445} \cdot (f_c)^{0.783}$ | $\gamma_3 = 0.008733$ |
| 4 | $\tau_4 = 0.3 \cdot \tau_{max}$ | $\gamma_4 = 0.048820$ |

It is evident the greater accuracy in matching the experimental behavior when the "optimal solution" is implemented in the numerical simulations, but the "proposed model" also provides satisfying results.

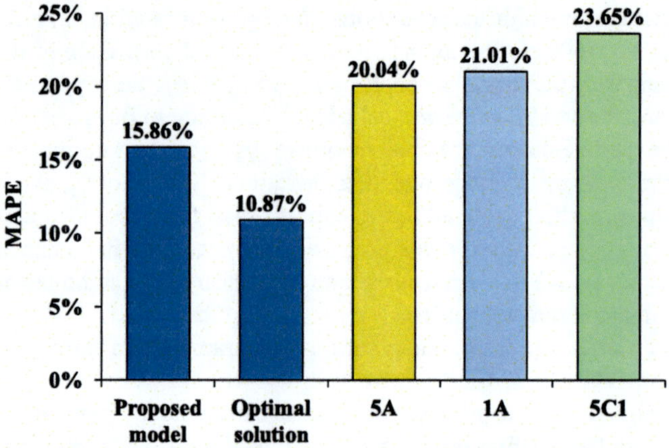


**Fig. 3.** Comparison between the proposed model, the optimal solution and the three best literature models in terms of MAPE.

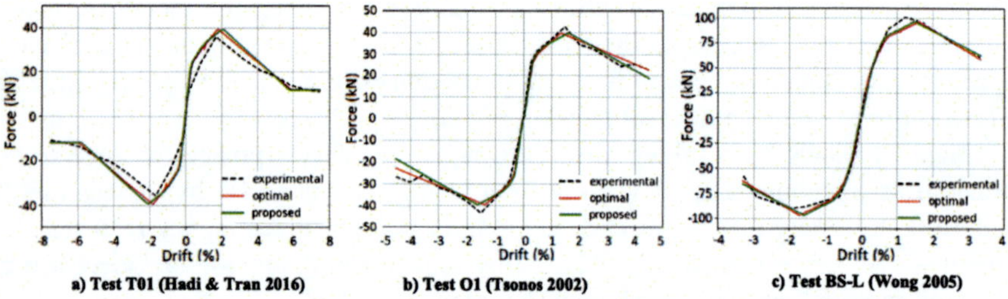


**Fig. 4.** Numerical vs experimental comparison of the monotonic envelopes considering the "optimal solution" and the "proposed model": test T01 (a), Test O1 (b), and test BS-L (c).

## 4 Modeling of Cyclic Response

The modeling of the cyclic response of joints represents another key aspect, characterized by a greater complexity with respect to the monotonic behavior. To this purpose, the "Pinching4" uniaxial material model, available in the OpenSees library (McKenna et al. 2010), was still used to set the parameters governing the hysteresis rule and pinching effect.

The cyclic degradation of the model is simulated by three damage rules: unloading stiffness degradation, reloading stiffness degradation and strength degradation. Each of the three hysteretic damage modes is characterized by the following damage index formulation, dependent on the displacement history and the accumulated energy:

$$\delta_i = \left(\alpha_1 (d_{max})^{\alpha_3} + \alpha_2 \left(\frac{E_i}{E_{monotonic}}\right)^{\alpha_4}\right) \le \delta_{lim} \tag{5}$$

In Eq. 5: $\delta_i$ is the damage index (0 = no damage, 1 = maximum damage); $\alpha_1$, $\alpha_2$, $\alpha_3$ and $\alpha_4$ are parameters required to fit the damage rule to experimental data; $E_i$

is the accumulated hysteretic energy; $E_{monotonic}$ is the energy required to achieve under monotonic loading the deformation at failure; $d_{max}$ is the maximum historic deformation demand; $\delta_{lim}$ is the maximum possible value of the damage index.

The four parameters ($\alpha_j$) in Eq. 5 must be calibrated for each of the three damage indices on the basis of experimental data. In the "Pinching4" model, these $\alpha_j$ parameters are replaced by proper degradation model parameters (gKj, gFj, gDj) to be defined for each damage index. Moreover, the model requires the definition of other parameters, i.e.: gE, a multiplication factor used to define the maximum energy dissipation under cyclic loading, and three couples of parameters (rDispP, rDispN; rForceP, rForceN; uForceP, uForceN) describing the pinching behavior of the unload-reload path, for a total of 22 parameters to set at proper values for a correct numerical simulation of the joint cyclic response.

Table 3 shows the values of the pinching parameters proposed by five selected studies, in which the calibration procedure was based on a fitting process of experimental cyclic responses of RC joints available to the authors. These studies were performed by: Lowes and Altoontash (2003), Theiss (2005), Hassan (2018), De Risi et al. (2016) and Jeon et al. (2015).

**Table 3.** Values of the "Pinching4" material parameters selected from the literature.

| | Parameter ID | Lowes & Altoontash | Theiss | Hassan | De Risi | Jeon |
|---|---|---|---|---|---|---|
| Pinching parameters | rDispP | 0.25 | 0.11 | 0.15 | 0.16 | 0.2 |
| | rForceP | 0.15 | 0.25 | 0.35 | 0.23 | 0.2 |
| | uForceP | 0 | 0 | −0.1 | −0.22 | 0 |
| | rDispN | −0.25 | −0.11 | 0.15 | 0.16 | 0.2 |
| | rForceN | −0.15 | −0.25 | 0.15 | 0.23 | 0.2 |
| | uForceN | 0 | 0 | −0.4 | −0.22 | 0 |
| Unloading stiffness degradation parameter | gK1 | 1.3 | 0.42 | 0.5 | 0.85 | 0.95 |
| | gK2 | 0 | 0.35 | 0.2 | 0 | 0 |
| | gK3 | 0.24 | 0.2 | 0.1 | 0.12 | 0.1 |
| | gK4 | 0 | 0.028 | 0.4 | 0 | 0 |
| | gKLim | 0.89 | 0.99 | 0.99 | 0.96 | 0.95 |
| Reloading stiffness degradation parameter | gD1 | 0.12 | 0.046 | 0.1 | 0.38 | 0.35 |
| | gD2 | 0 | 0.005 | 0.4 | 0 | 0 |
| | gD3 | 0.23 | 1.385 | 1 | 0.34 | 0.15 |
| | gD4 | 0 | 0 | 0.5 | 0 | 0 |
| | gDLim | 0.95 | 0.99 | 0.99 | 0.96 | 0.95 |

*(continued)*

**Table 3.** (*continued*)

| | Parameter ID | Lowes & Altoontash | Theiss | Hassan | De Risi | Jeon |
|---|---|---|---|---|---|---|
| Strength degradation parameter | gF1 | 1.11 | 1 | 0.05 | 0 | 0.05 |
| | gF2 | 0 | 0 | 0.02 | 0 | 0 |
| | gF3 | 0.32 | 2 | 1 | 0 | 0.32 |
| | gF4 | 0 | 0 | 0.05 | 0 | 0 |
| | gFLim | 0.13 | 0.99 | 0.99 | 0 | 0.25 |
| Energy | gE | 10 | 2 | 10 | 10 | 10 |

The finite element model geometry used to simulate the monotonic envelope of the 15 experimental joints, was again considered to carry out cyclic analyses. The multilinear moment-rotation law in Fig. 1 was adopted to describe the envelope backbone of the shear rotational spring, of which the key parameters were derived from the $\tau$ and $\gamma$ values previously calibrated by the authors (see Table 2).

For each of the 15 tests, a set of 5 cyclic analyses were carried out by implementing the corresponding 5 sets of cyclic degradation parameters listed in Table 3. To this purpose, it is worth highlighting that no strength degradation was considered in all the 5 models, meaning that the strength damage index parameters (gF1, gF2, gF3, gF4, gFlim) were always set equal to zero whatever the accounted model. Indeed, like the numerical study performed by De Risi et al. (2016), even in this work the monotonic envelope of each joint has been separately calibrated by performing specific monotonic analyses.

## 4.1 "PInching 4" Model Parameters: Assessment of the Five Proposals

In order to identify the model which best approximate the cyclic behavior of the tests, an assessment was carried out on the basis of the difference between the numerical and the experimental response in terms of dissipated energy and stiffness degradation at each cycle. In fact, these are the properties mainly controlling the shape of the hysteretic loops and their evaluation is a relatively simple issue in the post–processing phase of the force–drift cyclic response.

To compare the accuracy of the numerical simulations, resulting from the implementation of the five sets of "Pinching4" model parameters, the mean absolute percentage error was calculated between the experimental and the numerical values for both energy ($E_{rr,E_d}$) and stiffness ($E_{rr,K}$) at each cycle, expressed by the following expressions:

$$E_{rr,E_d} = \frac{\sum_{i=1}^{n} \frac{|E_{d,i}^{exp} - E_{d,i}^{num}|}{E_{d,i}^{exp}} \cdot 100}{n} \quad (6)$$

$$E_{rr,K} = \frac{\sum_{i=1}^{n} \frac{|K_{i}^{exp} - K_{i}^{num}|}{K_{i}^{exp}} \cdot 100}{n} \quad (7)$$

where n is the total number of measures in terms of energy and stiffness available from the experimental curve of each specimen, $E_{d,i}^{exp}$ and $E_{d,i}^{num}$ are, respectively, the experimental and numerical dissipated energy calculated at the i-th displacement cycle; $K_i^{exp}$ and $K_i^{num}$ are, respectively, the experimental and numerical stiffness calculated at the i-th displacement cycle according to:

$$K_i = \frac{\left|F^+_{max,i}\right| + \left|F^-_{max,i}\right|}{\left|D^+_{max,i}\right| + \left|D^-_{max,i}\right|} \tag{8}$$

where $F^+{}_{max,i}$ and $F^-{}_{max,i}$ are the peak lateral forces applied to the beam in the two directions of loading; $D^+{}_{max,i}$ and $D^-{}_{max,i}$ are the respective displacements.

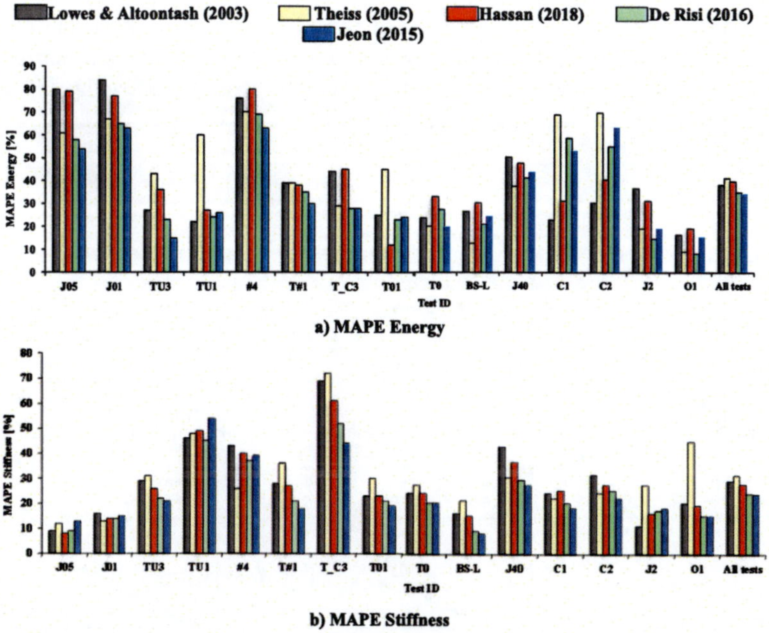


**Fig. 5.** Mean absolute percentage errors for energy dissipation (a) and stiffness degradation (b) calculated for the single test and on all the tests together.

The bar charts in Fig. 5 show the MAPE values calculated separately for the five models in Table 3 and for each test both in terms of energy (Fig. 5a) and stiffness (Fig. 5b) parameters; the mean value calculated per each model by considering all the experimental tests together is also plotted in Fig. 5.

As shown, the model by Jeon et al. (2015) (blue bars) generally yields the lowest MAPE values both in terms of energy and stiffness on the single tests; by considering all tests together, it provides MAPE values equal to 37.67% and 23.40%, respectively. In the ranking, the model by De Risi et al. (2016) follows (green bars), providing overall mean values equal to 38.13% for the dissipated energy and 23.73% for the stiffness degradation.

Figures 6a and 6b depict, respectively, the response in terms of stiffness degradation and cumulative dissipated energy obtained for three representative tests included in the database, the same already considered in Fig. 4. For these tests, the best proposals in terms of *"Pinching 4"* model parameters are taken into account in the numerical simulations, i.e., the model by Jeon et al. (2015) and by De Risi et al. (2016); the numerical simulations are, hence, compared with the experimental trends.

From Fig. 6 it is observed that, the two proposals provide very good simulations in terms of stiffness degradation whereas a higher scatter is noted in terms of dissipated energy, being this parameter very sensitive to the shape and size of the hysteretic cycles. This evidence is also confirmed from the experimental-numerical comparison reported in Fig. 7 where, with reference to the test O1 (Tsonos 2002) only, the numerical curves, even though are in an overall good agreement with the experimental response in terms of load-drift ratio curves, highlight the need of an ad-hoc calibration procedure of the "Pinching4" model parameters. The numerical hysteretic cycles appear to be narrower than the experimental one; the highest scatter is noted in terms of unloading stiffness degradation for which the various proposal reported in Table 3 suggest different values of its defining parameters.

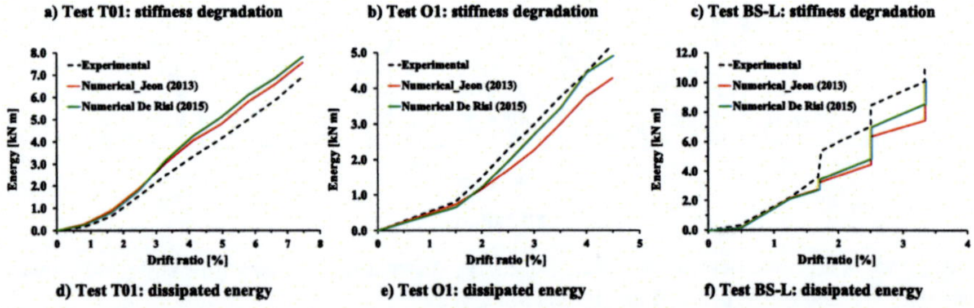


**Fig. 6.** Stiffness degradation (a,b,c) and cumulative dissipated energy (d,e,f): comparison between experimental results and numerical simulations obtained by considering the values of the "Pinching 4" model parameters proposed by Jeon et al. (2015) and De Risi et al. (2016).

Discrepancy in the shape of hysteretic cycles is even more evident from Fig. 8 where the numerical-experimental comparison available for the test O1 (Tsonos 2002) is focused on two cycles only, the first related to the attainment of the peak force and the second immediately following. As noted, in light of a good matching in terms of maximum force at each cycle, the comparison in terms of shape of hysteresis cycles

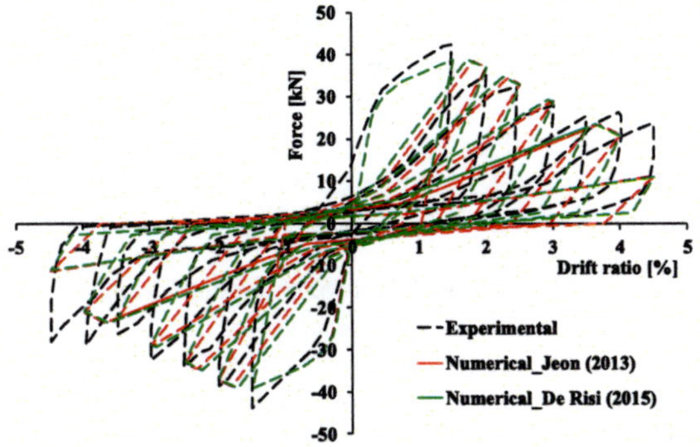


**Fig. 7.** Force – drift cyclic response using the pinching parameters proposed by Jeon et al. (2015) and by De Risi et al. (2016): test O1 (Tsonos 2002).

shows the weakness of the proposals taken under consideration and, therefore, a robust calibration procedure should be implemented to better set the values of the "Pinching4" material parameters.

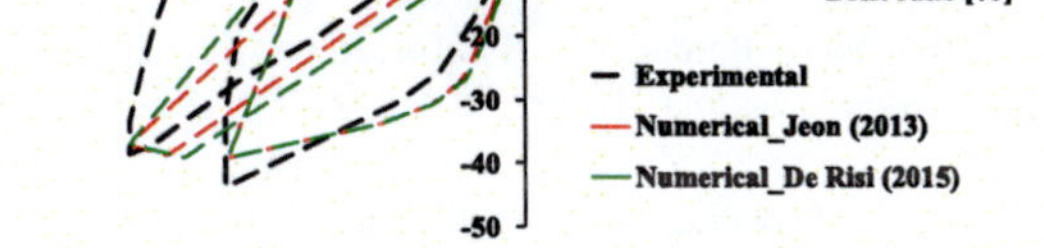


**Fig. 8.** Experimental-numerical comparison focused on a couple of force – drift hysteretic cycles available for the test O1 (Tsonos 2002).

## 5 Conclusions

A numerical modeling able to simulate the monotonic and cyclic behavior of exterior RC beam – column joints, based on the well-known scissor model, has been presented in the present paper. The monotonic analyses have been preliminary finalized to identify among the literature models simulating the joint shear response the ones that better reproduce the overall experimental behavior of the selected specimens. Subsequently, the calibration process has been carried out to improve the accuracy of the constitutive model in simulating the monotonic response of exterior RC joints. Finally, cyclic analyses have been carried out considering some literature proposals for defining the set of parameters governing the hysteresis rules and pinching effect. The obtained results have provided useful indications on how to further improve the cyclic modeling of the considered RC joints.

## References

Celik, O.C., Ellingwood, B.R.: Modeling beam column joints in fragility assessment of gravity load designed reinforced concrete frames. J. Earthquake Eng. **12**, 357–381 (2008)

De Risi, M.T., Ricci, P., Verderame, G.M.: Modelling exterior unreinforced beam-column joints in seismic analysis of non – ductile RC frames. Earthq. Eng. Struct. Dynam. **46**(6), 899–923 (2016)

Grande, E., Imbimbo, M., Napoli, A., Nitiffi, R., Realfonzo, R.: A macro-modelling approach for RC beam-column exterior joints: first results on monotonic behavior. J. Building Eng. **39**, 1–16 (2021)

Grande, E., Imbimbo, M., Napoli, A., Nitiffi, R., Realfonzo, R.: A nonlinear macro-model for the analysis of monotonic and cyclic behaviour of exterior RC beam- column joints. Front. Mater. **8**, 719716 (2021)

Hadi, M.N.S., Tran, T.M.: Seismic rehabilitation of reinforced concrete beam-column joints by bonding with concrete covers and wrapping with FRP composites. Mater. Struct. **49**, 467–485 (2016)

Hassan, W., Al-Zahraa Refaie, F., Belal, A.: Seismic vulnerability of concrete beam-column joints in older construction under high axial loads. In: Proceeding of the 16th European Conference on Earthquake Engineering, Thessaloniki, Greece, June 18–21 (Springer), pp. 1–10 (2018)

Hwang, S., Lee, H.: Analytical model for predicting shear strengths of exterior reinforced concrete beam-column joints for seismic resistance. ACI Struct. J. **96**(5), 846–857 (1999)

Jeon, J.S., Lowes, L.N., DesRoches, R., Brilakis, I.: Fragility curves for non-ductile reinforced concrete frames that exhibit different component response mechanisms. Eng. Struct. **85**, 127–143 (2015)

Kim, J., LaFave, J.M.: Joint Shear Behavior of Reinforced concrete Beam-Column Connections Subjected to Seismic Lateral Loading. Report No. NSEL-020, University of Illinois at Urbana-Champaign (2009)

Lowes, L.N., Altoontash, A.: Modeling reinforced-concrete beam-column joints subjected to cyclic loading. J. Struct. Eng. **129**, 1686–1697 (2003)

McKenna, F., Fenves, G.L., Scott, M.H.: OpenSees: Open System for Earthquake Engineering Simulation. Pacific Earthquake Engineering Research Center. University of California, Berkeley, CA, USA (2010)

Reyes de Ortiz, I.: Strut-and-Tie Modeling of Reinforced Concrete Short Beams and Beam-Column Joints. Ph.D. Thesis, University of Westminster, London (1993)

Sharma, A., Eligehausen, R., Reddy, G.R.: A new model to simulate joint shear behavior of poorly detailed beam–column connections in RC structures under seismic loads, part I: exterior joints. Eng. Struct. **33**(3), 1034–1051 (2011)

Shin, M., LaFave, J.M.: Modeling of cyclic joint shear deformation contributions in RC beam-column connections to overall frame behavior. Struct. Eng. Mech. **18**(5), 645–669 (2004)

Theiss, A.G.: Modeling the Earthquake Response of Older Reinforced Concrete Beam-Column Building Joints. Thesis, University of Washington, M.Sc (2005)

Tsonos, A.G.: Seismic repair of exterior R/C beam-to-column joints using two-sided and three-sided jackets. Struct. Eng. Mech. **13**(1), 17–34 (2002)

Uzumeri, S.M.: Strength and ductility of cast-in-place beam-column joints, American concrete institute annual convention. Symp. Reinf. Concr. Struct. Seismic Zones **53**, 293–350 (1977)

Vollum, R.L., Newman, J.B.: Strut and tie models for analysis/design of external beam-column joints. Mag. Concrete Res. **51**(6), 415–425 (1999)

Wong, H. F.: Shear Strength and Seismic Performance of Non-seismically Designed Reinforced concrete Beam-Column Joints. Ph.D. Thesis. The Hong Kong University of Science and Technology, Department of Civil Engineering (2005)

# On the Verification of Discontinuity Regions in Existing RC Structural Elements Using Strut and Tie Models

Giovanni Menichini(✉), Federico Gusella, and Maurizio Orlando

Department of Civil and Environmental Engineering, University of Florence, Florence, Italy
giovanni.menichini@unifi.it

**Abstract.** Reinforced concrete (RC) discontinuity regions (D-regions), such as dapped ends and corbels, can represent highly vulnerable zones in existing RC structures, since they are characterized by brittle failure mechanisms. In common design practice, the ultimate strength of D-regions is assessed through Strut and Tie models (STM). The present work shows that, if on the one hand, the resistant lattice trusses suggested by construction standards like Eurocode 2 are on the safe side in the design phase, on the other hand, they can provide non-conservative results in the assessment of the ultimate strength of existing D-regions, as they cannot be appropriate to reproduce the actual distribution of steel reinforcement. Within this work, some indications are provided to identify the resistant mechanisms of D-regions in existing RC members together with some preliminary indications how to increase their ultimate strength and ductility.

**Keywords:** RC discontinuity regions · Strut and Tie models · Existing RC structures

## 1 Introduction

### 1.1 RC Discontinuity Regions

Reinforced concrete (RC) discontinuity regions (D-regions), like dapped ends and corbels, exhibit small span-to-depth ratios so conventional sectional analysis procedures, based on the assumption that plane sections remain plane, are not appropriate for their design. For D regions, European standard code (Eurocode 2, 2004) recommends using Strut-and-Tie Models (STM): stress fields within the D region are discretized through a lattice truss composed of concrete compression elements (struts) and steel tension elements (ties), connected at pinned joints (nodes). The use of STM dates back to at least the late of 18th century, when Ritter (1899) and Mörsch (1901) presented a truss analogy for the shear design of RC beams. Schlaich et al. (1987) and Muttoni et al. (1996) significantly contributed to the modern use of STM providing information in identifying the trusses distribution within D-regions. Strut and Tie modelling is based on the lower bound theorem of limit analysis, which holds if the material has a rigid plastic constitutive law. This law is certainly acceptable for steel reinforcement, but not


M. A. Aiello and A. Bilotta (Eds.): ICC 2022, LNCE 435, pp. 442–456, 2024.
https://doi.org/10.1007/978-3-031-43102-9_34

for compressed concrete, which has a reduced ductility. Then STMs are reliable if the appropriate deformation capacity of concrete struts is provided to allow for the plastic capacity of all struts and ties to be attained. However, the accuracy of this method is highly dependent on the proper verification and localization of the concrete regions (nodal regions and struts).

An additional challenge arises when the STM approach is adopted to evaluate the load-bearing capacity of existing RC D-regions. In the assessment procedure, it is not possible to choose reinforcement layout, but the existing reinforcement layout should be considered, which may not fit the theoretical STM. In addition, shear provisions in the codes of the 1960s and 1970s were generally less severe than today. In some cases, minimum shear reinforcement ratios were not required at all. For these reasons, theoretical STM provided by Eurocode 2, when applied to evaluate the load-bearing capacity of existing D-regions within RC members can provide unconservative results. In the present work, indications for the identification of the resistant mechanisms of discontinuity regions of existing reinforced concrete dapped ends and corbels are provided, together with some preliminary indications on how to reinforce those elements in terms of strength and ductility.

## 1.2 Dapped End

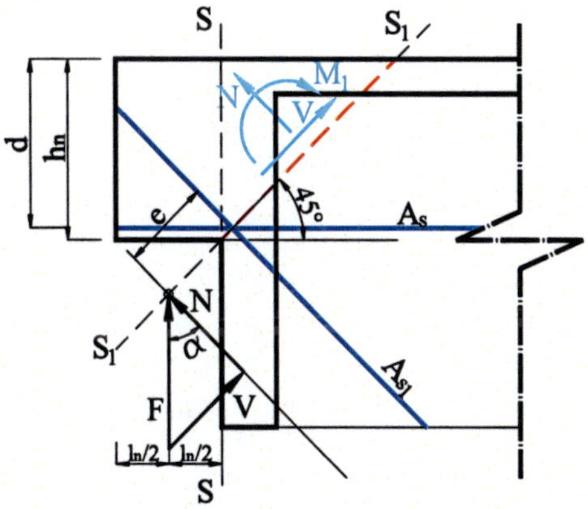


**Fig. 1.** Sections S and S1 in the dapped-end beam.

Before the introduction of strut-and-tie modelling, RC dapped-ends were designed checking the structure at the level of the vertical section S and the inclined section $S_1$, where the angle $\alpha$ is taken as 45° (Del Giudice, 1967) (Fig. 1):

Section S was verified to shear taking design shear stresses $\tau$ as

$$\tau = \frac{F}{0.9 \cdot d \cdot b} \tag{1}$$

where $F$ = the vertical reaction, $b$ = the beam width, $d$ = the effective height of section, $h_n$ = the nib height).

Moreover, S is also subject to the following bending moment:

$$M = F \cdot l_n/2 \tag{2}$$

where $l_n$ = length of the nib.

The following area of the horizontal reinforcement $A_s$ was evaluated:

$$A_s = \frac{M}{0.9 \cdot d \cdot \sigma_s} \tag{3}$$

Section $S_1$ is subject to a tensile axial force N equal to:

$$N = F \cdot \cos\alpha \tag{4}$$

acting at a distance *e* from the main inclined reinforcement.

The area of the inclined reinforcement $A_{s1}$ was evaluated assuming:

$$M_1 = N \cdot e; \quad r = \frac{\sigma_s}{\sigma_c}; \quad h = 0.675\sqrt{\frac{M_1}{b}} \tag{5}$$

and the bending reinforcement was calculated as:

$$A_{s1,b} = \frac{n}{2r \cdot (r+n)} \cdot bh \tag{6}$$

where $n$ = the modular ratio and $h$ = the height of section $S_1$. The formulation for $h$ and $A_{s1,b}$ are retrieved from Santarella (1944).

Once the tensile reinforcement is calculated as

$$A_{s1,t} = \frac{N}{\sigma_s} \tag{7}$$

the amount of total reinforcement $A_{s1}$ can be evaluated as:

$$A_{s1} = A_{s1,b} + A_{s1,t} \tag{8}$$

### 1.3 Corbels

An illustration of corbel design can be found in Prereswiet-Soltan (1982).

First, it is necessary to check that shear stresses in the corbel cross-section $h_p \times b$ (Fig. 2) do not exceed the allowable value $\tau_{c0}$:

$$\tau_c = \frac{P}{h_p \cdot b} \leq \tau_{c0} \tag{9}$$

where $h_p$ = the height of the corbel along the action line of the applied force and $b$ = the corbel width.

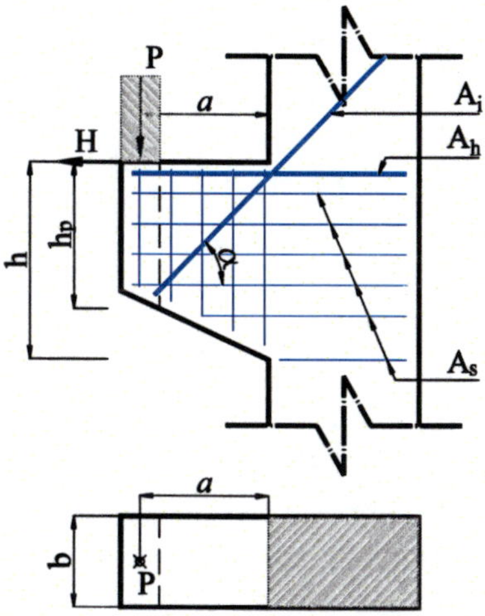


Fig. 2. Layout of corbel reinforcement.

The total area $A_h$ of the main horizontal tensile reinforcement is obtained from the following relationship:

$$A_h = \frac{1}{1.25 \cdot \sigma_s} \cdot \left( \frac{P \cdot a}{0.7 \cdot h} + H \right) \tag{10}$$

where $\sigma_s$ = the allowable tension of reinforcement steel, $P$ = the maximum force acting on the corbel (at a distance $a$ from the column), $h$ = the total height of the corbel, and $H$ = the applied horizontal force.

The total area $A_i$ of the main inclined tensile reinforcement is obtained as:

$$A_i = \frac{P}{1.80 \cdot \sigma_s} \tag{11}$$

Finally, the total area of horizontal stirrups $A_s$ is assumed not lower than 25% of $A_i$:

$$A_s \geq 0.25 \cdot A_i \tag{12}$$

## 2 Numerical Modelling

### 2.1 Dapped End

To study the reliability of the STM approach, a finite element (FE) model was developed using DIANA FEA (2019) code. The NS-REF specimen from the experimental campaign carried out by Desnerck et al. (2016) on dapped-end beams was taken as case study. Figure 3 illustrates the dimensions of the specimen and the amount of reinforcement. The width of the specimen is 400 mm.

Within the DIANA software, the concrete compressive behaviour was modelled using Thorenfeld's constitutive law, where the main parameters are the ultimate compressive

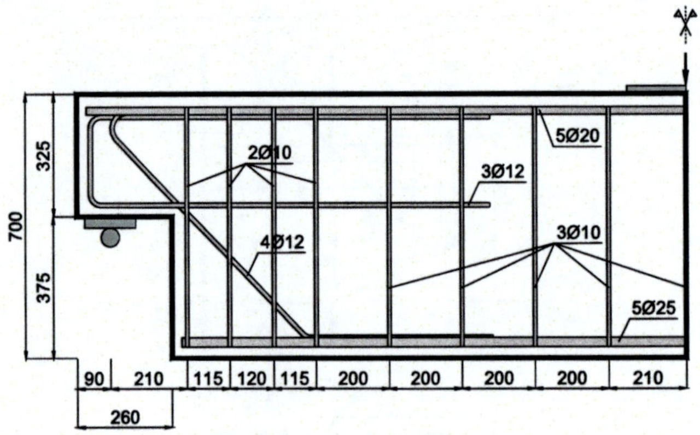


**Fig. 3.** Dapped-end beam specimen from Desnerck et al. (2016).

strength $f_c$ of concrete assumed to be 36.4 MPa and the elastic modulus $E_c$ equal to 34.6 GPa. Tensile behaviour was modelled using Hordijk's constitutive law, where the two main parameters are: concrete ultimate tensile strength $f_t$ assumed equal to 4.84 MPa and the fracture energy, calculated according to Model Code 2010, $G_f = 0.139$ N/mm$^3$. Concrete constitutive laws are based on a plane stress state and use smeared material properties, including the development of cracks using a rotating crack approach.

The reinforcement was modelled with discrete reinforcement lines consisting of embedded truss elements; the uniaxial stress state of the reinforcement was modelled by bilinear stress-strain relationship with hardening. The geometric and mechanical properties of each bar were taken from the work of Desnerck et al. (2016). Steel bearing plates were included in the FE model to prevent the development of local bearing failures and to simulate testing conditions more accurately. The bearing plates were modelled as linear elastic with an elastic modulus of 200 GPa. Analyses were performed in a displacement-controlled manner using a prescribed displacement increment of 0.1 mm for each loading stage. The element sizes considered in the FE model were 25 mm (Fig. 4).

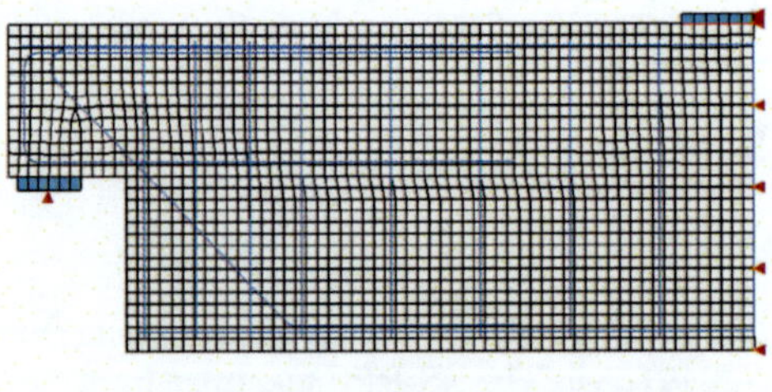

**Fig. 4.** Dapped-end beam numerical model.

To validate the reliability of the modelling approach, the load-displacement curves obtained numerically were compared with experimental results (Fig. 5).

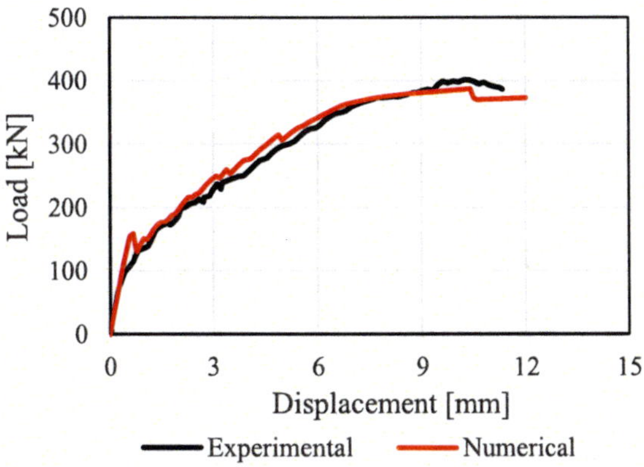


**Fig. 5.** Numerical and experimental load-displacement curves comparison.

## 2.2 Corbel

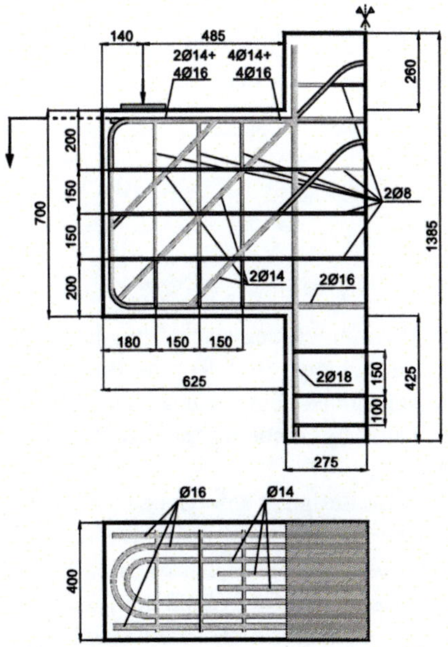


**Fig. 6.** Geometry of the corbel.

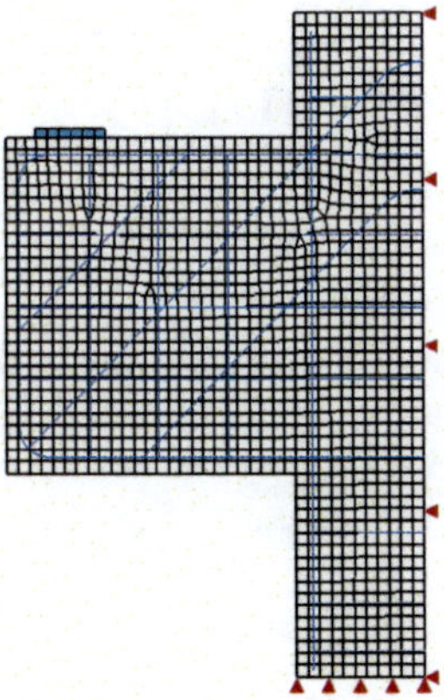

**Fig. 7.** Numerical model of the corbel.

The geometry of a corbel with inclined reinforcement was taken from the textbook by Prereswiet-Soltan (1982) on concrete and precast concrete structures. The dimensions and reinforcement layout of the analysed corbel are shown in Fig. 6. The study was limited to a numerical parametric analysis of the corbel at varying the position of inclined reinforcement. The same mechanical parameters and the same constitutive laws as the dapped end were assumed for concrete and steel.

A plane numerical model was also developed in DIANA. The material properties and the modelling parameters are the same as those given in Sub-Sect. 2.1 for the dapped-end beam. The element sizes considered in the FE model were 25 mm (Fig. 7), as in the FE model of the dapped end.

The numerical load-displacement curve obtained from the FE model for the corbel is shown in Fig. 8.

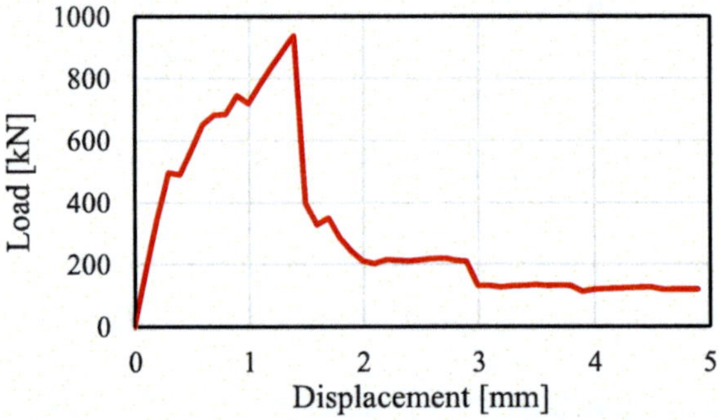


**Fig. 8.** Numerical load-displacement curve of the corbel.

# 3 STM Application

## 3.1 Dapped End

For the evaluation of the dapped end load-bearing capacity, according to Eurocode 2 (2004), two different strut-and-tie models are adopted (Fig. 9). The Eurocode leaves the possibility of using only one of the two models, but the evaluation of the load-bearing capacity can be more properly performed by combining the two trusses together, assigning each of them with 50% of the load.

Eurocode STMs are more appropriate for the design of new dapped ends than for the safety assessment of existing ones, where the reinforcement layout is already defined and cannot be chosen by the designer. In particular, for Truss 1, it is required to determine how many vertical stirrups the $T_{1,1}$ tie consists of, and for Truss 2 how many diagonal bars the $T_{1,2}$ tie consists of when more layers of diagonal bars are present, as is often the case for half-joints designed according to the procedure described in Sub-Sect. 1.2.

Concerning Truss 1, the number of stirrups involved in the $T_{1,1}$ tie can be calculated by summing the contribution of each stirrup until the strength of the horizontal tie $T_{2,1}$ is reached.

With regard to Fig. 10, the force $T_j$ in the $j$-th stirrup could be lower than its ultimate strength $T_{j,max}$:

$$T_j \leq T_{j,\max} = A_j f_{ju} \tag{13}$$

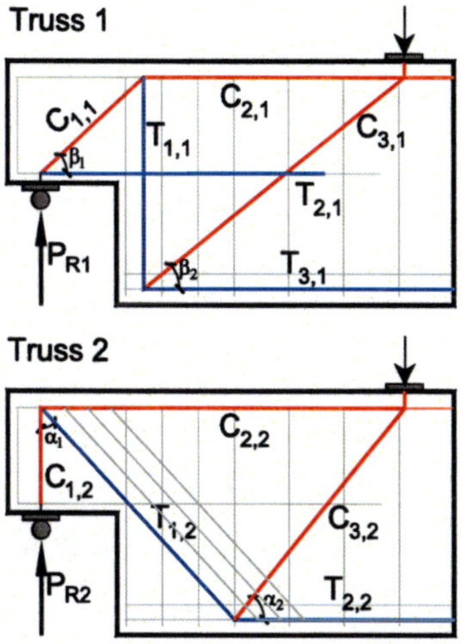


**Fig. 9.** EC2 strut-and-tie trussed for dapped end beams.

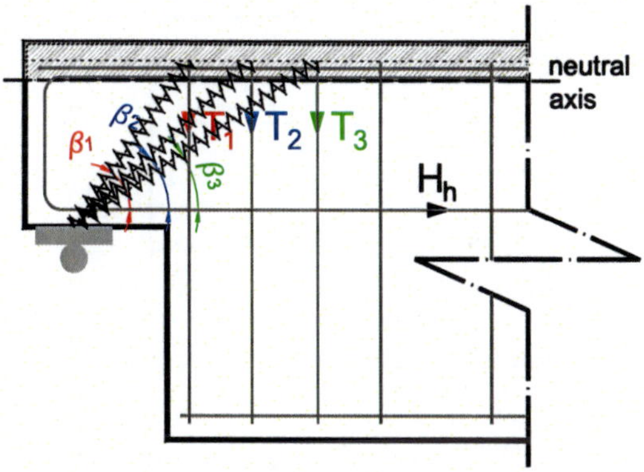


**Fig. 10.** Variable inclination of Truss 1.

where $A_j$ is the cross-sectional area of the stirrup and $f_{ju}$ is the ultimate strength of steel.

The horizontal force required to equilibrate $T_{j,max}$ is:

$$H_j \le T_{j,\max} \cdot \cot(\beta_j) \tag{14}$$

where $\beta_j$ is the angle identified in Fig. 10 (angle between the horizontal axis and the line connecting the external support to the intersection point of the compressive force on the beam and the stirrup position). The maximum force provided by the horizontal reinforcement is:

$$H_h = A_h f_{hu} \tag{15}$$

where $A_h$ = the cross-sectional area of the horizontal reinforcement and $f_{hu}$ = the ultimate strength.

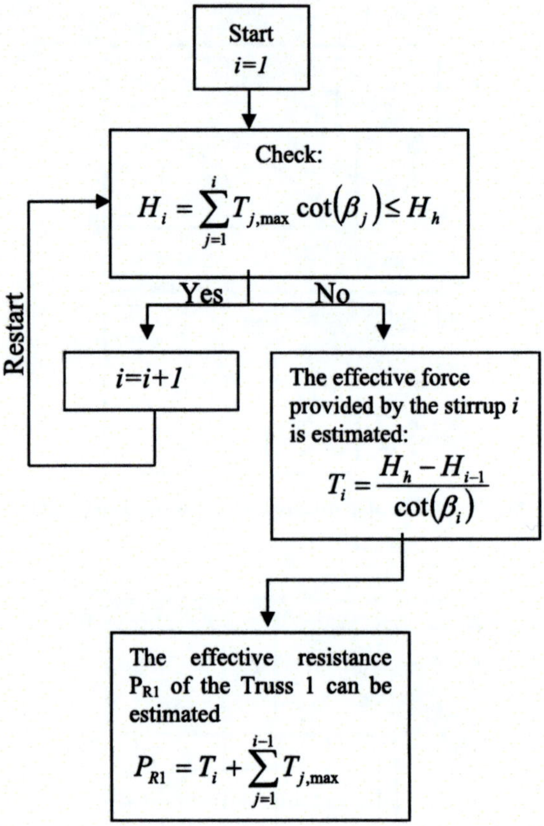


**Fig. 11.** Procedure for identifying the number of stirrups in truss 1.

The procedure in Fig. 11 is proposed to estimate the effective number of stirrups involved in the resistant mechanism and to evaluate the load-bearing capacity of the dapped end. With the aim to highlight the accuracy of the procedure, the NS-REF specimen (Desnerck et al. 2016) has been analysed as case study.

*Start - i = 1.*

The stirrup ultimate strength and its cotangent are evaluated as:

$$T_{1,max} = 157 \cdot 596 = 93572 \text{ N} = 93.6 \text{ kN}$$

where 157 $mm^2$ is the cross-sectional area of the first stirrup; $f_u = 596$ N/$mm^2$ is the experimental tensile strength.

The maximum horizontal force provided by the horizontal reinforcement is $H_h =$ 189.6 kN and it is greater than $H_1 = T_{1,max} \cot(\beta_1)$:

$$H_1 = 93.6 \cdot 0.8249 = 77.2 \text{ kN} \leq H_h = 189.6 \text{ kN}$$

where $\cot(\beta_1)$ is equal to *0.8249*. Therefore, the contribution of the second stirrup should be added and the procedure should be restarted.

*Restart - i = 2.*

$$T_{2,max} = 93.6\ \text{kN}$$
$$\cot(\beta_2) = 1.2766$$
$$H_2 = 77.2 + 93.6 \cdot 1.2766 = 196.7\ \text{kN}$$

The maximum horizontal force provided by the horizontal reinforcement is now lower than $H_2$

$$H_2 = 196.7\ \text{kN} > H_h = 189.6\ \text{kN}$$

where $H_2 = T_{1,max} \cot(\beta_1) + T_{2,max} \cot(\beta_2)$.

The effective contribution provided by the second stirrup is then lesser than its tensile strength and it is evaluated using the expression given in Fig. 11:

$$T_2 = \frac{H_h - H_1}{\cot(\beta_2)} = 88.1\ \text{kN}$$

Finally, the base reaction is assessed:

$$P_{r1} = 181.6\ \text{kN}$$

The method needs to be completed with the verification of the node where the reaction $P_R$ is applied to consider that more struts enter the node and not only one strut as in the traditional STM geometry.

The load-bearing capacity of the NS-REF specimen only considering the contribution of stirrups has also been evaluated through the FE model. Starting from the model described in Sect. 2 and removing the diagonal bars and carrying out the analysis by gradually increasing the number of vertical stirrups (2ϕ10) from 1 up to 4, it is obtained that after the addition of the second stirrup, the load no longer increases because the yield strength of the horizontal U-shaped bars has been reached (Fig. 12). The ratio between the theoretical and numerical value of the base reaction $P_{r1,th}/P_{r1,num}$ is 0.92 confirming the accuracy of the proposed approach, which can estimate the number of stirrups which participate in the resistant mechanism. The highest value of the numerical ultimate load is due to the same contributions considered in the STM.

For Truss 2, starting from the numerical model described in Sect. 2, both vertical stirrups (2ϕ10) and U-shaped horizontal bars consisting of 3ϕ14 were removed. The position of the diagonal bar was changed by horizontal steps of 75 mm (Fig. 13).

The results show that if the diagonal bar no longer involves the width of strut $C_{1,2}$ but gradually moves away from it (Pos. B, C, D in Fig. 13), the half-joint strength rapidly decreases because Truss 2 is only slightly activated (Fig. 14). This result well agrees with the mechanical model assumed in Del Giudice (1967): as the diagonal bar moves inwardly, the effective depth of the resisting diagonal cross-section reduces.

To better approximate numerically the geometry of the STM, a final analysis of the half-joint was performed neglecting the tensile strength of concrete. This analysis gave the principal stress trajectories shown in (Fig. 15a): it is evident that if the top end of the diagonal bar lies within the width of the vertical strut, Truss 2 can arise according to the geometry proposed by Eurocode 2. If the diagonal bar is moved toward the centre of the beam, a vertical pure shear crack is formed at the cross-section of the nib inner corner (Fig. 15b).

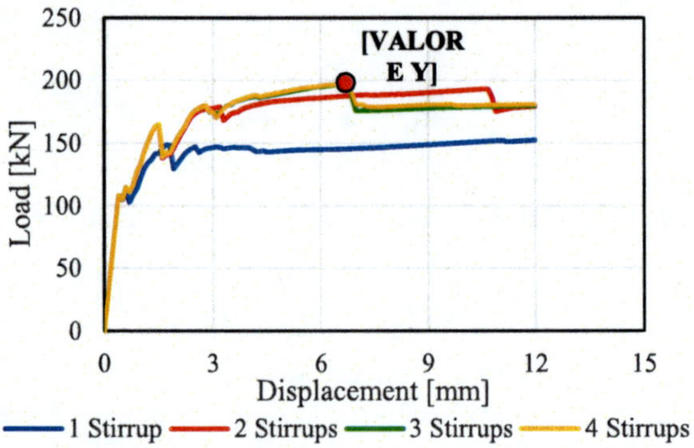


**Fig. 12.** Strength increasing by adding vertical stirrups for Truss 1.

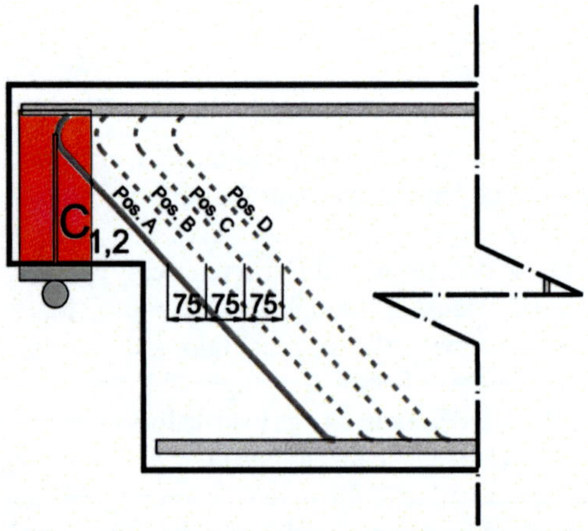


**Fig. 13.** Considered diagonal bar positions in the numerical model and $C_{1,2}$ strut width.

## 3.2 Corbel

For the evaluation of the corbel load-bearing capacity, according to Circolare (2009), two different strut-and-tie models are adopted (Fig. 16). The Circolare leaves the possibility of using only one of the two models, but the evaluation of the load-bearing capacity can be more properly performed by combining the two trusses together (Model A + Model B). The contribution of each resisting mechanism can be taken into account as:

$$V_{Rd} = V_{Rd,A} + 0.8V_{Rd,B} \quad (16)$$

considering the additional contribution of the inclined reinforcement from Model B reduced by 20%. Further details can be found in Angotti et al. (2022).

Starting from the model described in Sect. 2 (Fig. 2) and removing the diagonal bars, the FE analysis is repeated by considering three different reinforcement layouts:

1. the main horizontal reinforcement only ($A_s$)

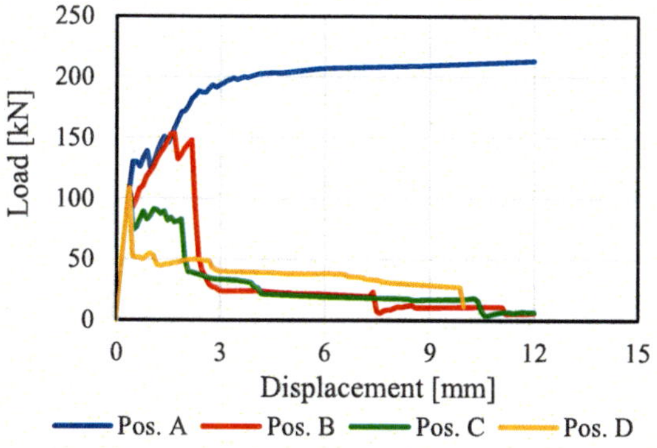


**Fig. 14.** Strength decreasing as the diagonal bar position changes.

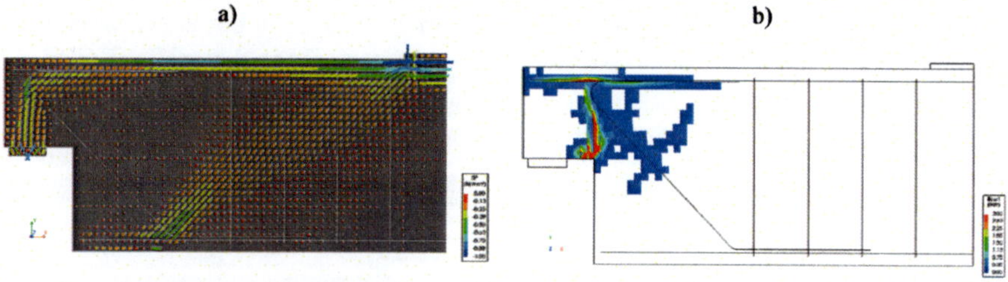


**Fig. 15.** Compressive principal stresses for diagonal reinforcement in Pos. A (a). Crack width for diagonal reinforcement in Pos. C (b).

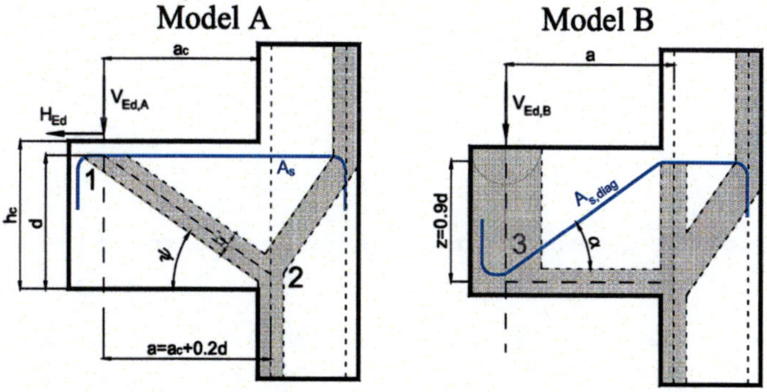


**Fig. 16.** Circolare (2009) strut-and-tie models for the corbel.

2. the main horizontal reinforcement and the horizontal stirrups ($A_s + H$)
3. the main horizontal reinforcement, the horizontal and vertical stirrups ($A_s + H + V$)

The results show that the largest contribution to the strength is provided by the main reinforcement $A_s$, and the addition of the horizontal and vertical stirrups makes a limited contribution to the overall strength of Model A (Fig. 17).

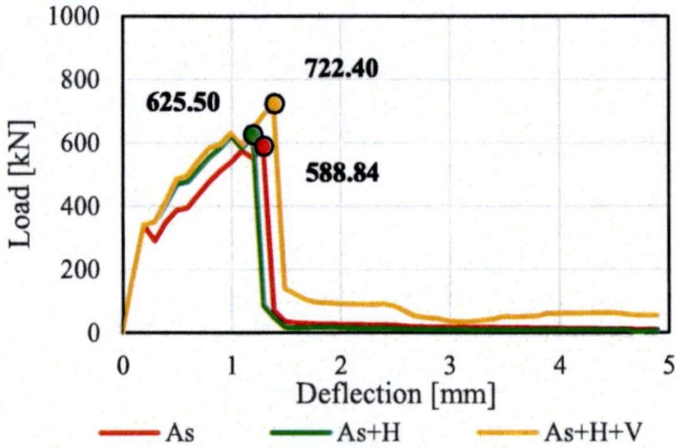


**Fig. 17.** Strength increasing by adding stirrups for the Model A.

For Model B, both vertical and horizontal stirrups and main horizontal bars were removed. One layer of diagonal bars at a time was considered in the analyses (Pos. A, B, C in Fig. 18).

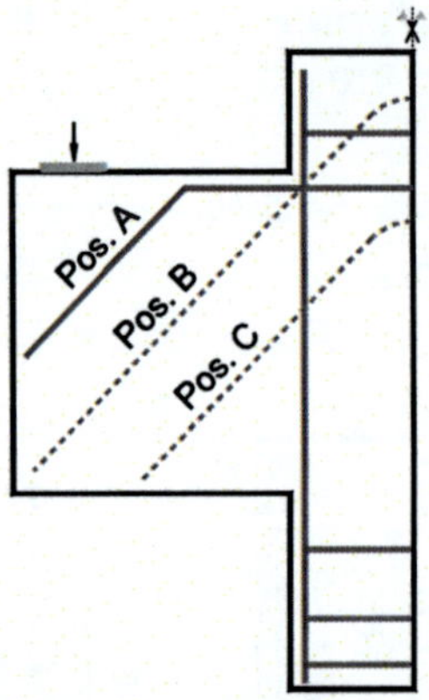


**Fig. 18.** Considered diagonal bar positions in the numerical model.

The results show that the highest resistance is given by the diagonal bar at position A then the contribution in terms of resistance decreases as the diagonal bars are moved to positions B and C. Similarly, to the dapped-end beam, the resistant contribution is about 50% lower than the horizontal reinforcement (Fig. 19).

Again, to better approximate numerically the geometry of the STM, some final analyses were performed neglecting the tensile strength of concrete. Observing principal stress trajectories from the numerical model, it can be seen that if the diagonal bar is in position A, the truss differs from the theoretical Model B (Fig. 20a) and the resistance cannot be accurately evaluated. Whereas for positions B and C the truss is very similar to the theoretical one (Fig. 20b, c).

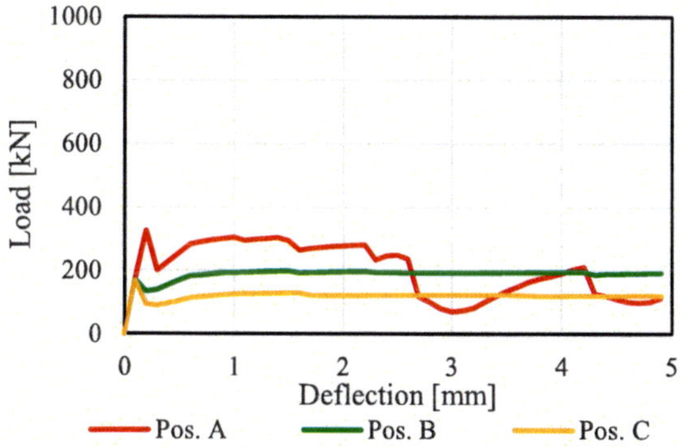


**Fig. 19.** Strength decreasing as the diagonal bar position changes.

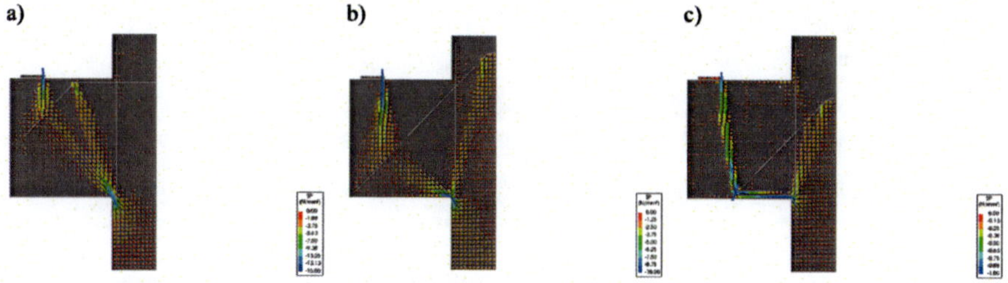


**Fig. 20.** Compressive principal stresses for diagonal reinforcement in Pos. A (a), in Pos. B (b) and in Pos. C (c).

## 4 Conclusions

The structural response of reinforced concrete (RC) discontinuity regions (D-regions) of dapped ends and corbels have been investigated. It has been observed that for existing RC structures, the Strut and Tie Models (STM) proposed by Eurocode can be not appropriate to reproduce the actual reinforcement layout. For dapped-end beams, a simple approach is proposed to estimate the number and percentage of resistance provided by vertical stirrups. Moreover, it is observed that the resistance provided by diagonal reinforcement reduces as its distance from the external support increases. FE modelling of the corbel has confirmed that the horizontal main reinforcement gives the highest contribution to strength, while diagonal bars are less effective, and stirrups, as expected, play a role in limiting the cracking pattern. Results provide useful information for a reliable estimation of the resistance of existing D-regions.

## References

Angotti, F., Guiglia, M., Marro, P., Orlando, M.: Strut-and-tie models. In: Angotti, F., Guiglia, M., Marro, P., Orlando, M. (eds.) Reinforced Concrete with Worked Examples, pp. 587–691. Springer, Cham (2022). https://doi.org/10.1007/978-3-030-92839-1_10

Circolare 2009. Circolare 2 febbraio 2009, n. 617 Istruzioni per l'applicazione delle «Nuove norme tecniche per le costruzioni». Coniglio Superiore dei Lavori Pubblici, Italy (2009)

Desnerck, P., Lees, J.M., Morley, C.T.: Impact of the reinforcement layout on the load capacity of reinforced concrete half-joints. Eng. Struct. **127**, 227–239 (2016)

DIANA FEA 2019. DIANA Finite Element Analysis User's Manual Release 10.3. Delft, The Netherlands (2019)

EN 1992-1-1 2004. Eurocode 2: Design of concrete structures. Part 1-1: General rules and rules for buildings. European Committee for Standardization, Brussels, Belgium (2004)

Del Giudice, G.: Ponte a travata Gerber in cemento armato. Vitali e Ghianda (Ed), Genova (1967)

Muttoni, A., Schwartz, J., Thürlimann, B.: Design of Concrete Structures with Stress Fields. Springer, Heidelberg (1996). https://doi.org/10.1007/978-3-0348-9047-2

Prereswiet-Soltan, S.: Guida alla progettazione di strutture tradizionali e prefabbricare in calcestruzzo armato. I.T.E.C. (Ed), Milano (1982)

Santarella, L.: Prontuario del Cemento Armato. HOEPLI (Ed), Milano (1944)

Schlaich, J., Schäfer, K., Jennewein, M.: Toward a consistent design of structural concrete. PCI J. **32**(3), 74–150 (1987)

# Non-Destructive Testing of a Cable-Stayed Bridge in Lisbon

Emanuele Codacci-Pisanelli[1] and Anna Reggio[2](✉)

[1] CONTEST Rilievi, Prove e Controlli nell'Ingegneria, Roma, Italy
ecp@contest.rm.it

[2] Department of Structural, Geotechnical and Building Engineering, Politecnico di Torino, Torino, Italy
anna.reggio@polito.it

**Abstract.** This paper discusses objectives, methodologies and results from a campaign of non-destructive investigations on a cable-stayed bridge in Lisbon (Portugal). Among the investigations held, the campaign encompassed dynamic testing of the stay cables in order to identify their natural vibration frequencies and to estimate their tension force. A frequency-domain output-only approach was carried out to identify the natural frequencies of the cables. Subsequently, the cable tension force was estimated on the basis of the flat taut string model. In an advantageous and cost-effective way, vibration tests were carried out without the interruption of the vehicular traffic. Results of the testing campaign were finally incorporated in the bridge condition assessment and used to provide both short- and long-term maintenance prescriptions.

**Keywords:** Cable-stayed bridge · Prestressed concrete · Non-destructive testing · Dynamic testing · Cable tension

## 1 Introduction

The present paper discusses objectives, methodologies and results from a campaign of non-destructive investigations on a cable-stayed bridge in Lisbon (Portugal). Non-destructive investigations were aimed at the health monitoring of the bridge and were finally used to formulate its maintenance plan, both short- and long-term.

Among the discussed methodologies, attention is given to the dynamic tests on the stay cables. Dynamic testing of the stay cables is, indeed, of paramount importance for cable-stayed bridges, during both construction and maintenance phases (Walther et al. 1999, Ryall et al. 2000). As pointed out by Cunha et al. 2001, different objectives can be pursued:

(i) the estimation of the cables tension force, either as a verification of the correct distribution of internal loadings in the bridge or as an indicator for structural health monitoring;

(ii) the identification of the cables damping properties (Codacci-Pisanelli & Reggio 2008);


M. A. Aiello and A. Bilotta (Eds.): ICC 2022, LNCE 435, pp. 457–465, 2024.
https://doi.org/10.1007/978-3-031-43102-9_35

(iii) the assessment of the cables fatigue phenomena, possibly caused by long-term traffic loads;

(iv) the evaluation of the level of importance of cables vibration;

(v) the experimental identification of global and local natural frequencies, for validating or updating finite element models.

In this work, dynamic testing was used to identify the natural vibration frequencies of the stay cables and subsequently estimate their tension force.

# 2 Santa Apolonia Cable-Stayed Bridge

## 2.1 General Description of the Bridge

The bridge under consideration is a crossing on railway tracks located in the nearby of the Santa Apolonia train station in Lisbon (Fig. 1). It was built between November 1997 and August 1998 at the initiative of *Caminos de Ferro Portugeses, E.P.*

The cable-stayed system is asymmetrical with a single pylon and two spans.

The main span is 80 m long; the shortest span is 50 m long. The pylon is made of reinforced concrete with a total height of 40 m above the foundation plane. The deck consists of a prestressed concrete box girder supported by way of 8 pairs of stay cables.

Stay cables are arranged in a single vertical plane at the centre of the deck and are connected to the pylon in a harp design (Fig. 2). At the side (shortest) span, stay cables are anchored to ground. Total deck width is 19 m, including two side platforms.

## 2.2 Characteristics of the Stay Cables

Santa Apolonia bridge was built by adopting the *TSR Stay Cable System* by company TENSA (formerly known as Tensacciai).

The stay cable system is composed of seven-wire steel strands with a nominal diameter of 15.7 mm and a nominal cross section of 150 mm$^2$. Low relaxation steel with a characteristic tensile strength of 1860 MPa is used. Strands are protected from corrosion using galvanised wires, a thin layer of corrosion inhibitor (wax) and a co-extruded high-density polyethylene (HDPE) coating.

Each stay cable consists of a compact bundle of parallel strands in a variable number: 19, 27, 31, 37, 46 or 55 strands, corresponding to linear mass density ranging from 29 kg/m to 78.28 kg/m. Stay cables are protected from atmospheric agents by means of a HDPE circular duct, along the entire free length, and are equipped with anti-vandalism tubes at the anchorages to both the pylon and the deck.

In the original project drawings of the bridge, stay cables anchored to the main (longest) span are denominated by letter "R", while stay cables anchored to ground are denominated by letter "T". The single cable is therefore identified by a letter and a number between 1 and 8 (Fig. 2). The same nomenclature is adopted in the present paper.

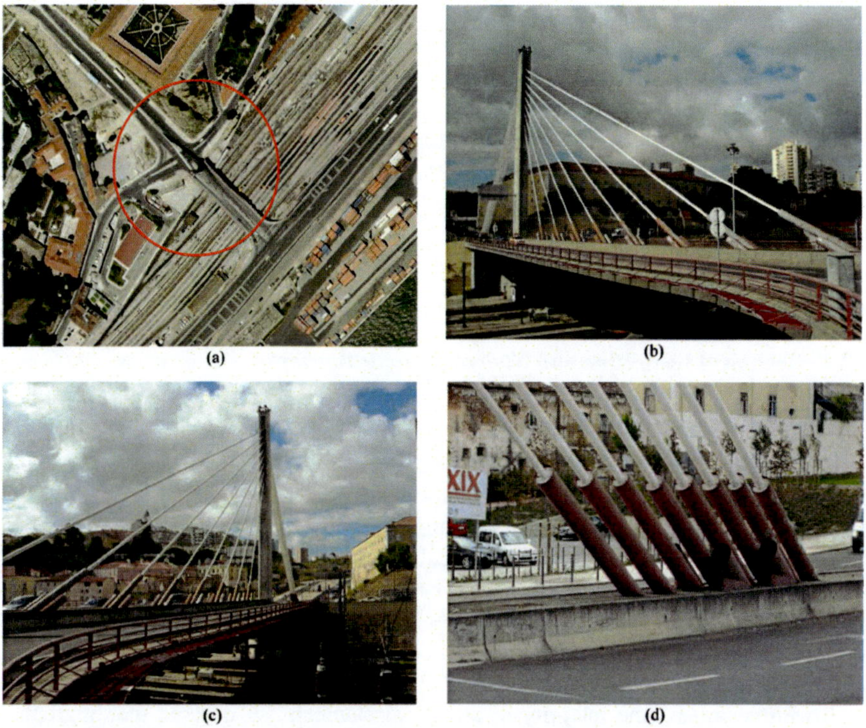


**Fig. 1.** Santa Apolonia cable-stayed bridge in Lisbon (Portugal): (a) aerial photogrammetric view; (b) and (c) overall views; (d) details of the ground anchorages of stay cables at side span.

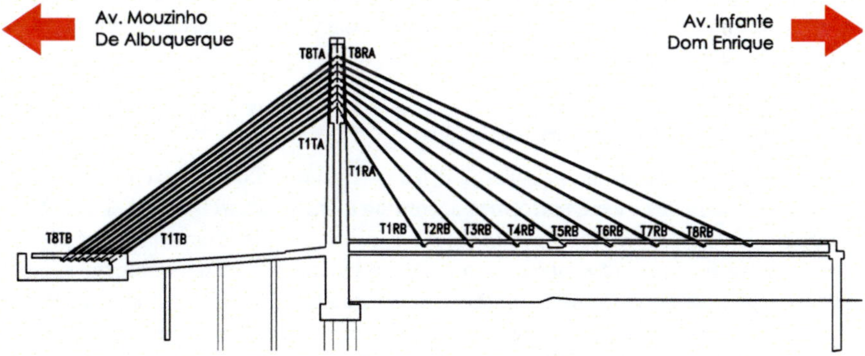


**Fig. 2.** Lateral view of the Santa Apolonia cable-stayed bridge.

# 3 Non-Destructive Investigations

## 3.1 Experimental Setup

The following *in situ* non-destructive investigations were performed on Santa Apolonia bridge: (i) technical inspection; (ii) thermographic survey; (iii) magnetometric tests; (iv) ultrasonic tests; (v) Time Domain Reflectometry tests; (vi) acoustic spectrum analysis tests.

## 3.2 Methodologies and Results

### 3.2.1 Technical Inspection and Thermographic Survey

During the first field survey, a technical inspection of the bridge was carried out, to assess the preservation condition of the structure and to detect the presence of any visible signs of damage, materials degradation, and unusual structural behaviour. For each stay cable, the inspection covered: anchor heads, anchor plates, and anchor wedges; the efficiency of the permanent load cells placed over the anchorages; bearing tubes and anti-vandalism tubes; the HDPE sheath covering the stay cable length.

The inspection has shown a good preservation condition of the bridge, except for minor anomalies, such as: presence of water in the anchoring chamber of the grounded stay cables (Fig. 3a); clogging of wax and cement grout in a few anchor heads of the stay cables anchored to the main span (Fig. 3b).

To further assess the integrity of structural elements, an infrared thermographic survey was carried out on stay cables, anchor heads and pylon. Thermographic survey was used to detect anomalous thermal bridges or anomalous thermographic signals caused by invisible material discontinuities and cracking. The images were acquired by means of a high-resolution infrared camera and processed by a dedicated software. Results confirmed the generally good preservation condition of the bridge, with a limited extent of degradation areas (Fig. 4a, b).

### 3.2.2 Magnetometric and Ultrasonic Tests

Magnetometric and ultrasonic tests were carried out on the reinforcement concrete pylon.

Magnetometric tests were performed according to the standard BS 1881–204:1988 – *Testing concrete. Recommendations on the use of electromagnetic covermeters* and used to measure the thickness of concrete cover and the spacing between reinforcement bars. Ultrasonic tests were performed according to the standard UNI EN 12504–4:2005 and used to evaluate the concrete compressive strength, correlated to the propagation velocity of the ultrasounds in the material (central frequency of the ultrasonic probe 54 Hz). Results were consistent with the original executive project of the bridge.

### 3.2.3 Time Domain Reflectometry and Acoustic Spectrum Analysis Tests

The integrity of the stay cables was investigated by way of Time Domain Reflectometry (TDR) tests and acoustic spectrum analysis tests.

(a) (b)

**Fig. 3.** Technical inspection: (a) presence of water in the anchoring chamber of the grounded stay cables; (b) wax clogging in a stay cable anchor head.

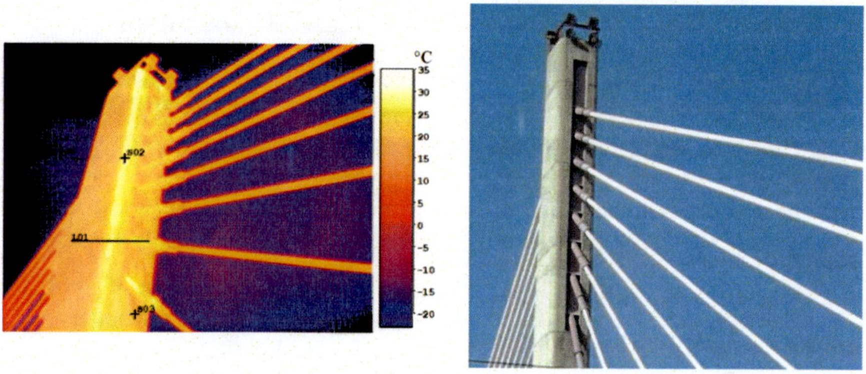


(a) (b)

**Fig. 4.** Thermographic survey: (a) thermographic image; (b) visible light image.

TDR tests were performed on strands, in correspondence with the anchor heads of the stay cables (Fig. 5). TDR testing consists in the application of electrical impulses, at one end of the strand, and in the subsequent measurement of properties and time delay of the reflected impulses. Any anomalies of the cable, such as cross-section reductions, cracks, short circuits, produce a variation in the electrical impedance and a reflection, which are detected by the TDR testing. The tests were performed using a portable equipment that shows results in real time through a graphical scheme on LCD display, identifying the defects and their position along the length of the cable.

Acoustic spectrum analysis tests were performed on strands, in correspondence with the anchor heads of the stay cables, and on the antivandalism tubes. The acoustic signal, generated by means of an instrumented metallic hammer, was acquired and analysed in the frequency band between 20 Hz and 20 kHz.

Results from TDR and acoustic spectrum analysis tests confirmed the integrity of the stay cables.

**Fig. 5.** Time Domain Reflectometry (TDR) tests on a stay cable, detail of the executive phases.

# 4 Dynamic Testing of the Stay Cables

Dynamic tests were carried out on all the stay cables of Santa Apolonia bridge, in order to identify their natural vibration frequencies and to evaluate their tension force.

## 4.1 Experimental Measurements and Identification of Natural Vibration Frequencies

Dynamic tests concerned the measurement of the stay cable response to the ambient excitation given by vehicular traffic and wind. Ambient vibration tests of this kind are comparatively cheaper than other dynamic tests because they do not require artificial excitation and can be advantageously performed without the interruption of the vehicular traffic.

Time histories of the stay cables accelerations were measured. The cables were equipped with piezoelectric uniaxial accelerometers characterized by high sensitivity and low electrical noise, as suited to small vibration testing of Civil Engineering structures (sensitivity 500 mV/g, where g is the gravity acceleration; maximum acceleration 14 g; shock acceleration 5000 g; band-pass frequencies 0.2 Hz $\div$ 6 kHz). Accelerometers were attached to the external surface of the stay cables and arranged to measure vibration in the vertical plane. One accelerometer per stay cable was used, located at half length of the cable. The output signals were conditioned and acquired by an acquisition unit at sampling frequency $f_s = 200\,\text{Hz}$. The total time $t_f$ of acquisition was set up as not smaller than 100 s, in order to have a frequency resolution $\Delta f = 1/t_f$ not greater than 0.01 Hz in the spectral analysis (Ewins 2000).

The acquisition system encompassed multiple recording channels and a custom software for managing and storing the measured data. Acceleration time histories were detected and stored continuously, starting from the activation time of the unit. The activation time was governed by a trigger event (i.e., a threshold level of the signal) and a pre-event memory was also included.

The measured acceleration time histories were subsequently processed in frequency domain to obtain the corresponding power spectra. Natural vibration frequencies of the stay cables were identified according to the classical approach based on the peak picking technique. This approach provides estimates of the natural frequencies that are reasonably accurate when the vibration modes of the system are well separated, as in the case of a stay cable (Irvine 1981).

Tables 1 and 2 show the values of the fundamental vibration frequency $f_1$ identified experimentally for each stay cable of the Santa Apolonia bridge. Tables also report, for each stay cable, the following data: length $L$; diameter $d$; the number $r$ of strands composing the cable; the mass per unit length $\rho$.

### 4.2 Evaluation of the Tension Force

The tension force in the stay cables was evaluated on the basis of the cables' natural vibration frequencies, previously identified experimentally. This indirect method of estimating the cable tension force has found favour in Civil Engineering practice because of its relative simplicity, cost-effectiveness, and reliability (Casas 1994, Fang & Wang 2012).

In the applied methodology, the dynamic behaviour of the stay cable is described according to the flat taut string model. This idealisation is sufficiently accurate when: (i) the sag of the cable is small compared to the span; (ii) the elastic bending stiffness of the cable is neglected; (iii) the cable vibrates in-plane.

**Table 1.** Stay cables anchored to the main span (cables' denomination ends with letter R): geometric properties, mass per unit length $\rho$, fundamental vibration frequency $f_1$ identified experimentally, estimated value $T$ and design value $T_d$ of the tension force.

| Stay cable | Length $L$ (m) | Diameter $d$ (m) | Number $r$ of strands | Mass per unit length $\rho$ (kg/m) | Fundamental frequency $f_1$ (Hz) | Estimated tension force $T$ (kN) | Design tension force $T_d$ (kN) |
|---|---|---|---|---|---|---|---|
| T1R | 24.2 | 0.181 | 31 | 45.74 | 5.51 | 3247 | 2148 |
| T2R | 30.3 | 0.181 | 31 | 45.74 | 4.07 | 2776 | 3103 |
| T3R | 37.0 | 0.204 | 37 | 54.21 | 3.46 | 3544 | 2971 |
| T4R | 44.1 | 0.204 | 37 | 54.21 | 2.98 | 3743 | 3905 |
| T5R | 51.5 | 0.204 | 37 | 54.21 | 2.73 | 4296 | 3888 |
| T6R | 59.1 | 0.226 | 46 | 66.58 | 2.20 | 4502 | 4881 |
| T7R | 66.8 | 0.226 | 46 | 66.58 | 1.92 | 4367 | 4774 |
| T8R | 74.5 | 0.226 | 46 | 66.58 | 1.87 | 5152 | 5537 |

**Table 2.** Stay cables anchored to ground (cables' denomination ends with letter T): geometric properties, mass per unit length $\rho$, fundamental vibration frequency $f_1$ identified experimentally, estimated value $T$ and design value $T_d$ of the tension force.

| Stay Cable | Length $L$ (m) | Diameter $d$ (m) | Number $r$ of strands | Mass per unit length $\rho$ (kg/m) | Fundamental frequency $f_1$ (Hz) | Estimated tension force $T$ (kN) | Design tension force $T_d$ (kN) |
|---|---|---|---|---|---|---|---|
| T1T | 39.4 | 0.143 | 19 | 29.00 | 3.40 | 2081 | 2125 |
| T2T | 41.2 | 0.181 | 27 | 40.54 | 3.22 | 2854 | 2803 |
| T3T | 43.1 | 0.181 | 27 | 40.54 | 3.13 | 2951 | 2914 |
| T4T | 44.9 | 0.204 | 37 | 54.12 | 3.01 | 3961 | 4042 |
| T5T | 46.7 | 0.226 | 46 | 66.58 | 2.75 | 4392 | 4140 |
| T6T | 48.5 | 0.226 | 46 | 66.58 | 2.73 | 4669 | 5085 |
| T7T | 50.2 | 0.226 | 46 | 66.58 | 2.65 | 4713 | 4903 |
| T8T | 52.0 | 0.226 | 55 | 78.28 | 2.58 | 5636 | 5728 |

The flat taut string model considers a tightly stretched elastic string of length $L$, fixed at both ends, whose $n^{\text{th}}$ natural vibration frequency $f_n$, in Hz, is given by

$$f_n = \frac{1}{2}\frac{n}{L}\sqrt{\frac{T}{\rho}}, \quad n = 1, 2, \ldots \tag{1}$$

where $T$ is the tension force in the string and $\rho$ is the mass per unit length (Rao 2017). From Eq. (1), it follows that higher natural frequencies are integer multiples of the fundamental vibration frequency $f_1$, and that the tension force in the string can be obtained as

$$T = 4\rho L^2 \left(\frac{f_n}{n}\right)^2, \quad n = 1, 2, \ldots \tag{2}$$

Tables 1 and 2 show the values of the tension force $T$ estimated for each stay cable of the Santa Apolonia bridge. The tension force $T$ has been estimated by applying Eq. (2) and by substituting into it the fundamental vibration frequency $f_1$ identified experimentally for the considered cable. Tables also report, for each stay cable, the design value $T_d$ of the cable tension force according to the original executive project of the bridge.

## 5 Conclusions

In this paper, a campaign of non-destructive investigations performed on the Santa Apolonia cable-stayed bridge in Lisbon has been presented and discussed. The investigations aimed at the health monitoring of the bridge and were finally used to formulate the maintenance plan, both short- and long-term. The obtained results represented a substantial and unbiased knowledge base according to which the prioritisation of maintenance interventions was defined. One focus of the investigations campaign was on the dynamic testing of the stay cables. Ambient vibration tests were carried out to experimentally identify the natural vibration frequencies of the stay cables and subsequently estimate their tension force according to the flat taut string model. In this way, the distribution of the tension forces among the stay cables and the consistency between operation conditions and design values were verified.

## References


Casas, J.R.: A combined method for measuring cable forces: the cable-stayed Alamillo Bridge, Spain. Struct. Eng. Int. J. IABSE **4**(4), 235–240 (1994)

Codacci-Pisanelli, E., Reggio, A.: Analisi vibrazionali: applicazioni agli stralli da ponte. In: Proceedings of the 3rd Workshop on Vibration Problems in Civil and Mechanical Engineering, Perugia, Italy, September 11–12 (2008)

Cunha, A., Caetano, E., Dalgado, R.: Dynamic tests on large cable-stayed bridge. ASCE J. Bridge Eng. **6**(1), 54–62 (2001)

Ewins, D.J.: Modal Testing: Theory, Practice and Application, 2nd edn. John Wiley & Sons, New York (2000)

Fang, Z., Wang, J.Q.: Practical formula for cable tension estimation by vibration method. ASCE J. Bridge Eng. **17**(1), 161–164 (2012)

Irvine, H.M.: Cable Structures. The Massachusetts Institute of Technology Press, Cambridge (USA) (1981)

Rao, S.S.: Mechanical Vibrations, 6th edn. Pearson Education Inc., Hoboken, NJ, USA (2017)

Ryall, M.J., Parke, G.A.R., Hewson, N., Harding, J.E. (eds.): Manual of Bridge Engineering, The Institution of Civil Engineers. Thomas Telford Ltd., London (2000)

Walther, R., Houriet, B., Isler, W., Moia, P., Klein, J.F.: Cable Stayed Bridges. Thomas Telford Ltd, London (1999)

# Safety Check of Reinforced Concrete Viaducts According to Past and Actual Design National Codes: A Real Case Study

Marco Gallo(✉), Romeo Tomeo, and Emidio Nigro

Department of Structure for the Engineering and Architecture,
University of Naples Federico II, Naples, Italy
{marco.gallo,romeo.tomeo,emidio.nigro}@unina.it

**Abstract.** This work deals with the static safety check of a real case study concerning a viaduct with simply supported prestressed concrete beams and reinforced concrete piles, recurrent in Italy in the 1960s–70s. A review of the typical design methods used in the past for bridges is provided, starting from the load case definitions and their possible combinations to the verification criteria. Moreover, the typical assumptions made in the original design procedure are shown, also considering the simplified models used. These original results are compared to the ones obtained from the application of the actual design method proposed in the national technical codes and the FEM modelling of the bridge. The whole static safety check process of the viaduct is also illustrated, starting from the structural survey plan and dynamic identification of the viaduct up to the development of the bridge structural elements checks, highlighting the most important aspects in the assessment of the typical viaduct examined.

**Keywords:** Concrete viaducts · Assessment of existing bridges · bridges FEM model

## 1 Introduction

### 1.1 Design Methods

The static and seismic design methods of the structures are deeply changed over time as the national codes and calculation systems are constantly being updated. On one hand, past national codes proposed load cases, combinations of them and verification criteria much different from the current ones. On the other hand, more sophisticated calculations methods and tools, as FEM for instance, were not available, thus, simplified models were typically used to perform hand evaluations and verifications. Existing structures, and in particular bridges, could eventually be not verified by using modern codes approach. This work, thus, aims to provide a review of the safety check of a real case study of a concrete viaduct focusing on the typical design methods used in the past for bridges and compare those to the application of the actual ones.


M. A. Aiello and A. Bilotta (Eds.): ICC 2022, LNCE 435, pp. 466–480, 2024.
https://doi.org/10.1007/978-3-031-43102-9_36

### 1.2 The Case Study

The case study bridge, designed in 1973, is a viaduct with simply supported prestressed concrete beams and reinforced concrete piles, recurrent in the 1960s–70s in Italy. Different systems are detectable along the viaduct in terms of piles and deck beams; however this work is focused on the 37.8 m long span. The prestressed beams have a 2.30 m high cross-section and they are supported by a portal frame with two rectangular piers; the plinth foundation of each portal frame has two piles, 26.50 m long with diameter D = 1.50 m. The cross-section of the bridge is shown in Fig. 1, while longitudinal section is shown in Fig. 2.

The safety checks of the bridge are carried out according to the Italian Technical Code for the Constructions (Norme Tecniche per le Costruzioni – in the following NTC2018) and the "Linee guida per la classificazione e gestione del rischio, la valutazione della sicurezza ed il monitoraggio dei ponti esistenti" (2021 – in the following LG2021).

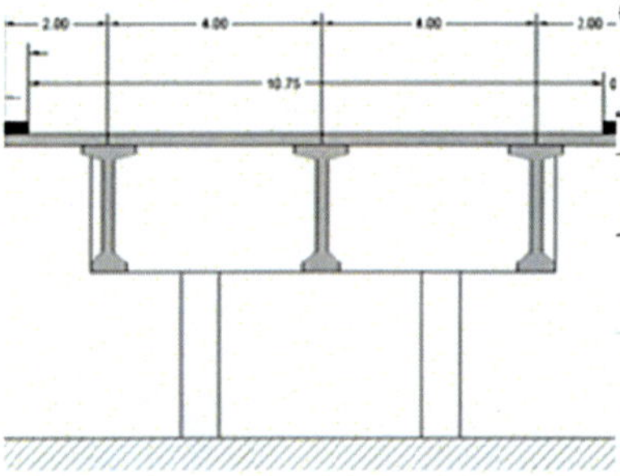

Fig. 1. Cross-section of the bridge (S-S)

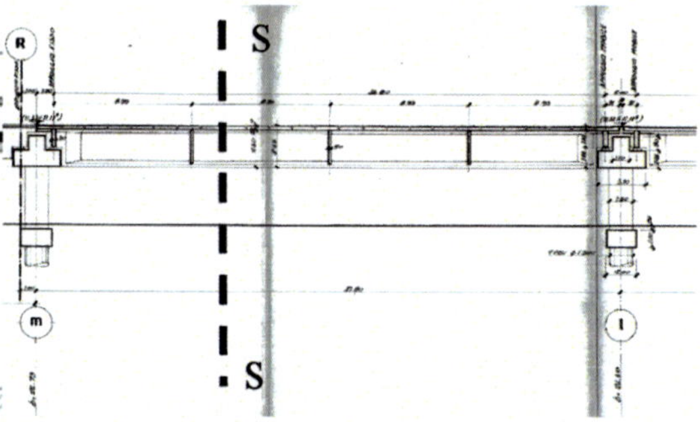


Fig. 2. Longitudinal section

## 2 Surveys Plan and Material Tests

### 2.1 Testing

The current Italian and European approach in existing buildings safety check preliminary foresees to put in place a set of testing, scheduled in a survey plan, in order to achieve one of the Knowledge Levels available in codes, from KL1 (limited knowledge) to KL3 (full knowledge), which also determines the confidence factor that must be used in design material resistance evaluation. For the purpose, the tests summarized in Table 1, for material properties investigation, were performed. In addition, some in situ surveys were performed to verify the original construction drawings for both structural geometry and rebars details, in order to achieve a KL3 (Knowledge Level 3 – full knowledge). The CF (confidence factor, partial factor for existing structures verification) equal to 1.00 will be used in the following to evaluate the material resistances.

Moreover, visual inspections of support devices (Fig. 3), transversal and longitudinal joints (Fig. 4), Gerber saddles and prestressing cables sheaths were conducted. These

**Table 1.** Survey plan for material properties investigation

| Test | Number |
|---|---|
| Compression test concrete | 48 + 16 |
| Carbonation test concrete | 48 + 16 |
| Rebar locator | 29 |
| Tensile test rebar (destructive) | 22 |
| Schmidt rebound hammer | 139 |
| Ultrasonic pulse velocity | 139 |

inspections confirmed the design information regarding the support devices and revealed a significant degradation state due to corrosion of anchor steel plates. A pronounced state of degradation was also evidenced for transversal and longitudinal joints, because of water percolation that caused concrete cover detachment.

**Fig. 3.** Support devices inspection

**Fig. 4.** Joints inspection

By contrast, Gerber saddles showed no appreciable cracks as well as the prestressing cables sheaths (Fig. 5), which appeared to be in a good state of preservation.

## 2.2 Material Properties

The resistance of materials for the calculation were determined in accordance with paragraph 6.3.4 of the LG2021 as shown in Eq. 1.

$$f_d = \min\left(\frac{f_m}{CF \cdot \gamma_M}; \frac{f_k}{CF}\right) \quad (1)$$

where $f_m$ = average resistance; $f_m$ = characteristic resistance; $CF$ = confidence factor and $\gamma_M$ = partial safety factor for the material.

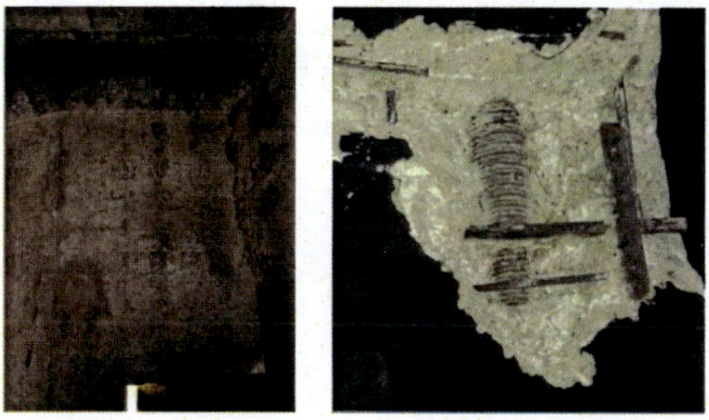

**Fig. 5.** Prestressing cables sheets inspection

The characteristic value of the resistance (or of a general value $x$) can be evaluated by assuming a log-normal distribution (Eq. 2).

$$\hat{x}_{0.05} = e^{\overline{\mu}_{0.16} - 1.64 \cdot \overline{\sigma}} \quad (2)$$

where $x$ = characteristic value of general value, $\mu$ = average value, $\sigma$ = standard deviation.

Thus, the design resistance values for concrete, reinforcing and prestressing steel bars are summarized in Table 2.

**Table 2.** Design material resistances (cubic resistance for concrete and yielding stress for rebars)

| Material | $f_m$ [MPa] | $f_k$ [MPa] | $f_d$ [MPa] |
|---|---|---|---|
| Concrete (piles) | 46.8 | 37.6 | 31.2 |
| Concrete (beams) | 48.6 | 32.4 | 32.4 |
| Concrete (slabs) | 41.9 | 27.9 | 27.9 |
| Reinforcing steel | 463 | 368 | 368 |
| Prestressing steel | 1852 | 1600 | 1600 |

## 2.3 Dynamic Identification

The identification of the modal parameters of the structure was conducted relying on the measurement of its dynamic response under operating conditions. The test aimed to deepen the knowledge of its dynamic behaviour in service situations in the current state, focusing on the fundamental vibration modes. It was useful to estimate natural frequencies, damping ratio and modal shapes. Specifically, the structure's measured dynamic response has been elaborated using reliable methods available in scientific literature (Brincker et al., 2001; Peeters & De Roeck, 2001).

The sensors were arranged in mutually orthogonal directions in order to ensure that the fundamental modes of shape of the structure were observable. The test measured the acceleration response of the structure at the level of the deck on two adjacent spans, highlighted in Fig. 6 and indicated as A-C and D-H, respectively.

The adopted test layout is shown in the same figure.

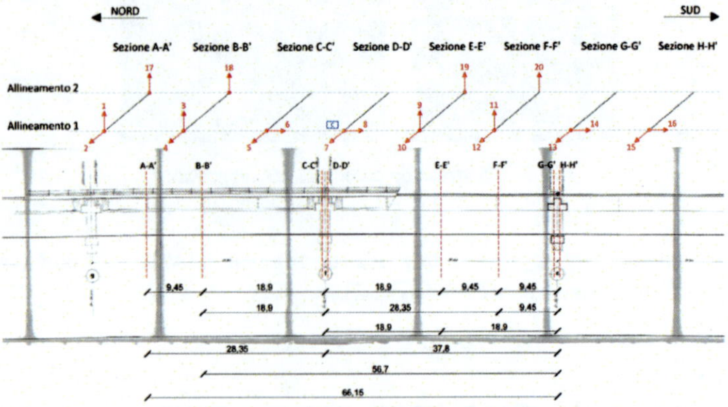


**Fig. 6.** Dynamic identification test set up

The analysis of the data allowed the reliable identification of the fundamental vibration modes of the structure. Table 3 shows the values of natural frequencies and damping ratios for the first two of them.

The detected first two modes (frequencies equal to 2.45 and 2.67 Hz) are global ones as the membrane behaviour of the slab is effective and they correspond to transversal and longitudinal global translational motions. In the interval between 3.5 Hz and 3.7 Hz several dominant frequencies were detected in the vertical degrees of freedom.

The lack in coherence between the signals measured on different spans suggests that the vertical response of each span is not correlated with that of adjacent ones because of the transversal joint that interrupts the structural continuity.

Considering the same span, instead, there was no perfect coherence between the detected signals, as the transversal connection between the span beams seems not completely efficient (Fig. 7).

**Table 3.** Natural frequencies and damping ratios

| Mode | f [Hz] Med. Value | f [Hz] Standard Dev. | ξ [%] Med. Value | ξ [%] Standard Dev. |
|---|---|---|---|---|
| 1 | 2.45 | 0.0096 | 3.0 | 0.274 |
| 2 | 2.67 | 0.0050 | 2.7 | 0.090 |

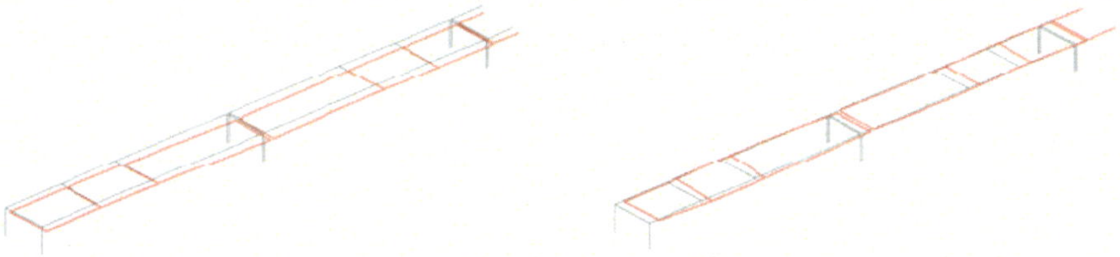

**Fig. 7.** First modal shapes

# 3 Original Design and Simplified Models

## 3.1 Original Design

About the case study under consideration, the original design drawings and technical reports are available. This documentation allowed to verify the load analysis and the criteria used for the safety checks of the structural elements. As usual in '70, the method of "allowable stress" was used for the safety checks considering the service loads.

## 3.2 Load Analysis

First of all, it is useful to numerically compare the values of loads assumed in original design with those used for the current safety evaluation. For instance, in Table 4 the permanent structural and non-structural loads assumed in the past for the deck beams analysis are shown and compared to the current ones.

**Table 4.** Permanent load analysis for beams

| Load | | Past | Actual |
|---|---|---|---|
| Beam (self-weight) | [kN/m] | 19.25 | 19.25 |
| Slab (self-weight) | [kN/m] | 20.40 | 20.40 |
| Non-structural permanent loads | [kN/m] | 10.83 | 10.75 |

About the vehicle load, the Italian technical reference n. 384/1962 provided for 6 load schemes which may be used for the two defined bridge categories. The scheme 5 (Fig. 8) provides for a 32 tons military load on the first lane, whereas on the other two lanes a civilian 12 tons load. Finally, equivalent distributed loads were proposed as a function of span length.

Vehicle distributed loads had different meanings as a military load was located on the lateral lane, the others (two in this case) was interested by a civilian load. However, loads analysis leads to quite similar results for past and actual design. Then, the loads were combined and distributed for each deck beam by using the well known Courbon's method (rigid deck hypothesis) as shown in Eq. 3:

$$r_i = P\frac{1}{n}\left(1 \pm \frac{ned_i}{\sum d_i^2}\right) = Pk_i \tag{3}$$

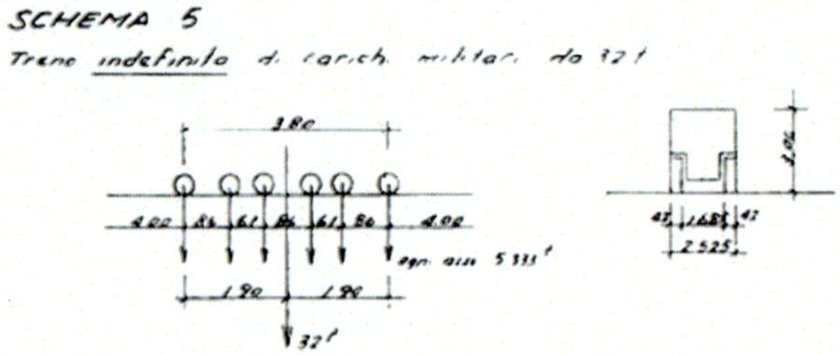


**Fig. 8.** Load scheme from original reference

where $n$ = number of beams, $e$ = eccentricity of the load, $d_i$ = eccentricity of beam axis.

The beam loads are multiplied by the dynamic factor $\phi$, according to Eq. 4:

$$\phi = 1 + \frac{(100 - L)^2}{[100(250 - L)]} \tag{4}$$

From Courbon formula application, the original design evaluated an equivalent distributed load, only due to the vehicles, of 4.97 tons/m (48.7 kN/m) for bending moment verification and 5.47 tons/m (53.66 kN/m) for shear verification. However, the application of NTC2018, considering the accidental load due to traffic, 9 kN/m$^2$ with a tandem of concentrated forces of 300 kN on the first lane, 2.5 kN/m$^2$ and 200 kN/100 kN concentrated forces on the others two, leads to a value of bending moment equal to 10310 kNm and 1165 kN of shear force, taking into account the load with no amplification due to the partial factors. It is possible to evaluate an equivalent load for bending verification and the shear one obtaining 64.35 kN/m and 65.08 kN/m respectively. As a result it is worth to underline how NTC2018 provides for a 30% bigger equivalent distributed load for bending moment verification (64.35 vs 48.7 kN/m) and 21% bigger for shear one (65.08 vs 53.66).

### 3.3 Simplified Models for the Stress Analysis of the Main Bridge Structural Members

In the original design, modern calculation methods and tools were not available, then a series of simplified models were needed to evaluate stresses even in the case of complex structures such as bridges (Fig. 9). As said, Courbon's method allows to relate all the loads on the slab with a distributed load on each beam. As a consequence of that, the beams can be analysed as simply supported, thus, manual calculation could be carried out. This makes it very easy to evaluate bending and shear stresses.

On the other hand, using the modern calculation capabilities, a FEM model can be built in SAP2000 software. In this case the base hypothesis of Courbon's method about the torsional stiffness of the deck is an output of the model behaviour. The comparison between the results obtained through the FEM model and those obtained in the original design (see Table 5) shows that the simplified Courbon's method is still valid in order to quickly evaluate the internal forces for simple situations as the case study is.

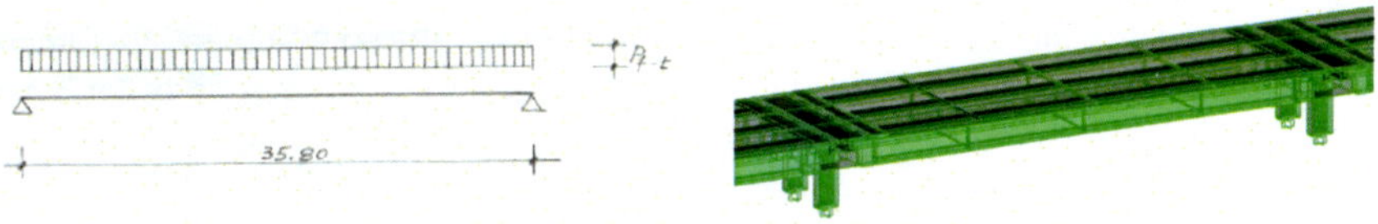


**Fig. 9.** Comparison between the deck beam model

The FEM model was implemented in SAP2000 software. The model involves three adjacent spans of the viaduct. Deck beams, cross beam, pulvinus and piers were modelled using frame elements. The cross section of the beams take into account the collaboration with the slab, in particular, a 4.0 m portion of the slab. In order to account for concrete cracking, a reduced E modulus has been used for pulvinus (50% reduction) and piers (60% reduction).

**Table 5.** Stress comparison

| | Courbon | FEM | Δ [%] |
|---|---|---|---|
| Bending moment [kNm] | 23303 | 22686 | 2.6 |
| Shear [kN] | 2620 | 2691 | −2.7 |

The reaction obtained from deck beams analysis are useful to evaluate the bending and shear stresses on the pulvinus. As for the deck beams, a simplified method is necessary for the purpose. In fact, as shown in Fig. 10, the original model for the pulvinus analysis was a simply supported beam with two lateral cantilever parts according to the geometry of the transversal portal frame. The static scheme, related to the in-plane behaviour of the portal, also includes three vertical forces, which are the reactions of the deck beams. However, the three forces are not applied on the pulvinus barycentric axes as the real supporting system has a longitudinal eccentricity.

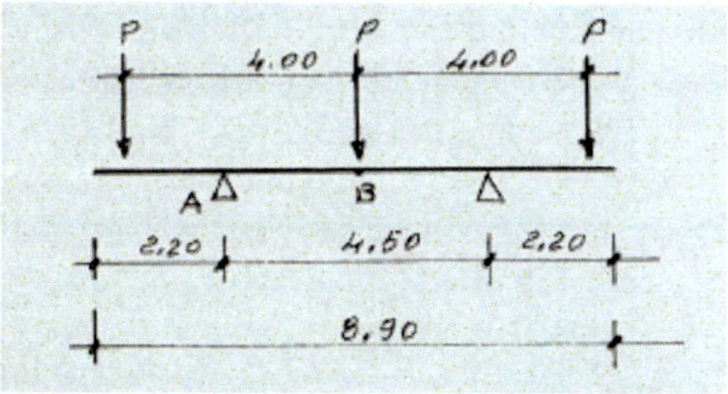


**Fig. 10.** Continuous beam for "pulvinus" modelling

For this reason a double verification is needed in this case. Firstly, as shown in Fig. 10, the pulvinus has a beam behaviour in his own plane, thus the verification are in terms of shear forces and bending moment applied on the inverted "T" cross section.

However, considering the eccentricity between the load application point and the axis of the pulvinus cross-section (Fig. 11), a simple cantilever static scheme is considered in the original design in order to evaluate the internal forces and carry out the related safety checks against bending moment and shear.

As can be noted in Fig. 11, the check of the pulvinus was performed in two phases. In the first phase (construction) the upper part of the pulvinus is not yet realized, in order to allow the tensioning of the prestressing cables, and the concentrated load (*P*) is due to the self-weight of the beam. In the second phase (operating conditions) the pulvinus is fully realized and the increase of the concentrated load ($\Delta P$) is due to the self-weight of deck slab and pavement and to the traffic loads. For the piles verification, the structural self-weight and the non-structural loads coming from the deck are considered. For the traffic load two combination are defined (see Fig. 12), one of them is useful to maximize the bending moment in the longitudinal direction of the bridge (Case A), while the other one is defined in order to maximize the bending moment in the transversal direction and the axial load on the piles (Case B). Figure 12 and 13 show the two simplified models that were used to evaluate stresses for "pulvinus" and piles.

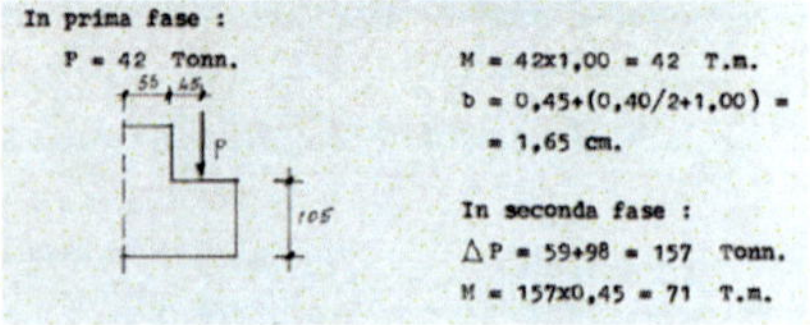


Fig. 11. Pulvinus behaviour in transversal direction

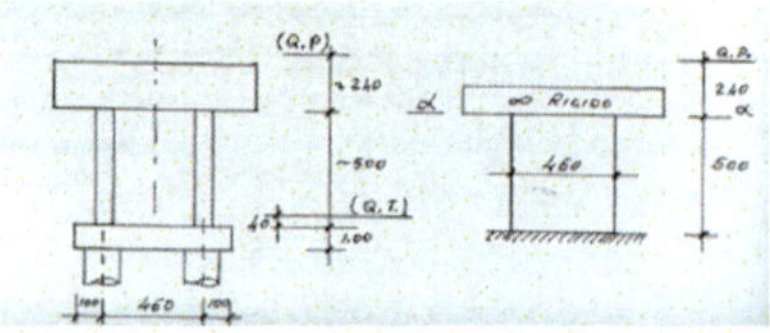


Fig. 12. Model for piles verification

Some considerations also must be done about the structural behaviour and verification of the slab deck. In this case the simplified model is a continuous beam supported at three points. Several load conditions must be taken into account, but for the method comparison it is sufficient one of them, shown in Fig. 14.

The slab resistance is also related to its collaborating width, especially in the case of relevant concentrated loads, as provided by NTC 2018. About that, an easy way to compute this width is to add L/2 to the directly loaded area, being L the span length of the slab. This procedure can be validated by a more sophisticated FEM model shown in Fig. 15. Using an effective width equal to 3.50 m, obtained by considering the load dimension increased by diffusion up to middle thickness of slab plus an half of length as shown in Fig. 16, a very good agreement between beam model and FEM 2-D model in terms of bending moments is reached (Table 6).

## 4 Safety Check of the Bridge

### 4.1 FEM Model

FEM model was built with the software SAP2000 for the structural analyses. Three adjacent spans, 37.80 m long, were modelled. Beams and piles were modelled via beam elements while the slab was taken into account in beam cross-section. A diaphragm

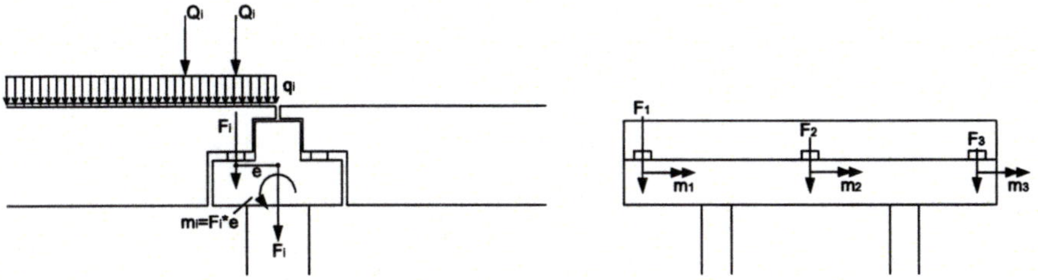


Case A

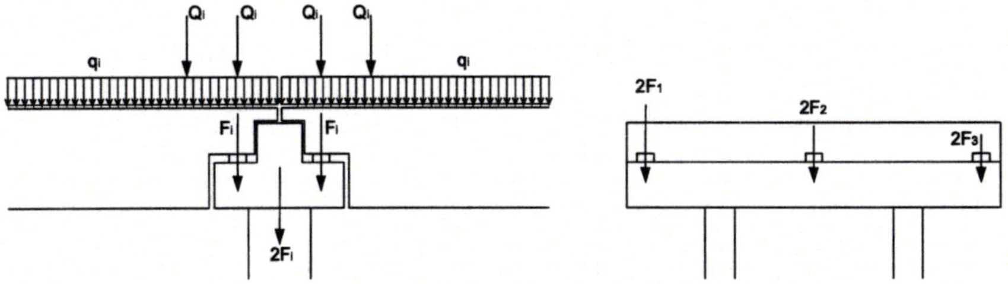


Case B

**Fig. 13.** FEM model of piles

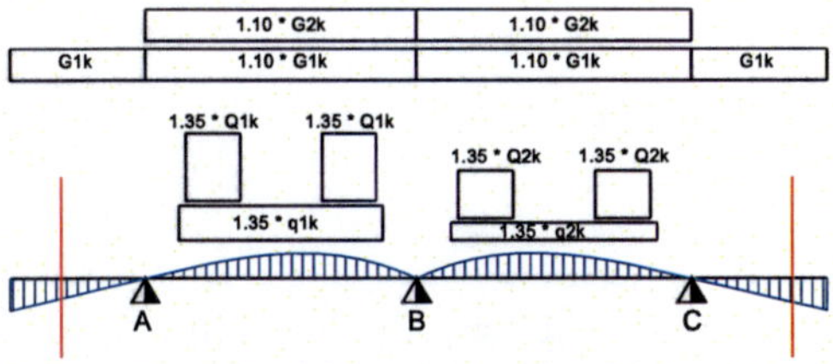


**Fig. 14.** Slab load condition

**Table 6.** Comparison between slab models

| Beam model | | | Shell model | |
|---|---|---|---|---|
| $b_{eff}$ m | M kNm | M/$b_{eff}$ kNm/m | $M_{max}$ kNm/m | $M_{ave}$ kNm/m |
| 3.50 | −185 | −53 | −55 | −53 |

constraint is considered for the deck joints, due to the rigid membrane behaviour of the 22 cm thick r.c. slab. In addiction, the transverse beams were also modelled in order to take into account their contribution to the load repartition. The connection between beams and pier was modelled via "Equal" internal constrains which affects X, Y and Z direction for fixed bearing and allows the longitudinal displacement for sliding ones.

**Fig. 15.** Slab FEM model

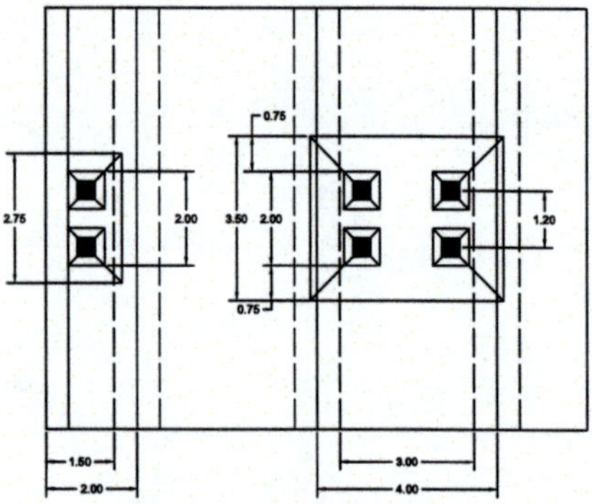


**Fig. 16.** Slab load condition

**Fig. 17.** Bridge FEM model

## 4.2 Model Validation

In order to validate the FEM model of the bridge in the presence of vertical loads, a comparison between the displacements measured for the load testing of the bridge during its construction and the displacements obtained from the FEM model for the same load condition on the deck was carried out. Load testing was performed by loading the deck with trucks defining three load schemes, applied in three steps.

In Table 7 the displacement comparison is shown. It is possible to underline how the numerical model provides values of the deflections at the midspan in good agreement with the experimental ones regarding points F7 and F8 for the different load schemes.

**Table 7.** Vertical displacements - experimental and numerical results comparison

| $f_{exp}$ [mm] | | | $f_{FEM}$ [mm] | | | Ratio $f_{FEM}/f_{exp}$ | | |
|---|---|---|---|---|---|---|---|---|
| F7 | F8 | F9 | F7 | F8 | F9 | F7 | F8 | F9 |
| 9.60 | 4.55 | −0.45 | 9.70 | 3.43 | −1.31 | 1.01 | 0.75 | 2.92 |
| 16.00 | 9.05 | 2.35 | 14.82 | 7.16 | 0.97 | 0.93 | 0.79 | 0.37 |
| 18.80 | 13.75 | 7.68 | 11.14 | 11.14 | 6.78 | 0.89 | 0.81 | 0.89 |

Model validation can be confirmed, moreover, by comparing the bending moment obtained applying the Courbon simplified method with the results of the FEM model. The Courbon method allows to evaluate how the vertical load of the different lanes load is divided on each deck beams. As the beams are simply supported, the value of the bending moment at the midspan can be easily evaluated in order to compare it with the numerical value of the FEM model (Fig. 18).

$$\begin{aligned} M_{Ed,Courbon} &= 23303\ \text{kNm} \\ M_{Ed,FEM} &= 22686\ \text{kNm} \end{aligned} \tag{5}$$

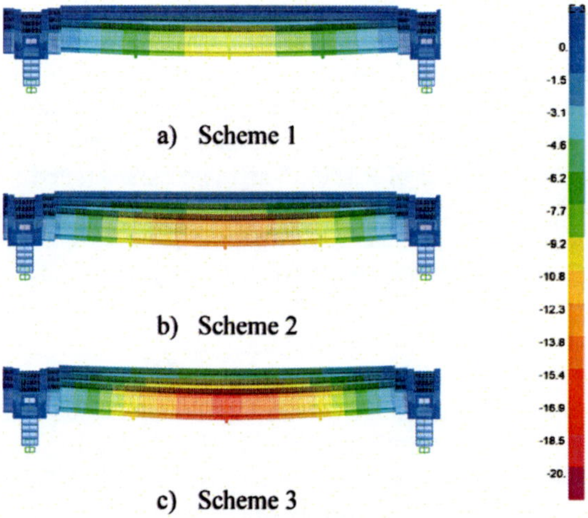


**Fig. 18.** FEM model displacements

### 4.3 Load Cases and Combinations

According to Italian technical references (NTC2018), and more generally to Eurocodes, the load that must be taken into account in the bridges design are:

- The total self-weight of structural and non-structural members ($G_{1k}$ and $G_{2k}$);
- Road traffic actions ($Q_{1k}$) sum of a uniformly distributed load and double-axle concentrated load (tandem system) as shown in Fig. 5;
- Braking and acceleration forces;
- Wind load (Figs. 19 and 20).

The loads are combined according to NTC 2018:

$$\gamma_{G1}G_1 + \gamma_{G2}G_2 + \gamma_P P + \gamma_{Q1}Q_1 + \gamma_{Q2}Q_2 \tag{6}$$

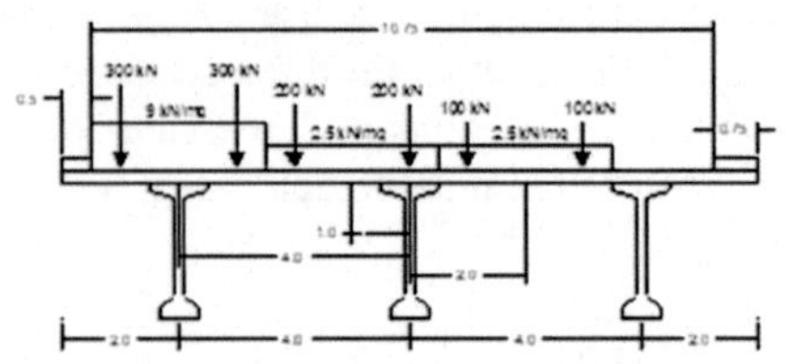


**Fig. 19.** Load lanes

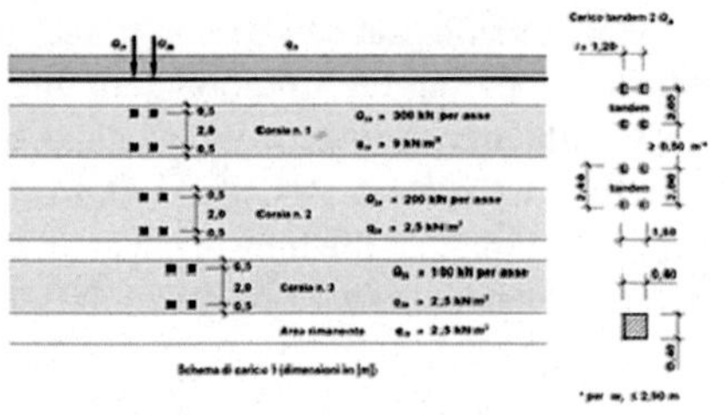

**Fig. 20.** Load Lanes NTC2018

## 4.4 Codes Comparison

As the FEM model was validated, it can be useful to underline the possible differences, in terms of stresses in the structural elements, coming from different vehicle load definition suggested in the past and actual codes. In fact, on one hand the NTC2018 suggest for the first lane a distributed load of 9 kN/m$^2$ and two concentrated forces of 300 kN. However, other technical references are also available such as the Italian "Codice della Strada" (CdS), where a vehicle of 44 tons is defined and a 9 kN/m$^2$ distributed load is also applied out of the vehicle footprint. Two load schemes are suggested where the 44 tons vehicle is applied only on the first two lane or on the third too. Figure 17 show how the load is defined and modelled in software SAP2000.

Table 8 shows the differences of the three load schemes, one from NTC2018 and the other from CdS, in which each result is affected by the application of the partial factors. CdS load schemes provides internal forces smaller than NTC2018, as a consequence of the value of the different loads defined according to CdS and NTC2018. The internal forces due to the Circ. 384/1962 are also shown.

**Table 8.** Stresses for different load schemes

| Code | Bending moment [kNm] | Shear [kN] |
|---|---|---|
| NTC2018 | 22686 | 2695 |
| 2 × 44 tons vehicle CdS | 19428 | 2273 |
| 3 × 44 tons vehicle CdS | 20382 | 2336 |
| Circ. 384/1962 | 16168 | 1889 |

## 4.5 Element Verification

As the loads are now given, it is possible to perform all strength verifications for vertical loads. The most stressed element are to be verified. Since the FEM model is available, it is not necessary to use simplified methods, and maximum stresses in structural elements can be obtained very easily. The elements interested by verification are:

- deck beams;
- pulvinus;
- Gerber saddles;
- support devices;
- piers.

According to current technical references the verification must be performed by comparing the acting internal forces and the ultimate resistance of the structural members. The latter, is evaluated with the ULS principles. The following Table 9 shows the maximum value of the ratio between resistance (R) and internal force (S) in terms of bending moment and shear. It results that the safety checks are satisfied.

It is worth noting that Gerber saddles were checked using a simplified strut-and-tie method, according to the provisions of italian code (Circolare n. 617/2009). The safety checks of the support devices, not shown here for sake of brevity, showed that they are almost capable to sustain the vertical loads prescribed by NTC2018. However, their apparent degradation (see Fig. 3) and the fact that they were designed for braking (or acceleration) forces generally smaller than those prescribed by modern codes suggest evaluating the substitution of the same.

**Table 9.** Safety checks of the structural members

| Element | R/S Bending moment | R/S Shear |
|---|---|---|
| Deck beams | 1.33 | 3.04 |
| Gerber paddle | – | 1.41 |
| Transversal beams | 1.46 | 3.09 |
| Pulvinus | 1.34 | 1.02 |
| Piles | 1.12 | 1.93 |

## 5 Conclusions

The following conclusions can be drawn:

- The traffic loads which must be considered for the analysis have became bigger of 30% approximately in NTC2018 with respect to the past technical codes.
- As it can be seen in Table 7, the value of the stresses in deck structural element, as beams, in terms of bending moment and shear are generally bigger using the NTC2018 as a consequence of the different load schemes defined.
- The FEM model calculation demonstrates that simple methods, as Courbon's method is, are correct whether their base hypothesis are respected. That is also the case for the collaborating width for slab verification;
- Finally, it is worth to underline that the safety checks of a typological bridge, as the one of the case study, could be satisfied according to the modern technical codes and calculation methods, if it has been correctly designed according to past technical references, relatively to the vertical load safety check.

**Acknowledgements.** This research activities is within the framework of the Research Agreement between Reluis and the Department of Civil Protection of the Italian Government.

## References

[D.M.II.TT., 2018]: Decreto Del Ministero Delle Infrastrutture E Dei Trasporti del 17 Gennaio 2018: Aggiornamento delle "Norme Tecniche per le Costruzioni", supplemento Ordinario n. 42 della Gazzetta Ufficiale della Repubblica Italiana del 20 Febbraio 2018

Circolare 21 gennaio 2019, n. 7 - Istruzioni per l'applicazione dell' <<Aggiornamento delle "Norme tecniche per le costruzioni">> di cui al D.M. 17 gennaio 2018

Circolare 2 febbraio 2009, n. 617 - Istruzioni per l'applicazione delle "Nuove norme tecniche per le costruzioni" di cui al D.M. 14 gennaio 2008

Linee Guida per la classificazione e gestione del rischio, la valutazione della sicurezza ed il monitoraggio dei ponti esistenti - Allegate al parere del Consiglio Superiore dei Lavori Pubblici n.96/2021

"Linee guida e manuale applicativo per la valutazione della sicurezza sismica e il consolidamento dei ponti esistenti in c.a." - Progetto DPC-Reluis 2005–2008 - Linea 3: Valutazione e riduzione del rischio sismico di ponti esistenti

RILEM. NDT 4 Recommendations for in situ concrete strength determination by combined non-destructive methods, Compendium of RILEN Technical Recommendations, E&FN Spon, London (1993)

SAP 2000. Vers. Advanced 14.0.0. Documentation and Help. 2005

Brincker, R., Zhang, L., Andersen, P.: Modal identification of output-only systems using frequency domain decomposition. Smart Mater. Struct. **10**, 441–445 (2001)

Peeters, B., De Roeck, G.: Stochastic system identification for operational modal analysis: a review. ASME J. Dyn. Syst. Measur. Control **123**, 659–667 (2001)

Pucinotti R. (2003). Diagnostica nella Prevenzione dei Crolli: Applicazione ad un Caso di Studio. In Atti del Convegno Internazionale "Crolli e Affidabilità delle Strutture", Napoli, 2003

Paulay, T., Priestley, M.J.N.: Seismic Design of Reinforced Concrete and Masonry Buildings. Wiley, New York (1992)

Gasparik, J.: Prove non distruttive nell' edilizia. Quaderno didattico AIPnD, Brescia (1992)

Di Leo, A., Pascale, G.: Prove non distruttive sulle costruzioni in c.a. Il giornale delle prove non distruttive, n. 4 (1994)

Min. LL.PP. n.384/1962. Norme relative ai carichi per il calcolo dei ponti stradali

# Structural Fire Safety Assessment of RC Parks for Cars and Motor Scooters Using the FSE Approach

Donatella de Silva(✉), Enrico Cardellino, Margherita Autiero, and Emidio Nigro

Department of Structures for Engineering and Architecture, University of Naples Federico II, Naples, Italy
donatella.desilva@unina.it

**Abstract.** During the last years, the use of performance-based approach according to the Fire Safety Engineering (FSE) is becoming ever more widespread. This approach is particularly used for the definition of the potential fires and of the structural fire behaviour of buildings with intended use of car parks. However, if the park is used for motor scooters, the international scientific framework offers very few information about its fire safety, both for fluid dynamic and thermomechanical aspects.

In this work, several fire scenarios correlated to the two different park intended uses were investigated, starting from experimental Heat Release Rate curves. These curves were used for CFD analyses to define the time-temperature distributions during selected fire scenarios. Several thermomechanical analyses of a reinforced concrete structure were performed using both natural and nominal fire curves. A benchmark between the different cases were performed, underlining the importance of the FSE to reduce the structural fire demand.

**Keywords:** Reinforced concrete structures · Thermo-mechanical analyses · Structural fire safety

## 1 Introduction

In the last decades, many underground car and motor scooters parking are constructed at the basement of residential, administrative and office buildings, because the usable land available is increasingly limited, especially in big cities. The main feature of these underground parks is the enclosed condition, which can become particularly critical, in case of fire (Autiero 2022). Indeed, the limited ventilation inside the compartment can cause large amount of smoke, high temperature with difficulty in evacuation with great human suffering and property loss. The fire involving cars and heavy goods vehicles are well known in literatures, both from the experimental and numerical point of view. In Johan (2004) and TGS8 (2008–2011) realistic scenarios of car fires in open car parks, by using ordinary medium-size passenger cars are studied. In Lecocq et al. (2012) also the comparison between the fire test results conducted on the electrical vehicles and internal combustion engine vehicles have been reported. This report showing that the


M. A. Aiello and A. Bilotta (Eds.): ICC 2022, LNCE 435, pp. 481–492, 2024.
https://doi.org/10.1007/978-3-031-43102-9_37

fires due to the E.V. are less dangerous than ones due to the ICEV. Gas temperatures were calculated using rate of heat release for the model plumes, and compared to experimental temperatures, finding a good agreement. Zhang (2007), design the development of car fire by letting surface densities of the fuel over the car. Fire spread and movement of smoke are then investigated under different ventilation conditions. Tohir and Spearpoint (2014) discusses a simplified approach for obtaining heat release rate curves for multiple vehicle fires. These letters were obtained starting from experimental single vehicle design fire curves. The good agreement with the experimental results confirms the robustness of the simplified method.

However, the fire of motor scooters is little investigated in the literature and one of the most relevant works has been proposed by Chen et al. (2005), in which a series of burning tests were conducted on motor scooters by using a 10 MW large-scale fire products collector. The emissions characteristics and the curves of heat release rate of burning motor scooters were identified. They found that as the number of scooters increased, the duration of the fully developed burning period became shorter and the ignition and fire spreading patterns affected the maximal heat release rate considerably.

Even if the potential fire generated by a motor scooter is less dangerous than a vehicle one for the structural collapse, it could become dangerous if more motor scooters burn at the same time; moreover, the fires of motor scooters in an underground parking can also be significant for the safe people evacuation. Starting from this lack of literature indications, the aim of this work is to investigate several fire scenarios correlated to the two different intended uses, to quantify the natural fire curves and the temperature levels reached in each fire scenario. These fire scenarios were considered for analysing the fire behaviour of a real reinforced concrete structure, underling all its criticalities and peculiarities, useful for the future fluid-dynamic and thermo-mechanical analyses, especially considering the motor scooters parks, which are not contained in any technical regulation (Miano 2020).

## 2 Fire Modelling

To study the fire resistance of structures according to the performance-based approach, the first step is the modelling of possible fire scenarios that lead to the design natural fire curves. First of all, the Heat Release Rate (HRR) curves that characterize the fire scenarios have to be defined. In this work the HRR curves of cars and motor scooters are considered and described in the following.

### 2.1 Cars Heat Release Rate (HRR) Curves

In the international scientific literature, vehicles are classified according to the thermal energy that can be released during the fire; an example is the classification reported in Kruppa et al. (2002).

The belonging to a certain category is a function of the car tank volume and the engine size. In relation to the most commercially available cars, at the Category 3 belong the ones with capacity less or equal to 2000 cc. However, the cars with a higher engine size belong to "Categories 4 and 5".

In this paper Category 3 were considered, in particular the best-selling cars with a calorific value of 9500 MJ were chosen and the HRR curves were provided from experimental fire test data (Kruppa et al. 2002). Category 3 may also include commercial vehicles, named "VANs", to which should be added a fire load derived by 250 kg of flammable material with a calorific value of 40 MJ/kg. The envelope of the experimental curves allowed the derivation of input HRR curves (Fig. 1) where peaks of 8000 kW and 18000 kW are reached for car and VAN respectively.

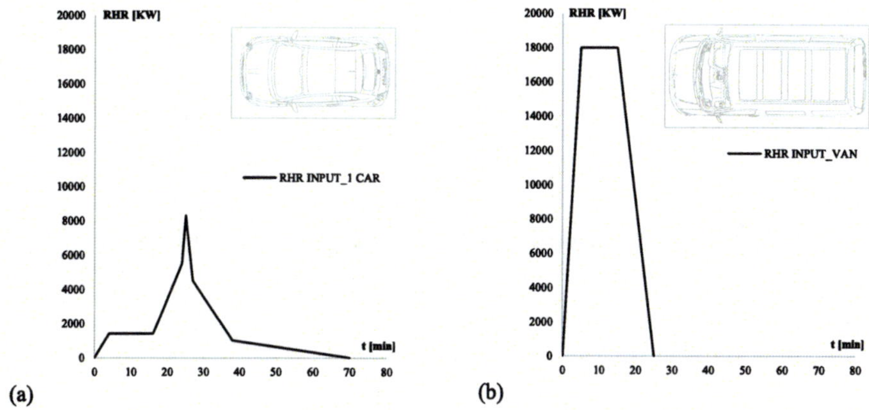


**Fig. 1.** The HRR curves for one car (a) and one VAN (b).

## 2.2 Motor Scooters Heat Release Rate (HRR) Curves

To determine the HRR curves for the motor scooters, as mentioned, one of the most relevant works has been proposed by Chen et al. (2005). The tests were conducted with one, two and three scooters to obtain the different thermal release. Also, the influence of the position of ignition has been investigated, as well as the fire spreading pattern, the duration of fully developed burning and the peak of heat release rate.

The test results show that one scooter has a peak heat release rate equal to 1 MW, two scooters have 2 MW, and three scooters have 4 MW, showing that the maximal heat release rate does not grow linearly with the number of the motor scooters (Chen et al. 2005). However, the total heat release is approximately equal to the combustion energies sum of all available flammable materials. On the contrary, the duration of full developed burning becomes shorter increasing the number of scooters. As results the available HRR curves has been obtained with a $HRR_{max}$ equal to about 1300 kW and a total duration of about 17 min, as shown in Fig. 2. Furthermore, the Fig. 3 show that the total energy released by the motor scooter is considerably less than the VAN and car ones.

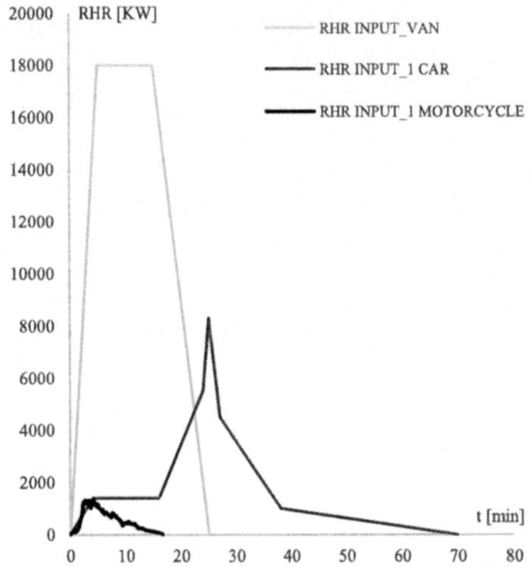


**Fig. 2.** The HRR curves for Motor scooters compared to VAN and car ones.

### 2.3 Thermo-Fluid-Dynamic Analyses

The HRR curve described in the previous paragraph were used as input for thermo-fluid dynamic analyses performed by Cfast software (Jones et al. 2006) to obtain the natural fire curves of each fire scenario. To better understand the effect of various parameters on the natural fires, parametric analyses were carried out; the results were described in the following.

#### 2.3.1 Parametric Analysis

Many thermo-fluid dynamic analyses were performed by varying the ignition time delay between each motor scooter, the number of motor scooters, the influence of the openings and finally the size of the compartment. The variation of motor scooters number, from 1 to 25, lead to observe an increase of temperature peak until about 600 °C shown in Fig. 3(a).

The opening factor (O), was varied from 0.025 to 0.096 m$^{-1}$, obtaining a variation of both peak and duration of the fire curve (see Fig. 3(b)). In particular, as the opening factor increases, the peak temperature decreases, and the fire duration grows. Finally, bigger is the compartment dimension (from 25 × 46 × 2,5 m to 40 × 60 × 2,5 m), smaller is the peak temperature, as shown in Fig. 3(c). The results obtained from this parametric analysis, which aims to define the fire scenarios, underlined that motor scooters fires are susceptible to all analysed parameters, therefore, to establish certain specific criteria become complex.

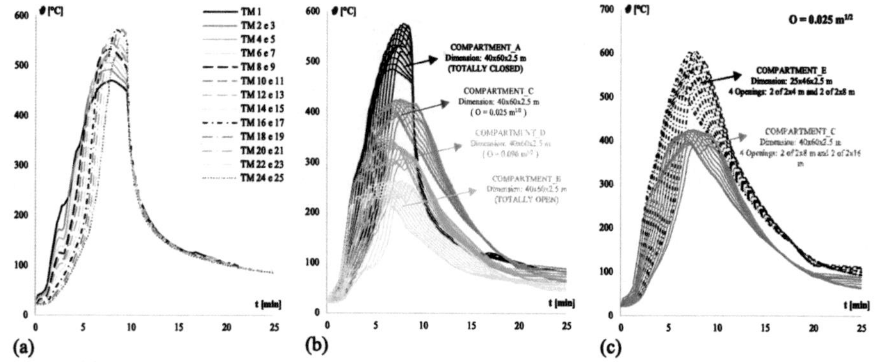


**Fig. 3.** Fire curves by varying: the number of motor scooters (a), the opening factor (b) and the compartment dimension (c).

# 3 Case Study Description

## 3.1 Generality

In the following paragraph the structural fire resistance assessment of an existing reinforced concrete structure, not designed for fire action, is described. In particular, the fire scenario regarding the parking with 366 motor scooters spaces available, located at the basement of the building, was analyzed. The other floors of the building are intended for offices and residential use. The parking is characterized by two underground levels; the first one, with a height of 4.9 m, is directly connected to the outside through two entrances and it is partly covered by a skylight. The second level, with a height of 3 m, doesn't have a direct connection with the outside; it is connected to the first level through an internal ramp.

## 3.2 Thermo-Fluid-Dynamic Analyses

The first step of the performance-based approach consists of the assessment of thermal action through the selection of the design fire scenarios. The design fire scenario is a qualitative description of the fire development during the time, identifying key events that characterise the fire and distinguish it from other possible fires. It typically defines the ignition and fire growth process, the fully developed stage and the decay one. As shown in the previous paragraph, the fire scenario is largely influenced by different conditions: internal ventilation, external environmental conditions, active protection system effectiveness, type, position and dimension of fire ignition, type and distribution of fire load, fire load density, windows and door state (close or open), conditions of the occupants. For this reason, the number of possible fire scenarios is usually too large and the analysis of each one is not practicable. Therefore, the design fire scenarios should be chosen, in order to analyse the most severe cases for the structure, the development and spread of the fire, the safety of the occupants and rescue teams. In general, the choice of the design fire scenarios should be carried out based on the risk (R), associated to each fire scenario, equal to the probability of occurrence multiplied by the consequence of each fire scenario. For each selected fire scenario, the designer has to define the natural fire curve, determined on the basis of fire models. In this study the identification of possible fire scenarios was based on maximising the fire effects related to the resistance of the structural elements and evacuation, thanks to engineering judgments. In particular, five fire scenarios were analysed and the fire temperature was estimated by placing the targets near each motor scooter. In the following, the Target of the Motor scooter are TM while the Target of the Car are TC. In each scenario, the ignition time delay between each motor scooter was considered equal to 24 s.

In the first (Scenario 1) and second (Scenario 2) scenarios, 5 motor scooters were involved. Scenario 1 aimed to maximise the fire effects on the evacuation and on the columns and floor located near to the ac-cess ramp, where there is a reduction in height from 4.9 m to 3m. Scenario 2 aimed to maximise the fire effects on the column where the height is 4.9 m, since this column has a smaller section and was surrounded by several motor scooters. Scenario 3 aimed to maximise the effects on the floor at a height of 4.9 m: therefore, it was considered the area with a lower ventilation because of the external grids, and it was considered a greater number of motor scooters, indeed 7 motor scooters were involved with a symmetrical fire spreading which started from the central one. Scenario 4 and Scenario 5 aimed to maximise the fire effects on both the displacement and bending moments of the beams. In Scenario 4, 5 motor scooters were concentrated at the central span, while in Scenario 5, 4 motor scooters were placed at the end span. Since the park structure has only exposed beams, the beam with the greatest span was considered. In addition, different scenarios were analysed also in the case of car parking, in order to compare the temperatures and consequently the performances in terms of strength and stiffness of the structural elements in the two different intended uses. Unlike for motor scooters parking, in the case of car parking, the national and international technical standards indicate how to apply the fire safety engineering methods, identifying several fire scenarios that have to be analysed.

Therefore, the code provides the experimental Heat Release Rate (HRR) curves which are described in the previous part of this paper and that are going to be used for the following structural analyses, as well as the ignition time delay between each car, that can be assumed to be equal to 12 min. The scenarios analysed in this case were similar to the ones described for motor scooters with the aim to maximise the effects on structural elements and evacuation but considering the arrangement of car spaces in this specific parking.

From the results, it is possible to observe that in case of cars, fluid-dynamic analyses show significantly higher temperatures than the ones reached in case of motor scooters (Table 1).

Therefore, it was deemed necessary to carry out structural thermo-mechanical analyses in the case of cars, because in the case of motor scooters the temperatures don't significantly affect the thermal properties of the concrete and so the performance in terms of strength and stability of the structural elements. However, the national fire code establishes five performance levels for assessing the structural fire resistance. In the case of motor scooters some of these performance levels can be considered satisfied without any check for the reason explained before (e.g. Performance Level III: maintaining the fire resistance requirements for the whole fire duration). In the case of Performance

**Table 1.** Maximum temperatures obtained from the fluid-dynamic analysis of fire scenarios for motor scooters and cars.

| $\theta_{max}$ [°C] | motor scooters | cars |
|---|---|---|
| Scenario 1 | 342.4 | 1143.2 |
| Scenario 2 | 163.6 | 940.4 |
| Scenario 3 | 440.6 | 956.6 |
| Scenario 4 | 165.1 | 921.2 |
| Scenario 5 | 161.5 | 956.7 |
| Scenario 6 | – | 813.6 |

Level IV (limited damage of the structure after fire duration) more specific checks has to be carried out. This may be also justified by the fact that the fire loads densities of the motor cycles is much smaller than both car and VAN ones:

$$q_{moto} = 181\,\text{MJ/m}^2 < q_{auto} = 543\,\text{MJ/m}^2 < q_{van} = 1512\,\text{MJ/m}^2 \quad (1)$$

However, the scenarios with motor scooters are useful for safe evacuation checks, which, for brevity reasons, are not shown in this paper.

## 3.3 Thermo-Mechanical Analyses for Natural Fire Scenarios

### 3.3.1 Selection of the Substructure

The fluid dynamic analysis of each scenario provides the natural fire curve which represent the thermal inputs to carry out the advanced mechanical analyses. According to Eurocodes, the structural analysis in case of fire should be carried out also for significant parts of the structure. These substructures differ in extension and boundary conditions; these latter should be chosen based on potential thermal expansion and deformation so that their interaction with the other parts of the structure can be approximated with constant boundary conditions during fire exposure (EN1994-1-2).

Therefore, each scenario corresponds to a different substructure and the substructure which might better simulate the behavior of the entire structure cannot be defined uniquely. Consequently, the designer, based on his own experience, must identify the extension of the substructure and its surrounding conditions, without modifying excessively the structural fire response. Among all analyzed scenarios, in the following, the results of most severe scenario in terms of maximum temperatures is shown.

This one corresponds to the first fire scenario, in which four cars are involved and it is characterized by a fire spreading of a VAN, to maximize the effects on the beam in the area with a height of 3 m. Figure 4 (a) shows the results obtained from the fluid-dynamic analysis in terms of temperatures recorded by the thermocouples placed near each car.

In particular, in the analyzed substructure (Fig. 6), the constraints at the basis joints of the columns A, B, C and D were considered fixed. The vertical load of the nodes I, L, M, and N was assumed constant and equal to the axial stress at time t = 0 ($N_{fi,d,0}$),

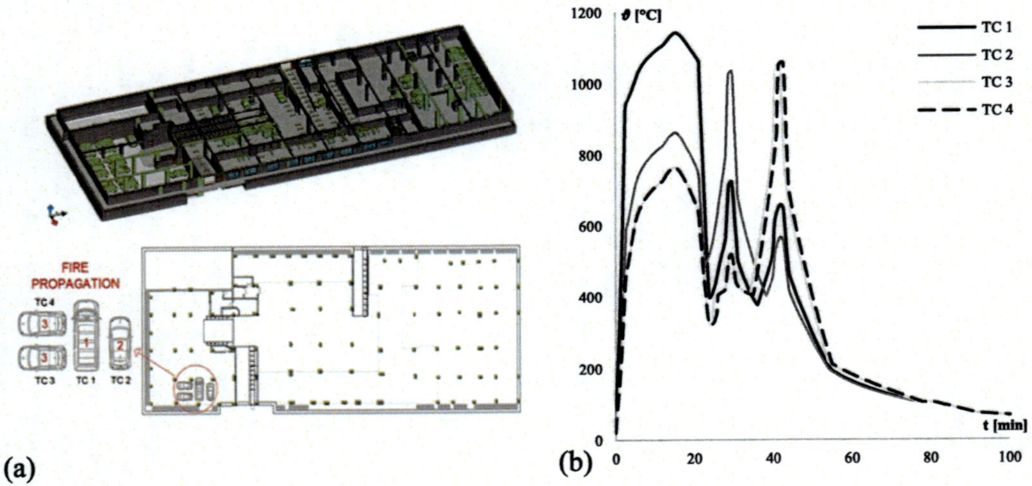


**Fig. 4.** (a) Fire propagation and arrangement of the Target Car (TC) in the fire scenario; (b) Thermal results of the Fire Scenario 1.

because the floor fire condition should not produce significant differential elongations in the heated columns, such as to activate the shear stiffness of the floor beams ILMN (including ILMN). Indeed, the beam ILMN remains cold, and it can be assumed that it has sufficient axial stiffness to resist the elongation induced by the thermal expansion of the heated beam EFGH: therefore, the horizontal displacement in nodes I, L, M and N are considered fixed.

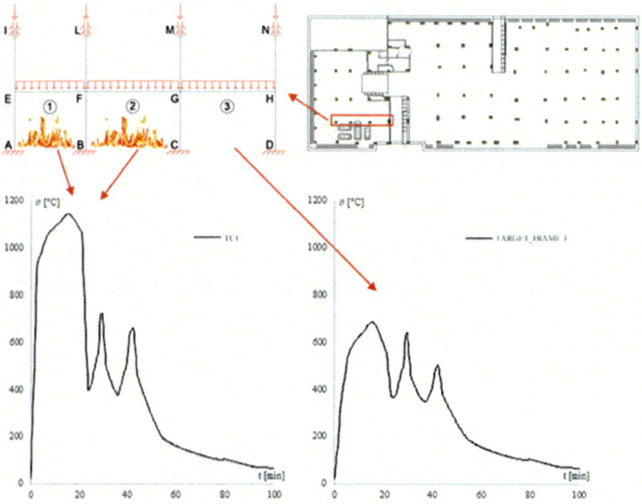


**Fig. 5.** Selected substructure and related fire curve.

On the contrary, the real rotational restrain condition at nodes I, L, M and N, is certainly intermediate between the one simulated by a constant bending moment and by a zero rotation. In this specific case, the rotational constraint condition of these nodes does not significantly change the result of the analysis, therefore, four fixed rotation conditions were simulated for safety reasons. In order to simulate the presence of the remaining structure in elevation, distributed loads were used on the EFGH beam. As said before the natural fire curves considered to carry out the mechanical analyses, are the ones obtained from the first fire scenario (see Fig. 5). In particular, among these curves, for the beams EF and FG, the heavier fire curve, i.e. TC1, was chosen, while for the GH beam was considered another curve obtained from an additional target located in the middle.

## 4 Results and Discussion

Once the substructure was identified, it was possible to carry out the thermal analyses of the structural RC members (column 55 × 75 cm with a concrete cover c = 6 cm and 10ϕ18 reinforcement steel bars; beam 40 × 55 cm with a concrete cover c = 4 cm and 5 + 5ϕ18 reinforcement steel bars) and consequently the mechanical analyses of the structure by using Safir software (Franssen and Gernay 2017). The thermo-mechanical analysis was conducted also using the nominal fire curves to provide a comparison between the two different approaches, prescriptive and performance-based one. In Fig. 6 (b), the thermal fields obtained for the column and the beam after 15 min, which is the natural fire curve peak, are represented by varying the fire curves. In the contest of the Italian fire code and according to a prescriptive approach, the required fire resistance time for the analyzed parking is 90 min. By comparing the heating curves Fig. 6 (b) and 6 (c) obtained for the two sections and for each fire curve (i.e., the two approaches), it is possible to observe that in the first stage of the heating, both for the column and the beam, the average temperatures, reached in the concrete cover, with the natural fire curve are higher than ones obtained with nominal fire. However, with the increase of the heating time, the trend is reversed.

The mechanical analysis was carried out by using the previous thermal fields as inputs. Figure 7 compares the bending moment obtained for the most stressed span ($M_{Sd,t}$) with continuous curves, and the resistant bending moment ($M_{Rd,t}$), with dashed curves. Therefore, it is possible to assume that the fire resistance is verified for the parking considered in this study, both using prescriptive and performance-based approach.

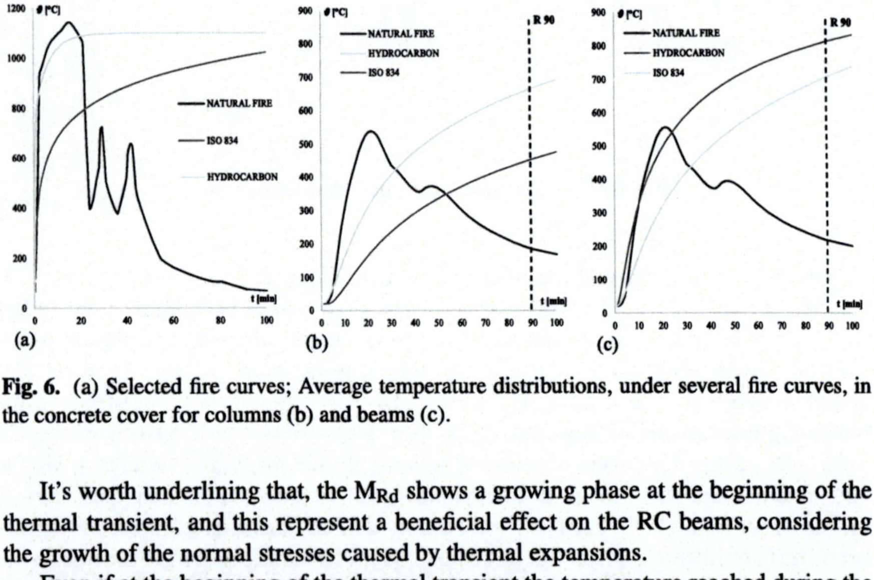


**Fig. 6.** (a) Selected fire curves; Average temperature distributions, under several fire curves, in the concrete cover for columns (b) and beams (c).

It's worth underlining that, the $M_{Rd}$ shows a growing phase at the beginning of the thermal transient, and this represent a beneficial effect on the RC beams, considering the growth of the normal stresses caused by thermal expansions.

Even if at the beginning of the thermal transient the temperature reached during the natural fire are greater than the nominal ones, the duration of these high temperatures is not sufficient to cause the collapse of the RC structure.

## 5 Conclusions

The fire resistance of reinforced concrete structures can be guaranteed using two different approaches, prescriptive or performance-based ones according to the Fire Safety Engineering (FSE) criteria. In the international research field, the application of the FSE

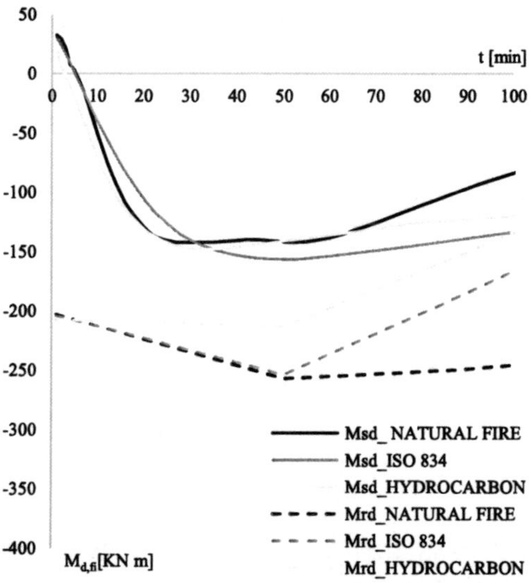


**Fig. 7.** Bending moments variation in the midspan of the beam E-F.

approach to study car parks fire behaviour is a relevant and a well-investigated topic. However, if the park is intended for motor scooters, no much information is provided neither in literature nor in the codes. Thus, the aim of this work is to investigate several fire scenarios related to the two different intended uses, to quantify the natural fire curves and the temperature levels reached in each fire scenario. Starting from a deep literature review about Heat Release Rate (HRR) curves measured in experimental fire tests, different fire curves were obtained thanks to thermo-fluid-dynamic zone models. In particular, many thermo-fluid dynamic analyses were performed by varying the ignition time delay between each motor scooter, the number of cycles, the influence of the openings and finally the size of the compartment. The results obtained from this parametric analysis, that allowed defining the fire scenarios, can be summarized as following:

- motor scooter fire scenarios are susceptible to the number of motor scooters and therefore it is difficult to establish a certain number of motor scooters to be considered, like for cars;
- as the opening factor increases, the maximum temperatures in terms of peak decreases, but the fire duration increases;
- decreasing the size of the compartment, the temperature increases for the same opening factors.

In the second part of this paper the fire behaviour of a real reinforced concrete structure was analysed, underling all its criticalities and peculiarities, which can be a useful reference for designers of parking intended for motor scooters. Preliminary results showed that the temperatures reached in the case of motor scooters motor scooter scenarios are less than the ones reached for car ones. Once the fire curves were chosen, advanced thermo-mechanical analyses were carried out on substructures selected, according to FSE criteria. The thermo-mechanical analyses carried out, both using natural and nominal fire curves, have shown that the collapse is not reached in both cases. Indeed, although the natural fire curve for a representative car fire scenario shows temperatures particularly high, the duration of these temperatures is not capable to lead at the collapse of the structure, as it is made of reinforced concrete. Indeed, given the high thermal inertia, concrete structures are not sensitive only to the temperature peak reached during the fire, but also to the duration of each temperature level. This underlines the importance of evaluating the fire resistance not only with prescriptive approaches, but also with performance-based ones, allowing to study on one hand the fires that can really happen and on the other hand, the real behaviour of the whole structure where indirect actions, redistribution of the stresses and internal actions according to the different stiffness of the structural members, play a fundamental role.

## References


Autiero, M., Cardellino, E., de Silva, D., Nigro, E.: Application of FSE approach for structural fire safety assessment of car and motor scooter parks. In: 14th fib International PhD Symposium in Civil Engineering, 8p. (2022)

Chen, C.J., et al.: Burning analysis of motor scooters. In: Eighth International Symposium on Fire Safety Science, Tsinghua University, Beijing, China, 18–23 September, pp. 671–680 (2005)

Dell'interno, I.M.: Approvazione di norme tecniche di prevenzione incendi, ai sensi dell'articolo 15 del decreto legislativo 8 marzo 2006, n. 139, GU n. 192 del 20/8/2015 – S.O. n. 51

EN 1994-1-2, Eurocode 4 2005: Design of composite steel and concrete structures – Part 1–2: General rules - Structural fire design (2005)

Franssen, J.M., Gernay, T.: Modeling structures in fire with SAFIR®: Theoretical background and capabilities. J. Struct. Fire Eng. **8**(3), 300–323 (2017). https://doi.org/10.1108/JSFE-07-2016-0010

Johan, M.: On the fire dynamics of vehicles and electrical equipment. Academic dissertation, University of Helsinki (2004)

Jones, W., Forney, G., Peacock, R., Reneke, P.: A technical reference for CFAST: an engineering tool for estimating fire and smoke transport, Technical Note (NIST TN), National Institute of Standards and Technology, Gaithersburg, MD (2003). https://doi.org/10.6028/NIST.tn.1431, 2 May 2022

Kruppa, J., Joyeux, D., Cajot, L.: Demonstration of real fire tests in car parks and high buildings: final report. European Commission, Directorate-General for Research and Innovation (2002)

Lecocq, A., Bertana, M., Truchot, B., Marlair, G.: Comparison of the fire consequences of an electric vehicle and an internal combustion engine vehicle. 2. In: International Conference on Fires In Vehicles - FIVE 2012, Chicago, United States, pp. 183–194 (2012)

Miano, A., De Silva, D., Compagnone, A., Chiumiento, G.: Probabilistic seismic and fire assessment of an existing reinforced concrete building and retrofit design. Struct. Eng. Mech. **74**(4), 481–494 (2020). https://doi.org/10.12989/sem.2020.74.4.481

Mohd Tohir, M.Z., Spearpoint, M.: Simplified approach to predict heat release rate curves from multiple vehicle fires in car parking buildings. In: 3rd International Conference on Fire in Vehicles (FIVE), Berlin, Germany (2014)

TGS8 "Steel products and applications for buildings, construction and industry", "Robustness of car parks against localised fire", 2008–2011

Zhang, X.G., Guo, Y.C., Chan, C.K., Li, W.Y.: Numerical simulations on fire spread and smoke movement in an underground car park. Build. Environ. **42**, 3466–3475 (2007)

# Assessment of Existing Bridges: The Swiss Experience of the Last Three Decades

Aurelio Muttoni[1,2,3](✉), Franco Lurati[1], Duarte Viúla Faria[2], João Simões[2,4], and Miguel Fernández[2,5]

[1] Lurati Muttoni Partner, Studio d'Ingegneria SA, Mendrisio, Switzerland
aurelio.muttoni@epfl.ch
[2] Muttoni & Fernández Consulting Engineers, Ecublens VD, Switzerland
[3] Swiss Federal Institute of Technology (EPFL), Lausanne, Switzerland
[4] Strutlantis Engineering, Lisbon, Portugal
[5] Universidad Politécnica de Madrid, Madrid, Spain

**Abstract.** Many of the existing reinforced and prestressed concrete infrastructures, such as bridges, subways or overpasses have reached or will reach, in this decade, half a century of existence, making it timely to assess their structural safety accounting for their state. It is desirable that the assessment of existing structures is carried out following a design philosophy by Levels-of-Approximation (LoA) as described in Model Code 2010. This design philosophy consists in starting with a lower LoA corresponding to simple and fast calculations with some safety margin (corresponding to an approach typically used in the design of new structures), refining in the following stages (higher LoAs) only the calculations associated with the governing structural verifications. This work aims at addressing the fundamental aspects of this methodology, based on the experience in Switzerland in the last three decades, discussing briefly the calculation methods that can be applied in higher LoA and that allow considering some reserves of structural resistances typically neglected for design of new structures.

**Keywords:** Assessment · existing structures · levels-of-approximation · advanced methods · cost reduction

## 1 Introduction

A large part of existing reinforced concrete infrastructures throughout Europe have reached half a century of age, raising concerns about their structural safety. Such concerns are in many cases increased by signs of structural degradation (reinforcement corrosion, concrete degradation, alkali-aggregate reaction). Other possibly harmful effects for the level of structural safety are related to the increase of actions or to the poor performance of typical details adopted in the past (half-joints, expansion joints,...). This subject has been on the agenda due to the recent collapses of several bridges or buildings in Europe (as for example the collapse of the Polcevera viaduct in Genoa or the punching failure of a slab of an underground car park in Santander, Spain).


M. A. Aiello and A. Bilotta (Eds.): ICC 2022, LNCE 435, pp. 493–506, 2024.
https://doi.org/10.1007/978-3-031-43102-9_38

An approach based on Levels-of-Approximation (LoA), which is quite suitable for the assessment of existing structures, was first introduced in the Swiss code for concrete structures in 2003 (SIA 262 2003; Muttoni 2003) and later adopted by the *fib* Model Code 2010 (fib 2013) for concrete structures (Muttoni and Férnandez Ruiz 2012a; Muttoni and Fernández Ruiz 2012b). Additional higher levels of approximation were subsequently introduced in the Swiss code for the assessment of existing concrete structures in 2011 (SIA 269/2 2011). As the name states, and as depicted in Fig. 1, this design philosophy consists in performing the structural verifications by increasing LoAs. Thus, in a first level, actions and resistances are calculated with simple and fast methods, with some level of conservatism. Only the governing verifications are then analysed with higher LoAs (which allows more detailed calculation of actions and resistances, requiring, however, a longer analysis time).

This design philosophy allows a more rational use of resources, directing the focus to the governing verifications to be analysed with methods associated to higher LoAs. It is worth mentioning that, for existing structures, a more detailed analysis of the structure should always be accompanied by a better knowledge of the structure (which requires geometric surveys, inspections, material testing and structure diagnostics).

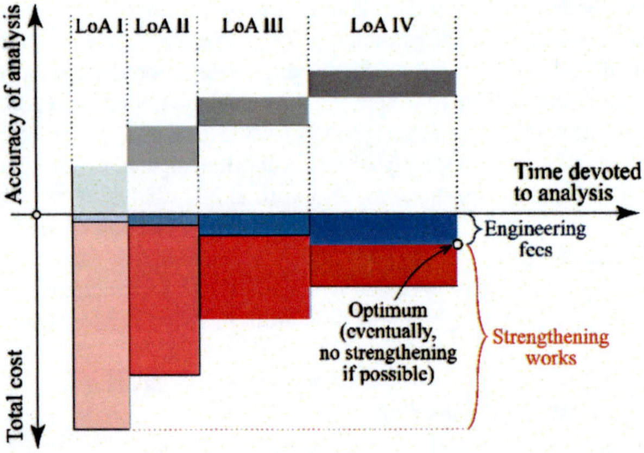


**Fig. 1.** Accuracy of the verification and costs associated with required strengthening works as a function of the required time for analyses for different levels of approximation (LoA).

The analyses with higher LoAs should be done through the application of calculation methods that allow considering resistances and favourable effects which are typically neglected for design of new structures (which tend to be conservative as they shall cover multiple cases in a simplified manner). Such reserves of resistance consist for example in the redistribution of internal actions or in the consideration of the membrane effect (compressions that develop in the highly stressed regions of a slab in response to the expansion due to cracking). The consideration of these physical effects, which can be considered through advanced calculation methods, allows, in most cases, to justify higher resistances than those obtained by applying the expressions usually recommended for design of new structures. Although not stated explicitly, this approach with different LoAs has been incorporated in the second-generation Eurocode 2 (prEC 2:2021).

A LoA approach can allow for an important reduction of the required structural strengthening, or, in some cases, even to avoid any strengthening, thus having a significant impact on the cost associated to the interventions (Fig. 1, a practical real case with a quantification of the savings in strengthening works is shown in Sect. 5). This paper aims at sharing the knowledge and experience of the authors in the application of the methods associated with higher LoAs to existing bridges in Switzerland since the introduction of the first code for the assessment of existing structures in 1994, discussing the potential gains, both in terms of the required intervention, as well as in terms of the associated costs.

## 2 Fundamental Concepts in the Asessment of Existing Structures

The assessment of existing structures is fundamentally different from the conceptual design and analysis of new structures because interventions in existing structures require in many cases a more significant economic effort when compared to new structures (structures in use usually have constraints that limit the intervention and the strengthening solutions). Thus, the main objective of the assessment and analysis of existing structures is to allow for the quantification of the actions and the resistances in a more accurate manner, having as target the reduction of costs associated to strengthening works. The code provisions used for design of new structures are of general application, tending to be conservative. These expressions can therefore be used in a first or second LoA, but the experience of the authors shows that, in many situations, there are non-negligible resistance reserves that can be activated when advanced models are used. These models, rationally-sound and validated against experimental evidence, allow analysing existing structures in a more general way. In Switzerland, the potentially different approaches for new and existing structures have been considered for the first time in 1994 with the introduction of a standard for the assessment of existing structures (SIA 462 1994).

It is important to mention that these analyses should be done by engineers experienced in this field. According to Walraven (2016), this type of analysis should be done by engineers with sufficient experience, allowing them to:

- Understand the actual behaviour of the structure;
- Determine the current state of the structure by determining the properties of the materials used;
- Make a quick and reliable decision on the state of the structure in case of a potential failure;
- Have the ability to deal with the effects of structural degradation on structural safety;
- Finally, to determine the structural safety.

In this work, special attention will be given to the determination of the flexural, punching and shear strength of existing structures. According to the authors' experience, these verifications are typically governing in prestressed reinforced concrete structures that have been designed with former generations of codes, have suffered changes of use or where significant degradations have occurred.

## 3 Flexural Resistance of Slabs

In general, engineers use finite element models that consider a linear-elastic response of the materials for the determination of the internal actions. The verification for bending is subsequently performed following a sectional-analysis by direct comparison of the acting moment and resisting moments (and in some cases, considering local internal forces redistribution). This methodology, suitable for design of new structures (being indirectly associated to a satisfactory behaviour for serviceability limit states), reveals to be very conservative, since it fully disregards the potential redistribution of internal actions that can occur in slabs. It thus neglects a reserve of resistance that can and should be considered in the analysis of existing structures.

For the determination of the flexural resistance of a slab panel, the application of limit analysis, namely, the yield lines method (Ingerslev 1923; Johansen 1943), can be applied. This method allows the determination of the global load-carrying capacity of the slab (instead of the local sectional resistance), since it considers the redistribution of internal forces that occurs after cracking (and yielding) of the sections until a failure mechanism is formed. Thus, the flexural strength is not determined by a local condition, but by the activation of the resisting moment along the yield lines. The geometry of the mechanism depends on the support conditions, loading and reinforcement arrangement.

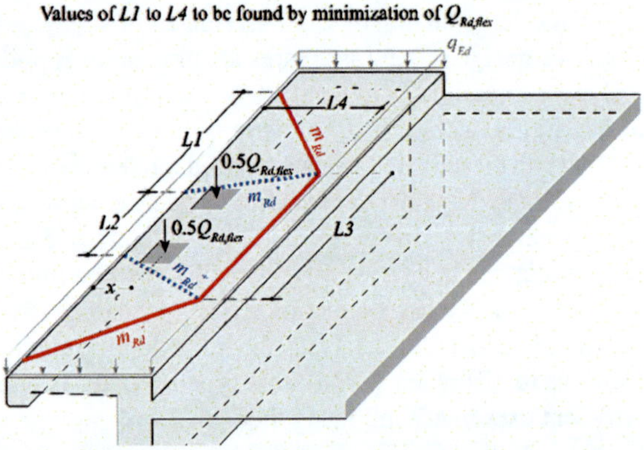


**Fig. 2.** Example of the application of the failure lines method in a cantilever of a bridge deck slab.

An example of the application of the failure lines method is shown in Fig. 2.

In the application of the yield lines method for evaluation of the load-carrying capacity of slabs, it is necessary to take into consideration that:

- the sections where yield lines are formed should present a sufficient deformation capacity, allowing the activation of the considered mechanism;
- since it is an upper limit of the plasticity theory, several mechanisms shall be analysed and/or optimised in order to identify the governing one (with the least load-carrying capacity);

- the verification with this methodology still requires to verify the resistance of other failure modes, as for instance those associated to brittle failures (such as punching, shear failures as discussed in following chapters, or anchorages, laps, and other potentially critical details).

According to the experience of the authors of this paper, the application of the yield-line method allows avoiding, in most cases, the need for flexural strengthening, with the obvious benefits in terms of costs associated to a strengthening intervention.

# 4 Shear Resistance

## 4.1 Elastic-Plastic Stress Fields for Members with Shear Reinforcement

For the verification of the shear resistance of webs in bridge girders, different LoA can also be used. In a first step (LoA I), the resistance associated to yielding of the shear reinforcement (assuming the minimum strut inclination defined in the standards for new structures) and the resistance associated to crushing of the concrete struts in the web can be verified separately. If such verification provides insufficient resistance, a more refined calculation can be conducted by calculating the inclination of the compression struts at failure assuming both yielding of the shear reinforcement and crushing of the concrete web. This is to be performed without accounting for the minimum compression field inclination defined for new structures, but determining the strength reduction factor $\nu$ for the concrete strength on the basis of the strain state in the web (LoA II). In case of post-tensioned girders, a more refined calculation can be conducted by estimating also the stress increase in the inclined tendons accounting for the strain state (LoA III). For these verifications, it is implicitly assumed that the concrete stress and the stress in the shear reinforcement is constant over the member's depth. Refined analyses have shown that in real structures, this is not the case and significant differences in terms of stresses and compression field inclinations can be observed near to the tension and compression chords (Rupf et al. 2013). This effect can be accounted for by analysing reinforced concrete members with the nonlinear finite element method where the strain compatibility is considered in all elements. A simple and robust approach consists in neglecting the concrete tensile strength and by modelling the concrete and steel behaviour with elastic-plastic constitutive laws (Fernández Ruiz and Muttoni 2007). This approach has shown to be suitable, accurate and very robust for members with sufficient reinforcement for crack control.

This allows generating elastic-plastic stress fields (EPSF) where the loads are carried by a system of ties (reinforcement) and compression fields (concrete). With such approach, not only reinforced and prestressed concrete girders can be analysed, but also members with a particular geometry, with complex reinforcement shapes and loads as for instance diaphragms, half-joints and complex framed piers (so-called D-type regions (Schlaich et al. 1987)).

In the case of the assessment of existing structures, the use of elastic-plastic stress fields (corresponding to a LoA IV) is very efficient (fib 2021). The main advantages are listed below:

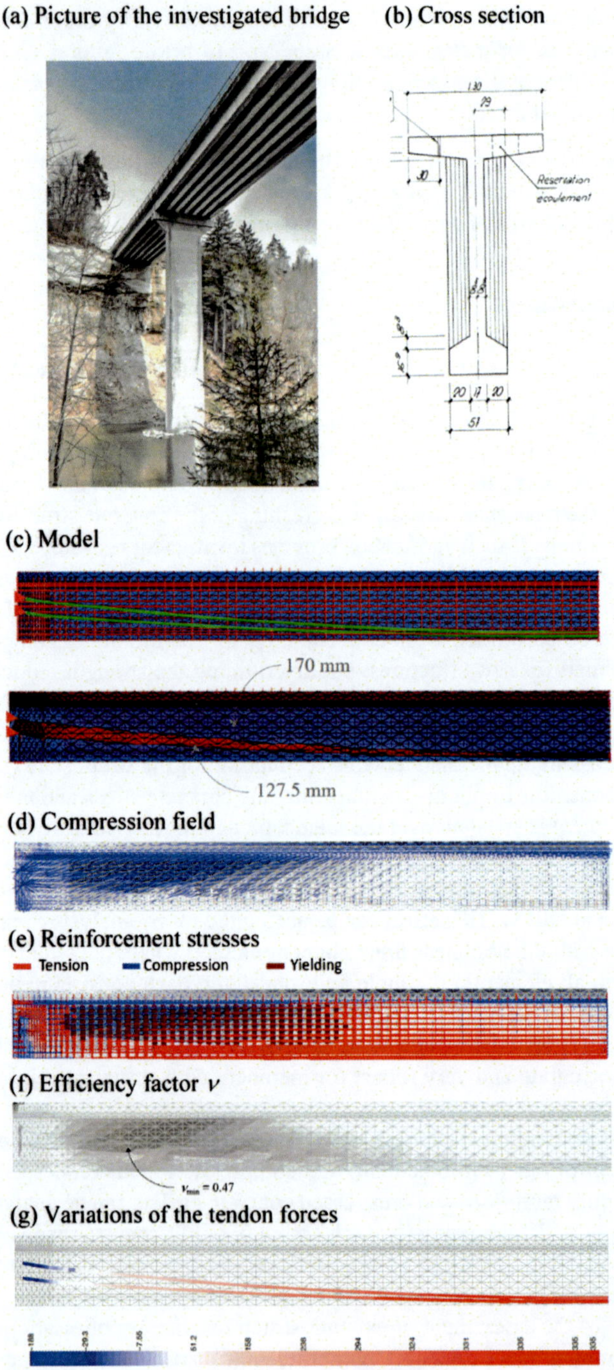


**Fig. 3.** Analysis of a precast prestress girder with elastic-plastic stress fields (LoA IV).

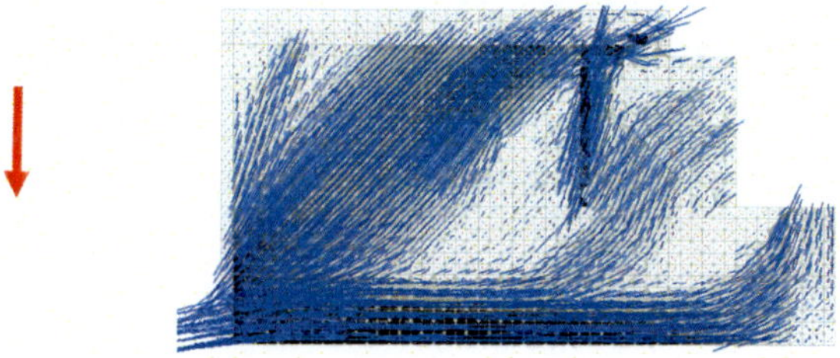

(a) Compression field

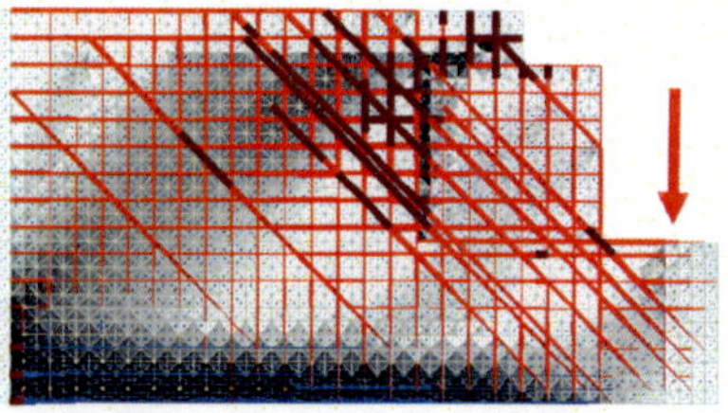

(b) Reinforcement stresses

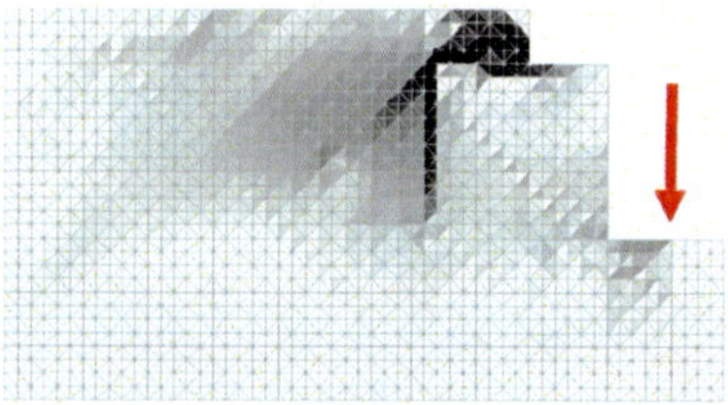

(c) Concrete efficiency factor $\nu$

**Fig. 4.** Analysis of a half-joint with elastic-plastic stress fields (LoA IV, actual case near Lausanne, Switzerland).

- Limited number of parameters (modulus of elasticity of concrete and reinforcement; compressive strength of concrete and yield strength of reinforcement), without the need to use an extremely large number of parameters whose physical meaning is complex to explain;
- Local calculation of the strength reduction factor ν for concrete compressive strength in presence of cracking (Vecchio and Collins 1986));
- Consideration of the actual arrangement of reinforcement;
- Additional activation of post-tensioning tendons (as a function of strains);
- Redistribution of internal forces and consideration of the membrane effect;
- Quantification of the shear force carried by the compression zone.

It should be noted that the use of elastic-plastic stress fields is proven by comparison with more than 200 tests published in specialized literature.

The authors have a large experience in the application of this method and were able to show in certain situations increases in resistance up to twice as high as those obtained by the provisions used for the design of new structures (these values can however vary significantly from case to case. Consequently, significant strengthening works, that would be necessary if only the provisions normally used for new structures were applied, were avoided with obvious benefits for the Owner.

Two examples resulting from the application of elastic-plastic stress fields are presented in Figs. 3 and 4, respectively for the case of a precast prestressed girder with additional post-tensioning tendons and a half-joint. In the first case, it was possible to optimise the strengthening works whereas in the case of the half-joint, it was possible to demonstrate that no strengthening was necessary despite the poor detailing with a construction joint near to the half-joint (see strain localization, yielding of the reinforcement and significant reduction of the efficiency factor ν shown in Fig. 4) and the large cracks observed in that region.

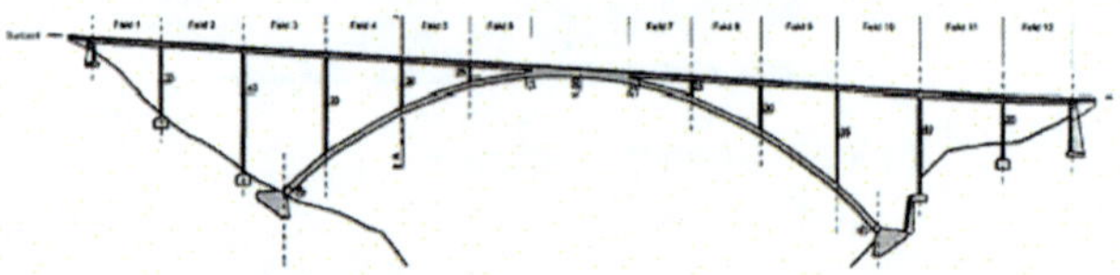

(a) elevation of the arch bridge

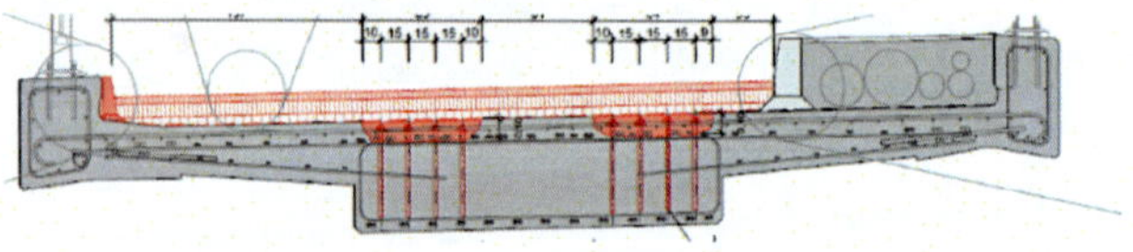

(b) cross-section of the deck with the new post-installed shear reinforcement

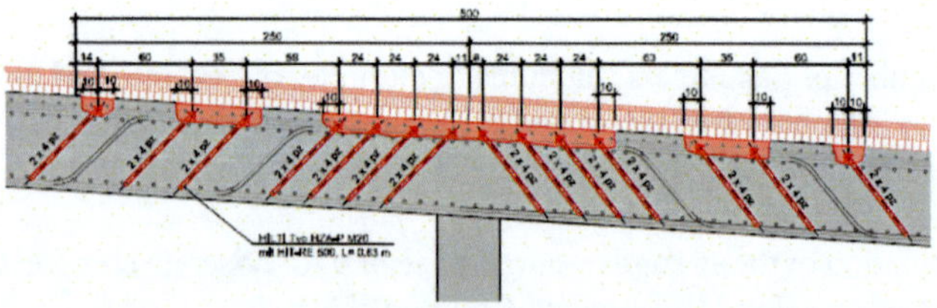

(c) distribution of the post-installed shear reinforcement complementing the existing bent-up bars

**Fig. 5.** Arch bridge built in 1962 and strengthening works in 2022.

In some cases, the strut-and-tie models (LoA I), the rigid-plastic stress fields (LoA II-III) and the elastic-plastic stress fields (LoA IV) allow verifying the critical details not complying with current codes of practice (too large spacings between shear reinforcements, unsuitable arrangement and shape of reinforcement, missing suspension reinforcement between longitudinal girders and cross beams in case of indirect supports,..). Figure 5 shows for instance the case of an arch bridge built in 1962 in the Swiss Alps where the deck is currently strengthened with post-installed shear reinforcement complementing the existing bent-up bars. In this case, the post-installed shear reinforcement specially developed for such strengthening works (Hilti 2009) has been optimized based on these approaches. In addition, the length of the zone to be strengthen has been reduced by using advanced tools for members without shear reinforcement as shown in the following subsection.

### 4.2 Cases of Bridge Deck Slabs and Other Members Without or With Insufficient Shear Reinforcement

The verification of the compliance with respect to the out-of-plane shear forces in existing bridge deck slabs is also a topic widely discussed among engineers since, in the past, bridges have been built with relatively thin deck slabs. Several parameters shall be accounted for, such as the position of the loads, slab of variable or constant thickness, presence or not of post-tensioning ducts (in case of bridges erected with the free cantilever method) and quality of injection, amongst others (Vaz Rodrigues et al. 2008; Natário et al. 2014). The calculation of the internal forces can be conducted using elastic tools (LoA I), elastic tools accounting for the redistribution of internal forces with simple rules (LoA II) or with Non-Linear Finite Element Analyses (NLFEA, LoA III). In this case, not only a more realistic distribution of the shear forces can be obtained (Fig. 6), but also the beneficial effect of the compressive membrane action in the critical zone can be accounted for. In any case, the study of the shear field (Vaz Rodrigues et al. 2008) is a very efficient tool to better understand the flow of forces to the supports and to detect the potentially critical zones (Fig. 6). With respect to the shear resistance of members without shear reinforcement, the Critical Shear Crack Theory (Muttoni et al. 2008; Cavagnis et al. 2020) has been adopted by several standards for different LoAs by adapting the general equation accordingly (see Table 1.). It allows at LoA I to estimate the shear resistance without knowing the amount of flexural reinforcement (but assuming that the latter is sufficient to fulfil the flexural resistance), to verify the shear resistance at LoA II just knowing the utilization ratio of the flexural reinforcement (assumed to be proportional to the ratio $m_{Ed}/m_{Rd}$) and calculating the shear resistance at LoA III as a function of the strain $\varepsilon_v$ in the flexural reinforcement. The latter can be obtained based on a sectional analysis as a function of the bending moment and the axial force or directly from a Nonlinear Finite Element Analysis (NLFEA) where the flexural and shear deformations are considered (Cantone et al. 2022), but the shear resistance is verified according the CSCT (this approach has the advantage of being significantly more robust that conventional NLFEAs since the concrete tensile strength can be neglected, or considered only with respect to the tension stiffening effects).

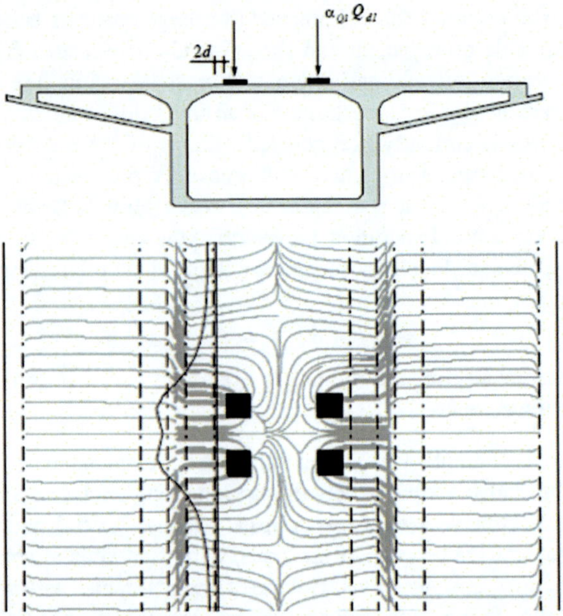


**Fig. 6.** Analysis of a bridge deck slab with location of concentrated loads and resulting shear field with distribution of the shear force along the control section.

**Table 1.** Equations for determining the shear resistance of members without shear reinforcement according to the CSCT for different LoA.

| LoA | Equation | Code |
|---|---|---|
| I | $V_{Rd} = \frac{0{,}3\sqrt{f_{ck}}}{\gamma_C} \frac{b_w \cdot d}{1+0.022 \cdot d}$ | Linee guida 2019, SIA 262 LoA I |
| II | $V_{Rd} = \frac{0{,}3\sqrt{f_{ck}}}{\gamma_C} \frac{b_w \cdot d}{1+0.022 \cdot d \cdot m_{Ed}/m_{Rd}}$ | SIA 262 LoA II |
| III | $V_{Rd} = \frac{0{,}33\sqrt{f_{ck}}}{\gamma_V^*} \frac{b_w \cdot d}{1+24\gamma_{def} \cdot \varepsilon_v \cdot d / d_{dg}}$ | prEN 1992-1-1:2021, Annex I (prEC 2:2021)) SIA 269/2:2011 |

## 5 Strengthening and Assessment Costs Associated to Different Levels of Approximation

With respect to the cost associated to the interventions, in the frame of a project related to the assessment and the strengthening against punching of a flat slab, the authors made a comparison of the costs associated to strengthening works and engineering studies related to three LoAs (II, III, IV). The assessment has been conducted according to *fib* MC2010 (see also Muttoni 2008) and the dimensioning of the post-installed reinforcement against punching has been conducted according to (Hilti 2009, compatible with *fib* MC2010).

It can be noted in fact that the same approach has been adopted by Annex I for existing structures in the draft of the second generation of Eurocode 2 (prEC 2 2021). For each level, the cost associated to the engineering study required to establish the strengthening solution to ensure the structural compliance and the cost associated to the strengthening works on-site were evaluated. The results of this analysis are presented in Fig. 7 and Table 2.. Figure 7 sketches the required structural strengthening solutions to ensure the structural safety compliance with respect to punching by applying LoA II, III and IV.

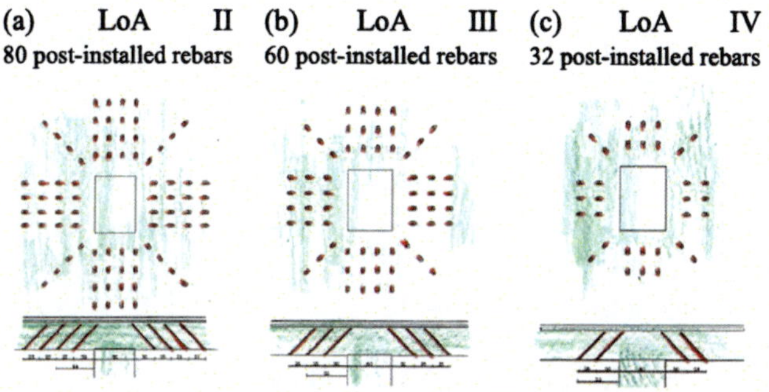


**Fig. 7.** Strengthening necessary to ensure the safety level with respect to punching shear in a slab-column connection, applying the LoA (a) II, (b) III and (c) IV.

It can be seen that, as the LoA increased, the number of post-installed bars to be introduced decreased from 80 to 32. The relative costs of the engineering studies are presented in Table 2..

**Table 2.** Relative engineering and strengthening costs.

| | Level II | Level III | Level IV |
|---|---|---|---|
| Engineering costs | reference | 1.5x | 4x |
| Strengthening costs | reference | −25% | −60% |

It can be seen that the application of LoA IV requires an analysis time approximately 4 times the one associated with LoA II (level usually used for the design of new structures). On the other hand, the costs in the strengthening works per connection (see Table 2.) are reduced by about 60% when comparing LoA II to LoA 4. Taking into account the balance between the costs associated (1) to the intervention and strengthening works and (2) to the fees for the engineering studies, it can be quickly concluded that the Owner has a positive balance. In this specific case, the engineering office fees were largely covered by the reduction in the costs associated with the strengthening works of a single connection. Thus, bearing in mind that in buildings or bridges, strengthening details are typically repeated several times, the costs in engineering studies using a LoA IV will be relatively small.

# 6 Effects of Structural Degradation and Application of Reliability Concepts

## 6.1 Material Degradation

Despite the fact that the effects of degradation of the materials that constitute the structure can be an important aspect to be accounted for when checking existing structures, this aspect is treated here only in a concise manner due to space limits. For these considerations, the engineer will have to have a sound knowledge of the degradation mechanisms (the most common examples of degradation are corrosion of reinforcement (resulting from carbonation of concrete and chloride attack) and the alkali-aggregate reaction. Note that the "effect" of degradation has been mentioned to highlight the fact that a degradation mechanism can have associated effects in addition to the more obvious ones. For example, in the case of corrosion, a reinforcement bar loses cross section and simultaneously the increase in volume of the bar can give rise to cracking and/or spalling of the concrete cover, with consequences at the level of the available concrete section, bond or deformation capacity of the reinforcement.

It is thus evident that the effect of the degradation of the materials which constitute the structures should be integrated in the verifications previously mentioned. However, this aspect is complex and is still the subject of several research works. The *fib* Model Code 2010 (fib 2013) already provides some information about the main mechanisms of degradation of materials, but detailed provisions will be presented in the next generation of fib Model Code (Matthews et al. 2018). It is strongly advised to accompany the studies by evaluations of the structure in situ, in order to obtain a detailed knowledge of the state of the structure, limiting the uncertainties at stake. Regular monitoring of the structure is also advised.

## 6.2 Reliability Analyses

Together with the use of advanced verification models that integrate the effect of degradation of structural materials, reliability analyses can also be performed in order to integrate and account in a rational manner for the uncertainties of all the variables (whether those associated with the models, the geometry, the mechanical properties or the actions). This type of analysis can eventually allow reducing the partial safety coefficients of actions and/or materials. Alternatively, an analysis of this type can allow determining the remaining lifetime for the structure.

# 7 Conclusions

This paper presents briefly advanced calculation methods for the assessment of existing concrete structures and the experience acquired in the last three decades based on the Swiss standards for the assessment of existing structures published since 1994. The main messages of this communication are summarized below:

- The design expressions recommended in the standards for the design of new structures are generally the result of simplifications of detailed mechanical models. More refined equations which directly result from the original detailed models can, however, be used for the assessment of existing structures. The use of such models makes it possible to obtain higher compliance values (ratios of resistances / internal actions), allowing for an important reduction of the necessary structural strengthening, or, in some cases, even justifying the absence of the need for strengthening.
- Some of the more detailed analyses that can be done are briefly presented in this paper. This type of analyses allows the consideration of some reserves of hidden structural resistances typically neglected for design of new structures. All the advanced analyses presented in this paper are already used in some Countries and will be allowed and provisioned in the second-generation Eurocode 2 (prEC2:2021).
- The additional fees associated with the engineering studies when more complex models are considered can in most cases be largely compensated by the reduction of costs associated with the strengthening intervention.
- These analyses with more complex mechanical models should, however, be conducted by experienced engineers with solid knowledge of the behaviour of structures.
- The effect of degradation is an important aspect to be accounted for when assessing existing structures (treated in this paper only in a concise manner due to space constrains).

## References


Cantone, R., Setiawan, A., Fernández Ruiz, M., Muttoni, A.: Characterization of shear deformations in reinforced concrete members without shear reinforcement. Eng. Struct. **257**, 16 (2022)

Cavagnis, F., Simões, J.T., Fernández, R.M., Muttoni, A.: Shear strength of members without transverse reinforcement based on development of critical shear crack. ACI Struct. J. **117**(1), 103–118 (2020)

Fernández, R.M., Muttoni, A.: On Development of suitable stress fields for structural concrete. ACI Struct. J. **104**(4), 495–502 (2007)

fib. Fédération internationale du béton, Model Code for Concrete Structures 2010, Ernst & Sohn, Germany, p. 434 (2013)

fib. fib Bulletin 80 – Partial factor methods for existing concrete structures - Recommendation Task Group 3.1, Fédération Internationale du Béton, Ernst & Sohn, Germany, p. 144 (2016)

fib. fib Bulletin 100 - Design and assessment with strut-and-tie models and stress fields: from simple calculations to detailed numerical analysis, Fédération Internationale du Béton , Ernst & Sohn, Germany, p. 235 (2021)

Ingerselv, A.: The strength of rectangular slabs. J. Inst. Struct. Eng. **1**(1), 4–14 (1923)

Johanssen, K.W.: Yield-Line Theory. Cement and Concrete Association, London, p. 181 (1943)

Hilti.: Post-Installed shear reinforcement Hilti HZA-P, Principles and design, Hilti, Fastening Technology Manual B2.6, p. 32 (2009)

Guida, L.: Ministero delle Infrastrutture e dei Trasporti, Consiglio Superiore dei Lavori Pubblici, Linee Guida per la classificazione e gestione del rischio, la valutazione della sicurezza ed il monitoraggio dei ponti esistenti, p. 90 (2019)

Matthews, S., Bigaj-van-Vliet, A., Walraven, J., Mancini, G., Dieteren, G.: fib model code 2020: towards a general code for both new and existing concrete structures. Struct. Concr. p. 969–979 (2018)

Muttoni, A.: Introduction à la norme SIA 262, Documentation SIA, Société suisse des ingénieurs et des architectes, Zurich, p. 5–9 (2003)

Muttoni, A.: Punching shear strength of reinforced concrete slabs without transverse reinforcement. ACI Struct. J. **105**(2), 163–172 (2008)

Muttoni, A., Fernández, R.M.: Shear strength of members without transverse reinforcement as function of critical shear crack width. ACI Struct. J. **105**(2), 163–172 (2008)

Muttoni, A., Fernández, R.M.: Levels-of-approximation approach in codes of practice. Struct. Eng. Int. **2**, 190–194 (2012)

Muttoni, A., Fernández, R.M.: The levels-of-approximation approach in MC 2010: application to punching shear provisions. Struct. Concr. **13**(1), 32–41 (2012)

Natário, F., Fernández, R.M., Muttoni, A.: Shear strength of RC slabs under concentrated loads near clamped linear supports. Eng. Struct. **76**, 10–23 (2014)

prEC2, 2021, prEN 1992-1-1, Eurocode 2-design of concrete structures- Part 1-1: general rules and rules for buildings, CEN, EN 1992-1-1, Brussels, Belgium, pp 383 (2021)

Rupf, M., Fernández, R.M., Muttoni, A.: Post-tensioned girders with low amounts of shear reinforcement: shear strength and influence of flanges. Eng. Struct. **56**, 357–371 (2013)

Schlaich, J., Schäfer, K., Jennewein, M.: Toward a consistent design of structural concrete. PCI J. **32**(3), 74–150 (1987)

SIA 262. Costruzioni di calcestruzzo, Società svizzera degli ingegneri e architetti, Zurich, p. 44 (2003)

SIA 269/2. Maintenance des structures porteuses – Structures enbéton, Société suisse des ingénieurs et des architectes, Zurich, p. 94 (2011)

SIA 462. Valutazione della sicurezza strutturale delle costruzioni esistenti, Società svizzera degli ingegneri e architetti, Zurich, p. 16 (1994)

Vaz, R.R., Fernández, R.M., Muttoni, A.: Shear strength of R/C bridge cantilever slabs. Eng. Struct. **30**, 3024–3033 (2008)

Vecchio, F.J., Collins, M.P.: The modified compression-field theory for reinforced concrete elements subjected to shear. ACI J. Proc. **83**(22), 219–231 (1986)

Walraven, J.: Forensic engineering: need for a new professional profile, fib symposium 2016, South Africa, pp. 97–110 (2016)

# Durability

# Experimental Evaluation of the Non-linear Behavior of Existing Gerber Half-Joints in Presence of Corrosion

Filippo Molaioni, Paolo Isabella(✉), Fabio Di Carlo, Zila Rinaldi, and Alberto Meda

Department of Civil Engineering and Computer Science Engineering (DICII), University of Rome Tor Vergata, Rome, Italy
beatrice.belletti@unipr.it

**Abstract.** The gerber half-joints assessment is today a topical problem since these are widespread in the italian infrastructure heritage. These elements, classified as "critical" by the recent italian guidelines (LG20), are often affected by chloride corrosion phenomena due to their positioning under the deck joint. The research program, aimed at evaluating the non-linear behaviour of gerber half-joints, includes the cast of sixteen specimens, which were designed and tested until failure according to different design models, reinforcement amount, and increasing corrosion levels of the reinforcement. This paper presents the first results of two un-corroded and two corroded specimens. The experimental outcomes show that both design details and reinforcement corrosion could compromise the strength and ductility requirements of the saddles. Furthermore, the authors believe that the results can represent a strong scientific reference for the analytical and numerical structural assessment of existing corroded gerber half-joints.

**Keywords:** Gerber half-joints · Existing bridges · Reinforced concrete · Chloride corrosion

## 1 Introduction

Reinforced concrete (RC) Gerber half-joints or saddles are a key connection between the structural elements of the bridges deck, allowing the realization of suspended beams. In Italy, this construction scheme has been widely adopted in the realization of the roads and highways network, between the 1930s and 1980s, given its numerous advantages, among which speed of construction, simplification of calculations, and optimization of stresses. The Gerber scheme is nowadays considered improper and no longer used because of critical issue related to maintenance and durability. Nevertheless, many existing bridges and viaducts still feature Gerber half-joints details. Among the most frequently encountered problems: structural design inadequacy; design loads lower than those provided by actual Codes; degradation phenomena due to seepage of water containing de-icing salt from the deck expansion joints; difficulties or impracticability of joint inspections due to geometry peculiarities.


M. A. Aiello and A. Bilotta (Eds.): ICC 2022, LNCE 435, pp. 509–521, 2024.
https://doi.org/10.1007/978-3-031-43102-9_39

Since most modern existing bridges have reached or are going to reach the end of their service life, the set-up of efficient strategies for safety assessment and maintenance emerges as a crucial research challenge in structural engineering. In this framework, in 2020, the Italian National Ministry of Infrastructures and Transports published the "Guidelines for risk management, safety assessment and monitoring of existing bridges" (C.S.LL.PP. 2020). For these reasons, the Guidelines identify the Gerber half-joint as a critical element of the bridge, highlighting the need to perform a structural assessment, which may result in a retrofit intervention to guarantee the required safety and durability conditions.

In Giannetti et al. (2020) a classification of existing Gerber bridges in Italy is made, within a focus on their recurrent degradation phenomena and a discussion on the historical sources which can be used to draw useful information for the safety assessment.

Experimental tests on un-corroded Gerber half-joints can be found in literature. Mattock et al. (1979) proposed a design procedure for reinforced or prestressed dapped-end beams, based on the results of an experimental investigation on eight dapped ends, four subjected to vertical load only, and four to a combination of vertical and horizontal loads. In Lu et al. (2012), the test results of 24 RC dapped-end beams are reported, varying the compressive strength of concrete, the shear span-to-depth ratio and the horizontal load. Desnerck et al. (2016) investigate the contribution of the internal reinforcement layout on the load capacity of RC saddles through an experimental survey on four full-scale half-joint beams. In Mata-Falcón et al. (2019) the results of an experimental program consisting of twenty-eight tests on fifteen different reinforcement configurations with and without inclined reinforcement are described, proposing a simplified procedure for defining the geometry of STMs considering spalling failures. Rajapakse et al. (2022) investigates shear and flexural failures of Gerber half-joints through an experimental campaign consisting of eight large-scale dapped-end tests. A historical review of the theoretical and experimental studies carried out to study the behaviour of RC half-joints beams can be found in Shakir (2020).

Corrosion induced by chloride-rich water can degrade the mechanical properties of the concrete and steel. Therefore, reductions in the bearing capacity of the Gerber half-joint and changes in the failure mechanism may be triggered by corrosion. Calculation methods, commonly used when dealing with the assessment of the remaining load carrying capacity of RC Gerber half-joint, are based on the Strut-and-tie method (STM). In Desnerk et al. (2017, 2018), an experimental survey on RC half-joint beams, incorporating different deterioration defects and reinforcement detailing, was performed to discuss the impact of reinforcement layout, anchorage, and concrete cracking on the structural capacity, evaluated by means of STMs. Santarsiero et al. (2021) propose a procedure for the durability analysis of Gerber saddles, considering the simultaneous presence of permanent loads and environmental actions under the form of chloride ions, through nonlinear FEM numerical simulations. In Di Prisco et al. (2018), a full-scale test on a saw-extracted beam-end from the Annone overpass on SS.36 is performed to check the effectiveness of the STM adopted for the evaluation of the failure mode of the element. Nevertheless, the problem of the assessment of the structural safety of corroded Gerber saddles is still far from a satisfactory solution. The paper presents and discusses the first results of a wider experimental survey carried out at the Laboratory of the University of Rome "Tor Vergata" within the framework of an Italian ReLUIS-2021 project for the evaluation of the non-linear behavior of existing reinforced concrete half-joints in presence of corrosion of the steel reinforcement.

The experimental survey is presented in Sect. 2. Section 2.1 is devoted to the description of the specimen preparation, with particular reference to geometry, details of the steel cage, and material properties of concrete and steel reinforcement. The description of the artificial corrosion process through electrolytic cells is reported in Sect. 3, giving information in terms of adopted values of current density, expected and measured mass loss for each corroded specimen. A description of the set-up and of the used measure devices is given in Sect. 4. Finally, a discussion on the obtained results for both un-corroded and corroded specimens, is made in Sect. 5 in terms of load-deflection diagrams, failure mode, and cracking pattern.

## 2 Test Specimens

A wide experimental survey for the evaluation of the behaviour of un-corroded and corroded RC Gerber half-joints has been planned and conducted at the Laboratory of the University of Rome "Tor Vergata". The design of the specimens is based on the Strut and Tie Models (STMs) proposed in EN 1992–1-1, named here Model A and Model B, respectively and reported in Fig. 1. Three main components can be identified in the STM: the ties and the struts, characterized by tensile and compressive stresses, respectively, and the nodes, representing the joining element between ties and struts. It is worth to highlight the existence of two zones for the half-joint: the "B-region" in which the hypothesis of Bernoulli holds and the "D-region", area of discontinuity in which the hypothesis of Bernoulli is no longer valid (Fig. 1). Two reinforcement layouts have been adopted in the experimental campaign, studied by considering the activation of the Model A only or a combination in parallel of Model A and Model B. With the aim to investigate different collapse mechanisms of the half-joints, two different steel reinforcement amounts are adopted, "low" and "high", respectively. The paper presents the first outcomes of the experimental survey related to four specimens, designed according to Model A + B, cast and tested up to failure. Two of the four specimens were kept un-corroded (UC) for reference. The remaining two specimens were subjected to an artificial corrosion process of the steel ties to investigate the effects of chlorides corrosion phenomena on the structural behaviour of the element. Table 1 shows the layout of the experimental program, together with the expected value of the mass loss of the two corroded specimens.

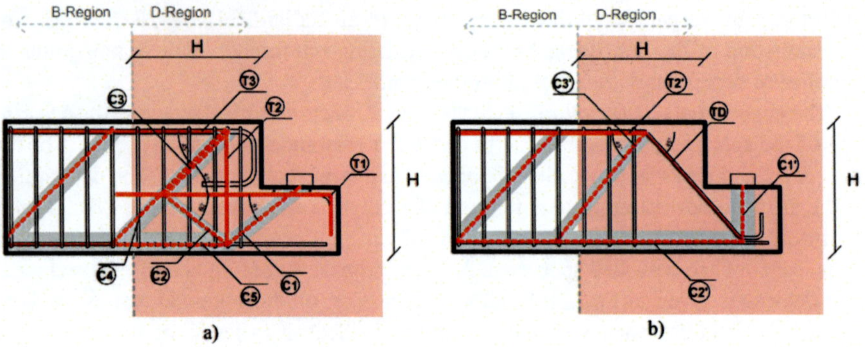


**Fig. 1.** Strut and Tie Models (STMs) proposed in EN 1992–1-1: a) Model A and b) Model B.

**Table 1.** Specimens layout.

| Specimen name | G1-NC | G2-NC | G1-C | G2-C |
|---|---|---|---|---|
| Strut and tie model | A + B | A + B | A + B | A + B |
| Reinforcement amount | low | low | low | low |
| Expected mass loss | - | - | 15% | 15% |

## 2.1 Geometry and Materials of the Specimens

The specimens are made by casting RC beams that have Gerber half-joint details at the end. Therefore, each cast beam allows the testing of two specimens. The geometry and the reinforcement details of the specimens are illustrated in Fig. 2. The beams are overall 3000 mm long, the nibs are 300 mm long with a 200 x 250 mm cross section, while the internal part of the beam has a 200 x 500 mm cross section.

As mentioned before, the reinforcement layout has been set to investigate STM A + B. The specimens are characterized by four types of steel ties: the longitudinal rebars located at the extrados of the nib and anchored in the inner part of beam (T1); the vertical stirrups close to the nib having the role of transferring the vertical actions from the bottom of the nib to the top of the beam (T2); the longitudinal reinforcement at the extrados of the beam, bringing the tensile stresses from the "D-region" to the "B-region", allowing the development of the Ritter-Morsch truss model in the beam (T3); the diagonal tie, which crosses both the nib and the beam, transferring the tensile stresses from the "D-Region" to the "B-Region" (TD). The tensile actions carried by the diagonal rebars are balanced by a lower compressed strut and struts inclined at 45°.

The layout also depends on the reinforcement amounts: low reinforced specimens and high reinforced specimens are distinguished. Table 2 summarizes the layout for the ties of each specimen.

As concern the other reinforcement, four Ø10 mm lower longitudinal rebars (compression) were used. The nibs are equipped with two Ø8 mm closed stirrups 200 mm spaced. The internal beam (B-Region), designed to lead the collapse in the half-joints,

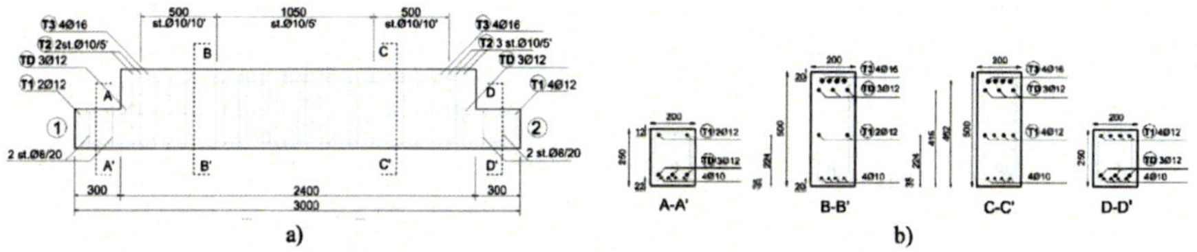

**Fig. 2.** Geometry and reinforcement layout of the specimens.

**Table 2.** Steel reinforcement layout of the specimens.

| Reinforcement amount | low | low | low | low | high | high | high | high |
|---|---|---|---|---|---|---|---|---|
| Tie | T1 | T2 | T3 | TD | T1 | T2 | T3 | TD |
| Diameter | 12 | 10 | 16 | 12 | 12 | 10 | 16 | 12 |
| Number | 2 | 2x2 | 4 | 3 | 4 | 3x2 | 4 | 3 |

is equipped with a transverse reinforcement able to resist the high shear actions attained during the execution of the tests. In particular, Ø10 mm closed stirrups, 100 mm spaced, are used in the two outermost parts of the beam, while Ø10 mm closed stirrups, 50 mm spaced, are used in the central portion. The concrete cover thickness is equal to 20 mm, while the distance between the first stirrup belonging to the tie T2 and the initial section of the nib is 75 mm.

The average concrete strength measured on seven 150 mm side cubes is equal to about 52.6 MPa. Furthermore, tensile tests have been performed on the steel rebars (Italian B450C type), considering all used diameters, obtaining values of the yielding strength equal to 503.0 MPa, 507.0 MPa and 455.0 for the steel rebars having diameter equal to 10 mm, 12 mm and 16 mm, respectively.

## 3 Artificial Corrosion Process

The specimens G1-C and G2-C were subjected to an artificial corrosion process of the steel ties. Corrosion has been provided with an accelerated process through electrolytic cells, by dipping the specimens in a 3% saline solution placed in a suitable pool, up to cover the upper surface of the saddle (Fig. 3). For each edge of the beam, the steel rebars belonging to the T1, T2 and TD ties have been connected to the positive pole of a power supply (anode). The cathodes have been realized with two Ø10 mm diameter steel bars placed inside the pool near each edge of the beam. A current intensity equal to 1.16 A and 1.53 A has been adopted for specimens G1-C and G2-C, respectively, in agreement with the values suggested in El Maaddawy & Soudki (2003). All other rebars of the steel cage have been epoxy coated against corrosion, as shown in detail in Fig. 4.

The time to obtain the desired corrosion level was evaluated with the Faraday's law, suitably modified in order to account for the concrete presence:

$$time[sec] = \frac{\lambda \cdot m_{loss} \cdot n_{specimen} \cdot C_{Faraday}}{current[A] \cdot M_{specimen}} \quad (1)$$

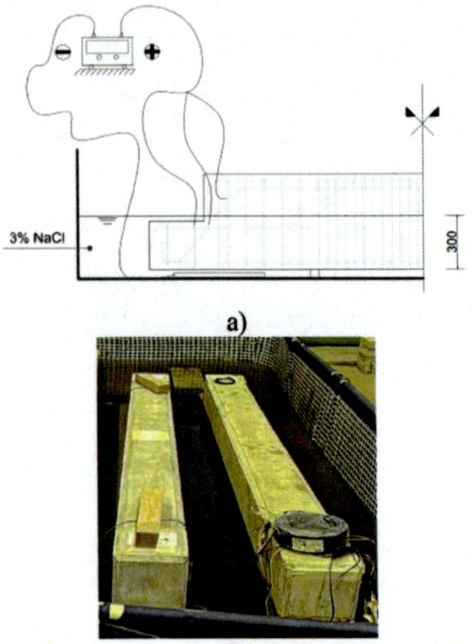


**Fig. 3.** Artificial corrosion process: a) scheme; b) view of the specimens inside the pool.

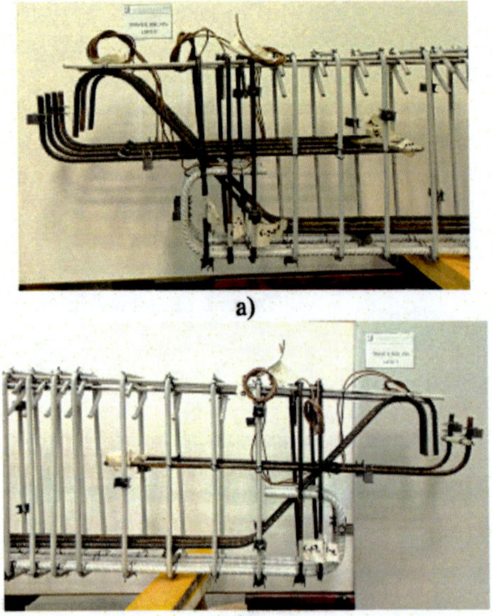


**Fig. 4.** Steel cage of the corroded specimens: a) detail of specimen G2-C; b) detail of specimen G1-C.

where $M_{specimen}$ is the molar mass of the reinforcing rebar (equal to 55.8 mol), $n_{specimen}$ is the valence equal to 2, $C_{Faraday}$ is the Faraday constant (equal to 96485 C/mol) and $\lambda$ is the constant accounting for the possibility that the corrosive process does not start immediately due to the concrete cover, assumed equal to 1.6, based on previous studies (Rinaldi et al. 2010; Meda et al. 2014, Rinaldi et al. 2022).

After the tests, the effective corrosion amount and morphology were evaluated by extracting the steel rebars from the specimens (Fig. 5). In particular, the actual corrosion is evaluated in terms of mass loss, measured by weighing the reinforcement after the rebar cleaning in agreement with (ASTM G1–90, 2002). The average mass loss values are summarised in Table 3 for each specimen. It is worth highlighting that the various steel ties belonging to each specimen show different values of the actual corrosion, as reported in Table 4. For the specimen G1-C (G2-C, respectively), the steel tie T1 (T2, respectively) presents a higher corrosion level with respect to the other two ties. The corrosion degrees of the ties have been evaluated by considering their overall length, except for the diagonal ties TD for which the length of the extracted portion of the rebar, as shown in Fig. 5, has been considered. Finally, Fig. 6 shows the crack pattern of the beam, including the two specimens, G1-C and G2-C, due to the corrosion process. A maximum observed crack width of 2 mm and 0.8 mm was observed, respectively, for specimens G1-C and G2-C.

**Table 3.** Corrosion degrees of the specimens.

| Specimen | G1-C | G2-C |
|---|---|---|
| Foreseen corrosion | 15 | 15 |
| Actual corrosion | 13 | 14 |

**Table 4.** Corrosion degrees of the steel ties.

| Specimen | G1-C | | | G2-C | | |
|---|---|---|---|---|---|---|
| Tie name | T1 | T2 | TD | T1 | T2 | TD |
| Actual corrosion [%] | 21 | 15 | 6 | 15 | 18 | 8 |

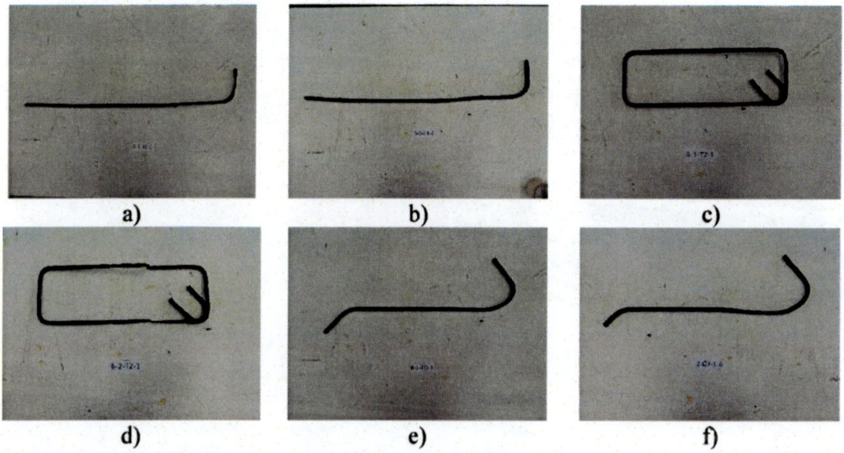


**Fig. 5.** Corroded embedded bars extracted after failure: a) T1/G1-C, b) T1/G2-C, c) T2/G1-C; d) T2/G2-C, e) TD/G1-C, f) TD/G2-C.

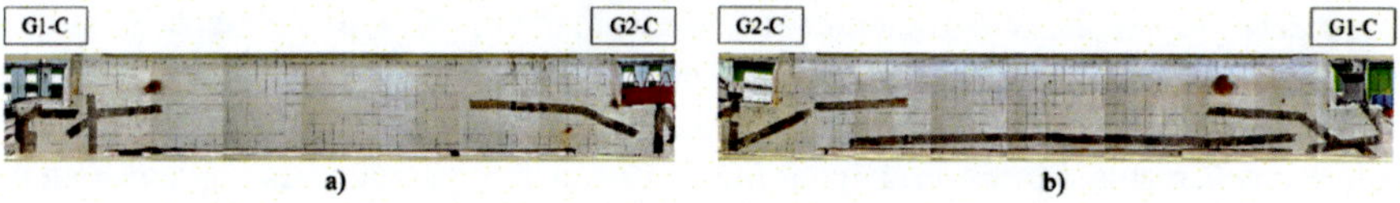


**Fig. 6.** Crack pattern due to corrosion expansion of the specimen G1-C (a) and G2-C (b).

## 4 Test Set-Up

The test set-up, shown in Fig. 7a with a three-dimensional sketch, it has been specifically defined, aiming to simulate the actual loading and restrain condition of an existing Gerber half-joint, since a RC Gerber bridge is characterized by a suspended span resting on the nibs of pier or adjacent beams (Fig. 7b). The beam is anchored to the laboratory strong floor by means of two steel beams and two pre-tensioned high strength rebars, each one connected to a hydraulic 300 kN jack, allowing the simulation of the existing structural continuity.

The load is applied through a 4000 kN hydraulic jack placed at a distance equal to 1500 mm from the axis of the steel beams (800 mm from the axis of the cylindrical support). The jack has been fixed to the load frame of the laboratory and is equipped with a 1000 kN load cell and a hinged joint. A view of the test set-up is reported in Fig. 8.

Figure 9 shows the sensors used to measure the displacements and deformation of the specimens. Five wire encoders (ENC) were used to measure the vertical saddle displacements. Two of these (ENC 1 and ENC 2) were placed at the load application section, on both sides of the specimen, with the aim of measuring the displacement at the loading point and catching the occurrence of torsional phenomena of the nib. To check the displacements at the constraints, a wire encoder was placed in correspondence with the cylinder support (ENC 3) and other two encoders were placed in correspondence with the steel beams (ENC 4 and ENC 5).

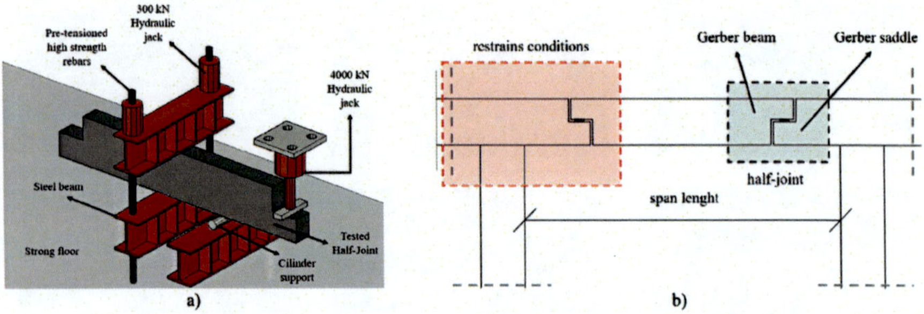


**Fig. 7.** a) 3D sketch of the test set-up, b) Scheme of the half-joint.

Four potentiometer transducers were placed, three of them in correspondence with the three steel ties of the STM (POT 1, POT 2 and POT 3); while POT 4 was placed at the extrados of the cylindrical support section to check the occurrence of inelastic behavior, as it is expected to be the area with the major bending stress. Finally, LVDT 1 was placed near to the re-entrant corner to evaluate the opening of the crack for shear-bending. LVDT 2 and LVDT 3 were placed orthogonally to the expected concrete struts with the aim to measure the compressive cracks and assess the role of concrete in the failure mechanisms. Moreover, Digital Image Correlation technique is implemented in one of the lateral surface, details and results can be found in Molaioni et al. (2022).

Fig. 8. Test setup.

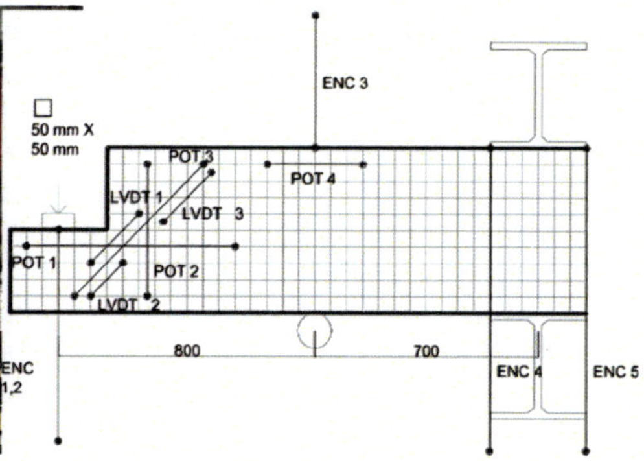


Fig. 9. Scheme of the measuring instrumentation.

## 5 Experimental Results

The experimental results of both the uncorroded, namely G1-NC and G2-NC, and the corroded twins Gerber half-joints specimens, namely G1-C and G2-C are presented and compared in Fig. 10, in terms of Load-Displacement curves. The load relates to the measurement of the 1000 kN load cell, while the displacement refers to the average displacement measured by ENC 1 and ENC 2.

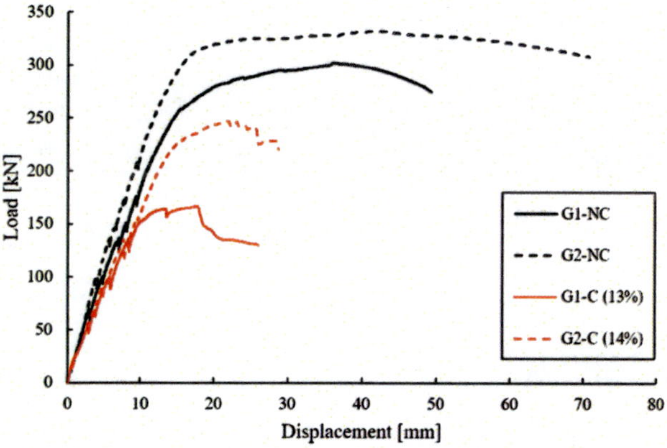


Fig. 10. Force-Displacement curves of the specimens.

### 5.1 Uncorroded Half-Joints

For the specimen G1-NC, characterized by the lower reinforcement amount, the maximum vertical load was equal to 301.4 kN for a vertical displacement of 36.3 mm. The first crack appeared at the nib re-entrant corner for a load of 69 kN and a displacement of 2.94 mm. An almost linear behavior followed up to a load of about 260 kN and a displacement of about 16 mm. After the maximum load, a softening behavior is observed until the attainment of the complete failure for a vertical load of 279 kN and a vertical displacement of 48.4 mm. The failure mechanism, shown in Fig. 11a, is governed by the rupture of the diagonal rebars in the nib (TD) and by the spalling of the concrete cover under the nib. A bending-shear failure is observed.

Relatively to the specimen G2-NC, the maximum measured load was equal to 331.3 kN for a displacement of 40.9 mm. The first crack appeared in the re-entrant corner for a load of 47.7 kN and a displacement of 1.9 mm. The structural behavior was almost linear up to a load of 308 kN and a displacement of 16.5 mm. At the failure condition, the measured load was equal to 308.8 kN, with a vertical displacement of 70.1 mm. The greater bearing capacity of about 10% observed for the specimens G2-NC with respect to the specimen G1-NC is due to the presence of a higher reinforcement amount. The failure mechanism is governed by the collapse of the compression strut in the nib and the yielding of the steel ties. As a higher reinforcement amount characterizes specimen G2-NC with respect to specimen G1-NC, the concrete struts have undergone greater stresses. It is remarked that, despite the shear-compression failure, which highlights that the weakest element in the STM is the concrete strut in the nib, the nonlinear behavior is characterized by the existence of a certain degree of ductility due to the yielding of the steel ties.

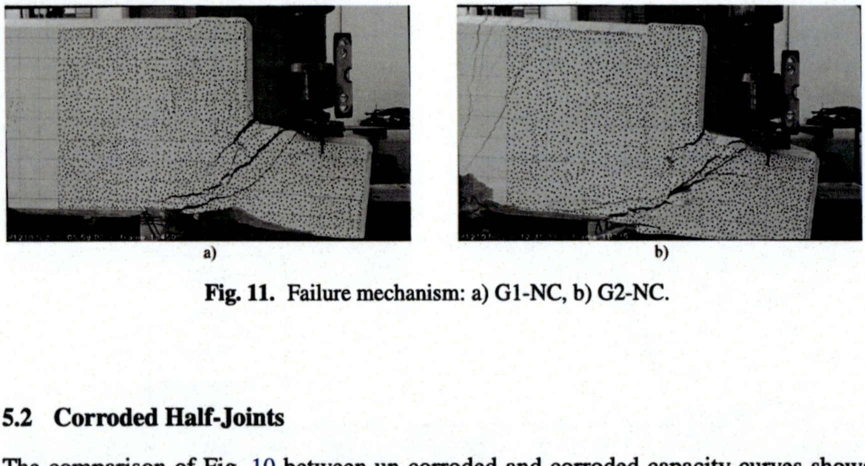

a) b)

**Fig. 11.** Failure mechanism: a) G1-NC, b) G2-NC.

## 5.2 Corroded Half-Joints

The comparison of Fig. 10 between un-corroded and corroded capacity curves shows reductions in both capacity and ductility. More specifically, a reduction of the capacity equal to 45% and 25% is observed for the specimens with low (G1-NC and G1-C) and high (G2-NC and G2-C) reinforcement amount, respectively. It is worth to highlight that an average mass loss equal to 13% and 14% was found in the corroded specimens G1-C and G2-C, respectively. Relatively to the specimen G1-C, a maximum load equal to 166 kN is observed for a displacement of 17.2 mm. Furthermore, the rupture of one of the corroded diagonals rebars was observed after the execution of the test. In Fig. 12a, the collapse mechanism of the specimen G1-C is reported. It could be observed how, due to the corrosion effect, the bending-shear crack in the nib governs the failure of the half-joint. It is also noted that the concrete in the beam suffered fewer cracks compared with the un-corroded twin G1-NC, as the corrosion of the tie acted as a fuse, preventing the development of the cracking of the struts. Specimen G2-C is the corroded twin of specimen G2-NC regarding the reinforcement amount. A maximum load of 247 kN for a displacement of 20.9 mm was observed. The failure mechanism shown in Fig. 12b features a shear-traction collapse due to the rupture of the corroded stirrups belonging to the tie T2. It must be remarked that the collapse mechanism of the un-corroded twin G2-NC was governed by the compression of the struts in the nib. Therefore, in addition to the reduction in both capacity and ductility, a change in the collapse mechanism can also be achieved due to the corrosion effect.

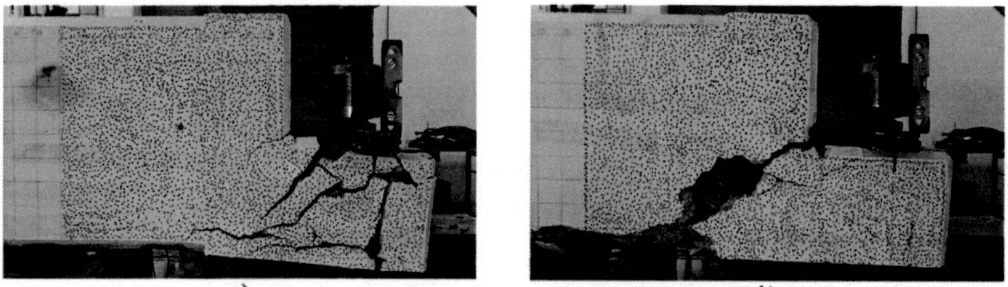

a) b)

**Fig. 12.** Failure mechanism: a) G1-C, b) G2-C.

## 6 Conclusions

The paper investigates the influence of steel reinforcement amount and corrosion on the non-linear behavior of RC half-joints, through experimental tests on two un-corroded and two corroded elements. The tested specimens are part of a larger research program, including the realization of sixteen specimens, designed, cast and tested until failure according to different design models, reinforcement amount, and increasing corrosion levels of the reinforcement. The corrosion was artificially induced with an accelerated process through electrolytic cells, by dipping the specimens in a 3% saline solution placed in a suitable pool, up to cover the upper surface of the saddle, in order to reach a mass loss of about 15%.

The corroded steel ties, extracted from the two corroded saddles, showed different values of the actual corrosion, with an average value of 13% and 14%, respectively.

The obtained results show that both construction details and reinforcement corrosion could compromise the non-linear behavior of the saddles. The following remarks can be made:

- the failure mechanism of the un-corroded specimen with the low reinforcement amount is governed by the rupture of the diagonal rebars (TD) and by the spalling of the concrete cover under the nib;
- the failure mechanism of the un-corroded specimen with the high reinforcement amount is governed by the collapse of the compression strut in the nib and by the yielding of the steel ties. The greater bearing capacity of about 10% with respect

to the specimen G1-NC is due to the presence of a higher reinforcement amount. A certain degree of ductility is still found;
- for both corroded specimens, a reduction in both capacity and ductility is observed;
- for the corroded specimens with the low reinforcement amount, the rupture of one of the corroded diagonals rebars was observed after the execution of the test. A bending-shear failure can be observed;
- for the corroded specimens with the high reinforcement amount, a change of the collapse mechanism is found, with a shear-traction collapse, due to the rupture of the corroded stirrups belonging to the tie T2.

Finally, the authors believe that the results can represent a strong scientific reference for the analytical and numerical structural assessment of existing corroded Gerber half-joints.

**Acknowledgements.** The paper presents some of the results obtained in the framework of the ReLUIS project (accordo attuativo DM 578/2020).

## References

ASTM G1–90. Practise for preparing, cleaning and evaluating corrosion test specimens. ASTM International, West Conshohocken, Pa (2002)

C.S.LL. Linee guida per la classificazione e gestione del rischio, la valutazione della sicurezza ed il monitoraggio dei ponti esistenti (2020)

Desnerck, P., Lees, J.M., Morley, C.T.: Impact of the reinforcement layout on the load capacity of reinforced concrete half-joints. Eng. Struct. **127**, 227–239 (2016)

Desnerk, P., Lees, J.M., Morley, C.T.: The effect of local reinforcing bar reductions and anchorage zone cracking on the load capacity of RC half-joints. Eng. Struct. **152**, 865–877 (2017)

Desnerk, P., Lees, J.M., Morley, C.T.: Strut-and-tie models for deteriorated reinforced concrete half-joints. Eng. Struct. **161**, 41–54 (2018)

Di Prisco, M., Colombo, M., Martinelli, P., Coronelli, D.: The technical causes of the collapse of Annone overpass on SS.36. In: Proceedings of Italian Concrete Days 2018, Lecco, Italy (2018)

El Maaddawy, T.A., Soudki, K.A.: Effectiveness of impressed current technique to simulate corrosion of steel reinforcement in concrete. J Mater Civ Eng **15**(1), 41–47 (2003)

EN 1992–1-1 "Eurocode 2: Design of Concrete Structures"

Giannetti, I., Mornati, S., Coccia, S., Di Carlo, F., Rinaldi, Z.: Safety assessment of existing post-war reinforced concrete bridges. The case study of 'Gerber Girders' bridges in Italy. In: 12th International Conference on Structural Analysis of Historical Constructions SAHC 2020 P. Roca, L. Pelà and C. Molins (Eds.), 1–12 (2020)

Lu, W.Y., Lin, I.J., Yu, H.W.: Behaviour of reinforced concrete dapped-end beams. Mag. Concr. Res. **64**(9), 793–805 (2012)

Mata-Falcón, J., Pallarés, L., Miguel, P.F.: Proposal and experimental validation of simplified strut-and-tie models on dapped-end beams. Eng. Struct. **183**, 594–609 (2019)

Mattock, A.H., Chan, T.C.: Design and Behavior of Dapped-End Beams. PCI J. 28–45 (1979)

Meda, A., Mostosi, S., Rinaldi, Z., Riva, P.: Experimental evaluation of the corrosion influence on the cyclic behaviour of RC columns. Eng. Struct. **76**, 112–123 (2014)

Molaioni, F., Talledo, D.A., Bartoli, M., Di Carlo, F.: Analysis of failure mechanisms of Gerber half-joint specimens through digital image correlation technique. In: Italian Concrete Conference, (October 2022) (2022)

Rajapakse, C., Degée, H., Mihaylov, B.: Investigation of shear and flexural failures of dapped-end connections with orthogonal reinforcement. Eng. Struct. **260**, 114233 (2022)
Rinaldi, Z., Imperatore, S., Valente, C.: Experimental evaluation of the flexural behavior of corroded P/C beams. Constr. Build. Mat. **24**(11), 2267–2278 (2010)
Rinaldi, Z., Di Carlo, F., Spagnuolo, S., Meda, A.: Influence of localised corrosion on the cyclic response of reinforced concrete columns. Eng. Struct. **256**, 114037 (2022)
Santarsiero, G., Masi, A., Picciano, V.: Durability of Gerber saddles in RC bridges: analyses and applications (Musmeci Bridge, Italy). Infrastructures **6**(2), 25 (2021)
Shakir, Q.M.: A review on structural behavior, analysis and design of RC dapped end beams. In: 3rd International Conference on Recent Innovations in Engineering (ICRIE 2020), IOP Conference Series: Materials Science and Engineering **978**, 012003 (2020)

# Effects of Stirrups Corrosion on the Shear Strength of RC Beams

Antonino Recupero[1], Pier Paolo Rossi[2], and Nino Spinella[2](✉)

[1] Department of Engineering, Università Di Messina, Messina, Italy
[2] Department of Civil Engineering and Architecture, Università Di Catania, Catania, Italy
nino.spinella@unict.it

**Abstract.** Based on different theories (plasticity theory, modified compression field theory and compression chord capacity theory), this paper investigates the effects of corrosion of the transverse reinforcement on the shear strength of reinforced concrete beams. The effects of corrosion of the transverse reinforcement are firstly described and, then, adequately introduced in the formulations of the plasticity and modified compression field theories. The considered formulations are applied to a set of RC beams tested in laboratory by other researchers and a comparison is first drawn between the shear strength values resulting from the considered models and the shear strength resulting from the laboratory tests. Then, the effects of corrosion on the mechanical parameters involved in the ultimate response of the examined reinforced concrete beams are investigated.

**Keywords:** Shear strength · RC beams · Corrosion · Stirrups

## 1 Introduction

In the past, many laboratory tests were conducted to investigate the effects of corrosion of the longitudinal reinforcement on the flexural strength of RC (Cheng et al., 2021; Dai et al., 2020; Rajput & Sharma, 2018; Zhu et al., 2013) and Prestressed Concrete (PC) structural members (Recupero & Spinella, 2019a, 2019b; Zhang et al., 2017). However, few tests were carried out to investigate the effects of stirrups corrosion on the response of such members. Nevertheless, it is of utmost importance for the assessment of the strength of any RC member because, owing to the minor protection of the transverse reinforcement compared with that of the longitudinal one, severe corrosion of the transverse reinforcement is more likely to occur in existing structures. In addition, the diameter of stirrups is smaller than that of the longitudinal steel bars. Owing to this, the percentage loss of the stirrup cross-sectional area is always higher than that of the longitudinal steel bar. Corrosion of steel stirrups can negatively affect the performance of RC members and cause, even in members designed to fail in flexure, the attainment of the shear capacity prior to any yielding of the longitudinal steel bars.

This work focuses on the prediction of the effects of corrosion of the transverse reinforcement on the shear strength of RC beams. Based on the results of laboratory tests, this work evaluates the accuracy in the prediction of the shear strength of both


M. A. Aiello and A. Bilotta (Eds.): ICC 2022, LNCE 435, pp. 522–533, 2024.
https://doi.org/10.1007/978-3-031-43102-9_40

slender and non-slender corroded RC beams by means of three models. The considered models are: the Compression Chord Capacity Model (CCCM) (Marí et al., 2014), a model based on the plasticity theory (PM) (Rossi, 2013), and a model based on the Simplified Modified Compression Field Theory (SMCFT) (Bentz et al., 2006). All these models have been proposed in the past with reference to members with un-corroded rebars (Recupero et al., 2003, 2005, 2018; Rossi, 2013). The CCCM (Cladera et al., 2021) has recently been modified to simulate those effects of steel corrosion and applied to beams with corroded reinforcements. Similarly, Rossi and Spinella (2022) has proposed a theoretical model based on the MCFT. Hence, the results obtained by means of this latter model are used as the term of comparison for the other two models. Further, the modifications introduced in the CCCM to consider the effects of corrosion are adopted in the other two models.

The selected set of beams, extracted from previous databases (Cladera et al., 2021; Lu et al., 2018), has been tested in laboratory by other researchers and includes both slender and non-slender beams failing in a shear mode. Prior to any numerical analysis, the data of the above previous databases have been revised and, where necessary, amended. While the PM and the CCCM are suggested for both slender and non-slender beams, the SMCFT is recommended for only slender beams (Bentz et al., 2006). This limitation will be considered in the analysis of the numerical results.

## 2 Corrosion-Damage Structural Effects

Corrosion of steel bars is distinguished between localized and uniform. The former is a cross-section loss concentrated in a small part of the bar (it is labelled as pitting). The latter is, instead, almost uniformly spread along the bar. Generally, pitting is the most dangerous, because alters the steel bar's load-deformation behavior by several mechanisms: (1) higher stresses in the corroded cross-section to resist tensile forces (2) varying strain rate due to the different stiffnesses of the corroded and non-corroded parts. Therefore, corrosion of steel bars involves some complex damage-mechanisms that must be taken into account in a simple way.

In this work, the effects on both steel and concrete are considered by means of an effective width of the beam and by means of reduced values of both cross-sectional area and strain capacity of steel bars.

### 2.1 Cross-section loss of steel bar due to corrosion

The reduction of the cross-sectional area of corroded steel bars is often evaluated by means of the cross-sectional area loss ratio, $\eta_a$, which can be expressed as function of the cross-sectional area, $A_v = \pi \varphi_v{}^2/4$, or, directly, of the rebar diameter, $\varphi_v$:

$$\eta_a = 100\frac{A_{v0} - A_v}{A_{v0}} = 100\frac{\phi_{v0}^2 - \phi_v^2}{\phi_{v0}^2} \tag{1}$$

where $A_{v0}$ and $\varphi_{v0}$ are the cross-sectional area and diameter of a steel bar before corrosion, respectively; $A_v$ and $\varphi_v$ are the cross-sectional area and diameter in a cross-section of a steel bar after corrosion. The cross-sectional area loss ratio, $\eta_a$ is a localized measure of corrosion, thus its maximum value ($\eta_{am}$) along the length of a steel bar is an estimate of the maximum corrosion level. The maximum corrosion level can significantly influence the axial force capacity of the entire rebar. For this reason, $\eta_{am}$ is used in this work to calculate the effective cross-sectional area of the steel bars.

When only the weight loss ratio, $\eta_w$, is available by test report it is converted into cross-section loss ratios, $\eta_{am}$, as recently proposed by Cladera et al. (2021):

$$\eta_{am} = 1.36\eta_w \tag{2}$$

### 2.2 Effective beam width after spalling of concrete

Concerning the effects of steel corrosion on spalling of the concrete cover, Higgins et al. (2003) have observed that these effects mainly depend on stirrup spacing ($s_v$) and concrete cover ($c_v$).

Higgins et al. (2003) assumed that the angle of discrete spalls is 20° and proposed two relations, mainly based on empirical data, to estimate the effective width of the beam ($b_{w,eff}$) for the calculation of the shear strength of the beam. These equations have been recently adjusted by Cladera et al. (2021), as follows:

$$b_{w,eff} = b_w - 2\left(c_{v+\phi_v}\right) + s_v/5.5 \tag{3}$$

$$b_{w,eff} = b_w - (5.5/s_v)\left(c_{v+\phi_v}\right)^2 \tag{4}$$

where the Eq. (4) is valid for $s_v \leq 5.5\,(c_v + \varphi_v)$ and the Eq. (5) in the remaining cases.

As suggested by Higgins et al. (2003), the beam width ($b_w$) should be reduced only if the average cross-sectional area loss ratio of stirrups is equal or higher than 10%.

## 3 Description of the Considered Models

In this Section, the three considered models – PM, SMCFT and CCCM - are described. The corrosion effects are introduced into these models as described in the previous section.

### 3.1 Model 1 (PM)

This model considers both beam and arch actions, and requires the application of the static theorem of limit analysis (Rossi, 2013).

The mathematical programming problem resulting from the application of the static theorem is nonlinear and defined by means of equilibrium equations and constraint conditions for geometric and mechanical parameters. The member is characterised by a rectangular cross-section and endowed with longitudinal and transverse reinforcements. The longitudinal steel bars are distinguished into flange ($A_{slf1}$ and $A_{slf2}$) and web ($A_{slw}$) bars (see Fig. 1a). The flange bars are those on the opposite sides of the cross-section -

either in tension or in compression owing to the bending moment - and close to the lateral surface of the member. These bars are considered to be concentrated in the centroid of their cross-section and located on the two opposite sides of the cross-section. The cross-section of the member is divided into three parts, named F1, F2 and F3, as shown in Fig. 1b. In each of these parts, the response of concrete and steel is defined by means of simplified stress fields. The stress-strain constitutive behaviour of concrete and steel is considered to be perfectly plastic. However, while longitudinal and transverse steel bars are assumed to resist both compression and tension, concrete is assumed to resist compression only. The geometry of the zones $F_1$, $F_2$ and $F_3$ of the cross-section is identified by the separation lines of the central part $F_3$ at abscissa $y = y_1$ and $y = y_2$, respectively. Arch mechanisms are also considered in the absence of any external axial load, i.e. in beams. However, this type of arch action can occur only if members are endowed with longitudinal tension reinforcement able to resist the transverse component of the diagonal compression force. The shear force is the objective function of the programming problem and is maximized to obtain the maximum value of this internal force. Because of the non-linearity of the relationships, this problem must be solved more than once with reference to different starting values of the variables of the problem. Of the different resulting shear forces, the highest is assumed as the shear strength of the member. An optimization algorithm is required to solve the mathematical programming problem.

In the past, the PM was applied to evaluate the shear strength of columns with rectangular cross-section and with both longitudinal and transverse reinforcements (Rossi, 2013).

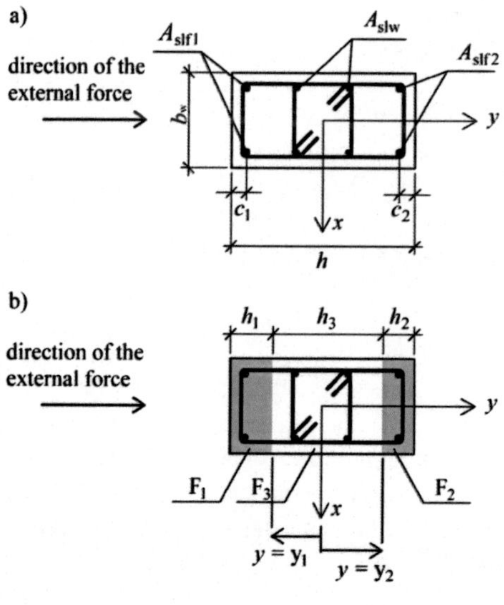


**Fig. 1.** Plastic Model: identification of (a) longitudinal flange and web bars and (b) parts F1, F2 and F3 of the cross-section.

### 3.2 Model 2 (SMCFT)

The Modified Compression Field Theory (MCFT) is based on a smeared rotating crack concept, and treats cracked RC as a new orthotropic material with unique constitutive relationships. Iterative procedures are needed to satisfy three groups of different equations (equilibrium, compatibility and constitutive equations).

Bentz et al. (2006) proposed a simplification of the MCFT (labelled as SMCFT) capable of predicting the shear strength of a wide range of RC members, with almost the same accuracy as the full theory.

In the SMCFT, the shear stress capacity ($\tau_R$) of a RC beam is calculated as the sum of concrete ($\tau_c$) and stirrup ($\tau_v$) contributions. The shear strength $V_R$ is given by the following relations:

$$V_R = (\tau_c + \tau_v)b_w d = \left(\beta\sqrt{f_{cm}} + \rho_v f_{yv} \cot\theta\right)b_w d \tag{5}$$

The inclination of the concrete strut in the web with respect to the longitudinal axis of the beam ($\theta$) and the factor $\beta$, which takes into account both tension stiffening and aggregate interlock effects, depend on the longitudinal strain at the centroid level of the cross-section ($\varepsilon_x$) by means of the following relations:

$$\beta = \frac{0.4}{1 + 1500\varepsilon_x}\frac{1300}{1000 + S_{xe}} \tag{6}$$

$$\theta = \left(29^\circ + 7000\varepsilon_x\right)\left(0.88 + S_{xe}/2500\right) \tag{7}$$

where $S_{xe} = 35\ S_x/(16 + d_g)$. $S_x$ is the vertical distance between the longitudinal steel bars and $d_g$ is the aggregate size, which is taken as zero if $f_{cm} > 70$ MPa.

To calculate the shear strength of the member, the value of $\varepsilon_x$ corresponding to the maximum capacity of the member is first estimated (e.g., one might assume $\varepsilon_x$ equal to $1 \times 10^{-3}$). Then, Eqs. (6) and (7) are used to calculate $\beta$ and $\theta$. If the longitudinal reinforcement is elastic, the new value of the longitudinal strain corresponding to the maximum shear stress ($\tau_R = \tau_c + \tau_v$) is estimated as $\varepsilon_x = (\tau_R \cot\theta - \tau_c/\cot\theta)/(\rho_l\ E_s)$, where $\rho_l$ is the geometric ratio of the longitudinal steel bars and $E_s$ is the elastic modulus of steel. Calculations have to be repeated until convergence on $\varepsilon_x$ has been reached. If the longitudinal reinforcement has yielded, the longitudinal strain $\varepsilon_x$ has to be adjusted until convergence.

### 3.3 Model 3 (CCCM)

The CCCM (Cladera et al., 2016) is a shear design-oriented mechanical model derived by a more complex analytical model referred to as the Multi-Action Shear Model (MASM) (Marí et al., 2014). Subsequently, Cladera et al. (2021) proposed to simplify the MASM into the CCCM, assuming that the contributions given by residual tensile stresses and dowel action are negligible.

The shear strength of a RC beam is estimated as the sum of concrete and stirrups contributions:

$$V_c = 0.25(\xi K_c + 20d_0)f_{cm}^{2/3}b_w d \tag{8}$$

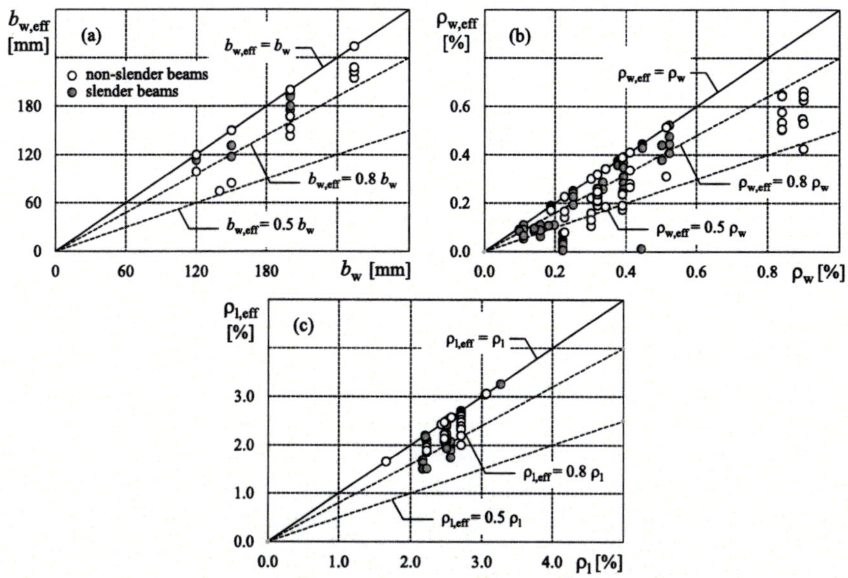


**Fig. 2.** Effects of corrosion on geometric properties of steel and concrete cross-sections: effective to original ratios for (a) width of the beam cross-section, (b) geometric ratio of the transverse reinforcement and (c) geometric ratio of the total longitudinal reinforcement.

$$V_v = 1.4(A_{sv}/s_v)f_{yv}(d - x)\cot\theta \tag{9}$$

In addition, the shear strength cannot be higher than the shear strength produced by strut crushing.

The shear strength of non-slender beams is governed by arch action. The CCCM take into account that the longitudinal strain gradient is not linear and that the depth of the neutral axis increases with the decrease of *a/d*.

## 4 Database of Laboratory Tests

The tests of the database herein considered are those considered by Cladera et al. (2021), based on two previously published databases (Lu et al., 2018; Soltani et al., 2020). The database collects 136 corrosion-damaged beams failing in shear, including specimens with shear span ratio *a/d* in the range from 1.0 to 4.7, namely 62 slender beams ($2.5 \leq a/d \leq 4.7$), and 74 non-slender beams ($1.0 \leq a/d \leq 2.2$). All the beams have rectangular cross-section and are over-reinforced in bending ($1.65\% \leq \rho_l \leq 3.02\%$). By contrast, the geometric ratio of (un-corroded) stirrups is fairly low ($0.10\% \leq \rho_v \leq 0.90\%$). The compression strength of concrete is always evaluated in terms of cylindrical compression strength. Therefore, if the only cubic compression strength ($R_{cm}$) is available in the original paper, it is transformed into a cylindrical compression strength by $f_{cm} = 0.83\ R_{cm}$. The $f_{cm}$ is in the range from 20.75 to 44.4 MPa. Further, the yielding strength of stirrups, $f_{yv}$, is in the range 275 ÷ 626 MPa.

The level of corrosion is reported in the original papers in terms of weight loss ratio, cross-sectional area loss ratio or corrosion depth penetration. In the present work, owing to the recognized great importance of the cross-sectional area loss ratio (Higgins & Farrow, 2006; Rodriguez et al., 1997; Xue et al., 2013), the level of corrosion is herein reported in terms of the cross-sectional area loss ratio by means Eq. (2).

### 4.1 Corrosion effects on the geometric properties of reinforcements and specimens' size

The effects of corrosion on the geometric properties of reinforcements and size of the specimens (as assumed in the numerical analyses) are plotted in Fig. 2, where the effective values of the width of the beam cross-section ($b_{w,eff}$) and the effective values of the transverse and longitudinal reinforcements are plotted as a function of the values before corrosion. The effective width of the beam cross-section is calculated by means of Eqs. (3) and (4). If the only cross-sectional area loss ratio of stirrups is reported in the original paper, the inverse of Eq. (2) is used to obtain the average section loss ratio.

The effective width of the beam cross-section may be as low as about half of $b_w$ in some specimens tested by Imam and Azad (2016), even though it is often in the range from 0.8–1 of $b_w$. The effective geometric ratio of the transverse reinforcement $\rho_{w,eff}$ is calculated as $\rho_v \times (1-\eta_w)$, $\rho_v$ being the original (un-corroded) value. The values of $\rho_{v,eff}$ vary from the original ratios to almost zero [e.g. in some specimens tested by Rodriguez et al. (1997)]. In most cases, the effective geometric ratio of stirrups $\rho_{v,eff}$ is lower than 80% of $\rho_v$, hence the reduction of the cross-sectional area of the transverse reinforcement is significant. Similarly, the effective geometric percentage of the longitudinal reinforcement $\rho_{l,eff}$ is calculated as $\rho_l \times (1-\eta_l)$. The values of $\rho_{l,eff}$ are in the range 100% ÷ 55% of the original value $\rho_l$.

## 5 Numerical Analysis

The numerical analysis is carried out considering both the reduction of the cross-sectional area of the reinforcements (longitudinal and/or transverse), and the effective width of the beam cross-section due to corrosion of stirrups.

In Fig. 3, the experimental shear strength ($V_{exp}$) of the 62 corroded slender specimens is plotted versus the shear strength resulting from the numerical models ($V_{num}$). The mean values of the ratio $R_V = V_{exp}/V_{num}$ are equal to 0.92, 1.00 and 0.93 in the case of the PM, CCCM and SMCFT, respectively. The Coefficients of Variation (CoV) of the ratios $R_V$ range from 0.13 to 0.26, showing appreciable differences. The difference between the laboratory and numerical shear strength values is also apparent from Fig. 3. Further, it is interesting to note that the plot referring to the results of the CCCM strongly resembles that referring to the SMCFT. The plot obtained from the results of the PM appears fairly different from the others and leading, in general, to more accurate predictions of the laboratory shear strength.

The shear strength resulting from the laboratory tests of the 74 corroded non-slender specimens is plotted in Fig. 4 versus the shear strength resulting from the PM and CCCM. The SMCFT is not suggested for non-slender beams: thus, results are not provided.

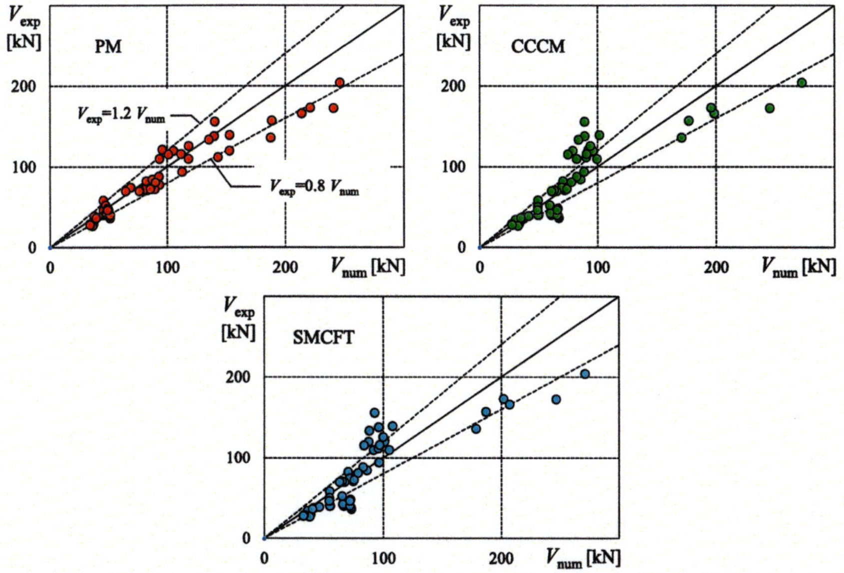


**Fig. 3.** Experimental versus numerical shear strength of slender beams for the different models: PM, SMCFT and CCCM.

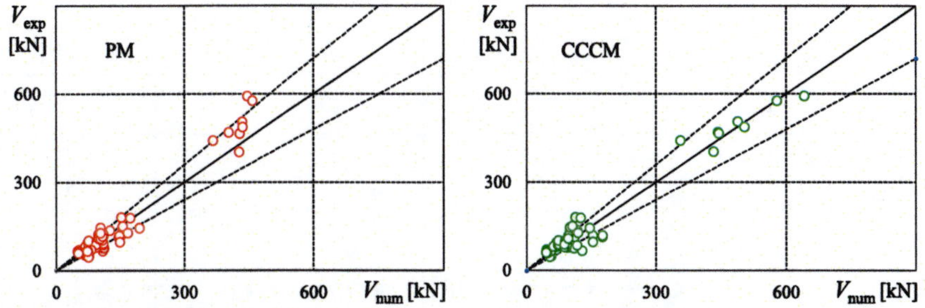


**Fig. 4.** E Experimental versus numerical shear strength of non-slender beams for the different models: PM and CCCM.

PM and CCCM lead to conservative predictions of the shear strength of the non-slender beams. The mean values of $R_V$ are equal to 1.04 and 1.12 in the case of the PM and CCCM, respectively. The CoVs of $R_V$ are quite similar and equal to 0.20 to 0.22 in the case of the PM and CCCM, respectively.

To validate the considered models further, the ratios $R_V$ obtained by the PM, CCCM and SMCFT are plotted versus the shear span ratio (Fig. 5), and the (un-corroded) transverse reinforcement ratio (Fig. 6).

The Fig. 5 highlights that, in the case of the PM, the ratio $R_V$ slightly increases with the decrease of *a*/*d*. The authors believe that the reason of the underestimate of the shear strength of the deep beams (*a*/*d* ~ 1) may be due to the presence of plates under the point load and at support. These plates reduce the length of the concrete strut. These effects are not considered in the present numerical analyses, and can be responsible for some underestimate of the shear strength.

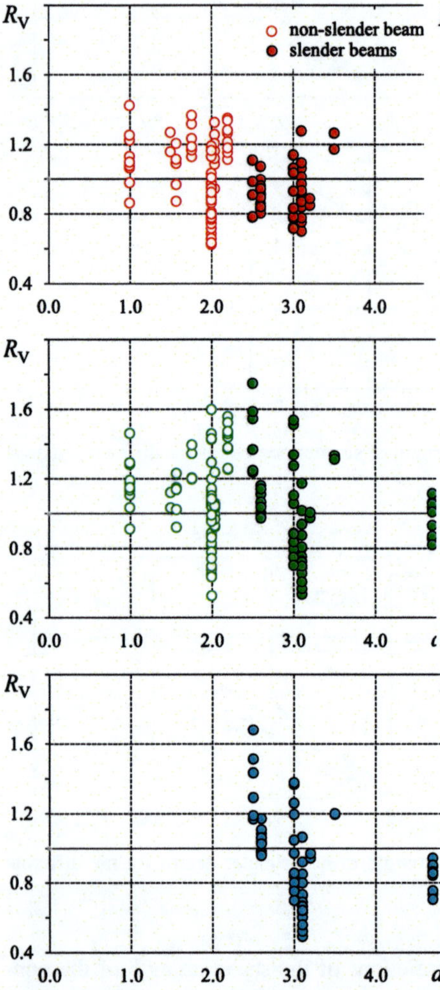


**Fig. 5.** Shear strength ratio RV versus shear span ratio.

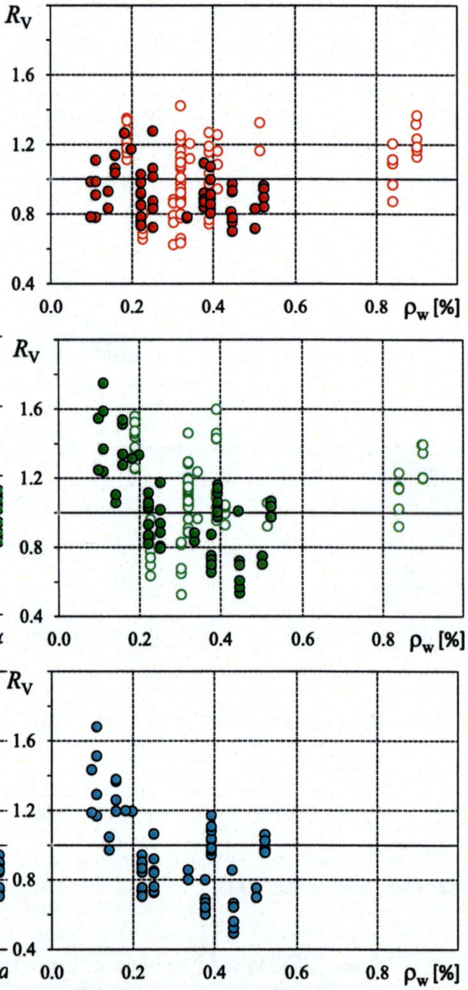


**Fig. 6.** Shear strength ratio RV versus geometric ratio of the transverse reinforcement.

In the case of the PM model, no particular correlation may be observed between the shear strength ratio $R_V$ and the geometric ratio of the transverse reinforcement $\rho_v$ (see Fig. 6). In the case of the CCCM and SMCFT, instead, the values of the ratio $R_V$ appear to decrease with the increase of $\rho_v$, particularly in slender beams.

## 6 Conclusions

This work investigates the accuracy of different models in the prediction of the shear strength of corroded reinforced concrete beams.

The considered models are the compression chord capacity model, a plastic model previously proposed by one of the authors, and the simplified modified compression field theory. Only the compression chord capacity model has recently been embedded with modifications that simulate those effects of steel corrosion and applied to beams with corroded reinforcements. The modifications introduced in the compression chord capacity model to consider the effects of corrosion are first embedded into the other two models. Then, all the models are used to calculate the shear strength of a set of reinforced concrete beams with corroded reinforcements.

The main conclusions of the numerical analyses are:

- the reduction of the steel reinforcement cross-section and concrete beam width appears to be appropriate to lead to sufficiently accurate predictions of the shear strength by means of the considered models.
- the compression chord capacity model leads to sufficiently accurate predictions of the shear strength of both slender and non-slender reinforced concrete beams.
- the model based on the plasticity theory, proposed by one of the authors in a previous paper for un-corroded reinforced concrete elements, leads to accurate predictions of the shear strength of both slender and non-slender reinforced concrete beams. The CoV of the shear strength ratios obtained for slender beams is lower than that obtained by means of the other examined models.
- the simplified modified compression field theory leads to satisfactory predictions of the shear strength of slender RC beams. The CoV of the shear strength ratios is only slightly lower than the value obtained by means of the compression chord capacity model for slender beams.
- even in the case of slender beams, the reduction of the cross-sectional area of the transverse reinforcement because of corrosion leads to a significant contribution of concrete of the compression chord to the shear strength of the member.
- the distinction between slender and non-slender members, as members with negligible or significant contribution of concrete of the compression chord to the shear strength of the member (beam or arch action), should be made in view of the shear span and effective transverse reinforcement ratios, and not on the basis of the only shear span ratio.

**Acknowledgements.** This research study was supported by the University of Catania within the research project "DU.SO.CA.P. – Durabilità e SOstenibilità delle strutture in Cemento Armato e Precompresso", the research program "PIAno di inCEntivi per la RIcerca di Ateneo 2020/2022 (PIA.CE.RI.) – Starting Grant – Linea di intervento 3", the research project "Definizione e validazione di procedure di progetto di interventi di adeguamento sismico di edifici in c.a mediante pareti oscillanti" and the research project "PIAno di inCEntivi per la RIcerca di Ateneo 2020/2022 (PIA.CE.RI.) - Linea di intervento 2". Special thanks are due to the Italian the Superior Council of Public Works (CC.SS.LL.PP.), and the Network of University Laboratories of Seismic Engineering (RELUIS). The results were achieved in the national technical agreement for implementing the agreement pursuant to art. 15 law 7 August 1990, No. 241 between the Superior Council of Public Works and RELUIS.

## References


Bentz, E.C., Vecchio, F.J., Collins, M.P.: Simplified modified compression field theory for calculating shear strength of reinforced concrete elements. ACI Struct. J. **103**(4), 614–624 (2006). https://doi.org/10.14359/16438

Cheng, H., Li, H.N., Biondini, F., Wang, D.S., Zou, Y.: Strain penetration effect on cyclic response of corroded RC columns. Eng. Struct. **243**, 112653 (2021). https://doi.org/10.1016/J.ENGSTRUCT.2021.112653

Cladera, A., Marí, A., Bairán, J.M., Ribas, C., Oller, E., Duarte, N.: The compression chord capacity model for the shear design and assessment of reinforced and prestressed concrete beams. Struct. Concr. **17**(6), 1017–1032 (2016). https://doi.org/10.1002/SUCO.201500214

Cladera, A., Marí, A., Ribas, C.: Mechanical model for the shear strength prediction of corrosion-damaged reinforced concrete slender and non slender beams. Eng. Struct. **247**, 113163 (2021). https://doi.org/10.1016/J.ENGSTRUCT.2021.113163

Dai, K.Y., Liu, C., Lu, D.G., Yu, X.H.: Experimental investigation on seismic behavior of corroded RC columns under artificial climate environment and electrochemical chloride extraction: a comparative study. Constr. Build. Mater. **242**, 118014 (2020). https://doi.org/10.1016/J.CONBUILDMAT.2020.118014

Higgins, C., et al.: Shear capacity assessment of corrosion-damaged reinforced concrete beams (2003)

Imam, A., Azad, A.K.: Prediction of residual shear strength of corroded reinforced concrete beams. Int. J. Adv. Struct. Eng. **8**(3), 307–318 (2016). https://doi.org/10.1007/S40091-016-0133-X

Lu, Z.H., Li, H., Li, W., Zhao, Y.G., Dong, W.: An empirical model for the shear strength of corroded reinforced concrete beam. Constr. Build. Mater. **188**, 1234–1248 (2018). https://doi.org/10.1016/J.CONBUILDMAT.2018.08.123

Marí, A., Bairán, J., Cladera, A., Oller, E., Ribas, C.: Shear-flexural strength mechanical model for the design and assessment of reinforced concrete beams. Struct. Infrastruct. Eng. **11**(11), 1399–1419 (2014). https://doi.org/10.1080/15732479.2014.964735

Rajput, A.S., Sharma, U.K.: Corroded reinforced concrete columns under simulated seismic loading. Eng. Struct. **171**, 453–463 (2018). https://doi.org/10.1016/J.ENGSTRUCT.2018.05.097

Recupero, A., D'Aveni, A., Ghersi, A.: N-M-V interaction domains for box and I-shaped reinforced concrete members. Struct. J. **100**(1), 113–119 (2003). https://doi.org/10.14359/12445

Recupero, A., D'Aveni, A., Ghersi, A.: Bending moment-shear force interaction domains for prestressed concrete beams. J. Struct. Eng. **131**(9), 1413–1421 (2005). https://doi.org/10.1061/(ASCE)0733-9445(2005)131:9(1413)

Recupero, A., Spinella, N.: Preliminary results of flexural tests on corroded prestressed concrete beams. In: Proceedings of the Fib Symposium 2019: Concrete - Innovations in Materials, Design and Structures, pp. 1323–1330 (2019a). https://www.scopus.com/inward/record.uri?eid=2-s2.0-85066067217&partnerID=40&md5=a66559dd71b9d55256f8e8320af77a10

Recupero, A., Spinella, N.: Experimental tests on corroded prestressed concrete beams subjected to transverse load. Structural Concrete, suco.201900242 (2019b). https://doi.org/10.1002/suco.201900242

Recupero, A., Spinella, N., Tondolo, F.: Failure analysis of corroded RC beams subjected to shear-flexural actions. Eng. Fail. Anal. **93**,(2018). https://doi.org/10.1016/j.engfailanal.2018.06.025

Rodriguez, J., Ortega, L., Casal, J.: Load carrying capacity of concrete structures with corroded reinforcement. Constr. Build. Mater. **11**(4), 239–248 (1997). https://doi.org/10.1016/S0950-0618(97)00043-3

Rossi, P.P.: Evaluation of the ultimate strength of R.C. rectangular columns subjected to axial force, bending moment and shear force. Eng. Struct. **57**, 339–355 (2013). https://doi.org/10.1016/J.ENGSTRUCT.2013.09.006

Rossi, P.P., Spinella, N.: Corroded stirrups effects on the shear behavior of reinforced concrete slender beams. In: 6th International Conference on Civil Engineering, ICOCE 2022, Singapore, 30–37 (2022). https://doi.org/10.1007/978-981-19-3983-9_3

Soltani, M., Abu-Abaileh, A., Rowe, B.S.: Statistical approach to modeling reduced shear capacity of corrosion-damaged reinforced concrete beams. Pract. Period. Struct. Des. Constr. **26**(2), 04020073 (2020). https://doi.org/10.1061/(ASCE)SC.1943-5576.0000564

Zhang, W., Ye, Z., Gu, X.: Effects of stirrup corrosion on shear behaviour of reinforced concrete beams. Struct. Infrastruct. Eng. **13**(8), 1081–1092 (2017). https://doi.org/10.1080/15732479.2016.1243563

Zhu, W., François, R., Coronelli, D., Cleland, D.: Effect of corrosion of reinforcement on the mechanical behaviour of highly corroded RC beams. Eng. Struct. **56**, 544–554 (2013). https://doi.org/10.1016/J.ENGSTRUCT.2013.04.017

# Shear Response Behavior of Slender RC Beams with Corroded Stirrups

Antonino Recupero[1], Pier Paolo Rossi[2], and Nino Spinella[2(✉)]

[1] Department of Engineering, Università di Messina, Messina, Italy
[2] Department of Civil Engineering and Architecture, Università di Catania, Catania, Italy
nino.spinella@unict.it

**Abstract.** The corrosion of steel reinforcement in reinforced concrete beams is cause of several structural-damage effects. When corrosion distresses stirrups, the shear response is drastically affected. The corrosion of stirrups produces rust all around causing reduction of cross-section steel rebar and spalling of concrete cover. This study presents a theoretical model to reproduce the shear response of reinforced concrete slender beams with corroded stirrups. The formulation extends an established procedure, based on the Modified Compression Field Theory, and, introducing the structural effects of stirrups corrosion, provides the entire load-displacement curve. Information about crack condition and compression field inclination at each stage of loading are also obtained. Firstly, the proposed model is assessed against a database of sixty-two specimens collected in the literature; then, it is used for a parametric analysis to highlight the role of the parameters involved.

**Keywords:** Shear strength · RC beams · corrosion · stirrups · MCFT

## 1 Introduction

Reinforced concrete (RC) structures represent a significant portion of the Italian building heritage. During the first half of the last century, the long-term durability of RC structures was not considered during their design. However, many RC buildings have been found to exhibit corrosion of the steel reinforcing bars. Corrosion of steel reinforcement is one of the main causes of deterioration of existing RC structures, and it can result in the reduction of bar cross sections, degradation of the bond between reinforcing bars and surrounding concrete, and cracking of concrete (Campione et al. 2017; Fernandez et al. 2016; Recupero & Spinella 2019a, 2019b). A large number of experimental tests have been conducted to investigate the influence of corrosion on the mechanical characteristics of reinforcing steel bars. A considerable amount of these researches has been directed to investigate the corrosion effects on the flexural behaviour of RC beams, and analytical studies have also been conducted to model the corrosion effects on the flexural and bond behaviour of RC beams with longitudinal corroded rebars (Rodriguez et al. 1997; Tondolo 2015). By contrast, a few studies have been focused on the shear strength of RC beams with corroded transverse reinforcement (Cladera et al. 2021; Lu et al. 2018;


M. A. Aiello and A. Bilotta (Eds.): ICC 2022, LNCE 435, pp. 534–546, 2024.
https://doi.org/10.1007/978-3-031-43102-9_41

Recupero et al. 2018). Stirrups play a key role for the shear strength of RC beams. They restrict the inclined crack width and improve the dowel action. Further, stirrups are located in the external thickness of the beam, then they are the most vulnerable reinforcement to the environmental conditions and, consequently, corrosion. In addition, diameter of stirrups is usually smaller than one of longitudinal steel bars. Thus, the loss of the cross-sectional area in the formers is expected to be more significant than in the latters.

It follows that stirrups corrosion is a critical issue. It affects the shear strength and the deflection capacity of beams. However, many of the models presented in the literature are able to predict only the load-bearing capacity at the Ultimate Limit State (ULS) of RC beams with corroded reinforcement, and, moreover, the equations are empirical in nature.

Recently, Cladera et al. (2021) adapted their own Compression Chord Capacity Model (CCCM) to predict the shear strength of corrosion-damaged RC beams. The predictions of 62 slender and 84 deep beams failing in shear, where the steel reinforcement was subjected to corrosion, led to achieve accurate results.

This paper aims to propose a theoretical model to predict the load-deformation curve of corroded RC slender beams by extending a previously established procedure, proposed by one of the authors, based on the Modified Compression Field Theory (MCFT) (Vecchio & Collins 1986).

A crack element containing longitudinal and transverse smeared steel bars is considered. In addition, both reduced rebar cross-section and effective beam width, due to corrosion, are taken into account. The proposed model allows the calculation of strain and stress fields by respecting all the equilibrium and compatibility equations. A corroboration of the proposed model, against the same experimental tests reported in Cladera et al. (2021), is presented.

Finally, a simple relation for the prediction of the shear strength of slender RC beams with low values of the effective geometric percentage of stirrups has been obtained.

## 2 Corrosion-Damage Structural Effects

Corrosion of steel bars is distinguished between localized and uniform corrosion. The former is a cross-section loss concentrated in a small part of the bar (it is also labelled as pitting). The latter, instead, is almost uniformly spread along the bar. Generally, pitting is the most dangerous, because alters the steel bar's load-deformation behavior by means of several mechanisms: (1) the higher stresses in the corroded cross-section to resist equal tensile forces as the non-corroded part, (2) a varying strain rate due to the different stiffnesses of the corroded and non-corroded parts. Therefore, the corrosion of steel bars involves some complex damage-mechanisms that must be taken into account in a simple way.

In the proposed model, the effects on both steel and concrete by means of an effective width of the beam and by means of reduced values of both the cross-sectional area and strain capacity of steel bars as described in the following sections.

### 2.1 Cross-section Loss of Steel Bar Due to Corrosion

The reduction of the cross-sectional area of corroded steel bars is often evaluated by means of the cross-sectional area loss ratio, $\eta_a$:

$$\eta_a = 100\frac{A_{v0} - A_v}{A_{v0}} = 100\frac{\phi_{v0}^2 - \phi_v^2}{\phi_{v0}^2} \tag{1}$$

where $A_{v0}$ and $\phi_{v0}$ are the cross-sectional area and diameter of a steel bar before corrosion, respectively; $A_v$ and $\phi_v$ are the cross-sectional area and diameter in a cross-section of a steel bar after corrosion. The cross-sectional area loss ratio $\eta_a$ is a localized measure of corrosion, then its maximum value ($\eta_{am}$) along the length of a steel bar is an estimate of the maximum corrosion level. The maximum corrosion level can significantly influence the axial force capacity of the entire rebar. For this reason, $\eta_{am}$ is used in this work to calculate the effective cross-sectional area of the steel bars.

When only the weight loss ratio is available by test report, $\eta_w = 100 \times (m_0 - m)/m_0$ ($m_0$ = mass of a length of the steel rebar before corrosion; and $m$ = mass after corrosion), it is converted into the cross-section loss ratio, $\eta_{am}$, as recently proposed by Cladera et al. (2021):

$$\eta_{am} = 1.36\eta_w \tag{2}$$

### 2.2 The Effective Beam Width After the Concrete Spalling

Concerning the effects of steel corrosion on spalling of concrete in the cover, Higgins et al. (2003) have observed that these effects mainly depend on stirrup spacing ($s_v$) and concrete cover ($c_v$).

Higgins et al. (Higgins et al. 2003) assumed that the angle of discrete spalls is 20°, then they have proposed two relations, mainly based on empirical data, to estimate the effective width of the beam ($b_{w,eff}$) for the calculation of the shear strength of the beam. These equations have been recently adjusted by Cladera et al. (2021), as follows:

$$b_{w,eff} = b_w - 2\left(c_{v+\phi_v}\right) + s_v/5.5 \tag{3}$$

$$b_{w,eff} = b_w - (5.5/s_v)\left(c_{v+\phi_v}\right)^2 \tag{4}$$

where the Eq. (4) is valid for $s_v \leq 5.5\,(c_v + \phi_v)$ and the Eq. (5) in the remaining cases.

As suggested by Higgins et al. (2003), the beam width ($b_w$) should be reduced only if the average cross-sectional area loss ratio of stirrups is equal or higher than 10%.

## 3 Formulation of the Proposed Model

The proposed model provides the shear-deflection curve, until peak load, of slender RC beams with corroded steel bars. It is based on the MCFT equations, and the flexural problem is preliminary solved with respect to the shear problem (Spinella 2019).

### 3.1 Flexural Model

To calculate the bending moment at a generic step, a top axial strain ($\varepsilon_{ct}$) value is assumed, then the longitudinal strain of the bottom reinforcement ($\varepsilon_{lb}$) is adjusted until the axial load equilibrium is achieved. This analysis provides the bending moment (M), the curvature ($\chi$), and the average axial strain from flexure ($\varepsilon_{xf}$) at the centroid level of the cross-section (Spinella 2013).

### 3.2 Shear Model

The MCFT, at the basis of the proposed model, assumes the following equilibrium conditions:

$$\sigma_{cv} = \sigma_{c1} - \tau/\cot\theta \tag{5}$$

$$\sigma_{c2} = \sigma_{c1} - \tau(1/\cot\theta + \cot\theta) \tag{6}$$

where $\sigma_{cv}$ = concrete stress along the vertical direction; $\sigma_{c1}$ and $\sigma_{c2}$ = concrete principal tensile and compressive stresses, respectively; $\tau$ = shear stress; and $\theta$ = strut angle (see Fig. 1). Clamping stress ($\sigma_v$) can be obtained as $\sigma_v = \sigma_{cv} + \rho_{sv}\,\sigma_{sv}$; where $\rho_{sv} = A_{sv}/(b_w\,s_v)$ is the geometric ratio of the transverse reinforcement; $A_{sv}$ is the total area of stirrups; and $\sigma_{sv}$ is the stirrup stress.

In slender beams, the clamping stress along a cross-section far enough from the point load can be neglected, and then Eq. (6) allows calculation of the shear stress as follows:

$$\tau = (\sigma_{c1} + \rho_{sv}\sigma_{sv})\cot\theta = (\sigma_{c1} - \sigma_{c2})\cot\theta/\left(1 + \cot^2\theta\right) \tag{7}$$

$$\cot^2\theta = (-\sigma_{c2} - \rho_{sv}\sigma_{sv})/(\sigma_{c1} + \rho_{sv}\sigma_{sv}) \tag{8}$$

Using compatibility equations, a dual form of the cotangent of the inclined strut can be obtained:

$$\varepsilon_x + \varepsilon_v = \varepsilon_1 + \varepsilon_2 \tag{9}$$

$$\cot^2\theta = (\varepsilon_v - \varepsilon_2)/(\varepsilon_x - \varepsilon_2) \tag{10}$$

where $\varepsilon_1$ and $\varepsilon_2$ are the principal tensile and compressive stresses, respectively and $\varepsilon_x$ and $\varepsilon_v$ are the horizontal and vertical strains, respectively.

Equations (8) and (9) return the same result. Therefore, equating the two expressions of cot $\theta$ gives the vertical strain, $\varepsilon_v$, that can be calculated independently of cot $\theta$ and depends of the steel stirrups behaviour (elastic or plastic).

If steel stirrups are elastic, the solution of $\varepsilon_v$ is:

$$\varepsilon_v = \left(-B_e + \sqrt{B_e^2 - 4A_eC_e}\right)/(2A_e) \tag{11}$$

and the coefficients $A_e$, $B_e$ and $C_e$ are:

$$A_e = \rho_{sv} E_s \tag{12}$$

$$B_e = \sigma_{c1} + A_e(\varepsilon_x - 2\varepsilon_2) \tag{13}$$

$$C_e = \sigma_{c2}(\varepsilon_x - \varepsilon_2) - \sigma_{c1}\varepsilon_2 \tag{14}$$

Alternatively, if steel stirrups are yielding, the solution of $\varepsilon_v$ is:

$$\varepsilon_v = -C_y/B_y \tag{15}$$

where $B_y$ and $C_y$ are:

$$B_y = \sigma_{c1} + \rho_{sv} f_{yv} \tag{16}$$

$$C_y = \sigma_{c2}(\varepsilon_x - \varepsilon_2) - \sigma_{c1}\varepsilon_2 + \rho_{sv} f_{yv}(\varepsilon_x - 2\varepsilon_2) \tag{17}$$

### 3.2.1 Constitutive Behaviour of Materials

The cracked concrete is subjected to a plane stress state. Therefore, the uniaxial constitutive laws of concrete in compression and tension are appropriately modified in the MCFT to take into account both the detrimental effect of the tensile strain ($\varepsilon_1$) field orthogonal to the principal compression stress ($\sigma_{c2}$), and the tension stiffening effect due to the interaction between concrete and rebar.

To reduce the compression strength of cracked concrete, the following compression softening coefficient is used:

$$\beta = f_{ce}/f_c = 1/[0.8 + 0.34(\varepsilon_1/\varepsilon_{c0})] \leq 1 \tag{18}$$

where $f_{ce}$ is the effective compressive strength of concrete, and $\varepsilon_{c0}$ is the strain at peak stress.

To calculate the compression stress, the Hognestad's parabola is used:

$$\sigma_{c2} = f_{ce}\left[2(\varepsilon_2/\varepsilon_{c0}) - (\varepsilon_2/\varepsilon_{c0})^2\right] \tag{19}$$

Tension stiffening effect is taken into account as suggested by the MCFT:

$$\sigma_{c1} = f_{ct}/\left(1 + \sqrt{c_t\varepsilon_1}\right) \tag{20}$$

where $c_t = 2.2\, m_t$ and $m_t$ is given by the following relation:

$$1/m_t = 4\left[(\rho_v/\phi_v)\left|\sin(90^\circ - \theta)\right| + (\rho_{lb}/\phi_{lb})\left|\sin(0^\circ - \theta)\right|\right] \tag{21}$$

where $\rho_{lb}$ is the geometric percentage of the longitudinal tension reinforcement.

Due to the capacity of cracked concrete to bridge forces across the crack sides, the tensile stress cannot be greater than a limit value, $\sigma_{c1,max}$. It is the minimum of the equilibrium of forces across the crack in the longitudinal and transverse directions, respectively:

$$\sigma_{c1,\max} = \min\left\{ \begin{array}{l} \tau_i \cot\theta + \rho_{sl}(f_{yl} - \sigma_{sl}) \\ \tau_i/\cot\theta + \rho_{sv}(f_{yv} - \sigma_{sv}) \end{array} \right\} \tag{22}$$

$$\begin{aligned} \tau_i &= 0.18\sqrt{f_c}/[0.31 + 24w/(d_g + 16)] \\ &\leq |\rho_{sv}(f_{yv} - \sigma_{sv}) - \rho_{sl}(f_{yl} - \sigma_{sl})| \sin\theta\cos\theta \end{aligned} \tag{23}$$

where $f_{yl}$ and $f_{yv}$ = yielding strength of the longitudinal and vertical rebar, respectively; $\tau_i$ = local shear stress; $d_g$ = maximum coarse aggregate size (in mm); $w = \varepsilon_1 S_{m\theta}$ is the average crack width (in mm); and $f_c$ is in MPa. In Eqs. (21) and (22), the residual reinforcement stresses at the crack are taken as zero when the yield stress is reached.

The average diagonal crack spacing, $S_{m\theta} = 1/(\sin\theta/S_{ml} + \cos\theta/S_{mv})$, depends on the average crack spacing along the two orthogonal directions $S_{ml}$ and $S_{mv}$. They are assumed equal to the stirrups spacing ($s_v$) and the effective depth ($d$) of the cross-section, respectively.

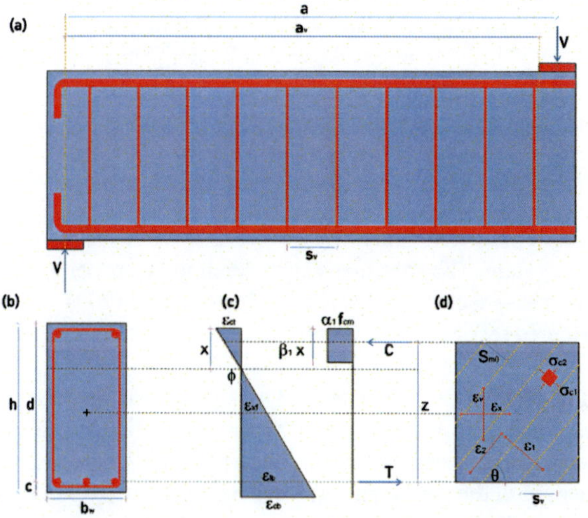


**Fig. 1.** Model of the RC beam: (a) lateral view, (b) cross-section, (c) flexural model and (d) shear model.

## 3.3 Analytical Procedure

### 3.3.1 Flexural Model

Preliminarily, the flexural model is solved for an assigned value of the compression strain at the top of the cross-section, $\varepsilon_{ct}$. Then, the flexural axial strain ($\varepsilon_{xf}$) at the centroid level of the cross-section, the curvature ($\chi$) and the bending moment (M) are calculated. The corresponding shear stress at the critical cross-section is obtained as depending on the static scheme. In a simply supported beam it is $\tau_{flex} = (M/a_v)/(b_w\, d)$, $a_v$ being the shear span length.

### 3.3.2 Shear Model

The shear analysis is performed by the following these steps:

1. Set a value of $\varepsilon_2$.
2. Set a value of $\varepsilon_1$, and then calculate β from Eq. (18) and $\sigma_{c2}$ from Eq. (19).
3. Use the stored values of θ and $\varepsilon_v$ to calculate $w$ and $\sigma_{c1}$.
4. Calculate $\varepsilon_v$ from Eq. (11) or (15).
5. Calculate $\sigma_{sv} = \min\{E_s\, \varepsilon_v; f_{yv}\}$.
6. Calculate the new θ by Eq. (10) and $\sigma_{c1,max}$ from Eqs. (22) and (23). If $\sigma_{c1} > \sigma_{c1,max}$, then set $\sigma_{c1} = \sigma_{c1,max}$ and return to Step 3.
7. Check if the value of $\varepsilon_2$ obtained by Eq. (9) is equal to the assumed value. If this is not the case, return to Step 2 and adjust $\varepsilon_1$.
8. Calculate τ from Eq. (7). If $\tau \neq \tau_{flex}$ return to Step 1 and adjust $\varepsilon_2$.

In the flexural model, the ULS can be achieved because of excessive deformation of either longitudinal tension steel bars or concrete. In the shear model, the ULS of the beam can be achieved because of excessive crack width, stirrups rupture ($\varepsilon_v > \varepsilon_{vu}$) or concrete crushing.

## 4 Corroboration

The proposed model is corroborated by comparison with results of laboratory tests available in the literature. The specimens considered in this paper are the same as those reported in the database recently collected by Cladera et al. (2021) As previously mentioned, the considered set of specimens consists of 62 corrosion-damaged slender beams endowed with rectangular cross-section and failed in shear. The examined beams are over reinforced in bending and characterized by a geometric percentage of the longitudinal tension reinforcement $\rho_{lb} = A_{lb}/b_w\, d$ (where $A_{lb}$ is the cross-sectional area of the longitudinal tension reinforcement) in the range from 1.77 to 3.02%. By contrast, the geometric percentage of the transverse reinforcement $\rho_{sv} = A_v/b_w\, s_v$ is in the range from 0.10 to 0.52%. The cylindrical compression strength, $f_c$, ranges from 20.8 to 44.4 MPa.

The yielding strength of stirrups is in the range from 300 to 626 MPa. Corrosion is present in the transverse reinforcement, and sometimes also in the longitudinal reinforcement. In some specimens tested by Rodriguez et al. (1997) the cross-sectional area loss level of the transverse reinforcement is as high as 97.2%.

### 4.1 Shear Strength Comparison

To assess the effectiveness of the proposed model against other mechanical models, the CCCM, recently adapted by Cladera et al. (2021) to predict the shear strength of corrosion damaged RC beams, is used here as benchmark. The shear strength resulting from the CCCM is calculated as the sum of concrete ($V_{cu}$) and transverse reinforcement ($V_{su}$) contributions and is not higher than the shear strength corresponding to concrete crushing ($V_{c,max}$).

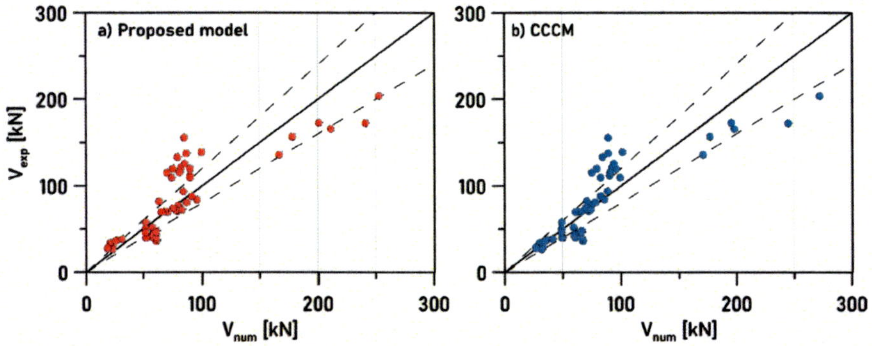


**Fig. 2.** Comparison between experimental and numerical shear strength: (a) proposed model and (b) CCCM.

As shown in Fig. 2, the shear strength of the examined specimens is predicted by the proposed model with satisfying accuracy. Similar results are also obtained from the use of the CCCM.

The ratios between the experimental and the numerical shear strength ($V_{exp}/V_{num}$), resulting from the application of the proposed model, range from 0.58 to 1.68, with a mean value equal to 1.07 and a Coefficient of Variation (CoV) equal to 0.30. Similarly, the results obtained from the CCCM show minimum and maximum values equal to 0.53 and 1.59, a mean value equal to 1.00 and a CoV equal to 0.26.

## 4.2 Shear Force - Deflection Curves Comparison

In this section, the experimental shear force – deflection curves obtained by means of the proposed model for some RC beams tested by Xue et al. (2016) are presented and discussed.

The specimens have a rectangular cross-Sect. (120 × 240 mm) and are strongly reinforced in bending (two D19 steel bars) to fail in shear. The specimens are classified into two series (B-39 and B-52) and are characterized by different geometric percentages of steel stirrups, i.e. 0.39% and 0.52%, respectively.

At the generic step, the mid-span deflection is estimated as the sum of the flexural and shear contributions, i.e. $\delta = \delta_f + \delta_s$. The flexural deflection is assumed as the sum of elastic and plastic contributions, i.e. $\delta_{fe}$ and $\delta_{fp}$, respectively. The elastic part is calculated as $\delta_{fe} = \chi\ [a\ (a/3 + a_m) + a_m^2/2]$, where $a_m$ is half of the length of the central part of the beam subjected to pure moment (= 0 for three-points test). The plastic part is calculated as $\delta_{fp} = (\chi - \chi_y) \times [(a - L_p/2) + (a_m - L_p/2)] \times L_p$, where $\chi_y$ is the curvature at yield and $L_p$ is the effective plastic hinge length (assumed equal to the cross-section depth). The shear deflection is obtained from the shear analysis as $\delta_s = 2a \times (\varepsilon_x - \varepsilon_2) \cot\theta$.

Figure 3 shows the experimental and numerical shear force - deflection curves for the specimens belonging to the B-39 and B-52 series tested by Xue et al. (2016) The peak load of the specimens is predicted with satisfactorily accuracy by the proposed model. However, the ascending branch of the curve obtained by the numerical model is much stiffer than that reported by the laboratory test. To evaluate the accuracy of the deflections estimated by means of the proposed model, an additional numerical analysis has been performed by means of the program Response2000 (r2k) (Bentz 2000) and plotted in Fig. 3 by means of black square dots. The numerical results of the two models (proposed and r2k) are comparable, in terms of both loads and deflections.

## 5 Simplified Model

In this Section, the proposed model is simplified in a formula to predict the shear strength of RC beam with corroded stirrups.

With this aim, a parametric analysis of a RC beam with rectangular cross-section has been carried out. The selected beam is representative of slender RC beams of existing structures. The *a/d* ratio is assumed equal to three and the cross-section is 300 × 500 mm in size. The beam is reinforced with a geometric percentage of the longitudinal tension and compression reinforcements equal to $\rho_{lb} = 1.94\%$ and $\rho_{lt} = 0.48\%$, respectively. The transverse reinforcement consists of stirrups with two legs.

The cross-sectional area loss ratio is uniform on the stirrups and it is in the range 0–90%. The cylindrical compression strength of concrete is $f_{cm} = 35$ MPa and the yield strength of steel is $f_y = 450$ MPa.

In Fig. 4, the response parameters that appear in Eq. (7) as obtained at peak load, i.e. cotangent of the angle of inclination of the diagonal compression stress of concrete (cotθ), and principal tensile stress to tensile strength ratio ($\sigma_{c1}/f_{ct}$), are plotted as a function of the mechanical percentage of (corroded) stirrups ($\omega_v = \rho_v f_{yv}/f_{cm}$) for different values of the geometric percentage of (uncorroded) stirrups ($\rho_{v0} = 0.10\% \div 0.30\%$).

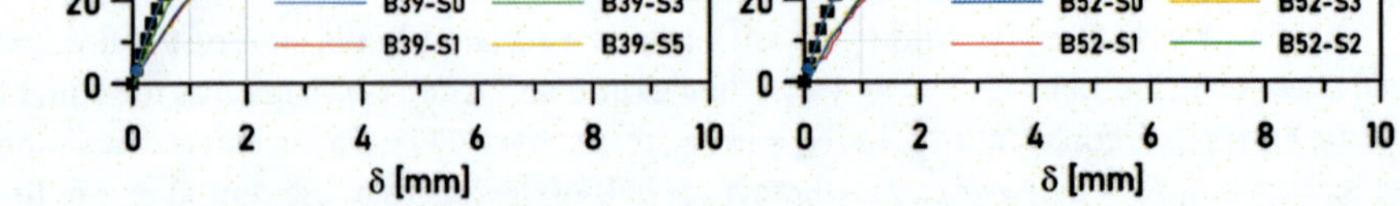


**Fig. 3.** Comparison between numerical and experimental responses of beams tested by Xue et al. (2016): experimental (solid) and numerical (circle/square) shear-displacement curves for (a) series B-39 and (b) series B-52.

The value of cotθ increases with the mechanical ratio $\omega_V$. In the cases where steel stirrups remain elastic (due to very low geometric percentages of stirrups and/or to medium-high levels of corrosion) the angle of inclination of the diagonal compression stress of concrete is about 33.2° (cot θ ≤ 1.53) (Fig. 4a), and negligible re-orientation of the diagonal compression stress of concrete takes place. The relationship between cot θ and $\omega_V$ follows a non-linear trend. In particular, the value of cot θ first increases with $\omega_V$ until a peak value has been reached. Then, slowly decreases with the increase of $\omega_V$. This final trend of cot θ is in accord with the results reported by Walraven et al. (2013) for beams with higher values of the geometric percentage of stirrups.

The first trend of the relationship is not reported by other researchers and is useful for a more in-depth comprehension of the response of slender beams with low geometric percentages of transverse reinforcement.

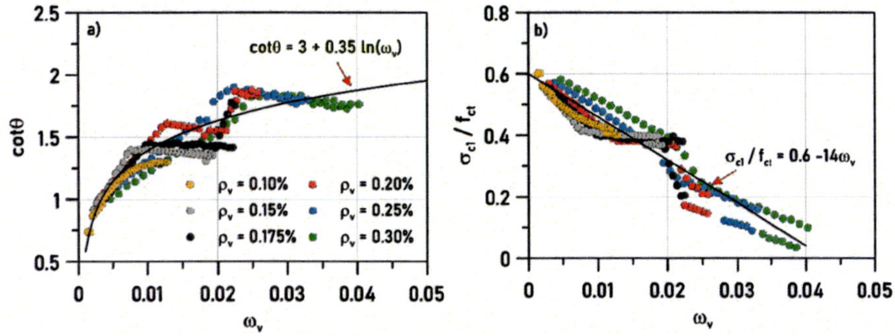


**Fig. 4.** Results of the parametric analysis: (a) cotangent of the angle of inclination of the diagonal compressive stress field of concrete and (b) principal tensile stress to tensile strength ratio versus the mechanical percentage of corroded stirrups.

A numerical regression analysis of the data reported in the figure leads to the simple relation:

$$\cot\theta = 3 + 0.35 \ln \omega_v \tag{24}$$

At the same manner, the principal tensile stress to tensile strength ratio ($\sigma_{c1}/f_{ct}$) almost linearly decreases with the increase of $\omega_V$ (Fig. 4b). This relationship may be analytically described by the following equation:

$$\sigma_{c1}/f_{ct} = 0.6 - 14\omega_v \tag{25}$$

Equations (24) and (25) suggest a simple formula for the prediction of the shear strength of RC beams with low values of $\rho_V$ (≤0.30%). Indeed, introducing these two equations into Eq. (7), and considering that stirrups yield at peak load - as also assumed in many codes - the shear strength can be easily calculated as follows:

$$V_{prop} = \left[(0.6 - 14\omega_v)f_{ct} + \rho_v f_{yv}\right][3 + 0.35 \ln \omega_v]b_w d \tag{26}$$

The simplified proposed formula is, then, validated against the previously considered database of 62 experimental tests. The results are plotted in Fig. 5.

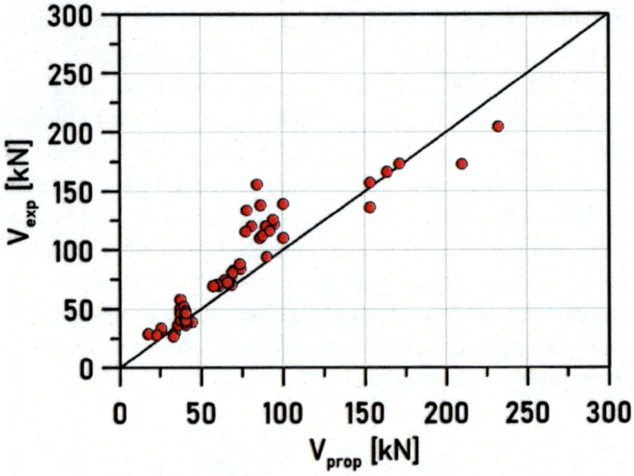


**Fig. 5.** Comparison between experimental shear strength ($V_{exp}$) and numerical shear strength obtained by means of the simplified proposed formulation ($V_{prop}$).

The $V_{exp}/V_{prop}$ ratio resulting from the application of the simplified proposed relation ranges from 0.80 to 1.86, with a mean value equal to 1.16 and a CoV equal to 0.19. The mean value indicates a slightly conservative estimate of the shear strength. The CoV obtained by the proposed relation is significantly lower than those obtained by means of the other models.

## 6 Conclusions

This paper presents an analytical procedure - based on the Modified Compression Field Theory and applied at the cross-section level - for the evaluation of the shear force - deflection response of slender RC beams with corroded stirrups. The proposed model considers the reduction of both the cross-sectional area and the strain capacity of the transverse reinforcement because of steel corrosion, along with the reduction of the beam width because of spalling of the concrete cover. The numerical model is validated by comparison with the results of 62 laboratory tests on slender RC beams with corroded stirrups and with the results deriving from the application of the Compression Chord Capacity Model to the same set of beams. Finally, a simple relation for the prediction of the shear strength of RC beams with low values of the effective geometric percentage of stirrups is derived.

The research findings are summarized as follows:

- the proposed model is able to consider the effects of the strain capacity of corroded steel bars, recognizing the failure modes governed by either stirrup yielding or rupture.
- the proposed model provides satisfying predictions of the shear strength of RC beams with corroded stirrups.
- the mean value of the numerical to experimental shear strength ratio is equal to 1.07 while the coefficient of variation is equal to 0.30. The response at peak load is in

good agreement with that resulting from the application of the Capacity Compression Chord Model.

- in beams with a geometric percentage of stirrups lower than about 0.10%, transverse steel remains elastic and the cotangent of the angle of inclination of the diagonal compression stress of concrete is lower than about 1.5. In such cases, a negligible re-orientation of the diagonal compression stress of concrete occurs.
- a simple formula has been derived for the estimation of the shear strength of reinforced concrete beams with corroded stirrups. The application of this proposed simple relation to the beams of the considered database highlights that it is able to provide accurate estimates of the load-bearing capacity of such RC beams. The mean value of the numerical to experimental shear strength ratio is equal to 1.16, whereas the coefficient of variation is equal to 0.19.

**Acknowledgements.** This research study was supported by the University of Catania within the research project "DU.SO.CA.P. – Durabilità e SOstenibilità delle strutture in Cemento Armato e Precompresso", the research program "PIAno di inCEntivi per la RIcerca di Ateneo 2020/2022 (PIA.CE.RI.) – Starting Grant – Linea di intervento 3", the research project "Definizione e validazione di procedure di progetto di interventi di adeguamento sismico di edifici in c.a mediante pareti oscillanti" and the research project "PIAno di inCEntivi per la RIcerca di Ateneo 2020/2022 (PIA.CE.RI.) - Linea di intervento 2". Special thanks are due to the Italian the Superior Council of Public Works (CC.SS.LL.PP.), and the Network of University Laboratories of Seismic Engineering (RELUIS). The results were achieved in the national technical agreement for implementing the agreement pursuant to art. 15 law 7 August 1990, No. 241 between the Superior Council of Public Works and RELUIS.

## References

Bentz, E.: Sectional analysis of reinforced concrete members [National Library of Canada = Biblio-thèque nationale du Canada] (2000). http://hdl.handle.net/1807/13811

Campione, G., Cannella, F., Cavaleri, L., Ferrotto, M.F.: Moment-axial force domain of corroded R.C. columns. Mater. Struct. **50**(1), 21 (2017). https://doi.org/10.1617/s11527-016-0930-5

Cladera, A., Marí, A., Ribas, C.: Mechanical model for the shear strength prediction of corrosion-damaged reinforced concrete slender and non slender beams. Eng. Struct. **247**, 113163 (2021). https://doi.org/10.1016/J.ENGSTRUCT.2021.113163

Fernandez, I., Bairán, J.M., Marí, A.R.: Mechanical model to evaluate steel reinforcement corrosion effects on σ–ε and fatigue curves. Exp. Calibration Validation. Eng. Struct. **118**, 320–333 (2016). https://doi.org/10.1016/J.ENGSTRUCT.2016.03.055

Higgins, C., et al.: Shear capacity assessment of corrosion-damaged reinforced concrete beams (2003)

Lu, Z.H., Li, H., Li, W., Zhao, Y.G., Dong, W.: An empirical model for the shear strength of corroded reinforced concrete beam. Constr. Build. Mater. **188**, 1234–1248 (2018). https://doi.org/10.1016/J.CONBUILDMAT.2018.08.123

Recupero, A., Spinella, N.: Preliminary results of flexural tests on corroded prestressed concrete beams. In: Proceedings of the Fib Symposium 2019: Concrete - Innovations in Materials, Design and Structures (2019a)

Recupero, A., Spinella, N.: Experimental tests on corroded prestressed concrete beams subjected to transverse load. Struct. Concr. suco.201900242 (2019b). https://doi.org/10.1002/suco.201900242

Recupero, A., Spinella, N., Tondolo, F.: Failure analysis of corroded RC beams subjected to shear-flexural actions. Eng. Fail. Anal. **93**, 26–37 (2018). https://doi.org/10.1016/j.engfailanal.2018.06.025

Rodriguez, J., Ortega, L., Casal, J.: Load carrying capacity of concrete structures with corroded rein-forcement. Constr. Build. Mater. **11**(4), 239–248 (1997). https://doi.org/10.1016/S0950-0618(97)00043-3

Spinella, N.: N-M-$\chi$ interaction for arbitrary cross section under biaxial bending and axial load. Pollack Periodica **8**(3), 87–100 (2013). https://doi.org/10.1556/Pollack.8.2013.3.9

Spinella, N.: Modeling of shear behavior of reinforced concrete beams strengthened with FRP. Compos. Struct. **215**, 351–364 (2019). https://doi.org/10.1016/j.compstruct.2019.02.073

Tondolo, F.: Bond behaviour with reinforcement corrosion. Constr. Build. Mater. **93**, 926–932 (2015). https://doi.org/10.1016/J.CONBUILDMAT.2015.05.067

Vecchio, F.J., Collins, M.P.: The modified compression-field theory for reinforced concrete elements subjected to shear. ACI J. Proc. **83**(2), 219–231 (1986). https://doi.org/10.14359/10416

Walraven, J., Belletti, B., Esposito, R.: Shear capacity of normal, lightweight, and high-strength concrete beams according to model code 2010. I: experimental results versus analytical model results. J. Struct. Eng. **139**(9), 1593–1599 (2013). https://doi.org/10.1061/(ASCE)ST.1943-541X.0000742

Xue, X., Seki, H., Song, Y.: Shear behavior of RC beams containing corroded stirrups. Adv. Struct. Eng. **17**(2), 165–178 (2016). https://doi.org/10.1260/1369-4332.17.2.165

# Cyclic Behaviour of Grouted Duct Connections in RC Precast Structures

Lorenzo Hofer[1(✉)], Mariano Angelo Zanini[1], Flora Faleschini[1], Klajdi Toska[1], Marco Nucci[2], and Carlo Pellegrino[1]

[1] Department of Civil, Environmental and Architectural Engineering, University of Padova, Padova, Italy
lorenzo.hofer@dicea.unipd.it
[2] NUOVA TESI SYSTEM S.r.l., Casale Sul Sile, Treviso, Italy

**Abstract.** The paper investigates the cyclic behaviour of a column-to-foundation joint for precast concrete structures. The connection investigated in the experimental campaign is realized using corrugated steel ducts in which column protruding longitudinal rebars are an-chored by grouted high performance mortar. The research program includes six full-scale RC columns having a square section, tested under an increasing cyclic lateral load and a constant axial load, with the aim to investigate the influence of the connection typology (cast-in-place or grouted connection), of different bar anchorage lengths and of different rebars diameters. Results are analysed in terms of hysteretic behaviour, energy dissipation, ductility values and plastic hinge location, and show a comparable structural behaviour between specimens with the precast joint and reference cast-in-place ones.

**Keywords:** column-to-foundation connection; grouted duct connection · precast concrete structures · cyclic test; seismic behaviour

## 1 Introduction

Reinforced concrete precast structures are widely diffused structural typologies, especially adopted to realize industrial plants, factories and warehouses [1]. Especially for this structural typology, connections play a key role since they allow the structural elements to be prefabricated and then only assembled on site. The common configuration of this kind of structures adopts cantilever columns connected with simply supported beams. Consequently, the structural behavior of the column base highly influences the overall structural response. When precast frame structures are subject to horizontal forces, the ductility demand is concentrated at the base of the column, which becomes the most critical part of the structure. Consequently column-to-foundation connections have to ensure an overall strength, ductility and dissipation comparable with those of the common cast-in-place (CIP) connections [2]. In last years, several column-to-foundation connection types have been developed and tested and can be summarized into two main typologies: *i.* traditional connection, where the entire column is inserted into a pocket foundation; *ii.* Connections aimed to directly connect the column's longitudinal bars or


M. A. Aiello and A. Bilotta (Eds.): ICC 2022, LNCE 435, pp. 547–559, 2024.
https://doi.org/10.1007/978-3-031-43102-9_42

the bottom of the column to the foundation. The latter ones that can be furtherly subdivided into two groups: *iia.* Connections adopting bar couplers or bolted connections; *iib.* Connections with high performance mortar grouted in corrugated ducts. Between the two typologies *i* and *ii*, pocket foundations represent the older solution. In this construction technique, the temporary bearing and the vertical alignment is ensured by the crane and by provisional props. Once positioned, the joint is filled with in-situ concrete pouring. Typically, such solution implies large excavations and difficult transportation. More information and extensive experimental test on this construction technique can be found in [3], and [4]. In recent years, scientific research mainly focused on the second connections typology, aiming to develop construction techniques able to speed up the on-site construction process. Regarding the connection type *iia*, [5] adopted mechanical bar couplers for connecting the reinforcement bars at the base of the column, while [6] investigated the behaviour of two other couplers, consisting of up-set headed couplers and grouted sleeve couplers. Further studies and experimental campaign can be found in [7] and [8]. Finally, an extensive experimental campaign aimed to investigate the structural behavior of four different mechanical column-foundation connections devices that can be included in the *iia*, is reported in [9]. In the last approach *iib*, reinforcement bars protruding from one precast element are grouted into corrugated steel ducts encased in the other unit. Usually, high performance mortar is adopted. This connection type is mainly applied in column-to-column and column-to-foundation connections. For the former case, readers are referred to [10] and [11]; both works showed the overall good behavior of such kind of connection. Regarding the column-to-foundation connections, two main solutions were studied in scientific literature. The first, in which bars protruding from the foundation are anchored in the column and then the opposite one, in which longitudinal rebars protruding from the column are anchored in the foundation. While the former connection was tested in several experimental campaigns ([12–14]), the latter was little investigated. In particular, [15] tested a one column with protruding bars anchored in foundation. Although the overall response turned out to be promising and satisfying, the steel tube used to center the column, influenced the column behavior. In the above context, this work wants to enlarge the current knowledge on the latter type of connection in which longitudinal bars protruding from the column base are anchored in the foundation by high performance mortar grouted in corrugated steel ducts. Further details on this work can be found in [16].

## 2 Anchorage Length Determination

For estimating the anchorage length values to adopt in the full-scale columns, a preliminary experimental campaign consisting in a series of pull-out test was carried out. Two rebar diameters were adopted among those mostly adopted in precast concrete structures, i.e., 24 mm and 30 mm, respectively $d_{b1}$ and $d_{b2}$. Grouted specimens were realized with both the diameter bars, whereas only 24 mm reinforcing bars were used to cast reference samples, these latter realized with reinforcing bars directly embedded in concrete. For $d_{b1}$grouted specimens, five anchorage lengths $a_L$ were analyzed: 10, 14, 17, 22 and 26$\cdot$ $d_{b1}$, thus varying between 240 and 624 mm, while for the reference $d_{b1}$ specimens only three $a_L$ values were considered (17, 22 and 26$\cdot db_1$). Instead, for

$d_{b2}$, only grouted specimens were considered with only the three highest values of $a_L$ (17, 22 and 26· $d_{b2}$corresponding to 510, 660 and 780 mm). For each configuration, three specimens were tested. Figure 1 shows a summary of the tested specimens, while Table 1 list the mechanical property of concrete (compressive cubic strength $f_{c,cube}$ and tensile strength $f_t$) and mortar (compressive cubic strength $f_{c,cube}$ and flexural strength $f_f$ evaluated on 40 × 40 × 160 mm prism according to EN 1015 series [17]) at the day of the test. According to the Italian Code for Construction (NTC 2018, [18]), C25/30 concrete strength class and B450C steel class are adopted.

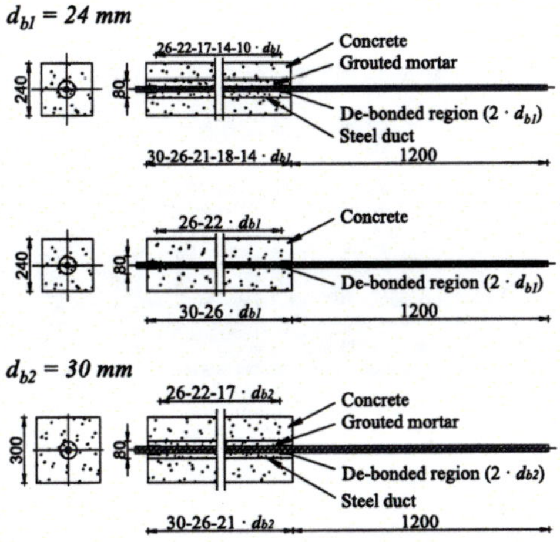


**Fig. 1.** Geometry of the pull-out specimens.

**Table 1.** Mechanical properties of the materials used in the pull-out specimens

| Connection Type | $a_L$ | Concrete | | Mortar | |
|---|---|---|---|---|---|
| | | $f_{c,cube}$ [MPa] | $f_t$ [MPa] | $f_{c,cube}$ [MPa] | $f_f$ [MPa] |
| Grouted | $10 \cdot d_{bl}$ | 40.1 | 4.08 | 73.8 | 7.41 |
| Grouted | 14 and $17 \cdot d_{bl}$ | 50.8 | 3.44 | 86.1 | 7.15 |
| Grouted | 22 and $26 \cdot d_{bl}$ | 44.2 | 4.16 | 83.1 | 7.55 |
| Reference | 22 and $26 \cdot d_{bl}$ | 42.3 | 3.83 | – | – |
| Grouted | 17, 22 and $26 \cdot d_{b2}$ | 50.3 | 4.28 | 63.7 | 7.16 |

Lastly, the corrugated ducts adopted are typical post-tensioning ducts, made with corrugated galvanized strip steel with an 80 mm internal diameter, an 84 mm external diameter, and a 0.6 mm thickness. Figure 2a shows the test layout, consisting of the concrete specimen located over the moving head of the universal machine with the steel bar gripped by two steel jaws. The load was applied by lifting the upper moving head at a velocity of 1.5 mm/min. Finally, at the top of the specimen, on the unloaded protruding bar, a 25 mm linear-variable displacement transducer (LVDT) was placed for recording the relative slip between concrete and the bar. Results shows that in the grouted duct specimens with an embedment length between 26·$d_b$ and 17·$d_b$, tensile failure of the steel bar occurred, for both the investigated diameters $d_{b1}$ and $d_{b2}$ (Fig. 2b). Specimens with anchorage length of 14·$d_{b1}$ and 10·$d_{b2}$ showed a satisfying response too, characterized by the bar yielding, but in some cases the splitting failure of the concrete cover occurred before the bar tensile failure (Fig. 2c). On the contrary, in the specimens where bars were directly immersed in concrete, even the highest anchorage length of 26·$d_{b1}$ was not enough for ensuring the bar failure in all the three specimens (Fig. 2d). Results are summarized in Table 2.

**Fig. 2.** Pull-out test setup (a); bar fracture for specimens with $a_L$ between 26·$d_b$ and 17·$d_b$ (b); splitting failure for specimens with $a_L$ of 10·$d_{b1}$ and 14·$d_{b1}$ (c); splitting failure for reference specimens (d) (adapted from [16]).

**Table 2.** Results of pull-out test (BF = bar failure, SY = splitting failure with bar yielding, S = splitting failure without bar yielding).

| Grouted duct | | | |
|---|---|---|---|
| $d_{b1}$ = 24 mm | | $d_{b2}$ = 30 mm | |
| $a_L$ | Failure mode | $a_L$ | Failure mode |
| 26·$d_{b1}$ | 3 BF | 26·$d_{b2}$ | 3 BF |
| 22·$d_{b1}$ | 3 BF | 22·$d_{b2}$ | 3 BF |
| 17·$d_{b1}$ | 3 BF | 17·$d_{b2}$ | 3 BF |
| 14·$d_{b1}$ | 2 BF + 1 SY | – | – |
| 10·$d_{b1}$ | 1 BF + 2 SY | – | – |
| Reference only concrete specimens | | | |
| 26·$d_{b1}$ | 2 BF + 1 SY | | |
| 22·$d_{b1}$ | 1 BF + 1 SY + 1 S | | |

In all cases where the bar failure was reached, there was no visible damage on the grouted mortar and the slip values at the unloaded end were in the order of 10–2 mm.

This preliminary experimental campaign showed that the grouted duct solution is able to provide a satisfying bar anchorage, higher than the one of the reference specimens, even at low embedment length values. According to the above results, the anchorage length of 26·$d_b$ and 17·$d_b$ were selected for anchoring the longitudinal bar of the full-scale RC columns.

## 3 Experimental Campaign

The experimental campaign presented in this work includes six full-scale RC columns, having a square section and tested under a combined cyclic horizontal load and a constant compressive axial load. The parameters investigated are:

- connection type (cast-in-place or grouted, respectively indicated as CIP and G);
- bar diameter ($d_{b1} = 24$ mm and $d_{b2} = 30$ mm);
- anchorage length ($a_{L1} = 26 \cdot d_b$ and $a_{L2} = 17 \cdot d_b$, respectively indicated as F for the full anchorage length and R for the reduced one).

### 3.1 Specimen Geometry

The columns tested in this experimental campaign have a square section of 400 mm side and 2900 mm height, while the foundations footprint is in all cases of 1400 × 1800 mm. Two different foundation hights of 700 mm and 850 mm are adopted for ensuring the anchorage length of 26·$d_{b1}$ and 26·$d_{b2}$ respectively for the column with 24 and 30 mm protruding rebars. Figure 3a shows the details of longitudinal and transverse reinforcement of the six columns, while Fig. 3b show the lateral view of the investigated specimens. Stirrups have a 100 mm spacing at the column base, and 150 mm in the upper part.

### 3.2 Material and Fabrication

Two different concrete classes were adopted for columns and foundations, respectively C45/55 and C25/30. The choice of adopting two strength classes is owned to the current engineering practice, where foundations have typically a reduced strength class (unless for peculiar durability issues) than elevation structures, and particularly, less than pre-cast elements. For each element, a single concrete batch was realized for the column and one for the foundation: concrete properties are listed in Table 3 in terms of compressive strength at the time of specimen testing (≈ 40 days, $f_{c40,cube}$), tensile strength via the Brazilian test ($f_t$) and the secant modulus of elasticity ($E_c$). According to [18], reinforcement bars were realized with B450C steel: Table 3 shows the steel property in terms of average yield strength $f_y$, average ultimate tensile strength fu and average percentage elongation at maximum force $\varepsilon_t$. Finally, the grouted mortar has, at the day of the test, a cubic compressive strength of 63.2 MPa and a flexural strength evaluated on 40 × 40 × 160 mm prism of 9.4 MPa.

Finally, Fig. 4 shows some photos of the realization process. First the foundation was realized with the eight embedded corrugated ducts (Fig. 4a), together with the column with the protruding longitudinal bars (Fig. 4b). Figure 4c, d show how the connection is

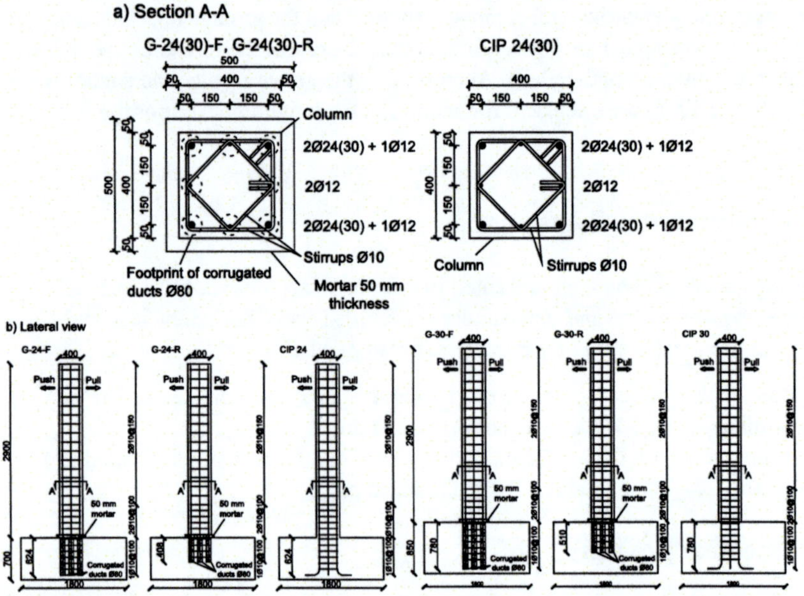


**Fig. 3.** Reinforcement details (a) and lateral view (b) of the six columns (dimension in mm) (adapted from [16]).

**Table 3.** Properties of hardened concrete and steel.

| Specimen | Column | | | Foundation | | | Steel | | |
|---|---|---|---|---|---|---|---|---|---|
| | $f_{c40,cube}$ | $f_t$ | $E_c$ | $f_{c40,cube}$ | $f_t$ | $E_c$ | $f_y$ | $f_t$ | $\varepsilon_t$ |
| | MPa | MPa | GPa | MPa | MPa | GPa | MPa | MPa | % |
| G-24-F | 71.4 | 4.4 | 42.0 | 50.6 | 4.1 | 34.1 | | | |
| G-24-R | 67.5 | 4.1 | 41.5 | 60.5 | 4.5 | 37.1 | 537 | 648 | 17.4 |
| CIP-24 | 62.3 | 5.5 | 38.2 | 63.8 | 5.5 | 39.8 | | | |
| G-30-F | 73.5 | 4.2 | 40.2 | 57 | 5.6 | 37.4 | | | |
| G-30-R | 78.0 | 5.1 | 39.4 | 60.6 | 4.4 | 35.6 | 572 | 673 | 19.8 |
| CIP-30 | 60.5 | 4.4 | 36.7 | 60.5 | 4.9 | 38.0 | | | |

completed through the addition of the high-performance mortar: first, some small steel plates are used to space the column from the foundation, and at the same time allowing the mortar to seal the connection without any voids, the mortar is injected.

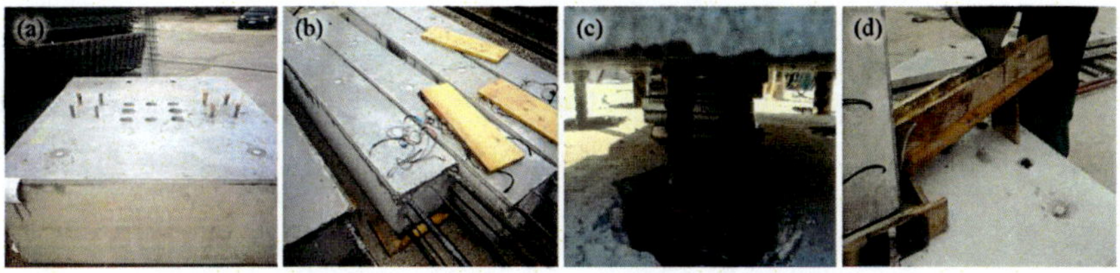


**Fig. 4.** Realization of the grouted connections (adapted from [16]).

### 3.3 Test Setup

As Fig. 5 shows, a cantilever scheme was adopted for testing the columns. The lateral displacement was applied at the top of the column by a hydraulic horizontal actuator, while the axial load was applied at the top of the column by a hydraulic jack that allows to tension two tie roads hinged at the base of the column, thus inducing a compression in the column. The axial load was kept constants for all the testing time. For the set of columns reinforced with db1 bars, a compressive load of 350 kN was kept constant during the test, while an axial force of 300 kN was adopted for specimens with $d_{b2}$ longitudinal rebars. Tests were performed by imposing a horizontal displacement in a quasi-static way, with a lateral loading rate varying from 0.1 mm/s for the lowest displacements cycles, to 0.8 mm/s for the largest ones. The test sequence was designed following the ACI 374.1–05 recommendations [19]. The same Fig. 5 shows the imposed displacement history in terms of drift ratio $\delta$, defined as the ratio between the imposed top displacement $\Delta$ and the column span $h$, equal to 2500 mm for columns reinforced with $d_{b1} = 24$ mm bars, and 2550 for columns with $d_{b2}$. The maximum imposed drift was 5.5 and 5%, respectively for $d_{b1}$ and $d_{b2}$ columns. Finally, details of the adopted instrumentation can be found in [16].

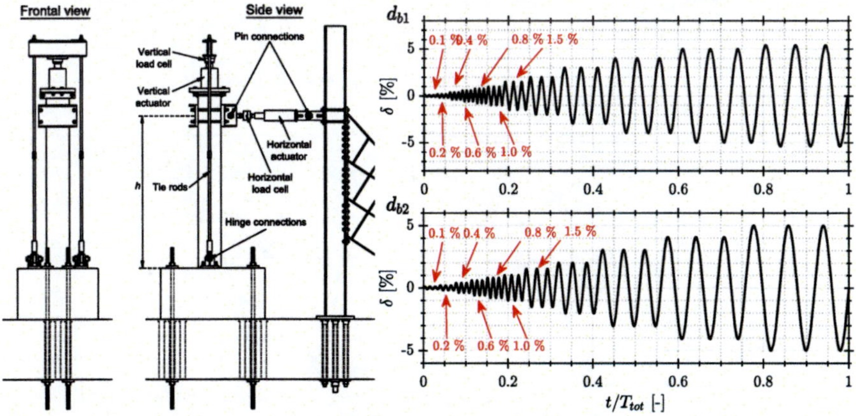


**Fig. 5.** Test setup and lateral imposed displacement (adapted from [16]).

# 4 Results and Discussion

## 4.1 Cracking Pattern and Hysteresis Curve

All the six columns showed a similar cracking pattern, characterized by the typical horizontal cracks of the flexural failure. At the lowest drift value ($\delta = 0.1\%$), columns behave elastically, with no cracks occurrence. Cracks developed starting from $\delta = 0.2\%$ and gradually increased along the height of the column. At $\delta = 1.5\%$ the reinforcing bars started yielding and only few new cracks appeared, whereas the existing ones grew. From $\delta = 2.0\%$, damage mainly increased and concentrated at the base of column (i.e., $100 \div 150$ mm from the bottom). Figure 6 shows the damage on the columns at the end of the test. In the grouted specimens the mortar collar crushed in analogy to the CIP specimens. In any grouted column no macro-damage was visible in the connections. Furthermore, the intermediate longitudinal bar $\Phi 12$ in one side of column cross-section of specimen G-24-R and CIP-24, broke during the last drift level. The damage extension was totally comparable for the two series of specimens, the first with $d_{b1}$ and the second with $d_{b2}$. Figure 7a and Fig. 7b show the load F vs drift $\delta$ curve respectively for the $d_{b1}$ and $d_{b2}$ specimens.

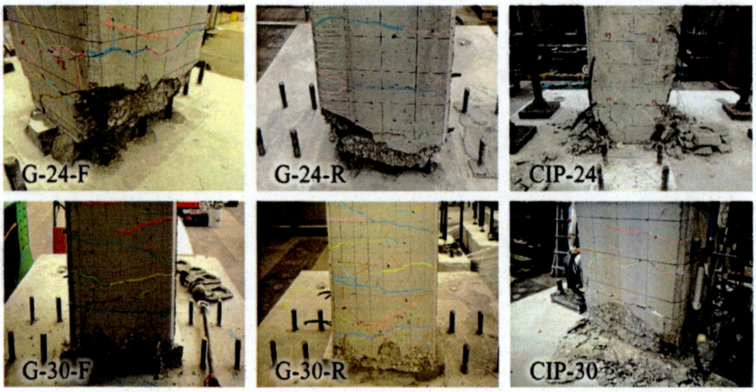


**Fig. 6.** Damage extension (adapted from [16]).

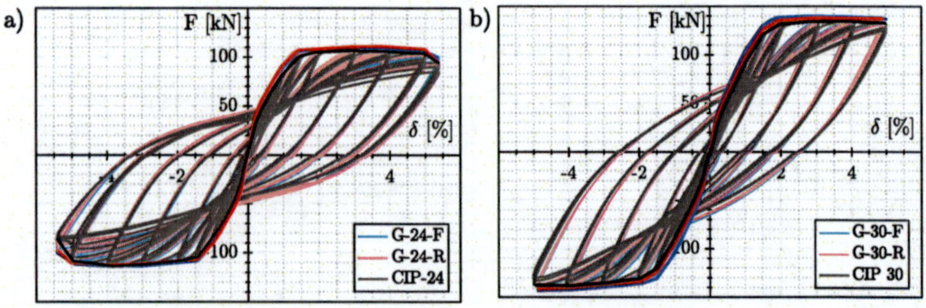


**Fig. 7.** Load $F$ vs drift $\delta$ curve for the specimens with $d_{b1}$ (a) and $d_{b2}$ (b) (adapted from [16]).

For the specimens with $d_{b1}$ rebars, both analyzing the load-drift curves and the relevant points (cracks opening, yielding and ultimate load), it is possible to state that the overall structural behavior is comparable. They display almost symmetric behavior in the push and pull conditions and are almost overlap-ping. Similar considerations can be made for columns reinforced with $d_{b2}$ longitudinal bars. Compared to previous samples, they display slightly larger unsymmetric results. However, overall, the behavior is considered sufficiently symmetric.

### 4.2 Energy Dissipation

Measurement of the dissipated energy is a meaningful index, independently from the ductility of a structural system [20]. In this section, the energy dissipation capacity is computed for evaluating the capacity of the columns to absorb energy during the entire loading history. For each $i^{th}$ cycle, the dissipated energy $E_{d,i}$ can be computed as:

$$E_{d,i} = \oint F(\Delta) d\Delta \quad (1)$$

Figure 8a and Fig. 8b show the dissipated energy specimens with $d_{b1}$ and $d_{b2}$ rebars, respectively, where both the dissipated energy for each $i^{th}$ cycle computed according to Eq. (1). Figure 8 shows that most of the energy dissipation is concentrated at the end of the test, when the bars exceed the yielding point. For both specimens with $d_{b1}$ and $d_{b2}$ rebars the energy dissipation was stable for every repetition of each drift level, thus evidencing a good energy dissipation capacity. Even in terms of dissipated energy the grouted specimens behave similarly to the cast-in-places ones, for both the considered diameters $d_{b1}$ and $d_{b2}$.

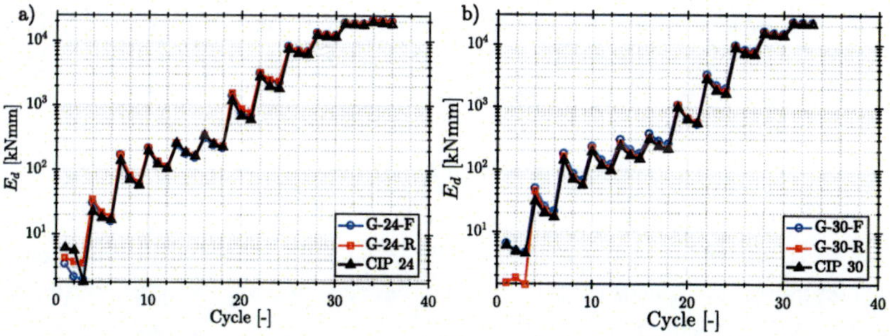


**Fig. 8.** Dissipated energy for the specimens with $d_{b1}$ (a) and $d_{b2}$ (b) (adapted from [16]).

### 4.3 Plastic Hinge Localization

Finally, columns are compared also in terms of plastic hinge localization, since the plastic hinge length is an important parameter to numerically simulate the nonlinear seismic response of the structural system. According to Priestley and Park (1987), the schematic curvature distribution along a RC member at ultimate stag can be assumed as in Fig. 9. This distribution is composed by the yield curvature ($\varphi_p$) linearly distributed along the column height and the rectangular plastic curvature ($\varphi_y$), concentrated in the plastic zone. Typically, in the initial stages, ($M < M_{cr}$), the response is elastic and linear. For higher load levels, longitudinal bars begin to yield ($\varphi_y$ and $M_y$, Fig. 9b), followed by concrete crushing ($\varphi_u$ and $M_u$). Beyond the yield limit, a large increase of curvature commonly occurs at the base of the column. According to Fig. 9c, the total rotation can be subdivided into elastic ($\theta_e$) and plastic ($\theta_p$) rotations.

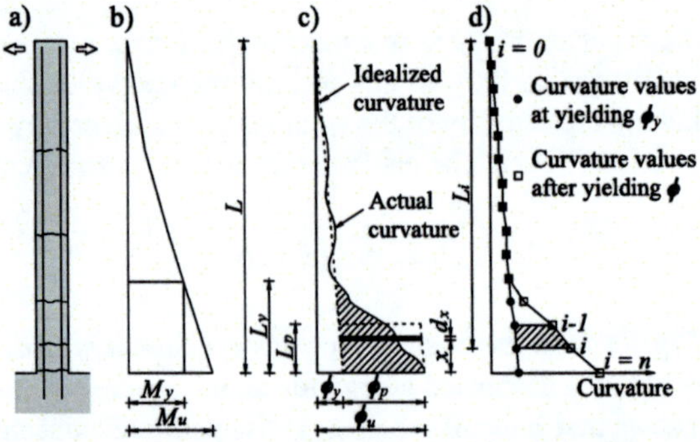


**Fig. 9.** Schematic curvature distribution along a RC column: cracks distribution (a), bending moment diagram (b), curvature diagram (c) and model for $L_p$ computation (d).

In practical applications, it is convenient to refer to an equivalent plastic hinge $L_p$, derived by computing $\theta_p$ as the rectangular area $\varphi_p \cdot L_p$. For computational purposes, the equivalent plastic hinge $L_p$ can be computed according to [21]. The computed plastic hinge lengths are compared with the one proposed by the Eurocode 8 (EC8), Eq. (A.9), developed for members with detailing for earthquake resistance and without lapping of longitudinal bars in the vicinity of the section where yielding is expected:

$$L_p = \frac{L_v}{30} + 0.20h + 0.11\frac{d_{bL}f_y}{\sqrt{f_c}} \tag{2}$$

Figure 10 shows for both specimens with $d_{b1}$ and $d_{b2}$ there are no significant discrepancies among the specimens with full and reduced anchorage lengths. Furthermore, the plastic hinge length computed with Eq. (2) well predicts the obtained values. For both rebars diameters, $L_p$ of the cast-in-place specimens is slightly shorter than the one of the precast specimens, maybe due to the stiff restrain introduced by the small steel plates. However, this difference seems to be negligible and not influencing the overall structural behaviour of the tested connections.

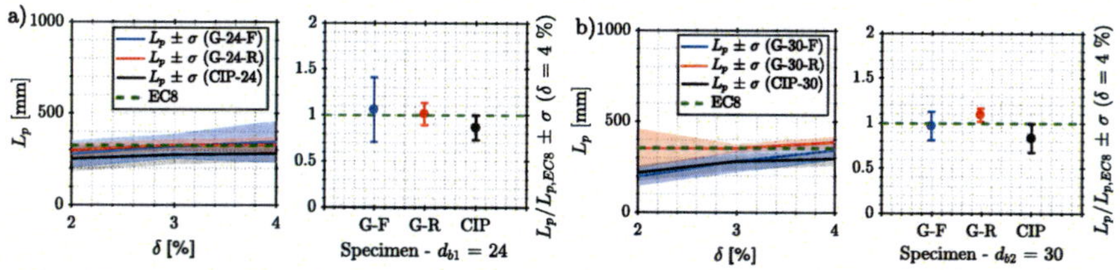


**Fig. 10.** Comparison of $L_p$ for $d_{b1}$ and $d_{b2}$ specimens with the one computed with the EC8 formula (adapted from [16]).

## 5 Conclusions

This paper presented the experimental campaigned aimed at investigating the structural performance of column-to-foundations connections realized with longitudinal protruding bars anchored in the foundation with high performance mortar grouted. This connection type seemed particularly efficient, since the steel ducts are well confined by the massive concrete cast of the foundation. Compared to similar technologies, this connection type avoids some specific detailing that are typical of other precast connection technologies (e.g. bar couplers, bolted and/or welded connections, detailing for confinement). According to the experimental evidences obtained in this work, it is possible to draw the following conclusions:

- a good performance of the connection was observed for all the grouted specimens with stable and symmetric hysteretic cycles. The cast-in-place and precast specimens' behaviours were comparable also in terms of ductility, dissipated energy and curvature profile;
- all the specimens exhibited a ductile flexural failure, with the formation of the plastic hinge at the base of the column. In the grouted specimens, mortar crushing occurred at the last drift levels, but no macro-damage was visible in the grouted connections.
- the two anchorage lengths adopted in this study seem sufficient to anchor the column longitudinal rebars into the foundation and to ensure a good cyclic behaviour. However, further studies are needed for investigating the joint seismic overstrength, even considering other failure modes and in presence of significant P-$\Delta$ effects.
- The plastic hinge length was evaluated, and its value was compared with that proposed by the Eurocode 8 formulation, showing a maximum discrepancy of about 16%.

**Acknowledgement.** The present work discusses the experimental behaviour of the column-to-foundation connection of the precast system developed by NUOVA TESI SYSTEM S.r.l. who realized the tested specimens and is gratefully thanked. The authors would like also to thank Eng. Giovanni Gobbi and Eng. Enrico Rovizzi.

**Funding.** This study was funded by NUOVA TESI SYSTEM S.r.l. who financed the realization of the specimens and supported costs for the execution of the experimental campaign.

## References


1. Buratti, N., Minghini, F., Ongaretto, E., Savoia, M., Tullini, N.: Empirical seismic fragility for the precast RC industrial buildings damaged by the 2012 Emilia (Italy) earthquakes. Earthq. Eng. Struct. Dynam. **46**, 2317–2335 (2017)
2. Negro, P., Toniolo, G.: Design guidelines for connections of precast structures under seismic actions. EUR - Scientific and Technical Research Reports. JRC Publication No. JRC71599. Publications Office of the European Union (2012)
3. Saisi, A., Toniolo, G.: Precast r.c. columns under cyclic loading: an experimental programme oriented to EC8. studies and researches, F.lli Pesenti Master school. Politecnico di Milano **19**, 373–414 (1998)
4. Xu, Y., Zeng, Z., Wang, Z., Ge, J.: Experimental studies of embedment length of precast bridge pier with socket connection to pile cap. Eng. Struct. **233**, 111906 (2021)
5. Philippi, D.J., Hegemier, G.A.: Use of mechanical couplers in concrete columns. In: Architectural Engineering Conference 2013, Pennsylvania State College, 3–5 April 2013, United States (2013)
6. Haber, Z.B., Saiidi, M.S., Sanders, D.H.: Seismic performance of precast columns with mechanically spliced column-footing connections. ACI Struct. J. **111**(3), 639–650 (2012)
7. Al-Jelawy, H., Haber, Z., Mackie, K.: Grouted splice pre-cast column connections with shifted plastic hinging. In: 16th World conference on earthquake - 16WCEE 2017, January 9–13, Santiago Chile (2017)
8. Wang, R., Ma, B., Chen, X.: Seismic performance of prefabricated segmental bridge piers with grouted splice sleeve connections. Eng. Struct. **229**, 111668 (2021)
9. Dal Lago, B., Toniolo, G., Lamperti Tornaghi, M.: Influence of different mechanical column-foundation connection devices on the seismic behaviour of precast structures. Bull. Earthq. Eng. **14**, 3485–3508 (2016)
10. Zheng, L.X.: Grouted precast concrete column connections under reversed cyclic bending and compression. ACI Struct. J. **93**(3), 247–256 (1996)
11. Tullini, N., Minghini, F.: Grouted sleeve connections used in precast reinforced concrete construction: experimental investigation of a column-to-column joint. Eng. Struct. **127**, 784–803 (2016)
12. Belleri, A., Riva, P.: Seismic performance and retrofit of precast concrete grouted sleeve connections. Pre-cast/Prestressed Concr. Inst. J. **57**(1), 97–109 (2012)
13. Popa, V., Papurcu, A., Cotofana, D., Pascu, R.: Experimental testing on emulative connections for precast columns using grouted corrugated steel sleeves. Bull. Earthq. Eng. **13**(8), 2429–2447 (2015)
14. Tullini, N., Minghini, F.: Cyclic test on a precast reinforced concrete column–to–foundation grouted duct connection. Bull Earthq. Eng. **18**, 1657–1691 (2020)
15. Buratti, N., Bacci, L., Mazzotti, L.: Seismic behaviour of grouted sleeve connections between foundations and pre-cast columns, In: 15th World conference of earthquake engineering (WCEE), 24–28 September 2012, Lisbon, Portugal, paper no. 1525 (2012)
16. Hofer, L., Zanini, M.A., Faleschini, F., Toska, K., Pellegrino, C.: Seismic behavior of precast reinforced concrete column-to-foundation grouted duct connections. Bull. Earthq. Eng. **19**, 5191–5218 (2021)
17. EN 1015–11:2019 Methods of test for mortar for ma-sonry. Determination of flexural and compressive strength of hardened mortar (2019)
18. NTC DM 17/01/2008: Norme Tecniche per le Costruzioni (in Italian) (2018)
19. ACI 374.1–05 Acceptance criteria for moment frames based on structural testing and commentary (2014)

20. Germano, F., Tiberti, G., Plizzari, G.: Experimental behavior of SFRC columns under uniaxial and biaxial cyclic loads. Compos. B **85**, 76–92 (2016)
21. Jiang, C., Wu, Y.F.: Variation of plastic hinge length for RC column. In: Proceedings of the 12th International Symposium on Fiber Reinforced Polymers for Reinforced Concrete Structures (FRPRCS-12) & The 5th Asia-Pacific Conference on Fiber Reinforced Polymers in Structures (APFIS-2015), Joint Conference, 14–16 December 2015, Nanjing, China (2015)

# Seismic Response of RC Frames Affected by Carbonation-Induced Corrosion

Francesco Nigro and Enzo Martinelli(✉)

Department of Civil Engineering, University of Salerno, Salerno, Italy
{fnigro,e.martinelli}@unisa.it

**Abstract.** In Europe, the vast majority of RC structures is characterized by a lack in both material quality and structural detailing, which significantly influences their seismic behaviour. Furthermore, a huge number of them may be affected by severe material degradation phenomena, since classical procedures adopted in concrete structures design have often failed to achieve sufficiently durable performances. Nowadays, code provisions towards durability rely on some "deemed-to-satisfy" rules concerning the values of concrete cover and of concrete strength to be adopted. The present paper reports a parametric study on the time-dependent variation of structural seismic reliability due to carbonation-induced corrosion. To this aim, a simple and explicit procedure has been applied. Such procedure involves a mechanically-consistent implementation of carbonation-induced effects and requires the use of nonlinear static (Pushover) analyses, although it is possible to extend its use to different degradation phenomena and to the use of nonlinear dynamic analyses.

**Keywords:** RC buildings · durability · seismic reliability · carbonation-induced corrosion · pushover analysis

## 1 Introduction

One of the most relevant issues of modern structural engineering consists in both designing durable RC structures and forecasting the influence of degradation phenomena on the behaviour of existing ones, as their seismic performance might be affected not only by a lack in structural detailing, but also by severe material degradation phenomena.

The introduction of practice-oriented criteria regarding the choice of both material properties and concrete cover thickness in EN 1992-1-1:2004 is the main results of a wide research activity aimed at quantifying the effects on reinforced concrete of different mechanisms, such as carbonation, chloride ingress, alkali-aggregate reaction and freeze-thaw cycles.

The DuraCrete Project (2000) brought to formulating statistical models for simulating the damage progression and semi-probabilistic checking methods applicable for service life design purposes.

As outlined by Francois et al. (2018), carbonation-induced corrosion represents the most common degradation phenomena affecting RC structures, as all RC structures are


M. A. Aiello and A. Bilotta (Eds.): ICC 2022, LNCE 435, pp. 560–574, 2024.
https://doi.org/10.1007/978-3-031-43102-9_43

exposed to atmospheric $CO_2$, even though it may induce milder degradation effects than other phenomena (e.g., chloride ingress).

Recent numerical studies of structural seismic behaviour, including the effects of carbonation-induced degradation, have clearly shown a pronounced capacity degradation, confirming that the observed cross sectional strength reduction reverberates its effects upwards at the structural scale (Erduran and Martinelli, 2020).

Well-established seismic reliability assessment methods, such as the "2000 SAC/FEMA" (Cornell et al. 2002), can be applied in different moments of the structural service life, in order to catch the time-variation of its seismic response, as such degradation phenomena develop, according to what has been suggested by Vamvatsikos and Dolšek (2011).

Although those methodologies are usually based on the results of Incremental Dynamic Analyses (IDAs), some alternative procedures have been explored by Dolšek and Fajfar (2004) and Faella et al. (2008) utilizing the N2 Method (Fajfar 1999) and response spectra derived from natural records.

Based on the aforementioned degradation models and reliability evaluation methods, Nigro et al. (2021) have recently proposed a general practice-oriented procedure for the evaluation of seismic reliability of RC structures affected by carbonation-induced degradation phenomena.

In the present paper, Sect. 2 offers an overview of the most relevant parameters concerning the modelling of carbonation-induced corrosion and the main steps of a procedure intended at determining the time-dependent seismic response. Section 3 describes the parametric study concerning the influence of water-cement ratio (*w/c*) and Relative Humidity (*RH*) based on the proposed framework. Finally, Sect. 4 remarks the main results.

## 2 Overview of Carbonation-Induced Degradation Phenomena

In principle, the influence of degradation phenomena should not be neglected, when assessing the safety and serviceability of RC structures. As a matter of fact, a significant reduction in structural reliability is expected as soon as the environmental condition become more severe, due to the combination of concrete degradation and reinforcement corrosion. To this aim, the "Model Code for Service Life Design – fib bulletin 34" (fib, 2006) defines the fundamentals for service life design approach, maintaining a code compatible structure.

### 2.1 Main Parameters Concerning Carbonation-Induced Corrosion

According to Tuutti (1982), reinforcement corrosion consists of two time phases:

- *initiation*, defined as the time until the reinforcement gets depassivated and, hence, corrosion may begin. Its duration, namely *depassivation time* $t_d$, is usually described by the Fick's law, which expresses the evolution of the carbonation depth:

$$t_d = \left[\frac{2k_e \cdot k_c \cdot (k_t \cdot R_{ACC,0}^{-1} + \varepsilon_t) \cdot C_S \cdot t_0^{2 \cdot W_t}}{(d' - \phi/2)^2}\right]^{\frac{1}{2 \cdot W_t - 1}} \tag{1}$$

where $t_d$ = depassivation time; $k_e$ = environmental function, which is related to *RH* [-]; $k_c$ = execution transfer parameter [–]; $k_t$ = regression parameter [–]; $R^{-1}_{ACC,0}$ = inverse effective carbonation resistance of concrete [(mm$^2$/years)/(kg/m$^3$)], which depends on the *w/c*; $\varepsilon_t$ is error term; $C_s$ = $CO_2$ concentration [kg/m$^3$]; $W_t$ is the weather function, intended at simulating the effects of driving rains on carbonation of the member under consideration [–];

- *propagation*, which is characterized by both reinforcement corrosion and concrete deterioration. In case of expanding corrosion, products of the reinforcement cracks along the reinforcing element are provoked, leading to spalling of the concrete cover. A linear assumption is often accepted to describe the time evolution of the bar radius loss:

$$x_{corr}(t) = V_{corr} \cdot w_t \cdot (t - t_d) \tag{2}$$

where $x_{corr}$ = bar radius loss; $V_{corr}$ = corrosion rate; $w_t$ = weather function; $t_d$ = depassivation time, defined by Eq. 1.

At the same time, it is possible to relate the crack width to the loss of bar radius as follows:

$$w = w_{cr} + \beta \cdot [x_{corr} - p_0] \tag{3}$$

where $w_{cr}$ = 0.05 mm, corresponding to visible crack opening; $\beta$ = parameter controlling propagation; $x_{corr}$ = bar radius loss; $p_0$ = parameter depending on the ratio between the initial concrete cover *d'* and the initial bar diameter $\phi$ and on the splitting strength of concrete.

Figures 1 and 2 point out the influence of water-cement ratio (*w/c*) and Relative Humidity (*RH*) on the depassivation time $t_d$.

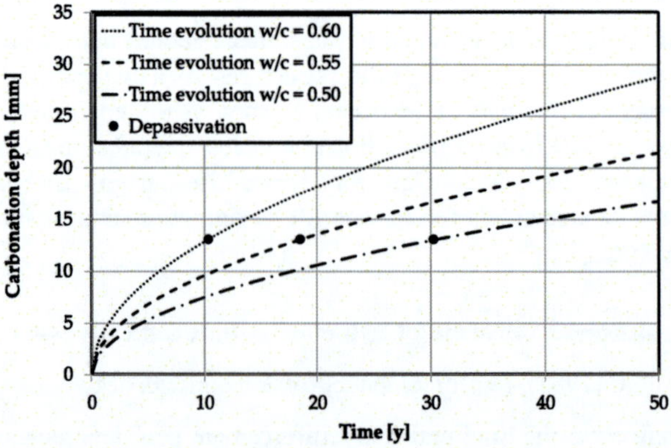


**Fig. 1.** Influence of water-cement ratio on the carbonation depth.

Although the previously outlined models have a probabilistic representation, in the following they are all used in a deterministic way in order to catch their influence on structural behaviour.

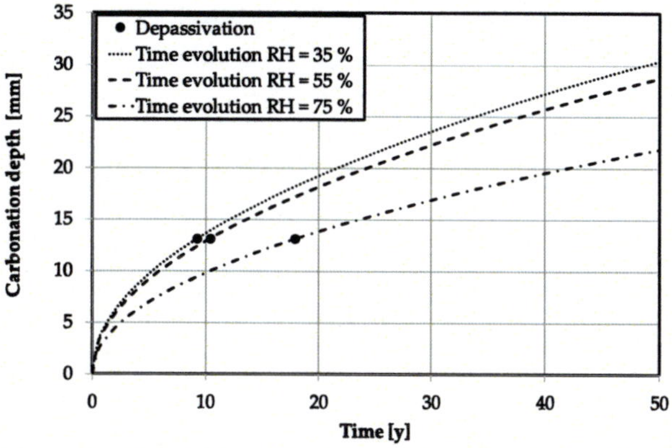


**Fig. 2.** Influence of RH on the depassivation time.

## 2.2 Effects of Carbonation-Induced Corrosion on Chord Rotation

Chord-rotation is the most widely adopted parameter to describe the displacement demand on structural members subjected to ductile mechanisms. Several models are currently in use to quantify the chord-rotation capacity of existing members depending on their main geometric and mechanical properties (Panagiotakos and Fardis 2001). Since it is clear that corrosion may imply a reduction in ductility and a reduction member capacity in terms of chord rotation, Vecchi and Belletti (2021) have recently proposed some corrective factors with the aim to include the effect of corrosion on the member capacity. Specifically, the following bilinear expression could be used to modify the standard expression of the ultimate chord-rotation capacity:

$$\alpha_{COR} = \frac{\theta_U(\psi)}{\theta_{EN}} = \begin{cases} a - b \cdot \psi, & \text{if } \psi \leq \psi_{\lim} \\ a - b \cdot \psi_{\lim} - c \cdot (\psi - \psi_{\lim}). & \text{if } \psi > \psi_{\lim} \end{cases} \quad (4)$$

where $\alpha_{corr}$ = correction factor; $\theta_U$ = chord rotation capacity accounting for corrosion; $\theta_{EN}$ = chord rotation according to EN 1998-3:2005; $\psi$ = reinforcement mass loss; *a,b,c* = calibration parameters proposed on the bases of the available experimental results (Vecchi and Belletti 2021).

## 2.3 Main Variables of the "2000 SAC/FEMA" Method

According to the "2000 SAC/FEMA" Method it is possible to express (through a closed-form expression) the probability of exceedance of a certain Limit State (LS) $P_{LS}$ for a structure exposed to a seismic hazard described by hazard curves, in order to assess its reliability. To this aim, it is necessary to define:

- an intensity measure ($I_M$) that normalizes each of the chosen seismic signals;
- a consistent engineering demand parameter ($E_{DP}$) aimed at quantifying the response of the structure.

Although many choices can be considered for the definition of both $I_M$ and $E_{DP}$, the proposed procedure adopts as $I_M$ the spectral acceleration $S_a(T_f)$ corresponding to the fundamental period of the structure $T_f$ and the chord rotation $\theta$ as $E_{DP}$.

Moreover, considering the seismic Demand "*D*" as a log-normal random variable, it is possible to express the median demand as a function of the intensity measure value:

$$\hat{D} = a \cdot s_a^b \tag{5}$$

where $\hat{D}$ = median demand;$s_a$ = spectral acceleration, assumed as intensity measure.

Then, it is possible to express the Hazard (*H*) as a function of $s_a$, by means of the Probabilistic Seismic Hazard Analysis (PSHA) of the construction site, according to a power law, as in Eq. 6:

$$H(s_a) = P(S_a \geq s_a) = k_0 \cdot s_a^{-k} \tag{6}$$

As clearly outlined by Pinto et al. (2004), based on applying the total probability theorem and assuming the capacity as deterministic, it is possible to express $P_{LS}$ by equalling the median demand to the capacity (in terms of $E_{DP}$) of the considered LS as follows:

$$P_{LS} = k_0 \cdot \left(\frac{E_{DP,LS}}{a}\right)^{-k/b} \cdot exp\left[\frac{k^2}{2b^2} \cdot \beta_D^2\right] \tag{7}$$

where $E_{DP,LS}$ = chord rotation capacity for the considered LS; $\beta_D{}^2$ = variance (namely, dispersion) of the random variable "*D*". As a matter of fact, in Eq. 7 it is possible to recognize that the exponential term is a magnification factor (accounting for demand dispersion around the median value) of the other factor which represents the probability for the structure to achieve the LS of interest considering both capacity and demand as deterministic variables.

## 2.4 The Equivalent Constant Rate (ECR)

The concepts of exceedance probabilities of "*p*% in $T_d$ years" integrated into our design methodologies may be modified as degradation phenomena occur, where *p* is expressed as follows:

$$p = 1 - \exp(-P_{LS} \cdot T_d) \tag{8}$$

where $p$ = probability of exceedance;$T_d$ = structural service life; $P_{LS}$ = constant rate of the homogeneous Poisson process.

Although the original formulation of the SAC/FEMA method does not take into account the time-dependent variation of the structural response, Vamvatsikos and Dolšek (2011) have recently proposed an extension of the method, with the aim of including the time-dependant nature of the structural response. Specifically, this evolution leads to defining an "equivalent constant rate" (ECR) of degradation (or $P_{LS}{}^{(ECR)}$) associated with the lifetime $T_d$, maintaining the following code-compatible format:

$$P_{LS}^{(ECR)} = \frac{\alpha}{1 - \exp(-\alpha \cdot T_d)} \cdot \int_0^{T_d} P_{LS}(t) \cdot \exp(-\alpha \cdot t)dt \tag{9}$$

where $P_{LS}(t)$ = regression law expressing the time evolution of $P_{LS}$; $\alpha$ = is the societal discount rate, which can be fixed for instance at 0.1%. In this case, the ECR will be approximately equal to the average (AVG, $P_{LS}^{(AVG)}$) rate:

$$P_{LS}^{(AVG)} = \frac{1}{T_d} \cdot \int_0^{T_d} P_{LS}(t)dt \tag{10}$$

## 2.5 Outline of the Procedure for Assessing the Time-Dependent Structural Performance

In the present Section the main points concerning the explicit durability assessment are summarized. Nigro et al. (2021) have provided further details concerning such procedure.

As soon as degradation phenomena starts developing, it is possible to evaluate their effects on member's cross section in terms of reduction of reinforcement area (due to corrosion) and, eventually, in terms of concrete cover loss (caused by spalling).

Consequently, it is possible to subdivide the structural service life $T_d$ in a proper number of intervals and determine $P_{LS}$ at different stages of the structural age, incorporating the aforementioned effects of material degradation phenomena. Thus, it is possible to evaluate the equivalent and average rates - $P_{LS}^{(ECR)}$ and $P_{LS}^{(AVG)}$ - and to compare them to the initial value of $P_{LS}$, namely $P_{LS,0\,y}$. From this comparison an explicit assessment of the structural durability can be performed. It is worth noticing that a better understanding of the results is gained if those comparisons are performed in terms of return period $T_R$, which is the inverse of $P_{LS}$.

### 2.5.1 Modelling of Members Affected by Degradation

Based on the degradation models outlined in Sect. 2.1, it is possible to explicitly take into account the effect of material degradation on the seismic behaviour of the structure.

Such operation is performed through the cross-section fibre model in the OpenSEES (Mazzoni et al. 2006) workspace, modifying the bar diameter due to corrosion and, eventually, eliminating some concrete fibres from the edges of the cross section.

### 2.5.2 Use of Non-linear (Pushover) Analyses

In order to enhance the cost-effectiveness of the procedure in terms of computational costs, the use of Pushover analyses is preferred to the IDAs.

Median Demand is evaluated according to Eq. 5. Even though in the first formulation of the method that relationship is determined through IDAs, in the following such power law is computed according to the results of the Incremental N2 Method, employing the natural spectra directly derived from the PEER Ground Motion Database Record considering the 22 ground shaking records suggested by FEMA (2009), represented in Fig. 3.

The dispersion $\beta_D{}^2$ of the seismic demand is accounted by considering the natural spectra, directly derived from the corresponding accelerograms. Each natural spectra is scaled with respect to the $I_M$, in order to cover the whole integration domain.

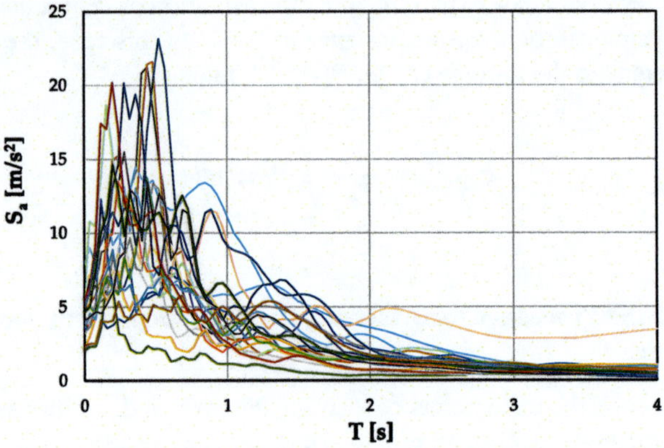


**Fig. 3.** 22 natural spectra (employed in the analyses) derived from the PEER Ground Motion Database Record.

## 3 Parametric Analyses

With the aim to point out the influence of carbonation-induced degradation on the seismic capacity, pushover analyses are carried out at different steps of structural age (from 0 to 50 years with steps of 10 years), varying in the first case the water-cement ratio (*w/c*) and the Relative Humidity (*RH*) in the second case. The (median) values of $R_{ACC,0}{}^{-1}$ and $V_{corr}$ are implicitly assumed, considering an exposure class XC2 and a cement CEM III/B 42.5. A design concrete cover value $d' = 20$ mm is adopted.

### 3.1 Structure Presentation

In the present study a 4 bay, 4 storey RC frame (Fig. 4) has been considered. It is ideally drawn from a GLD residential building, designed according to the provisions of "Regio Decreto 16/11/1939 n.2229". Its members are reinforced by 14 mm rebars. With the aim of reproducing typical material properties of existing RC buildings built in the 1970s and 1980s, C20/25 concrete and Aq42 rebars (characterized by a minimum yield stress equal to 230 MPa and by an ultimate stress included between 420 and 500 MPa) are implemented in the analysis.

### 3.2 Parametric Field and Reference PSHA

Table 1 summarizes the parametric field considered in the analyses.

Figure 5 shows the Hazard function concerning the site of Avellino (Italy) in terms of $S_a(T_f)$. The soil class is "B", while $T_f = 0.56$ s.

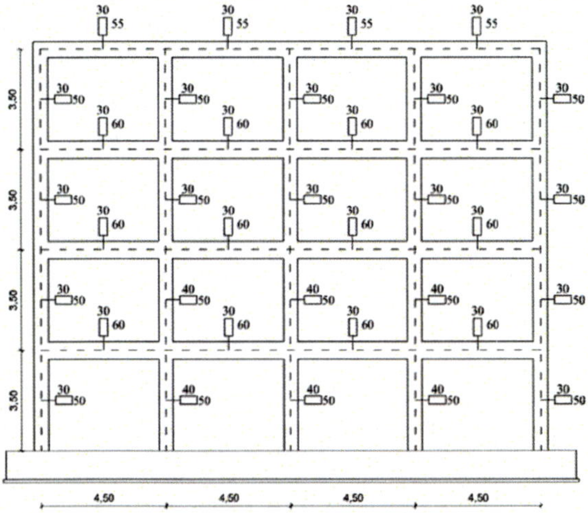

**Fig. 4.** Frame representation.

**Table 1.** Parametric field./Campo parametrico.

| Exposure | d' | w/c | RH |
|---|---|---|---|
| class | [mm] | [-] | [%] |
| XC2 | 20 | 0.50 | 55 |
| | | 0.55 | |
| | | 0.60 | |
| | | 0.60 | 35 |
| | | | 55 |
| | | | 75 |

## 3.3 Applications

The main results concerning the influence of *w/c* and *RH* are reported hereafter. For the sake of brevity, the results of two relevant Limit States (LSs) are reported: Damage Limitation LS and Life Safety LS. The values of chord rotation capacity ($E_{DP}$) for Life Safety LS has been defined according to "Equation A.1" of EN 1998-3:2005, whereas the capacity in the SLD is assumed constantly equal to 0.5%, as fixed by the Italian Code (2018). Such values are also reduced according to Eq. 4, as soon as the corrosion takes place.

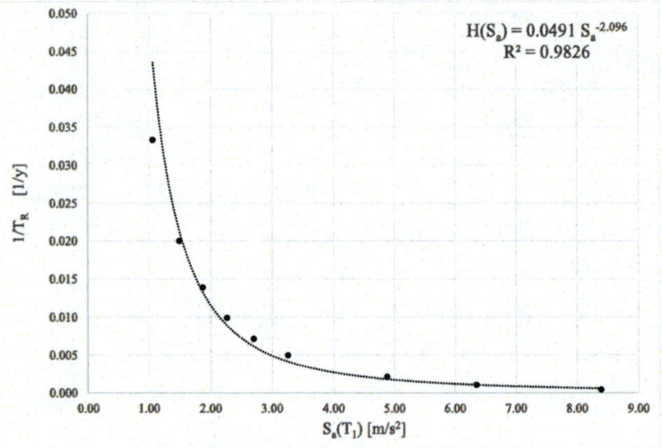


**Fig. 5.** Hazard function of the considered site (Avellino, soil class "B").

### 3.3.1 Variation of Water-Cement Ratio, w/c

In Fig. 6 a synthetic representation of the time-evolution (from 0 to 50 years) of pushover curves is presented, with the aim to show the influence of the water-cement ratio on its evolution.

Moreover, Fig. 7 represents the evolution of the return period $T_R$ for the Damage Limitation LS, while Fig. 8 represents the evolution of the return period $T_R$ for the Life Safety LS.

Finally, Table 2 and Table 3 summarize the comparison between the initial return period $T_{R,0\,y}$ and the equivalent $T_R^{ECR}$.

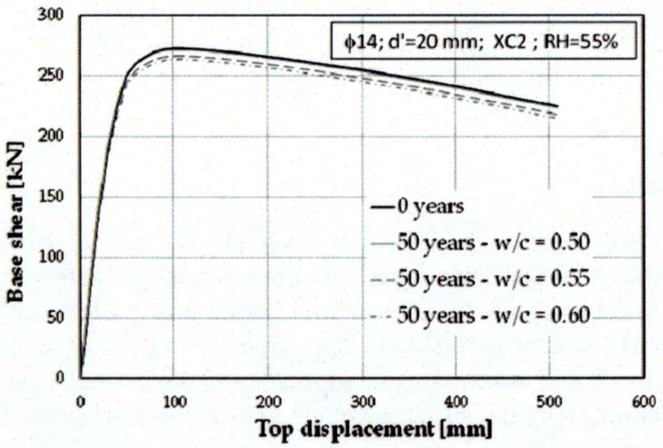


**Fig. 6.** Time evolution of the pushover curve varying the water-cement ratio (XC2, d' = 20 mm, RH = 55%).

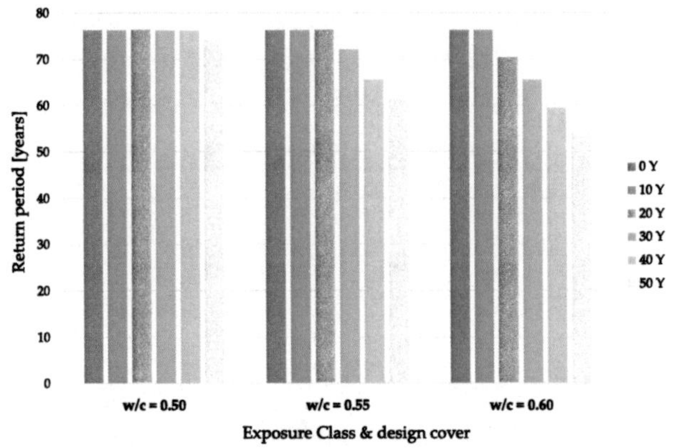


**Fig. 7.** Time evolution of return period for Damage Limitation LS varying the water-cement ratio (XC2, d' = 20 mm, RH = 55%).

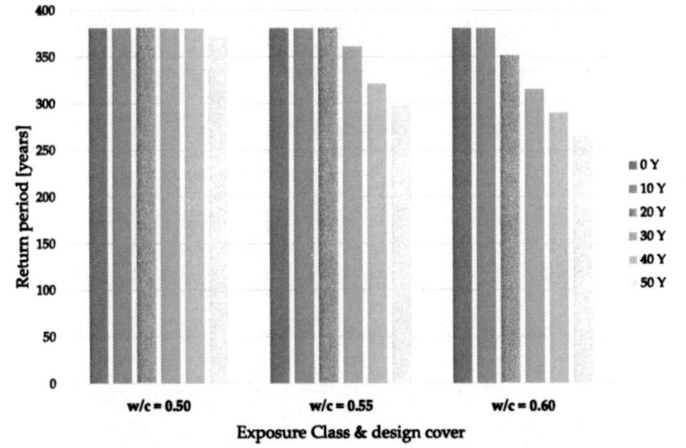


**Fig. 8.** Time evolution of return period for Life Safety LS varying the water-cement ratio (XC2, d' = 20 mm, RH = 55%).

**Table 2.** Influence of water-cement ratio. Difference betweeb $T_{R,0y}$, $T_R^{ECR}$ for Damage Limitation LS

| Water-cement ratio | $T_{R,0y}$ | $T_R^{ECR}$ | $T_R^{ECR} - T_{R,0y}$ | |
|---|---|---|---|---|
| | [years] | [years] | [years] | [%] |
| w/c = 0.50 | 76 | 76 | 0 | **0%** |
| w/c = 0.55 | 76 | 71 | 5 | **6%** |
| w/c = 0.60 | 76 | 67 | 10 | **12%** |

**Table 3.** Influence of water-cement ratio. Difference betweeb $T_{R,0y}$, $T_R^{ECR}$ for Life Safety LS.

| Water-cement ratio | $T_{R,0y}$ | $T_R^{ECR}$ | $T_R^{ECR} - T_{R,0y}$ | |
|---|---|---|---|---|
| | [years] | [years] | [years] | [%] |
| w/c = 0.50 | 381 | 380 | 2 | 0% |
| w/c = 0.55 | 381 | 355 | 28 | 7% |
| w/c = 0.60 | 381 | 331 | 52 | 13% |

### 3.3.2 Variation of Relative Humidity, RH

Figure 9 presents a synthetic representation of the time-evolution (from 0 to 50 years) of pushover curves, with the aim to show the influence of the RH on its evolution.

Moreover, Fig. 10 represents the evolution of the return period $T_R$ for the Damage Limitation LS, while Fig. 11 represents the evolution of the return period $T_R$ for the Life Safety LS.

Finally, Table 4 and Table 5 summarize the comparison between the initial return period $T_{R,0y}$ and the equivalent $T_R^{ECR}$.

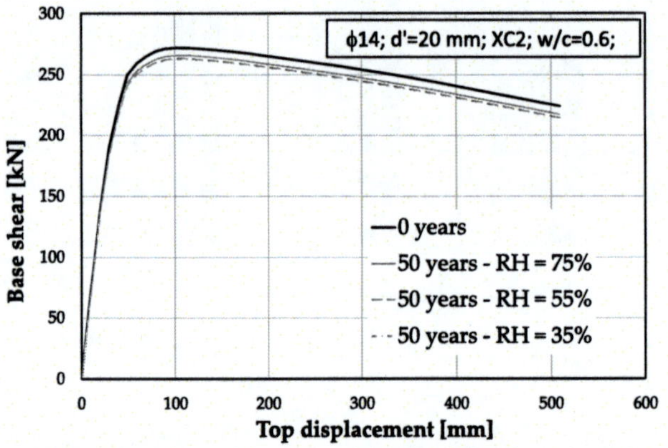


**Fig. 9.** Time evolution of the pushover varying the RH (XC2, d' = 20 mm, w/c = 0.60).

## 3.4 Result Discussion

The influence of the aforementioned parameter is highlighted by the comparison between the risk (in terms of return period) calculated neglecting the effects of environmental actions $T_{R,0y}$ and the equivalent risk $T_R^{ECR}$ faced by the structure during its service life. Such comparison is possible since the risk is computed at different stages of the theoretical structural service life $T_d$.

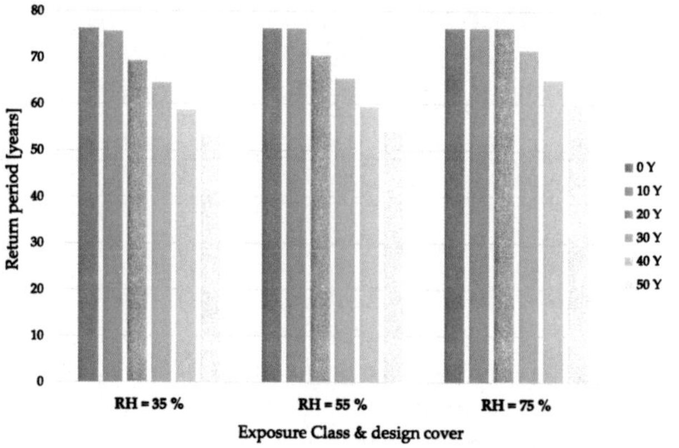


**Fig. 10.** Time evolution of return period for Damage Limitation LS varying the RH (XC2, d' = 20 mm, w/c = 0.60).

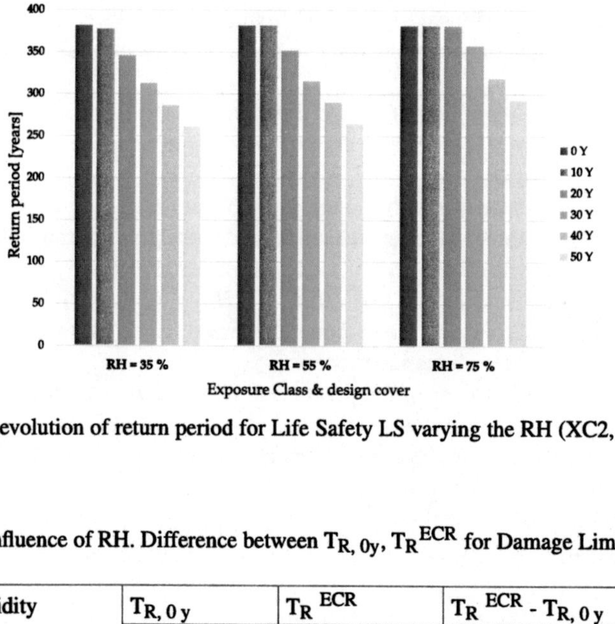


**Fig. 11.** Time evolution of return period for Life Safety LS varying the RH (XC2, d' = 20 mm, w/c = 0.60).

**Table 4.** Influence of RH. Difference between $T_{R,\,0y}$, $T_R^{ECR}$ for Damage Limitation LS

| Relative Humidity | $T_{R,\,0\,y}$ | $T_R^{\,ECR}$ | $T_R^{\,ECR} - T_{R,\,0\,y}$ | |
|---|---|---|---|---|
| | [years] | [years] | [years] | [%] |
| RH = 35% | 76 | 66 | 12 | **15%** |
| RH = 55% | 76 | 67 | 12 | **15%** |
| RH = 75% | 76 | 71 | 8 | **11%** |

**Table 5.** Influence of RH. Difference betweeb $T_{R,\,0\,y}$, $T_R^{ECR}$ for Life Safety LS.

| Relative Humidity | $T_{R,\,0\,y}$ | $T_R^{\ ECR}$ | $T_R^{\ ECR} - T_{R,\,0\,y}$ | |
|---|---|---|---|---|
| | [years] | [years] | [years] | [%] |
| RH = 35% | 381 | 328 | 64 | **16%** |
| RH = 55% | 381 | 331 | 63 | **16%** |
| RH = 75% | 381 | 353 | 46 | **12%** |

It is possible to summarize the main results (that result to be valid for each Limit State studied) in the following points:

- considering that the water-cement ratio (Sect. 3.3.1) is directly proportional to the inverse carbonation resistance $R^{-1}{}_{ACC,0}$, a value of *w/c* = 0.50 can thoroughly reduce the carbonation process in concrete elements. This effect is evident since, in the case of CEM III/B 42.5, passing from w/c = 0.60 to w/c = 0.50, the inverse carbonation resistance $R^{-1}{}_{ACC,0}$ is almost tripled; this circumstance can be easily explained by the fact that lower values of *w/c* imply lower concrete porosity, that delays the degradation phenomena.
- consequently, the usage of minor w/c is recommended in order to achieve sufficiently durable performances and allows, as well, to adopt a minor value of the design concrete cover;
- regarding the influence of relative humidity (*RH*)–outlined in Sect. 3.3.2-the results obtained using the three values of RH considered in the present study, although really distant, do not show a clear effect of the RH variation on Return Period reduction;
- nevertheless, it is worth saying that values of *RH* higher than 85%, if constant over time, can significantly delay the carbonation spread, since the presence of water in the pores inhibits the carbonation process;
- although the average values of the parameters describing material degradation phenomena may not usually lead to huge demand increments, corrosion spread can produce remarkable reductions of columns capacity in terms of chord rotation, conditioning the reliability assessment.

# 4 Conclusions

In the present paper the results of a preliminary approach for practice-oriented seismic reliability analysis of RC structures affected by carbonation-induced material degradation phenomena are presented.

Specifically, the influence of two relevant parameters, according to the degradation models (outlined in Sect. 2.1) on the reliability performance of a typical RC frame is explored. Through the aforementioned results, it should be observed that:

- when a maintenance plan is not properly implemented, the effects of material degradation phenomena should not be neglected in the seismic reliability assessment of RC structures;

- although carbonation-induced degradation may induce "milder" effects with respect to other degradation phenomena (e.g. pitting corrosion) it can actually reverberate its effect at the structural scale increasing the displacement demand for the structure and reducing members' chord rotation capacity, as well;
- the outcomes of the employed simplified procedure enable to predict explicitly the structural performance in terms of durability and are comparable with the deemed-to-satisfy provisions, given in EN 1992-1-1:2004, concerning the minimum values of concrete design cover;
- the variable $T_R^{ECR}$ is a good indicator of the structural ageing. As shown in the Tables of Sect. 3, if the difference between $T_{R,0\,y}$ and $T_R^{ECR}$ is up to 15%, the effect of ageing may be neglected, while higher values of this difference certainly imply a safety decrease (in terms of return period) higher than 25% (after 50 years);

Finally, it is worth highlighting that, although the actual rate of diffusion of degradation phenomena may lead to a certain variability of the structural behaviour, the manageable models currently available in literature give the possibility to obtain an estimation of structural reliability in a simple, but still comprehensible, way.

## References

Cornell, C.A., Jalayer, F., Hamburger, R.O., Foutch, D.A.: Probabilistic basis for 2000 SAC federal emergency management agency steel moment frame guidelines. ASCE J. Struct. Eng. **128**, 526–533 (2002)

DM 17/01/2018 (Italian Technical Code of constructions)

Dolšek, M., Fajfar, P.: IN2 – A simple alternative for IDA. In: Proceedings of the 13th World Conference on Earthquake Engineering, Vancouver, BC, Canada, 1–6 August 2004, no. 3353 (2004)

DuraCrete. Probabilistic Performance Based Durability Design of Concrete Structures: Statistical Quantification of the Variables in the Limit State Functions. Report No.: BE 95–1347. CUR, Amsterdam (2000)

EN 1998–3:2005. Design of structures for earthquake resistance. Part 3: Assessment and retrofitting of buildings. European Committee for Standardization, Bruxelles (2005)

EN 1992–1–1:2004. Design of Concrete Structures—Part 1–1: General Rules and Rules for Buildings. Comité Européen de Normalisation, Brussels, Belgium (2004)

Erduran, E., Martinelli, E.: Some remarks on the seismic assessment of RC frames affected by carbonation-induced corrosion of steel bars. CACRCS DAYS 2020: Capacity Assessment of Corroded Reinforced Concrete Structures (2020)

Faella, C., Lima, C., Martinelli, E.: Non-linear static methods for seismic fragility analysis and reliability evaluation of existing structures. In: The 14th World Conference on Earthquake Engineering, Beijing, China, 12–17 October 2008 (2008)

Fajfar, P.: Capacity spectrum method based on inelastic demand spectra. Earthq. Eng. Struct. Dynam. **28**, 979–993 (1999)

FEMA (Federal Emergency Management Agency). 2009. *Quantification of Building Seismic Performance Factors*

Fib. Model Code for Service Life Design; Fédération Internationale du Béton: Lausanne, Switzerland (2006)

François, R., Laurens, S., Deby, F.: Steel corrosion in reinforced concrete. In: Corrosion and its Consequences for Reinforced Concrete Structures, 1st edn., pp. 1–41. ISTE Press Ltd., London. Elsevier Ltd., New York (2018)

Mazzoni, S., McKenna, F., Scott, M.H., Fenves, G.L., et al.: Open System for Earthquake Engineering Simulation. User Command-Language Manual (2006)

Nigro, F., Zinco, A., Martinelli, E.: A practice-oriented procedure for seismic reliability assessment of rc structures affected by carbonation-induced degradation. Appl. Mech. **2**, 820–840 (2021). https://doi.org/10.3390/applmech2040047

Panagiotakos, T., Fardis, M.: Deformation of reinforced concrete members at yielding and ultimate. ACI Struct. J. **98**, 135–148 (2001)

Pacific Earthquake Engineering Research Center (PEER NGA DATABASE). http://peer.berkeley.edu/nga

Pinto, P.E., Giannini, R., Franchin, P.: Specialized methods for seismic problems. In: Seismic Reliability Analysis of Structures, 1st edn., pp. 215–218. IUSS Press, Pavia (2004)

Regio Decreto 16/11/1939, n.2229. (The eldest Italian Technical Code for Buildings)

Vamvatsikos, D., Dolšek, M.: Equivalent constant rates for performance-based seismic assessment of ageing structures. Struct. Saf. **33**, 8–18 (2011)

Vecchi, F., Belletti, B.: Capacity assessment of existing RC columns. Buildings **11**, 161 (2021)

# Winterization Methods in Post-tensioning Tendons

Tommaso Ciccone(✉), Luca Civati, and Giacomo Liberali

Tensacciai Srl, Milan, Italy
t.ciccone@tensainternational.com

**Abstract.** Prestressed concrete elements are nowadays used in several applications very different from the standard concrete structures built in the past. In such cases, post-tensioning systems are by far the best choice, considering their great flexibility. To achieve the most effective results for post-tensioning applications, a dedicated careful design is always required, even more when special boundary conditions are given to the project, such as construction of a post-tensioned structure in arctic climates. This case is nowadays more and more common, due to research in new energy sources like the use of liquefied natural gas.

In these conditions, special winterization measures for the post-tensioning system have to be designed, in order to manage properly extreme low temperature conditions. Among post tensioning activities, ones of the most important requiring special winterization measures are the grouting operation, i.e. how to fill tendons' ducts, and the pouring of special post-tensioning anchorages. In all these cases, heating of ducts and anchorages is required to guarantee proper grout curing.

The aim of this article is to provide a review of the possible winterization systems used in post-tensioning works, highlighting pros and cons for each one of them. Then, several applications used in job sites are presented.

**Keywords:** prestressed concrete · durability · post-tensioning tendons · winterization

## 1 Introduction

In recent years post-tensioning technologies have been involved in new projects requiring many times customized technical solutions to deal with constructability issues. Several problems are found when the post tensioned structure has to be built in very cold climates (ambient temperature less than 0 °C).

While tendons threading and stressing operations are not directly affected by cold temperature, the main problem raises with operations where cement material curing is involved. The most significant cases are:

- ducts injection with cement grout;
- bond anchorage injection with mortar or grout.


M. A. Aiello and A. Bilotta (Eds.): ICC 2022, LNCE 435, pp. 575–590, 2024.
https://doi.org/10.1007/978-3-031-43102-9_44

While grouting is required in all grouted post-tensioning tendons, a bond anchorage injection is a special application. In fact, this type of anchorage is generally used to prestress vertical elements like silos, tanks and walls, where there is no possibility to access to the bottom anchorage. These types of structures are often built in arctic zones, for example for liquefied gas containment.

Scope of this paper is to describe the issue and possible solutions, together with a case study where the solution was used and how it was numerically validated before the jobsite application.

## 2 Standards for Cold Temperature Operations

The main reference standard for duct grouting in cold climate is PTI M55.1-19 (2019), where the "cold climate" is defined as:

*"The ambient temperature shall be considered as the conditions for determining cold climate grouting conditions. Cold climate conditions shall be in effect whenever the ambient temperature is 4 °C (40 °F) and falling. Satisfactory mixing and protection conditions shall be employed."*

Indication to perform grouting in any way is:

*"Grouting operations shall be postponed if frost is expected within the next 2 days. However, if grouting has been performed, the temperature of the grout in the ducts shall be kept above 2 °C (35 °F) for three consecutive days after grouting, or until job-cured 50 mm (2 in.) cubes of grout reach a minimum compressive strength of 5.5 MPa (800 psi). To achieve this, the Contractor may use methods such as the installation of electric heating coils in the concrete, insulating the structure, or applying external heat."*

*fib* Bulletin 20 (2002) and EN 446 (2007) report almost the same recommendation with the second one saying that:

*"No grout shall be placed if the temperature of the structure adjacent to the tendon is below 3 °C, or is likely to fall below 3 °C during the following 48 h, or a higher temperature limit specified by the manufacturer for winch he has confirmed compliance of the grout with EN 447, unless the structure adjacent to the tendons is heated so as to maintain the temperature of the placed grout above 5 °C, or 2 °C above the temperature specified by the manufacturer, for at least 48 h."*

Both American and European standards indicate the minimum temperature below which injection operations must be avoided. However, special winterization measures are allowed. The aim of these measures is to provide a temperature of more than 5 °C around the duct to allow the correct grout curing.

No indication can be found in International standards for bond anchorages' injection; however, it is consistent to consider the same indications also in this case.

Several International standards, such as CWA 14646 (2003), allow to stress tendons only above a certain temperature.

Actually, by a technical point of view, temperature has to be taken in consideration in stressing operations for at least two reasons.

The first is that, when at the time of stressing temperature is very different from the one at the time of grouting, the thermal deformation occurring can increase or decrease the load within the tendon and such situation has to be considered in the design stage.

The second is that, if stressing is carried out at very low temperature, the following grouting operation very probably has to be managed in the same conditions and so winterizations measures, or just use of corrosion protection powders for tendons, have to be considered. However, threading and stressing operations in cold climates require in any case winterization measures on the equipment to be used (proper hydraulic oil, heaters).

## 3 Problems and Possible Solutions

The company Tensacciai S.r.l. is involved in important arctic projects for the construction of several lightweight concrete LNG tanks. In this application post tensioning (PT) tendons are installed in all structure walls in both directions (horizontal and vertical). Vertical tendons are made with the use of a dead bond anchorage (Tensacciai MTDB) on the lower end.

In the below paragraphs issues due to climate found in this project, solutions used, and how these were validated and installed, are presented.

### 3.1 Tendon Grouting

Grout used to fill tendons' duct is a cement-based grout composed by standard Portland cement, admixture and water. As for concrete, the curing required temperature has to be greater than 5 °C.

When this temperature on site falls below this value, winterization countermeasures can be:

- use of special admixtures (chloride and calcium nitrate, or chloride, calcium nitrite and nitrate): an issue is the validation of products which can be potentially aggressive for strands and could have lower compression strength. An application is described in Yang at al (2015);
- insulation and heating of concrete structure from the external: this method is often complex and expensive due to the necessity of covering large surfaces and unusual shapes;
- heating the duct and the concrete around it by mean of heating wires. Normally, this is the simplest and cheapest solution;
- a combination of the previous solutions.

### 3.2 Bond Anchorage Grouting (Tensacciai MTDB System)

A bond anchorage is a special post-tensioning anchorage that uses bond to transfer force between strands and concrete. In the past this type of anchorage (sometimes also known as "dead") was the simplest post-tensioning method and was done just by directly installing strands into the concrete. In recent years more effective post-tensioning anchorages have been developed. The Tensacciai MTDB system (Fig. 1) is a dead-end passive anchorage which is mainly designated for vertical or nearly vertical tendons and especially for the cases where the passive end presents no direct access. It can be found in many applications such as for example: bridge piers, silos, building cores, wind towers, reservoirs, retaining walls, dams, inclined structures, etc.

The system is composed by an anchorage made of a cast-iron block with relevant bottom closing plate embedded into concrete which contains the prestressing strands, each one provided with a compression grip extruded on the end. The anchorage is filled with high-performance grout in order to create the bond between the anchorage and strands, which allows to transfer the prestressing forces to the structure. The high-performance grout injection is made through an inlet vent placed at the bottom part of the anchorage.

The prestressing force is transmitted by the bond of the strands and the compression grips to the high-performance grout. The load is then transferred to the anchorage by interaction effect induced by the draw-in of the wedge-shaped hardened grout inside the cast-iron block.

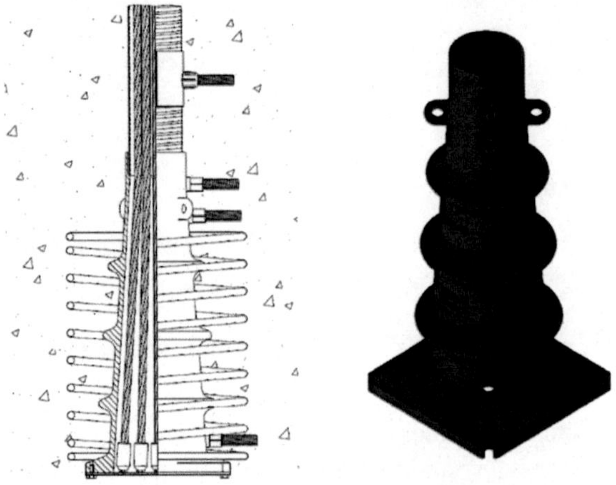

**Fig. 1.** Tensacciai MTDB system

The MTDB anchorage is often used in vertical PT cables in LNG tanks to prestress the walls or in circular silos to introduce vertical prestressing in the shell.

Winterization countermeasures, in order to allow the correct high-performance grout curing, can be:

- insulation and heating of concrete structure from external: this method is often complex and expensive;
- heating of the cast iron block and the surrounding concrete by means of heating wires. Normally, this is the simplest and cheapest option.

In the following paragraphs the winterization method with heating wires application is analyzed for both cases here below described. In case of ducts, two heating wires are installed along the entire duct's length and around the anchorages at the ends. While, in case of an MTDB anchorage, a single wire (with its specific length) is wrapped around the anchorage.

### 3.3 Adopted Solution

The solution proposed and used for both cases is the application (around the duct or around the MTDB anchorage) of a special heating cable (described in Fig. 2).

The cable used is a self-adjusting heating cable with a delivered emission power of 40 W/m at 10 °C and 55 W/m at −10 °C. The heating cable automatically varies heat emission without control thermostats being used. It consists of a conductor insulated by polymers, which varies its resistance according to the temperature it is subjected to.

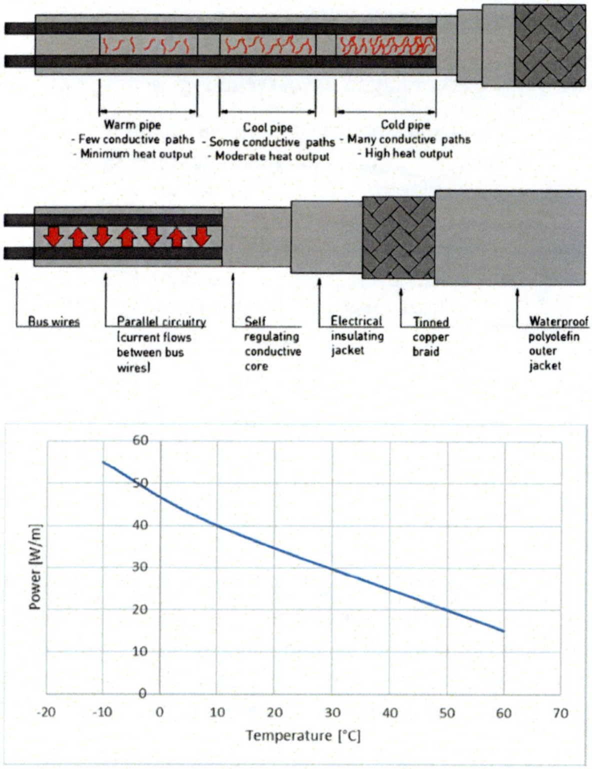


**Fig. 2.** Heating cable, Temperature-power curve

The main characteristics of the system are:

- cable can be cut to the desired length;
- total power depends on the length of the heating cable and the perceived temperature;
- cable is secured along the pipe with the relevant fixing;

- minimum bending radius: 25 mm;
- cable heats up along its entire length and can be wound in overlapping spirals as it limits its power at the overlapping points;
- temperature-power curve is shown in Fig. 2.

#### 3.3.1 Tendon's Heating

In order to evaluate the efficiency of the heating system described in the previous paragraph, several finite element analyses (with Straus7 software) have been carried out. The scope of the analysis is to understand if power emitted by the heating cable is enough to allow reaching a temperature greater than 5 °C around the duct. Moreover, the target was also to understand the time needed by the system to reach the above temperature.

Conditions considered are:

- temperature at external face of the wall: −10 °C;
- temperature at internal face of the wall: −5 °C;
- initial temperature on concrete/duct/strands: −5 °C;
- power emitted by wire according to the curve in Fig. 2.

All materials are considered isotropic and, in particular, reference values for steel, concrete and air were identified by common literature (refer to Table 1).

**Table 1.** Material properties

| Material | Density [kg/m$^3$] | Specific heat [J/kgK] | Conductivity [W/mK] |
|---|---|---|---|
| Light weight concrete | 2250 | 800 | 0.8 |
| Light weight concrete with steel bars | 2770 | 660 | 0.55 |
| Steel | 7850 | 500 | 51 |
| Air | 1.2 | 1005 | 0.026 |

A further study was dedicated to the evaluation of the parameters for lightened concrete in which reinforcements are present: they were estimated assuming a ratio between volumes of reinforcement/concrete equal to about $\nu = 20\%$. In particular, the followed approaches are summarized below regarding thermal properties. Specific heat $C_p$ was evaluated as mass-weighted average present in the volume unit:

$$C_p = \frac{(\rho_{cls} \cdot V_{cls} \cdot C_{p\_cls}) + (\rho_{arm} \cdot V_{arm} \cdot C_{p\_arm})}{(\rho_{cls} \cdot V_{cls}) + (\rho_{arm} \cdot V_{arm})} \tag{1}$$

where:

- $\rho_{cls}$, $V_{cls}$ and $C_{P\_cls}$ are concrete density, volume and specific heat respectively;
- $\rho_{arm}$, $V_{arm}$ and $C_{P\_arm}$ are rebar reinforcement density, volume and specific heat respectively.

Conductivity was instead assessed with reference to the proposal of Zhao et al (2013) which presents an analytical approach for the determination of the thermal conductivity $k$ in reinforced concrete according to the percentage and type of reinforcement present:

$$k = \frac{(1-a) \cdot k_{cls}^2 + a \cdot k_{cls} \cdot k_{arm}}{\left(1 + a^2 - a\right) \cdot k_{cls} + \left(a^2 - a\right) \cdot k_{arm}} \tag{2}$$

where:

- $a = \sqrt{\nu}$
- $k_{cls}$ and $k_{arm}$ are concrete and reinforcement conductivity respectively.

Two cases have been considered:

- case A, with empty duct;
- case B, with a central mass equal to the mass of strands bundle and air between duct and strands.

A non-linear analysis has been carried out in order to take into account the bi-linear law temperature-power of the cable. The concrete and steel have been modeled as isotropic materials with constant thermal characteristics (conductivity and specific heat).

CASEe A: EMPTY DUCT

Considering that the duct and heating wires run along the entire wall length (about 330 m), thermal plain conditions are considered.

The model represents a wall section with a height equal to ducts spacing (symmetry condition), the wall is 70 cm width and the steel duct has an internal diameter equal to 110 mm with a thickness about 0.6 mm. The FEM elements are plate type (mean dimensions 20 mm), the initial conditions are the nodes temperature while the external wall temperature are constant boundary conditions. The heating cable is modelled as heat flux on edge plates (refer to Fig. 3, light blue vectors are the flux while green vectors are the temperature boundary conditions).

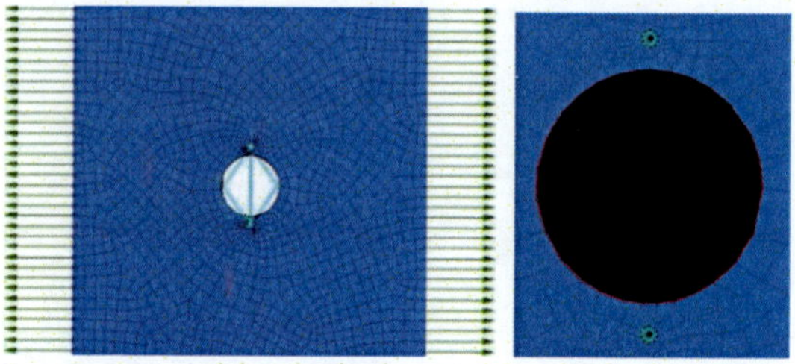

(a) (b)

**Fig. 3.** FEM model (a), duct and wire details (b)

Figure 4 shows duct and concrete temperature variation. The analysis simulates a duration of 72 h.

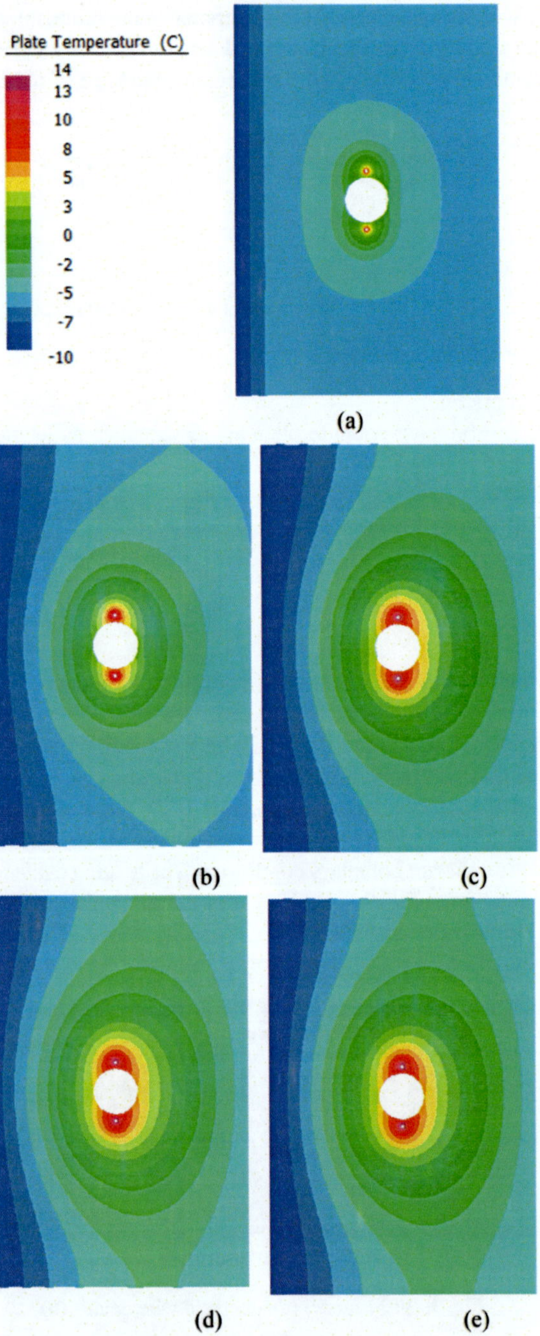


**Fig. 4.** Temperature at 1 h (a), 6 h (b), 24 h (c), 48 h (d) and 72 h (e)

As shown in Fig. 4, after 48 h duct and surrounding concrete temperatures are greater than 0 °C for all points with a value around 7 °C.

The diagram in Fig. 5a shows the time-temperature curve of the coldest duct point; after 48 h there is a steady situation with a temperature greater than 4 °C.

While in the diagram in Fig. 5b, the temperature of the wire and the electrical power adsorbed are shown.

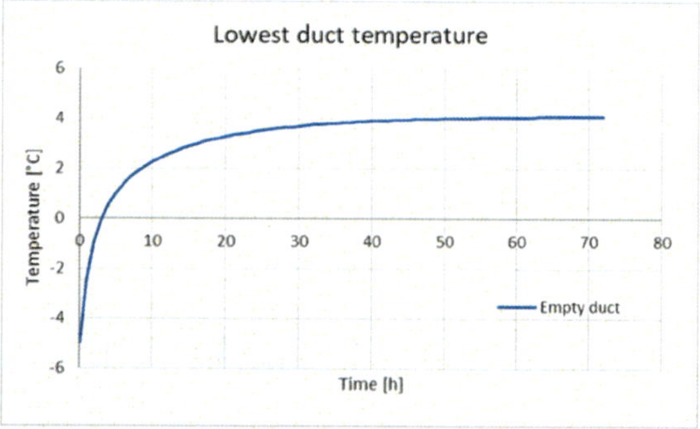


(a)

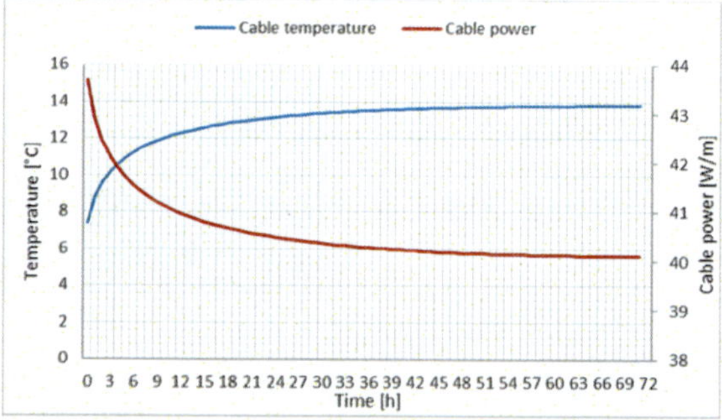


(b)

**Fig. 5.** (a) Duct temperature, (b) Cable power and temperature

CASE B: DUCT WITH STRANDS

The model represents a wall section with a height equal to ducts spacing (symmetry condition), the wall is 70 cm width and the steel duct has an internal diameter equal to 110 mm with a thickness about 0.6 mm. The FEM elements are plate type (mean dimensions 20 mm), the initial conditions are the nodes temperature while the external wall temperature are constant boundary conditions. Strands are modeled as a single element in the center of the duct with an area equal to the area of real strands bundle and the space between strand and duct is modelled as air element (refer to Fig. 6).

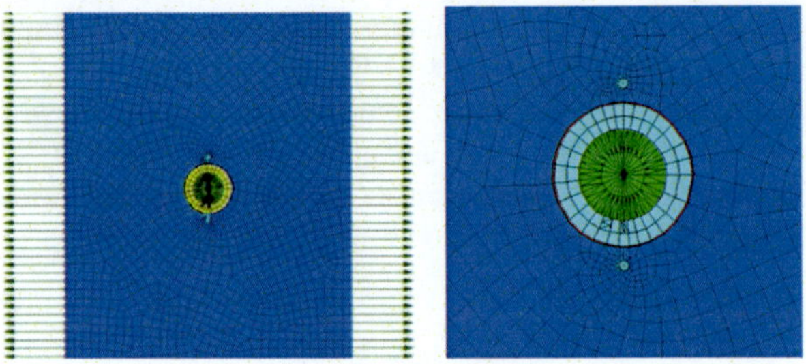

**Fig. 6.** FEM model and duct detail

Figure 7 shows duct and concrete temperature variation. The analysis simulates a duration of 72 h.

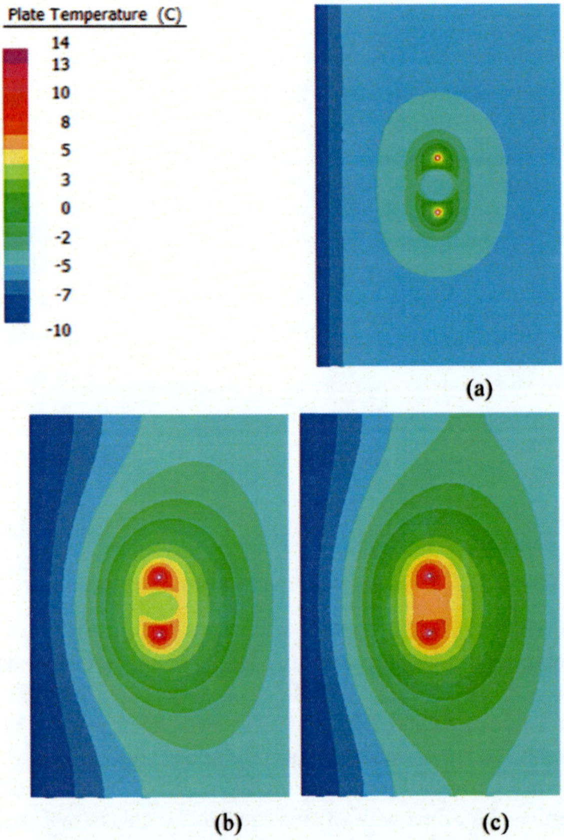


**Fig. 7.** Temperature at 1 h (a), 24 h (b) and 72 h (c)

Results are comparable with case A. In this case, it results that some more hours are required for reaching of the target temperature. But the temperature reached after 72 h by strands is greater than the previous case, due to the fact that steel has a greater thermal inertia.

Figure 8a shows the time-temperature curves of the coldest duct point and of strands; after 72 h there is a steady situation with a temperature greater than 4 degrees for the duct and 6° for the strand. In Fig. 8b the temperature of the wire and the electrical power adsorbed are shown. Values are similar to those resulting in case A.

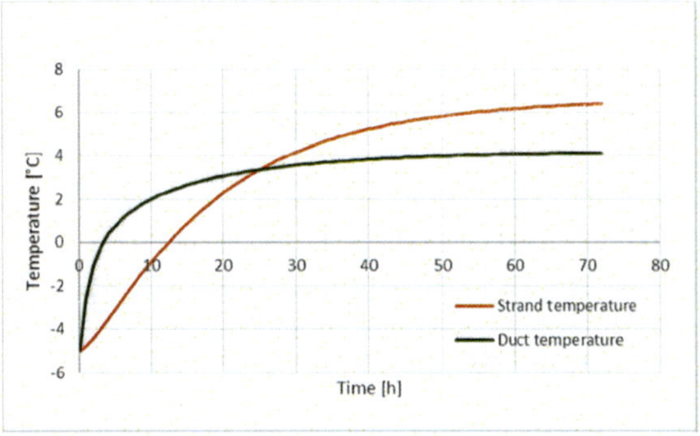


(a)

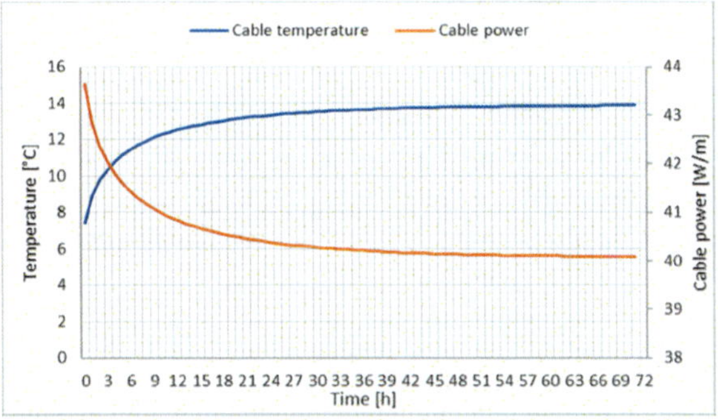


(b)

**Fig. 8.** (a) Duct and strands temperature, (b) Cable power and temperature

Figure 9 summarizes the FEM analysis results, i.e. duct temperature in case A, duct and strand temperature in case B.

It is therefore possible to define when it is time to turn on the system before performing grouting operations. Models show that, in both analyzed cases, the duct reaches in 48 h the average target temperature of 5 °C.

### 3.3.2 Bond Anchorage (Tensa MTDB System) Heating

In this case the 3D thermal analysis was carried out by Politecnico di Milano (under an agreement signed between Tensacciai and Politecnico). The model considers a portion

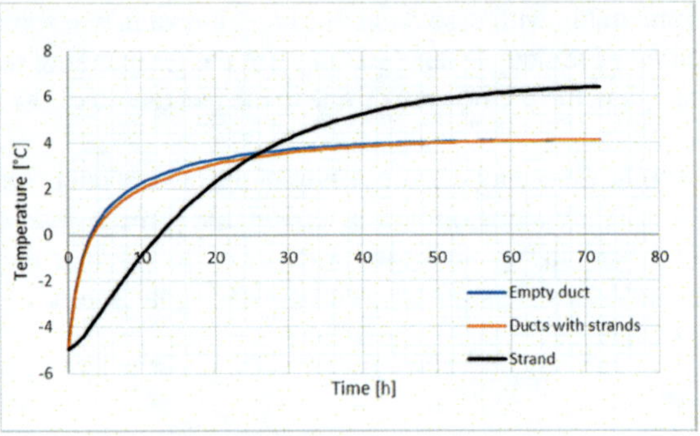


**Fig. 9.** Duct temperature case A and B

of wall, foundation and soil with a dead anchorage in the actual position (Fig. 10). The width of the modeled solid is the pitch between two tendons.

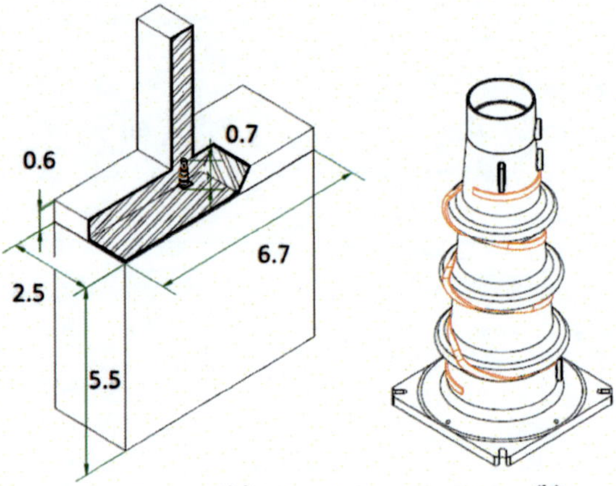


**Fig. 10.** (a) FEM model and (b) anchorage detail

The heating wire is idealized as a solid and can provide a heat flux in all directions.

Material data (density, conductivity, specific heat) used are the same computed for the plane analysis of the duct.

Considered temperature conditions are the following:

- Temperature at external face of the walls: −10 °C;
- Temperature at surrounding faces of the soil: 2 °C;
- Initial temperature on concrete/duct/strands: 5 °C;
- Power emitted by wire according to the curve shown in Fig. 2.

Detailed results are reported in Fig. 11 through some 3D views of the cast iron anchorage body.

A further post-processing check leads, first, to a temperature time history at 2 points inside cast iron block (Fig. 12).

Results confirm the solution suitability. After several hours (10/15 h) the temperature at point B (coldest point inside the anchorage) reaches the target value of 5 °C and it remains constant.

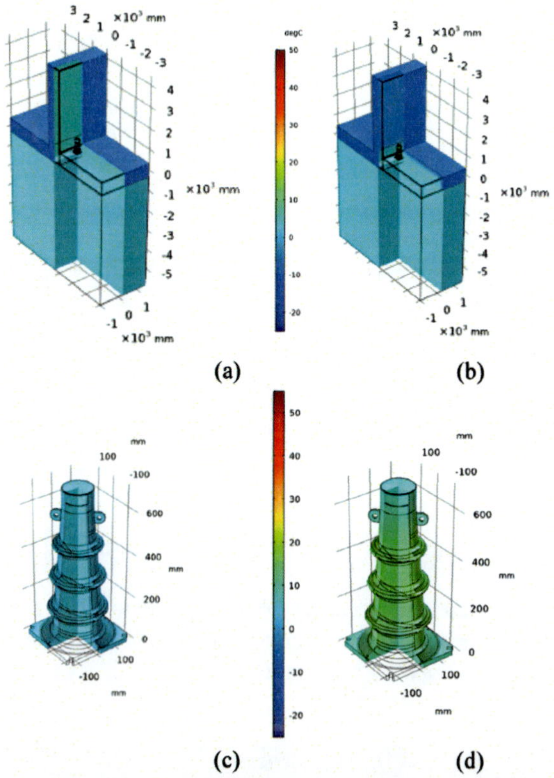

Fig. 11. Temperature at 0 h (a & c) and 11 h (b & d)

### 3.4 On Site Application

In the following paragraphs it is described how the system is installed. The heating system is supplied at the jobsite in small coils and ready for the installation. The connection between two pieces can be done in a simple way (like electrical wires), thus it is easy to install the system also when concrete is poured in segments.

#### 3.4.1 Heating Wire on Ducts

For the duct winterization the two wires are installed straight along the entire duct length and properly fixed with tape. At cable ends, it is important to extend the wire on

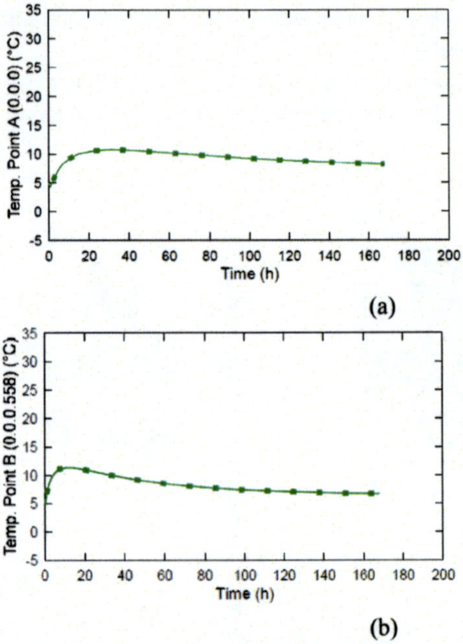


**Fig. 12.** Temperature at points A (a) and B (b)

the anchorages in order to guarantee proper heating also inside anchorage components (Fig. 13).

Then recesses are fixed to rebars or formwork in order to house sufficient extra wire to make the connection with the power cable (Fig. 14). The maximum wire length has to be chosen according to supplier's indication.

**Fig. 13.** Heating system fixed on duct and anchorage

(a)

(b)

**Fig. 14.** Recess before (a) and after (b) concrete cast

After concreting, when the heating system has to be turned on, the heating wires are connected to the power supply. The absorbed power depends on the total length of heating cable.

**Fig. 15.** Wire installation on anchorages

### 3.4.2 Bond Anchorage Grouting (Tensacciai MTDB System)

For the MTDB anchorage the wire is wrapped around the cast iron block (Fig. 15). A single wire can be used for several anchorages; however, it must be considered that the winterization system will then work contemporarily on these elements. It is therefore worth to choose tendons to be injected according to PT work schedule. Then, also in this case, recess is fixed to formwork to collect the wire end.

## 4 Conclusions

Post-tensioning is nowadays widely spread in new construction typologies and even in arctic climates where special projects require special applications. Adequate technological solutions are needed to address problems found in cold climates and those proposed here are effective both from the theoretical and field application points of view. The self-adjusting cables allow to guarantee the minimum temperature of the duct or dead anchorage, also with a lower ambient temperature. If the wires are embedded in a colder concrete, the emitted power increases automatically without any type of thermocouples.

The thermal analysis, carried out and described in the above paragraphs, confirm the efficiency of the proposed system. Also the experimental tests on site, through measures taken with thermocouples, showed temperatures fully compliant with those required for proper installation. The solution adopted confirms to be the most suitable one to deal with arctic conditions present in special projects with post-tensioning tendons.

## References

PTI M55.1-19. Specification for Grouting of Post-tensioned Structures (2019)

EN445. Grout for prestressing tendons – Test methods (2007)

EN446. Grout for prestressing tendons – Grouting procedures (2007)

fib Bulletin 20. Grouting of tendons in prestressed concrete (2002)

CWA 14646. Requirements for the installation of post-tensioning kits for prestressing of structures and qualification of the specialist company and its personnel (2003)

Yang, F., Ma, R., Zhang, Y., Zhang, Y.: Grouts for bridge post-tensioning tendons at below-freezing temperature. In: ACI SP303-20, pp. 21–32 (2015)

Zhao, S.B., Yang, S., Feng, X.Z., Lu, M.J.: Study on thermal conductivity of reinforced concrete plate. Appl. Mech. Mater. **438–439**, 321–328 (2013)

# Sustainable Concretes for the Offshore Wind Turbine Industry: Evaluation of the Durability of Innovative Materials in Offshore Structures

Vito Tarantino[1(✉)], Ilaria Ingrosso[1], Amaia Gomez San Martin[2], Valle Chozas Ligero[2], and Riccardo Angiuli[1]

[1] European Research Centre for Design Technologies and Materials, Brindisi, Italy
vito.tarantino@cetma.it

[2] Acciona Construcción Technology Centre, Madrid, Spain
amaia.gomez.sanmartin.ext@acciona.com

**Abstract.** Global warming derives from the emission of large quantities of greenhouse gases (ghg) into the atmosphere and cement production is one of the main processes responsible for environmental pollution. The production of eco-sustainable concretes with recycled aggregates can contribute to emissions reduction, but the development of new binders able to replace, at least in part, cement would be even more important. This paper presents some results on innovative concretes investigated in marewind project, an ambitious eu project that aims the development of a more durable concrete materials to achieve a sustainable foundation structure (lighter and enhanced durability) for an offshore wind turbine compared to the traditional solution. The results of the tests carried out on two different mixtures of eco-sustainable concretes, Alkali Activated Concrete (AAC) and Ultra High-Performance Concrete (UHPC), will allow to identify the best performing formulations in terms of mechanical and durability behaviour of the concretes in relevant environment.

**Keywords:** concrete · geopolymer · durability · recycled materials

## 1 Introduction

### 1.1 Background

Portland cement production is characterized by high energy demands, the consumption of non-renewable raw materials and the emission of greenhouse gases (essentially $CO_2$). For this reason it is necessary that the new frontiers of scientific research are focused on increasing the durability of structures made with Ordinary Portland Cement (OPC) and, at the same time, developing new binders capable of replacing or reducing, at least in part, the cement content.

The use of concrete in offshore structures is increasing in the last decade, and recent studies highlight the benefits of its use in this application, due to low installation cost, abundancy, well known manufacturing and placement process and better durability, are


M. A. Aiello and A. Bilotta (Eds.): ICC 2022, LNCE 435, pp. 591–604, 2024.
https://doi.org/10.1007/978-3-031-43102-9_45

some of the reasons that allow concrete (pre-stressed and reinforced concrete) to be a competitive and attractive material for the design of offshore structures [1, 2]. However, in the aggressive marine environment, traditional construction material generally undergo several durability threats during the service life, mainly related to corrosion and acid/sulphate attack of concrete and reinforced structures.

In the European MAREWIND project (Grant Agreement N° 952960), to which this study is involved to, particular attention has been paid to aspects of sustainability and durability of building materials, through an experimental program focused on Alkali Activated Concretes (AAC) as novel sustainable building material of substituting Portland cement-based concrete, and Ultra High Performance Concrete (UHPCs) to reduce maintenance and repairing costs, besides provide a more sustainable solution.

Referring to the Alkali Activated Materials AAM, Alkali activated slag (AAS) is the most renowned type of alkali activated material (AAM) due to its ability to consume less energy when used to replace OPC as binder [3]. Also, the production of 1 ton of slag requires about 1300 MJ of energy and produce only 0.07 ton of $CO_2$. While the production of equivalent amount of OPC requires 5000 MJ of energy and emits 1 ton of $CO_2$ into the environment [4]. In addition, AAS composites have been reported to enhance the mechanical and durability properties of concrete [5, 6].

However, major limiting factors to the use of AAS are high shrinkage [7, 8], and quick setting time [9].

Consequently, the present study on AAM is aimed to evaluate the feasibility of using glass powder combined with blast furnace slag waste activated with sodium hydroxide (NaOH) and sodium silicate ($Na_2SiO_3$), also known as waterglass, to obtain alternative cementitious binder, termed "geopolymer", to replace the cement.

In order to provide a general overview of the sustainability of the geopolymer mixtures developed, this part of the experimental program, therefore, first covered the development of the geopolymer binder and then the design and characterization of the rheological behavior in the fresh (density, workability, air content) and hardened (compressive strength) state of geopolymer mortars (AAMs) and concretes (AAC).

Referring to Ultra High Performance Concrete (UHPC), this advance material, is one of the recent developments in concrete technology that offers a structural behavior more similar to steel itself than to conventional concrete, especially due to its highly ductile behavior when subjected to stress. UHPC stands out for its high compressive strengths ($\pm$120–200 MPa) similar to the values get for steel. It has a substantial autogenous shrinkage within the first days, and restrained shrinkage could be the cause of micro and macro-cracking [10, 11].

This material exhibits a much denser microstructure than a conventional concrete achieving almost no capillary pores, which enhance its resistance to the diffusion of chloride ions compared to that of normal-strength concrete [12]. UHPC is characterized by high cement content (700–850 $kg/m^3$) and very low w/c ratio [10, 11].

Consequently, the present study on UHPC, will focus on the design of ultra-high concrete with SCC consistency, replacing as much clinker content as possible with SCM from wastes in order to reduce concrete $CO_2$ footprint and cost [13]. In order to improve material's early-age cracking resistance, superabsorbent polymers (SAP) are used as internal curing additive to limit self-desiccation and autogenous shrinkage.

# 2 Material and Methods

## 2.1 Components and Mix Design (AAM)

The materials used for the formulation of the AAM mixtures are natural aggregates (NA), recycled aggregates (RA), natural magnetite (NM from LKAB), glass powder, GGBS from Ecocem France, water and an alkaline activator (sodium silicate solution from Ingessil) mixed in different proportions. Different aggregates in three different size were used: natural or recycled sand (0–4 mm), natural magnetite (0–2 mm) and coarse aggregate (4–9 mm).

According to UNI EN 206-1 (2000), the GGBS was considered additional material for cement binders with effects on the technical performance of concrete (e.g., workability, mechanical properties). In particular, the adopted GGBS is defined as type 2 addition (with pozzolanic activity). Chemical compositional data of GGBS are reported in the technical data sheets that shows the following components CaO (43.4%), $SiO_2$ (36.4%) and $Al_2O_3$ (12.1%). According to the producer, the alkaline solution (Ingessil) has sodium silicate percentage 41.7–45.0% and $SiO_2/Na_2O$ mass ratio in the range 1.60–1.70.

In order to increase AAC mix sustainability, optimized binder-recycled aggregates compatibility was studied. Furthermore, very fine natural magnetite was used to try to reduce the cracking phenomena caused by the heat of hydration.

The materials density and absorbition was measured according to UNI EN 1097-6 and showed in Table 1.

**Table 1.** Material Density and Absorbition.

| Material | Density [kg/m$^3$] | Absorbition [w %] |
|---|---|---|
| GGBS | 2780 | / |
| GLASS | 1400 | 0.00 |
| NA (0–4 mm) | 2700 | 4.97 |
| NA (4–9 mm) | 2600 | 0.99 |
| RA (0–4 mm) | 2300 | 9.86 |
| RA (4–9 mm) | 2400 | 5.72 |
| NM (0–2 mm) | 5100 | 0.30 |

### 2.1.1 Paste Preparation and Lab Trials Conducted (AAM)

Since alkaline activated materials have a very viscous behaviour and generally have very fast setting times, AAM are hard materials to mix. In order to verify the compatibility of the materials the experimental program was divided into three different phases:

- in the first phase, the behaviour in the fresh and hardened state of the binder composed of glass powders, GGBS and activators such as water, sodium silicate and sodium hydroxide was characterized;

- in the second phase, previously optimized binder was mixed with four different types of sand for the realization and characterization (fresh/hardened state) of alkali activated mortars;
- in the last phase, based on previous binder and mortars, alkali activated concretes (AAC) mix design containing three different types of aggregates was performed.

After choosing the best performing mix design in terms of compressive strength, consistency and workability, a final proportion optimization and repeatability study was performed.

a) *Phase 1: Alkali Activated Binder design (AAB).*

   The binder design was carried out considering all the different parameters necessary for this step (water dosage, alkaline dosage, alkaline module). Moreover physical (i.e., density and workability) and mechanical (compressive strength) tests were carried out.

   Six different binder formulations were performed: three containing 70% blast furnace slag and 30% glass powder and three containing 80% blast furnace slag and 20% glass powder.

   The first step was to define all the parameters and establish the best proportions of water/binder, alkaline dosage, and alkaline modulus.

b) *Phase 2: Alkali Activated Mortar design (AAMs).*

   As regards the design of the mortar, the previously developed and optimized binder was mixed with the sands characterized in the laboratory in order to carry out a correct design of the mortars and subsequently of the concrete. Due to the high viscosity of the binder, the mortars made were tested to find out the compatibility, in terms of workability and mechanical strength, between the developed binder and the different types of sand previously characterized.

c) *Phase 3: Alkali Activated Concrete design (AAC).*

   In this phase of the experimental program, the activities were focused to optimize the water-binder ratio and improve the rheological behavior of the mixtures also using the coarse aggregates, with different dosage percentages.

To facilitate the sliding of the concrete between the prototype reinforcement bars, aggregates with a maximum diameter of 9 mm was chosen.

As was done for mortars, in the mixes natural aggregates were used to evaluate the behavior of the binder during the transition from mortar to concrete; recycled aggregates were used to increase the sustainability of the mixtures and very fine natural magnetite was used to try to reduce the cracking phenomena caused by the heat of hydration and to increase the density.

After choosing the best performing mix design in terms of compressive strength, consistency and workability, the mix was replicated five times to evaluate the repeatability of the result at 28 and 60 days, to optimize the proportions of different components and to produce the samples for durability tests (resistivity – chloride intrusion – resistance of the material to chloride penetration – depth of penetration of pressurized water) and adhesion tests (pull out test).

### 2.2 Ultra High Performance Concretes (UHPCs)

To achieve UHPC requirement stablished in the MAREWIND project for the use of this material in an offshore environment, first, a selection of the most suitable raw material was carried out. Once the cement, additives were selected, concrete dosage design was done.

#### 2.2.1 Paste Preparation and Lab Trials Conducted (UHPCs)

To achieve the best performance of the concrete mix, an initial selection of raw materials was carried out: Four different cements from different companies were analyzed. All the studied cements are CEM I 52.5R to achieve as high mechanical strength as possible. Three SCM (Supplementary Cementitious Materials) were tested: fly ash (FA), silica fume (SF) and blast furnace slag (BFS). Two commercial last generation additives were tested to define the most suitable one to improve concrete mix workability. For the aggregates portion, high performance materials were selected, such as quartzite sand and quartzite coarse aggregate together with fine silica sands.

Regarding the fibers, several products were tested to choose the best solution from a workability/mechanical performance point of view.

### 2.3 Durability Performance for AAC and UHPC

The durability performance of the concrete materials presented in this study, are being characterized by: chloride intrusion (NT Build NT-492 and AASHTO T358), water penetration resistance (EN 12390-8) and material freeze/thaw resistance (ASTM C666).

Required sample size was prepared for each test with the optimum dosage defined for each concrete material.

## 3 Results

### 3.1 Alkali Activated Materials Results

Glass powder combined with blast furnace slag activated with sodium hydroxide and sodium silicate were mixed to obtain an alkali activated binder. The binder developed has been in turn mixed with different sands to realize mortars, then, subsequently alkali activated concretes has been performed. All binders, mortars and concretes has been tested at the fresh and hardened state.

- Binders' characterization

  To optimize the formulations, the glass powder, GGBS, water and activators were mixed for fifteen minutes before the workability test. The alkaline dosage (M+ = 7.5) and the alkaline modulus (AM = 9) has been fixed while the water / binder ratio (w / b ratio) has been varied (Table 2).

  Alkali dosage (M+), defined as the mass ratio of total sodium oxide (Na2O) in the activating solution to PFA is a proxy for the concentration of the alkali activator solution:

$$M+=\frac{Na_2O}{PFA}(in\,mass) \quad (1)$$

Alkali modulus (AM), is the mass ratio of sodium oxide to silica in the activating solution and is a proxy for the amount of added silica in the activator solution:

$$AM = \frac{Na_2O}{SiO_2} \tag{2}$$

**Table 2.** Fresh state binder test results

| ID | Slag/Glass | | Workability | |
|---|---|---|---|---|
| | % | w/b | Consistency | Mm |
| Binder 1 | 70/30 | 0.50 | Fluid | >300 |
| Binder 2 | 70/30 | 0.40 | Fluid | 300 |
| Binder 3 | 70/30 | 0.35 | Workable | 180 |
| Binder 4 | 80/20 | 0.50 | Fluid | >300 |
| Binder 5 | 80/20 | 0.40 | Fluid | 300 |
| Binder 6 | 80/20 | 0.35 | Workable | 240 |

The binders showed a very fluid consistency and regarding workable, the best w/b ratio is equal to 0.35. With average density values of 1718.8 kg/mc and 1850.0 kg/m$^3$ respectively, the Binder 3 and Binder 6 showed also a good compressive strength (Table 3).

**Table 3.** Hardened state binder test results

| ID | Compressive strength | | | Density |
|---|---|---|---|---|
| | MPa | | | Kg/m$^3$ |
| | 2 days | 7 days | 14 days | 14 days |
| Binder 1 | 27.2 | 49.2 | 54.9 | 1631.9 |
| Binder 2 | 30.2 | 44.7 | 45.6 | 1640.6 |
| Binder 3 | 42.8 | 52.5 | 56.3 | 1718.8 |
| Binder 4 | 53.7 | 58.4 | 52.5 | 1691.0 |
| Binder 5 | 52.2 | 59.9 | 56.5 | 1769.1 |
| Binder 6 | 84.5 | 74.7 | 67.4 | 1850.0 |

Despite the good workability and the good compressive strength, it is necessary to underline the presence of shrinkage cracks (caused by the rapid setting time) and the consequent reduction of the compressive strength over time.

- Mortars characterization

Considering the density, workability and compressive strength values, the previously developed binders were used to make the mortars. In particular, the first four mortars were made with normalized sand (NS) while the others were made with natural sand (NA) recycled sand (RA) and natural magnetite (NM) reported in Table 1.

**Table 4.** Fresh state mortar test results

| ID | Aggregates | Slag/Glass | | Workability | |
|---|---|---|---|---|---|
| | % | % | w/b | Consistency | mm |
| Mortar 1 | 100 NS | 70/30 | 0.50 | Liquid | > 300 |
| Mortar 2 | 100 NS | 80/20 | 0.40 | Liquid | 300 |
| Mortar 3 | 100 NS | 80/20 | 0.35 | Workable | 280 |
| Mortar 4 | 100 NS | 80/20 | 0.35 | Dry | 240 |
| Mortar 5 | 100 NM | 80/20 | 0.35 | Fluid | 220 |
| Mortar 6 | 50 RA50 NM | 80/20 | 0.35 | Liquid | > 300 |
| Mortar 7 | 50 RA/50 NM | 80/20 | 0.35 | Workable | 220 |
| Mortar 8 | 50NA/50 NM | 80/20 | 0.35 | Workable | 220 |

From Table 4 it can be observed that, also in this case, the best w/b ratio is equal to 0.35 and that, as expected, the addition of sand reduces the spreading of the mortars. It should be emphasized that, despite the spreading values being reduced, the mortars were endowed with good rheological behavior in terms of workability. Furthermore, the addition of fine aggregate reduced the cracking phenomena and, consequently, the loss of compressive strength over time seen for the binders (Table 5).

**Table 5.** Hardened state mortars test results

| | Compressive strength | | | | Density |
|---|---|---|---|---|---|
| | Mpa | | | | Kg/m$^3$ |
| ID | 3 days | 7 days | 14 days | 28 days | |
| Mortar 1 | / | 22.8 | 34.1 | 31.0 | 2039.1 |
| Mortar 2 | / | 44.5 | 48.0 | 48.2 | 2246.0 |
| Mortar 3 | / | 46.2 | 46.2 | 51.1 | 2214.8 |
| Mortar 4 | / | 56.1 | 55.1 | 58.3 | 2242.2 |
| Mortar 5 | 49.2 | 58.8 | 72.6 | 78.6 | 2503.9 |
| Mortar 6 | 28.1 | 61.6 | 79.6 | 82.0 | 2304.2 |
| Mortar 7 | 59.8 | 66.3 | 69.3 | 83.7 | 2296.9 |
| Mortar 8 | 54.3 | 61.2 | 67.8 | 74.5 | 2257.8 |

**Table 6.** Percentages of materials contained in the mixtures

| | | AAC DESIGN | | | | |
|---|---|---|---|---|---|---|
| | | Aggregates | | | | |
| Mix ID | Binder | Natural (NA) | Recycled (RA) | Magnetite (NMG) | | |
| | | 0–2 mm | 4–8 mm | 0–5 mm | 5–9 mm | 0–2 mm |
| AAC_M1 | Binder 6 | 45% | 55% | | | |
| AAC_M2 | | | 50% | 25% | | 25% |
| AAC_M3 | | | | 25% | 50% | 25% |
| AAC_M4 | | | | 50% | 25% | 25% |
| AAC_M5 | | 50% | 25% | | | 25% |

The results of the compression tests confirmed that the best w/b ratio to be used is 0.35 with values over 70 MPa, regardless of the type of sand used.

- Concretes characterization

  A total of five dosages of concrete containing different percentages and types of aggregates were carried out (Table 6).

After choosing the best performing mix design in terms of compressive strength, consistency and workability, to achieve project specs, a final proportion optimization and repeatability study was performed (Table 7).

**Table 7.** Fresh state concrete test results

| | Density | Slump Test | | Air Content |
|---|---|---|---|---|
| ID | Kg/m$^3$ | Class/Slump mm | Spread mm | % |
| AAC_M1 | 2612 | S5/> 220 | 400 | 5–10 |
| AAC_M2 | 3100 | S2/70 | 110 | |
| AAC_M3 | 3026 | S5/> 220 | 500 | |
| AAC_M4 | 2974 | S5/< 220 | 500 | |
| AAC_M5 | 2940 | S5/> 220 | 500 | |

Such as for binders and mortars, all the concretes developed showed good workability, air content ranging from 5%–6% and no bleeding phenomena (Fig. 1).

The alkali activated concretes develops a compressive strength (60 MPa) lower than mortars (80 MPa), but acceptable for the project targets. Furthermore, unlike traditional concretes which reach full maturity after 28 days, the mixture reaches its maximum compressive strength about sixty days after casting (Table 8).

The results reported in Table 8 show that the best performing mix design is mix 5, having better characteristics in the fresh and hardened state than the other mixtures.

**Fig. 1.** Slump test on alkaline activated concretes (optimization mix 1 to mix 5)

**Table 8.** Hardened state concrete test results

| | Compressive strength | | | | | Density |
|---|---|---|---|---|---|---|
| | Mpa | | | | | Kg/m$^3$ |
| ID | 3 days | 7 days | 14 days | 28 days | 58 days | Average |
| AAC_M1 | 32.7 | 36.8 | / | 46.1 | / | 2262.2 |
| AAC_M2 | 20.0 | 42.6 | 44.0 | 60.8 | / | 2700.0 |
| AAC_M3 | 28.1 | 37.8 | 46.1 | 56.6 | 58.4 | 2646.0 |
| AAC_M4 | 22.2 | 19.4 | 36.2 | 44.5 | 48.1 | 2614.2 |
| AAC_M5 | 16.4 | 43.7 | 40.7 | 52.2 | 62.7 | 2613.0 |

Subsequently, the experimental program was focused on the optimization of mix 5 in order to evaluate the repeatability of the results by modifying the percentages of water and mixing times (Table 9).

The optimization of the mixture concerned the variation of the mixing order of the components, the percentage variation of water added as compared to that foreseen by the project and in the variation of the mixing time. For optimal curing, concretes activated with alkaline materials generally require temperatures above 20 °C.

For the purposes of the project, in which it is planned to build a large prototype, it will not be possible the curing in the oven, therefore, the samples were allowed to mature at temperature environment.

**Table 9.** Mix 5 Optimization results

| ID | Fresh state | | | | Hardened state | |
|---|---|---|---|---|---|---|
| | Density | Slump test | | | Compressive strength | Density |
| | $Kg/m^3$ | Class [mm] | Spread [mm] | MPa | | $Kg/m^3$ Average |
| | | | | 28 days | 58 days | |
| AAC_M5$^{I}$ | 2940 | S5 > 220 | 500 | 52.2 | 62.7 | 2600 |
| AAC_M5$^{II}$ | 2963 | | | 45.2 | 58.3 | |
| AAC_M5$^{II}$ | 2950 | | | 51.7 | 53.4 | |
| AAC_M5$^{IV}$ | 2967 | | | 48.6 | 59.9 | |
| AAC_M5$^{V}$ | 2964 | | | 55.3 | 60.1 | |

As also found in the early stages of the experimental program, temperatures below 20/22 °C increase the setting time in the fresh state, but, on the other hand, reduce the development of mechanical strength especially in the short term in the hardened state.

In conclusion the results obtained in the fresh and in the hardened state both provided very similar values, confirming the goodness of the mixture.

### 3.2 Ultra High Performance Concretes Results

Raw materials were combined to get the best performance that achieves compressive/durability performance required for the design of UHPC. Compressive strength (EN 12390-3), flexural tensile strength (EN 14651), setting time, Marsh cone and slump flow (EN 12350-8) were measured. Once the most suitable combination was defined, the UHPC mix design was carried out (Fig. 2).

**Fig. 2.** Slump flow test for UHPC (S27, left) and UHPC with SAP (S28, right)

A total of 30 dosages were tested being the one with almost a 1000 $kg/m^3$ of binder, w/b ratio = 0.16 the one that achieved self-compacting consistency and > 120 MPa compressive strength (S27). Aggregates maximum size is $\leq$ 6 mm so the manufacturing of slender structures is permitted. The main innovation is based in the reduction on the cement content of the UHPC up to 37% with the subsequent improvement in the concrete dosage sustainability (Fig. 3).

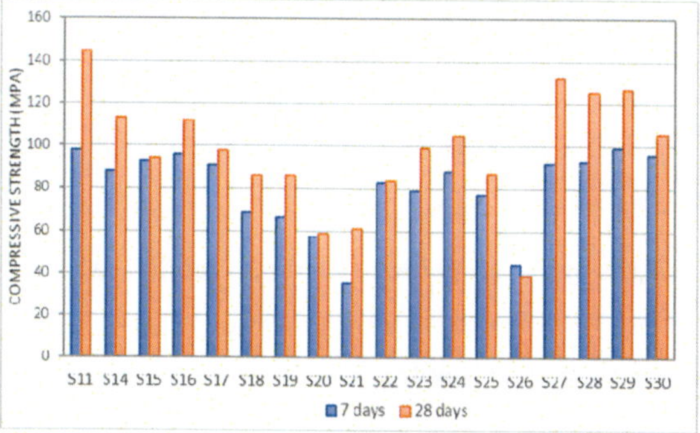


**Fig. 3.** UHPC studied concrete dosages compressive strength (Mpa) comparative results at 7 and 28 days.

Shrinkage characterisation of the selected concrete mixes was done by the "Ring test" according to ASTM C1581/C1581M (Fig. 4), to define the age of cracking under restrained conditions. The results of shrinkage under restrained conditions of HPC (Sample 10), UHPC without SAP (Sample 27) and UHPC with SAP (Sample 28) are shown in Fig. 5.

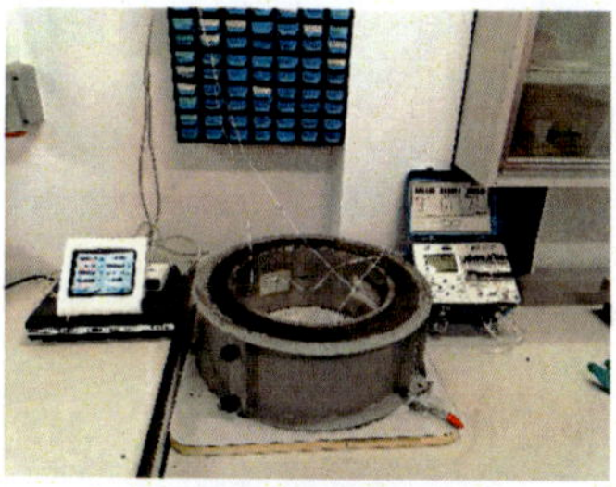

**Fig. 4.** Ring test equipment.

Regarding the achieved results and according to the Table 10 included in ASTM C1581/C1581M Standard, UHPC with PVA fibers presents has moderate-low potential for cracking (16 days) and cracking under restrained conditions and UHPC without SAP additive (9 days) has moderate-high potential, whereas the results for UHPC samples with SAP additive (4 days), and conventional concrete (7 days) have high potential for cracking under restrained conditions.

### 3.3 Durability Results

Durability results are part of the Marewind project work, and several ages will be tested. Tests are still on-going, so net table presents the available results to date.

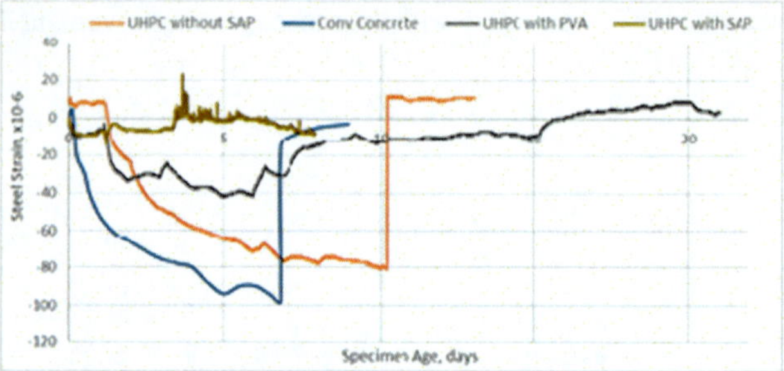


**Fig. 5.** Steel ring strain vs specimen age (days) test results according to ASTM C1581/C1581M

**Table 10.** Summary of the results obtained on durability tests carried out for Marewind concretes.

| | Age (month) | Corrosion | | | Chloride penetration resistance ($m^2/s$) | | Water penetration resistance | | |
|---|---|---|---|---|---|---|---|---|---|
| | | Rate (kΩ·cm) | | Likehood (kΩ·cm) | | | | | |
| | | Assess | | Assess | value | Assess | Max. (mm) | Avg. (mm) | Assess |
| UHPC No SAP | **28 days** | 63 | High | High risk | 1.41 | Extr. high | 7.0 | 1.0 | Very high |
| | **3** | 200 | Low-mod | Mod. risk | 0.37 | Extr. high | 0 | 0 | Very high |
| UHPC with SAP | **28 days** | 65 | High | High risk | 1.56 | Extr. high | 7.5 | 1.0 | Very high |
| | **3** | 189 | Low-mod | Mod. risk | 0.36 | Extr. high | 0 | 0 | Very high |
| AAC | **28 days** | 9.3 | Very high | High risk | 4.93 | Very High | 10.5 | 6.4 | High |
| | **3** | 17.0 | Very high | High risk | 1.76 | Extr. high | 23.1 | 67.2 | Very low |

From the previous table, durability results obtained for UHPC are very promising at 3 months age and improved a standard C60 offshore concrete durability performance.

AAC, present high corrosion rate and likehood. Regarding the chloride penetration resistance, it shows an extremely high resistance. It has better performance at early-ages to water penetration, whereas it seems to be likely to develop corrosion risk during it service life. A deeper analysis will be done when all the ages results are available, to check if this tendency keeps in time or not (Fig. 6).

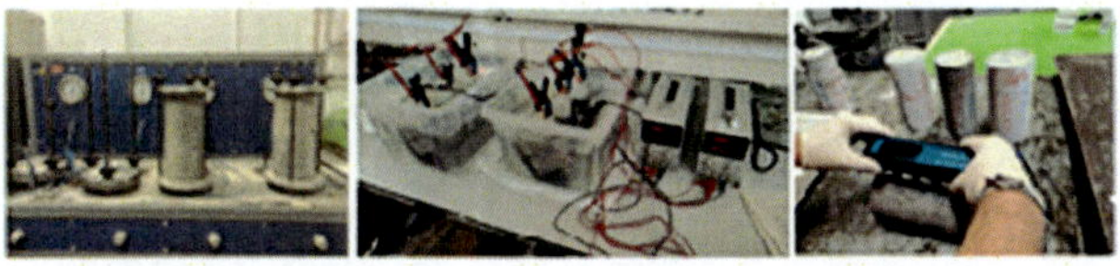

**Fig. 6.** Durability tests: water penetration (left), chloride intrusion (center), resistivity measurement (AASHTO T358) (right).

## 4 Conclusion

Summarizing the development activities of activated alkaline materials (AAM) it is possible to provide specific indications on the binders, mortars and concretes developed.

Due to its high viscosity, the preliminary activities on the binder helped to understand its rheological behavior in the fresh state and the mechanical characteristics in the hardened state, i.e. without the addition of the aggregates.

Since the first days of curing, the binders have shown good workability and an excellent ability to develop compressive strength. The same physical-mechanical performances were observed for the mortars, even if quite different sands were used. With reference to concrete (AAC) it was noted that the environmental temperature has a considerable influence on workability and compressive strength. The low ambient temperature, in fact, makes the mixture more workable, but reduces the resistance to compression and vice versa. Furthermore, unlike binders and mortars, the concrete specimens, probably due to the greater volume and the high percentages of sand contained, develop very low strengths in the first days of curing. This kind of problems, however, can be overcome during the activities envisaged in in the continuation of the project where upscaling tests in larger samples will be done to experimentally determine the heat evolution and the maximum Tª reached during the curing step.

As for the durability tests, they are still in progress, but the results obtained so far will be useful for a further phase of optimization of the concrete to be used for the prototype to be assessed in a relevant environment.

Finally, it can be concluded that:

- the developed cement-free binder is suitable for producing both mortar and concrete;
- the developed concrete has rheological behaviour comparable to self-compacting concrete (SCC), thus ideal for massive castings;
- the developed cement free concrete has mechanical performance comparable to Ordinary Portland Concrete (OPC), with further space for improvement.

UHPC dosage has been optimised and meets with design requirement of consistency (self-compacting concrete) and compressive strength. The best dosage taking into account the results of physical and durability characterization has been Sample 27. These characteristics allow the use of this material for the manufacturing of slender structures. Regarding durability results, UHPC with and without SAP and HPC, have very high resistance to water penetration and extremely high chloride penetration resistance at 3 months of age. In relation to corrosion rate, UHPC with and without SAP have low to moderate values and HPC has low values of corrosion rate. Results at 6 and 12 months of age are pending and will be presented on documents related to subsequent project tasks. In any case, the durability performance of both new concretes improves the performance of a standard C60 offshore concrete.

**Acknowledgement.** This research was supported by the Marewind project. This project has received funding from the European Union's Horizon 2020 research and innovation program under Grant Agreement N° 952960.

## References

1. Mathern, A., Von der Haar, C., Marx, S.: Concrete support structures for offshore wind turbines: current status, challenges, and future trends. Energies **14**(7), 1995 (2012). https://doi.org/10.3390/en14071995
2. Fernandes, J.F., Bittencourt, T.N., Helene, P.: A Review of the Application of Concrete to Offshore Structures, pp. 377–392. American Concrete Institute, ACI Special Publication, Farmington Hills (2008)
3. Jiang, M., Chen, X., Rajabipour, F., Hendrickson, C.T.: Comparative life cycle assessment of conventional, glass powder, and alkali-activated slag concrete and mortar. J. Infrastruct. Syst. **20** (2014)
4. Imbabi, M.S., Carrigan, C., McKenna, S.: Trends and developments in green cement and concrete technology. Int. J. Sustain. Built Environ. **1**, 194–216 (2012). https://doi.org/10.1016/j.ijsbe.2013.05.001]
5. Aydın, S., Baradan, B.: Effect of activator type and content on properties of alkali-activated slag mortars. Compos. Part B Eng. **57**, 166–172 (2014). https://doi.org/10.1016/j.compositesb.2013.10.001
6. Chi, M.: Effects of dosage of alkali-activated solution and curing conditions on the properties and durability of alkali-activated slag concrete. Constr. Build. Mater. **35**, 240–245 (2012). https://doi.org/10.1016/j.conbuildmat.2012.04.005
7. Puertas, F., et al.: Alkali-activated slag concrete: fresh and hardened behaviour. Cem. Concr. Compos. **85**, 22–31 (2018). https://doi.org/10.1016/j.cemconcomp.2017.10.003
8. Collins, F., Sanjayan, J.G.: Effect of pore size distribution on drying shrinking of alkali-activated slag concrete. Cem. Concr. Res. **30**, 1401–1406 (2000). https://doi.org/10.1016/S0008-8846(00)00327-6
9. Puertas, F., Varga, C., Alonso, M.M.: Rheology of alkali-activated slag pastes: effect of the nature and concentration of the activating solution. Cem. Concr. Compos. **53**, 279–288 (2014). https://doi.org/10.1016/j.cemconcomp.2014.07.012
10. Du, J., et al.: New development of ultra-high-performance concrete. Compos. Part B Eng. **224** (2021)
11. Marvila, M., de Azevedo, A., de Matos, P., Monteiro, S., Vieira, C.: Materials for production of high and ultra-high performance concrete: review and perspective of possible novel materials. Materials **14**(15), 4304 (2021). https://doi.org/10.3390/ma14154304
12. Li, J., Wu, Z., Shi, C., Yuan, Q., Zhang, Z.: Durability of ultra-high performance concrete – a review. Constr. Build. Mater. **255**, 119296 (2020). https://doi.org/10.1016/j.conbuildmat.2020.119296
13. Shah, H., Yuan, Q., Photwichai, N.: Use of materials to lower the cost of ultra-high-performance concrete – a review. Constr. Build. Mater. **327**, 127045 (2022). https://doi.org/10.1016/j.conbuildmat.2022.127045

# From In-situ Corrosion Detection to Structural Evaluation: A Simplified Protocol for the Assessment of Existing RC Structures

Elena Casprini[1(✉)], Chiara Passoni[1], Alessandra Marini[1], Gianni Bartoli[2], and Paolo Riva[1]

[1] Department of Engineering and Applied Sciences, University of Bergamo, Dalmine, Italy
elena.casprini@unibg.it
[2] Department of Civil and Environmental Engineering, University of Florence, Florence, Italy

**Abstract.** The actual level of deterioration, especially connected to corrosion, is a critical information to be included in the whole assessment process for existing RC structures, to allow for a reliable assessment of the structure as-is condition, an effective choice of the renovation strategy, and the prediction of the structural performances along the renovated life cycle. The DEMSA protocol, enabling the detection, evaluation, and modelling of corrosion effects on RC structures was recently proposed by the authors. The whole procedure guiding the engineer from the on-site inspection to structural evaluation is briefly described herein, and focus is made on the description of the new tools conceived to define such a simplified procedure available for the professional practice. In this paper, the validation of the procedure is presented, through the application of the protocol to some reference case-studies.

**Keywords:** existing RC structures · corrosion effects · assessment procedure

## 1 Introduction

Reinforced Concrete (RC) structures may be affected by deterioration processes evolving over time during their life cycle. Corrosion of steel embedded in concrete is acknowledged as one of the most common and impacting causes of deterioration, since it reduces the strength, stiffness, and ductility of RC structural members (Bertolini et al. 2013, Zhu & Francois 2014). Such variations influence the life cycle performances of RC buildings, in terms of safety and durability. Therefore, the evaluation of corrosion effects should be integrated in the diagnosis and assessment process, to investigate the actual capacity and the expected residual life of structures, and to select the most effective renovation strategy.

However, the actual level of deterioration of existing structures is rarely addressed in the current practice, with the risk to invest in great renovation projects on structures that are seriously compromised. Indeed, unless corrosion signs are clearly manifest on concrete surface, the investigation on possible causes and effects of corrosion hidden


M. A. Aiello and A. Bilotta (Eds.): ICC 2022, LNCE 435, pp. 605–616, 2024.
https://doi.org/10.1007/978-3-031-43102-9_46

in the structures may be disregarded due to the lack of validated and simplified tools available to professional engineers. Many formulations are available in the literature, addressing the effects of corrosion at material level (Andrade, 2019), or proposing either analytical models for the evaluation of the residual capacity of beams and columns (Coronelli, 2020), or more complex finite element models to account for the variation of the structural sub-assembly performance over time (Dizaj et al. 2018).

However, being the global structural assessment and the identification of structural vulnerabilities the main objectives of the assessment process, the formulations available in the literature need to be systematized, and a simplified procedure guiding the engineer from the building inspection to the structural modelling need to be introduced. In this perspective, the DEMSA (Deterioration Effect Modelling for Structural Assessment) Protocol was recently proposed by the authors (Casprini et al. 2022a). In this paper, the procedure is briefly presented, focusing on those new simplified tools which may help professional engineers in practical applications.

## 2 DEMSA Protocol: Simplified Tools

The assessment of corrosion effects on the structural behaviour requires a transversal and multidisciplinary approach. On-site diagnosis techniques need to be improved to detect either the risk of corrosion or the effects of active corrosion processes; the phenomenon of corrosion at material level is investigated by experts of electrochemistry and material engineering, while analytical and numerical models for the assessment of the single structural members or sub-assembly are formulated by structural engineers (Fig. 1). As a result, the current state of the art is often fragmented into not-interrelated disciplines and approaches, leading to some issues: for example, the level of accuracy at which the phenomenon is investigated at material level is often not applicable at the scale of the structure; also, reliable quantitative information about the corrosion process (measurable on the structure) should be defined as input data to be implemented in the available models.

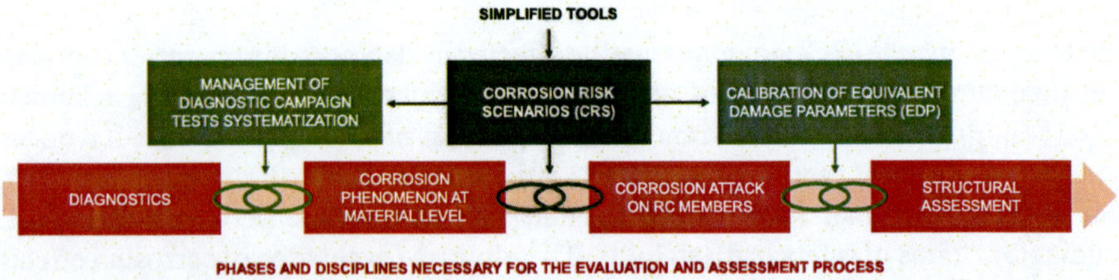


**Fig. 1.** Introduction of simplified tools to include corrosion effects in the structural assessment process

Furthermore, from a simple representation of the main stages of corrosion in concrete (Fig. 2), it is shown that the absence of evident signs of corrosion on the concrete surface does not entail absence of corrosion processes. An apparently sound concrete may hide different stages (Casprini et al. 2022b): absence of aggressive substances ($CO_2$ and

$Cl^-$) in the concrete cover (ST0); penetration of substances into the cover which do not reach rebar level (ST1) or depassivated bars but absence of conditions activating corrosion (ST2); activated corrosion which does not result yet in concrete cracking (ST3). It clearly emerges that the risk of corrosion cannot be disregarded based on visual inspection only. On the other hand, when signs of corrosion-induced deterioration, such as delamination (ST4) or cracking in correspondence of reinforcement bars are clearly manifest (ST5), the corrosion may be in a very advanced stage. The DEMSA protocol provides straightforward tools allowing detection of corrosion process, even at the early stages, based on the presence of specific environmental and aggressiveness conditions, classified in Corrosion Risk Scenarios and aggressiveness classes.

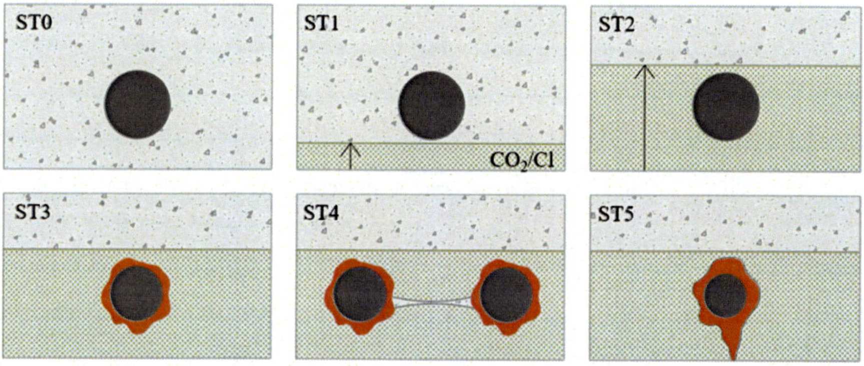


**Fig. 2.** Corrosion stages of steel bars in concrete

## 2.1 Definition of Corrosion Risk Scenarios

The Corrosion Risk Scenarios (CRS) represent not only the type of environment in which corrosion is likely to occur, but also the type of attack to which a structure may be subjected in such an environment (Table 1).

Scenario 0 includes all those exposure conditions inhibiting corrosion or leading to a negligible corrosion rate (lower than 1–2 μm/year, not significant for the structural performance), which means absence of aggressive conditions at rebar level able to depassivate the rebars (carbonation and/or chlorides), or absence of water and oxygen at rebar level (always saturated or dry environment, with relative humidity R.H. < 40%).

Scenario 1 is related to carbonation-induced corrosion in absence of chloride, in which the bars are depassivated when the carbonation front reaches the concrete cover depth; in this case, for corrosion processes to be significant, wet/dry cycles or high R.H. (>70%) need to be present.

Potentially, all the structures may be in this Scenario, if not sheltered from rain, or when defects in the impermeabilization or in the water collection system are present.

Scenario 2 is also introduced, characterized by the simultaneous presence of low chloride content in the cementitious matrix (below chloride threshold for pitting corrosion initiation, thus ranging between 0.1% and 0.4% with respect to cement weight)

**Table 1.** Corrosion Risk Scenarios (1-2-3) and representative values of the related corrosion attack characteristics ($v_{avg}$ and $R_p$)

| **Corrosion Risk Scenario (CRS)** | | 1 | 2 | 3 |
|---|---|---|---|---|
| Aggressiveness conditions on the reinforcement surface | | Carbonated concrete + R.H. > 70% | Carbonated concrete + 0.1%< $Cl^-$ < 0.4% R.H. > 50% | $Cl^-$ > 0.4%+ RH>40% |
| Corrosion phenomenon | | Carbonation-induced Corrosion | Carbonation-induced corrosion + $Cl^-$ | Chloride-induced corrosion |
| $v_{avg}$ [μm/year] | *CLASS O*: Ordinary R.H. (CRS 2) or marine atmosphere (CRS 3) | | 2-10 | 10-50 |
| | *CLASS H*: High R.H. (CRS 1-2) or chloride airborne (CRS 3) | 2-10 | 10-50 | 50-100 |
| | *CLASS E*: Alternation of wet/dry environment (CRS 1-2) or water stagnation, tidal zone (CRS 3) | 10-50 | 100-200 | 100-300 |
| Maximum to average attack ratio $R_p$ | | 1-2 | 3-7 | 4-10 |

and carbonation; in this case, chloride bounds to the hydrated phase, or in the form of calcium chloroaluminate hydrates, may be liberated, making the pore solution even more aggressive; corrosion rates become relevant also in moderate humidity conditions (R.H. > 50%), for the hygroscopic nature of chloride-contaminated concrete. Although in modern structures limits are imposed on the chloride content, this problem cannot be disregarded when dealing with existing structures, since chloride could have been introduced in RC elements through the use of contaminated raw materials, or the addition of accelerant admixtures based on calcium chloride (structures built in the '60s and '70s).

Finally, all the exposure conditions leading to chloride-induced corrosion are grouped in Scenario 3, with a chloride content at rebar level of $Cl^- > 0.4\%$ with respect to cement weight, typical of marine environment, or exposure to industrial brine or infrastructure where the use of deicing salts is frequent.

Once the Corrosion Risk Scenarios are defined, representative values of the fundamental characteristics of the corrosion attack can be associated to each Scenario. The parameters required to implement most of the models and formulations available in the literature are the bar's average and minimum residual cross-sections. Therefore, a representative value for the average corrosion rate $v_{avg}$, which is the rate of the corrosion attack penetration into the bar section, measured for practical purposes in μm/year, and the maximum to average attack ratio on the bar section $R_p$ are required. Starting from data proposed in the literature (Bertolini et al. 2013, Martinez & Andrade 2009, RILEM 1996, Tuutti 1982) associated to different aggressiveness conditions, and data measured on existing structures in different Scenarios (Casprini 2021), preliminary tentative ranges of values are introduced (Table 1). In order to reduce such intervals of values, three aggressiveness classes are included within each Scenario, defined as Ordinary, High and Extreme, characterized by different exposures to moisture and chloride.

## 2.2 Management of the Diagnostic Campaign

Since Corrosion Risk Scenarios are introduced to define the corrosion attack characteristics necessary to evaluate the corrosion effects at the structural level, the on-site diagnostic campaign is then planned and carried out with the aim of identifying the Corrosion Risk Scenario of interest.

Accordingly, the professional engineer is aware of the aggressiveness conditions which need to be checked, and guidance is provided in the protocol to measure such relevant information, thus reducing time and costs of diagnostic (following the flowchart in Fig. 3). After visual inspection (an abacus of visible deterioration conditions is provided in the protocol), the possible ingress of chloride from the outside into the concrete is evaluated by examining the surrounding environment.

If such exposure is likely, the presence of a chloride penetration profile is first assessed; when its presence is detected, Scenario 3 can be directly selected, and the profile depth is compared with the concrete cover.

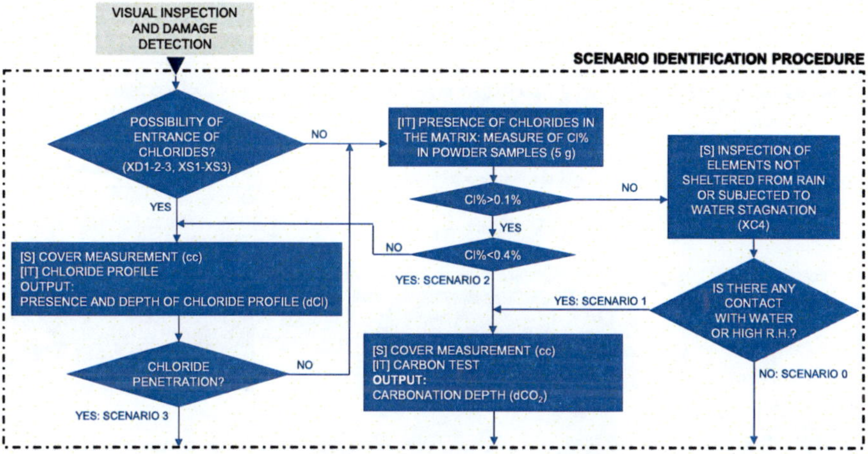


**Fig. 3.** Scenario Identification procedure: [S] refers to survey and [IT] to instrumental tests (from Casprini et al. 2022b)

If a chloride penetration profile is not found, or such exposure is a priori excluded, the presence of chloride in the concrete matrix from construction should be checked by measuring the chloride content in concrete powder samples from different locations. While for $Cl^- > 0.4\%$ (with respect to cement weight), the possibility to have Scenario 3 should be reconsidered, for $0.1\% < Cl^- \leq 0.4\%$, Scenario 2 is selected. In case $Cl^- \leq 0.1\%$, relevant corrosion rates may be found only for carbonation-induced corrosion in the case of contact with water or very high levels of relative humidity (R.H. $\geq 70\%$); in this case, the structure would be in Scenario 1. Therefore, the possibility of water leakage or infiltration is examined, and particular attention is paid to those elements which are not sheltered from rain. If all these conditions are excluded, Scenario 0 can be selected. When either Scenario 1 or 2 is identified for the structure, carbonation depth and concrete cover dimension should be measured and compared, to assess whether corrosion is likely to be already active or not. References to necessary tests are reported in Table 2. Two relevant aspects emerge from the proposed procedure. Firstly, despite carbonation tests are widely performed in the current practice to assess the building's state of preservation, they may be avoided when Scenarios 0 or 3 are recognized, thereby saving time and costs.

**Table 2.** Tests required for the Scenario Identification procedure: description and codes of reference

| Test | Code of reference | Instructions | Instruments required |
|---|---|---|---|
| Chloride Content Analysis | EN 14629, 2007 | Collect concrete powder samples at the locations of interest and execute chemical laboratory test. | Concrete powder (drill) |
| Chloride penetration depth | EN 14629, 2007 | Extract a core or collect concrete powder in separate depth increments from the outer surface to bar level and execute chemical laboratory tests. | concrete core or concrete powder (drill or core-drill) |
| | Collepardi, 1972 | Not codified – spray fluorescein on a freshly broken concrete surface and then silver nitrate. Chloride containing zone turns dark pink whereas free chloride zone turns to dark brown | fluorescein (1 g/L in a 70% solution of ethyl alcohol in water) + silver nitrate (0.1 mol/L AgNO3 solution) |
| Carbonation depth | EN 14630, 2006 | Extract the core or the powder according to the method adopted and spray the indicator phenolphthalein on fresh concrete. Carbonated concrete does not change its colour while non-carbonated concrete turns purple | concrete core + phenolphthalein or CARBONTEST© kit (Felicetti, 2009) |
| concrete cover measurement | ACI 222R-19, 2019 | Measure concrete cover at different locations to obtain an average value | Covermeter and/or calibre |

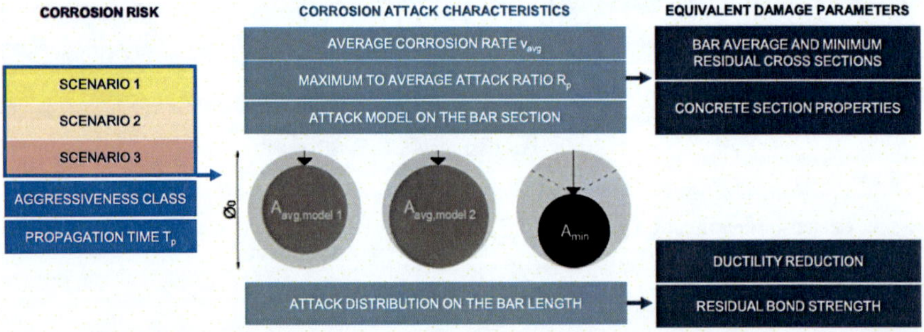


**Fig. 4.** Definition of the equivalent damage parameters to be implemented in the structural analyses (from Casprini et al. 2022b)

Secondly, although the check on the possible presence of chlorides in the concrete matrix is always significant when dealing with existing structures, this aspect is often disregarded since the standardized test for measuring the chloride penetration depth is costly and time-consuming, while the available rapid tests are not as reliable as the standardized one. In this perspective, an experimental programme is being carried out to adapt a commercial titrator for chloride, conceived for solutions, to concrete powder.

### 2.3 Calibration of the Equivalent Damage Parameters

Another critical aspect in the current assessment procedure consists in defining the input data for the structural models. Once the corrosion attack characteristics have been defined according to the Scenarios, equivalent damage parameters can be calibrated through the formulations proposed in the literature. In the protocol, some formulations among those proposed for the evaluation of corrosion effects at a sectional level are selected and systematized within the procedure, while a simplified method is proposed to define an equivalent distribution of the corrosion attack along the element length (Fig. 4).

In detail, the representative value of the average corrosion rate and the maximum to average attack ratio can be selected from Table 1, according to the Scenario and the aggressiveness class of interest; in order to preliminary estimate the average and minimum bars residual cross-sections, a regular circular residual cross-section may be considered, among the corrosion attack penetration models proposed in the literature.

It is also necessary to estimate the effective duration of the corrosion attack; following the simplified model proposed in the Contecvet manual (2001), based on the square-root formula, such time period is obtained by comparing the aggressive substances penetration depth and the concrete cover dimension.

All these data allow calculating the bars residual cross-sections, and the residual compressive strength of the cracked concrete cover material, according to Coronelli & Gambarova (2004).

As for bond strength and ductility reduction, further research is needed; although most of the available formulations are based on the attack penetration at a sectional level, theoretical modelling of these effects (Chen et al. 2020) and experimental evidence (Casprini 2021) show that also the attack pattern distribution along the bar length is critical. For this reason, it would be useful to relate simplified model of the attack pattern along the bar length to each Scenario: a simplified method is proposed in Casprini et al. (2022b), by defining a single equivalent defect in the bar characterized by the minimum residual cross-section, and a certain length that provide the same bar ductility reduction of the bar corroded with a possible natural corrosion pattern.

It should be emphasized that each single method or formulation adopted in the protocol, can be substituted as soon as new research achievements are available, or in some cases, if other formulations are more suitable to the specific structure under investigation. The protocol is conceived as a flexible procedure composed by different steps, in which the required input data are specified, such as the relevance of each consideration in relation to the global structural evaluation.

### 2.4 Impact of Deterioration Effects on Modelling/Retrofit Strategy

The information collected during the visual inspection, the Corrosion Risk Scenario Identification procedure and the calibration of the equivalent damage parameters are then used to implement the steps of the DEMSA protocol (Fig. 5).

The equivalent damage parameters (EDP) are first implemented both in analytical models for the estimation of the residual capacity of RC beams and columns, or in finite element models to assess the global structural behaviour and evaluate the capacity reduction both in terms of base shear and displacement due to corrosion effects. Such evaluations allow performing sensitivity analyses, in order to support the decision-making

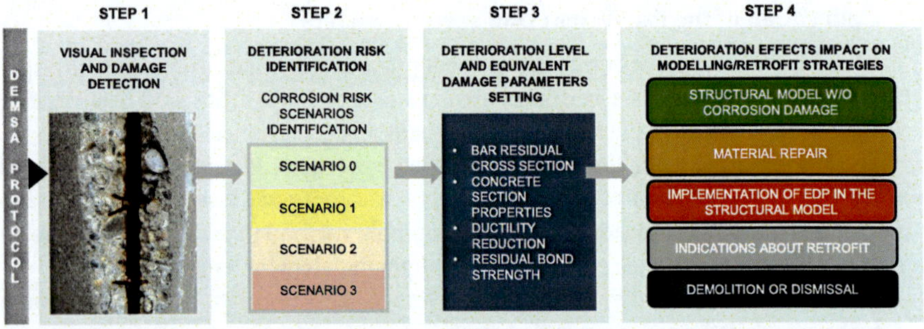


**Fig. 5.** Summary of the DEMSA Protocol guided procedure: from visual inspection to the actions to be implemented in the structural model and preliminary selection of the renovation strategy (adapted from Casprini et al. 2022a)

process for the structure of interest. Two detailed flow-charts (Casprini et al. 2022a) provide instructions on how to carry out the evaluation if either visible corrosion-induced signs of deterioration are present or absent; in the former case, the objective is to quantify the effects of corrosion (ST4-5 in Fig. 2), while in the latter, it is interesting to detect hidden corrosion processes or evaluate possible effects in the future life of the structure (ST0-3 in Fig. 2).

Several outcomes are possible depending on the results of the sensitivity analyses and the strategy of intervention. If Scenario 0 is selected, or corrosion is detected early enough to not have caused relevant effects (in this case removal of corrosion causes is required and material repair should be carried out in the concrete cover), structural modelling may be performed disregarding corrosion effects in the present and future life of the structure. Conversely, in some cases, corrosion may have affected structural elements so heavily to compromise their structural function.

Accordingly, indications about structural retrofit may be provided (when single structural elements or a group of them is compromised) or, when is unavoidable, demolition or dismissal of the structure may be preferred. In all the intermediate conditions, the protocol provide guidance on how to account for the EDP (constant or variable in time) in the structural modelling, to perform further evaluations. After this preliminary evaluation, the professional engineer is aware of the relevance of corrosion effects on the structure under investigation; when a more accurate quantification of corrosion effects is critical to determine the strategy of intervention, or when material repair is necessary, the support of a corrosion expert is fundamental, to obtain more detailed information on the corrosion processes. In these cases, the sensitivity analyses carried out in the context of the protocol allow identifying the location of thorough investigations and chemical analyses.

## 2.5 Application of the Protocol Procedure to Bridge Structures

The DEMSA protocol was mainly conceived for buildings; other structural typologies, such as bridges, are even more affected by deterioration processes and vulnerable to their effects. Recently, national guidelines aimed at the assessment of the state of preservation of bridges are being applied on a large scale. These guidelines (CSLPP, 2020), allow the definition of a level of attention for each structure. In case deterioration processes are manifest, it is required to consider their effects in the further structural evaluation.

The procedure defined in the protocol may be integrated within the framework of these guidelines, especially in two stages of the assessment process. First, rapid on-site tests able to detect the risk of corrosion in early stages based on the hypothesis of a possible Corrosion Risk Scenario would be useful to avoid neglecting corrosion effects when are not visible. Then, concerning structural modelling, the approach proposed in the protocol may be adapted to bridge structures to provide the input data to be implemented in the structural models. In this perspective, ongoing research is being carried out: the possibility to extend the classification of the Corrosion Risk Scenarios is being considered, to include also specific problems of preservation of different structural typologies, such as pre-stressed structures (in the framework of the project DPC - ReLUIS 2021/2023 WP4, task 4.1).

## 3 Demsa Protocol: Application to Case-Studies

The proposed protocol was validated through the application to several case studies, addressing structures in different Scenarios. For all these buildings, a possible Scenario and an expected corrosion attack were estimated following the DEMSA protocol (Casprini, 2021); then, aggressiveness conditions in the concrete cover were measured and the state of preservation of steel bars was assessed by removing some portions of the bars, which were cleaned from corrosion products and then subjected to computerized industrial tomographic scan (performed by *TecEurolab*©). The scans provided the accurate description of the residual cross-section distribution, from which the corrosion attack characteristics were measured; such information, along with an example of the corrosion patterns observed on bars from different Scenarios is reported in Fig. 6.

The investigation carried out allowed to validate the protocol since activated corrosion processes were detected by following the simple and slightly destructive instrumental tests and by checking the exposure conditions. Furthermore, the average corrosion rates and the maximum to average attack ratio measured in naturally corroded bars were in good agreement with the ranges proposed according to each Scenario.

Moreover, it was confirmed that, even within the same structural sub-assembly, the absence of aggressiveness conditions in a structural member leads to the absence of activated corrosion processes in that member, while in the adjacent member, corrosion effect may be significant: samples AB1 and AB3 (Fig. 6) were extracted from two adjacent beams in the same colonnade in an abandoned building, and in both cases the carbonation front had reached the rebar level (in absence of chloride). AB1 was in a beam sheltered from rain, while the beam of AB3 was subjected to leakage phenomena, and frequent contact with water.

For other buildings, the protocol provided indications able to support the decision-making process and guide the definition of the best renovation strategy. A 25-years-old industrial building abandoned after only 10 years of service, due to the deterioration of the waterproofing system, was investigated to be restored. While evident signs of deterioration (efflorescence and biological colonization) appeared on the concrete surface

| CRS - class | 0 | 1 - E | 2 - H | 3 – E |
|---|---|---|---|---|
| Bar tomographic scan | 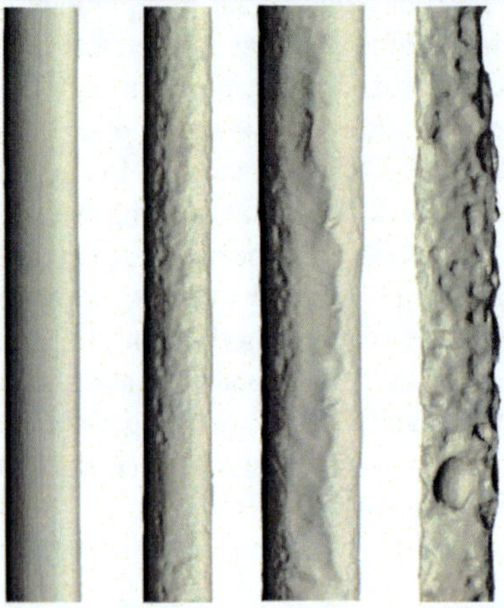 | | | |
| $v_{avg}$ [μm/year] | - | 22 | 24 | 220 |
| $R_p$ | - | 1.8 | 2.0 | 1.6 |

**Fig. 6.** Examples of corrosion attack characteristics in bars extracted from existing structures

of all the structural elements, the guided on-site survey allowed to assess that, despite problem of water infiltrations involved all the elements, the vertical resisting system was not yet affected by corrosion, since the carbonation front (25 mm) had not reached the rebars (35 mm). Conversely, heavy corrosion effects were found in the roof system, where the bars were depassivated (in absence of chlorides).

The time left to corrosion initiation in the columns could be estimated (24 years); since repairing of the impermeabilization system was planned, thus eliminating the problem of contact with water, the columns were not expected to be affected by corrosion effects, neither at the moment of the evaluation nor in the future service life. On the contrary, the roof was seriously compromised; Scenario 1E was selected (Table 1), which may have caused for the bars of interest (ϕ16) up to a cross-section reduction of 18%. A significant reduction of the load-bearing capacity may be expected, such as bond strength deterioration due to diffuse concrete cracking along the element length. Such condition was detected in almost 50% of the roof elements.

The evaluations carried out led to the following conclusion: if the roof was refurbished, the elements should have been verified against static load by assuming a reduced capacity, and structural retrofit could have been necessary by integrating reinforcing bars. Even if the elements were verified, the removal of the damaged concrete and the formation of a new cover would have been necessary. The cost for that operation was estimated to be relevant. For this reason, based on these considerations, the owner decided to also evaluate the solution of complete replacement of the roof and to compare the costs. Detailed reference of the mentioned case-studies, among others, can be found in Casprini (2021).

## 4 Concluding Remarks

The DEMSA Protocol represent a straightforward instrument to preliminary evaluate the effects of corrosion on existing RC structures. It is not conceived to provide an accurate description of the corrosion phenomena, but to help the professional engineers in practical applications of diagnosis and assessment.

The main advancement introduced with respect of the current assessment practice are: the definition of a relationship between specific environmental and aggressiveness conditions (herein classified in Corrosion Risk Scenarios) and corrosion attack characteristics; a guided diagnostic campaign allowing for the identification of a possible Scenario; specific guidance on how to implement the equivalent damage parameters in the structural analysis and how to interpretate the results for structural assessment.

More correlations among environmental conditions and actual corrosion damage found in existing structures is still needed to make the evaluation more reliable. However, a progressive updating of each step of the protocol is allowed by the flexible and rationalized nature of the procedure, in which new research achievements, such as advanced diagnostic techniques and more accurate formulations, can be easily implemented.

## References

Bertolini, L., Elsener, B., Pedeferri, P., Redaelli, E., Polder, R.: Corrosion of Steel in Concrete – Prevention, Diagnosis, Repair. Weinheim, Germany: Wiley VCH (2013)

Zhu, W., Francois, R.: Corrosion of the reinforcement and its influence on the residual structural performance of a 26-year-old corroded RC beam. Constr. Build. Mater. **51**, 461–472 (2014)

Andrade, C.: Propagation of reinforcement corrosion: principles, testing and modelling. Mater. Struct. **52** (2) 2019

Coronelli, D.: Resistance of corroded RC beams: extending fib Model Code 2010 models. Struct. Concr. **21**, 1747–1762 (2020)

Dizaj, E.A., Madandoust, R., Kashani, M.M.: Probabilistic seismic vulnerability analysis of corroded reinforced concrete frames including spatial variability of pitting corrosion. Soil Dyn. Earthq. Eng. **114**, 97–112 (2018)

Casprini, E., Passoni, C., Marini, A., Bartoli, G.: DEMSA protocol: deterioration effect modelling for structural assessment of RC buildings. Buildings **12**(5), 574 (2022). https://doi.org/10.3390/buildings12050574

Martínez, I., Andrade, C.: Examples of reinforcement corrosion monitoring by embedded sensors in concrete structures. Cem. Concr. Compos. **31**, 545–554 (2009)

RILEM: Durability Design of Concrete Structures; Report No. 14; E & FN Spon: London, UK (1996)

Tuutti, K.: Corrosion of Steel in Concrete. Swedish Cement and Concrete Research Institute: Stockholm, Sweden (1982)

Casprini, E.: A protocol for the assessment of corrosion effects in RC structures in a life cycle engineering framework. Ph.D. thesis, University of Bergamo, Dalmine, Italy (2021)

EN 14629: Products and Systems for the Protection and Repair of Concrete Structures—Test Methods—Determination of Chloride Content in Hardened Concrete. CEN: Brussels, Belgium (2007)

Collepardi, M., Marcialis, A., Turriziani, R.: Penetration of chloride ions in cement pastes and in concretes. J. Am. Ceram. Soc. **55**, 534–535 (1972)

EN 14630: Products and Systems for the Protection and Repair of Concrete Structures—Test Methods—Determination of Carbonation Depth in Hardened Concrete by the Phenolphtalein Method. CEN: Brussels, Belgium (2006)

Felicetti, R.: Improved Procedure for the Analysis of Construction Materials and Device to Implement This Procedure. Italian Patent Application MI2009A 001073, 17 June 2009

ACI Committee 222, ACI 222R-19: Guide to Protection of Reinforcing Steel in Concrete against Corrosion; ACI: Farmington Hills, MI, USA (2019)

IN30902I. CONTECVET: A Validated User Manual for Assessing the Residual Life of Concrete Structures. DG Enterprise, CEC (2001)

Coronelli, D., Gambarova, P.G.: Structural assessment of corroding R/C beams: modelling guidelines. ASCE J. Struct. Eng. **130**, 1214–1224 (2004)

Chen, E., Berrocal, C.G., Fernandez, I., Lofgren, I., Lundgren, K.: Assessment of the mechanical behaviour of reinforcement bars with localised pitting corrosion by digital image correlation. Eng. Struct **219**, 110939 (2020)

Casprini, E., Passoni, C., Marini, A., Bartoli, G.: Toward the definition of equivalent damage parameters for the assessment of corroded RC structures. Struct. Concr. (2022). https://doi.org/10.1002/suco.202200368

Ministero delle Infrastrutture e dei trasporti, CSLPP n. 88/2019: Linee guida per la classificazione e gestione del rischio, la valutazione della sicurezza ed il monitoraggio dei ponti esistenti (2020)

# Corrosion Effects on Seismic Vulnerability of Reinforced Concrete Structures from Different Periods

Antonio Bossio, Gian Piero Lignola(✉), and Andrea Prota

Department of Structures for Engineering and Architecture, University of Naples "Federico II", Naples, Italy
glignola@unina.it

**Abstract.** Clear knowledge of the seismic vulnerability of corroded reinforced concrete structures is very important. Reinforced concrete has been used from over a century and properties of materials have changed and evolved, so the evaluation of the seismic behaviour of existing structures (corroded or not) should consider such progress of used materials. The aim of the present research is to investigate about the seismic capacity of a simple corroded reinforced concrete structure, considering three different building periods (evaluated by considering properties of materials related to different periods), two different corrosion levels of the structure (evaluated by considering bar reduction and concrete cover cracking/delamination). Push-over analyses were performed in order to understand the influence of materials in terms of seismic capacity of structures and how much corrosion phenomenon contributes to this reduction. Results show that material properties have a limited influence on seismic behaviour, but older ones allow for a faster and earlier corrosion process leading to faster and earlier reduction of structural capacity.

**Keywords:** corrosion · seismic behaviour · construction period · push-over analysis

## 1 Introduction

Corrosion of bars has severe effects on seismic capacity of Reinforced Concrete (RC) structures, significantly of internal reinforcement, particularly when they are in aggressive environments (Capozzucca 1995, Biondini et al. 2011, Zhu et al. 2013, Caprili et al. 2015, Zanini et al. 2017, Bossio et al. 2018a). Corrosion yields to reduction of cross section of longitudinal and transverse reinforcement (Cabrera 1996, Rodriguez et al. 1997, El Maaddawy et al. 2005, Bossio et al. 2015a, Fernandez et al. 2016), to decrease of bond between concrete and steel reinforcement (Cairns et al. 2007, Berto et al. 2008, Desnerck et al. 2015, Çaliskan and Aras 2017, Coccia et al. 2016), to decrease of ductility (Cairns et al. 2005, Apostolopoulos & Papadakis 2007, Li et al. 2008, Imperatore et al. 2012 & 2017, Su and Zhang 2015, Fahy et al. 2017) and cracking of concrete cover leading to delamination (Pantazopoulou and Papoulia 2001, Bossio et al. 2017a).


M. A. Aiello and A. Bilotta (Eds.): ICC 2022, LNCE 435, pp. 617–628, 2024.
https://doi.org/10.1007/978-3-031-43102-9_47

The occurrence of these critical phenomena earlier than design service life (according to codes) is the reason for unforeseen costs in the retrofit of structural elements (Imperatore et al. 2016). Industrialized countries are spending unexpected large amount of money and these corroded RC structures are becoming dangerous for public safety other than expensive (Hays 2010, UK DTI 2015). Nowadays codes and guidelines are providing significant specification to reduce corrosion effects for new-built structures, but less is available for existing structures and their residual seismic capacity after corrosion is still debated in the scientific and technical field. The aim of this paper is to compare the seismic behaviour of 50 years old corroded RC structures, by means of push-over analyses. Three different periods of construction have been considered by means of different characteristics of materials. The numerical case study was already discussed by the authors in previous papers (Bossio et al. 2018a), however the focus and novelty are now on the differences in terms of construction materials.

## 2 Numerical Case Study

Properties of materials for RC structures changed and evolved during last century, consequently seismic behaviour of existing structures (corroded or not corroded) should be analysed considering the evolution of materials, in terms of properties and durability. Geometry of the simple structure, material properties and corrosion levels have been defined before the push-over analyses (Celarec et al. 2011, Bossio et al. 2018a).

### 2.1 Geometry of the Structure

A structure with (10.00 × 5.00) m$^2$ rectangular plan and two floors is considered with a total height of 6.40 m. There are two frames in the X-direction, each one with two spans 5.00 m long, while in the Y-direction there are three frames, each 5.00 m long. Roof is made of beams oriented in the X-direction. A square cross Sect. (30 × 30) cm$^2$ is assumed for the columns, reinforced by eight longitudinal D14 bars and two-legs D10 stirrups at 20 cm spacing. The ID of columns is given by letter "C" followed by letter "g" to indicate ground floor and number "1" to indicate first floor; second figure is the identification number (1 to 6). Rectangular cross section of beams is (30 × 50) cm$^2$ and they have eight longitudinal bars and stirrups similar to the reinforcement of columns. The ID of beams is given by letter "B" followed by letter "g" to indicate ground floor and number "1" to indicate first floor; a further couple of figures identifies the columns positioned at their ends (see Fig. 1). Minimum dimensions and minimum reinforcement ratios given by modern building codes were used for beams and columns, however they were replicated for structures built in elder periods only for comparison purposes. In fact, the main parameter is the material properties, and the interaction with other structural parameters was prevented.

### 2.2 Material Properties and Corrosion Levels

Evolution of materials used in RC building yields to different concrete and steel during time, hence different values of characteristic cubic compressive strength ($R_{ck}$) and characteristic yielding stress ($f_{yk}$), respectively. In order to simplify the comparison of simulations, the effects of corrosion were limited to the material constitutive behaviour of steel bar and to concrete cover delamination. The three considered building periods are related to two specific Italian Building Codes for Structures. The first is the Decree of Italian Ministry of 05/30/1972 establishing the ribbed bars in RC members, hence reducing the use of smooth bars. The second is the Decree of Italian Ministry of 01/14/2008 that requires different minimum values of properties of materials in RC structures. Table 1 shows the material properties in each building period. The push-over analyses were, in all cases, conforming to last Italian Building Code.

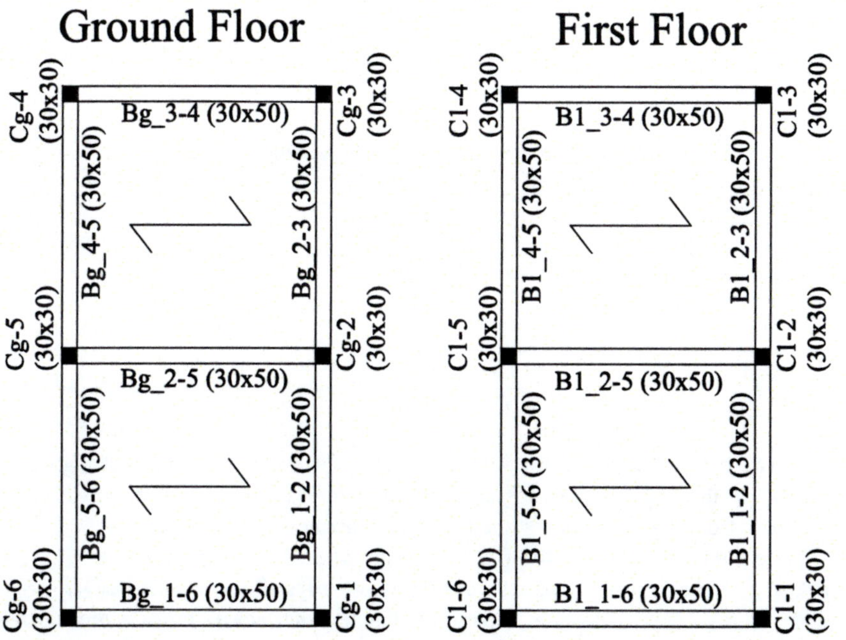


**Fig. 1.** Plans of ground and first floors with columns and beams identification labels and dimensions (in cm)

A generalized corrosion was considered for 50 years exposure (XC4 class) or service life. Carbon dioxide diffusivity is the key factor to assess carbonation rate for concrete and to compute durability. Two coefficients for carbon dioxide penetration were considered: $k = 6$ mm/year$^{1/2}$ or $k = 8$ mm/year$^{1/2}$. This yields to two different corrosion initiation periods (e.g. 25 and 11 years for longitudinal bars and stirrups respectively, with lower k), due both to different rates and different concrete covers for longitudinal bars and stirrups. Assuming also two different average rates for corrosion, respectively, 45 μm/year and 65 μm/year, accounts for different levels of (high) relative humidity (Pedeferri and Bertolini 2000, Bertolini et al. 2013). The likely higher permeability of older concretes can be correlated to higher k values. For lower k and 45 μm/year, the corrosion penetration for longitudinal reinforcements, $x_b$, (i.e. reduction of radius) is

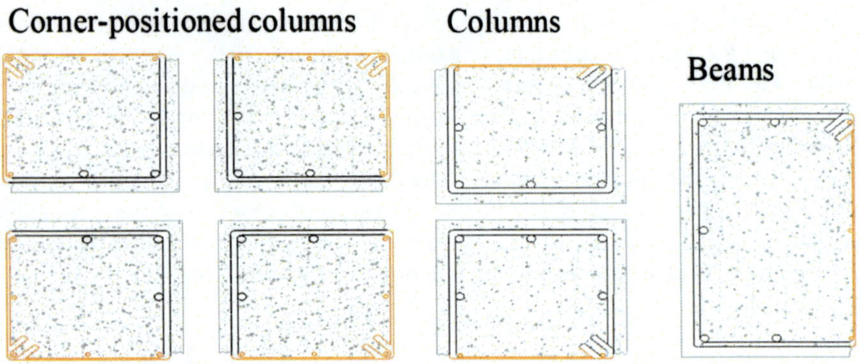


**Fig. 2.** Cross sections of columns and beams considered in the push-over analyses, accounting for corrosion effects

**Table 1.** Different combinations of material properties

| Construction ages | Concrete | Steel |
|---|---|---|
| | $R_{ck}$ (MPa) | $f_{yk}$ (MPa) |
| Before 1972 | 18 | 310 |
| 1972–2008 | 25 | 380 |
| After 2008 | 30 | 450 |

1,125 μm, but the reduction for stirrups, $x_s$, is 1,750 μm (same rate, but higher time interval). In the worst case of higher k and 65 μm/year, $x_s$ is 2,843 μm, while $x_b$ is 2,515 μm. Concrete cover is assumed to be delaminated in both cases.

Table 2 reports bar diameter, $\phi_b$, stirrups diameter, $\phi_s$, radial reduction of bars and stirrups, $x_b$ and $x_s$, residual diameter of bars and stirrups after corrosion, $\phi_{b,corr}$ e $\phi_{s,corr}$ and the 30 mm concrete cover, $c_c$, condition. Figure 2 outlines the cross sections considered in the push-over analyses including concrete cover delamination. Corrosion rate is similar in terms of mass loss with lower k, while for stirrups it is 1.5 times the mass loss compared to longitudinal bars for higher k, however it can be even higher in different conditions (Lin et al. 2019).

**Table 2.** Diameter of bars and stirrups related to corrosion level (mm)

| | $\phi_b$ | $\phi_s$ | $x_b$ | $x_s$ | $\phi_{b\ corr}$ | $\phi_{s\ corr}$ | $c_c$ |
|---|---|---|---|---|---|---|---|
| k = 6 | 14 | 10 | 1.125 | 1.750 | 11.8 | 6.5 | Delaminated |
| k = 8 | 14 | 10 | 2.515 | 2.843 | 9.0 | 4.3 | Delaminated |

The corrosion is present on all the external facades of the structures; the RC elements are not symmetrically corroded, but degradation (i.e. reduction of bars and concrete cover delamination) occurs on some sides only (two sides for corner-positioned columns and one side for all other columns and for beams, Fig. 2). In this way the behaviour of the cross sections under bending action is not symmetric. In particular, when delaminated concrete cover is compressed, the asymmetric behaviour is more evident for the concrete cross section with significant reductions of load capacity (Di Ludovico et al. 2010, Bossio et al. 2018b for P-M domains).

## 2.3 Modelling of Structural Elements Capacity

Structural elements have been modelled by means of concentrated plasticity. Plastic hinges model is based on the proposal by Italian Seismic Code NTC2008. Plastic hinges formation at the two ends and shear failure yield to nonlinear behaviour defined with reference to the chord rotation. Four hinges were inserted in the columns (i.e. two base-hinges and two top-hinges in each principal direction). Conversely only two hinges were inserted in the beams. The yield rotation was estimated according to Panagiotakos and Fardis (2001) assuming uni-axial hinges in both principal directions. In the numerical simulations only the reduction of performance (Imperatore et al. 2017) of corroded longitudinal bars and stirrups, cracking and delamination of concrete cover were included, but bond between concrete and steel was neglected (e.g. bent anchorage at bar ends prevent bond issues) and buckling effect (Vecchi and Belletti 2021), too.

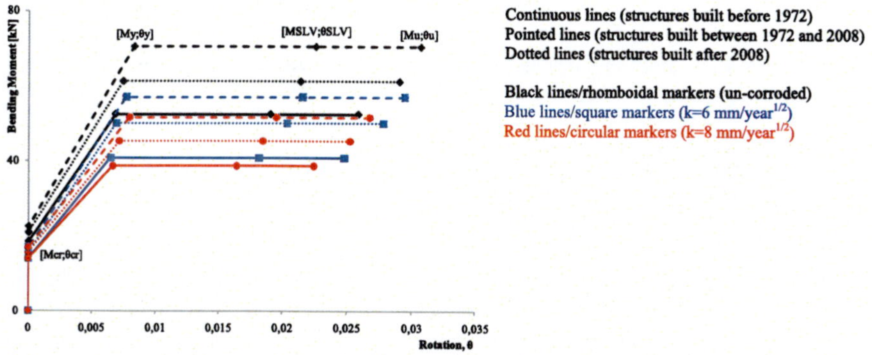


**Fig. 3.** Bending moment vs rotation of the plastic hinges at different corrosion levels for considered building periods

These assumptions were aimed to reduce uncertainties in understanding the influence of material properties on lateral capacity and behaviour of corroded structures. Figure 3 depicts the nonlinear relationship between bending moment and rotations for plastic hinges. In the figures, continuous lines correspond to structures built before 1972, pointed lines to structures built between 1972 and 2008, while dotted lines to structures built after 2008.

Black lines (rhomboidal markers) correspond to un-corroded structures, blue lines (square markers) and red lines (circular markers) correspond to corroded structures with $k = 6$ mm/year$^{1/2}$ and $k = 8$ mm/year$^{1/2}$, respectively. M and $\theta$ are bending moments and rotations, respectively, while subscripts cr, y, MAX and u correspond to cracking, yielding, peak and ultimate values of bending moment or corresponding rotation.

Shear capacity is crucial in the evaluation of seismic capacity, and it was included. The Biskinis et al. (2004) model is one of the best accepted shear models for existing structures including cyclic degradation. In fact, the shear capacity is reduced by the inelastic (ductility) demand on the structural element.

## 3 Push-Over Analysis and Discussion of Results

Push-over analyses both for X-direction and Y-direction (both positive and negative verses) and for each considered building period are presented, yielding to thirty-six analyses. Twelve of them depict the three building periods for un-corroded structure, the other twelve plus twelve depict the structures corroded with a coefficient for carbon dioxide, k, equal to 6 mm/year$^{1/2}$ and to 8 mm/year$^{1/2}$. For the sake of brevity results for X + -direction only are shown.

Figure 4 outlines the bi-linear curves of un-corroded and corroded structures for all considered building periods (X + -direction only). The same meaning for line stiles and markers of Fig. 3 is replicated for Fig. 4. Initial slope for all the curves is similar, being the effects of material properties and corrosion level not significant on the elastic portion. Clear differences occur in terms of maximum displacement capacity and ductility: they reduce with age (being older structures less performant) and corrosion level. A different trend can be found only for the oldest structure (built before 1972) with highest corrosion level, in terms of initial slope, (smallest) strength, displacement and ductility capacity. Nevertheless, this effect is emphasised by the bi-linearization procedure as discussed later.

### 3.1 Performance of Un-corroded Structure

Acceleration-Displacement Response Spectrum (ADRS) curves and corresponding bi-linear curves are shown in Fig. 5 for un-corroded structure. Structures built after the 2008 or built between 1972 and 2008, as expected, have similar capacity with negligible differences, while older structures (built before 1972) have lower capacity than the others.

All the structures have no deficiencies in the un-corroded state since their capacity is always higher than the demand (for the Italian site considered in the design phase).

### 3.2 Performance of Corroded Structure

ADRS curves and bi-linear curves for corroded structure (k = 6 mm/year$^{1/2}$) are depicted in Fig. 6. Structures built after 2008 and structures built between 1972 and 2008 have similar capacity, higher than demand, again. Conversely, structures built before 1972 have lower capacity, and capacity lower than demand.

The case of corrosion with k = 8 mm/year$^{1/2}$ is depicted in Fig. 7 and all the ADRS capacity curves are lower than the demand. The significant difference with the case k = 6 mm/year$^{1/2}$ confirms that material quality plays a fundamental role in terms of durability (w/c ratio, permeability, etc.). In this case there is a considerable drop of the capacity for the structure built before 1972.

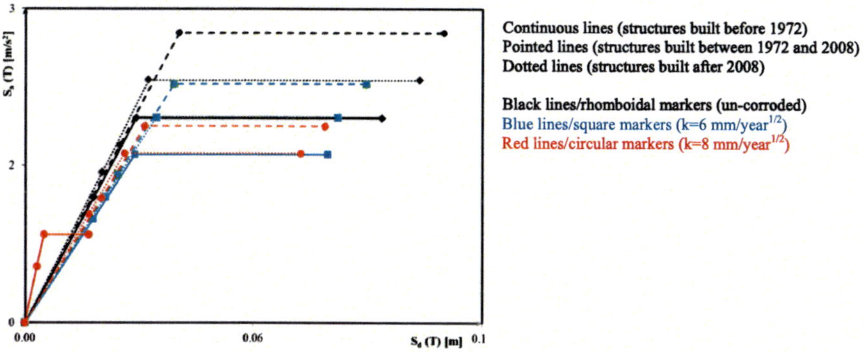


**Fig. 4.** Bi-linear curves of un-corroded and corroded structures for all considered building periods (X + direction)

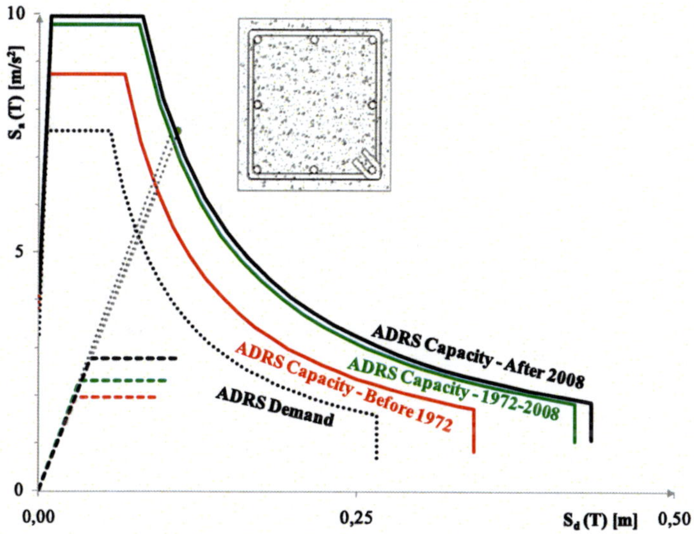


**Fig. 5.** ADRS and Bi-linear curves in the case of un-corroded structure

### 3.3 Discussion and Comparisons

In the worst case (i.e. $k = 8$ mm/year$^{1/2}$) there is a premature shear failure while plastic hinge of beams rotates, hence there is a limited ductility capacity. Early shear failure, according to Biskinis et al. (2004), occurs while rotational demand is still limited (Fig. 8).

Since flexural failure is not fully developed as in the other cases, structure built before 1972 has a significant higher slope in the bi-linearized curve (red curve in Fig. 7). In the bi-linearization the reference point at 60% of peak, to estimate elastic stiffness, is not on the flexural decay branch, but on the first stiffer branch.

Return periods, $T_R$, of seismic events able to induce collapse in structures built in different building periods are shown in Fig. 9, with corrosion or not.

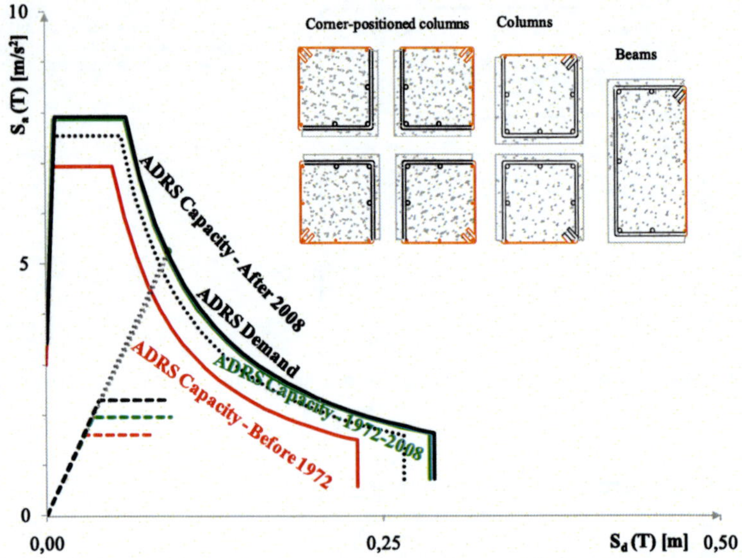


**Fig. 6.** ADRS and Bi-linear curves in the case of corroded structure with $k = 6$ mm/year$^{1/2}$

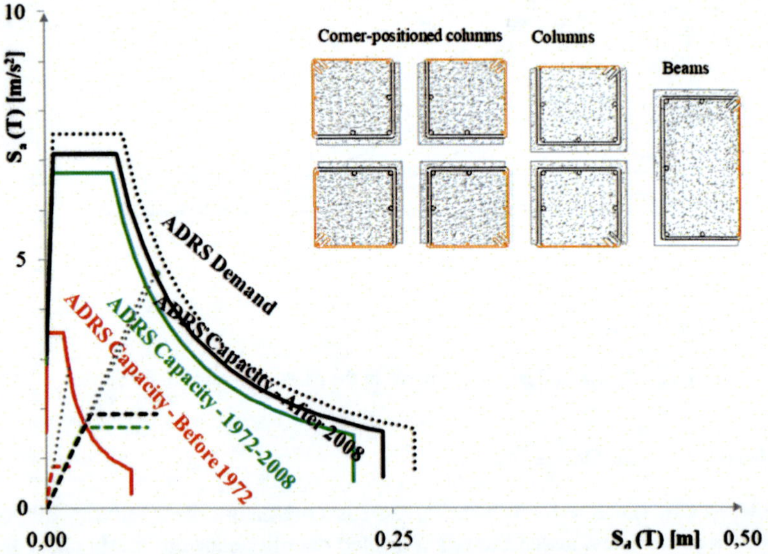


**Fig. 7.** ADRS and Bi-linear curves in the case of corroded structure with $k = 8$ mm/year$^{1/2}$

Given a return period, $T_R = 475$ years (thin red line), structures built before 1972 do not satisfy their safety level neither with $k = 6$ mm/year$^{1/2}$ nor with $k = 8$ mm/year$^{1/2}$. On the other hand, safety is ensured by structures built after 1972 at least in the case of better structures ($k = 6$ mm/year$^{1/2}$).

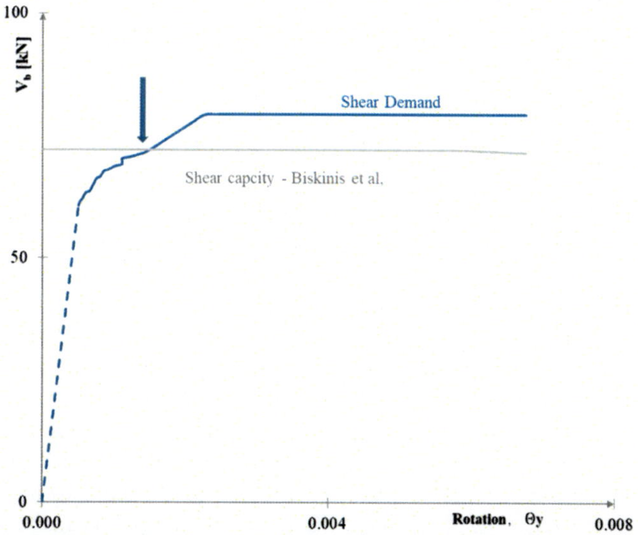


**Fig. 8.** Beam shear failure (arrow) in the case of structure built before 1972 and $k = 8$ mm/year$^{1/2}$

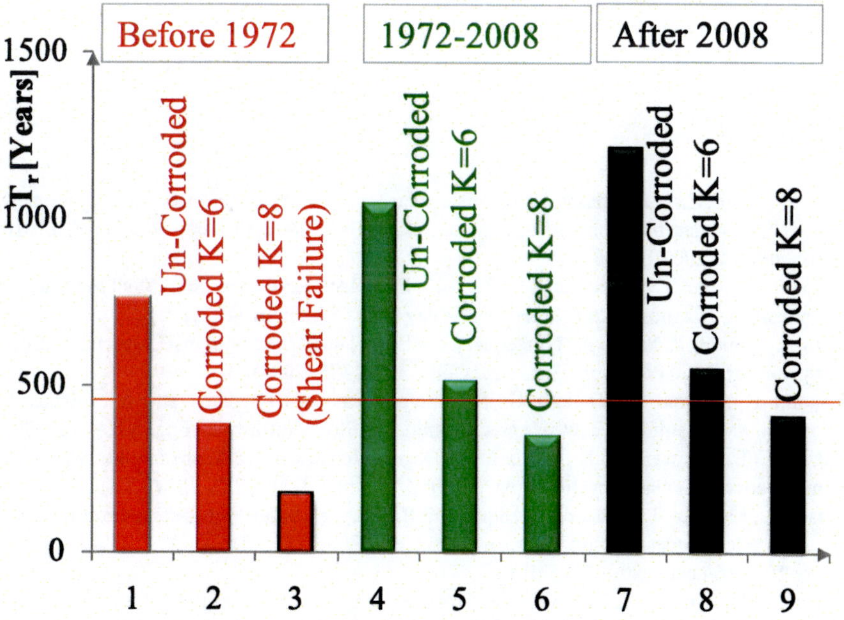


**Fig. 9.** Return periods comparison (k in mm/year$^{1/2}$)

# 4 Conclusions

Seismic behaviours of a simple corroded reinforced concrete (RC) 3D structure, in terms of push-over analyses, are compared focusing on different corrosion levels on structures built in different periods, hence accounting for the evolution of materials technology over time. Different building periods are considered by means of different material properties, while two different corrosion levels are considered for the structure after a typical service life of 50 years (steel bar corrosion and concrete cover cracking/delamination). Other potential modelling parameters (e.g. beam column joints behaviour, Bossio et al. 2015b and 2017b) are neglected to focus on previous points and will be focused in next studies.

At the beginning, un-corroded structures are safe in terms of seismic behaviour according to modern Italian Building Code. Material properties and quality of execution (e.g. permeability,...) are simulated by means of a variable coefficient of carbon dioxide penetration k. Older structures have potentially lower quality, hence higher k. Corrosion impacts the stirrups before longitudinal bars, since they are more exposed. Premature shear failure, for this reason could reduce the seismic capacity changing the failure mode from flexure, with very limited ductility. This is extremely sensitive in the evaluation of the seismic capacity of structures.

Material properties alone have a limited influence on seismic behaviour, if structures were designed according to relevant seismic codes, however older structures revealed a faster and earlier corrosion process leading to faster and earlier reduction of structural capacity.

## References

Apostolopoulos, C.A., Papadakis, V.G.: Consequences of steel corrosion on the ductility properties of reinforcement bar. Constr. Build. Mater. **22**, 2316–2324 (2008). https://doi.org/10.1016/j.conbuildmat.2007.10.006

Bertolini, L., Elsener, B., Pedeferri, P., Redaelli, E., Polder, R.: Corrosion of Steel in Concrete – Prevention, Diagnosis, Repair, 2nd edn. Wiley VCH, Weinheim (2013)

Berto, L., Simioni, P., Saetta, A.: Numerical modelling of bond behaviour in RC structures affected by reinforcement corrosion. Eng. Struct. **30**, 1375–1385 (2008)

Biondini, F., Palermo, A., Toniolo, G.: Seismic performance of concrete structures exposed to corrosion: case studies of low-rise precast buildings. Struct. Infrastruct. Eng. **7**, 109–119 (2011)

Biskinis, D.E., Roupakias, G.K., Fardis, M.N.: Degradation of shear strength of reinforced concrete members with inelastic cyclic displacement. ACI Struct. J. **101**, 773–783 (2004)

Bossio, A., Monetta, T., Bellucci, F., Lignola, G.P., Prota, A.: Modeling of concrete cracking due to corrosion process of reinforcement bars. Cem. Concr. Res. **71**, 78–92 (2015). https://doi.org/10.1016/j.cemconres.2015.01.010

Bossio, A., Fabbrocino, F., Lignola, G.P., Prota, A., Manfredi, G.: Simplified model for strengthening design of beam-column internal joints in reinforced concrete frames. Polymers **7**(9), 1732–1754 (2015). https://doi.org/10.3390/polym7091479

Bossio, A., Fabbrocino, F., Lignola, G.P., Prota, A., Manfredi, G.: Design oriented model for the assessment of T-shaped beam-column joints in reinforced concrete frames. Buildings **7**(4), 118 (2017). https://doi.org/10.3390/buildings7040118

Bossio, A., et al.: Corrosion effects on seismic capacity of reinforced concrete structures. Corros. Rev. **37**(1), 45–56 (2018). https://doi.org/10.1515/corrrev-2018-0044

Bossio, A., et al.: Nondestructive assessment of corrosion of reinforcing bars through surface concrete cracks. Struct. Concr. **18**(1), 104–117 (2017). https://doi.org/10.1002/suco.201600034

Bossio, A., Lignola, G.P., Prota, A.: An overview of assessment and retrofit of corroded reinforced concrete structures. Procedia Struct. Integr. **11**, 394–401 (2018). https://doi.org/10.1016/j.prostr.2018.11.051

Cabrera, J.B.: Deterioration of concrete due to reinforcement steel corrosion. Cement Concr. Compos. **18**(1), 47–59 (1996). https://doi.org/10.1016/0958-9465(95)00043-7

Cairns, J., Du, Y., Law, D.: Influence of corrosion on the friction characteristics of the steel/concrete interface. Constr. Build. Mater. **21**, 190–197 (2007)

Cairns, J., Plizzari, G.A., Du, Y., Law, D.W., Franzoni, C.: Mechanical properties of corrosion-damaged reinforcement. ACI Mater. J. **102**, 256 (2005)

Çaliskan, Ö., Aras, M.: Experimental investigation of behaviour and failure modes of chemical anchorages bonded to concrete. Constr. Build. Mater. **156**, 362–375 (2017)

Capozucca, R.: Damage to reinforced concrete due to reinforcement corrosion. Constr. Build. Mater. **9**(5), 295–303 (1995)

Caprili, S., Moersch, J., Salvatore, W.: Mechanical performance versus corrosion damage indicators for corroded steel reinforcing bars. Adv. Mater. Sci. Eng. **2015**, 19 (2015). https://doi.org/10.1155/2015/739625

Celarec, D., Vamvatsikos, D., Dolšek, M.: Simplified estimation of seismic risk for reinforced concrete buildings with consideration of corrosion over time. Bull. Earthq. Eng. **9**, 1137–1155 (2011)

Coccia, S., Imperatore, S., Rinaldi, Z.: Influence of corrosion on the bond strength of steel rebars in concrete. Mater. Struct. **49**(1–2), 537–551 (2016)

Desnerck, P., Lees, J.M., Morley, C.T.: Bond behaviour of reinforcing bars in cracked concrete. Constr. Build. Mater. **94**, 126–136 (2015)

Di Ludovico, M., Lignola, G.P., Prota, A., Cosenza, E.: Nonlinear analysis of cross sections under axial load and biaxial bending. ACI Struct. J. **107**(4), 390–399 (2010)

El Maaddawy, T., Soudki, K., Topper, T.: Analytical model to predict nonlinear flexural behavior of corroded reinforced concrete beams. ACI Struct. J. **102**, 550–559 (2005)

Fahy, C., Wheelera, S.J., Gallipoli, D., Grassla, P.: Corrosion induced cracking modelled by a coupled transport-structural approach. Cem. Concr. Res. **94**, 24–35 (2017)

Fernandez, I., Herrador, M.F., Marí, A.R., Bairàn, J.M.: Structural effects of steel reinforcement corrosion on statically indeterminate reinforced concrete members. Mater. Struct. **49**, 4959–4973 (2016)

Hays, G.F.: Now is the Time, World Corrosion Organization (2010)

Imperatore, S., Leonardi, A., Rinaldi, Z.: Strength decay of RC sections for chloride attack. Int. J. Struct. Integr. **7**(2), 194–212 (2016)

Imperatore, S., Leonardi, A., Rinaldi, Z: Mechanical behaviour of corroded rebars in reinforced concrete elements. Models Methods Civ. Eng. 207–220 (2012)

Imperatore, S., Rinaldi, Z., Drago, C.: Degradation relationships for the mechanical properties of corroded steel rebars. Constr. Build. Mater. **148**, 219–230 (2017)

Lin, H., Zhao, Y., Yang, J., Feng, P., Ožbolt, J., Ye, H.: Effects of the corrosion of main bar and stirrups on the bond behavior of reinforcing steel bar. Constr. Build. Mater. **225**, 13–28 (2019)

Li, S., Wang, M., Li, S.: Model for cover cracking due to corrosion expansion and uniform stresses at infinity. Appl. Math. Model. **32**, 1436–1444 (2008)

Ministero delle Infrastrutture: Decreto Ministeriale 14 gennaio 2008 – Norme Tecniche per le Costruzioni, NTC (2008)

Ministero per i Lavori Pubblici: Decreto Ministeriale 30 maggio 1972 – Norme tecniche alle quali devono uniformarsi le costruzioni in conglomerato cementizio, normale e precompresso ed a struttura metallica

Panagiotakos, T.B., Fardis, M.N.: Deformations of reinforced concrete members at yielding and ultimate. ACI Struct. J. **98**, 135–148 (2001)

Pantazopoulou, S.J., Papoulia, K.D.: Modelling cover-cracking due to reinforcement corrosion in RC structures. J. Eng. Mech. **127**(4), 342–351 (2001)

Pedeferri, P., Bertolini, L.: La durabilità del calcestruzzo armato. McGraw-Hill, Milano (2000)

Rodriguez, J., Ortega, L.M., Casal, J.: Load carrying capacity of concrete structures with corroded reinforcement. Constr. Build. Mater. **11**, 239–248 (1997)

Su, R.K.L., Zhang, Y.: A double-cylinder model incorporating confinement effects for the analysis of corrosion-caused cover cracking in reinforced concrete structures. Corros. Sci. **99**, 205–218 (2015)

U.K. Department of Trade and Industry. Effective Cost Analysis for Repairing of Corrosion Damaged Reinforced Concrete Structures (2015). http://projects.bre.co.uk/rebarcorrosioncost

Vecchi, F., Belletti, B.: Capacity assessment of existing RC columns. Buildings **11**(4), 161 (2021)

Zanini, M.A., Faleschini, F., Pellegrino, C.: Probabilistic seismic risk forecasting of aging bridge networks. Eng. Struct. **136**, 219–232 (2017)

Zhu, W., François, R., Coronelli, D., Cleland, D.: Effect of corrosion of reinforcement on the mechanical behaviour of highly corroded RC beams. Eng. Struct. **56**, 544–554 (2013)

# Self-monitoring Precast RC Beams Industrial Production with FBG Sensors for Quality Control and Real-Time Monitoring

Monica Capasso[1](✉), Reza Darban[1], Davide Lavorato[1], Carlotta Pia Contiguglia[1], Michele Arturo Caponero[2], Cristina Mazzotta[2], Paolo Clemente[3], Claudio Failla[4], Sergio Signorini[5], Francesco Sonzogni[5], and Camillo Nuti[1]

[1] Department of Architecture, University of Roma Tre, Rome, Italy
monica.capasso@uniroma3.it
[2] ENEA, Fusion and Nuclear Safety Department, C.R. Frascati, Frascati (RM), Italy
[3] ENEA, Department for Sustainability, C.R. Casaccia, Rome, Italy
[4] Municipality of Milan, Milan, Italy
[5] Magnetti Building Spa, Carvico (BG), Italy

**Abstract.** Fibre optical sensors technology is widely employed for structural health monitoring of civil engineering structures, mainly existing structures, to guarantee a proper structural safety level. This research proposes an industrial solution which leads to the production of self-monitoring prestressed, precast RC beams through optical fibre technology, for quality prebuilt beam control and to monitor structures when the implementation in the construction site is done. As a first step, steel strands have been instrumented with Fiber Bragg Grating sensors. It was feasible through the embedding of FBGs into fibreglass manufactured saddles for easy positioning and the fixing of optical sensors on rebars during the pre-cast production site of beams. The following step consists of performing executing tensile tests for comparing and validating FBG monitoring results to traditional measurement systems (extensometers) and Digital Image Correlation measurement system (DIC). Throughout this article, the manufacturing saddles process and preliminary thermal tests are presented to display the first monitoring parameters' results.

**Keywords:** Fiber Bragg Grating (FBG) sensors · self-monitoring element · prestressed precast reinforced concrete (RC) · beam quality control

## 1 Introduction

Structural health monitoring (SHM) of civil engineering structures plays a fundamental role in the knowledge of in-service buildings and infrastructures safety (Ferdinand 2014). Real-time monitoring is a useful solution for prompt defects and structural damages detection, to require focused routine maintenance, avoiding extraordinary structural repairs which would request higher economic costs. Nowadays, there are several technologies developed to monitor structural safety, locating, identifying, and quantifying


M. A. Aiello and A. Bilotta (Eds.): ICC 2022, LNCE 435, pp. 629–643, 2024.
https://doi.org/10.1007/978-3-031-43102-9_48

anomalies in different materials: however, most of them have high intrusiveness and they are sensitive to electromagnetic interferences. Moreover, more than one sensor should be embedded in buildings and infrastructures, to instal a proper detection system. Therefore, it is easier to provide a sensor monitoring system in pre-cast RC structures during their industrial production site. Embedding sensors in pre-cast beams, columns, etc. during construction allows a product's quality control, useful to detect preliminary defects which could be responsible for lower capacity loads and durability. Another point worth mentioning is that embedding sensors in pre-cast processes are useful for continuous monitoring structures and infrastructures during all their serviceability life.

Therefore, this study wishes to find a solution to provide for self–monitoring pre-cast RC beams, to control and notify when damage occurs. It is also useful for controlling the structural elements' quality, preliminarily. Furthermore, to avoid high intrusiveness and electromagnetic interference issues, this research develops its project through fibre optical systems (FOS), using quasi-distributed Fiber Bragg Grating (FBG) technology sensors (Fig. 1).

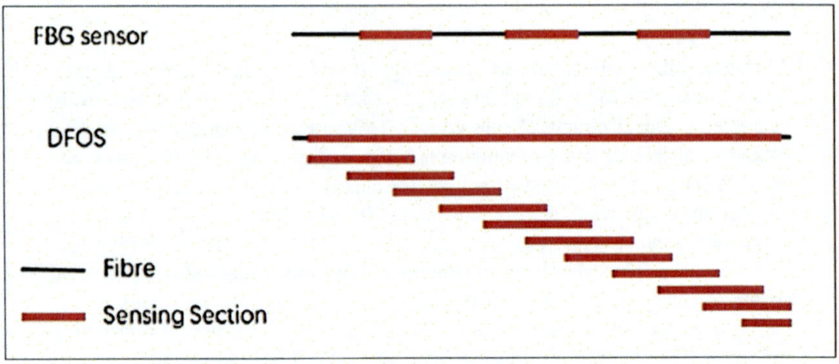


**Fig. 1.** Quasi-distributed and distributed fibre optical sensors (Soga et al. 2018).

**Table 1.** Fiber Bragg Grating (FBG) sensors and Distributed fiber optical sensors (DFOS) properties.

| **Fiber optical sensors properties** | |
|---|---|
| FBG & DFOS | Measurement parameters (temperature, strains,...) |
| DFOS | Detection (cracking,...) |
| FBG & DFOS | Small diameters (125μm) |
| FBG & DFOS | Flexibility |
| FBG & DFOS | Electromagnetic interference |

The FBG sensors are suitable for static and dynamic measurements, and they can measure strains, displacements, and temperature variations. In fact, thanks to their Bragg

Grating, the fibre starts to become sensitive to several parameters, corresponding to those punctual optical sensors (Table 1).

The 'self-monitoring' concept aims to underline the advantages of the embedding of the quasi-distributed FBG sensors into prebuilt RC beams. The only necessity for their SHM system is to connect the projected wiring of optical sensors to optical interrogators for temperature, static and dynamic measurements. In this way, the final precast elements are already set to be monitored at the construction site, eliminating their consequently instrumenting step.

### 1.1 Prefos Project

Prefos Project is a Regional Research Project (Latium Region, Italy). It is supported by the European Regional Development Fund POR 2014–2020 and it involves different Partners: the University of Roma Tre-Department of Architecture, Enea C.R. and Magnetti Building S.p.a.

The University of Roma Tre, (Dept. of Architecture) based in Rome, boasts a structural Laboratory that is fundamental for developing and testing some steps of this project.

Enea is the National Agency for New Technologies, Energy and Sustainable Economic Development, a public body aimed at research, technological innovation, and the provision of advanced services to enterprises, public administration and citizens in the sector of energy, the environment and sustainable economic development (article 4, Law no. 22 of December 2015).

Magnetti Building S.p.a. is a construction society, based in the north of Italy, close to the city of Bergamo. It aims to find solutions studying right down for a suitable one and also to create even more complex architectural structures. Moreover, Magnetti Building S.p.a.'s areas of intervention also regard existing structures that need to be recovered and redeveloped.

The involvement of different entities will nudge the project into a complete self-monitoring precast prestressed RC beam's design, testing the FBG optical technology and its application related to. The optical sensors will be embedded in the concrete section, and they will monitor also steel rebars, thanks to fibreglass saddles (instrumented with FBG sensors) fixed on them.

## 2 Fiber Bragg Grating

### 2.1 Fiber Bragg Grating Principles

A Fiber Bragg Grating is optical sensor technology, fabricated by exposing a short segment (100 ÷ 150 mm, typically) of the optical fibre's core to a particular pattern of ultraviolet (UV) light. During the exposure, the refractive index of the core is modified periodically, and a diffraction grating (Bragg grating) is written. When a broadband light is an input through the grating, a wavelength is reflected (Fig. 2). The wavelength reflected is called "Bragg Wavelength": the unexpired wavelengths keep travelling through the fibre and grating region (Keiser 2003; Kashyap 2009). The central Bragg wavelength follows the Bragg condition which can be expressed as:

$$\lambda_B = 2n_{eff}\Lambda \tag{1}$$

where $\lambda_B$ is the central wavelength of the FBG sensor, $n_{eff}$ is the effective refractive index and $\Lambda$ is the grating period. An FBG spectrum displayed by an optical interrogator is shown in Fig. 3.

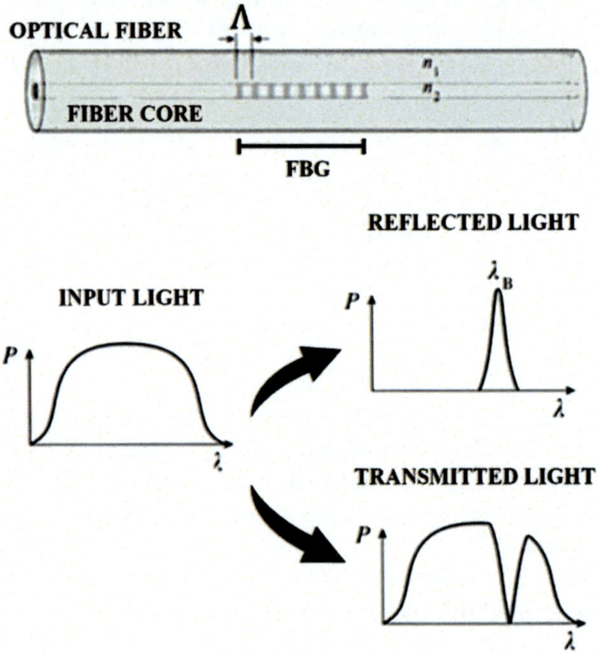


**Fig. 2.** FBG principles

The distance between each grating ($\Lambda$) changes when temperature and strain alterations occur on the fibre: consequently, there is a shift of the Bragg wavelength's FBG sensor. The variations of the Bragg wavelength ($\Delta\lambda_B$) let the monitoring and detection of strain and temperature parameters change.

Thanks to the multiplexing technology, more than one FBG sensor can be connected in series on the same optical cable: each sensor can be detected by its Bragg wavelength ($\lambda_{B1}$, $\lambda_{B2}$, ..., $\lambda_{Bn}$) by an interrogator. In Fig. 4, the spectrum of three FBGs connected in series is shown.

Moreover, the quasi – distributed Fiber Bragg Grating optical sensors can measure strains and temperature variations with high-speed acquisitions: they are suitable for static and dynamic measurements. Their strain resolution is about 0.1 μm/m and the temperature resolution is about 0.01 °C (Ferdinand P. 2014).

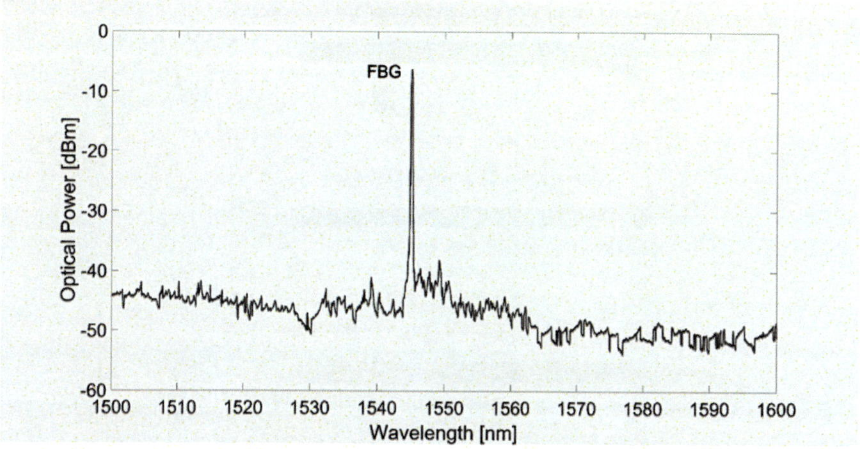


**Fig. 3.** FBG spectrum shown by interrogator.

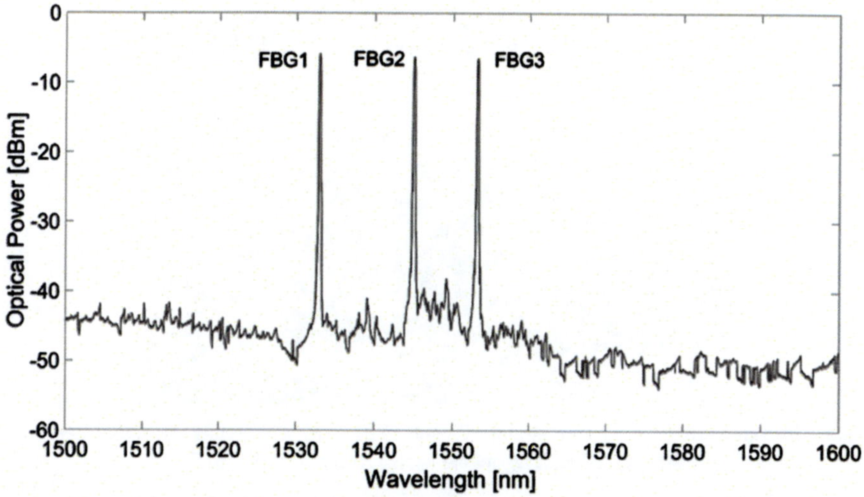


**Fig. 4.** Spectrum of three Fiber Bragg Gratings connected in series on the same optical fibre cable.

## 3 Steel Strand Monitoring with Fiber Bragg Gratings

Over the past few years, previous studies provided several solutions for instrumenting a 7 wires steel strand. However, all of them proposed a modified strand structure, (Figs. 5, 6, 7 and 8), to insert the optical cable into the central wire of the strand (Whanxu Z. et al. 2021, Kim S.T. et al. 2015, Kim J.M. et al. 2012)

Prefos project aims to propose and test some options for instrumenting a prestressed precast RC self – monitoring beam. As a first step, the project proposes a solution for

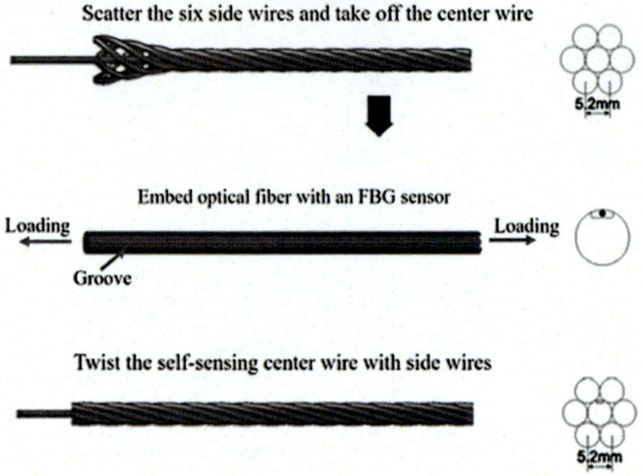


**Fig. 5.** Instrumenting a 7-wire steel strand with optical fibre, milling the central wire (Whanxu et al. 2021).

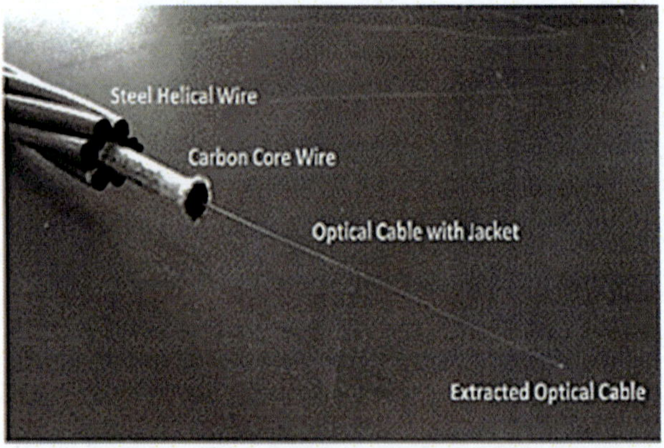


**Fig. 6.** Instrumenting a 7-wire steel strand with optical fibre, by substituting the central steel wire with a carbon wire instrumented with an optical cable (Kim S.T. et al. 2015).

monitoring prestressing steel rebars (7 wires steel strands): a fibreglass saddle instrumented with an FBG sensor helps the sensor's positioning procedure on the steel strands surface.

## 3.1 Materials and Methods

Instrumenting a steel strand during the precast site could require special care and warn for those who should deal with it. For this reason, an easy approach to solve and make the fixing-on-strand Fiber Bragg Grating sensors process easier is presented here.

Fibreglass saddles were instrumented with the FBG optical sensors: saddles have a quasi-cylindrical shape (Fig. 9), given to them using a 7-wire strand as mould.

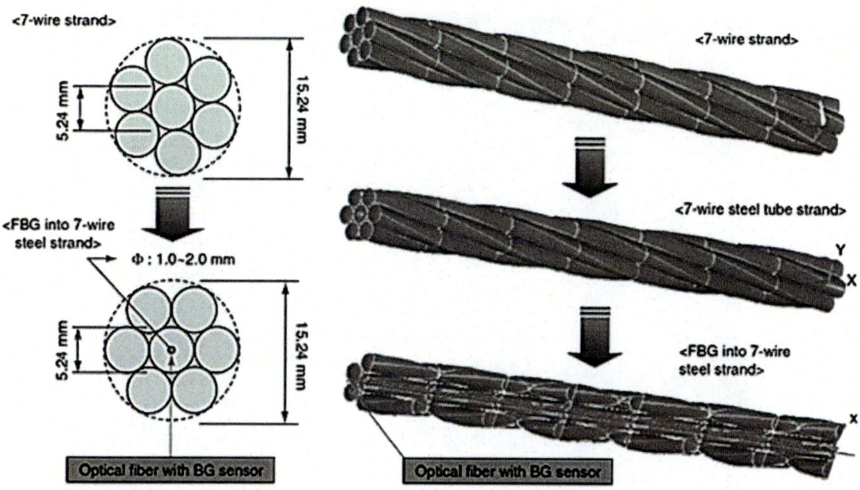


**Fig. 7.** Instrumenting a 7-wire steel strand with optical fibre, by substituting the central steel wire with a different steel wire instrumented with an optical cable (Kim J.M. et al. 2012).

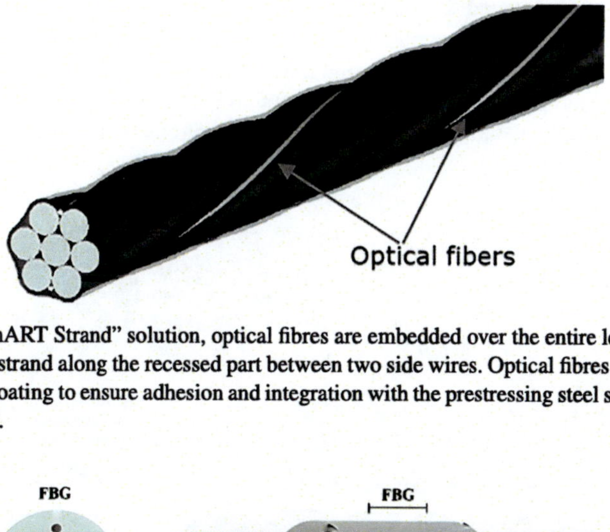


**Fig. 8.** In "SmART Strand" solution, optical fibres are embedded over the entire length of a prestressing steel strand along the recessed part between two side wires. Optical fibres are embedded into the resin coating to ensure adhesion and integration with the prestressing steel strand (Nakaue S. et al. 2021).

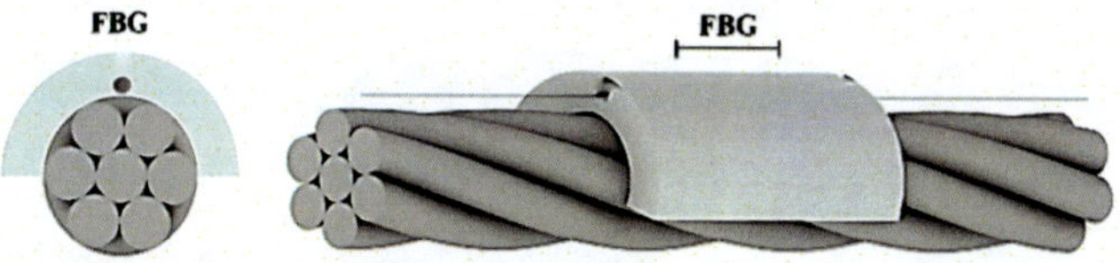


**Fig. 9.** Fibreglass saddle instrumented with a Fiber Bragg Grating.

First and foremost, a fibreglass paper (Fig. 10) was cut and saturated with epoxy resin (bicomponent, 4:1): Table 2 shows the mechanical properties of the resin.

**Fig. 10.** Fibreglass roll.

Before curing, the imbibed fibreglass was turned around the mould (a 7-wire steel strand used only for this preliminary step) for just one layer. Secondly, a Fiber Bragg Grating sensor was positioned between the first and the second layer and the remaining length of the paper covered the strand for the last layer of it.

Then, the saddle was covered by a film waiting for the curing (about 12 h); after that, it was cut in two longitudinal directions, just to remove it from the mold and, above all, for letting the saddle fix on the strand to be monitored (Fig. 11). Finally, the instrumented saddle was glued on the strand to be monitored with epoxy resin "Araldite 2011" (bicomponent, 1:1).

**Fig. 11.** Steps for making instrumented fibreglass saddles.

The above–mentioned process has been used t produce saddles with two different lengths, respectively about 30 mm and 110 mm (Fig. 12) and also two different diameters about 12.5 mm and 15.2 mm (for positioning them on two different steel strands about the same saddles' diameters).

**Table 2.** "Sikadur 330" epoxy resin mechanical properties.

| | | | | | |
|---|---|---|---|---|---|
| **Compressive Strength** | | 60 °F (16 °C) | 73 °F (23 °C) | 90 °F (32 °C) | (ASTM D-695) |
| | 8 hour | - | - | 8,000 psi (55.2 MPa) | 50 % R.H. |
| | 1 day | 8,100 psi (55.8 MPa) | 10,700 psi (73.7 MPa) | 10,600 psi (73.1 MPa) | |
| | 3 day | 11,200 psi (77.2 MPa) | 11,100 psi (76.5 MPa) | 11,000 psi (75.8 MPa) | |
| | 7 day | 11,600 psi (80.0 MPa) | 11,200 psi (77.2 MPa) | 11,800 psi (81.3 MPa) | |
| | 14 day | 12,400 psi (85.5 MPa) | 11,800 psi (81.3 MPa) | 11,900 psi (82.0 MPa) | |
| **Flexural Strength** | 8,800 psi (60.6 MPa) (7 days) | | | | (ASTM D-790) 73 °F (23 °C) 50 % R.H. |
| **Modulus of Elasticity in Flexure** | 5.06 x 105 psi (3,489 MPa) (7 days) | | | | (ASTM D-790) 73 °F (23 °C) 50 % R.H. |
| **Tensile Strength** | 4,900 psi (33.8 MPa) (7 days) | | | | (ASTM D-638) 73 °F (23 °C) 50 % R.H. |
| **Elongation at Break** | 1.2 % (7 days) | | | | (ASTM D-638) 73 °F (23 °C) 50 % R.H. |
| **Heat Deflection Temperature** | 120 °F (50 °C) (7 days) | | | | (ASTM D-648) [fiber stress loading=264 psi (1.8 MPa)] |

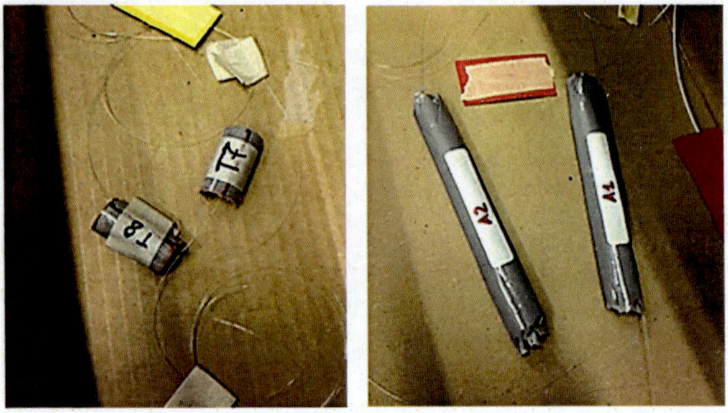


**Fig. 12.** Photo on the left: 30 mm (length) saddles: on the right: 110 mm (length) saddles.

### 3.2 Thermal Tests on Instrumented Saddles

After curing time, two different diameter steel strands have been inserted into a thermal machine, as shown in Fig. 13.

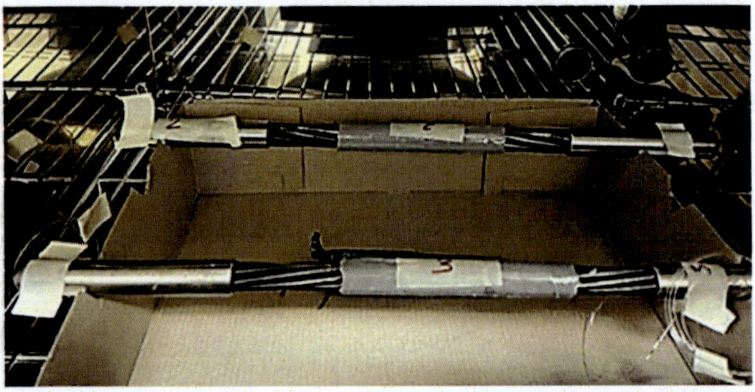

**Fig. 13.** Saddles bonded on two different diameters (12,5 mm; 15,2 mm) strands.

For monitoring the temperature variations (in degrees, unit of measurement), a temperature sensor was combined with the Fiber Bragg Grating sensors, during the thermal test.

First, the test was set up: several increasing temperature steps were configured from the machine, between 20 °C and 70 °C. Figures 14, 15 plot temperature variations monitored by the temperature sensor and FBG sensors wavelength shifts for two different diameter steel strands, respectively.

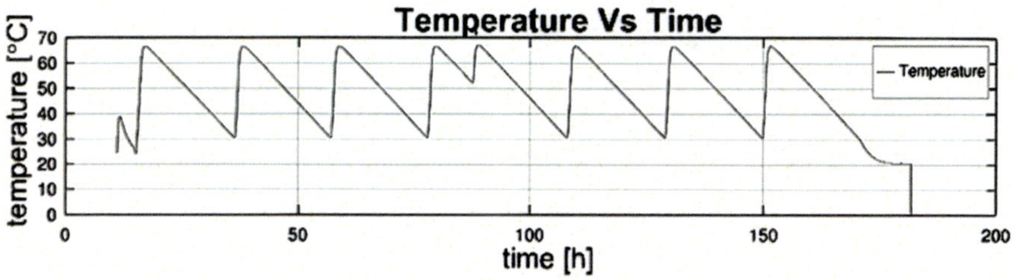


**Fig. 14.** Temperature variation in time, during thermal test.

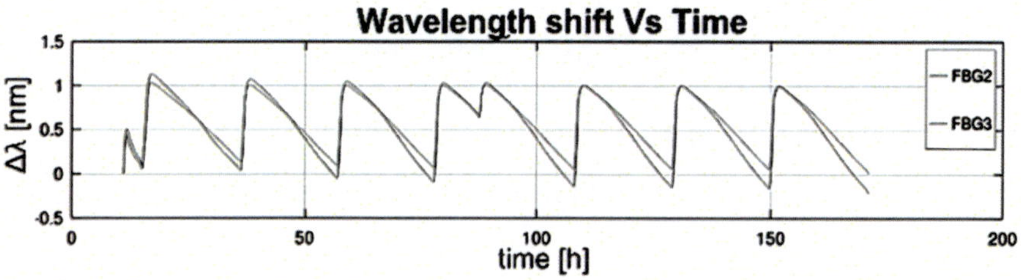


**Fig. 15.** Delta-lambda variation (on the FBG) in time, during thermal test.

### 3.3 Thermal Tests on Chains of Instrumented Saddles

Several cyclic tests were done to verify the instrumented saddles and different chains of FBG's behaviour. These were realized by connecting sensors previously embedded into the fibreglass saddles.

Four chains were prepared, jointing saddles through a fusion splicer: three of them consist of some 30 mm length saddle; one consists of two 110 mm FBG saddles connected with a fusion joint.

Figures 16–17 show the mentioned chain of FBG sensors embedded into saddles of different sizes.

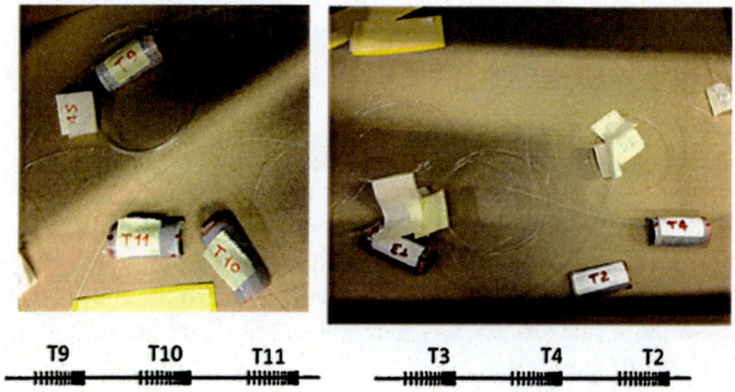


**Fig. 16.** Chains of FBG sensors embedded into fibreglass saddles.

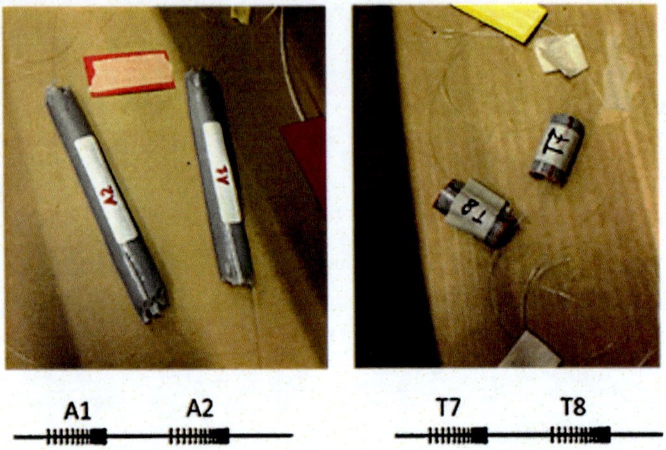


**Fig. 17.** Chains of FBG sensors embedded into fibreglass saddles.

First, each chain was connected to a single interrogator's channel for monitoring the wavelength shift of the FBG sensors, during the test. A temperature sensor inside the thermal machine was also used to monitor the temperature in time.

Then, the test was set up: several increasing temperature steps were configured from the machine, between 20 °C and 70 °C.

In Fig. 18 the temperature trend is displayed.

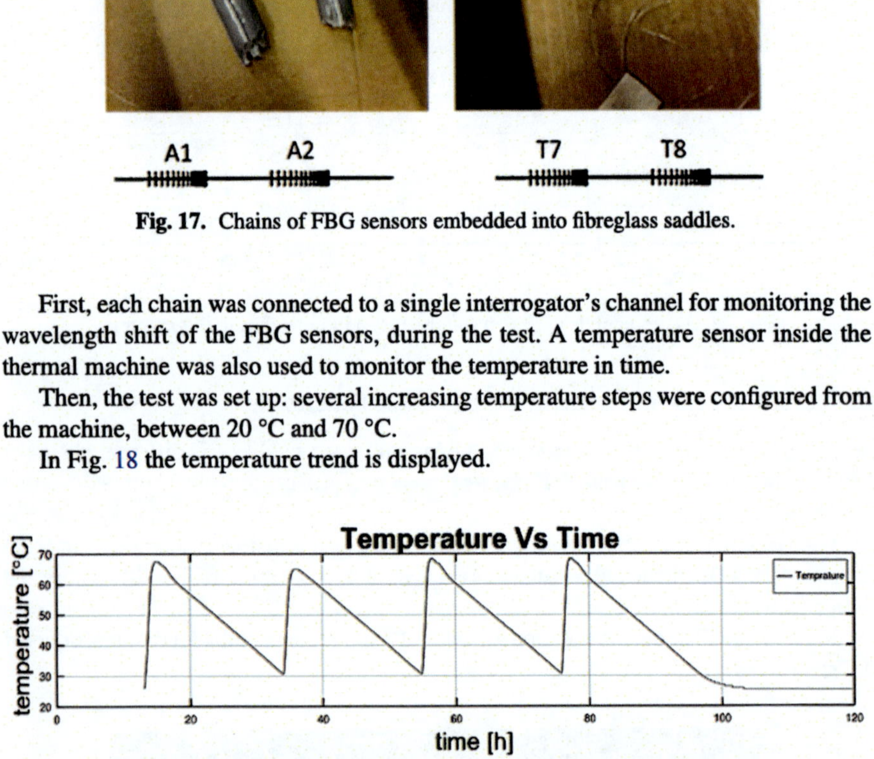


**Fig. 18.** Temperature variation in time, during thermal test on chains of instrumented saddles.

The FBG sensors' output is combined with temperature variations monitored by the temperature sensor: Figs. 19, 20, 21 and 22 plot the results from each FBG chain.

Reading the results, it is clear that the wavelength shift from chain number 3 (Fig. 21) is higher than the others. This result should be due to the different sizes of the saddles (saddles A1 and A2 are about 110 mm in length) and also the fibreglass paper used to manufacture those two saddles was longer than the pieces cut for the 30 mm ones (400 mm against 240 mm).

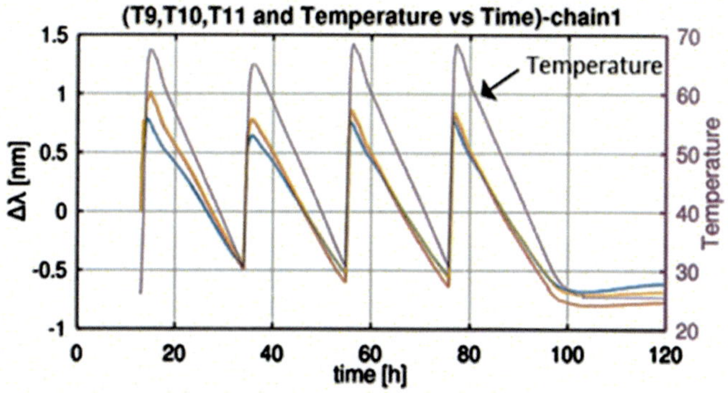


Fig. 19. Wavelength shift and temperature against time, chain n.1.

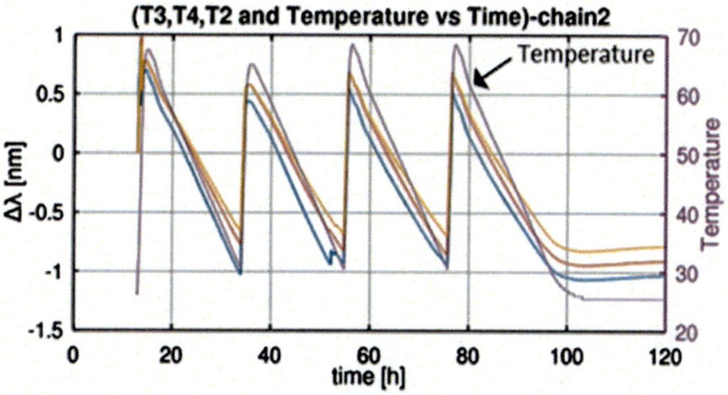


Fig. 20. Wavelength shift and temperature against time, chain n.2

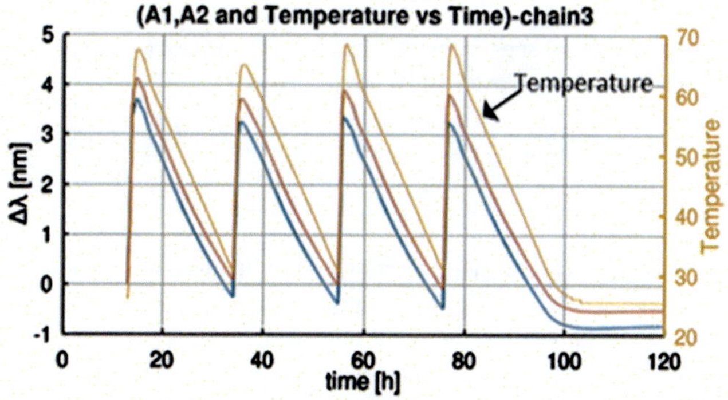


Fig. 21. Wavelength shift and temperature against time, chain n.3.

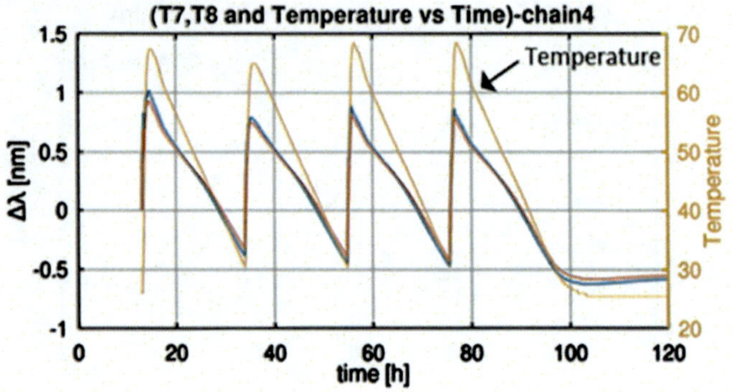


**Fig. 22.** Wavelength shift and temperature against time, chain n.4.

## 4 Conclusions

This article shows the preliminary results of a research project still in development. From the first results, it emerges as the proposed monitoring solution, that includes the manufacturing of fibreglass saddles which incorporate optical Fiber Bragg Grating sensors, can return data comparable with the traditional temperature sensor monitoring temperature variations due to external environment. Fibreglass saddles would also solve difficulties related to the instrumenting of pre-stressed cables during the industrial production of pre-cast beams: in this way, it will be possible to place the chain of saddles instrumented with optical sensors on the steel strand before their pre-stressing process. During future steps, tensile tests on strands will be conducted, instrumenting cables with saddles, which will be produced as described in this article, calibrating the system by comparing tensile tests results with traditional monitoring systems (extensometers) and Digital Image Correlation.

**Acknowledgments.** Financial support for this research was receipt by FESR Fondo Europeo di Sviluppo Regionale Programma Operativo Regione Lazio.

## References

Ferdinand, P.: The Evolution of Optical Fiber Sensors Technologies during the last 35 last years and their applications in Structural Health Monitoring. In: 7th European Workshop on Structural Health Monitoring, La Cité, 8–11 July 2014, Nantes (2014)

Keiser, G.: Optical communications essentials. McGraw-Hill (2003)

Kashyap, R.: Fiber Bragg Gratings. Academic press (2009)

Wang, H.P., Xiang, P., Jiang, L.Z.: Strain transfer theory of industrialized optical fiber-based sensors in civil engineering: a review on measurement accuracy, design and calibration. Sensors Actuators A: Phys., 414–426 (2019)

Wang, H.P., Dai J.G., Wang, X.Z.: Improved temperature compensation of fiber Bragg grating – based sensors applied to structures under different loading conditions. Opt. Fiber Technol. (2021)

Zhu, W., Shen, Q., Qin, H.: Monitoring of prestress and bond stress of self – sensing FBG steel strand. Measurement (2021)

Kim, S.T., Park, Y.H., Park, S.Y., Cho, K., Cho, J.R.: A sensor – Type PC strand with an embedded FBG sensor for monitoring prestress forces. Sensor (2015)

Kim, J.M., Kim, H.W., Park, Y.H., Yang, I.H., Kim, Y.S.: FBG Sensors encapsulated into 7 – Wire steel strand for tension monitoring of a prestressing tendon. Adv. Struc. Eng. (2012)

Lavorato, D., Nuti, C.: Pseudo – dynamic tests on reinforced concrete bridges repaired and retrofitted after seismic damage, Elsevier. Eng. Struc. (2016)

Lavorato, D., Nuti, C., Santini, S.: Experimental investigation of the shear strength of RC beams extracted from an old structure and strengthened by carbon FRP U-strips. Appli. Sci. (2018)

Lavorato, D., Bergami, A.V., Fiorentino, G., Fiore, A., Santini, S., Nuti, C.: Experimental tests on existing RC beams strengthened in flexure and retrofitted for shear by C-FRP in presence of negative moments. Inter. J. Adv. Struct. Eng. **10**(3), 211–232 (2018)

Xue, J., et al.: Severely damaged reinforced concrete circular columns repaired by turned steel rebar and high – performance concrete jacketing with steel or polymer fibers. Appli. Sci. (2018)

Xue, J., Lavorato, D., Tarantino, A.M., Brisighella, B., Nuti, C.: Rebar replacement in severly damaged RC bridge column plastic hinges: design criteria and experimental investigation. J. Struct. Eng. **149**(3), art. no. 04023007–1 (2023)

Soga, K., Luo, L.: Distributed fiber optics sensors for civil engineering infrastructure sensing. J. Struct. Integrity Maint. (2018)

Lupi, C., et al.: Critical isuues of double – metal layer coating on FBG for applications at high temperatures. Sensors (2019)

Caponero, M.A., Dell'Erba, D., Kropp, C.: Use of the fibre optic sensors for structural monitoring of temporary reinforcements of the church S. Maria delle Grazie in Accumuli. J. Civil Struct. Health Monit., 353–360 (2019)

Nakaue, S., Oshima, K., Oikawa, M., Nishino, M., Matsubara, Y., Yamada, M.: SmART strand prestressing steel strand with optical fiber for tensioning monitoring,. Sumitomo Electric Techn. Rev. **92**, 62–67 (2021)

# On the Influence of Corrosion on the Force-Displacement Behaviour of Steel Wires and Strands

Matteo Marra(✉), Michele Palermo, Stefano Silvestri, and Tomaso Trombetti

Department of Civil, Chemical, Environmental, and Materials Engineering DICAM, University of Bologna, Viale del Risorgimento 2, 40136 Bologna, Italy
{matteo.marra6,michele.palermo7,stefano.silvestri, tomaso.trombetti}@unibo.it

**Abstract.** Steel seven-wire strands are highly vulnerable to corrosion phenomena, which can strongly reduce their strength and ductility capacities. In the present work, the key parameters governing the mechanical response of corroded wires and strands are identified. They include parameters related to both the geometrical configuration of the corrosion (in terms of reduction of cross-sectional area and longitudinal extension) and to the material constitutive model (strength, hardening and ductility). In this respect, both elastic-perfectly plastic and bilinear with hardening models are considered for the steel material. A simplified mechanical model describing the tensile force-displacement behavior of corroded wires and strands is proposed. The corroded strand is modelled as a parallel system of corroded wires. Parametric simulations have been then carried out in order to highlight the effects of the identified parameters, in terms of force-displacement relationship. The objective of this work is to provide an estimation of the corroded strand maximum force under tensile load, which is of great interest from both design and safety assessment points of view.

**Keywords:** corrosion · corroded strands · simplified models · force-displacement relationship

## 1 Introduction

Cementitious grouting of the steel strands in post-tensioned beams is essential to provide their permanent protection against corrosion. In fact, they are highly vulnerable to corrosion phenomena, which can strongly reduce their strength and ductility capacities. The recent catastrophic collapse of the Polcevera bridge (Domaneschi et al. 2020) and subsequent extended bridge inspections on Italian bridges (Anania, Badalà, and D'Agata 2018) highlighted that several post-tensioned beams present lacks in the grout, which triggers the formation of corrosion in the steel strands. Although many studies have been conducted on corroded seven-wire strands (Chi-ho, Jae-bin, and Chang-su 2017; Darmawan and Stewart 2006; Haskins et al. 2016; Jeon et al. 2019; Jeon, Nguyen, and Shim 2020; Lee et al. 2017), a formulation for the estimation of their mechanical behaviour


M. A. Aiello and A. Bilotta (Eds.): ICC 2022, LNCE 435, pp. 644–653, 2024.
https://doi.org/10.1007/978-3-031-43102-9_49

is still missing. This paper aims at providing the basis for future analyses in a prospective development of approaches for the estimation of the residual strength capacity of corroded seven-wire strands. In this regard, a simplified mechanical model for corroded wires and strands is proposed based on key-parameters related to the corrosion geometry and mechanical behaviour of steel material. In detail, a new key-parameter describing the longitudinal extension of the corrosion is here introduced. The influence of the key parameters on the force-displacement behaviour of the strand is then discussed. The parametric study is carried out by means of several Monte Carlo simulations whose results could pave the way for future works currently underway.

## 2 Steel Material and Corrosion Geometry

The parameters that might influence the mechanical behaviour of corroded wires and strands are related to the mechanical behaviour of steel material and the geometrical configuration of the corrosion.

The mechanical behaviour of the steel material is here represented by two stress-strain relationships: (i) elastic - perfectly plastic behaviour and (ii) bilinear with hardening behaviour. These models are defined by the following parameters: the yield stress $f_y$, the ultimate stress $f_u$, the yield strain $\varepsilon_y$, the ultimate strain $\varepsilon_u$, the hardening ratio $r = k_2/k_1$ and the material ductility $\mu_\varepsilon = \varepsilon_u/\varepsilon_y$. The two constitutive relationships are reported in Fig. 1.

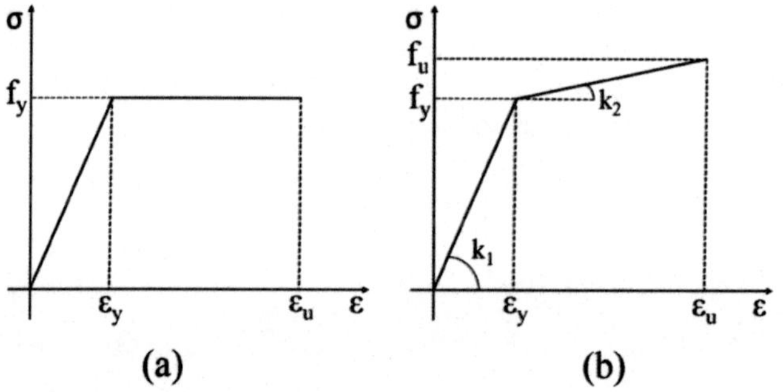


**Fig. 1.** Steel stress-strain relationships considered in the parametric study: (a) elastic - perfectly plastic, (b) bilinear with hardening.

The geometry of the corrosion in the wires is here described by two parameters: (i) the adimensional reduction of cross-section area $\rho = A_r/A_0$ and (ii) the here proposed adimensional corrosion length $\lambda = L_c/L_0$. $A_r$ and $A_0$ are the residual and initial cross-section area, respectively. $L_c$ and $L_0$ are the longitudinal extensions of the corroded zone and the reference length, respectively. The reference length represents the portion of wire or strand for which lack of grouting occurred, and therefore the portion where the corrosion can develop. Figure 2 reports the schematization of corrosion geometry in a corroded wire which can represent either a uniform cross-section reduction along the corroded length or an idealization of a more complex, or even random, cross-section reduction.

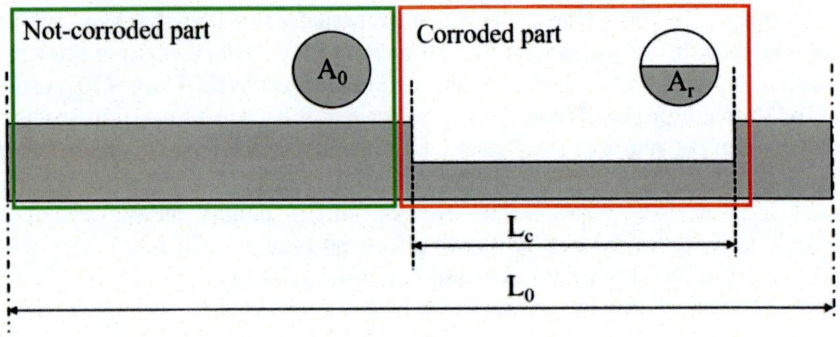


**Fig. 2.** The schematization of corrosion geometry in a corroded wire.

# 3 Mechanical Models

## 3.1 Corroded Single Wire

The proposed mechanical model of the corroded single wire is based on the following main assumptions: (i) the wire is straight and subjected to tension force only, (ii) the corrosion model provides cross-section reduction both in transversal and longitudinal direction, (iii) the wire is described as an equivalent series system composed by the corroded and not-corroded parts that can be identified along the reference length, (iv) the mechanical behaviour of the corroded wire is represented by a force-displacement relationship that also accounts analytically for the corrosion effect on the elastic stiffness. Figure 3 reports the mechanical model of a corroded wire and illustrative force-displacement relationships for the corroded and not-corroded parts.

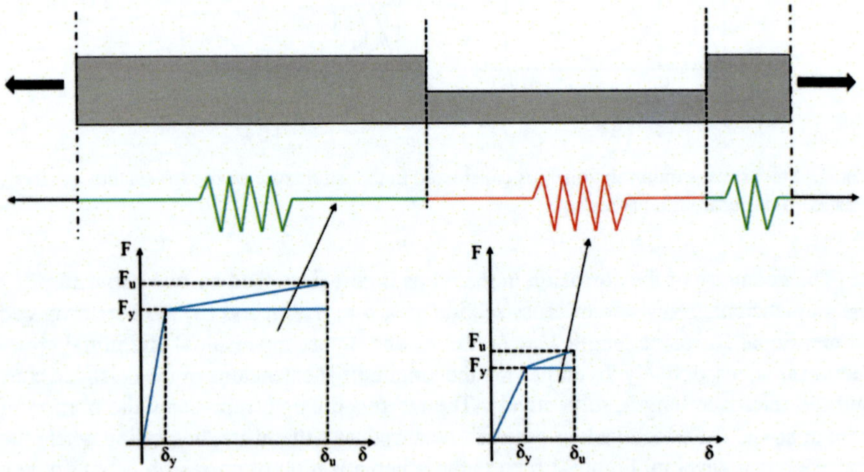


**Fig. 3.** Spring series system representing the mechanical model of a corroded wire.

From the properties of series systems, the following analytical expressions for the elastic stiffness $K$, yield force $F_y$ and yield displacement $\delta_y$, can be derived:

$$K = \frac{EA_0}{L_0}\left[\frac{1}{1+\lambda\left(\frac{1}{\rho}-1\right)}\right] \quad (1)$$

$$F_y = \rho \cdot f_y \cdot A_0 \quad (2)$$

$$\delta_y = \frac{F_y}{K} \quad (3)$$

In addition, considering an elastic - perfectly plastic stress-strain relationship for the steel material allows to obtain analytically also the ultimate displacement $\delta_u$ and the displacement ductility $\mu_\delta$ of the corroded wire:

$$\delta_u = \delta_y + \left(\varepsilon_u - \varepsilon_y\right) \cdot \lambda \cdot L_0 \quad (4)$$

$$\mu_\delta = \frac{\delta_u}{\delta_y} = 1 + \frac{\varepsilon_u - \varepsilon_y}{\varepsilon_y} \cdot \frac{\lambda}{\rho\left[1+\lambda\left(\frac{1}{\rho}-1\right)\right]} \quad (5)$$

### 3.2 Corroded Seven-Wire Strand

The proposed mechanical model of a corroded straight seven-wire strand is based on the following main assumptions: (i) the wires composing the strand are straight and subjected to tension forces only, (ii) any kind of interaction (e.g., friction) between the wires is neglected, (iii) the strand is described as a parallel system between the equivalent series systems representing the wires (according to the mechanical behaviour described in Sect. 3.1), (iv) the mechanical behaviour of a corroded seven-wire strand is represented by a force-displacement relationship. Figure 4 reports the schematization of the parallel model of a corroded seven-wire strand.

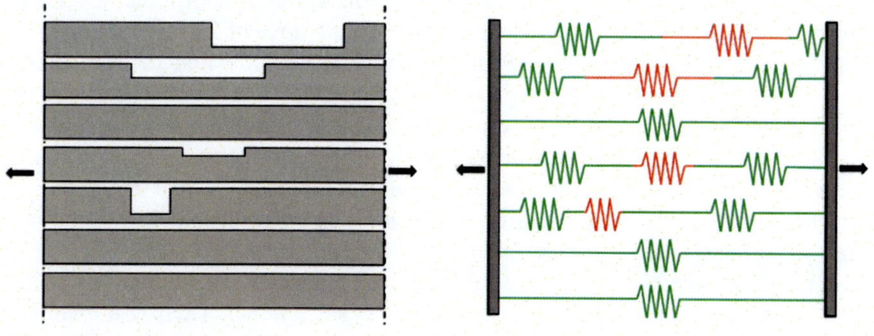

**Fig. 4.** Spring parallel system representing the mechanical model of a corroded seven-wire strand.

# 4 Parametric Study

In order to investigate the effects of the identified parameters on the force-displacement behaviour of corroded wires and strands, a parametric study has been carried out varying the values of the key-parameters $\rho$, $\lambda$, $r$ and $\mu_\varepsilon$ within usual values, as described in Sects. 4.1 and 4.2.

## 4.1 Corroded Single Wire

In this parametric study, two selected values of material ductility, namely 100 and 10 was considered representative of mild and harmonic (high-strength) steel, respectively. For each configuration of the material ductility, three values of hardening coefficient were considered: (i) $r = 0$ (i.e., elastic - perfectly plastic steel material), (ii) $r = 0.5\%$ and (iii) $r = 1.5\%$. The adimensional reduction of the cross-section area $\rho$ and the adimensional corrosion length $\lambda$ were assumed in the range 0–1.

The results of the parametric study are represented in terms of adimensional force-displacement relationships, i.e. normalized with respect to the yield point of the not-corroded wire, identified by $F_{y,w} = A_{0,w} \cdot f_y$ and $\delta_{y,w} = F_{y,w}/K$, where $A_{0,w}$ and $K_{0,w}$ are the cross-section and the elastic stiffness of the not-corroded wire, respectively. Figure 5 reports the force-displacement relationships of corroded wires characterized by $\rho = 0.9$ with $\lambda = 0.05$ and $\lambda = 0.10$, resulting from the parametrical simulations.

Figure 5 clearly shows that the different configurations (in terms of $\lambda$, $r$ and $\mu_\varepsilon$) leading to different results in terms of maximum force, yield and ultimate displacement. Figure 5 allows identifying two different behaviours of the wires characterized by $r > 0$. The first one is represented by a bilinear behaviour with the collapse due to the complete plasticization of the corroded part only, while the not-corroded part remaining within the elastic behaviour. The second one is represented by a trilinear-behaviour with two consecutive changes in the slope of the diagram after the yielding point. This behavior occurs because the collapse involves a partial plasticization of the not-corroded part in addition to the full plasticization of the corroded part, thus leading to a higher ductile behavior. In other words, an exploitation of the post-yielding branch response of the not-corroded part is exploited. This "post-yielding response exploitation" leads to larger ultimate displacement and, for $r > 0$, also to increased maximum force, describing the capacity of the corroded wire to exceed the yielding force of the not-corroded one. This depends on the specific combination of the values of $\lambda$, $r$ and $\mu_\varepsilon$. Accounting for the combination of these three parameters can provide a better description of the behaviour of corroded wires with respect to the usual estimation. Indeed, typically, as a first approximation, the force-displacement response of the corroded wire can be estimated with reference to the corroded part only: specifically, the maximum force of a corroded wire can be obtained with reference to the cross-section reduction of the corroded part only (i.e. $F_{\max} = \rho f_y A_{0,w}$).

Furthermore, the higher the material ductility, the higher the maximum force. The post-yielding response exploitation can also occur for small values of $\lambda$. In case of limited material ductility, which is the case of harmonic high-strength steel used in prestressed concrete structures, this effect strongly depends on the coupling between $\rho$ and $\lambda$. In particular, for small values of $\rho$, it can be observed only for high values of hardening

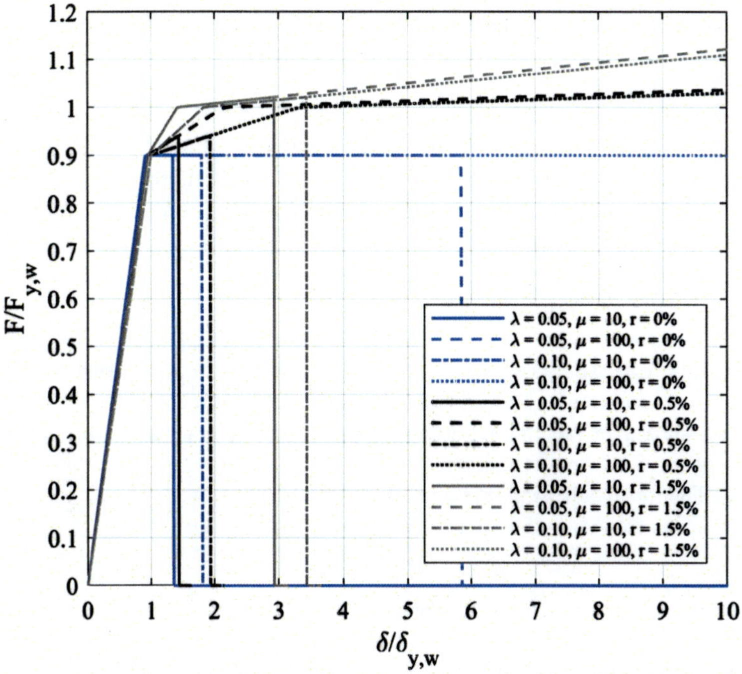


**Fig. 5.** Results of the parametric study in terms of force-displacement relationships of corroded wires characterized by $\rho = 0.9$.

coefficient (i.e. $r = 1.5\%$). In the other cases, the plastic deformation involve only the corroded part. This phenomenon leads to a very brittle behaviour for the corroded wire. This is clear from Figs. 6a and 6b which show the force-displacement relationships of corroded wires characterized by $\rho = 0.5$ and $\rho = 0.1$, respectively.

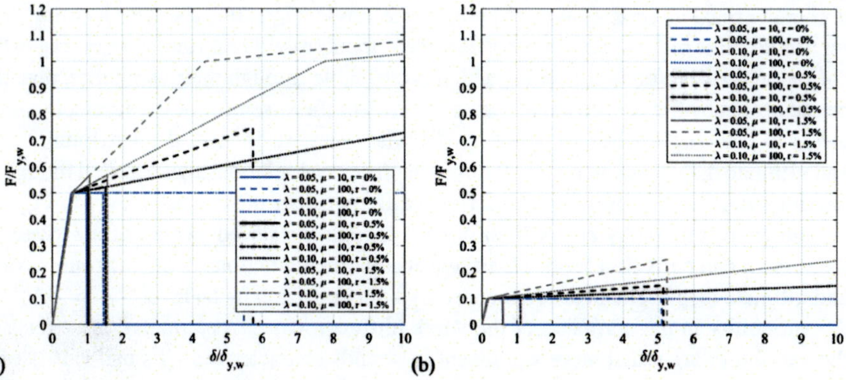


(a) (b)

**Fig. 6.** Results of the parametric study in terms of force-displacement relationships of corroded wires characterized by: (a) $\rho = 0.5$ and (b) $\rho = 0.1$.

From this first parametric study carried out on the single wire, the following considerations can be drawn: (i) a particular coupling of $\rho$ and $\lambda$ might lead to a post-yielding response exploitation of the not-corroded part that allows the capacity of the corroded wire to exceed the one of the corroded part, (ii) the maximum force and the ultimate displacement for the case of limited material ductility and small hardening coefficient are governed by the adimensional reduction of cross-section area $\rho$ and the adimensional corrosion length $\lambda$, respectively, (iii) accounting for $\lambda$ allows to model the wire embrittlement caused by the corrosion.

### 4.2 Corroded Seven-Wire Strand

A second parametric study has been carried out considering two scenarios characterized by different fixed corrosion levels in the wires and varying the main key parameters (S1 and S2 as described in Table 1). In particular, the two different scenarios are characterized by different corrosion levels for the single wires, but mantaining the same mean corrosion level $\rho_M$. The mean corrosion level is defined as:

$$\rho_M = \frac{\sum_{i=1}^{n_w} \rho_i}{n_w} \tag{6}$$

where $n_w$ is the number of the wires in the strand (typically, equal to 7). The seven corroded wires of the strands are supposed to be affected by the same adimensional corrosion length, i.e. by the same $\lambda$ value.

The results of the parametric study are represented in terms of force-displacement relationships with respect to the yielding point of the not-corroded seven-wire strand, identified by $F_{y,s} = A_{0,s} \cdot f_y$ and $\delta_{y,s} = F_{y,s}/K_{0,s}$, where $A_{0,s}$ and $K_{0,s}$ are the cross-section and the elastic stiffness of the not-corroded seven-wire strand, respectively.

Figures 7a and 7b report the results of the simulations for $\lambda = 0.05$ and $\lambda = 0.1$.

Figures 7a and 7b show that the proposed parallel model is able to simulate the typical force-displacement relationship of corroded wire strands (Jeon et al. 2020) that is characterized by progressive collapses of the single wires. It is also worth noting that, for a fixed value of $\lambda$, the behaviour of corroded strands might be different according to the corrosion distribution in the wires, which is evaluable through their variability, that can be quantified by the coefficient of variation (*cov*) of the ensemble of the seven corrosion levels of the single wires constituting the whole strand. In fact, the maximum force and ultimate displacement given by force-displacement relationships reported in Fig. 7a (strand characterized by $cov = 0.123$) are different than those ones shown in Fig. 7b (strand characterized by $cov = 0.606$), even though the strands were characterized by the same mean corrosion level. This suggests that the reductions in the maximum force and ultimate displacement are influenced by both corrosion mean level ($\rho_M = 0.5$) and the variability in the corrosion of the wires (different *cov*). In particular, the effects of the variability are much more significant in strands characterized by small values of $\lambda$ (localized corrosion), where the interaction with $\rho$ is higher. In other words, the ductility in the wires is not enough to enable the strand achieving the maximum force given by the mean corrosion level. Such first results, which requires further investigation, allows

**Table 1.** Values of the wires adimensional reduction of cross-section area for two scenarios of corroded seven-wire strands characterized by $\rho_M = 0.5$.

| *wire* | S1 | S2 |
|---|---|---|
| $\rho_1$ | 0.41 | 0.15 |
| $\rho_2$ | 0.44 | 0.20 |
| $\rho_3$ | 0.47 | 0.25 |
| $\rho_4$ | 0.50 | 0.50 |
| $\rho_5$ | 0.53 | 0.75 |
| $\rho_6$ | 0.56 | 0.80 |
| $\rho_7$ | 0.59 | 0.85 |
| *mean* | **0.5** | **0.5** |
| *Cov* | **0.123** | **0.606** |
| $\lambda$ | range 0–1 | range 0–1 |
| *r* | 0%, 0.5%, 1.5% | 0%, 0.5%, 1.5% |

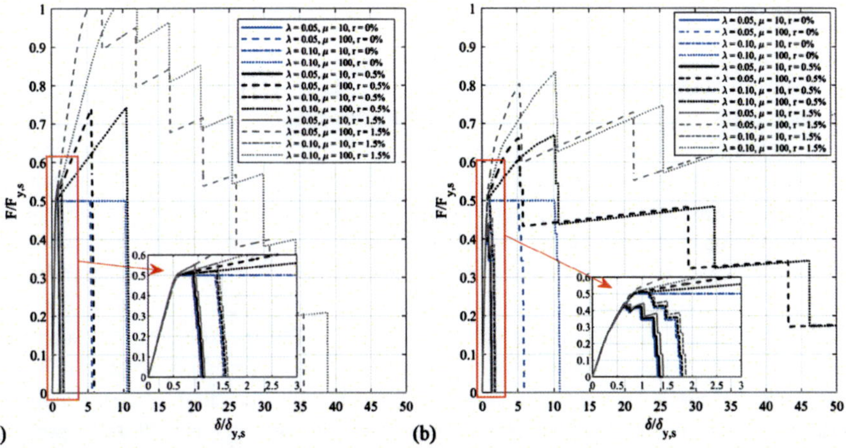


(a) (b)

**Fig. 7.** Results of the parametric study in terms of force-displacement relationships of the: (a) S1 corroded seven-wire strand and (b) S2 corroded seven-wire strand.

to highlight the importance of considering both parameters $\rho$ and $\lambda$ of the wires (and not only $\rho$), in the mechanical behaviour of corroded seven-wire strands. Furthermore, the hardening effect is limited for small material ductility (i.e. high strength steel with $\mu_\varepsilon = 10$), whilst it leads to higher maximum force and higher ultimate displacement for large material ductility (i.e. mild steel with $\mu_\varepsilon = 100$). In other words, the behavior of corroded seven-wire strands made up of high strength steel is well represented by an elastic - perfectly plastic constitutive model.

To sum up, the results of the parametric study allows to make the following consideration s: (i) the corrosion distribution in the wires provides a further reduction on the maximum force in addition to that one due to the reduction of the cross-section area and (ii) for a fixed value of $\lambda$, the higher the variability of the corrosion, the higher the reduction of the maximum force.

## 5 Conclusions

This paper investigates the effects of corrosion on the force-displacement response of both steel wires and strands. It is well known that corrosion can strongly reduce the strength capacity of wires and strands. However, a new description of the corrosion geometry that takes into account its longitudinal extension allows for further understanding of the phenomenon. The corrosion geometry is described by the adimensional reduction of cross-section area $\rho$ and the adimensional corrosion length $\lambda$. The importance of these parameters, together with the ones regarding the steel material, clearly emerges from the results of the parametric studies. The following conclusions can be drawn.

The results on the parametric study on the single wires allowed to clearly identify two different mechanical behaviour. The first one is represented by a bilinear force-displacement response whose collapse is due to the complete plasticization of the corroded part only. The second one is characterized by a trilinear-behaviour wih the collapse reached following a plasticization of the not-corroded part in addition to the full plasticization of the not-corroded part, leading to a higher ductility. In other words, taking into account the adimensional corrosion length allows for an exploitation of the post-yielding response of the not-corroded part, leading to higher maximum force and ultimate displacement (with respect to the ones of the corroded part, which can represent a first approximation of the mechanical behavior of a corroded wire).

The results of the parametric study conducted on the strands allowed to highlight the effect of a non-uniform corrosion (strands characterized by different levels of corrosion for the single wires) on their mechanical behaviour. The reductions in the maximum force and ultimate displacement with respect to the not-corroded strand are due to both the corrosion mean level and the corrosion variability in the single wires. The higher the variability, the higher the reduction of the strand maximum force. This represents an original result whose implications are currently under further investigation and would pave the way for code provisions.

Finally, the parametric study here presented lays the foundations for the development of engineering tools for the estimation of the strength capacity of corroded seven-wire strands, which is currently ongoing.

## References

Anania, L., Badalà, A., D'Agata, G.: Damage and collapse mode of existing post tensioned precast concrete bridge: the case of petrulla viaduct. Eng. Struct. **162**(January), 226–244 (2018). https://doi.org/10.1016/j.engstruct.2018.02.039

Chi-h, J., Jae-bin, L., Chang-su, S.: Tensile test of corroded strand and maintenance of corroded prestressed concrete girders. Int. J. Urban Civ. Eng. **11**(10), 1384–1388 (2017)

Darmawan, M.S., Stewart, M.G.: Effect of spatially variable pitting corrosion on structural reliability of prestressed concrete bridge girders. Aust. J. Struct. Eng. **6**(2), 147–158 (2006). https://doi.org/10.1080/13287982.2006.11464951

Domaneschi, M., et al.: Collapse analysis of the polcevera viaduct by the applied element method. Eng. Struct. **214**(April), 110659 (2020). https://doi.org/10.1016/j.engstruct.2020.110659

Haskins, R., Barry, W., Robert, E., James, E.: Relating Corroded Seven-Strand, Posttensioned Cable Cross-Sectional Properties to Load Capacity. Journal of Engineering (United Kingdom) **2016** (2016)https://doi.org/10.1155/2016/5719156

Jeon, C.H., Lee, J.B., Lon, S., Shim, C.S.: Equivalent material model of corroded prestressing steel strand. J. Market. Res. **8**(2), 2450–2460 (2019). https://doi.org/10.1016/j.jmrt.2019.02.010

Jeon, C.H., Nguyen, C.D., Shim, C.S.: Assessment of mechanical properties of corroded prestressing strands. Appl. Sci. (Switzerland) **10**(12), 1–20 (2020). https://doi.org/10.3390/APP10124055

Lee, B.Y., Koh, K.T., Ismail, M.A., Ryu, H.S., Kwon, S.J.: Corrosion and strength behaviors in prestressed tendon under various tensile stress and impressed current conditions. Adv. Mater. Sci. Eng. **2017**, 1–7 (2017) https://doi.org/10.1155/2017/8575816

# Resistance Against Calcium Chloride Attack of Alternative Binder-Based Sustainable Mortars

Denny Coffetti(✉) and Luigi Coppola

Department of Engineering and Applied Sciences, University of Bergamo, Dalmine, BG, Italy
{denny.coffetti,luigi.coppola}@unibg.it

**Abstract.** Nowadays, calcium chloride is commonly used as de-icer on roads and infrastructures to remove ice and snow, ensuring the safety of vehicles and pedestrian. The deterioration of Portland-based reinforced concretes exposed to $CaCl_2$ is well known while no data are available for alternative and innovatie binding materials. This paper focuses on the resistance to chemical attach of mortars manufactured with different low-carbon binders such as alkali activated slag cements and calcium sulphoaluminate (CSA) cement-based blends in presence of calcium chloride-based de-icing salts in cold weathers. Results indicated that alkali activated slag-based mortars are quasi-immune to $CaCl_2$ attack due to their mineralogical composition. On the contrary, calcium sulphoaluminate-based blends show strong deterioration, especially when CSA cement is used with gypsum and Portland cement. In this case, the total loss of binding capacity can be detected both in cold (4 °C) and hot (38 °C) climates.

**Keywords:** Calcium sulphoaluminate cement · alkali activated slag cement · calcium chloride resistance

## 1 Introduction

Road salts including sodium chloride (NaCl), calcium chloride ($CaCl_2$) and magnesium chloride ($MgCl_2$) are commonly used as de-icers on roads, infrastructures and service area to remove ice and snow and ensure the safety of vehicles and pedestrians. Even if sodium chloride is the most common road salt applied globally, today alternative chloride-based salts or chloride mixes are frequently used in many locations. Nonchloride-based salts such as calcium magnesium acetate and potassium acetate are also used but only marginally (Fay and Shi 2012). In the last decades, the amount of de-icing salt placed on the infrastructures has considerably increased worldwide. For example, in the United States the consumption of road salts was limited to about 0.5 million tons in 1950 while in 1994 about 15 million tons of chloride-based de-icing salts were used in USA to control the ice formation in the highways (US Geological Survey and US Department of Interior, n.d.). To date, every year the United States places about 20 million tons of road salts. On average, depending on weather conditions, from 5 to a maximum of 40 grammes of salt are actually spread per square meter. In contrast


M. A. Aiello and A. Bilotta (Eds.): ICC 2022, LNCE 435, pp. 654–663, 2024.
https://doi.org/10.1007/978-3-031-43102-9_50

to the noticeable improvements in preventing hazardous driving conditions, the use of de-icing salts is one of the main causes of the degradation of reinforced concrete-based infrastructures such as surface scaling (Amini et al. 2020, 2019; Sun and Scherer 2010) and steel reinforcement corrosion (Bertolini et al. 2013; Bohni 2005). However, another highly deleterious but often overlooked mechanism that affects concrete treated with de-icing salts is the formation of an expansive salts, mainly calcium oxychloride (CAOXY) but also Friedel's and Kuzel's salts. This mechanism generally occurs when calcium chloride-based de-icing salts were spread on concrete structures and it results in concrete scaling, cracking and spalling, impacting the structural integrity of cementitious materials.

CAOXY formation was observed and investigated in Portland cement-based mixtures since 1970's by Chatterji (Chatterji 1978) and Collepardi et al. (Collepardi et al. 1994; Monosi et al. 1989; Monosi and Collepardi 1990). Today, thanks to advanced testing techniques (Peterson et al. 2013; Qiao et al. 2018; Suraneni et al. 2018; Sutter et al. 2006; Traore et al. 2021), this mode of damage is comprehensively studied in traditional concretes even if the exact damage mechanism is still unknown as reported in a recent review by Jones et al. (Jones et al. 2020). However, in the last few years, growing attention has been paid to sustainable alternative binders to Portland cement such as alkali activated materials (Ameri et al. 2019; Luigi Coppola et al. 2020; Provis 2018), calcium sulphoaluminate (CSA) cements-based blends (Coppola et al. 2018; Koumpouri et al. 2021) or mixtures containing ultrafine supplementary cementitious materials (Mohan and Mini 2018). For these systems, the resistance against calcium chloride attack has not yet been investigated. Therefore, this paper presents the preliminary results of a research on the durability of different sustainable binders (alkali activated slag-based materials, calcium sulphoaluminate-based blends, binders where plain and/or ultrafine fly ash partially replace cement) exposed to cold solution containing de-icing salts based on $CaCl_2$.

## 2 Materials and Methods

### 2.1 Materials

Ten cementitious mortar compositions were tested in this study. Two different cements compliant with the EN 197–1 standard (CEM I 42.5 R Portland cement and type III/B 42.5 R blast-furnace cement), plain and ultrafine low calcium siliceous fly ash (type V according to EN 450–1), ground granulated blast-furnace slag (compliant with EN 15167–1), commercial calcium sulphoaluminate (CSA) clinker, technical grade anhydrite ($C\hat{S}$) and CL90-S hydrated lime (CH – according to EN 459–1) were used as binder. Moreover, tap water at room temperature and natural siliceous sand with maximum diameter equal to 2.5 mm were employed to manufacture mortars. The binder content was fixed at 500 kg/m$^3$ while water-to-binder ratio varies from 0.50 to 0.55 in order to obtain a 28-day compressive strength in the range of 40–55 MPa.

The composition of mortars is given in Table 1. A PCE superplasticizer (ester of acrylic or methacrylic acid monomer, compliant with EN 934–2) was added at 0.5 wt.% by binder mass to the reference (PC and BFC) and fly ash-based mortars (FA and UFFA) to ensure a consistency of fresh mixtures determined by flow table (EN 1015–3) close to 150 mm. The blend of solid activators to manufacture the alkali activated slag-based (AAS) mortar used in this research was optimized in previous works (L. Coppola et al. 2020; D'Alessandro et al. 2020), it is composed by sodium silicate, potassium hydroxide and sodium carbonate and its dosage was set at 16 wt.% by slag mass. Finally, the CSA clinker was used to produce a traditional OPC-CSA-C$\hat{S}$ ternary mixture (PC-CSA) and two low carbon quaternary mortars (S-CSA and FA-CSA) containing calcium sulphoaluminate clinker, calcium sulphate, supplementary cementitious materials (plain fly ash or ground granulated blast-furnace slag) and hydrated lime (Coppola et al. 2019, 2018). In these mixtures, proper setting time and workability were ensured by means of a tartaric acid-based admixture at 0.8 wt.% by binder mass (Cabrini et al. 2020).

**Table 1.** Composition of mortars

| | PC | BFC | FA5 | FA25 | UFFA5 | UFFA25 | AAS | PC-CSA | S-CSA | FA-CSA |
|---|---|---|---|---|---|---|---|---|---|---|
| CEM I | 500 | | 475 | 375 | 475 | 375 | | 200 | | |
| CEM III/B | | 500 | | | | | | | | |
| Plain fly ash | | | 25 | 125 | | | | | | 175 |
| Ultrafine fly ash | | | | | 25 | 125 | | | | |
| Slag | | | | | | | 500 | | 175 | |
| CSA clinker | | | | | | | | 200 | 200 | 200 |
| Anhydrite | | | | | | | | 100 | 100 | 100 |
| Hydrated lime | | | | | | | | | 25 | 25 |
| Sand | 1500 | 1500 | 1500 | 1500 | 1500 | 1500 | 1500 | 1500 | 1500 | 1500 |
| Water | 250 | 250 | 250 | 250 | 250 | 250 | 275 | 265 | 265 | 265 |
| Superplasticizer | 2.5 | 2.5 | 2.5 | 2.5 | 2.5 | 2.5 | | | | |
| Activators | | | | | | | 80 | | | |
| Tartaric acid | | | | | | | | 4 | 4 | 4 |

## 2.2 Experimental Methods

Mortars were mixed by adding the cementitious powder, sand, water and admixtures in an automatic mixer with a planetary motion with the following procedure:

1. Admixtures were dispersed in the mixing water;
2. Binders, solid activators (only for AAS) and mixing water were placed into the bowl;
3. The compound was mixed at low speed for 30 s;
4. The automatic mixer integrated the sand to the compound with a high-speed mixing;
5. Rest time of 90 s;
6. Further 60 s of mixing at high speed.

The prismatic samples (40x40x160 mm) were cured in a climatic chamber at 20 °C and R.H. 95% for 28 days. Firstly, the average compressive strength of mortars was evaluated on 3 specimens at 1, 7 and 28 days in accordance with EN 1015–11. After 28-day curing age, the mortar samples were immersed for 4 weeks in a 30 wt.% $CaCl_2$

aqueous solution (prepared by mixing appropriate masses of technical grade $CaCl_2$ with deionized water) at 38 °C to favor the calcium chloride diffusion into the cementitious mixtures. Then, the temperature was changed from 38 °C to 4 °C in order to promote the damaging effect caused by the calcium chloride chemical action. This procedure is aimed to speed up the formation of complex salt inside cementitious materials exposed to $CaCl_2$ solutions, it was followed by several authors (Collepardi et al. 1994; Monosi et al. 1989) and it was fully compatible with the findings of Farnan et al. (Farnam et al. 2015) that proposed a ternary phase diagram for a $Ca(OH)_2$-$CaCl_2$-$H_2O$ system to describe the behavior of a concrete exposed to road salts. During the storage of samples in cold $CaCl_2$ solution the deterioration of mortars was evaluated every 7 days (up to 42 days) by measuring the compressive strength (EN 1015–11, average between three specimens) and the specific mass (EN 1015–10, average between three specimens) in comparison with samples stored in cold deionized water. Moreover, visual observations were carried out to identify expansion, cracks formation or alteration in surface texture. Finally, X-ray diffractions (XRD) were conducted on alkali activated slag-based pastes to evaluate the mineralogical composition of mixtures cured in a climatic chamber at 20 °C and R.H. > 95%. XRD analysis were done by means of a Rigaku Miniflex diffractometer in a θ-2θ configuration using an incident beam monocromator employing the Cu Kα radiation (λ = 1.5418 Å) in a range of 5°–50° (2θ) with step size of 0.02° and a count time of 10 s for each step.

## 3 Results and Discussion

### 3.1 Traditional Binders

The exposure of Portland-based mortar (PC) to 30 wt.% $CaCl_2$ solution at 4 °C promotes a strong deterioration of cementitious matrix both in terms of specific mass reduction (about -4%) and strength loss (the compressive strength decreases of about 50% after 42 days of test) with respect to samples stored in water (Fig. 1). Moreover, considerable cracks on the edge of the specimens are macroscopically detectable already after few days in cold calcium chloride solution (Fig. 2), symptom of a severe attack by calcium oxychloride formation. On the contrary, mortar manufactured with blast-furnace cement exhibits negligible strength loss without any superficial degradation of samples. This behavior is in accordance with the experimental results summarized in (Jones et al. 2020) that evidenced, for Portland cement-based mortars exposed at 5 °C $CaCl_2$ solution with concentration in the range 22–30%, reduction in strength ranging from 34% to 67%.

The replacement of Portland cement with blast-furnace cement allows to strongly improve the resistance against CAOXY formation due to a dilution effect (less Portland cement clinker) combined with the pozzolanic nature of slag. In fact, Suraneni et al. (Suraneni et al. 2018) evidenced the linear relationship between CAOXY formation and calcium hydroxide amount in the cementitious materials, indicating that these two actions (dilution and pozzolanic nature) make blast-furnace and pozzolanic cement quasi-immune against calcium chloride attack.

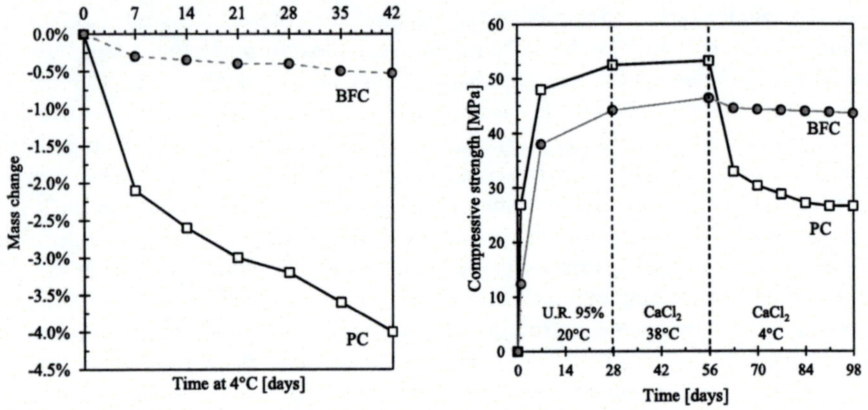


**Fig. 1.** Mass change (left) of traditional cementitious mortars exposed to 30 wt.% $CaCl_2$ solution at 4 °C in comparison with mortars stored in water. Strength development of mortars (right) in $CaCl_2$ solution.

**Fig. 2.** PC (up) and BFC (down) samples after 42 days in 30 wt.% $CaCl_2$ solution at 4 °C.

### 3.2 Fly Ash-Based Blended Binders

Similarly to the use of blast-furnace or pozzolanic cements, one of the most effective strategies to prevent the CAOXY formation in cementitious systems is the partial replacement of Portland cement with supplementary cementitious materials. However, Traore et al. (Traore et al. 2021) recently supposed that the beneficial effect deriving from the use of fly ash on cement pastes is evident only when its use is higher than 15% by mass. Results reported in Fig. 3 confirm the findings of Traore et al. and indicated that mortars containing 5% of fly ash (FA5 and UFFA5), regardless of its fineness, are poorly resistant to the chemical attack by $CaCl_2$, evidencing degradation similar to that of PC mortars. The influence of fineness is evident on UFFA25 mortar that shows a mass loss and strength reduction halved in comparison with FA25 mortar, -1.6% vs -0.9% and -23% vs -14%, respectively. In this case, the impact of the fineness of fly ash on CAOXY formation is not related to the dilution effect but it is due to the enhanced reaction of ultrafine fly ash with portlandite (from cement hydration) forming C-S-H gel, thus reducing both the porosity of mortars and the $Ca(OH)_2$ available to CAOXY formation (Chindaprasirt et al. 2007, 2005).

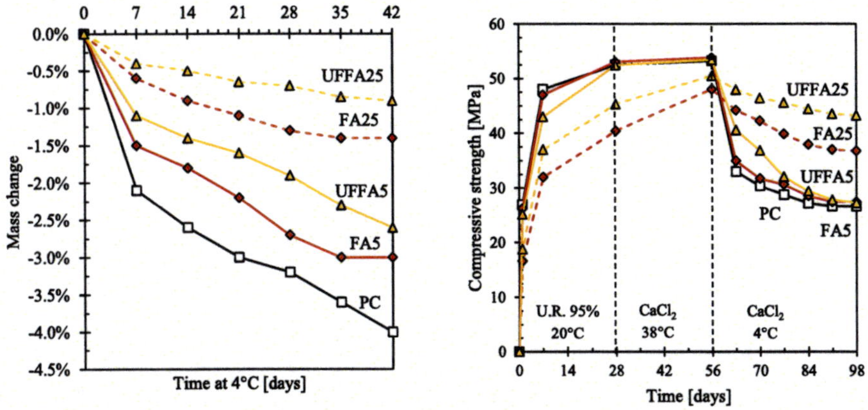


**Fig. 3.** Mass change (left) of fly ash-based mortars exposed to 30 wt.% $CaCl_2$ solution at 4 °C in comparison with mortars stored in water. Strength development of mortars (right) in $CaCl_2$ solution.

### 3.3 Alkali Activated Slag-Based Binders

Alkali activated AAS mortar seems to be immune to $CaCl_2$ attack because it shows negligible reduction in both specific mass and compressive strength after 42 days in cold de-icing salt-based solution as reported in Fig. 4. This excellent behavior is also confirmed by the visual observation that did not show any cracks or surface degradation and it could be ascribed to the phase composition of AAS mortar, mainly composed by C-(A)-S-H gel and totally free of $Ca(OH)_2$ as shown in the XRD patterns in Fig. 5 (the peak close to $2\theta = 18°$ is absent at all ages) and also confirmed by Myers et al. (Myers et al. 2017).

### 3.4 Calcium Sulphoaluminate Blends

The exposure of CSA-based mortars at $CaCl_2$ solution causes a strong deterioration of the samples, mass loss up to 20% and compressive strength reduction ranging from 60% to more than 80% with respect to reference specimens stored in water (Fig. 6). PC-CSA mixture shows the most severe damage, so much so as to make it impossible to measure the specific mass and the mechanical strength already after 14 days of storage at 4 °C. On the other hand, S-CSA and FA-CSA, even if roughly damaged, were tested up to day 42, evidencing a mass change close to 7% and a compressive strength loss equal to 63%, regardless of supplementary cementitious material used. The deterioration of the mortars can be highlighted not only when samples are stored in cold de-icing salt-rich solution but also at the end of the curing at 38 °C as reported in Figs. 6 and 7. This behavior is

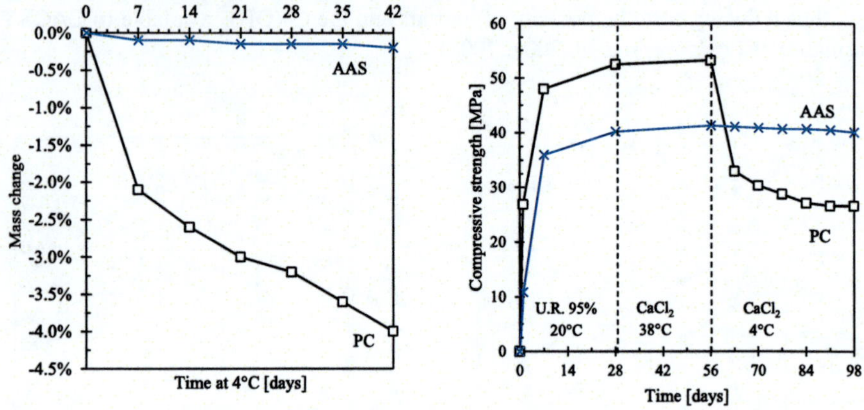


**Fig. 4.** Mass change (left) of alkali activated mortars exposed to 30 wt.% $CaCl_2$ solution at 4 °C in comparison with mortars stored in water. Strength development of mortars (right) in $CaCl_2$ solution.

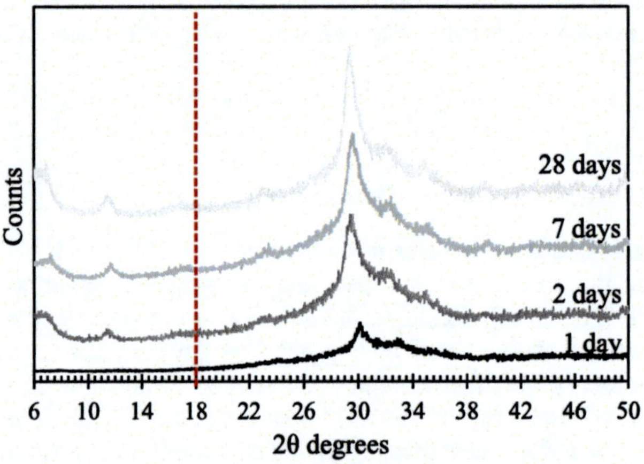


**Fig. 5.** XRD pattern of AAS mortars after 1,2,7 and 28 days from casting. Samples cured in dry environment.

not compatible with the formation of CAOXY which, in accordance with Farnam et al. (Farnam et al. 2015), is inhibited when the concentration of the $CaCl_2$ solution is equal to 30% and the temperature is above 35 °C. In this case, it is possible to hypothesize different degradation mechanisms from those detectable in Portland cement-based systems (i.e. calcium oxychloride, Friedel's and Kuzel's salts formation) which cause severe damage even when temperatures are strongly above 0 °C. Further investigations are underway to fully understand the mechanism that leads to the degradation of CSA systems in the presence of calcium chloride, thus jeopardizing the durability of the concrete elements manufactured with calcium sulphoaluminate cement and subjected to winter salting.

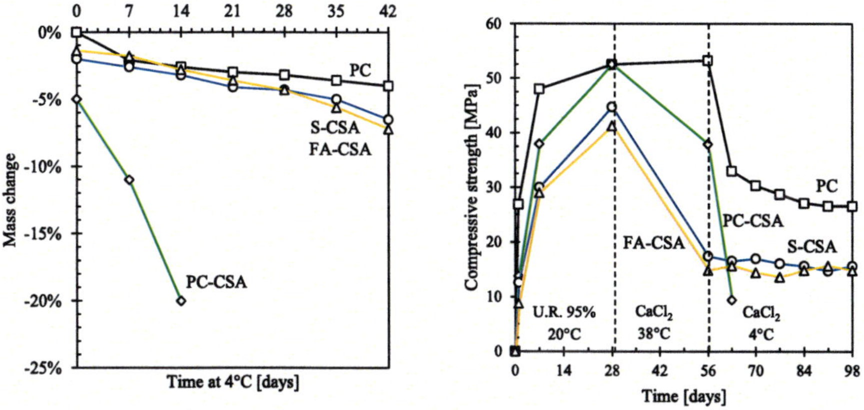


**Fig. 6.** Mass change (left) of CSA-based mortars exposed to 30 wt.% $CaCl_2$ solution at 4 °C in comparison with mortars stored in water. Strength development of mortars (right) in $CaCl_2$ solution.

**Fig. 7.** PC-CSA sample after 14 days immersed in cold calcium chloride-rich solutions.

## 4 Conclusions

The purpose of this work is to preliminarily evaluate the durability of mortars manufactured with different green binders to calcium chloride attack. The following conclusions can be drawn:

- The use of blast-furnace cement allows to strongly enhance the resistance against calcium oxychloride formation due to a dilution effect (less Portland cement clinker) combined with the pozzolanic reaction of slag.
- The partial substitution of Portland cement with plain or ultrafine fly ash reduces the mass reduction and strength loss of mortars exposed to $CaCl_2$ solutions only if the replacement rate is not lower than 25%.
- Alkali activated materials are quasi-immune to $CaCl_2$ attack due to the negligible amount of $Ca(OH)_2$ in its hardened phase. For this, the CAOXY formation is inhibited.
- CSA-based mortars suffer from severe damage when immersed in calcium chloride-rich solutions even at temperature close to 40 °C. The strength and mass loss are probably due to mechanisms other than CAOXY formation.

## 5 Future Developments

The topic of the durability of low-carbon binders-based mixtures against de-icing salts is an important issue to guarantee the sustainability of novel green concretes. In this paper the weakness of CSA-based mixtures in $CaCl_2$-rich solutions was highlighted but the degradation mechanism is still unclear. The authors are working to develop a model capable to explain the deterioration of these mixtures in contact with de-icing salts and, at the same time, are investigating strategies to improve their durability.

## References

Ameri, F., Shoaei, P., Zareei, S.A., Behforouz, B.: Geopolymers vs. alkali-activated materials (AAMs): a comparative study on durability, microstructure, and resistance to elevated temperatures of lightweight mortars. Constr. Build. Mater. **222**, 49–63 (2019). https://doi.org/10.1016/j.conbuildmat.2019.06.079

Amini, K., Cetin, K., Ceylan, H., Taylor, P.C.: A summary of factors affecting concrete salt-scaling performance. ACI Mater. J. **117**, 53–62 (2020). https://doi.org/10.14359/51724614

Amini, K., Ceylan, H., Taylor, P.C.: Effect of finishing practices on surface structure and salt-scaling resistance of concrete. Cem. Concr. Compos. **104**, 103345 (2019). https://doi.org/10.1016/j.cemconcomp.2019.103345

Bertolini, L., Elsener, B., Pedeferri, P., Redaelli, E., Polder, R.B.: Corrosion of steel in concrete: prevention, diagnosis, repair. Wiley WCH (2013)

Bohni, H.: Corrosion in Reinforced Concrete Structures. Woodhead Publishing Limited, Cambridge (England) (2005)

Cabrini, M., Lorenzi, S., Coppola, L., Coffetti, D., Pastore, T.: Inhibition effect of tartrate ions on the localized corrosion of steel in pore solution at different chloride concentration. Buildings **10**, 105 (2020). https://doi.org/10.3390/buildings10060105

Chatterji, S.: Mechanism of the CaCl2 attack on portland cement concrete. Cem. Concr. Res. **8**, 461–467 (1978). https://doi.org/10.1016/0008-8846(78)90026-1

Chindaprasirt, P., Jaturapitakkul, C., Sinsiri, T.: Effect of fly ash fineness on microstructure of blended cement paste. Constr. Build. Mater. **21**, 1534–1541 (2007). https://doi.org/10.1016/j.conbuildmat.2005.12.024

Chindaprasirt, P., Jaturapitakkul, C., Sinsiri, T.: Effect of fly ash fineness on compressive strength and pore size of blended cement paste. Cem. Concr. Compos. **27**, 425–428 (2005). https://doi.org/10.1016/j.cemconcomp.2004.07.003

Collepardi, M., Coppola, L., Pistolesi, C.: Durability of concrete structures exposed to CaCl2 based deicing salts. In: Proceedings of the 3rd CANMET/ACI International Conference. Nice, France, pp. 107–120 (1994)

Coppola, L., et al.: The combined use of admixtures for shrinkage reduction in one-part alkali activated slag-based mortars and pastes. Constr. Build. Mater. **248**, 118682 (2020). https://doi.org/10.1016/j.conbuildmat.2020.118682

Coppola, L., Coffetti, D., Crotti, E., Dell'Aversano, R., Gazzaniga, G., Pastore, T.: Influence of lithium carbonate and sodium carbonate on physical and elastic properties and on carbonation resistance of calcium sulphoaluminate-based mortars. Appl. Sci. **10**, 176 (2019). https://doi.org/10.3390/app10010176

Coppola, L., Coffetti, D., Crotti, E., Gazzaniga, G., Pastore, T.: The durability of one-part alkali activated slag-based mortars in different environments. Sustainability **12**, 3561 (2020)

Coppola, L., Coffetti, D., Crotti, E., Pastore, T.: CSA-based portland-free binders to manufacture sustainable concretes for jointless slabs on ground. Constr. Build. Mater. **187**, 691–698 (2018). https://doi.org/10.1016/j.conbuildmat.2018.07.221

D'Alessandro, A., Coffetti, D., Crotti, E., Coppola, L., Meoni, A., Ubertini, F.: Self-sensing properties of green alkali-activated binders with carbon-based nanoinclusions. Sustainability **12**, 9916 (2020). https://doi.org/10.3390/su12239916

EUSalt (European Salt Producers' Association). Environmental impact of winter maintenance with salt (2021)

Farnam, Y., Dick, S., Wiese, A., Davis, J., Bentz, D., Weiss, J.: The influence of calcium chloride deicing salt on phase changes and damage development in cementitious materials. Cem. Concr. Compos. **64**, 1–15 (2015). https://doi.org/10.1016/j.cemconcomp.2015.09.006

Fay, L., Shi, X.: Environmental impacts of chemicals for snow and ice control: state of the knowledge. Water. Air. Soil Pollut. **223**, 2751–2770 (2012). https://doi.org/10.1007/s11270-011-1064-6

Jones, C., Ramanathan, S., Suraneni, P., Hale, W.M.: Calcium oxychloride: a critical review of the literature surrounding the formation, deterioration, testing procedures, and recommended mitigation techniques. Cem. Concr. Compos. **113**, 103663 (2020). https://doi.org/10.1016/j.cemconcomp.2020.103663

Koumpouri, D., Karatasios, I., Psycharis, V., Giannakopoulos, I.G., Katsiotis, M.S., Kilikoglou, V.: Effect of clinkering conditions on phase evolution and microstructure of belite calcium-sulpho-aluminate cement clinker. Cem. Concr. Res. **147**, 106529 (2021). https://doi.org/10.1016/j.cemconres.2021.106529

Mohan, A., Mini, K.M.: Strength and durability studies of SCC incorporating silica fume and ultra fine GGBS. Constr. Build. Mater. **171**, 919–928 (2018). https://doi.org/10.1016/j.conbuildmat.2018.03.186

Monosi, S., Alvera, A., Collepardi, M.: Chemical attack of calcium chloride on the portland cement paste. Cem. **86**, 97–104 (1989)

Monosi, S., Collepardi, M.: Research on 3CaO.CaCl2.15H2O identified in concretes damaged by CaCl2 attack. Cem. **87**, 3–8 (1990)

Myers, R.J., Bernal, S.A., Provis, J.L.: Phase diagrams for alkali-activated slag binders. Cem. Concr. Res. **95**, 30–38 (2017). https://doi.org/10.1016/j.cemconres.2017.02.006

Peterson, K., Julio-Betancourt, G., Sutter, L., Hooton, R.D., Johnston, D.: Observations of chloride ingress and calcium oxychloride formation in laboratory concrete and mortar at 5 C. Cem. Concr. Res. **45**, 79–90 (2013). https://doi.org/10.1016/j.cemconres.2013.01.001

Provis, J.L.: Alkali-activated materials. Cem. Concr. Res. **114**, 40–48 (2018). https://doi.org/10.1016/j.cemconres.2017.02.009

Qiao, C., Suraneni, P., Weiss, J.: Flexural strength reduction of cement pastes exposed to CaCl2 solutions. Cem. Concr. Compos. **86**, 297–305 (2018). https://doi.org/10.1016/j.cemconcomp.2017.11.021

Sun, Z., Scherer, G.W.: Effect of air voids on salt scaling and internal freezing. Cem. Concr. Res. **40**, 260–270 (2010). https://doi.org/10.1016/j.cemconres.2009.09.027

Suraneni, P., Azad, V.J., Isgor, O.B., Weiss, J.: Role of supplementary cementitious material type in the mitigation of calcium oxychloride formation in cementitious pastes. J. Mater. Civ. Eng. **30**, 04018248 (2018). https://doi.org/10.1061/(asce)mt.1943-5533.0002425

Sutter, L., Peterson, K., Touton, S., Van Dam, T., Johnston, D.: Petrographic evidence of calcium oxychloride formation in mortars exposed to magnesium chloride solution. Cem. Concr. Res. **36**, 1533–1541 (2006). https://doi.org/10.1016/j.cemconres.2006.05.022

Traore, F., Jones, C., Ramanathan, S., Suraneni, P., Hale, W.M.: Using compressive strength and mass change to verify the calcium oxychloride threshold in cementitious pastes with fly ash. Constr. Build. Mater. **296**, 123640 (2021). https://doi.org/10.1016/j.conbuildmat.2021.123640

US Geological Survey, US Department of Interior, n.d. Mineral Yearbook - Metals and Minerals - From 1950 to 2017

# Effect of Composition on the Properties of Concrete Made with RAP Aggregate

Elena Redaelli(✉), Maddalena Carsana, Andrea Filippi, and Federica Lollini

Department of Chemistry, Materials and Chemical Engineering "Giulio Natta", Politecnico di Milano, Milan, Italy
elena.redaelli@polimi.it

**Abstract.** Replacing natural aggregate with recycled materials can decrease the environmental impact of concrete by reducing not only the use of natural resources, but also the disposal of waste materials. Recycled asphalt pavement (RAP) is a granular material obtained from the maintenance of road pavements whose size and distribution make it suitable as aggregate for concrete. This note presents the results of an experimental research aimed at characterising properties of concretes made with RAP as replacement of natural aggregate (in fractions ranging from 0 to 100%), considering the effect of compositional parameters such as the cement type (a limestone portland cement, CEM II A-LL 42.5R, and a pozzolanic cement, CEM IV/A (P-V) 42.5N-SR) and the water/cement ratio (0.45 and 0.65). Besides basic characterisation at fresh and hardened state, also durability properties were measured, such as resistance to carbonation, sorptivity, water absorption and electrical resistivity.

**Keywords:** circular economy · concrete · recycled aggregate · recycled asphalt pavement (RAP)

## 1 Introduction

Concrete as well as cementitious materials more generally have a strong environmental impact due to the use of portland cement and the consumption of natural resources such as water and aggregates. Because of this, most of the research related with such construction materials is oriented towards the use of alternative binders (e.g. low-clinker cements), together with the replacement of natural aggregates with materials coming from construction and demolition waste (C&DW) (Meyer 2009). In this regard, recycled asphalt pavement (RAP) potentially represents an interesting opportunity: it is made of natural stones or slags bound by asphalt and it is milled from road pavements; hence, it comes in granules with shape and size distribution that are similar to those of natural aggregate. Its use as recycled aggregate in concrete is less common compared to other C&DW, probably as a consequence of the organic fraction it contains and the potential contaminants. In the past, several researches investigated its use as partial replacement of natural aggregate in concrete (Coppola et al. 2016, Erdem et al. 2014), however some aspects still need to be verified, in particular with respect to the long-term behaviour.


M. A. Aiello and A. Bilotta (Eds.): ICC 2022, LNCE 435, pp. 664–680, 2024.
https://doi.org/10.1007/978-3-031-43102-9_51

This note presents results of an experimental research aimed at studying concretes made with replacements of natural aggregate with RAP in fractions ranging from 0 to 100%. Various concrete compositions were considered, differing in cement type (a limestone portland cement and a pozzolanic cement) and water/cement ratio (0.45 and 0.65). Besides the basic characterisation at fresh and hardened state (i.e., slump, density, compressive strength and dynamic modulus of elasticity), some durability properties were also measured, such as resistance to carbonation, sorptivity, water absorption and electrical resistivity.

## 2 Methodology

### 2.1 Mix Design of Concrete

Concrete mixtures were designed to investigate the role of RAP as a replacement of natural aggregates as well as other compositional parameters such as cement type and water/cement ratio. More specifically, cement types CEM II/A-LL 42.5R and CEM IV/A (P-V) 42.5N-SR were chosen: the former is the most used cement type in Italy, while the latter is more representative of a cement with increased performance thanks to the pozzolanic additions of natural pozzolana and fly ash. For water/cement ratio, values of 0.45 and 0.65 were chosen, which were considered representative of the requirements for concrete structures exposed to aggressive environments due to the presence of chlorides (e.g., exposure class XS3 or XD3 according to EN 206) and moderately aggressive environments with risk of carbonation (e.g., exposure class XC1).

Table 1 reports the compositions of the four reference mixtures, that were chosen by keeping the volume fraction of cement paste (and, hence, of aggregates) constant. Natural limestone aggregates were used (details of size distribution and physical properties are given in Filippi et al. 2022). A sulphonated naphthalene admixture was used as superplasticizer (SP) with the initial aim of achieving a consistence class of S3-S4.

Mixtures with RAP in partial or total replacement of natural aggregates have the composition shown in Tables 2, 3, 4 and 5: given that the density of RAP is lower if compared to natural aggregate, the dosage of RAP ($M_{RAP}$ in $kg/m^3$) was obtained by solving Eqs. 1 and 2:

$$\frac{M_{RAP}}{M_{RAP} + M_{NAT}} = \%RAP \tag{1}$$

$$V_{tot,aggr} = M_{NAT} \cdot \rho_{NAT} + M_{RAP} \cdot \rho_{RAP} \tag{2}$$

where $M_{NAT}$ is the dosage of natural aggregate ($kg/m^3$), %RAP is the mass % of RAP, $V_{tot,aggr}$ is the total volume of aggregates ($l/m^3$), $\rho_{NAT}$ and $\rho_{RAP}$ are the densities ($kg/m^3$) of natural and RAP aggregates, respectively. Since natural aggregate was available in six fractions, the combination of the fractions with RAP was obtained by optimising the combined size distribution with the least squares method, approaching a Fuller curve.

**Table 1.** Composition and designation of reference mixtures

| Concrete | L-a-N | L-b-N | P-a-N | P-b-N |
|---|---|---|---|---|
| Cement | CEM II/A-LL42.5R | | CEM IV/A(P-V)42.5N-SR | |
| w/c ratio | 0.45 | 0.65 | 0.45 | 0.65 |
| Composition ssd (kg/m$^3$) | | | | |
| Cement | 357 | 283 | 357 | 283 |
| Water | 161 | 184 | 161 | 184 |
| NAT aggr | 1980 | 1981 | 1980 | 1981 |
| SP adm | 6.5 | 4.5 | 9.8 | 3.8 |

**Table 2.** Composition and designation of L-a mixtures

| Concrete | L-a-R20 | L-a-R40 | L-a-R60 | L-a-R80 | L-a-R |
|---|---|---|---|---|---|
| Cement | CEM II A-LL 42.5R | | | | |
| w/c | | | 0.45 | | |
| %RAP | 20 | 40 | 60 | 80 | 100 |
| Composition ssd (kg/m$^3$) | | | | | |
| Cement | | | 357 | | |
| Water | | | 161 | | |
| NAT aggr | 1565 | 1153 | 756 | 371 | 0 |
| RAP aggr | 391 | 769 | 1134 | 1485 | 1826 |
| SP adm | 6.1 | 7.7 | 8.7 | 9.7 | 10.6 |

**Table 3.** Composition and designation of L-b mixtures

| Concrete | L-b-R20 | L-b-R40 | L-b-R60 | L-b-R80 | L-b-R |
|---|---|---|---|---|---|
| Cement | CEM II A-LL 42.5R | | | | |
| w/c | | | 0.65 | | |
| %RAP | 20 | 40 | 60 | 80 | 100 |
| Composition ssd (kg/m$^3$) | | | | | |
| Cement | | | 283 | | |
| Water | | | 184 | | |
| NAT aggr | 1566 | 1154 | 755 | 372 | 0 |
| RAP aggr | 391 | 769 | 1134 | 1486 | 1827 |
| SP adm | 3.1 | 3.6 | 2.8 | 4.7 | 4.7 |

**Table 4.** Composition and designation of P-a mixtures

| Concrete | P-a-R20 | P-a-R40 | P-a-R60 | P-a-R80 | P-a-R |
|---|---|---|---|---|---|
| Cement | CEM IV/A (P-V) 42.5N-SR | | | | |
| w/c | 0.45 | | | | |
| %RAP | 20 | 40 | 60 | 80 | 100 |
| Composition ssd (kg/m$^3$) | | | | | |
| Cement | 357 | | | | |
| Water | 161 | | | | |
| NAT aggr | 1565 | 1153 | 756 | 371 | 0 |
| RAP aggr | 391 | 769 | 1134 | 1486 | 1826 |
| SP adm | 9.2 | 9.5 | 9.9 | 12.3 | 7.1 |

**Table 5.** Composition and designation of P-b mixtures

| Concrete | P-b-R20 | P-b-R40 | P-b-R60 | P-b-R80 | P-b-R |
|---|---|---|---|---|---|
| Cement | CEM IV/A (P-V) 42.5N-SR | | | | |
| w/c | 0.65 | | | | |
| %RAP | 20 | 40 | 60 | 80 | 100 |
| Composition ssd (kg/m$^3$) | | | | | |
| Cement | 283 | | | | |
| Water | 184 | | | | |
| NAT aggr | 1566 | 1154 | 756 | 372 | 0 |
| RAP aggr | 392 | 769 | 1134 | 1486 | 1827 |
| SP adm | 2.8 | 3.7 | 4.1 | 5.0 | 4.0 |

## 2.2 Specimens and Tests

Concretes were mixed with the following procedure: 90 s blending of aggregates in the mixer, addition of cement and 90 s blending, addition of water admixed with initial amount of SP and 90 s mixing, further addition of SP and further 90 s mixing. Workability was measured with the Abrams' test. The target consistence class of S3-S4 could not be achieved in some cases; however, the amount of SP was kept under 3% vs. cement mass (except for mixture P-a-R80) to avoid risk of segregation. Figure 1 shows the dosage of SP added to the various mixtures: although the values cannot be directly compared, since the consistence was not always the same, in some cases it appears that the dosage of SP increases with the %RAP (e.g., for mixtures L-a and P-a, that have a lower dosage of water), however this effect was not systematic. Mixtures L-b and P-b, that have a higher dosage of water, were admixed with similar dosages of SP. It is worth noticing that, even for mixtures that had a low workability, the specimens could be easily compacted and no segregation was observed at demoulding.

Cubic specimens with side of 100 mm were cast in metallic moulds, that were compacted on a vibrating table. Specimens were demoulded the day after casting and then stored in a curing chamber with > 95% RH and T = 20 °C.

Compressive strength was measured at demoulding (1 day) and after 7, 28 and 90 days of wet curing. Before the test, non-destructive measurements of mass, electrical conductance and ultra-sonic (US) pulse velocity were also carried out on the same cubes (Redaelli et al. 2022). All measurements were carried out on duplicate specimens.

Some of the cubes were wet cured for 7 days and they were used for durability tests. Resistance to accelerated carbonation was measured on cubes that were coated with epoxy resin on all sides except two parallel faces and exposed to 3% $CO_2$ at 20 °C and 65% RH. After 77 days of exposure in the carbonation chamber they were split in halves and the fracture surface was sprayed with phenolphthalein indicator. The depth of carbonation was expressed as the average of the two faces.

Cylindrical specimens with nominal size of 100 mm in diameter and 50 mm in height were obtained by coring and cutting cubes that had been wet cured for 7 days. Such specimens were used to measure water sorptivity (S, in $kg/m^2 \cdot h^{0.5}$): after drying, one of their sides was placed in contact with water and their mass increase was measured in time. Then, they were completely immersed to measure the water absorption (W, as mass %).

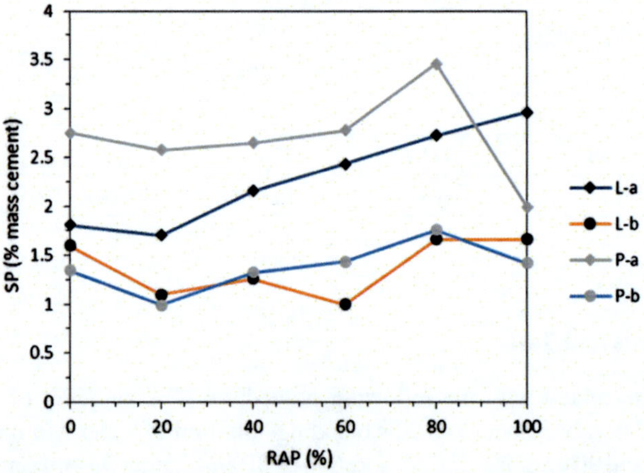


**Fig. 1.** Dosage of superplasticizer (SP) added to the mixtures as a function of %RAP

# 3 Results and Discussion

## 3.1 Physical and Mechanical Properties

Figures 2, 3, 4 and 5 show the evolution of compressive strength during wet curing. All concretes exhibited the typical increasing trend with time. After 1 d of wet curing, reference concretes with limestone cement (L) had values of 33 MPa for w/c = 0.45 and 22 MPa for w/c = 0.65, while concretes with pozzolanic cement (P) had values of 28 MPa for w/c = 0.45 and 12 MPa for w/c = 0.65. After 90 d, the strength of reference concretes reached values of 70 MPa and 50 MPa for limestone cement, and 75 MPa and 45 MPa for pozzolanic cement, for w/c = 0.45 and 0.65 respectively.

The trends clearly show that the presence of RAP aggregate reduced the compressive strength at all curing times, however it did not seem to affect the rate of strength development, as it is visible from Fig. 6 that shows, as an example, the ratio of compressive strengths at the various curing times on the compressive strength at 90 d for concretes made with pozzolanic cement and w/c = 0.45. The trends with the different %RAP tend to overlap. Comparable behaviour was obtained for all concretes, regardless of the type of cement and w/c ratio.

Figure 7 shows the effect of the %RAP on the compressive strength after 28 d of wet curing. For both cement types and w/c ratios, the strength decreased with the %RAP, however some irregularities were observed in some of the trends, such as P-a and L-b, that showed increased values for %RAP = 80% with respect to 60%. Anyway, all the trends showed an overall decrease, which occurred also at the other curing times. When the decreasing trends are plotted in terms of relative values, i.e. the ratio of the compressive strength of RAP concrete on the compressive strength of reference concrete, as it is done in Fig. 8, no systematic effect seemed to emerge related with cement type or w/c ratio.

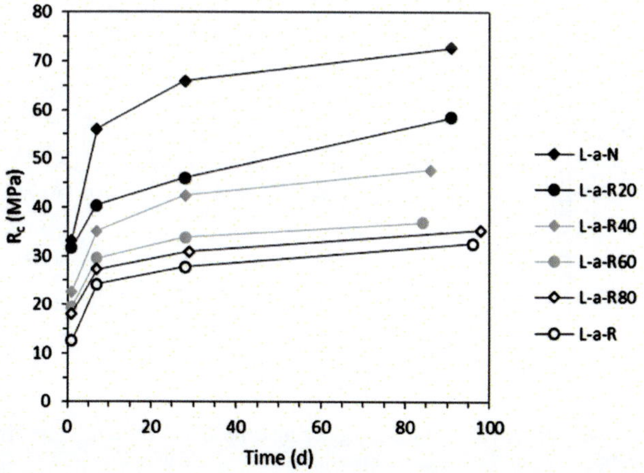


**Fig. 2.** Compressive strength as a function of curing time of concrete mixtures made with CEM II and w/c = 0.45

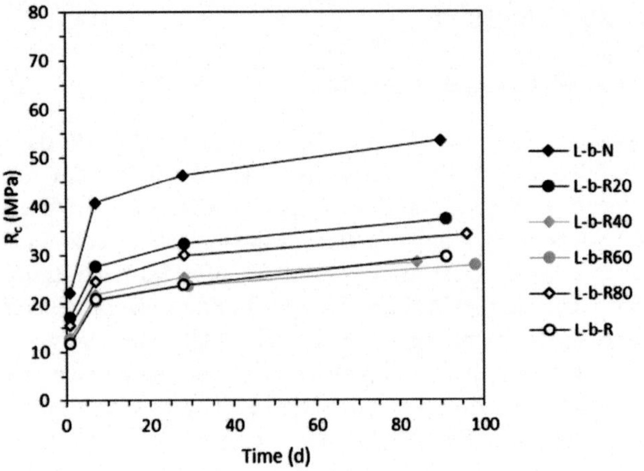


**Fig. 3.** Compressive strength as a function of curing time of concrete mixtures made with CEM II and w/c = 0.65

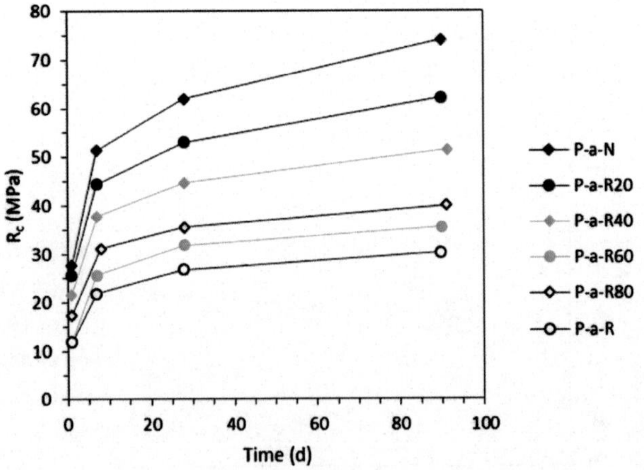


**Fig. 4.** Compressive strength as a function of curing time of concrete mixtures made with CEM IV and w/c = 0.45

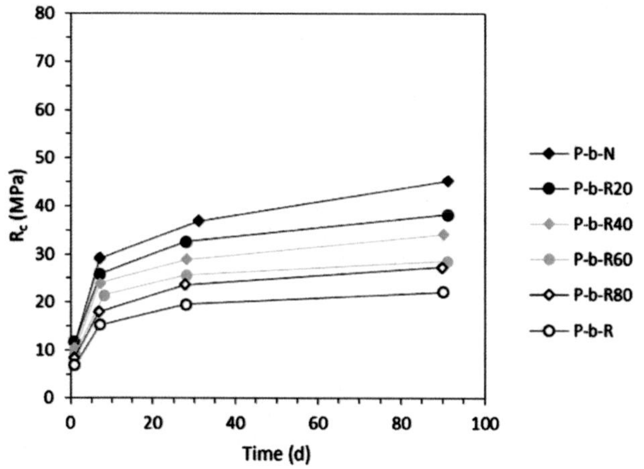


**Fig. 5.** Compressive strength as a function of curing time of concrete mixtures made with CEM IV and w/c = 0.65

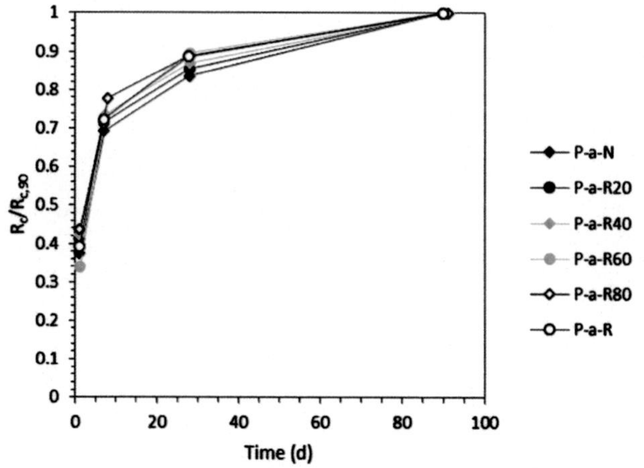


**Fig. 6.** Ratio of compressive strength on the compressive strength at 90 d for concrete mixtures made with CEM IV and w/c = 0.45

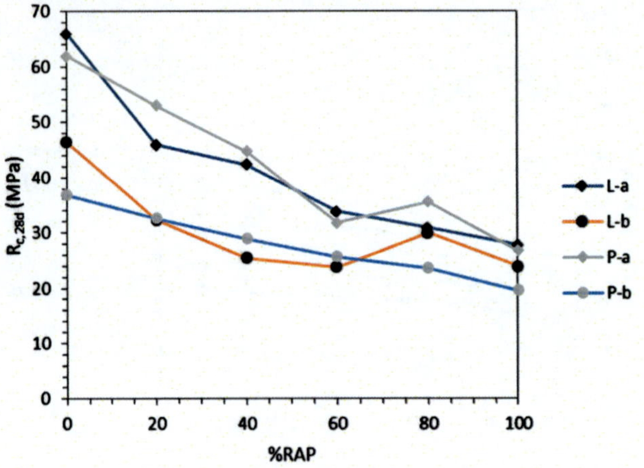


**Fig. 7.** Effect of %RAP on the compressive strength of concrete after 28 d of wet curing

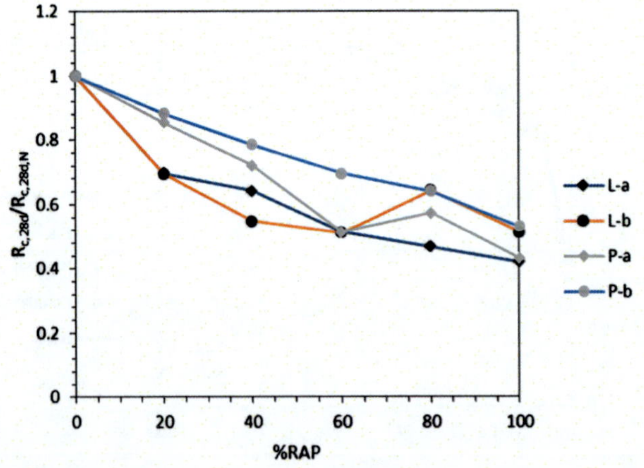


**Fig. 8.** Effect of %RAP on the ratio of compressive strength of RAP concrete on compressive strength of reference concrete

Density at hardened state is another property that showed a decreasing trend with %RAP, as it can be seen in Fig. 9. As previously mentioned, RAP has a lower density compared to natural aggregate: values of 2.58 kg/l and 2.81 kg/l were obtained for the former and the latter, respectively. Hence, the reduction of density is an expected side effect of partial replacement of natural aggregate with RAP. However, there are probably other factors that seem to play a role. Figure 10 compares the measured density (after 28 d) with the theoretical density for all concrete mixtures, and it shows that in spite of the good correlation between the two, the measured density is in most cases lower compared to the theoretical one. This effect is mainly present in concrete mixtures with RAP and it is likely due to an increased difficulty of compaction, although no visible signs of macroscopic entrapped air voids or segregation were observed.

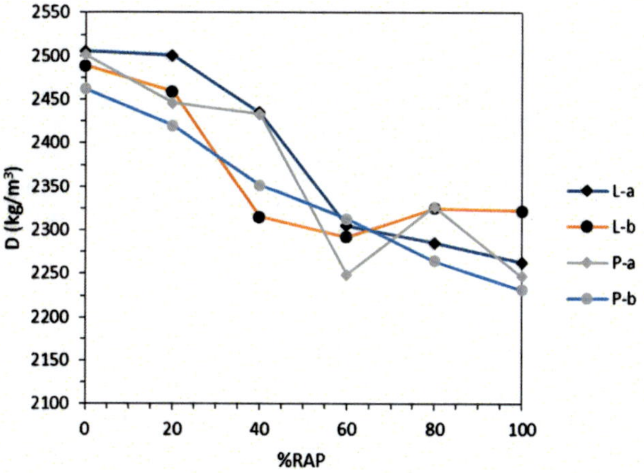


**Fig. 9.** Effect of %RAP on concrete density

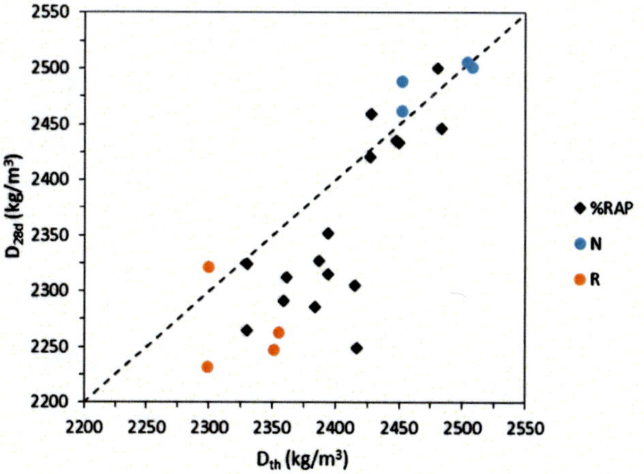


**Fig. 10.** Concrete density at hardened state ($D_{28d}$) vs. theoretical density ($D_{th}$)

Figure 11 plots the dynamic modulus of elasticity obtained after 28 days of wet curing: even for this parameter decreasing trends were observed, with values ranging from 40 GPa for reference concretes to 25 GPa for concretes with 100% RAP.

Finally, Figs. 12 and 13 show, as examples, the evolution of the electrical resistivity of concrete mixtures made with pozzolanic cement and w/c = 0.45 and 0.65, respectively: as a consequence of the progression of cement hydration during wet curing, the resistivity increased from values of few tens of Ω·m after 1 d to values around 200 Ω·m after 90 d. The effect of RAP aggregate on resistivity seemed to be negligible, as it is shown in Fig. 14 that compares values after 28 d for all concrete mixtures: no systematic effect emerged and the limited variations of this parameter do not appear to be linked to the %RAP.

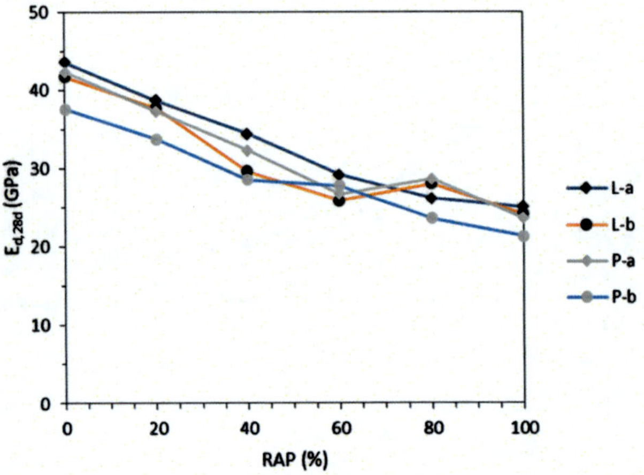


**Fig. 11.** Effect of %RAP on the dynamic modulus of elasticity of concrete after 28 d of wet curing

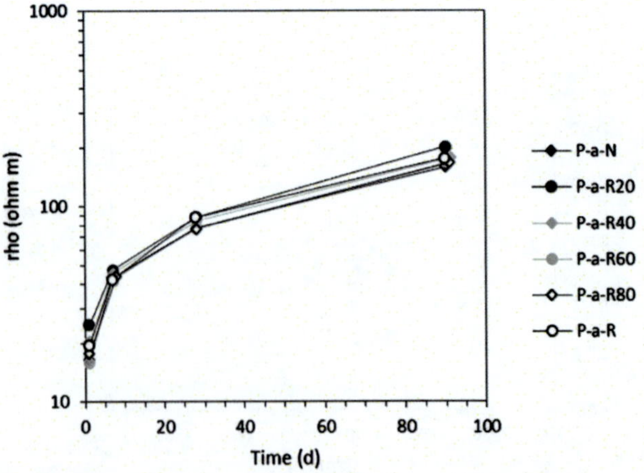


**Fig. 12.** Electrical resistivity as a function of curing time of concrete mixtures made with CEM IV and w/c = 0.45

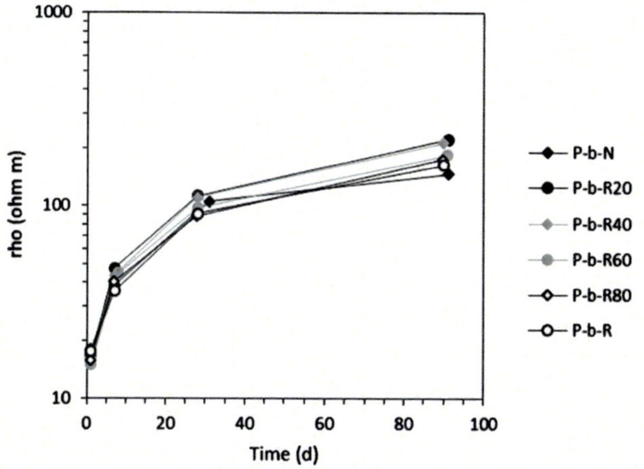


**Fig. 13.** Electrical resistivity as a function of curing time of concrete mixtures made with CEM IV and w/c = 0.65

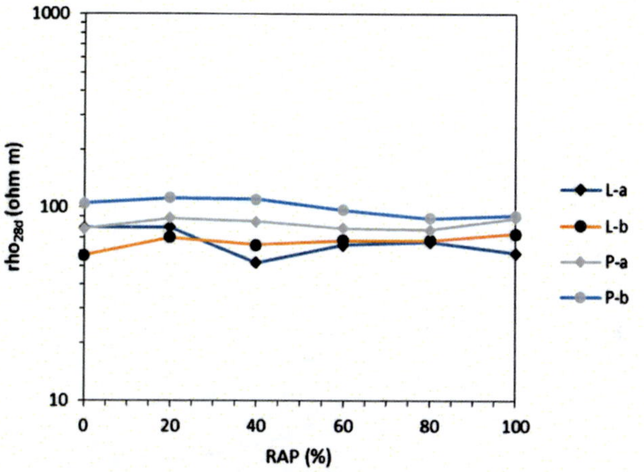


**Fig. 14.** Effect of %RAP on the electrical resistivity of concrete after 28 d of wet curing

## 3.2 Sorptivity and Water Absorption

Figure 15 shows, as an example, the evolution of absorbed water during a sorptivity test and the interpolation with the least squares method to obtain the sorptivity (S). The values of S for all concrete mixtures are reported in Fig. 16, while Fig. 17 reports the values of water absorption (W). The sorptivity decreases with the %RAP, from values around 0.8–1.4 $kg/m^2 \cdot h^{0.5}$ for reference concretes to values around 0.3–0.5 $kg/m^2 \cdot h^{0.5}$ for concretes with 100% RAP. Conversely, the water absorption does not seem to be affected by the %RAP and it shows values between 4–6%. Sorptivity is a measure of the initial rate of absorption of water when the concrete is initially dry, whilst water absorption is the total water absorbed after saturation. Results seem to indicate that S depends on %RAP, probably due to the hydrophobic nature of the bituminous fraction (Masi et al. 2022), while W is essentially independent on RAP, probably because after prolonged immersion in water RAP aggregates can reach a condition of saturation.

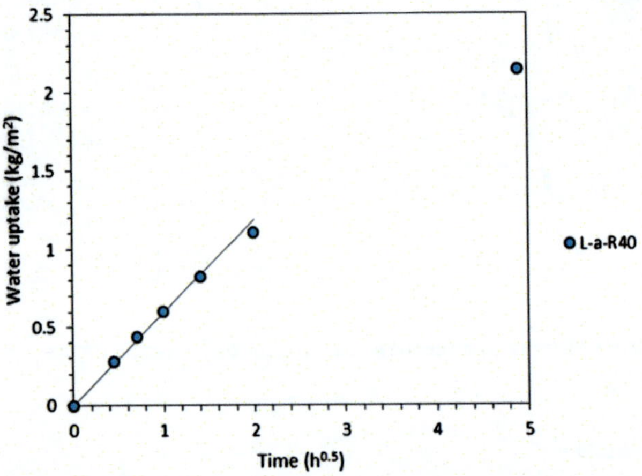


**Fig. 15.** Mass of water absorbed per unit surface as a function of square root of time

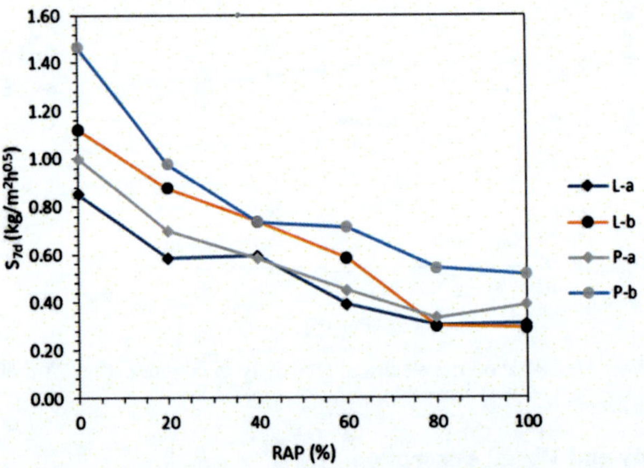


**Fig. 16.** Effect of %RAP on the sorptivity of concrete after 7 d of wet curing

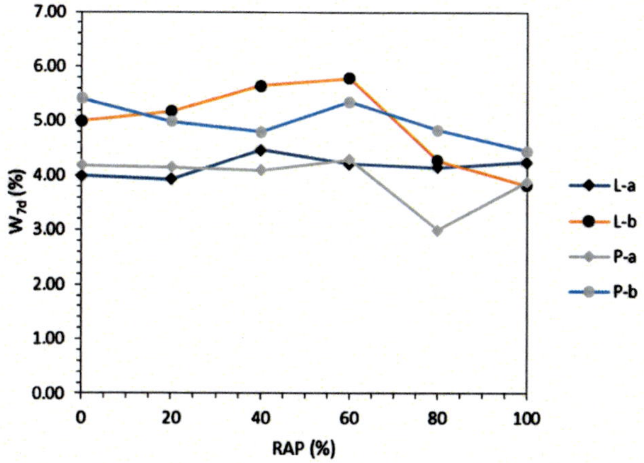


**Fig. 17.** Effect of %RAP on the water absorption of concrete after 7 d of wet curing

## 3.3 Carbonation

Figure 18 shows the coefficient of accelerated carbonation for the various concrete mixtures. Reference concretes had values between 20 and 40 mm/y$^{0.5}$. No systematic trend emerged as regards the effect of %RAP and the scatter of results was not linked to other compositional factors, either. These results probably indicate that the resistance to carbonation is not primarily affected by the nature of the aggregate, but it mainly depends on the cement paste, whose volume was kept constant in the different mixtures.

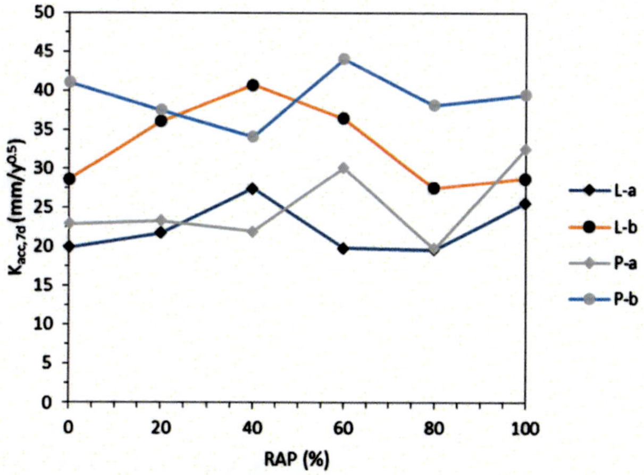


**Fig. 18.** Effect of %RAP on the coefficient of accelerated carbonation of concrete after 7 d of wet curing

### 3.4 Effect of w/c Ratio and Cement Type

The role of w/c ratio and cement type on strength properties will be analysed in relation with the possibility of compensating possible side effects related with the use of RAP. Figures 19 and 20 report the curves of $R_c$ at 28 d as a function of w/c, for mixtures with CEM II and CEM IV respectively. The curves were obtained by interpolating the experimental data with Abrams' law shown in Eq. 3:

$$R_c = \frac{k_1}{(k_2)^{w/c}} \tag{3}$$

where $k_1$ and $k_2$ are regression parameters. Both graphs highlight the detrimental effect of RAP on the compressive strength, however they also allow to notice that the compressive strength of concrete with RAP aggregates is still higher than 30 MPa for most concretes, making them suitable for structural applications, and also that the reduction of $R_c$ due to RAP can to some extent be compensated by reducing the w/c ratio. Also the selection of the pozzolanic cement mitigates the reduction of strength due to RAP, as it appears from Fig. 19 that shows that when this type of cement is used, values of $R_c$ higher than 30 MPa can be achieved even with the higher w/c ratio of 0.65.

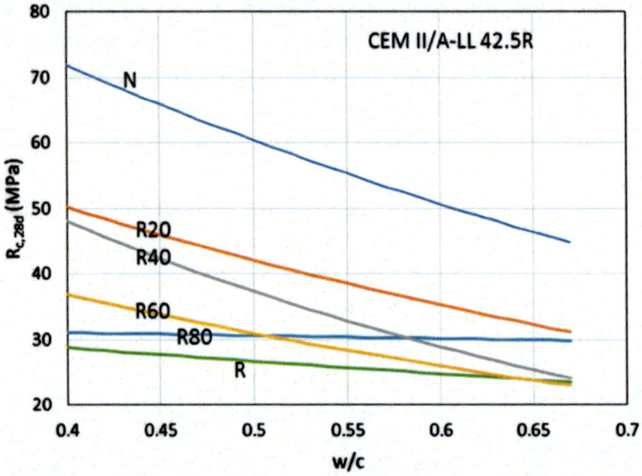


**Fig. 19.** Compressive strength after 28 d of wet curing versus w/c obtained from interpolation for concrete mixtures made with CEM II

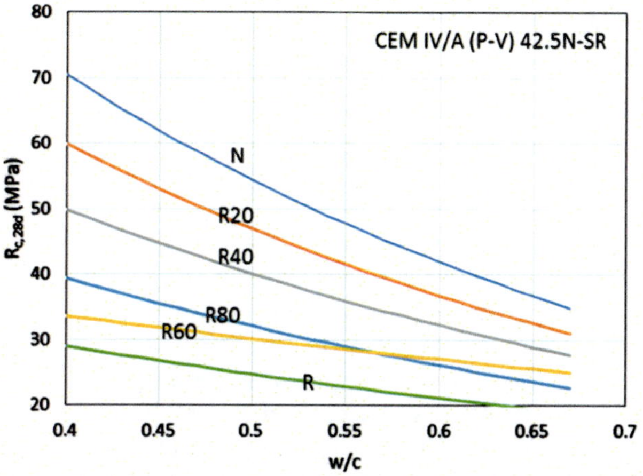


**Fig. 20.** Compressive strength after 28 d of wet curing versus w/c obtained from interpolation for concrete mixtures made with CEM IV

Finally, Fig. 21 plots the sorptivity versus the water absorption, both useful parameters to value durability-related properties. These parameters do not seem to be correlated as they normally are for ordinary concrete, for which an increase of the total porosity comes with an increase of the average pore size, and hence when S increases, W increases too. Anyway, although there are no specific requirements, the values of both S and W can be considered suitable for most applications; in fact, the reduction of S as a result of the addition of RAP can be considered beneficial, since it delays the increase of humidity content of concrete in the initial stages of direct contact with liquid water (e.g., wetting

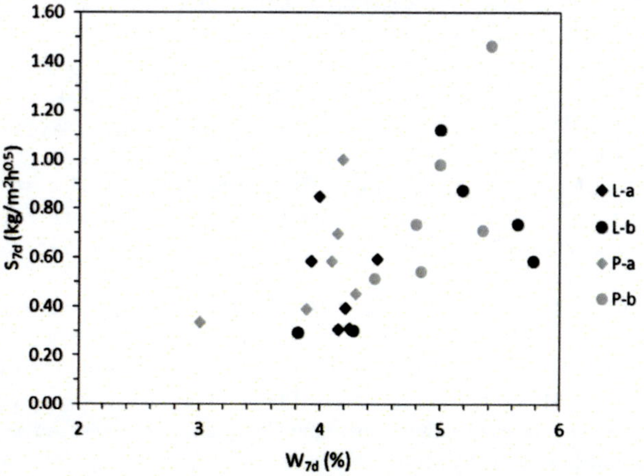


**Fig. 21.** Sorptivity versus water absorption of concrete, both after 7 d of wet curing

with rain). For prolonged times, the humidity content after saturation also plays a relevant role, however the values of W are very close to those typical of ordinary concrete mixtures.

## 4 Conclusions

An experimental research was carried out aimed at characterising the properties at fresh and hardened state of concrete mixtures made with RAP aggregate.

Compressive strength of concretes with RAP was lower compared to compressive strength of reference concretes. However, this detrimental effect could be mitigated by a reduction of w/c ratio and selecting the pozzolanic cement.

The rate of water absorption of concretes with RAP was lower compared to that of reference concretes, while the total water absorption was virtually unaffected by the presence of RAP. Also the resistance of carbonation in accelerated conditions was not substantially affected by the presence of RAP.

The preliminary results that we presented here seem to indicate that some of the properties that determine the durability behaviour of concrete are not as affected by RAP as the mechanical properties.

**Acknowledgements.** Fondazione Cariplo is gratefully acknowledged for supporting the research in the framework of the call "Circular Economy for a Sustainable Future" (RAPCON Project, ref. 2019–2286).

## References

Coppola, L., Kara, P., Lorenzi, S.: Concrete manufactured with crushed asphalt as partial replacement of natural aggregates. Mater. Constr. **66**(324), 1–7 (2016)

EN 206: Concrete. Specification, performance, production and conformity (2013)

Erdem, S., Blankson, M.A.: Environmental performance and mechanical analysis of concrete containing recycled asphalt pavement (RAP) and waste precast concrete as aggregate. J. Hazard. Mater. **264**, 403–410 (2014)

Filippi, A., Carsana, M., Lollini, F., Redaelli, E.: Characterization of recycled asphalt pavement as aggregate for concrete. In: Fifteenth International Conference on Recent Advances in Concrete Technology and Sustainability Issues, Milan, 13–15 July 2022, pp. 245–262. ACI SP-355 (2022)

Masi, G., Michelacci, A., Manzi, S., Bignozzi, M.C.: Assessment of reclaimed asphalt pavement (RAP) as recycled aggregate for concrete. Constr. Build. Mater. **341**, 1–9 (2022)

Meyer, C.: The greening of the concrete industry. Cement Concr. Compos. **31**(8), 601–605 (2009)

Redaelli, E., Carsana, M., Filippi, A., Lollini, F.: Preliminary characterization of concrete made with RAP aggregate. In: Fifteenth International Conference on Recent Advances in Concrete Technology and Sustainability Issues, Milan, 13–15 July 2022, pp. 201–212. ACI SP-355 (2022)

# Research on the High Temperature Behaviour of Ultra-High-Performance Concrete (UHPC) with Polypropylene (PP) and Steel Fibres

Francesca Sciarretta[1(✉)], Stefano Fava[2], Marco Francini[2], Luca Ponticelli[3], Mauro Caciolai[3], Bruno Briseghella[4], and Camillo Nuti[5]

[1] Dipartimento di Culture del Progetto, Università IUAV di Venezia, Venice, Italy
scifra@iuav.it
[2] Buzzi Unicem S. p. A., via Luigi Buzzi, 6 - 15033 Casale Monferrato, AL, Italy
[3] Corpo Nazionale dei Vigili del Fuoco, Rome, Italy
[4] College of Civil Engineering, Fuzhou University, Fujian, China
[5] Department of Architecture, University Roma Tre, Rome, Italy

**Abstract.** The research features a series of tests on two types of Ultra-High Performance Concrete (UHPC, 150 and 180 N/mm$^2$) with polypropylene (PP) fibres (0.27% of volume) and variable content of steel (S) fibres (0% to 1.92%), aimed at investigating the residual mechanical properties of the material after high temperature exposure. The results are compared to available research on small UHPC specimens exposed to high temperatures, with PP fibres from 0.03% to 2%, and S fibres from 0 to 3% of volume. The results demonstrate that UHPCs need hybrid fibre reinforcement (PP + S) to withstand high temperatures, and that the residual strength increases after 200 °C exposure, at all steel fibre dosages; this is in line with literature. Available research also shows that strength loss is possible in hot conditions, as found in the present research, while PP fibres alone do not always prevent the occurrence of spalling in small UHPC samples.

**Keywords:** Ultra-High-Performance Concrete · polypropylene fibres · fire behaviour · experimental tests · spalling

## 1 Introduction

### 1.1 UHPC and Fibre Addition

In the last 25 years, ultra-high-performance concrete (UHPC) has been gaining attention in the field of construction industry and research, due to its very high mechanical strength, high energy absorption capacity in tension and durability (Graybeal 2011, AFG/SETRA 2013, Lavorato et al. 2015, Lavorato et al. 2017, Xue et al. 2018, Xue et al. 2020).

There is a need of experimental research on UHPC mechanical properties under fire exposure (or 'hot' properties), as well as after heating and subsequent cooling (or 'residual'). As for the testing methods, no specific standards for mechanical tests on UHPC under/after high temperatures exist yet (ISO 2020). The compressive strength of


M. A. Aiello and A. Bilotta (Eds.): ICC 2022, LNCE 435, pp. 681–693, 2024.
https://doi.org/10.1007/978-3-031-43102-9_52

UHPCs exposed to high temperatures has been investigated since the 2000s especially in the residual condition, with tests on small samples subjected to pre-established target temperatures and slowly cooled down, to avoid thermal shock. The material strength of UHPC is inevitably investigated on small samples; the material properties must be considered along with the structure's size, loading and environmental conditions.

The paper presents experimental research on the hot and residual compressive strength of two UHPC mixes with polypropylene (PP) and steel (S) fibres at 200 °C. Tests at higher temperature will be carried out in the future on the basis of the present experience. In fact, explosive spalling occurred at 230–290 °C, indicating the need of changes in hybrid fibre composition with respect to what here investigated.

The use of PP and S fibres is frequent in UHPC mixes (Cree et al. 2017, Yermak et al. 2017, Xiong and Liew 2015, Li and Tan 2016, Larsen and Thorstensen 2020). The addition of PP fibres (Chen et al. 2020, Liang et al. 2018) is usual in concrete to withstand high temperatures. PP fibres melt at 165 °C, leaving a network of micro-channels which increases the material permeability, relieving the pore pressure and avoiding the development of high thermal gradients (Yang et al. 2019, Pimienta et al. 2012). As UHPC releases a high amount of energy at cracking, the addition of S fibres improves the tensile behaviour of the concrete and can reduce the risk of explosive spalling (Burke 2011).

Eurocode 2 (CEN 2004) recommends a minimum PP fibre content of 2 kg/m3 (about 0.2% of volume) for the highest grades of HSC (High Strength Concrete); such quantity can be effective for most types of PP fibres in UHPC (Lu and Fontana 2016). According to the American Concrete Institute, a 0.1% of volume in PP fibres is able to significantly reduce spalling (ACI 2018); for hydrocarbon fire exposure, the Canadian standard (CSA 2019) specifies a minimum of 0.3% in volume. On the other hand, research suggests that the quantity of PP fibres required to avoid explosive spalling depends on the UHPC mix design (ACI 2018); values from 0.03% to 2% of volume are reported.

However, despite the usually low heating rates, research reports the occurrence of explosive spalling during heating cycles on UHPC samples for mechanical testing under or after high temperatures (e. g. Cree et al. 2017, Yermak et al. 2017, Yang et al. 2019, Kahanji et al. 2016). Thus, spalling can impair the mechanical investigations on UHPC.

### 1.2 Available Studies on the Effect of PP, Steel and Hybrid Fibre Content

The available studies agree about the positive contribution of PP fibres in reducing explosive spalling phenomena in UHPC under fire, (Larsen and Thorstensen 2020, Ye et al. 2012, Bei and Zhixiang 2016), in line with experience on normal concrete and HSC. To be effective, fibres must have sufficient length and a minimal diameter not to affect too much the concrete volume (Bei and Zhixiang 2016). The escape of steam from the UHPC matrix can be facilitated by the micro-channels and/or micro-cracks generated by the melting of PP fibres, around 165 °C (Liu and Zhang 2020). Missemer et al. (2019) and Liu et al. (2008) enlighten the important role of micro-cracks in connecting the micro-channels network, and the good performance of PP fibres 6 mm long, with a diameter of 0.018 mm (aspect ratio 330). At 360–400 °C, other phenomena can affect the pore pressure (the melted PP releases various volatiles, e. g. pentane, propylene) (Liang et al. 2019), but such cases have not been experienced in this research.

The effect of S fibres on the compressive strength of UHPCs at normal temperature is not as substantial as for NSC and HSC (Burke 2011); in fact, the higher stiffness of UHPC impairs the transfer of energy from the matrix to the fibres before cracking, and thus reduces the confinement effect. For the same reasons, the effectiveness of S fibres in preventing explosive spalling is reduced in UHPCs (Ozawa et al. 2019, Huang et al. 2019). Fibres have a bridge effect across cracks, but they may have insufficient stiffness to ensure the integrity of the UHPC specimen; moreover, their strength reduces with temperature (Huang et al. 2019, Peng et al. 2016). Way and Wille (2016) observe embrittlement and weakening of S fibres after exposure to 500 °C and beyond, and full melting after 800 °C.

Hybrid reinforcement appears to be a need for UHPC exposed at high temperatures. PP and S fibres together have shown effectiveness in improving the residual strength of UHPC and reducing the explosive spalling hazard (Lee et al. 2012, Heinz et al. 2004). Liang et al. (2019) tested a UHPC with 2% in volume of PP fibres – a very high content, which may impair the material workability – and 1% in volume of S fibres, which retained the 70% of the compressive strength after exposure to 1000 °C.

The available tests in literature do not allow to draw exhaustive conclusions about the effects of different dosages and geometry of PP fibres; further research seems useful in this field.

# 2 Experimental Investigation

## 2.1 Materials and Testing Procedure

The UHCP mixes CLS-A and CLS-B (Table 1) were designed to attain compressive strength of 150 and 180 N/mm$^2$ respectively, after 28 days.

The materials, after samples production and curing, were brought to the Laboratory of Strength of Materials of the National Fire Brigade in Capannelle (Rome); there, the heating cycles and mechanical tests in hot conditions were performed.

A single percentage in volume of PP fibres was adopted for all mixes: 0,27%, according to available recommendations (CEN 2004, ACI 2018, Yang et al. 2021). The PP fibres have diameter of 0.05 mm and length of 12 mm (aspect ratio: 240).

CLS-A (strength 150 N/mm$^2$) has a single mix design and three different S fibre contents, i. e. 0%, 0.62% and 1.25%. CLS-B, which was designed to achieve 180 N/mm$^2$, has two different mixes. The mix indicated with B-Bt (where 'Bt' is for short term curing) reaches the desired performance in 28 days, while B-Lt (where 'Lt' is for long term curing) could reach the expected compressive strength only after 90 days. For both mixes, two S fibre dosages were adopted, i. e. 0% and 1.9% of volume.

The S fibres were of two types, i. e. straight fibres of diameter 0.4 mm and length 20 mm (aspect ratio: 50), and hooked fibres of diameter 0.7 mm and length 30 mm (aspect ratio: 43).

The settings of the thermal cycles were established on the grounds of the provisions published by the Italian Council for Research (CNR 2008); target temperatures were selected as multiples of 200 °C, with a very low heating rate (0.5 °C/min).

**Table 1.** Tested UHPC mix proportions (kg/m$^3$)

| ID<br>$f_c$ [N/mm$^2$] | CLS-A<br>150 | CLS-B<br>180 | |
|---|---|---|---|
| | | Lt* | Bt** |
| Nanodur compound 5941 | 800 | 1000 | 700 |
| Concrete I 52.5 R | - | - | 300 |
| Silica fume | 80 | 80 | 80 |
| Blast furnace slag | 100 | 100 | 100 |
| Metakaolin | 30 | 30 | 30 |
| Super-plasticiser | 40 | 40 | 40 |
| De-aerator | 0.2 | 0.5 | 1.0 |
| Sand Dolomia Genova 0–4 | 750 | 310 | 210 |
| Gravel Dolomia Genova 4–8 | 550 | 730 | 850 |
| Water | 156 | 156 | 165 |
| PP fibres | 2.5 | 2.5 | 2.5 |
| S fibres | 0 / 50 / 100 | 0 / 150 | 0 / 150 |

* Long-term curing
** Short-term curing

**Table 2.** Cold tests results.

| Material | Age (d) | PP | S | n. tests | Cold tests | |
|---|---|---|---|---|---|---|
| | | (% vol.) | | | $f_{cm}$ (N/mm$^2$) | CoV |
| CLS-A | 28 | 0.27 | 0 | 3 | 141.9 | 0.04 |
| | | | 0.62 | 3 | 144.3 | 0.02 |
| | | | 1.25 | 3 | 146.5 | 0.03 |
| CLS-A | 90 | 0.27 | 0 | 3 | 135.1 | 0.14 |
| | | | 0.62 | 3 | 149.2 | 0.04 |
| | | | 125 | 3 | 167.4 | 0.02 |
| CLS-B-Bt | 28 | 0.27 | 0 | 3 | 169.8 | 0.06 |
| | | | 1.90 | 3 | 185.1 | 0.01 |
| CLS-B-Lt | 90 | 0.27 | 0 | 3 | 171.6 | 0.08 |
| | | | 1.90 | 3 | 177.0 | 0.02 |

Three samples could be tested for each target. Concrete cubes of 100x100x100 mm were used as samples, according to the capacity of the testing equipment and the very high material strength.

**Table 3.** Hot tests results.

| Material | Age (d) | PP | S | n. tests | Hot tests | | |
|---|---|---|---|---|---|---|---|
| | | (% vol.) | | | $f_{cm}$ (N/mm$^2$) | CoV | %$f$ |
| CLS-A | 28 | 0.27 | 0 | 3 | 135.2 | 0.10 | –5 |
| | | | 0.62 | 3 | 135.5 | 0.17 | –6 |
| | | | 1.25 | 3 | 162.2 | 0.02 | +11 |
| CLS-A | 90 | 0.27 | 0 | 3 | 123.4 | 0.15 | –9 |
| | | | 0.62 | 3 | 152.4 | 0.03 | +2 |
| | | | 125 | 3 | 166.9 | 0.04 | –2 |
| CLS-B-Bt | 28 | 0.27 | 0 | 1 | 133.7 | - | –21 |
| | | | 1.90 | 3 | 160.8 | 0.02 | –13 |
| CLS-B-Lt | 90 | 0.27 | 0 | 0 | - | - | - |
| | | | 1.90 | 3 | 178.1 | 0.03 | +1 |

**Table 4.** Residual tests results.

| Material | Age (d) | PP | S | n. tests | Residual tests | | |
|---|---|---|---|---|---|---|---|
| | | (% vol.) | | | $f$cm (N/mm$^2$) | CoV | %$f$res |
| CLS-A | 28 | 0.27 | 0 | 3 | 166.8 | 0.15 | +18 |
| | | | 0.62 | 3 | 183.2 | 0.01 | +27 |
| | | | 1.25 | 3 | 199.0 | 0.01 | +36 |
| CLS-A | 90 | 0.27 | 0 | 3 | 183.0 | 0.05 | +36 |
| | | | 0.62 | 3 | 186.1 | 0.05 | +25 |
| | | | 125 | 3 | 198.1 | 0.02 | +18 |
| CLS-B-Bt | 28 | 0.27 | 0 | 3 | 190.0 | 0.01 | +12 |
| | | | 1.90 | 3 | 199.9 | 0.03 | +8 |
| CLS-B-Lt | 90 | 0.27 | 0 | 3 | 198.3 | 0.03 | +16 |
| | | | 1.90 | 3 | 124.6 | 0.03 | +27 |

To perform the hot tests, the samples were heated inside a muffle oven and immediately tested according to EN 12390-3 protocol (EN 2009). This procedure had been previously adopted at the Fire Brigade Laboratory; the oven and testing machine were moved close to each other and an insulating box was used to keep the sample at the attained temperature, once out of oven, before and during the mechanical test. The box was made of four refractory bricks and rockwool layers, and it was pre-heated in another oven before use, to the same temperature of the sample. The cold and residual tests were performed at the Material and Structures Research Laboratory of the University Roma Tre.

Attention was paid to the measurement of the temperature reached by the concrete; literature shows that for Ø100 mm cylindric samples, a less than 2-h hold does not allow reaching the target temperature at the core of the sample (Cree et al. 2017, Yermak et al. 2017, Xiong and Liew 2015, Ozawa 2019). Thermocouples were inserted in some samples to record the core temperature, through a hole of 4 mm diameter and 50 mm depth. The recordings verified that the samples reached the target temperature (Fig. 1a).

The research programme was interrupted after the cycles at target temperature 200 °C, due to explosive spalling of all the samples (Fig. 1b). In detail, the 28-days and the 90-days old samples exploded respectively at temperatures in the range 230–260 °C (CLS-A) and around 290 °C (CLS-B).

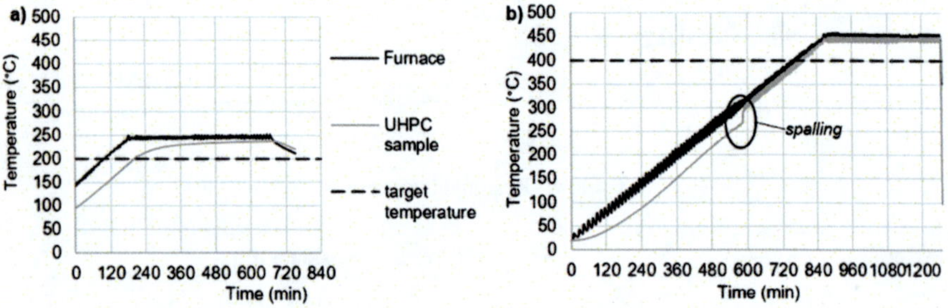


**Fig. 1.** Temperature recordings of cycles (a) 200 °C; (b) 400 °C

## 2.2 Results

The hot and residual compressive test results, for samples exposed to 200 °C, are commented below in comparison to cold tests (at 20 °C). Henceforth, the volume percentages in PP and S fibres are abbreviated into PP and S respectively.

Tables 2, 3 and 4 report the values of mean compressive strength $f_{cm}$, $f_{cm,\theta}$, and $f_{cm,\theta,res}$ respectively for cold, hot and residual tests, with the coefficient of variation (CoV, evaluated on the three samples) and the per cent difference of the property in hot or residual conditions versus the respective cold value ($\%f$ and $\%f_{res}$). The strength values of each sample are depicted in histograms in Figs. 2, 3 and 4.

Observing the original values (mean values in Table 2), the 28-days compressive strength of CLS-A shows no relevant variation with the fibre content. On the other hand, at 90-days it increases with S fibre content, i. e. from 135 to 149, and to 167 N/mm$^2$, meaning + 3% and + 14% respectively for S = 0.62% and S = 1.25%. The single specimen results in Fig. 2a show a good uniformity at 28 days for each S dosage, and at 90 days except for S = 0%. It may be assumed that the longer curing time increases the bond at the matrix-fibres interfaces, giving a positive contribution to strength.

For CLS-B (Fig. 2b), both mixes have the same basic strength. They show a moderate strength increase with S = 1.92% for the 28-day old samples (from 170 to 185 N/mm$^2$,

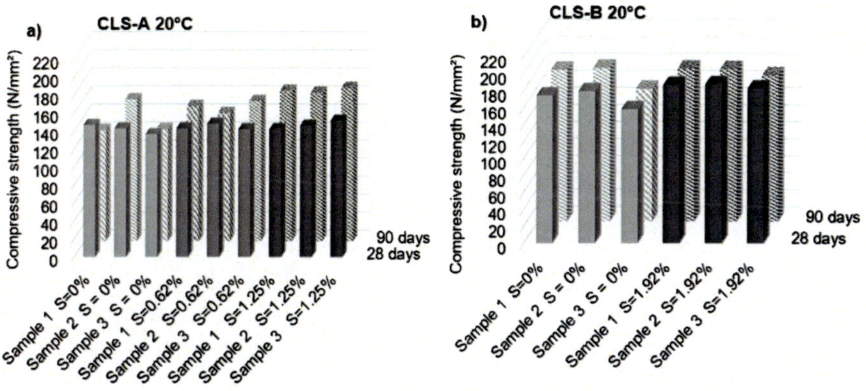


**Fig. 2.** Results of the single tests at 20 °C (cold).

+ 9%), while for the 90-day old the increase is very slight (from 172 to 177 N/mm$^2$, + 4%). In the latter case, the single specimens indicate some dispersion in the results for S = 0% (Fig. 2b); thus, the limited performance improvement may be disregarded.

As far as it concerns the 200 °C hot tests (Table 3), the S fibre dosage affects the strength of CLS-A. At 28 days of maturation, the samples with S = 0% and S = 0.62% undergo a very slight reduction in strength (about –5% of mean original strength), while the 1.25% content of S fibres brings on a moderate increase (+11%). For the former two S percentages, Fig. 3a shows some dispersion in the results of single samples, while data are quite compact in the latter case. At 90 days, the mean strength reduces by –9% for S = 0%, while for both dosages of S fibres the strength is almost equal to the original (+2 and –2%). Figure 3a points out a good uniformity in the results of S = 0.62% and S = 1.25%. Thus, at hot conditions (200 °C), the effect of curing time on the strength variation of CLS-A is very limited.

The CLS-B with PP fibres only (S = 0%) has shown a fragile behaviour both in cold and hot conditions. In particular, violent failures have occurred to the loaded specimens of both CLS-B mixes in hot conditions, so that the testing apparel was damaged. The only sample successfully tested, at 28 days (Fig. 3b), resulted in a 21% reduction in strength. It was then decided not to complete the hot tests on the CLS-B samples without S fibre. Such fragile behaviour was assumed to impair any applicability of the material without S fibre addition.

The hybrid fibre reinforced samples of CLS-B, S = 1.92%, have encountered more ductile failure modes; the strength decreases by –13% for the 28-day old samples and undergoes no relevant variation for the 90-day old. Again, Fig. 3b enlightens uniform results with a relevant content in S fibres.

The residual tests after exposure to 200 °C (Table 4) show increased compressive strength for all the samples of both mixes.

For CLS-A, the mean strength values at each steel fibre dosage are almost equal for 28- and 90-days old samples. Figure 4a indicates dispersion in the S = 0% tests at 28 days, while the data are quite compact for the hybrid fibre reinforcement and show

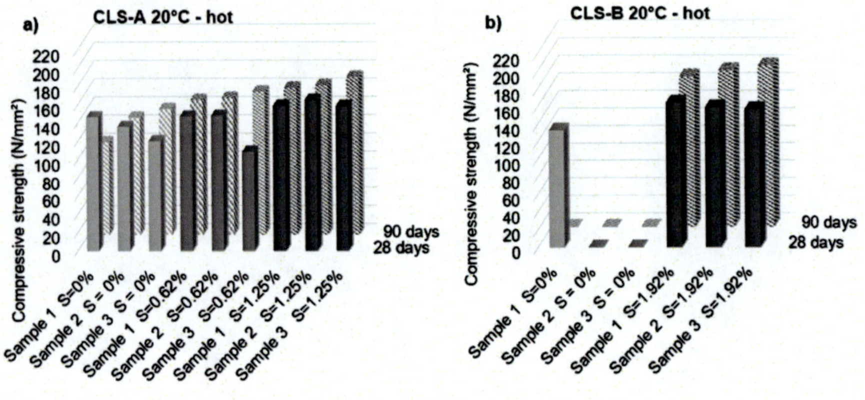


**Fig. 3.** Results of the single tests at 200 °C (hot)

a significant strength increase for both S fibre dosages (+27% and + 36% in the mean values for S = 0.62% and S = 1.25% respectively).

The residual strength of CLS-B with S = 1.92% is moderately higher than the original for the 28-days old samples (+8%) and significantly higher (+27%) for the 90-days old samples. This mix with the hybrid reinforcement shows uniform results (Fig. 4b).

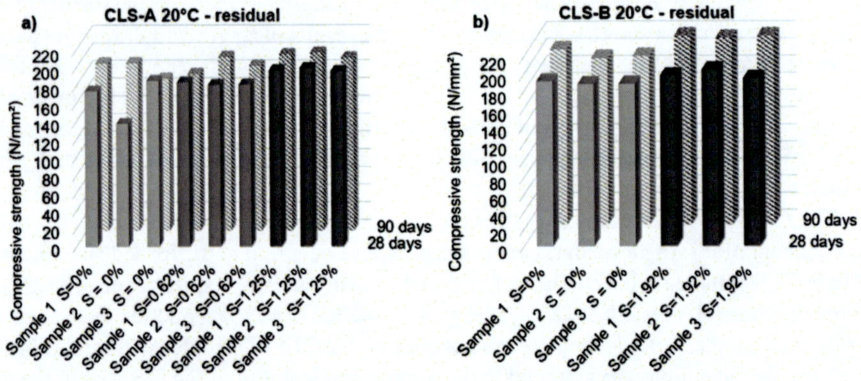


**Fig. 4.** Results of the single tests after 200 °C (residual).

## 3 Discussion

Comparisons with available literature data are presented for both the hot and residual conditions of the investigated mixes. More detailed analyses were published in the paper by Sciarretta et al. (2021).

### 3.1 Hot Behaviour

Figure 5 shows the plots of the present research, together with the data of CERIB and BSI®-fire concretes in hot conditions (Li and Tan 2016), both having PP = 0.33% in volume.

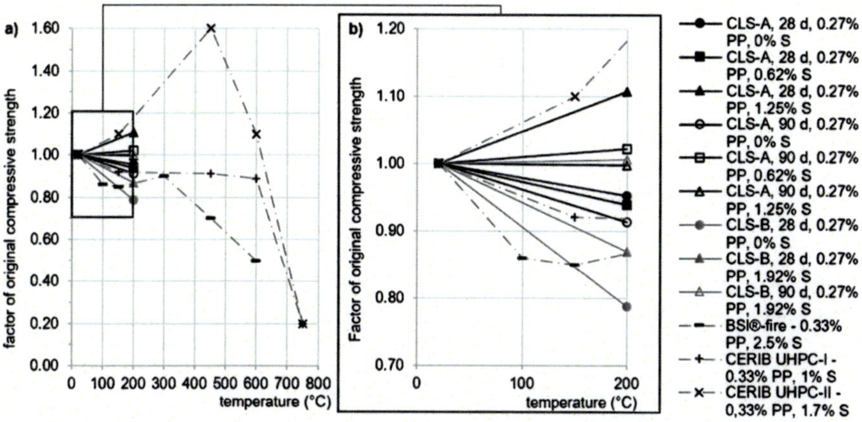


**Fig. 5.** (a) Trends of experimental hot compressive strength, compared to literature data (Li and Tan 2016); (b) zoom in.

CERIB UHPC-I and UHPC-II have compressive strength of 170 N/mm$^2$ with S = 1%, and 200 N/mm$^2$ with S = 1.7%. Both have PP fibres 12 mm long, aspect ratio 667, and S fibres 13 mm long, aspect ratio 81. In hot conditions, the latter has a higher strength gain than the former. BSI®-fire has a strength of 148–165 N/mm$^2$ with S = 2.5%, with the same PP fibres as the previous two, and S fibres 20 mm long with aspect ratio 67. It undergoes a slight reduction in strength in hot conditions.

At 200 °C (Fig. 5a), the behaviour of CLS-A is intermediate between the two CERIB concretes. As above noted, CLS-A shows a moderate strength increase for the highest dosage in S fibres (S = 1.25%) at 28 days of age; this is in line with the behaviour of CERIB UHPC II, which displays a strength increase of the same entity at 150 °C.

The decreasing strength of CLS-B at 28 days of age is in line with the behaviour of CERIB UHPC I (S = 1%) and BSI®-fire (S = 2.5%). As noted above, the different CLS-B mix with longer curing time and S reinforcement S = 1.92% withstands the heating without losing strength, which is an intermediate behaviour between CERIB UHPC-I and -II.

### 3.2 Residual Behaviour

Figure 6 shows the plots of the present research, together with the data of Ducorit®D4 concrete (Yang et al. 2019), of PP = 0.25%.

Ducorit®D4 has a strength of 167 N/mm$^2$ with S = 0%, and PP fibres 13 mm long with aspect ratio 433; it shows a drop (–50%) in the residual strength beyond 200 °C exposure.

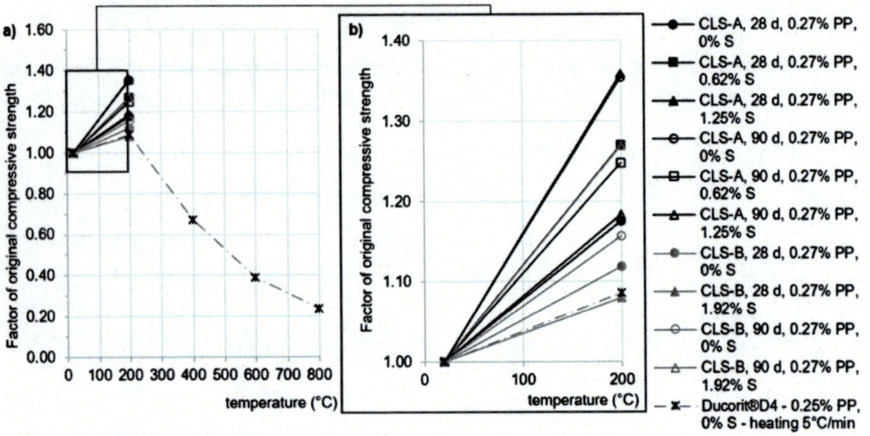


**Fig. 6.** (a) Trends of experimental residual compressive strength, compared to literature data (Yang et al. 2019) (b) zoom in

In residual conditions (Fig. 6a), CLS-A has a stronger increase in strength than the literature case Ducorit®D4 after 200 °C exposure; the effect of the S fibre reinforcement (which is absent in the literature case) is evident. As well, the performance of the CLS-B mix at 28 days of age is very similar to Ducorit®D4, regardless of the S reinforcement. The other mix, which attains the desired strength at 90 days, shows larger strength increases than the literature case, especially in presence of S fibres.

Generally, all the cases tested in the present research lie within the envelope of the literature cases. Beyond 200 °C, in hot conditions, the presented reference cases show a stable behaviour up to 600 °C. In residual conditions, the cited case undergoes mechanical decay (Figs. 5b and 6b).

## 4 Conclusions

The paper has presented experimental research aimed at assessing the high temperature behaviour of three UHPC mixes – one (CLS-A) of 150 and two (CLS-B) of 180 N/mm$^2$ basic strength – with polypropylene fibres (PP, 0.27% in volume) and steel fibre (S, various percentages from 0% to 1.92%). The compressive strength was measured under and after high temperature exposure. The aspect ratio of the PP fibres (12 mm long) is 240, while S fibres (of two shapes, length 20 and 30 mm) have aspect ratios of 50 and 43. Results during (hot) and after (residual) exposure to a cycle to 200 °C are given.

- At 200 °C, in hot conditions, CLS-A shows limited strength variations, in line with literature information. At this condition, the concrete age is not relevant for CLS-A, while the increasing S fibre dosage brings on a slight performance improvement.
- At the same condition, CLS-B undergoes small strength reductions at 28 days, both for PP-only and hybrid reinforcements; the strength remains at the original levels for the mix which attains the desired strength at 90 days of age.

- After 200 °C, the residual strength of both CLS-A and CLS-B is higher than the original values through all the S fibre dosages and concrete ages. The contribution of S fibre is relevant, in fact S = 1.25% and S = 1.92% increases the strength of CLS-A and CLS-B up to about 200 and 225 N/mm$^2$ respectively.
- After 200 °C, the effect of the S fibre content on CLS-A is more intense than in hot conditions at the same temperature.
- The benefit of hybrid (PP + S) fibre content is significant. In the present research, a S fibre content of 1.25% in volume for CLS-A has increased both hot (+11%) and residual strength (+36%) at 28 days of age. S = 1.92% for CLS-B-Lt (90 days) has increased the residual strength (+27%).
- The tested mixes withstood maximum temperatures of only 230–290 °C, in laboratory conditions. Although the issues of small heated samples would not affect real-scale applications, the next steps of the research will investigate the ways to improve the UHPC microstructural properties. To this purpose, the fibre type, geometry and content will be again key research parameters.

**Acknowledgements.** The authors warmly thank the Italian Fire Brigade, Buzzi Unicalcestruzzi, Federbeton (Italy) and ReLUIS for financial support. The research was also supported by the National Natural Science Foundation of China (Grant No. 51778148).

## References

ACI: Ultra-High-Performance Concrete: An Emerging Technology Report. PRC-239-18 (2018)

AFGC/SETRA: Betons fibrés à ultra-hautes performances – Ultra-high-performances fibre-reinforced concretes – Recommendations (2013)

Bei, S., Zhixiang, L.: Investigation on spalling resistance of ultra-high-strength concrete under rapid heating and rapid cooling. Case Stud. Construct. Mater. **4**, 146–153 (2016)

Burke, B.T.: Residual strength of Ultra-High Performance Concrete after exposure to elevated temperatures, MS thesis, University of Connecticut (2011)

CEN 2004. Eurocode 2: Design of concrete structures – Part 1–2: General rules – Structural fire design. EN 1992–1–2

CEN: Testing hardened concrete - Part 3: Compressive strength of test specimens. EN 12390-3 (2009)

Chen, H.-J., Yu, Y.-L., Tang, C.-W.: Mechanical properties of Ultra-High performance concrete before and after exposure to high temperatures. Materials **13**, 770 (2020)

CNR: Istruzioni per la progettazione, l'esecuzione e il controllo di strutture in calcestruzzo fibrorinforzato. CNR DT 204/2006 (2008)

Cree, D., Pliya, P., Green, M.F., Noumowé, A.: Thermal behaviour of unstressed and stressed high strength concrete containing polypropylene fibers at elevated temperature. J. Struct. Fire Eng. **8**(4), 402–417 (2017)

CSA: Concrete materials and methods of concrete construction/Test methods and standard practices for concrete. A23.1/A23.2 (2019)

Graybeal, B.: Ultra-High-Performance concrete. Federal Highway Administration Research and Technology. FHWA-HRT-11-038 (2011)

Heinz, D., Dehn, F., Urbonas, L.: Fire resistance of Ultra High Performance Concrete (UHPC) – testing of laboratory samples and columns under load. In: Schmidt, M., Fehling, E., Geisenhanslüke, C. (Eds.) Ultra High Performance Concrete, Proceedings of 3rd International Symposium, Kassel, 13–15 September 2004. Kassel University Press, Kassel (2004)

Huang, H., Wang, R., Gao, X.: Improvement effect of fiber alignment on resistance to elevated temperatures of ultra-high performance concrete. Compos. B **177**, 107454 (2019)

ISO: Fire resistance – Tests for thermophysical and mechanical properties of structural materials at elevated temperatures for fire engineering design. ISO/TR 15655 (2020)

Kahanji, Ch., Ali, F., Nadjai, A.: Experimental study of ultra-high performance fibre reinforced concrete under ISO 834 fire. In: Garlock, M., Kodur, V. (Eds.) SiF - Structures in Fire, Proceedings of 9th International Conference, Princeton University, 8–10 June 2016. DEStech, Lancaster (2016)

Larsen, I.L., Thorstensen, R.T.: The influence of steel fibres on compressive and tensile strength of Ultra-High-performance concrete: a review. Constr. Build. Mater. **256**, 119459 (2020)

Lavorato, D., Nuti, C., Santini, S., Briseghella, B., Xue, J.: A repair and retrofitting intervention to improve plastic dissipation and shear strength of Chinese RC bridges. In: IABSE Symposium Report, vol. 105, no. 9, pp. 1–6. International Association for Bridge and Structural Engineering, Zürich, Switzerland (2015)

Lavorato, D., et al.: Ultra-high-performance fibre-reinforced concrete jacket for the repair and the seismic retrofitting of Italian and Chinese RC bridges. In: Proceedings of the 6th International Conference on Computational Methods in Structural Dynamics and Earthquake Engineering, Rhodes Island, Greece, vol. 1, pp. 2149–2160 (2017)

Li, Y., Tan, K.H.: Effects of polypropylene and steel fibers on permeability of ultra-high performance concrete at hot state. In: Garlock, M., Kodur, V. (Eds.) SiF - Structures in Fire, Proceedings of 9th International Conference, Princeton University, 8–10 June 2016. DEStech, Lancaster (2016)

Liang, X., Wu, C., Yang, Y., Li, Z.: Experimental study on ultra-high performance concrete with high fire resistance under simultaneous effect of elevated temperature and impact loading. Cement Concr. Compos. **98**, 29–38 (2019)

Liang, X., Wu, C., Su, Y., Chen, Z., Li, Z.: Development of Ultra-High-Performance Concrete with high fire resistance. Constr. Build. Mater. **179**, 400–412 (2018)

Liu, J.-C., Zhang, Z.: Neural network models to predict explosive spalling of PP fiber reinforced concrete under heating. J. Build. Eng. **32**, 101472 (2020)

Liu, X., Ye, G., De Schutter, G., Yuan, Y., Taerwe, L.: On the mechanism of polypropylene fibres in preventing fire spalling in self-compacting and high-performance cement paste. Cem. Concr. Res. **38**, 487–499 (2008)

Lu, F., Fontana, M.: Effects of polypropylene fibres on preventing concrete spalling in fire. In: Garlock, M., Kodur, V. (Eds.), SiF - Structures in Fire, Proceedings of 9th International Conference, Princeton University, 8–10 June 2016. DEStech, Lancaster (2016)

Missemer, L., Ouedrago, E., Malecot, Y., Clergue, C., Rogat, D.: Fire spalling of ultra-high performance concrete: from a global analysis to microstructure investigations. Cem. Concr. Res. **115**, 207–219 (2019)

Ozawa, M., Parajuli, S.S., Uchida, Y., Zhou, B.: Preventive effects of polypropylene and jute fibers on spalling of UHPC at high temperatures in combination with waste porous ceramic fine aggregate as an internal curing material. Constr. Build. Mater. **206**, 219–225 (2019)

Peng, G.F., Yang, J., Long, Q.Q., Nin, X.J., Zeng, Q.P.: Comparison between Ultra-High-Performance Concretes with recycled steel fiber and normal industrial steel fiber. In: Ganjian, E., Ghafoori, N., Claisse, P. (Eds.), Sustainable Construction Materials and Technologies, Proceedings of 4th International Conference, Las Vegas, 7–11 August 2016 (2016)

Pimienta, P., et al.: Literature review on the behaviour of UHPFRC at high temperature. In: Schmidt, M., Fehling, E., Glotzbach, C., Frölich, S., Piotrowski, S. (Eds.) Hipermat – 3rd International Symposium on UHPC and Nanotechnology for High Performance Construction Materials, Proceedings of 3rd International Symposium, Kassel, 7–9 March 2012. Kassel University Press, Kassel (2012)

Sarwar, M.A.: Characterizing temperature-induced strength degradation and explosive spalling in ultra-high performance concrete. Civil Engineering MS thesis, Michigan State University (2017)

Sciarretta, F., et al.: Ultra-High-Performance Concrete (UHPC) with Polypropylene (PP) and steel fibres: investigation on the high temperature behaviour. Constr. Build. Mater. **304**, 124608 (2021)

Way, R.T., Wille, K.: Effect of heat-induced chemical degradation on the residual mechanical properties of ultrahigh performance fibre-reinforced concrete. J. Mater. Civ. Eng. **28**(4), 04015164 (2016)

Xiong, M.-X., Liew, J.Y.R.: Spalling behavior and residual resistance of fibre reinforced Ultra-High-Performance Concrete after exposure to high temperatures. Mater. Constr. **65**(320), e071 (2015)

Xue, J., et al.: Severely damaged reinforced concrete circular columns repaired by turned steel rebar and high-performance concrete jacketing with steel or polymer fibers. Appl. Sci. **8**(9), 1671 (2018)

Xue, J., Briseghella, B., Huang, F., Nuti, C., Tabatabai, H., Chen, B.: Review of ultra-high performance concrete and its application in bridge engineering. Constr. Build. Mater. **260**, 119844 (2020)

Yang, J., Chen, B., Nuti, C.: Influence of steel fiber on compressive properties of ultra-high performance fiber-reinforced concrete. Constr. Build. Mater. **302**, 124104 (2021)

Yang, J., Peng, G.-F., Zhao, J., Shui, G.-S.: On the explosive spalling behavior of Ultra-High-Performance Concrete with and without coarse aggregate exposed to high temperature. Constr. Build. Mater. **226**, 932–944 (2019)

Ye, H., et al.: Research on fire resistance of ultra-high-performance concrete. Adv. Mater. Sci. Eng. 530948 (2012)

Yermak, N., Pliya Beaucour, A.-L., Simon, A., Noumowé, A.: Influence of steel and/or polypropylene fibres on the behaviour of concrete at high temperature: spalling, transfer and mechanical properties. Construct. Build. Mater.**132**, 240–250 (2017)

# Road and Industrial Pavements

# Self-sensing Cementitious Pavements with Carbon Inclusions for Weigh-In-Motion and Monitoring of Infrastructures: Calibration and Field Tests

Antonella D'Alessandro(✉), Hasan Borke Birgin, and Filippo Ubertini

Department of Civil and Environmental Engineering, University of Perugia, Perugia, Italy
antonella.dalessandro@unipg.it

**Abstract.** Smart carbon-based cementitious composites possess multifunctional properties, adding novel capabilities to mechanical ones. Of great interest in civil engineering are self-sensing materials, which permit to carry out a continuous, diffuse and simple monitoring on structures and infrastructures. In the field of infrastructures, weigh-in motion, traffic and health monitoring are especially worthy of investigation, because they permit to control the safety of critical structures, such as bridges, and to identify states of damage or loss of structural integrity. The authors present the characterization and investigation of smart cementitious composites for road pavements doped with carbon microfibers and graphite, with self-monitoring capabilities. The optimal material was developed by carrying out laboratory tests on small-scale samples with different amounts of hybrid fillers, and the sensing properties were proved through fields tests. The results demonstrated the valuable capabilities of the developed cementitious pavement in weigh-in-motion and traffic monitoring of vehicles.

**Keywords:** Carbon-based fillers · Cement-based composites · Carbon microfibers · Sustainable concrete · Structural health monitoring · Earth materials

## 1 Introduction

Concrete is nowadays a versatile construction material for various applications. Its composite nature makes it suitable for addition of fillers and additive for enhancing its properties and behaviour (Metaxa et al. 2009, Yoo et al. 2018). As a matter of fact, cementitious matrices could constitute the core material for smart devices and ground breaking technologies (Han et al. 2015, Rainieri et al. 2013). In literature are present several ongoing researches on enhanced concrete materials which conjugate multifunctional capabilities with load bearing resistance, for applications in civil engineering (D'Alessandro et al. 2020, Han et al. 2011). In particular, recent studies report various multifunctionalities for concrete including self-sensing and structural assessment (Galao et al. 2017, Coppola et al.2011), strain-sensing (Loh et al. 2009, Azhari et al. 2012), deicing (Galao et al. 2016), and thermal-energy storing (D'Alessandro et al. 2017). More


M. A. Aiello and A. Bilotta (Eds.): ICC 2022, LNCE 435, pp. 697–707, 2024.
https://doi.org/10.1007/978-3-031-43102-9_53

advanced applications include concrete with computing abilities (Przyczyna et al. 2021). The authors in the last years are developing different experimental campaign and theoretical researches about smart composite materials for structural health monitoring, for the evaluation of the strain fields and the damage detection of buildings, or the weigh-in-motion (WIM) and monitoring of roads and infrastructures (Birgin et al. 2020a, 2020b, D'Alessandro et al. 2020a, D'Alessandro et al. 2020b, Meoni et al. 2018). This paper aims to demonstrate the effectiveness of the use of smart concrete sensors for WIM and traffic sensors, as embedded sensing elements in real infrastructures. For this purpose, small and medium-scale elements are first investigated, in order to identify the optimal composite material and setup. Field tests carried out on a full-scale setting with real transiting loads are then introduced.

## 2 Aim of the Research

The good monitoring results of previous research campaign carried out by the authors on samples made of smart cement-based materials and different types of carbon fillers, appear promising for the development of smart concrete applications. The current research is aimed at investigating the sensitivity of concrete based smart sensors for their multipurpose utilization for advanced monitoring systems.

The study started with the identification of the best carbon inclusion mix for a concrete-based composite with self-/strain- sensing abilities. For this purpose, small-scale cube samples with different percentages of carbon microfibers and graphite were subjected to electrical and electro-mechanical tests for the identification of the optimal hybrid carbon filler concentrations. Small concrete beams were subjected to tests for crack localization and identification. Then, medium scaled samples made of the most performing composite were produced and embedded in a full-scale pavement for field tests with real-scale loads. The final aim of the research is the WIM and traffic monitoring of infrastructures.

## 3 Materials and Devices

### 3.1 Matrix Material, Fillers and Preparation Procedure

The materials of the study were Portland Cement 42.5R, fine and coarse aggregates, Sigrafil® carbon short microfibers (CMF) with a length of 6 mm and thickness of 7 $\mu$m, with a single filament resistivity of 15 $\mu\Omega$m, designed specifically for water-based systems, and graphite powder (G). The fillers were added at various percentages. Figure 2 shows the preparation procedure of the samples. The carbon fillers were firstly mechanically mixed with water (Figs. 2(a) and (b)), and then cement, sand and gravel were added (Figs. 2(c) and (d)). After a homogeneous compound has been obtained, the material was poured into oiled molds. The samples were unmolded after 48 h, and cured at laboratory conditions. Fragments of hardened concrete were investigated through optical microscope and Scanning electron microscope. The micrographs show an acceptable distribution of the fillers within the matrix (Fig. 1).

### 3.2 Types of Samples

The samples of the study were small-scaled cubes, beams and medium-scale plates (Fig. 2(e)). The cubes had 5 cm side length, were equipped with embedded steel-net electrodes placed inside bulk composite material with a separation of 2 cm. The beams were square prism with dimensions of $4 \times 4 \times 16$ cm$^3$. The beams were equipped with copper wire electrodes placed 2 cm apart, in the centre of the greater side. The plate samples had the base dimensions of 35 cm × 25 cm and a thickness of 4 cm. The plate samples included 5 copper wire electrodes distributed at the half thickness with a mutual distance of 4 cm. Figure 3 shows the sketches of the samples, and the electrodes' configurations.

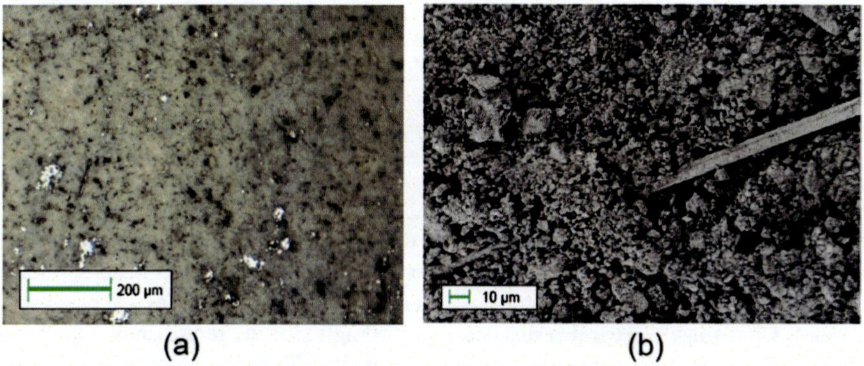


**Fig. 1.** Micrographs of concrete with carbon microfibers and graphite using (a) an optical microscope; (b) SEM

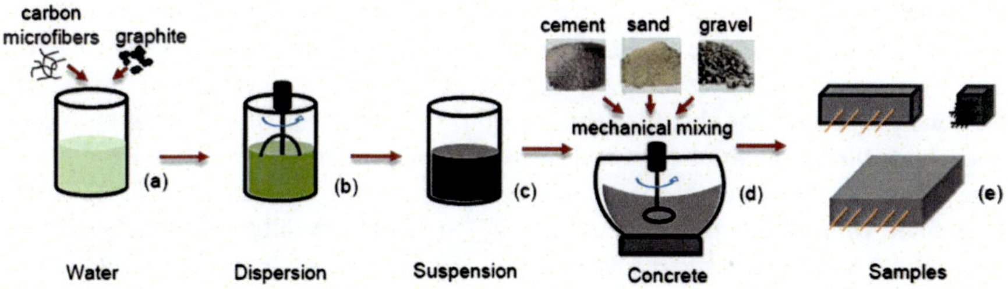


**Fig. 2.** Preparation procedures of the different types of smart concrete samples

### 3.3 Instrumentation and Setups

The electrical devices for the electromechanical tests were supplied through a National Instruments PXIe-1092 chassis equipped with modules PXIe-4138 and PXIe-4302, which had a voltage source with 60V and a 3A maximum output capacity and the

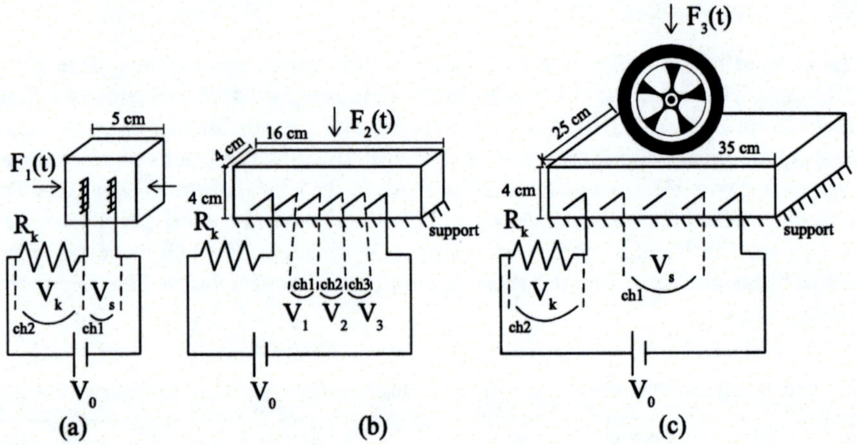


**Fig. 3.** Sketch of the dimensions, instrumentation and electrical setup of cubes, small beams and plate samples for the electromechanical and sensing tests

32-channel 24-bit voltage reader, respectively. The data acquisition procedure was developed under LABVIEW® environment. The electrical setup for the electromechanical and sensing tests on the three types of samples are illustrated in Fig. 3, which represents also the load application. The electrical resistance between the electrodes was calculated through Ohm's law electrical resistances. Accordingly, for the setups shown in Fig. 3 (a) and (c), for tests on cubes and plates, respectively, the general formulation for the sample resistance, $R_s(t)$, as a time history is:

$$R_s(t) = R_k \frac{V_s(t)}{V_k(t)} \tag{1}$$

where, $V_s(t)$, is the voltage reading through the sample acquired via channel 1 (ch1), and, $V_k(t)$, is the voltage reading through the shunt resistor via channel 2 (ch2), $R_k$ is the constant value of the shunt resistor, equal to 10 k Ω for the study. This setup allows to evaluate the variation of electrical resistivity and the gauge factors of the different types of mixes. The setup of the small beams was slightly different (Fig. 3(b)). In this case, three voltage time histories, namely $V_{1,2,3}(t)$, were simultaneously collected and compared. The three measurements analysed the behaviour of three adjacent sections of the plates. This setup was developed for investigating the capability of the smart concrete of detecting and localizing the transiting loads.

## 4 Experimental Tests

### 4.1 Compressive Tests on Cubes

The cubes were subjected to triangular cyclic load, as depicted in the Fig. 4. Accordingly, the compression load increased up to 5 kN with a rate of 1kN/s, followed by two cycles of load increase-decrease at the rate of 0.5 kN/s with the maximum load of 10 kN. During

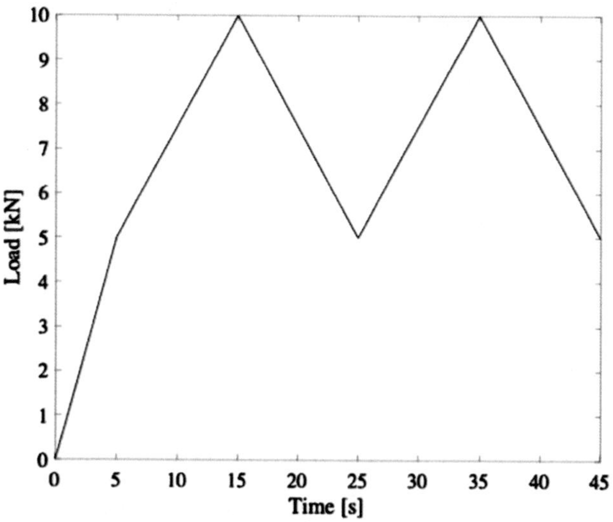


**Fig. 4.** Load history of cyclical triangular compressive loads on cubic samples

the tests the strain was recorded by the data acquisition system. A servo-controlled pneumatic universal dynamic testing machine model IPC Global UTM14P was used for load application. Nine different concrete mixes were tested: normal, with 5 and 10% in weight of G with respect to cement (5G0CMF and 10G0CMF), with 1/32 and 1/16% of CMF (0G32CMF and 0G16CMF), and all the combinations of them, following the same type of nomenclature (e.g. 5G32CMF).

### 4.2 Testing of Small Beams

The beams were tested through compressive loads on the lateral sides of the sample, as represented in Fig. 3(b). The loads were increased monotonically with a rate of 0.2 kN/s until the crack occurred.

### 4.3 Field Tests

The field tests were carried out on an internal road inside a factory in Ceneselli (RO), Italy, where vehicles and trucks ordinarily passed. The electrodes were cabled and connected to the external data acquisition system by electrical welded wires, and the slabs were covered by use of asphalt, remaining embedded within the pavement, along the transversal direction of the road. The sensitivity tests were carried out by use of a forklift with a weight of about 33 kN.

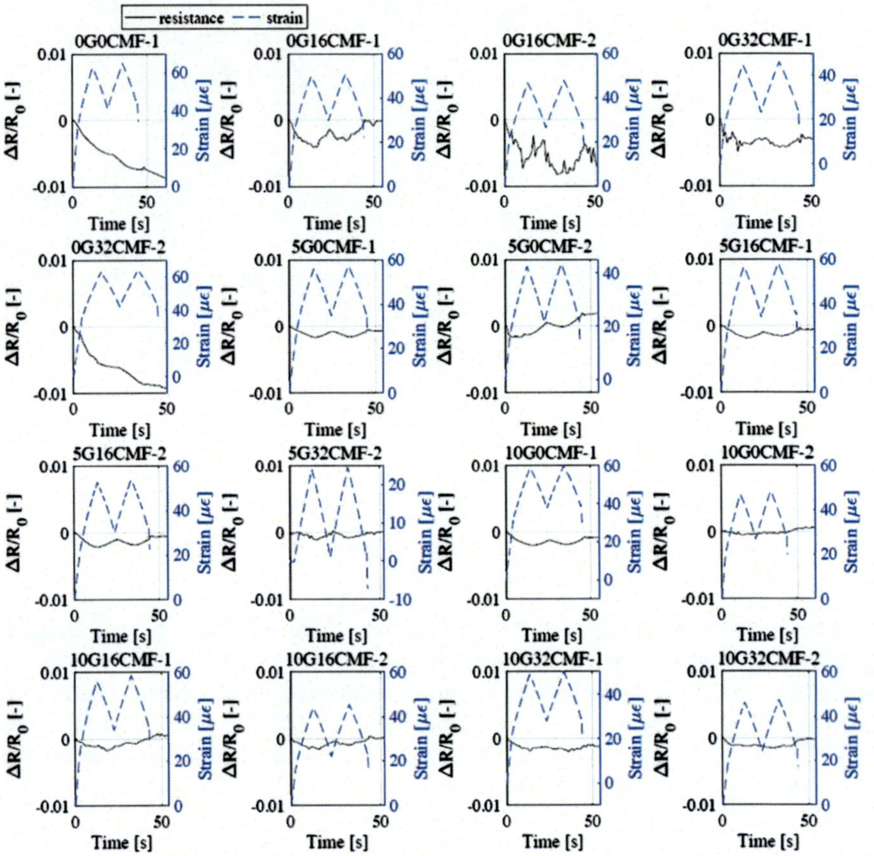


**Fig. 5.** Electrical and strain outputs of the concrete cubes with different fillers' content and typologies, subjected to cyclical compressive loads

# 5 Results

## 5.1 Compressive Tests on Cubes

The electromechanical tests carried out on the cubes with various types of fillers demonstrated different behaviour, as shown in Fig. 5. The signals clearly show that noise and drifts due to polarization effect are quite impactful for the quality of outcomes. An overall assessment of the collected signals yields into the conclusion that the addition of both graphite or carbon microfibers enhanced the signal. The hybrid composition determines a particularly performing electrical path. Consistent outputs were acquired from 5G16CMF and 10G32CMF, while for the other materials, huge differences were observed between two repetitions.

Figure 6 shows the linearity of reading, and the sensitivity. The results clearly demonstrated the different performance of the composites. The samples doped with only carbon microfibers generated signals with higher gauge factors; however, the linearity of the sensing models was found not adequate.

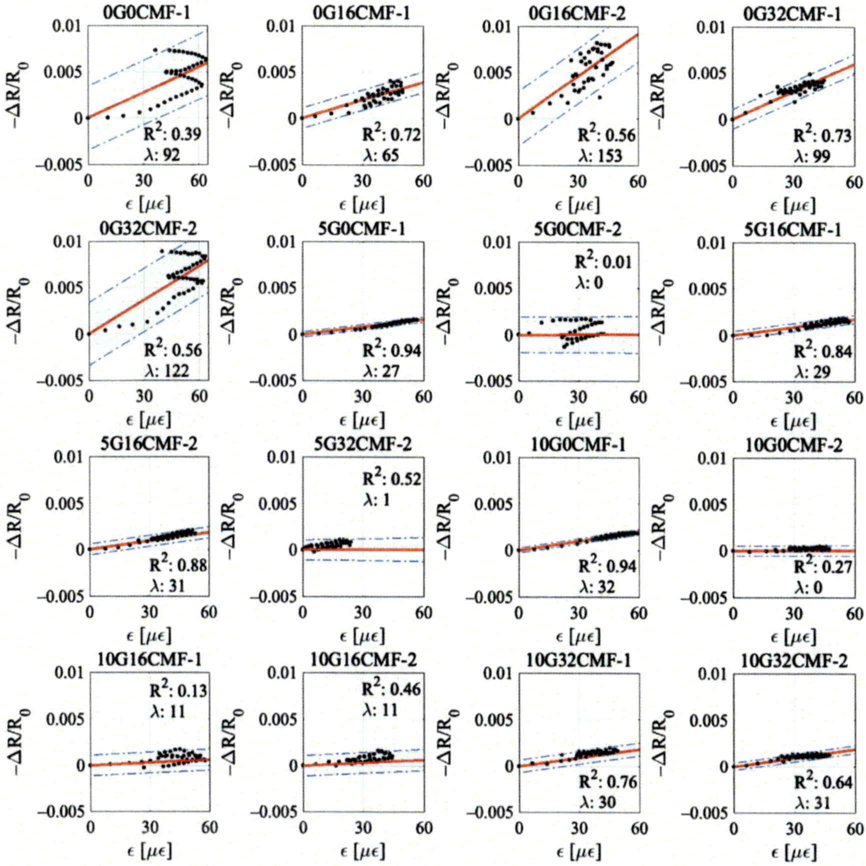


**Fig. 6.** Evaluation of linearity and sensitivity of concrete cubes with different fillers' content and typologies, subjected to cyclical compressive loads

Among the sample set, the best performing material combinations were found to be 5G0CMF and 10G0CMF. However, the results were not repetitive, due to the polarization effect. In general, the hybrid specimens showed a more feasible behaviour, with the exception of 5G32CMF which produced unreadable and affected signals. From the analysis of the signals, considering the sensitivity and the feasibility, 5G16CMF and 10G32CMF materials were qualified as the best ones for application. Analysing the overall behaviour, the best composite could be identified in the mix 10G32CMF.

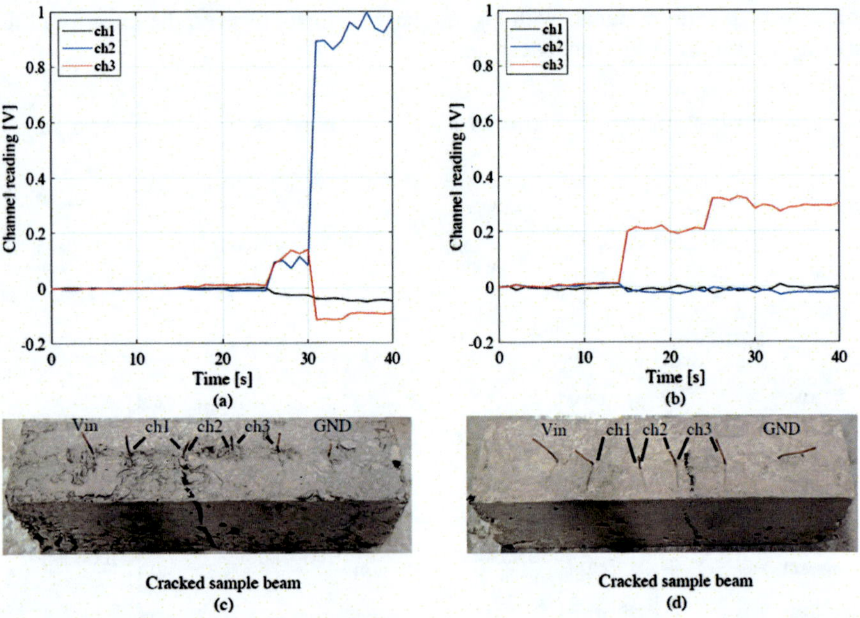


**Fig. 7.** Damage test results – electrical outputs and placement of the cracks - of small beams: (a and c) sample 10G16CMF, (b and d) 0G32CMF

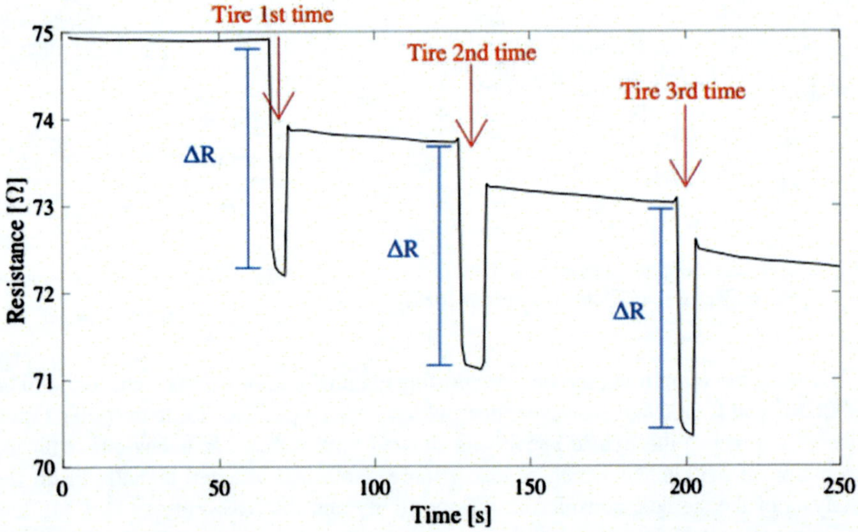


**Fig. 8.** The results of sensitivity tests on embedded slab concrete sensors in field: identification of the passing of the forklift

### 5.2 Tests on Small Beams

The damage tests performed on small beams with distributed electrodes showed that the cracks formed during the tests can be detected by observing the voltage-time histories collected through internal segments of the concrete sensors. As a matter of fact, the signal jumps visible in the real-time voltage acquisitions permitted to identify the crack occurrence. The readings and the cracked specimens with marked cracks are shown in Fig. 7. Accordingly, the voltage read through the segments where the cracks have been formed experienced a larger increase in ch2 for 10G16CMF and in ch3 for 0G32CMF than in other ones, due to a significant increase in the electrical resistance of the cracked zone.

### 5.3 Field Tests

Figure 8 reports the outputs of the electrical resistance obtained from the measurements in field occurring when an unload forklift passed three times over the concrete sensor embedded in the road pavement. The graph clearly demonstrates that the concrete sensor, although with a limited extension with respect to the transversal dimension of the read, is able to sense the presence of the load showing a proportional decrease in electrical resistance. The signals appear clear and repeatable.

## 6 Discussion

The hybrid carbon fillers selected for this study were graphite and carbon microfibers, and both have been found suitable for producing sensors with larger dimensions than the small laboratory ones, due to their capability of being successfully mixed through mechanical methods.

The electromechanical tests with small-scale smart cubes showed that under dynamic compression, the filler mixture of graphite and carbon microfibers generated repetitive results for strain sensing.

Tests on small beams demonstrated that the smart concrete is able of damage detection, allowing the identification and localization of cracks. The field tests on hybrid concrete plate sensors embedded in a road pavement and subjected to real loads revealed that such sensors could sense the loads and appeared feasible.

The results of the three types of tests were promising for further developments in real applications. However, the biphasic signal and the dimensional characteristics of the sensors should be tailored for the particular application of dynamic monitoring. Another important issue on such sensors is the impact of possible electrical interactions with the reinforcement in RC structures. A data-driven black-box model would be required in this case for conducting load and state identifications.

Further improvements on this topic will include the optimization of the sensor and setup design for enhancing the performance of the monitoring system for the specific vehicle characteristics, the structural configuration and the expected speeds.

## 7 Conclusion

The presented research investigated the self-sensing performance of concrete sensors doped with carbon-based fillers. The developed smart concrete sensor and sensing setup showed good monitoring capabilities at different scales, from laboratory one to real full scale. Possible applications could be the traffic monitoring, the self-evaluation of the structural integrity, and the WIM of infrastructures.

**Acknowledgements.** The authors would like to thank the European Union's Horizon 2020 research and innovation programme (Grant Agreement N. 765057 - SAFERUP!), and to acknowledge funding by the European Union – Next Generation EU under the Italian Ministry of University and Research (MUR) National Innovation Ecosystem Grant ECS 00000041 – Vitality. Padana Resine s.r.l. is also acknowledged by the authors for its contributions in field tests.

## References

Azhari, F., Banthia, N.: Cement-based sensors with carbon fibers and carbon nanotubes for piezoresistive sensing. Cem. Concr. Comp. **34**, 866–873 (2012)

Birgin, H.B., D'Alessandro, A., Laflamme, S., Ubertini, F.: Smart graphite–cement composite for roadway-integrated weigh-in-motion sensing. Sensors, **20**(16), 4518 (2020a). https://doi.org/10.3390/s20164518

Birgin, H.B., Laflamme, S., D'Alessandro, A., Garcia-Macias, E., Ubertini, F.: A weigh-in-motion characterization algorithm for smart pavements based on conductive cementitious materials. Sensors **20**, 659 (2020). https://doi.org/10.3390/s20030659

Coppola, L., Buoso, A., Corazza, F.: Electrical properties of carbon nanotubes cement composites for monitoring stress conditions in concrete structures. Appl. Mech. Mater. **82**, 118–123 (2011)

D'Alessandro, A., Materazzi, A.L., Ubertini, F.: Nanotechnology in Cement-Based Construction. Jenny Stanford Publishing, 424 Pages (2020a)

D'Alessandro, A., Coffetti, D., Crotti, E., Coppola, L., Meoni, A., Ubertini, F.: Self-sensing properties of green Alkali-activated binders with carbon-based nanoinclusions. Sustainability **12**(23), 1–13 (2020)

D'Alessandro, A., Fabiani, C., Pisello, A.L., Ubertini, F., Materazzi, A.L., Cotana, F.: Innovative concretes for low-carbon constructions: a review. Int. J. Low-Carbon Technol. **12**(3), 289–309 (2017)

Galao, O., Baeza, F.J., Zornoza, E., Garcés, P.: Carbon nanofiber cement sensors to detect strain and damage of concrete specimens under compression. Nanomaterials **7**(12), 413 (2017)

Galao, O., Bañón, L., Baeza, F.J., Carmona, J., Garcés, P.: Highly conductive carbon fiber reinforced concrete for icing prevention and curing. Materials **9**(4) (2016)

Han, B., Yu, X., Ou, J.: Self-Sensing Concrete in Smart Structures. Butterworth-Heinemann. Elsevier Inc. (2015). ISBN 978-0-12-800517-0

Han, B., Yu, X., Ou, J.: Multifunctional and smart nanotube reinforced cement-based materials. In: Gipalakrishnan, K., Birgisson, B., Taylor, P., Attoh-Okine, N. (Eds.) Nanotechnology in Civil Infrastructure. A Paradigm Shift, pp. 1–48. Springer (2011)

Loh, K.J., Hou, T.C., Lynch, J.P., Kotov, N.A.: Carbon nanotube sensing skins for spatial strain and impact damage identification. J. Nondestruct. Eval. **28**, 9–25 (2009). https://doi.org/10.1007/s10921-009-0043-y

Meoni, A., et al.: An experimental study on static and dynamic strain sensitivity of embeddable smart concrete sensors doped with carbon nanotubes for SHM of large structures. Sensors **18**(3), 831 (2018)

Metaxa, Z.S., Konsta-Gdoutos, M.S., Shah, S.P.: Carbon nanotubes reinforced concrete. ACI Symp. Publ. **267**, 11–20 (2009)

Przyczyna, D., Suchecki, M., Adamatzky, A., Szaciłowski, K.: Towards embedded computation with building materials. Materials **14**, 1724 (2021)

Rainieri, C., Song, Y., Fabbrocino, G., Markand, J.S., Shanov, V.: CNT-cement based composites: fabrication, self-sensing properties and prospective applications to structural health monitoring. In: Proceedings of the SPIE, Fourth International Conference on Smart Materials and Nanotechnology in Engineering, vol. 8793, 10pp. (2013)

Yoo, D.-Y., You, I., Youn, H., Lee, S.-J.: Electrical and piezoresistive properties of cement composites with carbon nanomaterials. J. Comp. Mater. **52**(24) (2018)

# Calcium Sulphoaluminate-Based Binders to Produce Expansive Concrete for Slabs on Ground

Denny Coffetti(✉) and Luigi Coppola

Department of Engineering and Applied Sciences, University of Bergamo, Bergamo, Dalmine (BG), Italy
{denny.coffetti,luigi.coppola}@unibg.it

**Abstract.** This article is aimed to the study of the influence of water/binder ratio, set retarding admixture dosage and curing condition on the properties of low environmental impact expansive concretes manufactured with calcium sulphoaluminate cement (CSA), gypsum, lime and supplementary cementitious materials instead of Portland cement (OPC). Experimental results indicated that tartaric acid-based set-retarding admixture influences the behaviour of concrete both in fresh and hardened state. In addition, according to Abram's model, results evidenced the water/binder ratio as a key factor in strength gain. Moreover, tartaric acid allows the production of shrinkage-compensating Portland-free concretes recommended for slabs on ground. Finally, by replacing OPC with CSA-based binders, it is possible to obtain, both for $CO_2$-emissions and energy consumption, a reduction up to 60% at equal strength class respect to an OPC-based concrete.

**Keywords:** Calcium sulphoaluminate cement · expansive concrete · jointless slabs on ground

## 1 Introduction

Nowadays, reinforced concrete ground floors are increasingly present both in infrastructures and in residential or industrial buildings. unfortunately, in many cases these concrete elements suffer from severe damage due to wrong materials selection, inaccurate concrete casting or poor design (Mynarcik 2013). Drying shrinkage is one of the common causes of cracking and curling of concrete slabs-on-ground, also because of the high ratio between the surfaces exposed to the air and the volume of concrete (Bissonnette et al., n.d.). in fact, an excessive shrinkage, in presence of internal and external constraints (such as reinforcing bars, floor foundation or other structural elements), determines notable internal tensile stress (Ababneh et al. 2017). However, expansive or shrinkage-compensating concrete (EC), although considerably more expensive than Portland cement-based mixtures, is valuable in concrete structures in which a reduction in cracking is of importance, not only pavement slabs but also bridge decks and liquid storage tanks. This technique is based on the early restrained expansion that occurs between the expansive agents and water (Han et al. 2016; Liu et al. 2016; Monosi et al.


M. A. Aiello and A. Bilotta (Eds.): ICC 2022, LNCE 435, pp. 708–719, 2024.
https://doi.org/10.1007/978-3-031-43102-9_54

2011). Generally, EC are manufactured with expansive agents that lead to the formation of ettringite ($C_3A \cdot 3C\overline{S} \cdot H_{32}$) or calcium hydroxide (CH) according to the following reactions:

$$C_4\,A_3\overline{S} + 6C + 8CS + 96H => 3C_3\,A \cdot 3C\overline{S} \cdot H_{32} \quad (1)$$

$$C + H = > CH \quad (2)$$

However, several authors (Collepardi et al. 2005; Sant et al. 2011) show that EC can be advantageously used to build reinforced concrete slabs-on grade without contraction joints only if an adequate wet curing is ensured. In particular, depending on the nature of expansive agent, 2- or 7-days wet curing is needed. Otherwise, the technique is totally unsuccessful.

Another effective method to produce EC involves the use of expansive cements, based on a controlled production of ettringite, instead of ordinary Portland cement. Between these special binders, ternary mixtures based on calcium sulphoaluminate cements (CSA), Portland cement (OPC) and gypsum (CSA:OPC:$C\overline{S}$) are certainly the most widespread (Monosi et al. 1996). Recently, Coppola et al. (Coppola et al. 2018) showed the possibility to manufacture environmentally friendly shrinkage-compensating mortars using CSA-based ternary mixtures in which OPC is totally replaced by supplementary cementitious materials (SCM's, such as fly ash and ground granulated blast furnace slag) and lime (CH).

The purpose of this paper is the evaluation of rheological, elasto-mechanical and physical performances of shrinkage-compensating Portland-free concrete manufactured with CSA:SCM:CH:$C\overline{S}$ mixtures at different water/binder ratios and tartaric acid-based set-retarding admixture dosage for concrete slabs on grade.

## 2 Materials and Methods

### 2.1 Materials

A commercial CSA clinker, ordinary Portland cement type I 52.5 R (EN 197-1 compliant), technical grade anhydrite ($C\overline{S}$) were used in this study to manufacture the reference shrinkage-compensating concretes (CSA:OPC:$C\overline{S}$ = 40:40:20). Ground granulated blast furnace slag (S: according to EN 15167-1), type V (according to EN 450-1 and EN 197-1) low calcium siliceous fly ash (FA) and hydrated lime (CH) CL90-S (according to EN 459-1) were employed to replace totally OPC in environmentally friendly mixtures (CSA:SCM:CH:$C\overline{S}$ = 40:35:5:20). The physical properties of binders were reported in Table 1. Furthermore, four different types of natural calcareous aggregates (maximum diameter equal to 32 mm) were combined to meet the Bolomey curve (Table 2).

Tartaric acid-based set-retarding admixture was added up to 0.6% with respect to binder mass in order to set the expansive behavior and the workability loss over time. Finally, the mixing water was fixed equal to about 200 kg/m$^3$ for ensure the fluid consistency and water/binder ratio was varied between 0.55 and 0.70. Composition of concretes are reported in Table 3.

**Table 1.** Physical and environmental properties of binders

| | OPC | CSA | $C\overline{S}$ | CH | S | FA |
|---|---|---|---|---|---|---|
| $D_{50}$ [μm] | 5.19 | 8.18 | 2.93 | 3.00 | 5.48 | 11.1 |
| Spec. Surface [$cm^2/g$] | 3175 | 2722 | 4837 | 4678 | 3049 | 2283 |
| Spec. Mass [$kg/m^3$] | 3150 | 2650 | 2670 | 2120 | 2730 | 2010 |
| GER [MJ/kg] | 5.50 | 2.70 | 1.30 | 4.50 | 0.31 | 0.10 |
| GWP [kg $CO_2$/kg] | 9.8 · 10–1 | $7.4 \cdot 10^{-1}$ | $2.4 \cdot 10^{-1}$ | $4.2 \cdot 10^{-1}$ | $1.7 \cdot 10^{-2}$ | $5.3 \cdot 10^{-3}$ |

**Table 2.** Physical and environmental properties of aggregates

| | Fine sand S | Fine gravel G1 | Coarse gravel G2 | Coarse gravel G3 |
|---|---|---|---|---|
| Diameter min/max [mm] | 0 / 6 | 6 / 12 | 10 / 20 | 20 / 30 |
| Water absorption [%] | 1.69% | 2.12% | 1.62% | 1.16% |
| Specific mass [$kg/m^3$] | 2550 | 2660 | 2680 | 2650 |

**Table 3.** Composition of concretes

| Composition [$kg/m^3$] | CSA | CEM I 52.5 R | C$ | CH | S | FA | Aggregates | Water | Tartaric acid |
|---|---|---|---|---|---|---|---|---|---|
| RC 0.55-0.4 | 142 | 142 | 72 | | | | 1788 | 196 | 2.20 |
| RC 0.60-0.4 | 132 | 132 | 66 | | | | 1818 | 197 | 2.22 |
| RC 0.65-0.4 | 122 | 122 | 61 | | | | 1845 | 199 | 2.23 |
| RC 0.70-0.4 | 113 | 113 | 57 | | | | 1857 | 199 | 2.23 |
| S 0.55-0.4 | 142 | | 71 | 18 | 123 | | 1776 | 195 | 2.18 |
| S 0.60-0.4 | 131 | | 65 | 16 | 115 | | 1806 | 196 | 2.20 |
| S 0.65-0.4 | 121 | | 60 | 15 | 105 | | 1821 | 197 | 2.21 |
| S 0.70-0.4 | 113 | | 57 | 14 | 98 | | 1845 | 198 | 2.22 |
| FA 0.55-0.4 | 142 | | 71 | 18 | | 124 | 1780 | 195 | 2.19 |
| FA 0.60-0.4 | 131 | | 65 | 16 | | 115 | 1802 | 196 | 2.20 |
| FA 0.65-0.4 | 121 | | 60 | 15 | | 106 | 1825 | 197 | 2.21 |
| FA 0.70-0.4 | 113 | | 57 | 14 | | 98 | 1849 | 198 | 2.22 |
| RC 0.55-0.6 | 142 | 142 | 72 | | | | 1788 | 196 | 3.35 |
| S 0.55-0.6 | 143 | | 72 | 18 | 125 | | 1799 | 197 | 3.37 |
| FA 0.55-0.6 | 142 | | 71 | 18 | | 124 | 1780 | 195 | 3.33 |

## 2.2 Tests on Concrete

Fifteen concretes were manufactured according to EN 12390-2; at the end of the mixing procedure, workability was measured over time (at 0, 30, 60, 90, and 180 min from mixing) by means of Abram's cone according to EN 12350–5 (Fig. 1). In addition, specific mass and entrapped air were evaluated on fresh concretes according to EN 12350-6 and EN 12350-7 standards. Specimens were produced and cured both under water at 20 °C and in a climatic chamber at 20 °C and R.H. 60%. Specific mass and compressive strength at 1, 7 and 28 days were also determined (EN 12390-3). In addition, only for mixture containing 0.6% of tartaric acid, free and restrained shrinkage/expansion were measured up to 56 days on specimens stored both under water at 20 °C and in dry environment (20 °C, R.H. 60%) according to EN 11307 and EN 8148, respectively. Finally, tensile strength on 28-day cured cylindrical specimens (according to EN 12390-6), elastic modulus (in accordance with method B, EN 12390-13) and water permeability under pressure (EN 12390-8) were measured.

**Fig. 1.** Slump test of expansive concrete

## 3 Results and Discussion

Workability at the end of the mixing procedure remains almost constant independently of the water/binder ratio by using 0.4% of tartaric acid dosage with respect to binder mass (Figs. 2 and 3). In particular, reference mixtures and concretes manufactured with S show an initial slump equal to 200 mm, reaching the consistency class S2 (100 mm slump) after about 60 min. On the contrary, FA-based concretes, at the same initial consistency, evidence a reduction in workability loss over time, achieving the consistency class S2 30 min later than the reference and S concrete (S2 after 90 min from casting).

According to Coppola et al. (Coppola et al. 2018) the tartaric acid dosage strongly influences the slump of concretes. A general increase in the initial workability (more marked in FA-based mixtures than those containing OPC and S) and a reduction in workability loss over time are observed by using 0.6% tartaric acid with respect to binder mass. In detail, reference concretes reach the consistency target (S2 class) after 120 min, those containing slag after about 90 min while mixtures based on fly ash show an excellent maintenance of workability over time, reaching the S2 consistency class in about 180 min. In general, it is possible to conclude that, for practical uses, OPC- or S-based concretes require greater set-retarding admixture dosage (0.6% by binder mass) than that (0.4% by binder mass) needed for FA mixtures.

On the other hand, the variation in water/binder and tartaric acid dosage do not determine substantial variations of entrapped air (between 0.8% and 1.5% by concrete volume) and specific mass in the fresh state. In particular, density is close to 2340 kg/m$^3$

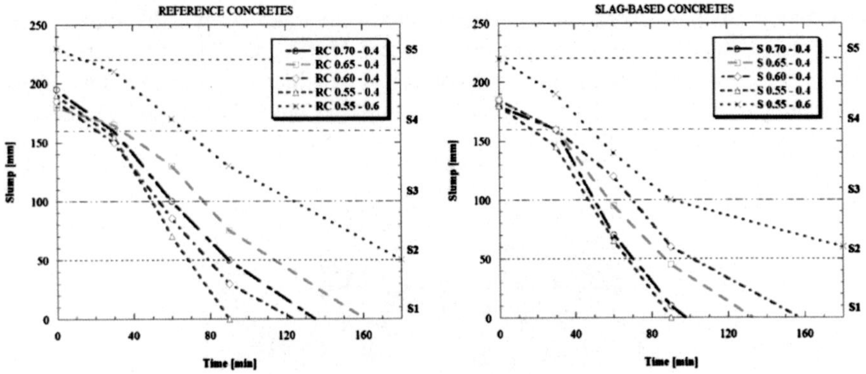


**Fig. 2.** Workability over time of OPC-based (left) and S-based (right) concretes

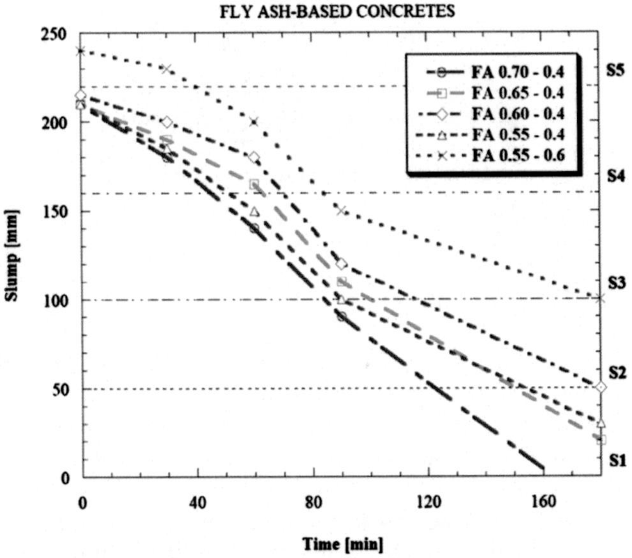


**Fig. 3.** Workability over time of FA-based concretes

for reference concretes while it attains values close to 2325 kg/m$^3$ for compounds in which SCMs have totally replaced Portland cement. On the contrary, the increase in water/binder ratio leads to a linear decrease in 28-day specific mass, independently of tartaric acid dosage and the type of binder (Fig. 4).

Concerning compressive strength measured on cubic specimens cured under water, it is possible to note that water/binder is a key factor (Figs. 5 and 6). Indeed, similarly to Portland cement concretes (Neville 1995), low w/b allows to obtain mixtures of excellent quality while increasing this parameter results in a general worsening of mechanical properties, regardless of binders employed and the age of concrete. Moreover, replacing OPC with hydrated lime and SCMs, negligible changes in 24-h strength are noted. On

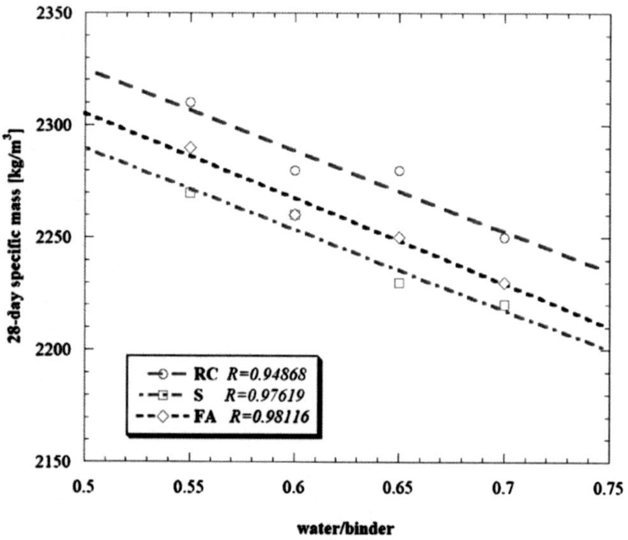


**Fig. 4.** 28-day specific mass vs water/binder ratio

the contrary, 30% reduction in compressive strength at 7 and 28 days were measured, independently of w/b. However, SCM-based concretes with w/b ratio from 0.55 to 0.70 exhibit perfectly compatible 28-day compressive strength for flooring (25–40 MPa).

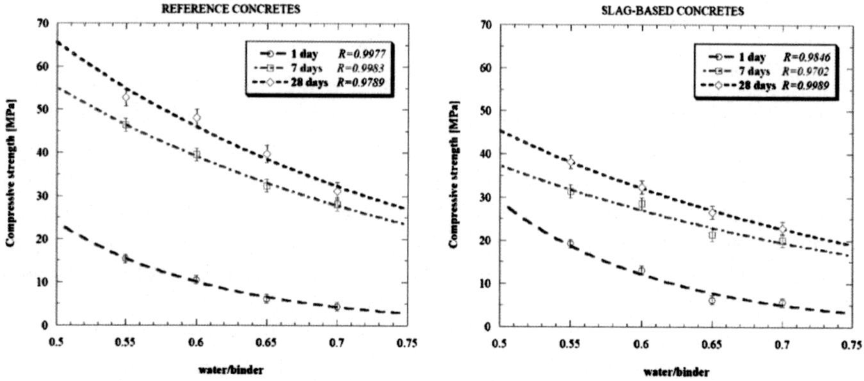


**Fig. 5.** Compressive strength of OPC-based (left) and S-based (right) concretes vs water/binder ratio

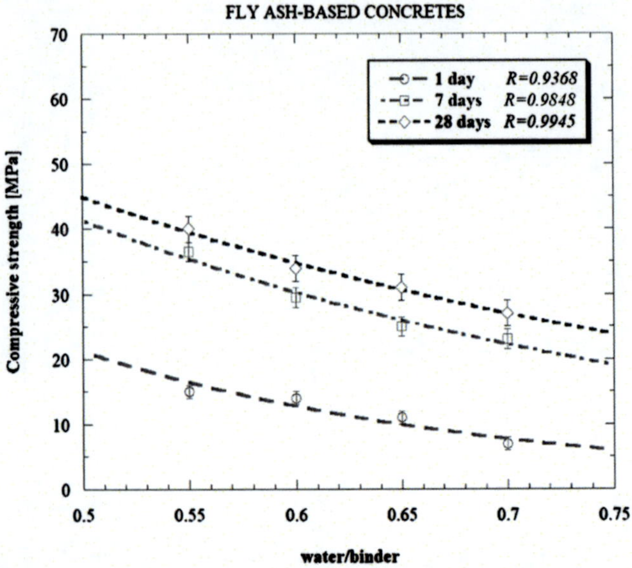


**Fig. 6.** Compressive strength of FA-based concretes vs water/binder ratio

Experimental data were used to determine the parameters A and B on Abram's model (III) for predicting the compressive strength at 28 days from casting of concrete manufactured with 0.4% vs binder mass of tartaric acid and cured underwater at 20 °C:

$$f_{c,28} = {A_{28}}/{B_{28}^{x}} \tag{3}$$

where ($f_{c,28}$) is the concrete compressive strength at 28 days, ($A_{28}$) and ($B_{28}$) are experimental parameters depending on composition and (x) is the water/binder ratio (Abd elaty 2014; Yeh 2006). Results reported in Table 4 and Fig. 7 show that concretes based on SCMs and lime have a mechanical behavior similar to that shown by traditional concretes manufactured with CEM I 52.5 R or CEM II/A-LL 42.5 R. On the contrary, reference mixture CSA:OPC:C$\overline{S}$ is more affected by w/b ratio, allowing to achieve, for low w/b ratios, compressive strength higher than those exhibited by CEM I 52.5 R-based composites. Finally, it should be noted that, by using sustainable CSA-based mixtures manufactured with FA or S, it is possible to reach similar mechanical strength to those obtainable, at equal w/c ratio, with a traditional limestone cement widespread on market (CEM II/A-LL 42.5 R).

**Table 4.** Coefficient of Abram's model for different mixtures

| | RC 0.4 | S 0.4 | FA 0.4 | CEM I 52.5 R | CEM II/A-LL 42.5 R |
|---|---|---|---|---|---|
| A28 | 386.61 | 255.31 | 261.47 | 261.25 | 263.32 |
| $B_{28}$ | 34.78 | 31.72 | 33.15 | 22.00 | 29.61 |

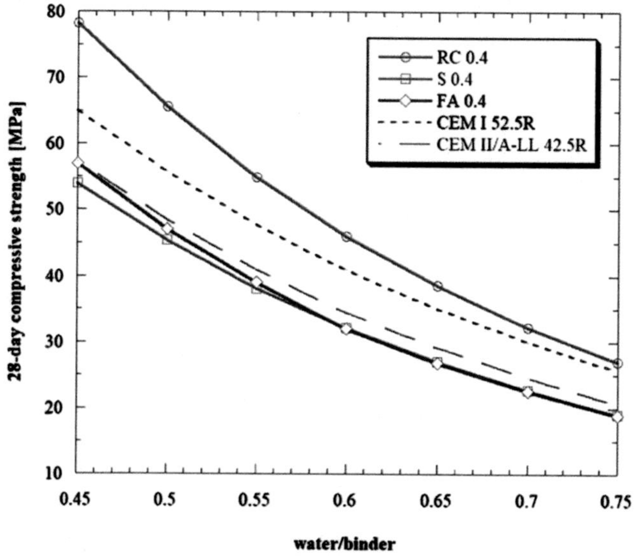


**Fig. 7.** 28-day compressive strength of concretes vs water/binder ratio (Abram's model, wet curing)

Also, the curing conditions strongly influence the mechanical properties of the concrete (Fig. 8). In fact, reference concrete cured in dry environment (T = 20 °C, R.H. 60%) exhibited compressive strength approximately 15% higher compared to that of the same mixture cured under water. Concrete manufactured with SCMs replacing OPC show more marked differences, up to 30%, between wet and dry cured specimens. Furthermore, increasing the tartaric acid dosage up to 0.6% vs binder mass, all concretes

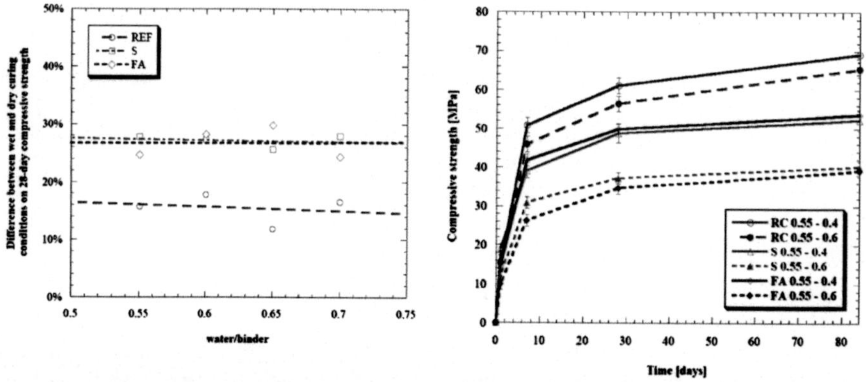


**Fig. 8.** Difference between wet and dry curing conditions on 28-day compressive strength vs water/binder ratio (left). Compressive strength development over time of concretes manufactured with different tartaric acid dosage (right)

(both references and those containing SCMs replacing OPC) exhibit a general reduction in mechanical performances up to 25% both at early and long ages (Fig. 8).

The total replacement of OPC with supplementary cementitious materials and the underwater curing conditions determine a general worsening of elasto-mechanical properties and water permeability of concretes. In FA and S based concretes, tensile strength decreases up to 40% compared to the reference mixture, independently of the curing conditions (wet or dry). However, tensile strength of CSA-based concretes (Fig. 9) follows the equation proposed by Eurocode 2 (EN 1992-1-1) for ordinary Portland cement concretes (strength class lower than C50/60):

$$f_{ctm} = 0.30 \cdot f_{ck}^{2/3} \tag{4}$$

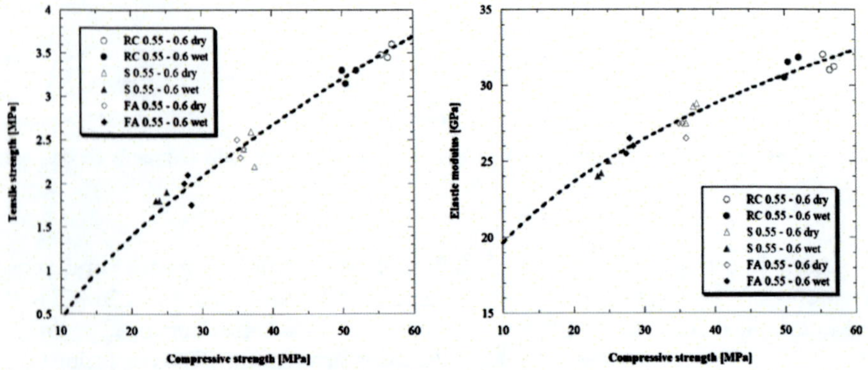


**Fig. 9.** Tensile strength (left) and elastic modulus (right) of concrete vs 28-day compressive strength. In dash line, the correlation proposed by EC2

Young's modulus decreases, at the same w/b, replacing Portland cement with SCMs due to the reduction of compressive strength caused by using FA or S (Fig. 9). Nevertheless, elastic modulus of concrete based on calcium sulphoaluminate cement can be well approximated by the following equation proposed by Eurocode 2:

$$E_{cm} = k \cdot \left(\frac{f_{cm}}{10}\right)^{0.30} \tag{5}$$

with k variable according to the mineralogical nature of aggregates used.

Regardless of binder used, shrinkage of CSA-based concretes is strongly influenced by curing conditions (Chen et al. 2012; Valenti et al. 2012). Indeed, free and restrained shrinkage tests show a stable behavior over time when specimens are stored in a climatic chamber at 20 °C and 60% R.H. On the other hand, in concretes cured under water an initial expansion was followed by a negligible shrinkage (Fig. 10). Total replacement of OPC with SCMs modifies the shrinkage of concretes; in fact, reference mixtures show more marked expansion underwater at early ages respect to Portland-free compounds, both in free and restrained conditions.

Figure 11 shows the environmental parameters GER (Gross Energy Requirement that correspond to the total energy expended in manufacture of 1 $m^3$ of concrete) and GWP (Global Warming Potential, related to the greenhouse gases emitted for producing 1 $m^3$ of mixture) for class C30/37 concrete manufactured with different type of binders calculated starting form raw materials data reported in Table 1 (CEM II/A-LL 42.5R: 3.60 MJ/kg, 8.8 • $10^{-1}$ kg $CO_2$/kg – aggregates: 0.13 MJ/kg, 2.4 • $10^{-3}$ kg $CO_2$/kg). If the aim is to increase sustainability, reducing both the emissions of $CO_2$ (GWP) and the primary energy required (GER) for the production of one cubic meter of concrete, the replacement of Portland cement type I with limestone Portland cement type II or ternary mixture, in which OPC and CSA are present in equal parts, is not a suitable solution to obtain a sharp reduction of environmental impact. In fact, the improvements

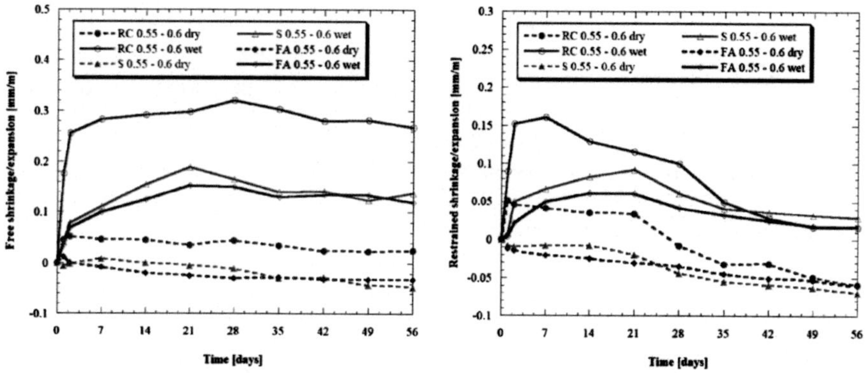


**Fig. 10.** Free (left) and restrained (right) shrinkage over time on different curing conditions

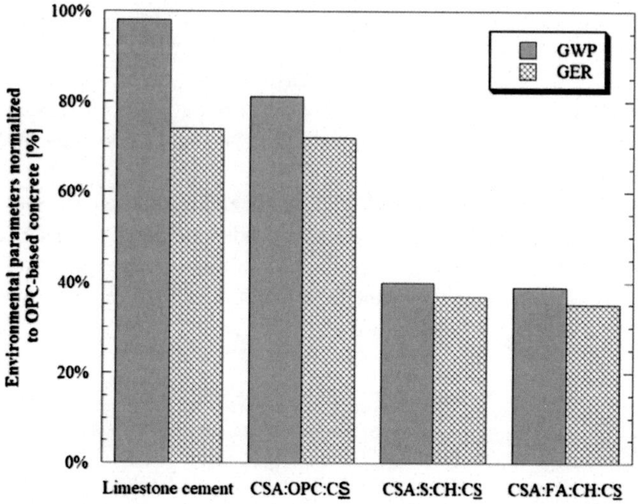


**Fig. 11.** GWP and GER parameters normalized to OPC-based concrete at equal strength class C30/37

obtained are rather limited, generally between 15% and 25%, due to both the high kiln temperatures required during Portland clinker production and the strong environmental impact of the extraction and grinding phase (Marroccoli et al. 2010).

The best route to achieve a clear improvement in terms of sustainability is to use mixtures based on sulphoaluminate cement in which OPC has been totally replaced by supplementary cementitious materials and hydrated lime. In this case, it is possible to obtain, both for GHG emissions and consumption of energy, a reduction that reaches 60% at equal strength class due to the nature of binders employed (that required limited processing before being used in mortars and concretes.

## 4 Conclusions

In this paper, the influence of water/binder ratio, curing conditions and set-retarding admixture dosage on the properties of environmentally friendly shrinkage-compensating CSA-based concretes was investigated. according to the experimental data, the following conclusions can be drawn:

- Workability at the end of the mixing procedure remains almost constant by varying water/binder ratio and binders employed while the tartaric acid-based set-retarding admixture acts as a superplasticizer.
- For using in job sites, OPC- or S- based mixtures require higher amount of tartaric acid-based set retarding admixture (0.6% vs binder mass) respect to that needed for concretes manufactured with FA (0.4% vs binder mass) in order to ensure a suitable workability retention.
- Portland-free concretes have mechanical behavior close to that shown by traditional concretes manufactured with Portland cement or limestone Portland cement;
- The total replacement of OPC with supplementary cementitious materials and the underwater curing conditions determine a general worsening of elasto-mechanical properties (compressive and tensile strength, Young's modulus) of concretes.
- Independently of binders employed, shrinkage of CSA-based concrete exhibit a stable behavior over time when specimens were cured at 20 °C and 60% R.H. while an underwater curing determines an initial expansion of concretes followed by a negligible shrinkage.
- CSA-based concretes manufactured with SCMs and hydrated lime in place of OPC are very promising from an environmentally point of view since GER and GWP parameters decrease about 60% at equal strength class compared to traditional OPC or CSA-OPC-C$ concretes due to the nature of binders that required low processing before being used in buildings industry.

## References

Ababneh, A.N., Al-Rousan, R.Z., Alhassan, M.A., Sheban, M.A.: Assessment of shrinkage-induced cracks in restrained and unrestrained cement-based slabs. (2017). https://doi.org/10.1016/j.conbuildmat.2016.11.036

Abd elaty, M. abd allah: Compressive strength prediction of Portland cement concrete with age using a new model. HBRC J. **10**, 145–155 (2014). https://doi.org/10.1016/j.hbrcj.2013.09.005

Bissonnette, B., Miltenberger, M.A., Fortin, C., Attiogbe, E.K.: Drying Shrinkage, curling, and joint opening of slabs-on-ground. Mater. J. **104** (2007). https://doi.org/10.14359/18671

Chen, I.A., Hargis, C.W., Juenger, M.C.G.: Understanding expansion in calcium sulfoaluminate-belite cements. Cem. Concr. Res. **42**, 51–60 (2012). https://doi.org/10.1016/j.cemconres.2011.07.010

Collepardi, M., Borsoi, A., Collepardi, S., Ogoumah Olagot, J.J., Troli, R.: Effects of shrinkage reducing admixture in shrinkage compensating concrete under non-wet curing conditions. Cem. Concr. Compos. **27**, 704–708 (2005). https://doi.org/10.1016/j.cemconcomp.2004.09.020

Coppola, L., Coffetti, D., Crotti, E.: Use of tartaric acid for the production of sustainable Portland-free CSA-based mortars. Constr. Build. Mater. **171** (2018). https://doi.org/10.1016/j.conbuildmat.2018.03.137

Han, J., Jia, D., Yan, P.: Understanding the shrinkage compensating ability of type K expansive agent in concrete. Constr. Build. Mater. **116**, 36–44 (2016). https://doi.org/10.1016/j.conbuildmat.2016.04.092

Liu, F., Shen, S.-L., Hou, D.-W., Arulrajah, A., Horpibulsuk, S.: Enhancing behavior of large volume underground concrete structure using expansive agents. Constr. Build. Mater. **114**, 49–55 (2016). https://doi.org/10.1016/J.CONBUILDMAT.2016.03.075

Marroccoli, M., Montagnaro, F., Telesca, A., Valenti, G.L.: Environmental implications of the manufacture of calcium sulfoaluminate-based cements. In: 2nd International Conference on Sustainable Construction Materials and Technologies, vol. 1, pp. 625–635 (2010)

Monosi, S., Troli, R., Coppola, L., Collepardi, M.: Water reducers for the high alumina cement-silica fume system. Mater. Struct. Constr. **29** (1996)

Monosi, S., Troli, R., Favoni, O., Tittarelli, F.: Effect of SRA on the expansive behaviour of mortars based on sulphoaluminate agent. Cem. Concr. Compos. **33**, 485–489 (2011). https://doi.org/10.1016/j.cemconcomp.2011.01.001

Mynarcik, P.: Technology and trends of concrete industrial floors. Proc. Eng. 107–112 (2013). https://doi.org/10.1016/j.proeng.2013.09.019

Neville, A.M.: Properties of Concrete. Longman, London (1995)

Sant, G., Lothenbach, B., Juilland, P., Saout, G.L., Weiss, J., Scrivener, K.: The origin of early age expansions induced in cementitious materials containing shrinkage reducing admixtures. Cem. Concr. Res. **41**, 218–229 (2011). https://doi.org/10.1016/j.cemconres.2010.12.004

Valenti, G.L., Marroccoli, M., Pace, M.L., Telesca, A.: Discussion of the paper Understanding expansion in calcium sulfoaluminate-belite cements by I.A. Chen et al., Cem. Concr. Res. **42**, 51–60 (2012). Cem. Concr. Res. (2012). https://doi.org/10.1016/j.cemconres.2012.08.002

Yeh, I.C.: Generalization of strength versus water-cementitious ratio relationship to age. Cem. Concr. Res. **36**, 1865–1873 (2006). https://doi.org/10.1016/j.cemconres.2006.05.013

# Author Index





M. A. Aiello and A. Bilotta (Eds.): ICC 2022, LNCE 435, pp. 721–723, 2024.
https://doi.org/10.1007/978-3-031-43102-9